Pavonia Cavanilles (Malvaceae)

Published by
The New York Botanical Garden Press
Bronx, NY 10458

International Standard Serial Number 0071-5794

The paper used in this publication meets the requirements of
the American National Standard for Information Sciences—Permanence of Paper for
Publications and Documents in Libraries and Archives, ANSI/NISO (Z39.48–1992).

Printed in the United States of America using soy-based ink on recycled paper

Library of Congress Cataloging-in-Publication Data

Flora neotropica. — Monograph no. 1 — New York: Published
for Organization for Flora Neotropica by The New York
Botanical Garden, 1968–
v.: ill.; 26 cm.

Irregular.
Each issue has distinctive title.
Separately catalogued and classified in LC before monograph no. 40.
ISSN 0071-5794 = Flora neotropica.

1. Botany — Latin America — Classification — Collected works.
2. Botany — Tropics — Classification — Collected works. 3. Botany —
Classification — Collected works. I. Organization for Flora Neotropica.
II. New York Botanical Garden.

QK205.F58 581.98'012—dc19 85-647083

Library of Congress [8508]
ISBN 0-89327-424-0

FLORA NEOTROPICA MONOGRAPH 76

PAVONIA CAVANILLES (MALVACEAE)

PAUL A. FRYXELL

Published for
The Organization for Flora Neotropica
by
The New York Botanical Garden Press
Bronx, New York

Issued 25 June 1999

Flora Neotropica

Published for
Organization for Flora Neotropica
by
The New York Botanical Garden Press

ORGANIZATION FOR FLORA NEOTROPICA
Founded and Conducted under the Auspices of UNESCO

Pavonia Cavanilles (Malvaceae)

Pavonia Cavanilles (Malvaceae)

PAUL A. FRYXELL

Contents

ABSTRACT

Fryxell, P. A. (Botany Department, University of Texas, Austin, TX 78713, U.S.A.). *Pavonia* Cavanilles (Malvaceae). Flora Neotropica Monograph 76: 1–285. 1999.—The genus *Pavonia* (Malvaceae) is probably the largest genus of its family, containing approximately 250 species, of which 224 are here recognized from the Americas. The New World species are subdivided into five subgenera, which are further subdivided into 15 sections and seven subsections to better portray relationships within this very diverse genus. *Pavonia* subgen. *Pavonia* includes 114 species; *P.* subgen. *Asterochlamys*, 32 species; *P.* subgen. *Malache*, 18 species; *P.* subgen. *Typhalea*, 34 species; and *P.* subgen. *Goetheoides*, 26 species.

As in Malvaceae generally, the flowers in *Pavonia* are monadelphous, with a central staminal column and usually numerous anthers, with five petals inserted at the base of the staminal column and five sepals forming a more or less gamosepalous calyx. The flower is subtended by an involucel (or epicalyx) made up of from 4 to 24 bracts. Fruit characters are important in distinguishing the subgenera, including spines, rugosities, and other ornamentation; the relative length of the carpel and the column to which it is attached; and whether the mature carpel is dehiscent or indehiscent. An unusual feature of *Pavonia* (in common with the other seven genera of the tribe Malvavisceae) is the possession of 10 styles and stigmas in the flower but only 5 carpels in the fruit.

Although *Pavonia* is predominantly tropical, a few species reach into the temperate zones in the southern United States and in northern Argentina. Many species are of limited distribution, whereas a few are widespread and sometimes weedy. They occupy a wide variety of habitats, ranging from mangrove associations to rain forests to seasonally dry deciduous forests to more open (often dry) environments.

The base chromosome number for *Pavonia* is x = 7, with reported counts ranging from $2n = 28$ (tetraploid) to $2n = 112$ (16-ploid), with known ploidy levels of 4x, 6x, 8x, and 16x.

The following species are herein described as new: *Pavonia aschersonioides*, *P. atlantica*, *P. aurelii*, *P. clarkii*, *P. cristobaliae*, *P. decora*, *P. formosa*, *P. froesii*, *P. heterotricha*, *P. hirticalyx*, *P. immitis*, *P. krapovickasii*, *P. longitricha*, *P. longitudinalis*, *P. nayarensis*, *P. neei*, *P. paraënsis*, *P. pedersenii*, *P. pseudolaxifolia*, *P. quadrifida*, *P. rudis*, *P. sepioides*, *P. spuria*, and *P. vanii*. In addition, the following new combinations are made: *P. benadirensis* (based on *Marconia benadirensis* Mattei), *P. cauliflora* (based on *Goethea cauliflora* Nees), *P. glaziovii* (based on *Codonochlamys glaziovii* Ulbrich), *P. goetheoides* (based on *Blanchetiastrum goetheoides* Hassler), *P. mogotensis* (based on *Hibiscus mogotensis* Carabia), *P. ramosissima* (based on *Pavonia sepium* var. *ramosissima* Arechavaleta), *P. setifer* (based on *Hibiscus setifer* Presl), *P. strictiflora* (based on *Goethea strictiflora* Hooker), and *P. tiliifolia* (based on *Codonochlamys tiliifolia* Ulbrich). New names in infrageneric rank are proposed as follows: *Pavonia* sect. *Albae*, *P.* subgen. *Asterochlamys*, *P.* sect. *Carcerariae*, *P.* sect. *Collicolae*, *P.* subsect. *Conditiflorae*, *P.* subsect. *Cordifoliae*, *P.* sect. *Diathericae*, *P.* subsect. *Hastifoliae*, *P.* sect. *Laminares*, *P.* subgen. *Malache*, *P.* subsect. *Rubriflorae*, and *P.* sect. *Varians*.

RESUMEN

Fryxell, P. A. (Botany Department, University of Texas, Austin, TX 78713, U.S.A.). *Pavonia* Cavanilles (Malvaceae). Flora Neotropica Monograph 76: 1–285. 1999.—El género *Pavonia* probablemente representa el género más grande de la familia Malvaceae, abarcando alrededor de 250 especies, de las cuales 224 aquí reconozco para las Américas. Las especies del Nuevo Mundo se colocan en cinco subgéneros, los cuales se dividen en 15 secciones y siete subsecciones, con el fin de mejor dilucidar las interrelaciones dentro de este género tan diverso. *Pavonia* subgen. *Pavonia* abarca 114 especies; *P.* subgen. *Asterochlamys*, 32 especies; *P.* subgen. *Malache*, 18 especies; *P.* subgen. *Typhalea*, 34 especies; y *P.* subgen. *Goetheoides*, 26 especies.

Al igual que las malváceas en general, las flores de *Pavonia* son monadelfas, con una columna estaminal central y usualmente con anteras numerosas, con cinco pétalos insertos en la base de la columna estaminal, y con cinco sépalos formando un cáliz más o menos gamosépalo. Un calículo (o epicáliz) abraza cada flor y contiene desde cuatro hasta 24 brácteas. Los caracteres de los frutos, tales como las espinas, rugosidades, y otros tipos de ornamentación; la longitud del carpelo en relación a la de la columna adjunta; y la naturaleza dehiscente o no del carpelo maduro, son importantes para distinguir a los subgéneros. Una peculiaridad notable de *Pavonia* (junto con los siete otros géneros de la tribu Malvavisceae) es la presencia de diez estilos y estigmas en la flor pero solamente cinco carpelos en el fruto.

Aunque la mayoría de las especies de *Pavonia* son de distribución tropical, unas pocas llegan hasta las zonas templadas del sur de los Estados Unidos y el norte de la República Argentina. Muchas especies son de distribución restringida, mientras que unas pocas son de distribución amplia y a veces llegan a ser malezas. Se encuentran en diversos habitats, desde manglares y selvas perennifolias hasta bosques deciduos y en ambientes más abiertos (a menudo áridos).

El número cromosómico base de *Pavonia* es x = 7; los recuentos reportados, desde $2n = 28$ (tetraploide) hasta $2n = 112$ (16-ploide), representan niveles de ploidia de 4x, 6x, 8x, y 16x.

Se describe las siguientes especies como nuevas: *Pavonia aschersonioides*, *P. atlantica*, *P. aurelii*, *P. clarkii*, *P. cristobaliae*, *P. decora*, *P. formosa*, *P. froesii*, *P. heterotricha*, *P. hirticalyx*, *P. immitis*, *P. krapovickasii*, *P. longitricha*, *P. longitudinalis*, *P. nayarensis*, *P. neei*, *P. paraënsis*, *P. pedersenii*, *P. pseudolaxifolia*, *P. quadrifida*, *P. rudis*, *P. sepioides*, *P. spuria* y *P. vanii*. Además, se proponen las siguientes combinaciones nuevas: *P. benadirensis* (basado en *Marconia benadirensis* Mattei), *P. cauliflora* (basado en *Goethea cauliflora* Nees), *P. glaziovii* (basado en *Codonochlamys glaziovii* Ulbrich), *P. goetheoides* (basado en *Blanchetiastrum goetheoides* Hassler), *P. mogotensis* (basado en *Hibiscus mogotensis* Carabia), *P. ramosissima* (basado en *Pavonia sepium* var. *ramosissima* Arechavaleta), *P. setifer* (basado en *Hibiscus setifer* Presl), *P. strictiflora* (basado en *Goethea strictiflora* Hooker) y *P. tiliifolia* (basado en *Codonochlamys tiliifolia* Ulbrich). Nombres nuevos en rango infragenérico se proponen como en siguiente: *Pavonia* sect. *Albae*, *P.* subgen. *Asterochlamys*, *P.* sect. *Carcerariae*, *P.* sect. *Collicolae*, *P.* subsect. *Conditiflorae*, *P.* subsect. *Cordifoliae*, *P.* sect. *Diathericae*, *P.* subsect. *Hastifoliae*, *P.* sect. *Laminares*, *P.* subgen. *Malache*, *P.* subsect. *Rubriflorae* y *P.* sect. *Varians*.

Introduction

The genus *Pavonia* Cavanilles is found in tropical and subtropical regions, sometimes reaches temperate climates, and is probably the largest genus of Malvaceae. It contains from 100 to more than 200 species by various estimates (Airy Shaw, 1966; Garcke, 1896; Hutchinson, 1967; Maas & Westra, 1993; Mabberley, 1987; Wielgorskaya, 1995). The present study recognizes 224 species from the New World. The genus is most richly developed and diverse in South America (Esteves, 1996; Gürke, 1892b; Kearney, 1958) and is well developed in Central America, the West Indies (Liogier, 1981), and Mexico (Fryxell, 1988a; Kearney, 1954). Ulbrich (1921) recognized 47 species from Africa, although in my opinion the actual number of African species is appreciably smaller when synonymies are considered (see Vollesen, 1995); there are fewer than 10 additional species from Asia and Australia (Fryxell, 1988b; Paul & Nayar, 1988; Philcox, 1997). I am not aware of any species that are common to the New World and the Old World, except for certain adventives such as *Pavonia hastata* and ornamentals such as *P. multiflora*. Thus, the genus on a global basis contains on the order of 250 species. Table I summarizes the historical development of views on the size of the genus.

Taxonomic History

Delimitation of the Genus

Pavonia is a representative of (and the largest genus in) the tribe Malvavisceae, which is clearly characterized by having twice as many styles and stigmas (10)

Table I

Historical summary of the number of species included in *Pavonia* as circumscribed in the present treatment

Author	Year	No. of species
Cavanilles	1787	14
Willdenow	1800	15
Link	1822	7
de Candolle	1824	31
Sprengel	1826	31
Don	1831	57
Steudel	1841	63
Walpers	1842	43
Dietrich	1847	69
Bentham & Hooker	1863	60+
Baillon	1872–1873	73
Gürke	1892b	90+
Ulbrich	1921	143
Kearney (Americas only)	1954, 1958	140
Airy Shaw	1966	210
Hutchinson	1967	200+
Mabberley	1987	150
Maas & Westra (Americas only)	1993	130
Wielgorskaya	1995	200
Fryxell (Americas only)	Present treatment	224 (33 new)

in the flower as it has mericarps (5) in the fruit. This feature has been examined developmentally by van Heel (1978), as described below (see morphology and terminology). There seems to be little doubt that the genera traditionally included in the Malavisceae are members of the same, presumably monophyletic lineage, but it is as yet unclear if other genera need to be added to the lineage (see La Duke & Doebley, 1995). As considered here, the tribe includes eight genera, of which three are monotypic: *Anotea* (de Candolle) Kunth, *Phragmocarpidium* Krapovickas, and *Rojasimalva* Fryxell. All eight of the genera occur in the New World, and three of them (*Malachra* Linnaeus, *Pavonia*, and *Urena* Linnaeus) extend to the Old World as well.

The question of the delimitation of *Pavonia* has received attention from various workers, and certain segregate genera have been removed from it, especially among South American plants. Some that have been removed in the past but that are generally included in *Pavonia* by contemporary interpretations include *Asterochlaena* Garcke, *Lebretonia* Schrank, and *Typhalea* (de Candolle) C. Presl. The genus *Peltaea* (C. Presl) Standley is generally accepted as distinct from *Pavonia* (Krapovickas & Cristóbal, 1965), as is the case for the genera *Anotea*, *Phragmocarpidium*, and *Rojasimalva*, which, although never formally included in *Pavonia*, nevertheless deserve mention here because of strong similarities and clear alliance to *Pavonia* as members of the tribe Malvaviscеae. A base chromosome number of x = 25 (Krapovickas & Cristóbal, 1965) characterizes *Peltaea*; it is the only member of the tribe Malvaviscеae known to have a base number other than x = 7.

The segregate genera *Codonochlamys* Ulbrich, *Blanchetiastrum* Hassler, *Goethea* Nees, *Lopimia* Martius, and *Triplochlamys* Ulbrich are returned to *Pavonia* in the present work. The genus *Lopimia* was not included earlier (Fryxell, 1988a), but a consideration of the larger group indicates that the two species concerned (*P. dasypetala* and *P. malacophylla*) are better included in *Pavonia* than excluded from it. The two species described in *Codonochlamys* by Ulbrich are here recognized to pertain to *P.* sect. *Malvaviscoides*. The other three segregate genera named are here included in *P.* subgen. *Goetheoides*, and the reasons for doing so are discussed in greater detail under that subgenus.

The delimitation of *Urena* from *Pavonia* presents additional problems. Some workers (e.g., Borssum Waalkes, 1966: 133–135) suggest that *Urena* be submerged in *Pavonia*, but most workers (e.g., Gürke, 1892a; Hutchinson, 1967; Kearney, 1951) distinguish *Urena* on the basis of glochidiate fruits and the presence of foliar nectaries. It should be noted, however, that the Old World species of *Urena* (Fryxell & Craven, 1989; Gürke, 1892a) lack the glochidiate spines on the fruits, and only the pantropical *U. lobata* Linnaeus and *U. sinuata* Linnaeus have these distinctive structures. Hochreutiner (1901) and Paul and Nayar (1988) restricted *Urena* to the glochidiate-fruited species, relegating the Old World species to *Pavonia*, but this view has not been followed by others (e.g., Sivarajan & Pradeep, 1996) and is not followed here. Thus, the principal character separating *Urena* from *Pavonia* is the presence in the former of the distinctive and well differentiated foliar nectaries (generally with yellow margins); the possession of subsessile anthers is a supporting character for maintaining these genera distinct. In my opinion, *Urena* appears to be a natural taxon by any standard and is the only genus in the tribe Malvaviscеae to possess such foliar nectaries. The data presented by La Duke and Doebley (1995) strongly support the distinctiveness of *Urena*.

KEY TO THE GENERA OF THE TRIBE MALVAVISCEAE

1. Leaves with one or more prominent nectaries abaxially on the principal nerves near their base; involucellar bracts 5; anthers subsessile. . . . . *Urena* Linnaeus
1. Leaves lacking nectaries; involucellar bracts usually >5 (4–24) or absent; anthers usually borne on manifest filaments.
    2. Fruits fleshy, succulent and smooth.
        3. Petals yellowish, lacking auricle at base; mature fruits blue-black. . . . . *Anotea* (de Candolle) Kunth
        3. Petals usually red (sometimes white, never yellow), prominently auriculate at base; mature fruits usually red (sometimes white). . . . . *Malvaviscus* Fabricius
    2. Fruits not fleshy but instead chartaceous or woody and sometimes spined, rugose, or otherwise ornamented.
        4. Flowers solitary or in inflorescences, but not subtended or enclosed by specialized foliaceous bracts. . . . . *Pavonia* Cavanilles
        4. Flowers or groups of flowers (heads) subtended and more or less enclosed by cordate-ovate, subsessile bracts.
            5. Mericarps 2-celled, the lower cell empty, with irregular internal partitions; flowers solitary. . . . . *Phragmocarpidium* Krapovickas
            5. Mericarps 1-celled; flowers solitary or grouped in heads.
                6. Mericarps with a single, apical spine longer than the mericarp body; involucel present, the bracts linear or filiform. . . . . *Rojasimalva* Fryxell
                6. Mericarps spineless; involucel present or absent, if present then the bracts often spatulate or geniculate (seemingly bifurcate).
                    7. Plants stellate-pubescent; leaves usually unlobed; involucellar bracts spatulate or geniculate (seemingly bifurcate); flowers solitary or grouped in heads; floral bracts ovate. . . . . *Peltaea* (C. Presl) Standley
                    7. Plants often setose; leaves often more or less lobed; involucel absent (or present in one species and the bracts linear); flowers grouped in heads; floral bracts often dissected at base. . . . . *Malachra* Linnaeus

SUBDIVISION OF THE GENUS PAVONIA

Previous workers have variously subdivided *Pavonia* into infrageneric groups, and the genus has had segregate genera removed from it, as previously discussed. The most elaborate system of subdivision is the one presented by Ulbrich (1921), who included 4 subgenera, 21 sections, and 12 subsections. Many of Ulbrich's groupings are clearly artificial, and his names for non-African subgroups are *nomina nuda*; his system will not be followed here. Esteves (1996) also used both subgeneric and sectional ranks in subdividing the genus.

Other workers (e.g., Fuertes, 1989; Gürke, 1892b; Kearney, 1951, 1958; Krapovickas, 1977, 1982a, 1982b, 1983; Krapovickas & Cristóbal, 1962; Vollesen, 1995) have generally subdivided the genus at a single level into sections and (as did Esteves, 1996) have usually found more natural groupings than did Ulbrich, although Gürke's (1892b) and Kearney's (1958) treatments include some ambiguities. The present treatment builds on and extends those prior treatments, recognizing five subgenera, and it follows Ulbrich (1921) and Esteves (1996) in using a range of infrageneric ranks, which permits a more sensitive method of presentation of relationships. A synoptic treatment of the subgenera follows, including a key to the 5 subgenera, 15 sections, and 7 subsections that are recognized here.

Morphology and Terminology

The descriptive terminology used here generally follows the discussion given for the family Malvaceae in Fryxell (1988a: 5–11) but with consideration of the following particulars in application to the genus *Pavonia*. Character states that are invariant among the species of *Pavonia* are as a rule omitted from the species descriptions. Such characters include:

- alternate, stipulate leaves;
- calyx gamosepalous and usually about half-divided (exceptions are noted: e.g., sepals nearly distinct in *P. flavispina*, *P. aschersoniodes*, and others; calyx shallowly dentate in some species of *P.* sect. *Malvaviscoides*);
- petals 5;
- androecium monadelphous;
- staminal column with a crown of 5 apical teeth at the summit that are comparable to (homologous

with?) similar features, called staminodes, found in members of the Sterculiaceae (e.g., *Byttneria*, *Theobroma*, a few species of *Ayenia*), and that are found in all tribes of the Malvaceae except the Malveae (Fryxell, 1988a: 5);

- styles and stigmas always 10;
- stigmas capitate;
- schizocarpic fruits with 5 carpels and solitary seeds.

As noted above, a distinctive feature of the gynoecium of *Pavonia*, in common with other genera of the tribe Malvavisceae, is that it has twice as many styles and stigmas (10) in the flower as it has carpels (5) in the fruit. This curious arrangement was studied developmentally by van Heel (1978), who found that the very young ovary has 10 carpels, but that alternate carpels abort during their development, except for their upper portions (the styles and stigmas), which persist. Thus the flower has 10 styles and stigmas, but the resulting fruit has only 5 carpels, herein called mericarps in conformity to the existing literature on Malvaceae (= "camaria," according to the terminology of Spjut, 1994). Because these numbers are constant in all of the species treated here, they will not be further mentioned, except in the generic description.

In some cases, species of *Pavonia* exhibit staminodes at the base of the staminal column. These take two forms. In most (all?) species of *P.* subgen. *Asterochlamys*, they appear as "filaments without anthers" but are sometimes thicker in structure than the filaments of the fertile stamens. They are more easily studied in fresh flowers than in dried specimens. The other more specialized form is found only in the two varieties of *P. oxyphylla*. These laminar structures are illustrated in Figure 71 and are further described in the species description. They seem to have no parallel elsewhere in the genus.

Fruit morphology has a strong bearing on classification within *Pavonia*, as it tends to have throughout the Malvaceae. Indeed, the subgenera and sections of *Pavonia* can be more or less defined on characters of mericarp morphology, although other characteristics of the plants are also involved. Because fruit and mericarp morphology are of basic taxonomic importance in *Pavonia*, a series of comparative drawings (see Figs. 4, 30, 36, 53, 65, 73) is presented to enable the reader to better visualize the distinctions among the generic subdivisions as well as among the species.

One feature of fruit morphology that has not previously been emphasized is here regarded as an important character in defining the subgenera. In three subgenera (*P.* subgen. *Asterochlamys*, *P.* subgen. *Malache*, and *P.* subgen. *Goetheoides*), the carpels are adnate to the column or connate above the column for their entire length. By contrast, in the other two subgenera (*P.* subgen. *Pavonia* and *P.* subgen. *Typhalea*), the carpels are adnate to the column only in the basal half (more or less), and the mericarps are essentially free distally and approximately twice as long as the column. An alternative description of this difference is to describe the fruits as either obtuse or retuse. This feature is readily discerned in the isolated mericarp (see Figs. 4, 73), because the zone of attachment along the ventral side is clearly defined. There is some variation around this norm within the subgenera, but it is appropriate to draw attention to this ventral attachment zone in order to define unambiguously the apex of the fruit, so that the position of spines and other ornamentation can be clearly stated.

For example, in *Pavonia multiflora* (see Fig. 30) the apical point is at the apex both of the mericarp and of the fruit. By contrast, in *P. geminiflora* (see Fig. 53) the point may be regarded as at the apex of the mericarp, but in relation to the fruit as a whole it is subapical or dorsal. In this frame of reference, thus, one sees that in *P.* subgen. *Pavonia* and *P.* subgen. *Typhalea* the distal points or spines (where present) are always subapical rather than apical. Truly apical points are less commonly present, primarily in species of *P.* subgen. *Goetheoides* and in *P. rhizophorae* (of *P.* subgen. *Malache*).

Without elaborating in any greater detail, but leaving it for the reader to comprehend from scanning the drawings, it is evident that there is a certain amount of variation on a common theme within each of the subgenera and sections, the least variation being found in *P.* subgen. *Malache*, where relatively unornamented mericarps are the rule (see Fig. 65).

Although seeds are often not evident in *Pavonia* specimens, whether because a given specimen is nonfruiting or because the mericarps are indehiscent, the seeds can provide taxonomically useful characters. The most notable examples are the striate seeds that characterize most species in *P.* subgen. *Asterochlamys* and the seeds with tufts of hairs at opposite ends of the hilum that are found only in *P.* sect. *Lebretonia*. Although other supporting characters are involved, these seed characters are essentially diagnostic of the two groups named. Chauhan (1987) describes some features of seed structure for six species of *Pavonia*, three of which are neotropical: *P. sepium*, *P. spinifex*, and *P. xanthgloea*.

Descriptive statements concerning stem pubescence may be taken to apply also to the pubescence of the petioles and pedicels, unless otherwise noted.

The distinctive narrow lines of fine (often recurved) pubescence that in many species are found running longitudinally along the stem, decurrent from the stipules, are generally also found on the adaxial side of the petioles and pedicels, with the curious exception of *Pavonia decora* in which they occur on the *ad*axial (upper) side of the petioles and the *ab*axial (under) side of the pedicels (see Fig. 29). These lines of fine pubescence may occur alone or in combination with other types of pubescence and are found in many other genera as well.

Measurements given for pedicels, involucels, and calyces are for anthesis or later, not for structures still in bud. If these structures are known to be accrescent post-anthesis (i.e., in fruit), it is so stated and additional measurements are given for the fruiting condition.

Growth Form and Habitat

Species of *Pavonia* range the gamut of growth form from prostrate herbs to shrubs to arborescent plants. Prostrate or decumbent herbs include *P. cancellata, P. cochensis, P. cymbalaria, P. glechomoides, P. humifusa, P. renifolia, P. repens, P. subrotunda,* and *P. vannii.* Rarely these plants are repent; that is, they root at the nodes (e.g., *P. flavispina, P. repens*). Most species of *Pavonia* are erect shrubs or subshrubs and range from relatively small plants (*P. aspera, P. cracens, P. intermixta, P. hoehnei, P. kleinii, P. nana,* and *P. nepetifolia*), which are often semi-decumbent, to robust shrubs several meters tall that may even be subarborescent (e.g., *P. malacophylla, P. dasypetala*). The fully arborescent species are primarily in *P.* subgen. *Goetheoides*, some species to 8 m tall, but subarborescent growth habits (or scandent shrubs) occur in *P. paludicola, P. troyana, P. bahamensis,* and *P. rhizophorae* (all of which are in *P.* sect. *Malache*), among others.

The habitats occupied by species of *Pavonia* are also varied. At one extreme are mangrove plants, such as *P. kearneyi, P. paludicola, P. rhizophorae,* and *P. bahamensis.* At another extreme are plants of arid habitats, including *P. submutica, P. ekmanii, P. arachnoidea,* and numerous others. Members of *P.* subgen. *Typhalea* sect. *Typhalea* and some members of *P.* sect. *Urenoideae* characteristically grow in the shaded understory of deciduous and evergreen forests. Some species grow in seasonally inundated localities along rivers or in otherwise wet soil (e.g., *P. piptocalyx, P. gracilis, P. opulifolia, P. restiaria, P. missionum, P. vitifolia,* and *P. patuliloba*). Members of *P.* subgen. *Goetheoides* are mostly arborescent and are confined to the tall rain forest along the easternmost coast of Brazil (a habitat that unfortunately is rapidly disappearing). A few species (e.g., *P. cancellata, P. humifusa, P. intermixta,* and *P. subrotunda*) prefer open habitats with deep sandy soils. Most species of *Pavonia*, however, are found in less extreme habitats, including deciduous or evergreen forests, savannas, cerrado formations, shrublands, and on roadsides or in otherwise disturbed habitats.

Pollination Syndromes

Several floral adaptations may be noted in *Pavonia* that evidently are related to pollination. Some of the more striking examples are the following.

In a few species (*Pavonia geminiflora, P. harleyi,* and *P. oxyphylla*) the flowers are presented horizontally at anthesis, with the staminal column declined and the anthers secund and with up-turned styles and stigmas, thus making a flower zygomorphic. This adaptation presumably provides a "landing platform" for a nectar feeder, which picks up and deposits pollen as a result of its visits to the flowers. Similar horizontal presentation (but with less marked zygomorphy) is found in other species, including *P. lasiopetala, P. purpusii, P. sapucayensis,* and *P. mollis*, for example. Gottsberger (1972) reported observations on insect pollination in *P. mollis* and illustrated the horizontal presentation of the flower (loc. cit.: fig. 18).

Various species have exserted genitalia: most notably *Pavonia luetzelburgii, P. macrostyla,* and *P. candida* (see Figs. 23, 68) but also including *P. hassleriana, P. pleuranthera, P. quadrifida* (see Fig. 62), *P. windischii* (see Fig. 26), *P. schwackei*, most members of *P.* subsect. *Exsertae* (e.g., *P. coccinea*; see Fig. 38), *P. bahamensis*, and some members of *P.* subgen. *Goetheoides* (e.g., *P. ducke-limae*; see Fig. 91, and *P. multiflora*; see Fig. 93). Species with exserted genitalia generally have tubular corollas. The specific pollinators of these various species are not known in any detail, although hummingbird pollination has been suggested for some of these species by Porsch (1929) and has been observed for the red-flowered *P. viscosa* (as *P. montana*) and *P. malvaviscoides* by Sazima (1981) and for *P. dasypetala* by Roubik (1982) and McDade and Davidar (1984). Hummingbird pollination was also reported by Gottsberger (1972) in *P. schrankii*, in which the corolla is yellow. Pollination by Bananaquits (*Coereba flaveola*: Coerebidae) and Bahama Woodstars (*Calliphlox evelynae*: Trochilidae) has been observed for *P. bahamensis* (Rathcke, 1996, 1999).

In a few species of *Pavonia* sect. *Malvaviscoides* (e.g., *P. glazioviana*) and of *P.* subgen. *Goetheoides* (e.g., *P. makoyana*) the involucel or calyx or both have taken on the pigmentation of the corolla and presumably have replaced its function of attracting pollinators. The size of the corolla is generally reduced in such species.

Most species of *Pavonia* are thought to be relatively generalized with regard to attracting pollinators. The corollas exhibit various colors (white, yellow, red, lavender, purple) and are with or without dark centers. Where present, these dark centers (dark red, maroon, purple) presumably function as "nectar guides" to attract floral visitors. In *P. oxyphylla* (see Fig. 71) distinctive laminar, darkly pigmented staminodes, which are an alternative kind of nectar guide, are found at the base of the staminal column.

Pollen

Published palynological information on the New World species of *Pavonia* is scant, with reports being found (Cervantes Aceves, 1992, Chaudhuri, 1965, Hashmi, 1970, Salgado-Labouraiu and Guzmán, 1967) for only four of the 224 species: *Pavonia fryxellii, P. lasiopetala, P. sagittata,* and *P. strictiflora* (as *Goethea strictiflora*). These reports agree in describing the grains as spheroidal, panporate, and relatively large (88–109 μm diam. for *P. strictiflora*; >130 μm for the other three species). The pollen of *P. strictiflora* lacks spines, a condition that is not otherwise reported in the Malvaceae. The other three species are spinose, with 30–60 apertures per grain and with pore diameters of 3.66 μm (*P. lasiopetala*) or 5.2 μm (*P. sagittata*). In *P. lasiopetala* the pores alternate with single spines, whereas in *P. sagittata* the pores alternate with pairs of spines. Clearly, sufficient information is not at hand to evaluate whether the palynological data have taxonomic significance and support the infrageneric divisions presented here, beyond the obvious hypothesis that the unusual condition of spineless pollen grains (in *P. strictiflora*) in fact characterizes *P.* subgen. *Goetheoides*.

However, pollen grains are sufficiently large in the Malvaceae that useful observations can often be made with relatively low-power magnification on unprepared pollen grains, without becoming involved in a full-scale palynological study. Thus, simple observation of flowering herbarium specimens reveals that in *Pavonia* subgen. *Goetheoides*, some species have spinose pollen grains (*P. alnifolia, P. calyculosa, P. crassipedicellata, P. longipedunculata, P. makoyana, P. multiflora,* and *P. stipularis*) and some species have pollen grains that lack spines (*P. cauliflora* and *P. strictiflora*). The subgenus, thus, is polymorphic for this character, and, although spineless pollen grains are highly unusual in the family, it is not appropriate to use this character as a basis for generic distinction.

Seed Dispersal

The fruits of *Pavonia* are uniformly composed of five mericarps, each containing a solitary seed. It is commonly the case that the mericarp itself is the disseminule, as a result of fruit maturation involving septicidal but not loculicidal dehiscence. The exception is found in *P.* subgen. *Asterochlamys*, in which most species exhibit some level of loculicidal dehiscence (sometimes with more or less suppressed septicidal dehiscence). In such plants the seeds are released directly from the mature fruits, and the seeds themselves serve as disseminules, generally without any obvious adaptation for dispersal over distance. Mericarp walls are relatively fragile in these loculicidally dehiscent fruits. It was on this character of loculicidal dehiscence that Hassler (1909) based the genus *Pseudopavonia* Hassler; however, virtually no subsequent workers have followed this view. It should be noted that in Figure 53 the mericarps of *Pavonia* subgen. *Asterochlamys* are drawn as if loculicidal dehiscence were suppressed (as in a slightly immature fruit) in order to make the drawings comparable to those of the other sections.

In *Pavonia* subgen. *Typhalea* each mericarp (see Fig. 73) bears three barbed spines (or only one in *P. stellata*) that readily adhere to the fur, feathers, clothing, or even skin of passing animals or humans. The mericarps are widely dispersed by such vectors, as is discussed by Mori and Brown (1998).

In *Pavonia* sect. *Pavonia* and *P.* sect. *Lebretonia* the mericarps are indehiscent and variously ornamented (see Figs. 4, 36). The ornamentations, although sometimes elaborate (e.g., *P. schrankii, P. rosa-campestris*), do not seem to have any obvious relationship to dispersal (except in *P. cancellata* and *P. humifusa*; see Fig. 4, where sometimes barbed projections have developed in a parallel adaptation to the spines of *P.* subgen. *Typhalea*). These various kinds of ornamentation (ribbing, rugosities, excrescences) tend to involve indurate tissue that is protective of the enclosed seed, so presumably this adaptation confers a type of dormancy on the seed that permits it to survive for a long time within the mericarp. Thus it can be passively disseminated over a considerable distance (the longer the dormancy, the greater the potential distance).

Table II

Chromosome counts in New World species of *Pavonia*

Taxon	2*n* count	Ploidy level	Species reported
subgen. *Pavonia*			
sect. *Malvaviscoides*	—	—	—
sect. *Pavonia*	56	8	*cancellata, laxifolia, luetzelburgii*
	112	16	*paniculata*
subsect. *Cordifoliae*	28	4	*mattogrossensis*
subsect. *Hastifoliae*	56	8	*aurigloba, consobrina, cymbalaria, distinguenda, hastata, lanata, missionum, revoluta, rosengurtii, sagittata, xanthogloea*
subsect. *Lebretonia*	56	8	*schrankii*
subsect. *Exsertae*	—	—	—
subgen. *Asterochlamys*	56	8	*garckeana, hexaphylla, sidifolia*
subgen. *Malache*	56	8	*ecostata, melanommata*
subgen *Typhalea*			
sect. *Typhalea*	28	4	*fruticosa*
sect. *Urenoideae*	56	8	*sepium*
	112	16	*communis*
sect. *Carcerariae*	42	6	*arachnoidea*
subgen. *Goetheoides*	56	8	*multiflora*

Sources: Bates, 1967, 1976; Davie, 1933; Fernández, 1974, 1981; Ford, 1938; Fryxell & Stelly, 1993; Krapovickas & Cristóbal, 1962, 1965; Skovsted, 1935, 1941.
Note: Some of the reports are unvouchered or are for species of questionable identity. A long dash indicates data unavailable.

A similar argument can be presented for the species of *Pavonia* subgen. *Malache*. The mericarps of these species are not ornamented in the same way, although a few of them have marginal wings (see Fig. 65), but they are characterized by having woody (even bony) mericarp walls as well as by having completely indehiscent mericarps. The extreme case is *P. candida*, whose seed can be extracted from the mericarp only with the aid of a hammer because the mericarp walls are 1–2 mm thick and bony. Presumably, the physical or bacteriological degradation of such a mericarp wall takes a significant period of time (possibly years), during which the mericarp can be passively disseminated over an appreciable distance. Despite these stout walls however, bruchid beetles can breach the mericarp walls and destroy the seeds. Neat emergence holes in the mericarps of *P. candida* led to the discovery and description of the bruchid *Acanthoscelides pavoniestes* C. D. Johnson (Johnson, 1983: 151).

A special case is presented by *Pavonia dasypetala* and *P. malacophylla* of *P.* subgen. *Pavonia*, which have viscid mericarps, a characteristic that clearly has implications for dispersal, perhaps by birds. Observational data are not available on the dispersal of these viscid mericarps, but such observations are much to be desired.

No information, even conjectural, is availablecon-cerning seed dispersal in species of *Pavonia* sect. *Malvaviscoides* or in those of *P.* subgen. *Goetheoides*.

Chromosomes

Table II presents data on chromosome numbers known for New World representatives of *Pavonia*, arranged according to infrageneric groupings recognized in this treatment. The reported counts indicate several ploidy levels based on a base number of x = 7. No instances of aneuploid numbers are known, and diploids ($2n = 14$) have not been reported. The most common number is $2n = 56$ (8-ploid), found in *P.* subgen. *Pavonia*, *P.* subgen. *Asterochlamys*, *P.* subgen. *Typhalea*, and *P.* subgen. *Goetheoides*. The lowest (tetraploid) value ($2n = 28$) is reported for one species of *P.* sect. *Typhalea* and for one species of *P.* subsect. *Cordifoliae* of *P.* subgen. *Pavonia*. The highest report (the 16-ploid level of $2n = 112$) is reported for the type of the genus, *P. paniculata*, and for *P. communis*. One can conclude that available data on chromosome numbers do not contradict the taxonomic subdivision presented here but give only imperfect support for it. Clearly, additional data are needed on chromosome numbers for a full cytotaxonomic interpretation of *Pavonia*.

Uses

Bertoni (1911) describes *Pavonia restiaria* as a potential crop plant, of interest both for its bast - fiber content, described as of high quality, and for its cellulose, noted for lacking gums and other contaminants, and thus of potential use in the paper pulp industry. In this respect it is similar to *Hibiscus cannabinus* Linnaeus ("kenaf"), which is grown commercially for these purposes. Evidently, Bertoni's suggestion of domesticating *P. restiaria* was never implemented. *Pavonia malacophylla*, on the other hand, is occasionally used in Brazil (and perhaps elsewhere) for its bast fibers to produce cordage.

Lentz (1993) cites *Pavonia schiedeana* (as *P. rosea*) as used for "tea from ground roots and hot water given to women to ease childbirth pain" in Honduras. Similar medicinal uses for this species (also as *P. rosea*) are noted from eastern Nicaragua by Coe and Anderson (1996).

Medicinal use has been reported for *Pavonia paniculata* (*Sántiz-Cruz 356*, NY), and the same species has been reported (*González 320*, WIS) as a source of tea derived from the leaves and used to treat whooping cough ("tos ferina"). A collector's note (*Krukoff 1085*) states "leaves boiled and used for stomach disease" for *P. cancellata.*

Plants grown as ornamentals include *Pavonia lasiopetala* in central Texas (Nokes, 1986); *P. hastata* in Australia (Mitchell, 1982); *P. candida* in various parts of Mexico; *P. alnifolia, P. cauliflora, P. multiflora, P. ×gledhillii* (see Cheek, 1989), *P. makoyana, P. multiflora* (probably the most widely cultivated species), *P. strictiflora*, and *P. tricalycaris* in tropical botanical gardens and sometimes elsewhere.

Systematic Treatment

Pavonia Cavanilles, Diss. **2:** unpaged addendum. 1786; **3:** 132. 1787, nom. conserv. Type. *Pavonia paniculata* Cavanilles. The generic name honors José Antonio Pavón (1754–1844) of Madrid, a physician and traveler in Peru and one of the authors of *Flora Peruviana.* (Non *Pavonia* Dombey *ex* Lamarck [Boraginaceae], Tabl. Encycl. **1:** 421. 1792; nec *Pavonia* Ruiz and Pavón [Monimiaceae], Fl. Peruv. Prodr. 127, t. 28. 1794; nec *Pavonia* H. F. A. de Roussel [Dictyotaceae], Fl. Cavados, ed. 2, 99. 1806.)

Lass Adanson, Fam. **2:** 400, 568. 1763, nom. rejic. vs. *Pavonia* Cavanilles. *Lassa* Kuntze, Revis. Gen. Pl. **3(2):** 19. 1893, orth. mut. Type. *Hibiscus spinifex* Linnaeus.

Malache B. Vogel in Trew, Pl. Select. **9:** 50, t. 90. 1772, nom. rejic. vs. *Pavonia.* Type. *Malache scabra* B. Vogel.

Prestonia Scopoli, Intr. Hist. Nat. 281. 1777, nom. rejic. vs. *Prestonia* R. Brown (Apocynaceae), On Asclepiad. 58. 1810, nom. conserv. Type. *Hibiscus spinifex* Linnaeus.

Schouwia Schrader, Goett. Gel. Anz. **1:** 717. 1821 (5 May). Type. *Schouwia semiserrata* Schrader [non *Schouwia* de Candolle (Cruciferae) Mém. Mus. Hist. Nat. **7:** 244. 1821 (20 Apr)].

Goethea Nees, Flora **4:** 304. 1821 (21 May). Lectotype (Gürke, 1892b). *Goethea cauliflora* Nees.

Lebretonia Schrank, Pl. Rar. Hort. Monac. 90, t. 90. 1822. Type. *Lebretonia coccinea* Schrank (≡ *Pavonia schrankii* Sprengel).

Lopimia Martius, Nova Acta Phys.-Med. Acad. Caes. Leop.-Carol. Nat. Cur. **11:** 96. 1823. Type. *Lopimia malacophylla* (Link & Otto) Martius (≡ *Sida malacophylla* Link & Otto).

Brehmia Schrank, Syll. Pl. Nov. **1:** 85. 1824. Type. *Brehmia arborescens* Schrank (= *Pavonia malacophylla* [(Link & Otto) Martius] [non *Brehmia* Harvey (Strychnaceae) *in* Hooker, London J. Bot. **1:** 25. 1842].

Typhalea (de Candolle) C. Presl, Abh. Königl. Böhm. Ges. Wiss., ser. 5, **3:** 449. 1845. Basionym: *Pavonia* sect. *Typhalea* de Candolle, Prodr. **1:** 442. 1824. Type. *Pavonia typhalea* (Linnaeus) Cavanilles, ICBN, art. 22.5 (non *Typhalea* Necker, nom. illegit., *ex* Britton & Wilson [see below]).

Asterochlaena Garcke, Bot. Zeitung (Berlin) **8:** 666. 1850. Type. *Asterochlaena cuspidata* Garcke (= *Pavonia mutisii* Humboldt, Bonpland & Kunth) (non *Asterochlaena* Corda [fossil fern], Beitr. Fl. Vorwelt 81. 1845–n.v.; nec *Astrochlaena* Hallier fil. [Convolvulaceae], Ann. Ist. Bot. Roma **7:** 228. 1894–n.v.).

Greevesia F. Mueller, Trans. Proc. Victorian Inst. Adv. Sci. **1854–1855:** 115. 1855. Type. *Greevesia cleisocalyx* F. Mueller (= *Pavonia hastata* Cavanilles).

Diplopenta Alefeld, Österr. Bot. Zeitschr. **13:** 10. 1863. Lectotype, here designated. *Diplopenta odorata* (Willdenow) Alefeld (≡ *Pavonia odorata* Willdenow).

Luederitzia Schumann, Bot. Jahrb. Syst. **10:** 45. 1889. Type. *Luederitzia pentaptera* K. Schumann (= *Pavonia rehmannii* Szyszylowicz).

Pseudopavonia Hassler, Feddes Repert. Spec. Nov. Regni Veg. **7:** 74. 1909. Type. *Pseudopavonia tenax* Hassler (= *Pavonia balansae* Gürke).

Blanchetiastrum Hassler, Feddes Repert. Spec. Nov. Regni Veg. **8:** 28. 1910. Type. *Blanchetiastrum goetheoides* Hassler [≡ *Pavonia goetheoides* (Hassler) Fryxell].

Codonochlamys Ulbrich, Notizbl. Bot. Gart. Mus. Berlin-Dahlem **6:** 329. 1915. Lectotype, here designated. *Codonochlamys tiliifolia* Ulbrich [≡ *Pavonia tiliifolia* (Ulbrich) Fryxell].

Triplochlamys Ulbrich, Notizbl. Bot. Gart. Mus. Berlin-Dahlem **6:** 333. 1915. Based on: *Pavonia* sect. *Tricalycaris* Gürke *in* Martius, Fl. Bras. **12(3)**: 479. 1892. Type (ICBN, art. 22.5). *Pavonia tricalycaris* St.-Hilaire.

Cancellaria (de Candolle) Mattei, L'Umbria Medica **2(14):** 215. 1921. Basionym: *Pavonia* sect. *Cancellaria* de Candolle, Prodr. **1:** 444. 1824. Type. *Pavonia zeylanica* Willdenow (non *Cancellaria* Schultz Bipontius [Asteraceae] *ex* Oliver, Fl. Trop. Afr. **3:** 321. 1877).

Marconia Mattei, L'Umbria Medica **2(14):** 215. 1921. Type. *Marconia benadirensis* Mattei. [≡ *Pavonia benadirensis* (Mattei) Fryxell, comb. nov.].

Pteropavonia Mattei, L'Umbria Medica **2(14):** 215. 1921. Type. *Pteropavonia kotzschii* (Hochstetter) Mattei (≡ *Pavonia kotzschii* Hochstetter).

Typhalea Necker, Elem. Bot. **2:** 412. 1790 (illegit., see ICBN, art. 32.8 and App. 5) *ex* Britton & Wilson, Bot. Porto Rico **5:** 560. 1924 (non *Typhalea* (de Candolle) C. Presl). Type. *Hibiscus spinifex* Linnaeus.

Prostrate perennial herbs, erect subshrubs, shrubs, or small trees, often stellate-pubescent, sometimes with glandular hairs, long simple hairs, scales, or other pubescence in various combinations, sometimes glabrous or glabrescent. *Leaves* alternate, usually petiolate, sometimes subsessile, distichous or spirally disposed, the blades ovate, elliptic, lanceolate, oblanceolate, deltoid, or reniform, sometimes lobed (hastately, sagittately, or palmately), sometimes asymmetrical, dentate or crenate, less often entire; stipules filiform, subulate, lanceolate, falcate, or spatulate, erect or reflexed. *Flowers* solitary or paired in the leaf axils or variously disposed in cymes, racemes, panicles, or heads; involucel whorled at the base of the calyx, the bracts 4–24, broadly foliaceous, ligulate, oblanceolate, spatulate, or filiform, persistent, sometimes colored (pink or red), usually distinct, sometimes basally connate or gamophyllous, sometimes seemingly 2seriate (alternately erect and reflexed) in subgen. *Goetheoides*; calyx gamosepalous, campanulate or tubular, shallowly to deeply 5-lobed, persistent (except deciduous in *P. piptocalyx*); corolla rotate, campanulate, or cylindric, white, yellow, rose, lavender, red, or purple, with or without a dark center, the petals 5, usually broadly obovate, sometimes narrowly ligulate, the claw pubescent or glabrous; androecium monadelphous, included or exserted, sometimes declined with secund anthers, surmounted by a crown of 5 apical teeth; styles 10, exceeding the androecium, more or less free distally, the 10 stigmas capitate. *Fruits* schizocarpic, pubescent or glabrous; mericarps 5, septicidally dehiscent, usually not loculicidally dehiscent (except in subgen. *Asterochlamys*), chartaceous or ligneous, sometimes winged, spined, rugose, carinate, or otherwise ornamented, or sometimes unornamented. *Seeds* solitary, glabrous or slightly pubescent.

Synopsis of Subgenera of Pavonia

I. **Pavonia** subgen. **Pavonia**

III. **Pavonia** subgen. **Asterochlamys** Fryxell, nom. nov. et stat. nov. Replaced basionym: *Asterochlaena* Garcke, Bot. Zeitung (Berlin) **8:** 666. 1850 (non *Asterochlaena* Corda, 1845). *Pavonia* sect. *Asterochlaena* (Garcke) Ulbrich, Bot. Jahrb. Syst. **57:** 64. 1921. Type. *Asterochlaena cuspidata* Garcke (= *Pavonia mutisii* Humboldt, Bonpland & Kunth).

Synonym: *Pseudopavonia* Hassler, Feddes Repert. Nov. Spec. Regni Veg. **7:** 74. 1909. *Asterochlaena* sect. *Pseudopavonia* Hassler, Feddes Repert. Nov. Spec. Regni Veg. **7:** 76. 1909, nom. provis. *Pavonia* subsect. *Pseudopavonia* Hassler *ex* Ulbrich, Bot. Jahrb. Syst. **57:** 64. 1921, nom. nud. Type. *Pseudopavonia tenax* Hassler (= *Pavonia balansae* Gürke).

III. **Pavonia** subgen. **Malache** (Vogel) Fryxell, stat. nov. Basionym: *Malache* Vogel *in* Trew, Pl. Select. **9:** 50. 1772, nom. rejic. *Pavonia* sect. *Malache* (Vogel) de Candolle, Prodr. **1:** 443. 1824. *Pavonia* subsect. *Malache* (Vogel) Endlicher, Gen. Pl. 982. 1840. Type. *Malache scabra* Vogel (≡ *Pavonia paludicola* Nicolson).

IV. **Pavonia** subgen. **Typhalea** (de Candolle) Ulbrich, Bot. Jahrb. Syst. **57:** 62. 1921. Basionym: *Pavonia* sect. *Typhalea* de Candolle, Prodr. **1:** 442. 1824. *Pavonia* subsect. *Typhalea* (de Candolle) Endlicher, Gen. P. 981. 1840. *Typhalea* (de Candolle) C. Presl, Abh. Königl. Böhm. Ges. Wiss. ser. 5, **3:** 449. 1845 (non *Typhalea* Necker, Elem. Bot. **2:** 412. 1790, illegit. Art. 32.8 *ex* Britton & Wilson, Bot. Porto Rico **5:** 560. 1924). Type. *Urena typhalea* Linnaeus [≡ *Pavonia typhalea* (Linnaeus) Cavanilles = *P. fruticosa* (Miller) Fawcett & Rendle].

Synonym: *Typhalea* Necker *ex* Britton & Wilson, Bot. Porto Rico **5:** 560. 1924. Type. *Hibiscus spinifex* Linnaeus [≡ *Pavonia spinifex* (Linnaeus) Cavanilles].

V. **Pavonia** subgen. **Goetheoides** (Gürke) Ulbrich, Bot. Jahrb. Syst. **57:** 66. 1921. Basionym: *Pavonia* sect. *Goetheoides* Gürke *in* Martius, Fl. Bras. **12(3):** 479. 1892. Type. *Pavonia goetheoides* (Hassler) Fryxell.

Key to the Subgenera, Sections, and Subsections of Pavonia

1. Mericarps with 1, 2, or (usually) 3 retrorsely barbed spines (see Fig. 73), the central spine subapical, erect (suppressed in *P. kearneyi*), the two lateral spines erect or divaricate (suppressed in *P. stellata*); subgen. *Typhalea*
   2. Mericarps 2-spined, the central spine suppressed (see Fig. 88); leaves small, spatulate, obtuse, penninerved, 5 times as long as wide. . . . . sect. *Diathericae* (p. 228)
   2. Mericarps 3-spined (or 1-spined in *P. stellata*), the central spine well developed; leaves various, often elliptic or oblanceolate, usually acute, mostly 1–3 times as long as wide.
      3. Leaves palmately lobed, with undulate margins (see Fig. 89); corolla with red center; calyx lobes broadly cordate, plicate in bud. . . . . sect. *Varians* (p. 234)

3. Leaves unlobed; corolla lacking red center; calyx lobes lanceolate or ovate.
    4. Internodes of equal length, the flowers solitary in the leaf axils; involucellar bracts 5–16; corolla usually yellow, white, or red. . . . . sect. *Urenoideae* (p. 205)
    4. Internodes sometimes apically shortened, the flowers solitary or aggregated into definite inflorescences, if solitary then bracts 10 or more; corolla yellow, white, or pink.
        5. Bracts of the involucel 10–15, arcuate, filiform, setose; internodes of equal length, the flowers solitary in the leaf axils; petals yellow. . . . . sect. *Carcerariae* (p. 231)
        5. Bracts of the involucel 5–10; internodes usually apically shortened, the flowers in terminal leafless inflorescences (usually racemes or heads; see, e.g., Figs. 72, 77) more or less above the leaves; petals pink or white. . . . . sect. *Typhalea* (p. 187)

1. Mericarps variously ornamented or unornamented, but lacking barbed spines.
    6. Mericarps ca. twice the length of the column; subgen. *Pavonia*
        7. Calyx tubular, the lobes usually ca. ¼ as long as tube (see Fig. 31). . . . . sect. *Malvaviscoides* (p. 64)
        7. Calyx more or less campanulate, the lobes subequal to the tube.
            8. Seeds with tufts of hairs at each end of hilum; mericarps usually carinate; involucellar bracts usually 5 (occasionally 6–7); . . . . sect. *Lebretonia*
                9. Leaves lance-ovate or elliptic, basally cuneate, never lobed (see Fig. 36). . . . . subsect. *Lebretonia* (p. 77)
                9. Leaves ovate or deltoid, basally cordate or truncate, sometimes hastately or sagittately lobed.
                    10. Genitalia usually exserted (see Fig. 38); involucel sometimes basally gamophyllous; leaves often discolorous. . . . . subsect. *Exsertae* (p. 84)
                    10. Genitalia included; involucellar bracts distinct; leaves concolorous or discolorous.
                        11. Leaf blades 6–16 cm long, almost as wide, broadly cordate-ovate, often distally lobulate (see Figs. 41, 43); corolla 2–5 cm long, yellow, white, rose, or purple. . . . . subsect. *Cordifoliae* (p. 95)
                        11. Leaf blades 1–7 cm long, often 2–3 times as long as wide, ovate-triangular, often hastately (see Fig. 46) or sagittately (see Fig. 53) lobed basally; corolla 1–4.5 cm long, yellow, rose, or red. . . . . subsect. *Hastifoliae* (p. 102)
            8. Seeds glabrous or pubescent but never tufted; mericarps generally not carinate; involucellar bracts 5–24.
                12. Plants procumbent; leaves sometimes asymmetrical (see Fig. 27); bracts of the involucel ca. 15, filiform, arcuate; mericarps with 3 erect apical points, these sometimes barbed (see Fig. 4). . . . . sect. *Cancellaria* (p. 59)
                12. Plants erect; leaves usually symmetrical; bracts of the involucel 5–24, spreading, erect, or arcuate; mericarps lacking apical points.
                    13. Stipules erect (see Fig. 36); involucellar bracts 10–24; corolla 2.5–5.5 cm long (if smaller, then red or rose), rose, red, yellow, or cream. . . . . sect. *Lopimia* (p. 32)
                    13. Stipules reflexed, usually more or less falcate (see Fig. 1); involucellar bracts 5–15; corolla 0.5–3 cm long, yellow. . . . . sect. *Pavonia* (p. 14)
    6. Mericarps subequal to the column.
        14. Leaves penninerved, elliptic or oval to oblong or lance-elliptic, never lobed, often large, up to 36 cm long (see Figs. 90–93); stipules often erect and appressed to stem, sometimes of 2 or more elements (see Fig. 90); pedicels usually solitary in the axils, often elongated, articulated at base leaving axillary scars, less often in terminal racemes; involucel prominent, sometimes colored; fruits twice as long as wide, glabrous and shiny, with a small apical mucro but otherwise unornamented; shrubs or trees 1–8 m tall. . . . . subgen. *Goetheoides* (p. 235)
        14. Leaves usually palminerved, often ovate, sometimes lobed, seldom >15 cm long; stipules various, often inconspicuous; pedicels solitary in the axils or in racemes, usually relatively short; involucel various, seldom colored; fruits oblate, glabrous or pubescent, usually 5- or 10-ribbed (rarely unribbed), sometimes winged, woody or even bony (see Fig. 65); plants shrubs, seldom >3 m tall; subgen. *Malache*
            15. Mericarps indehiscent, often ligneous; seeds variously pubescent or glabrous but not striate.
                16. Leaves subentire or obscurely crenate or denticulate; corolla 1–2 cm long, greeenish yellow; fruits sometimes crowned with 5 crests or wings; plants usually scandent shrubs of mangrove habitats (except *P. troyana*, occurring in Jamaican uplands); West Indies, Central America, NW South America. . . . . sect. *Malache* (p. 164)

16. Leaves manifestly crenate or dentate; corolla 1–8 cm long, white, yellowish, salmon, reddish, or dark red; fruits 5- or 10-ribbed or smooth but not winged; plants erect shrubs of upland Mexican habitats.
    17. Corolla rotate; staminal column more or less declined (flowers somewhat zygomorphic), with prominent purplish laminar staminodes at base (see Fig. 71). . . . . sect. *Laminares* (p. 182)
    17. Corolla more or less tubular (flowers actinomorphic), the staminal column erect (central, relative to the petals), lacking basal staminodes; sect. *Collicolae*
        18. Flowers solitary in the axils, the corolla salmon or reddish, narrowly tubular; genitalia exserted; fruits weakly 10-ribbed or smooth, semi-woody.. . . . . subsect. *Conditiflorae* (p. 180)
        18. Flowers more or less aggregated in leafless terminal racemes (sometimes axillary in *P. firmiflora,* but then corolla yellowish and genitalia included); corolla white, yellow, reddish, or dark red, broadly tubular; genitalia included or exserted; fruits often 5- or 10-ribbed, strongly woody.
            19. Corolla dark red, 1.2–2.5 cm long; genitalia included; styles connate nearly to apex, the stigmas subsessile (see Fig. 70). . . . . . subsect. *Rubriflorae* (p. 177)
            19. Corolla white or yellowish, 1–8 cm long; genitalia included or exserted; styles apically distinct (see Figs. 68, 69). . . . . . . subsect. *Collicolae* (p. 171)
15. Mericarps more or less loculicidally dehiscent, chartaceous (see Fig. 53); seeds usually striate; subgen. *Asterochlamys*
    20. Seeds glabrous, striate; corolla white, rose, or yellow. . . . . sect. *Asterochlamys* (p. 132)
    20. Seeds minutely pubescent, non-striate; corolla white. . . . . sect. *Albae* (p. 162)

Each subgenus will be treated separately in sequence, with further subdivisions as indicated in the preceding key.

Notes. In the following treatment, references to specimens in "herb. Pedersen" are references to the private herbarium of Troels Myndel Pedersen in Estancia Santa Teresa, Paraguay.

In citations of illustrations, an asterisk (*) with a plate number indicates that the illustration referred to is in color.

PAVONIA SUBGEN. PAVONIA

Shrubs, subshrubs, or herbs with simple, stellate, or glandular hairs; leaves ovate or elliptic, sometimes lobed; stipules simple, erect or reflexed; flowers solitary or in inflorescences, the involucellar bracts 4–24; corolla of various colors; mericarps twice length of column, variously ornamented or unornamented.

Type. *Pavonia paniculata* Cavanilles.

Synopsis of Pavonia subgen. Pavonia

A precise delimitation of the nominate subgroup (in whatever rank) of the genus *Pavonia* has not previously been clearly made. It has often been treated simply as a residue after the other better defined subgroups (e.g., *P.* subgen. *Typhalea*) have been removed. Kearney (1958: 226, couplet 15) referred to "section *Eupavonia*, including *Asterochlaena*" but did not subsequently distinguish the two and indeed included all of *P.* sect. *Lebretonia* as well. Gürke (1892b) similarly compounded these groups under "sect. *Eupavonia*." The interpretation here proposed treats *P.* subgen. *Pavonia* as a complex comprising the following five sections and four subsections.

1. **Pavonia** sect. **Pavonia** (15 species).
2. **Pavonia** sect. **Lopimia** (Martius) Endlicher, Gen. Pl. 982. 1840. Type. *Sida malacophylla* Link & Otto (≡ *Pavonia malacophylla* (Link & Otto) Garcke) (23 species).
3. **Pavonia** sect. **Cancellaria** de Candolle, Prodr. **1:** 444. 1824. Type. *Pavonia cancellata* Cavanilles [although *P. zeylanica* (Linnaeus) Willdenow was named lectotype by Mattei (1921), the type is *P. cancellata* according to ICBN, art. 22.5] (4 species).
4. **Pavonia** sect. **Malvaviscoides** St.-Hilaire, Fl. Bras. Merid. **1:** 237. 1827. *Pavonia* subsect. *Malvaviscoides* (St.-Hilaire) Endlicher, Gen. Pl. 982. 1840. *Pavonia* subgen. *Malvaviscoides* (St.-Hilaire) Ulbrich, Bot. Jahrb. Syst. **57:** 65. 1921. Type. *Pavonia malvaviscoides* St.-Hilaire (12 species).
5. **Pavonia** sect. **Lebretonia** (Schrank) Endlicher, Gen. Pl. 982. 1840. Basionym: *Lebretonia* Schrank, Pl. Rar. Hort. Monac. pl. 90. 1822. Type. *Lebretonia coccinea* Schrank (≡ *Pavonia schrankii* Sprengel).
    i. **Pavonia** subsect. **Lebretonia** (6 species).
    ii. **Pavonia** subsect. **Exsertae** Krapovickas, Trabalh. XXVI Congr. Nac. Bot. (Brasil)

332. 1977. Type. *Pavonia exserta* Krapovickas & Cristóbal (= *P. commutata*) (14 species).

iii. **Pavonia** subsect. **Cordifoliae** Fryxell, subsect. nov. Type. *Pavonia fryxellii* Krapovickas (7 species).

iv. **Pavonia** subsect. **Hastifoliae** Fryxell, subsect. nov. Type. *Pavonia hastata* Cavanilles (33 species).

Pavonia subgen. *Pavonia* sect. *Pavonia*

Erect *shrubs* or *subshrubs* with simple, stellate, or glandular hairs; *leaves* broadly ovate, sometimes weakly 3-lobed to narrowly hastate; stipules reflexed, usually more or less falcate, relatively persistent; *inflorescences* terminal, condensed or open racemes or panicles; involucel of 5–16 bracts, these linear to filiform, erect to arcuate; corolla yellow; *mericarps* obovate, reticulate-veined, blackish.

This group of species (the "*Pavonia paniculata* complex") is clearly delimited, and is most clearly characterized by the reflexed stipules and yellow corollas. The form of the mericarps (see Fig. 4) is also characteristic, although it is more easily illustrated than described.

Within this section, however, the individual species are harder to define and characterize. A first step is to obtain a sharper definition of the common and widespread *Pavonia paniculata*: a much clearer view results when *P. corymbosa* and *P. restiaria* are recognized as distinct species and segregated out of a broadly conceived *P. paniculata*. Although these three species are closely allied, their recognition as distinct seems justified on geographic and ecological grounds, as well as for morphological reasons.

Most of the remaining species, as well as *Pavonia restiaria*, are riverine species, in contrast to *P. corymbosa* and *P. paniculata*, which occur in dryer habitats. Interestingly, *P. restiaria* occurs immediately on riverbanks (or even in riverbeds), directly associated with flowing water (Krapovickas, pers. comm.), whereas the other species are plants of stagnant water, away from the river currents. Delimiting these species is more difficult. Some of the species (e.g., *P. gracilis, P. harleyi*) are easily recognized morphologically, but others are not. Two courses present themselves. One might lump them all together and use the oldest available name (*P. opulifolia*), or one might make a best effort to sort out distinct entities on the basis of available material, recognizing that there will be morphological and geographic anomalies that will need to be corrected in future studies. The latter course is adopted here, in part to provide a framework for such future studies.

Key to the New World species of *Pavonia* sect. *Pavonia*

1. Involucellar bracts 10–16.
  2. Involucellar bracts 12–16; mature leaves more or less hastate.
    3. Stems minutely and finely glandular-puberulent. . . . . 2. *P. cristobaliae*
    3. Stems setose and with underlying glandular hairs.
      4. Stipules broadly falcate; pedicels 2–4.5 cm long; upper leaf surface with appressed simple hairs 0.2–0.5 mm long. . . . . 14. *P. subhastata*
      4. Stipules narrowly lanceolate; pedicels 1–3 cm long; upper leaf surface with appressed simple hairs 0.5–1 mm long. . . . . 7. *P. laetevirens*
  2. Involucellar bracts 10–12(–13); mature leaves usually ovate, sometimes palmately lobed (*P. patuliloba*).
    5. Upper leaf surface with abundant simple hairs 0.5–1 mm long, erect, recurved, or appressed.
      6. Involucel 11–19 mm long; stems stellate-pubescent (hairs 1 mm long) and with narrow longitudinal rows of shorter pubescence; leaves commonly 3-lobed; staminal column not bearded at base. . . . . 11. *P. patuliloba*
      6. Involucel 6–8 mm long; stems setose (hairs 2–4 mm long) and densely and evenly glandular-pubescent; leaves ovate; staminal column densely bearded at base at juncture with corolla, otherwise glabrous. . . . . 4. *P. harleyi*
    5. Upper leaf surface stellate-pubescent or glabrate (if simple hairs present, then <0.5 mm long).
      7. Corolla 2.5 cm long; upper leaf surface with often bifurcate appressed hairs 0.3 mm long. . . . . 9. *P. opulifolia*
      7. Corolla 1–2 cm long; upper leaf surface otherwise.
        8. Inflorescence congested terminally; staminal column 3–4 mm long; calyx 4–5 mm long; petals 12–15 mm long; stems sparsely pubescent. . . . . 1. *P. corymbosa*
        8. Inflorescence open; staminal column ca. 1 cm long; calyx 5–8 mm long; petals ca. 2 cm long; stems strigillose. . . . . 15. *P. vitifolia*

1. Involucellar bracts 5–9.
 9. Involucellar bracts 7–9.
  10. Involucel 5–10 mm long; fruits 6–7 mm diam. . . . . 10. *P. paniculata*
  10. Involucel 10–15 mm long; fruits 8–9 mm diam. . . . . 13. *P. restiaria*
 9. Involucellar bracts 5–6.
  11. Hairs of upper leaf surface simple, appressed, 0.5–1.5 mm long; involucellar bracts up to 2 mm wide; fruits 6–7 mm diam.
   12. Stems pilose (hairs 2–3 mm long) and with underlying minute glandular hairs. . . . . 8. *P. laxifolia*
   12. Stems with scattered multi-armed stellate hairs, lacking both pilose and glandular hairs. 12. *P. pseudolaxifolia*
  11. Upper leaf surface stellate-pubescent or glabrate (if simple appressed hairs present, these 0.2–0.6 mm long); involucellar bracts 0.5–1 mm wide; fruits 4–6 mm diam.
   13. Involucellar bracts 5–7 mm long (subequal to calyx); staminal column 4–5 mm long, the filaments 2–3 mm long.
    14. Leaves hastate, the upper surface with simple curved hairs 0.3–0.6 mm long. . . . . 5. *P. heterotricha*
    14. Leaves ovate-cordate, the upper surface with simple, bifurcate, or stellate hairs 0.2–0.3 mm long. . . . . 3. *P. gracilis*
   13. Involucellar bracts 10–12 mm long (2–3 times as long as calyx); staminal column 1–2 mm long, the filaments 3–6 mm long. . . . . 6. *P. insperabilis*

1. Pavonia corymbosa (Swartz) Willdenow, Sp. Pl. **3:** 836. 1803. *Althaea corymbosa* Swartz, Prodr. 101. 1788. *Pavonia paniculata* var. *corymbosa* (Swartz) Gürke *in* Martius, Fl. Bras. **12(3):** 506. 1892. Type. Jamaica: *Swartz s.n.* (holotype, S).

Malachra cordata Poiret, Encycl. Suppl. **3:** 578. 1814. Type. Puerto Rico: *Riedlé s.n.* (P as photo CTES).

Shrubs 1–2 m tall, the stems with prominent longitudinal lines of dense recurved pubescence, otherwise with very sparse stellate hairs or with prominent simple hairs 1.5–3 mm long. *Leaf* blades ovate to somewhat 3-lobed, to 10 cm long, about as wide, basally cordate, the margin serrate, the apex acute or acuminate, palmately 7–9-nerved, the upper surface with extremely minute bifurcate hairs or glabrate on lamina (sometimes with simple hairs 1 mm long on veins, rarely with appressed simple or bifurcate hairs 1 mm long), the lower surface with scattered 3- and 4-armed stellate hairs; petioles to 9 cm long, densely pubescent on adaxial side, otherwise very sparsely stellate-pubescent, sometimes prominently setose; stipules falcate, reflexed, 7–10 mm long. *Flowers* in dense terminal (corymbiform?) inflorescences, the pedicels 1 cm long or less, setose (hairs 2–3 mm long); involucellar bracts 10–12, 12–15 mm long, narrowly linear, setose; calyx 4–5 mm long, ciliate; petals yellow, 12–15 mm long, densely pubescent on claw; staminal column 3–4 mm long, basally pubescent, the filaments 3 mm long. *Fruits* 8–9 mm diam.; mericarps 3 mm long, reticulate-angulate.

Distribution (Fig. 2). Widely distributed in South and Central America and the Greater Antilles, more or less sympatric with *Pavonia paniculata*, with which it is often confused.

Representative specimens examined. HONDURAS. W of Tela River, *Wilson 216* (F, NY, US).

NICARAGUA. Vic. of San Juan del Norte (Greytown), *C. L. Smith 6* (GH, MO, NY, US). **Zelaya:** Caño Montecristo, *Moreno 15099* (HNMN, NY).

COSTA RICA. Cartago: Turrialba River, *J. D. Smith 6452* (GH, NY, US). **Heredia:** Finca La Selva, just E of junction of Río Puerto and Río Sarapiquí, *Mowbray & Wilbur 1361* (DUKE). **Limón:** Tortuguero National Park, *Robles 2629* (MO, NY).

PANAMA. Colón/Canal Zone: Rd. between Trans-isthmian Hwy. and Madden Dam, *A. H. Gentry 6690* (MO, NY). **Darién:** Along the Sambú River, above the tide limit, *Pittier 5535* (NY, US). **Panamá:** Between Capira and Potrero, *Dodge & Hunter 8646* (C, F, GH, MO, NY, S, U, US)

HAITI. Massif de Cahos, near Las Caobas, Morne Das Bois-Rouge, *Ekman 5561* (NY, S, TEX, US); Massif de la Hotte, Fond-des-Nègres, Miragoane, *Ekman 7183* (S); Dépt. du Nord: Vic. of Dondon, *E. C. Leonard 8586* (GH, NY, US).

DOMINICAN REPUBLIC. Villa Riva (Almacén), Prov. Pacificador, *Abbott 609* (GH, NY, US); along Yabon River, ca. 12 km S of Sabana de la Mar, *A. H. Liogier 18439* (F, NY); Prov. Peravia: La Cienega, 11 km from the hwy. from San José de Oca to Río Abajo, *Mejía & Pimentel 9979* (JBSD, NY). Río San Juan, *G. S. Miller 1140* (US).

PUERTO RICO. Without locality, *Ventenat s.n.* (C).

COLOMBIA. Chocó: Pichimá River region, Waunana, *Forero 730* (MO); valley of the San Juan River, *Forero et al. 4186* (MO, NY); area of Baudó, 13 km upstream from estuary, *Fuchs & Zanella 21720* (K, MO, US). **Valle del Cauca:** Mun. Buenaventura, San Isidro, *van Rooden et al. 330* (CTES, MO, U).

ECUADOR. Esmeraldas: Near Borbon, *Játiva & Epling 2159* (MO, NY, S–2, US). **Napo:** Armenia Vieja at Río Napo, ca. 12 km SW of Coca, *Lugo 2647* (A, F, GB, NY,

S); ca. 38 km S of Coca, *Lugo 2612* (GB, K, NY, QCA, U, US). **Pichincha:** Río Palenque Biological Station, *Webster et al. 22755* (DAV, TEX), *Zarucchi & Carlosama 2395* (GH, QCA, S, US).

PERU. **Huánuco:** Tingo María, *Asplund 12167* (S, US), *Asplund 12994* (NY, S); Prov. Marañon, Monzón River, near Tingo María, *Ferreyra 4234* (CTES, US, USM); Panguana, Yupapichis River affluent to Pachitea River, *Seidenschwarz 273/1* (NY). **Loreto:** Prov. Ucayali, Contumana, *Fosberg 29004* (NY, US); Prov. Maynas, Padre Isla just below Iquitos, *A. H. Gentry et al. 22165* (F, MO, NY); Florida, Putumayo River, *Klug 2044* (A, GH, K, MO, NY, S, US); Prov. Requena, Cocha Paucate, Ucayali River near Pto. Perú, *R. Vásquez & N. Jaramillo 902* (MO). **Madre de Dios:** Prov. Manu, Manu National Park, Cocha Cashu Station, *Foster 9850* (MO, NY).

BRAZIL. **Acre:** Juruá River, on the edge of lake Juruá Miry, *Ule 5529* (K). **Amazonas:** Basin of Rio Demeni, vic. of Totobí, *Prance et al. 10336* (CTES, K, NY, S, US); Ilha Aramaçá almost opposite Tabatinga, *Prance et al. 16708* (CTES, K, MO, NY, S, U, US); Manaus, 16 Jan 1975, *Vogt s.n.* (CTES). **Pará:** Belem, *Black 48-2926* (NY, U, US); Belém, Igapó, *Murça-Pires & Black 692* (GH, US); Rio Giamá, Fazenda Vilha, *Murça-Pires & Black 1549* (CTES); Belem, *N. T. da Silva 92* (NY).

BOLIVIA. **Santa Cruz:** Prov. Ñuflo de Chávez, shores of the Río Negro, 10 km downstream from Perseverancia, *I. G. Vargas 642* (MO, NY, TEX).

Pavonia corymbosa and *P. paniculata* are often confused. They can be easily distinguished by their inflorescences, which are very crowded and compact in the former and very open in the latter. This distinction, however, is sufficiently subjective that it cannot be used in a diagnostic key, and other characters must be relied upon.

2. Pavonia cristobaliae Fryxell, sp. nov. Type. Brazil. Mato Grosso: 80 km W of Miranda, region of Morro do Aceite, *Krapovickas & Cristóbal 34443* (holotype, CTES; isotype, TEX). Fig. 1

Frutices, caulibus minute glanduloso-pubescentibus; laminis foliorum ovatis vel triangularibus subtiliter puberulentibus, stipulis reflexis; floribus solitariis in axillis foliorum; bracteis involucellorum 13–16 ciliatis; corollis 2.5–3 cm longis flavidis; fructibus ca. 9 mm diametro glabris rugulosis.

Shrubs 1.5 m tall, the stems minutely and finely glandular-puberulent. *Leaf* blades ovate to hastate-triangular, basally truncate, serrate, acute, palmately 5–7-nerved, 4–8 × 3–5 cm, nearly concolorous, finely puberulent above and beneath and with widely scattered larger stellate spicules beneath; petioles 1–3 cm long, glandular-puberulent; stipules lanceolate to falcate, reflexed, 5–6 mm long. *Flowers* solitary in the leaf axils, the pedicels 1–2.5 cm long, finely puberulent like stem and in addition with pilose hairs 1–1.5 mm long distally; involucellar bracts (11-)13–16, linear, 15–20 mm long, copiously ciliate, with underlying glandular puberulence; calyx 1 cm long, ciliate and with underlying glandular puberulence; corolla 2.5–3 cm long, yellow; androecium included in corolla, the column ca. half the length of the petals; styles emergent from the column, the stigmas subequal to the petals. *Fruits* ca. 9 mm diam., glabrous; mericarps 5 mm tall, rugulose.

Distribution (Fig. 2). Known from S Brazil (Mato Grosso). The type locality is at approximately 19°55′S, 56°55′W.

Representative specimens examined. BRAZIL. Mato Grosso do Sul: Mun. Corumbá, Fazenda Acorizal, *Pott 3061* (CTES); Mun. Miranda, Fazenda São Francisco, *Pott & Ferreira 2648B* (CTES).

The specific epithet is chosen to honor the co-collector of the type, Carmen L. Cristóbal of Corrientes, Argentina, knowledgeable specialist with the Sterculiaceae. As is indicated by its position in the key, *Pavonia cristobaliae* is evidently a vicariad of *P. subhastata*, the two species sharing many characters, yet being clearly distinct, both geographically and morphologically.

3. Pavonia gracilis R. E. Fries, Kongl. Svenska Vetenskapsakad. Handl., ser. 3, **24(2):** 27. 1947. Type. Brazil. Mato Grosso: Santa Ana de Chapada, *Malme II-3415* (holotype, S; isotypes, CTES [fragment], S). Figs. 3, 4

Subshrubs 1.5–2 m tall, the stems scabridulous with sparse stellate hairs and with minute glandular hairs on younger growth. *Leaf* blades broadly ovate (with reduced upper leaves sometimes hastately or palmately lobed), truncate or cordate, serrate, acute or acuminate, palmately 7–9-nerved, minutely pubescent (the hairs of the upper surface 0.2–0.3 mm long, simple, bifurcate, or stellate), those of the lower surface somewhat larger and with scattered 3–4-armed spicules, mostly 4–13 × 3–10 cm; petioles 1–4(–7) cm long, stellate-scabridulous and adaxially bearded; stipules triangular or falcate, 3–7 × 2 mm, often reflexed. *Flowers* generally in a terminal panicle; pedicels 0.5–3 cm long, slender; involucellar bracts 5(–6), linear, 5–7 × 1 mm, sparsely pubescent; calyx 4–6 mm long, short-ciliate; corolla 8–11 mm long, yellow, pubescent on margins of claw; staminal column 4–5 mm long (ca. half the length of the petals), glabrous, pallid, the filaments 2–3 mm long; styles

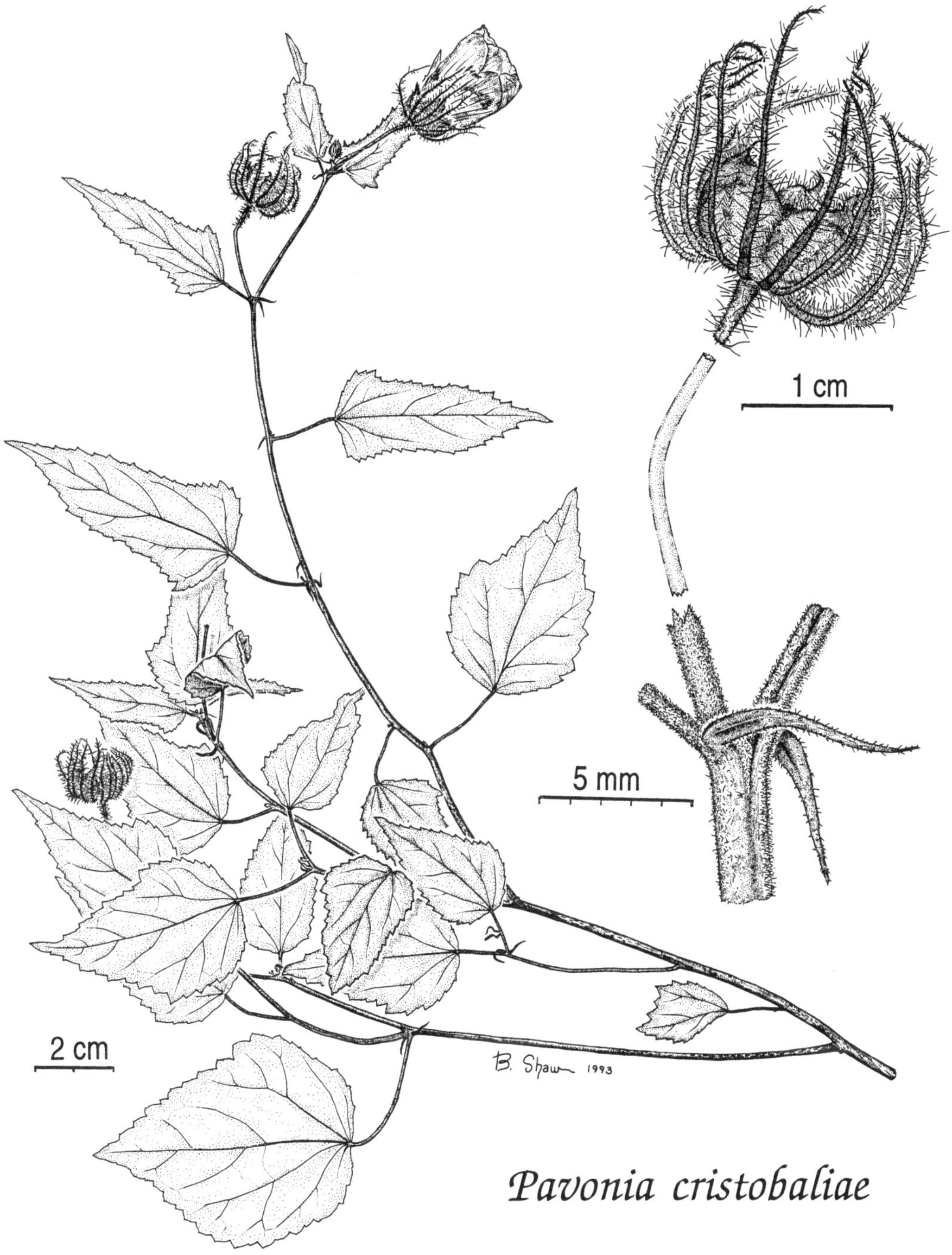

FIG. 1. *Pavonia cristobaliae*. Flowering branch (left) with enlargements of fruit (upper right) and node with stipules (center right) (*Krapovickas & Cristóbal 34443*).

exserted from column, subequal to petals, connate nearly to apex, the stigmas scarcely separated. *Fruits* 4 mm diam., glabrous, unornamented.

Distribution (Fig. 5). From S Brazil, Bolivia, and Paraguay, growing at elevations of 150–1000 m, but with one collection (*J. Steinbach 5730*) from

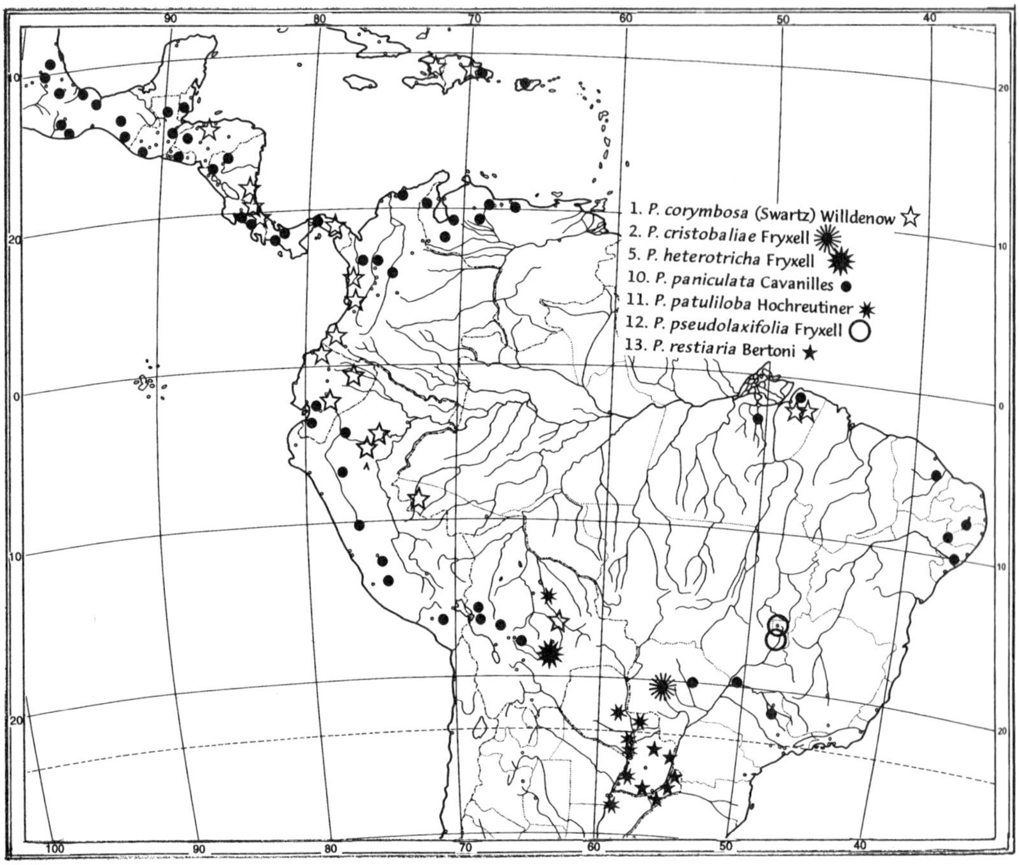

FIG. 2. Distribution map of seven neotropical species of *Pavonia* sect. *Pavonia*.

2500 m. This species is often found growing in wet habitats.

Additional illustrations. Fries (1947: pl. 3, figs. 5–9).

Representative specimens examined. **BRAZIL**. **GOIÁS:** Without locality, *Glaziou 20713* (K); Caiaponia-Aragarças rd., 75 km from Aragarças, *Hunt 6100* (CTES, K); 10 km from Goiânia on the rd. to Guapó, *Rizzo & Barbosa 48* (CTES); as edge of the João Leite River, which in 400 m empties into the Meia Ponte River, *Rizzo & Barbosa 598* (CTES). **MATO GROSSO:** Mun. Cáceres, 9 km ENE of Porto Esperidiɿo, *Krapovickas et al. 40117* (BM, C, CEN, CTES, LIL, NY, RB, UC); 15 km N of Rosario Oeste, *Krapovickas et al. 40219* (CEN, CTES, F, G, K, NY); without precise locality, Jul 1892, *Kuntze s.n.* (F, NY, US). **MINAS GERAIS:** Mun. Ituiritaba, Fundas, *Macedo 786* (US).

BOLIVIA. COCHABAMBA: Sacaba, *J. Steinbach 5730* (GH, K, MO). **SANTA CRUZ:** Caranda, 20 km SW of Portachuelo, *Krapovickas 34643* (CTES); Santa Cruz, vic. of El Zoológico, *Krapovickas & Schinini 36147* (CTES); Santa Cruz, near the Universidad, *Krapovickas & Schinini 36134* (CTES); Prov. Gutiérrez, 10 km W of Portachuelo, *Krapovickas & Schinini 31586* (CTES); Prov. San Andrés Ibañez, 7 km SE of Naranjillos, *M. Nee 38930* (MO, NY, TEX); Prov. Lara, *J. Steinbach 2090* (GH, NY).

PARAGUAY. In the Apa River region, *Hassler 11250* (A, CTES [fragment], K, MO, NY, S); Sierra de Amambay, *Hassler 11951* (A).

Local name. Vassoura do mata (Brazil: Minas Gerais).

Pavonia gracilis is easily recognized within *Pavonia* sect. *Pavoniae* by its flowers which are reduced in size and in number of parts. It is similar to *P. pseudolaxifolia* but differs from it in the characters indicated in the key.

4. Pavonia harleyi Krapovickas, Bonplandia **5:** 266. 1983. Type. Brazil. Bahia: Serra do Sincorá by Rio Cumbuca, ca. 3 km N of Mucugê, on the Andarai rd. (13°0′S, 41°23′W), *Harley, Renvoize, Erskine, Brighton, & Pinheiro 15997* (holotype, CEPEC; isotypes, CTES, K). Fig. 4

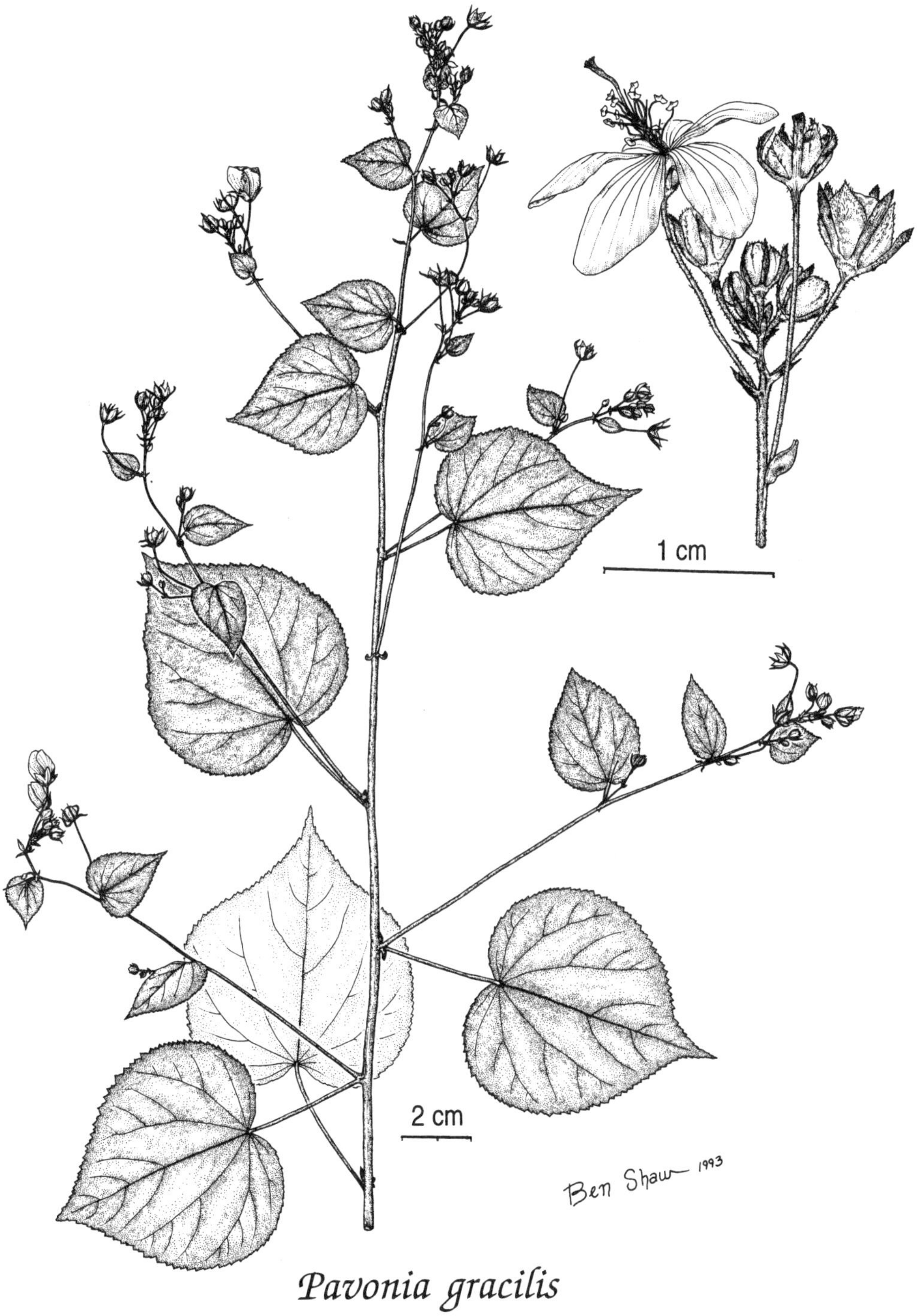

FIG. 3. *Pavonia gracilis*. Flowering branch with enlargement of flower and buds (*Krapovickas et al. 40117, 40219*)

Shrubs 1 m tall, the stems setose (hairs 2–4 mm long) and densely glandular-pubescent. *Leaf* blades ovate or triangular, mostly 3–7 × 1.5–3.5 cm, basally truncate, serrulate, acute, somewhat discolorous, palmately 5-nerved, abundantly glandular-puberulent and with appressed setae (0.5–1 mm) on both surfaces;

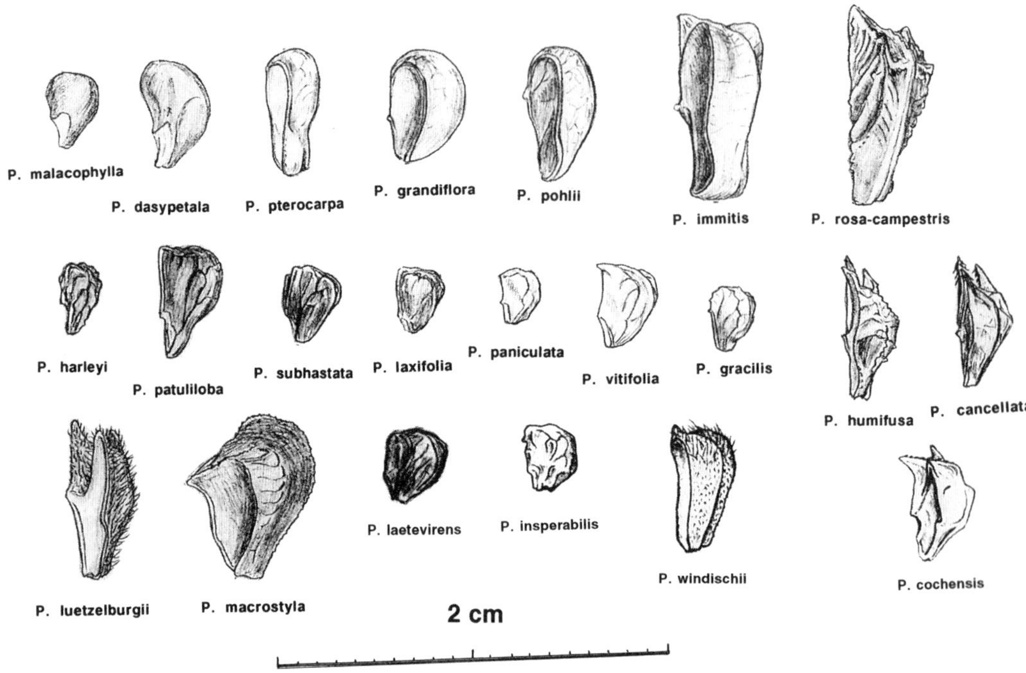

FIG. 4. *Pavonia* subgen. *Pavonia* mericarps in left lateral view. Top row (left to right): *P. malacophylla* (*Mexia 9297*), *P. dasypetala* (*Molina et al. 18198*), *P. pterocarpa* (*Damazio 1772*), *P. grandiflora* (*Hatschbach et al. 36281*), *P. pohlii* (*Heringer 14453*), *P. immitis* (*Anderson et al. 6741*), *P. rosa-campestris* (*Cid et al. 4482*). Center row (left to right): *P. harleyi* (*Lewis et al. 7654*), *P. patuliloba* (*Krapovickas et al. 13053*), *P. subhastata* (*Silverstone-Sopkin 2063*), *P. laxifolia* (*Solomon 13715*), *P. paniculata* (*King 4414*), *P. vitifolia* (*Woolston 1012*), *P. gracilis* (*Macedo 786*), *P. humifusa* (*Krapovickas 12901*), *P. cancellata* (*Haber & Zuchowski 8903*). Bottom row (left to right): *P. luetzelburgii* (*Fryxell 1935*), *P. macrostyla* (*Harley et al. 26953*), *P. laetevirens* (*Malme 3071a*), *P. insperabilis* (*Klug 4008*), *P. windischii* (*Toledo 1*), *P. cochensis* (*Johnston 18*).

petioles 1–6 cm long; stipules 3–4 mm long, subulate, viscid. *Flowers* solitary in the leaf axils or sometimes forming paniculate inflorescences, the pedicels 1–2.5 cm long; involucellar bracts 10–12, linear, 6–8 mm long, viscid and ciliate; calyx 7 mm long, ciliate; corolla 2 cm long, white, pale lavender, or yellowish fading lavender, densely white-bearded on claw in a ring at juncture with staminal column; staminal column 1.5 cm long, glabrous, more or less declined and curved with secund anthers, the filaments 1.5–2 mm long; styles exserted, free for distal 3 mm. *Fruits* 4.5 mm diam.; mericarps subglabrous, ca. 3 mm high, somewhat rugulose.

Distribution (Fig. 8). From central Brazil (Bahia), growing at elevations of 500–1250 m.

Illustration. Krapovickas (1983: fig. 3).

Representative specimens examined. BRAZIL. BAHIA: Ca. 2 km on the rd. from Lençois to BR-242, *de Carvalho & Saunders 2921* (TEX); Mun. Jacobina, vic. of Hotel Serra do Ouro, *Coradin et al. 6134* (CTES, K); 1 km E of the rd. from Lençois to BR-242, at km 8, *Coradin et al. 6485* (CTES); 34 km from Andaraí in the direction of Mucugê, *Esteves 2358* (CTES); Mun. Morro do Chapeu, Cachoeira do Ferro Doido, 29 km from Morro de Chapeu, *Esteves & Lyra-Lemos 2523* (NY, SPF); Ibyquara, Serra Sincorá region, *Fróes 20145* (F, K, NY, US); rd. from Mucugê to Guiné 5 km from Mucugê, *Furlan et al. 1959* (CTES, NY, SPF); Serra da Jacobina, Morro do Cruzeiro, *Furlan et al. 7551* (NY, SPF); Serra da Jacobina, W of Estiva, 12 km N of Senhor do Bonfim, *Harley et al. 16559* (K); about 3 km from Mucugê, *Harley et al. 20646* (CTES, K); Mun. Morro do Chapeu, 3 km SE of Morro do Chapeu, *Harley et al. 22933* (CTES, K); 10–12 km NW of Mucugê, *Hatschbach 48244* (C, CM, CTES, F, RB, S, US); Senhor do Bonfim, Serra de Santana, *G. P. Lewis et al. 7654* (F, NY, SPF); rd. between Carnaiba and Itapicuru, 22 Oct 1991, *Menezes et al. s.n.* (SPF); Mun. Mucugê, 10–12 km NW of Mucugê on the rd. to Andaraí, *Mori et al. 12695* (CTES, NY, US); vic. of Lençois, *Morim et al. 1686* (NY, TEX); Lençois, *Noblick 1770* (CTES); rd. between Andaraí and Mucugê, *Noblick & Pinto 2876* (CTES); Mucugêzinho-Palmeiras, *Noblick & Pinto 2831* (CTES); between Lençois and Itaberaba, *J. Pereira 2042* (CTES); Serra da Saúde, *Pinto 314/81* (CTES); Mun. Lençois, *Pinto 20978* (NY); rd. from Andarai to Mucugê, 4–5 km from Andarai, *Pirani et al. 2079* (K); Mun. Campo Formoso, Morro do Cruzeiro, *Thomas et al. 9687* (NY).

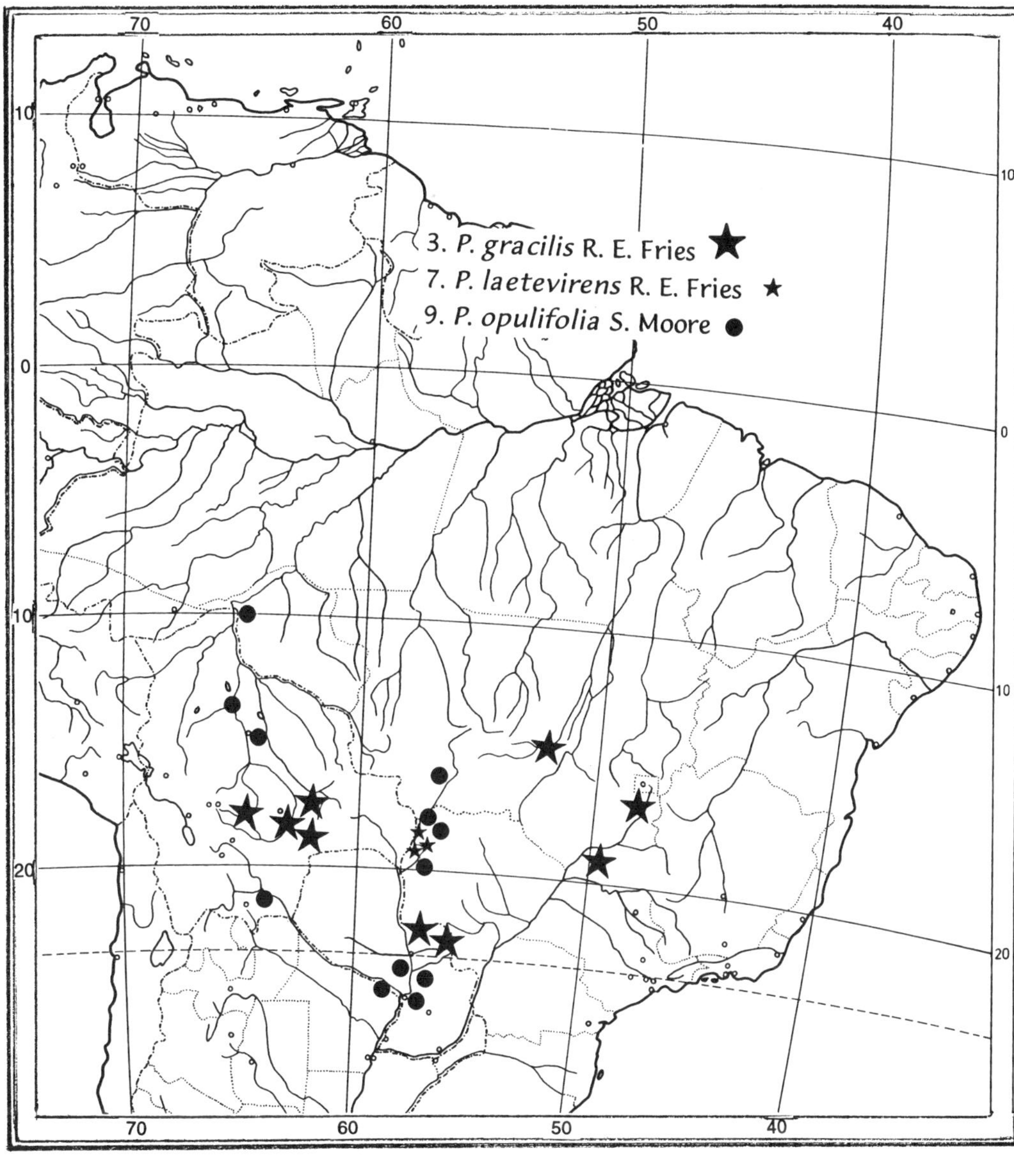

FIG. 5. Distribution map of three South American species of *Pavonia* sect. *Pavonia*.

Pavonia harleyi is similar to *P. geminiflora* in general aspect and differs from it in being more densely viscid, having truncate, narrower leaves, smaller fruits, and rounded mericarps. Differences in mericarp morphology (see Figs. 4, 53) indicate that the two species pertain to different subgenera despite their superficial resemblance. Both species have a horizontal presentation of flowers with a declined staminal column (bearded at the base) and secund anthers—i.e., a slightly zygomorphic flower—which suggests a morphological convergence as a result of a common pollinator syndrome.

5. Pavonia heterotricha Fryxell, sp. nov. Type. Bolivia. Santa Cruz: Prov. Andrés Ibáñez, La Guardia, ca. 15 km W of Santa Cruz, 1 Feb 1958, *Krapovickas 9047* (holotype, CTES).

Frutices, caulibus petiolisque heterotrichibus; laminis foliorum hastatis atque 3-lobis, in superficie superioris pilis simplicibus curvatis; floribus aggregatis in apicibus; bracteis involucellorum 5, brevi-ciliatis; calycibus 5 mm longis; corollis ca. 1 cm longis, flavidis; fructibus ignotis.

Shrubs 1.5 m tall, the stems covered with stellate hairs of two size classes (heterotrichous), the larger 0.6–0.9 mm diam., yellowish, scabridulous, scattered, the smaller 0.2–0.3 mm diam., whitish, densely overlapping. *Leaf* blades hastately 3-lobed, up to 9 × 7 cm (across basal lobes; the central lobe 2–3 cm wide), basally cordate, serrulate, acute, palmately 7–9-nerved, discolorous, the upper leaf surface with curved simple hairs 0.3–0.6 mm long, the lower surface with stellate hairs ca. 1 mm diam.; petioles 2–5 cm long, heterotrichous like stem, the larger yellowish hairs very dense distally; stipules triangular-falcate, to 7 × 3–4 mm (shorter and wider in the inflorescence), sparsely pubescent. *Flowers* aggregated at tips of branches, the pedicels 1.5–2.5 cm long, slender, stellate-pubescent and with underlying glandular puberulence; involucellar bracts 5, lanceolate, 5–7 × 1 mm, ciliate (hairs 0.5 mm); calyx 5 mm long, more densely pubescent than involucel; petals ca. 1 cm long, yellow. *Fruits* unknown.

Distribution (Fig. 2). From Bolivia, at present known only from the type collection. The type locality is at 17°55′S, 63°10′W.

As the name indicates, *Pavonia heterotricha* is distinguished by its strikingly heteromorphic indument on stems and petioles.

6. Pavonia insperabilis Fryxell, Contr. Univ. Michigan Herb. **17:** 168. 1990. Type. Peru. San Martín: environs of Chazuta forest along the Huallaga River (6°36′S, 76°11′W), 5 Apr 1986, *Knapp & Mallet 7018* (holotype, MO; isotypes, CTES, F, NY, TEX). Fig. 4

Shrubs 1–2 m tall, the stems with dense longitudinal rows of twisted hairs ca. 1 mm long and with scattered to dense spreading simple hairs (slightly longer), also with glandular hairs on very youngest growth. *Leaf* blades subhastate to ovate, mostly 4–12 × 1–8 cm (progressively reduced upward), cordate to truncate, crenate, acute, palmately 5–9-nerved, minutely pubescent; petioles 1–6.5 cm long; stipules falcate, 3–7 × 1–1.8 mm, often recurved. *Flowers* solitary in the leaf axils or forming a terminal panicle, the pedicels 1–3 cm long, scabridulous; involucellar bracts 6, narrowly ligulate, 10–12 × 0.5 mm, ciliate (hairs 1.5–2 mm long); calyx 3–5 mm long, sparsely ciliate; corolla 8–10 mm long, pale yellow, pubescent on claw; staminal column 1–2 mm long, the erect filaments 3–6 mm long; styles exserted from androecium, united below, free for distal 1–2 mm. *Fruits* oblate, ca. 4 mm diam., subglabrous; mericarps 3.5 mm high, reticulately veined.

Distribution (Fig. 8). Known from the Dept. San Martín in Peru, from elevations of approximately 250–300 m.

Illustration. Fryxell (1990: fig. 3).

Representative specimens examined. PERU. San Martín: Chazuta, Río Huallaga, *Klug 4008* (CAS, F, GH, K, MO, NY, S, US); Prov. Mariscal Cáceres, Puerto Pizana, *Schunke-Vigo 6989* (MO, NY, US), *Schunke-Vigo 4971* (F, GH, K, MO, NY, US).

Pavonia insperabilis has involucellar bracts (6 in number) that are twice the length of the calyx. It is also notable for having filaments that are longer than the staminal column.

7. Pavonia laetevirens R. E. Fries, Kongl. Svenska Vetenskapsakad. Handl. **42(12):** 47. 1908. *Pavonia subhastata* var. *laetevirens* (R. E. Fries) Hassler, Feddes Repert. Spec. Nov. Regni Veg. **8:** 30. 1910. Type. Brazil. Mato Grosso: Corumbá, *Malme II-3071* (lectotype, here designated, S; isolectotype, S). Figs. 4, 6

Pavonia opulifolia var. *major* S. Moore, Trans. Linn. Soc., ser. 2, **4:** 315. 1895. Type. Paraguay, near Corumbá [Brazil], *Moore 913* (holotype, BM as photo CTES; isotypes, B as photo F-9456, CTES [fragment]).

Annual *subshrubs* to 1.5 m tall, the stems with dense longitudinal lines of minute recurved hairs, densely to sparsely setose with simple hairs 0.5–1 mm long and with underlying minute glandular hairs especially on younger stems. *Leaf* blades ovate to more or less hastate, 3–9 × 2.5–8 cm, narrower upward, basally cordate to truncate, crenate-dentate, acute, palmately 5–7-nerved, the upper surface with appressed or suberect simple hairs 0.5–1 mm long especially on the nerves, the lower surface similar and with scattered 3-armed stellate spicules; petioles 1–4 cm long (shorter upward), densely bearded on adaxial side, otherwise sparsely setose; stipules 3–7 mm long, narrowly lanceolate, more or less reflexed, ciliate. *Pedicels* solitary in the leaf axils, 1–3 cm long, setose and with underlying minute recurved hairs; involucellar bracts 12–14, arcuate, 10–20 × ca. 0.5 mm, prominently ciliate (hairs 2–3 mm long) and with minute glandular hairs; calyx 5–7 mm long, shallowly 5-lobed, ciliate and with minute glandular hairs; petals yellow, 1.5–2 cm long(?), the genitalia included. *Fruits* 8–9 mm diam., subglabrous; mericarps 3.5–4 mm tall, obovoid, blackish, weakly reticulate-veined.

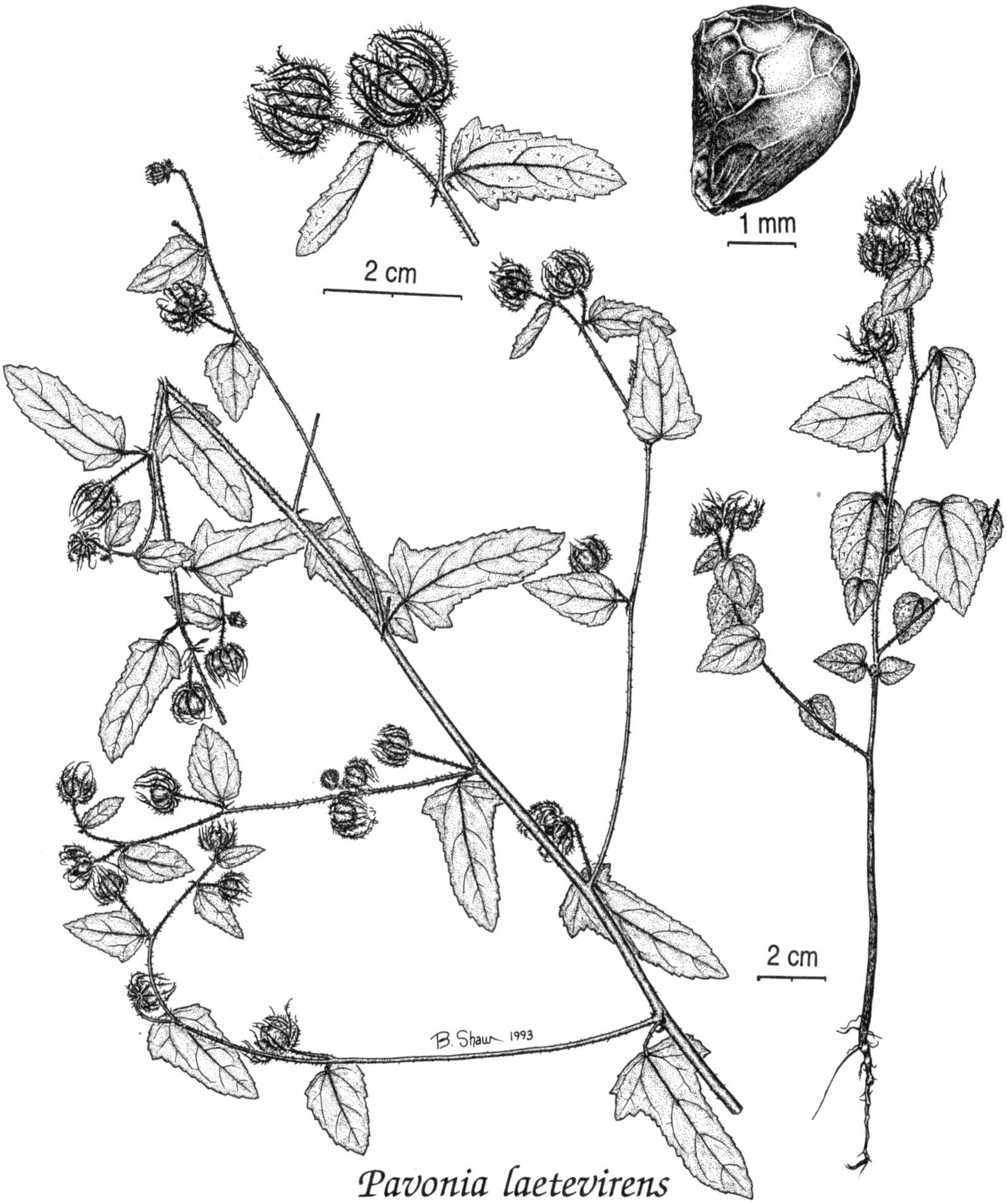

FIG. 6. *Pavonia laetevirens*. Flowering branch and small plant, with enlargements of two flowers and of mericarp (*Krapovickas & Cristóbal 34451*, *Gregory et al. 10004*).

Distribution (Fig. 5). From the general vicinity of Corumbá, Mato Grosso do Sul, Brazil. The type locality is at 19°00′S, 57°38′W.

Representative specimens examined. BRAZIL. MATO GROSSO DO SUL: Corumbá, rd. to Ladario, *Gregory et al. 10004* (CTES, NY, TEX); Fazenda Nova Campina, 30 km E of Corumbá, *Krapovickas 29807* (CTES, NY, TEX); 90 km W of Miranda, Morro do Aceite, *Krapovickas & Cristóbal 34451* (CTES, NY, TEX); Corumbá, *Malme II-3071a* (S-syntype).

Pavonia laetevirens is notable for its combination of axillary flowers, large number (12–14) of involucellar bracts, and appressed setae on the upper leaf surface.

8. Pavonia laxifolia St.-Hilaire, Fl. Bras. Merid. **1:** 226. 1827. Type. Brazil. Minas Gerais: Near S.

María de Baependy, *St.-Hilaire 550* (holotype, P as photo CTES; isotype, CTES). Fig. 4

Shrubs to 1.5 m tall, the stems pilose (hairs 2–3 mm long) and with a sparse understory of minute glandular hairs and narrow lines of dense, recurved hairs. *Leaf* blades ovate (reduced upward to narrowly triangular), sometimes hastately lobed, cordate, serrate, acute, up to 12 × 9 cm (narrower and smaller upward), both surfaces with appressed simple hairs 0.5–1.5 mm long (and also stellate hairs beneath); petioles 0.5–8 cm long; stipules 7–15 × 2–5 mm, asymmetrically ovate or falcate, reflexed. *Flowers* usually in a terminal raceme or panicle more or less above the leaves; pedicels 2–3 cm long; involucellar bracts 5–6(–8), lanceolate, 7–9 × 0.5–2 mm, ciliate; calyx 5–7 mm long, ciliate; corolla 8–10 mm long, yellow; androecium shorter than to subequal to corolla; styles glabrous, exserted from staminal column and from corolla, proximally united, distally free for 2 mm. *Fruits* 7 mm diam., minutely scabridulous, obscurely rugose. Chromosome number, $2n = 56$ (Fernández, 1981).

Distribution (Fig. 8). Found scattered across S Brazil, N Argentina, and parts of Bolivia, at elevations up to 1750 m. The type locality is at 21°58′S, 44°54′W.

Representative specimens examined. BRAZIL. Minas Gerais: Paraisópolis, *Hoehne 20058* (CTES); Belo Horizonte, Botanical Garden, *Barreto 8020* (F), *Barreto 8024* (F); 6 km NE of Ipuiuna, *Krapovickas & Schinini 35499* (CTES, MO). **Paraná:** Mun. Londrina, Leroville, Campo das Pedras, *Hatschbach & Guimarães 24848* (CTES); Mun. Arapoti, Fazenda Chapada Sto. Antonio, *Hatschbach & Ribas 54087* (CTES). **Rio de Janeiro:** São Cristóbao, km 45 between Volta Redonda and Angra dos Reis, *Krapovickas & Cristóbal 33554* (C, CTES, F, G, HAS, K, MO, NY, SPF). **São Paulo:** Ityrapina, *Gehrt 8327* (NY); near Jundiai, edge of Via Anhanguera, *Grotta 5823* (CTES, NY), *Hoehne 10735* (CTES).

BOLIVIA. La Paz: Polo-Polo, near Coroico, 1100 m, *Buchtien 3844* (US); Nor Yungas, 0.5 km below Coroico, 1750 m, *Solomon 13715* (MO, NY). **Santa Cruz:** Prov. Andrés Ibañez, Campus Universitario, *Saldias 165* (NY).

ARGENTINA. Misiones: Posadas, on streambanks of the Alto Paraná River near Santa Ana, *Ekman 174 v. 175* (MO).

Pavonia laxifolia is notable for the combination of a small number (5–6) of involucellar bracts and long, appressed setae on the upper leaf surface. It is similar to *P. pseudo-laxifolia* but is distinguished from the latter by the pubescence characters given in the key.

9. Pavonia opulifolia S. Moore, Trans. Linn. Soc., ser. 2, **4:** 314. 1895. *Pavonia subhastata* forma *opulifolia* (S. Moore) Hassler, Feddes Repert. Nov. Spec. Regni Veg. **8:** 31. 1910. Type. Brazil. Near Cuyabá, *Moore 3* (holotype, BM–n.v.).

Shrubs 1.5–3.5 m tall, the stems scabridulous with minute stellate hairs and narrow rows of fine recurved hairs. *Leaf* blades ovate to narrowly triangular to more or less 3–5-lobed, basally truncate to cordate, acute or acuminate, 5–8 × 6–8 cm, smaller and narrower upward, the upper surface with minute appressed hairs (0.3 mm long, frequently bifurcate), the lower surface densely villous on main veins and with widely scattered 3-armed stellate hairs; petioles 2–10 mm long; stipules ca. 10 × 2–4 mm, falcate or triangular, reflexed. *Flowers* solitary in the leaf axils or forming terminal panicles by apical congestion, the pedicels 1–2 cm long; involucellar bracts 10–13, linear, 10–13 mm long, ciliate (hairs 1–2 mm); calyx 6–9 mm long, hirsute; corolla yellow, 2.5 cm long. *Fruits* 7–8 mm diam., glabrous, the mericarps 5–7 mm long, strongly rugulose, often carinate, the base elongated.

Distribution (Fig. 5). Known from S Brazil (Mato Grosso, Paraná, Rondônia), Paraguay, and Argentina in the drainage of the Río Paraguay, and from Peru and Bolivia, growing in stagnant water away from the currents (Krapovickas, pers. comm.). The type locality is at 15°40′S, 56°06′W.

Representative specimens examined. PERU. San Martín: Tarapoto, *Woytkowski 35126* (F, MO).

BRAZIL. Mato Grosso: Mun. Corvurleá, Paraguai River, 14 Apr 1985, *Gasparini s.n.* (CTES); Corumbá, near Ilha do Sargento, *de Paula & Conceição 1568* (CTES); Mun. Corumbá, Jacadigo, *Pott et al. 5610* (CTES); Mun. Caceres, Retiro Jatoba, *Pott et al. 5661* (CTES); Mun. Corumbá, Fazenda Acurizal, Nabileque, *Pott et al. 4070* (CTES); Mun. Corumbá, Beira do Corixo Mutum, *Pott & Tavares 3195* (CPAP, CTES, NY). **Paraná:** San Carlos do Ivai, *Hatschbach 14337* (F, NY). **Rondônia:** Trail N of Rio Madeira, 2 km below confluence with Rio Abuña, *Prance et al. 8311* (CTES, F, K, NY, S, U, US).

BOLIVIA. Beni: Prov. Ballivián, Espíritu in the zone of the Yacuma River, *S. Beck 5251* (CTES); Prov. Moxos, San Ignacio de Moxo, *Krapovickas & Schinini 34928* (CTES); Prov. Yacuma, Espíritu, Carnaval Creek and Yacuma River, *de Moraes & Sarmiento 1306* (US). **Tarija:** San Telmo (Río Bermejo), rd. from Bermejo to Tarija, *Krapovickas et al. 18680* (CTES).

PARAGUAY. Central: Tavarory, *Zardini & C. Velásquez 27500* (CTES, MO); *Zardini & C. Velázquez 22419* (MO). **Chaco:** Riacho Negro, *Lindman A2049* (S–2). **Cordillera:** Tobatí, "Ybytú Silla" mesa, *Zardini & C. Velásquez 27351* (CTES, MO).

ARGENTINA. Formosa: Dept. Pilcomayo, Estancia Bouvier, Puesto Santa Rosa, Reserva de Flora y Fauna, *Fortunato et al. 2484* (CTES).

Pavonia opulifolia is distinguished by its relatively large, solitary flowers (corolla 2.5 cm long) with 10–13 involucellar bracts and by its usually bifurcate hairs on the upper leaf surface.

10. Pavonia paniculata Cavanilles, Diss. **3:** 135. t. 46, f. 2. 1787. *Malache paniculata* (Cavanilles) Kuntze, Rev. Gen. Pl. **1:** 70. 1891. *Lassa paniculata* (Cavanilles) Kuntze, Rev. Gen. Pl. **3(2):** 20. 1898. *Lebretonia paniculata* (Cavanilles) Britton, Bot. Porto Rico **5:** 561. 1924. Type. Peru. Near the Marañon River, *Dombey s.n.* (types, P as photo CTES, P-LA as photo CTES and as microfiche). Garilleti (1993) indicates that the fragments in the Cavanilles herbarium (MA-476073) cannot be identified as type material. Fig. 4

Hibiscus cyanogynus de Candolle, Prodr. **1:** 455. 1824. Type. *Icones Florae Mexicanae s.n.* (Torner Collection acc. no. 6331.844, Hunt Institute) (cf. photo F-30499).

Pavonia scabra C. Presl, Reliq. Haenk. **2:** 129. 1835 [non *Pavonia scabra* Bentham ex Garcke, 1881; nec *Pavonia scabra* (Vogel *in* Trew) Ciferri, 1936]. *Malache scabra* (C. Presl) Kuntze, Revis. Gen. Pl. **1:** 71. 1891 (non *Malache scabra* Vogel *in* Trew 1772). Type. Mexico. "in serris occidentalibus," *Haenke s.n.* (lectotype, here designated, PR-197901B; isolectotype, PR).

Pavonia caracasana Turczaninow, Bull. Soc. Imp. Naturalistes Moscou **31:** 188. 1858. *Malache heterophylla* (Turczaninow) Kuntze, Revis. Gen. Pl. **1:** 70.1891. Type. Venezuela. Caracas near Galipan, 1845, *Funck & Schlim 191* (holotype, KW; isotypes, BM, CTES [fragment], OXF, P as photo CTES).

Pavonia heterophylla Turczaninow, Bull. Soc. Imp. Naturalistes Moscou **31:** 188. 1858. Type. Mexico. Oaxaca: Pacific coast, 1840, *Galeotti 4092* (holotype, KW; isotype, K). The type is cited in error as "4192" by Turczaninow (Fryxell & Krapovickas, 1990).

Pavonia ageratoides Rusby, Mem. New York Bot. Gard. **7:** 297. 1927. Type. Bolivia. Canamina, *White 276* (holotype, NY).

Pavonia subtriloba Rusby, Mem. New York Bot. Gard. **7:** 298. 1927. Type. Bolivia. Canamina, *White 512* (holotype, NY).

Subshrubs or shrubs to 3 m tall, the stems stellate-pubescent with denser lines of pubescence longitudinally, and with glandular hairs especially in the inflorescence, sometimes pilose with simple hairs 2–3 mm long. *Leaf* blades mostly 6–12 cm long, as wide as long, ovate, oblong, or shallowly 3-lobed, cordate, serrate, acute or acuminate, discolorous, minutely pubescent; petioles ⅓–⅔(–1) times the length of the blades; stipules 5–10 mm long, falcate, recurved. *Inflorescence* a widely branched viscid panicle, at least the upper parts leafless; pedicels 1–2.5 cm long; involucellar bracts usually 7–9, linear, 5–10 mm long, viscid and ciliate, sometimes dark-pigmented; calyx ca. 6 mm long; corolla 1–2 cm long, pale yellow to yellow-orange; staminal column 4–5 mm long, pallid, glabrous, the filaments 4–5 mm long; styles pallid, glabrous, the stigmas reddish. *Fruits* 6–7 mm diam., subglabrous, the mericarps smooth or reticulate-veined. Chromosome number, $2n = 112$ (Krapovickas & Cristóbal, 1965; Bates, 1967).

Distribution (Fig. 2). Widely distributed from Mexico and the West Indies to Argentina, including most of the intervening countries. It is found at elevations of less than 100 m to 2500 m. It occurs in a wide variety of habitats.

Illustrations. Cavanilles (1787: pl. 46, fig. 2); Dodson & Gentry (1978: pl. 93C); Fryxell (1977: fig. 13F; 1979: figs. 3G, 6A); Fuertes (1989: pl. 13); Gürke (1892b: pl. 96, fig. 1); Robyns (1965: fig. 6B).

Representative specimens examined. **MEXICO. CHIAPAS:** Sibanil Ha', *Breedlove 9432* (CAS, F, LL, US); Mun. Ocozocuautla, 26–28 km N of Ocozocuautla, *Breedlove 23879* (LL, MO, RSA); Mun. Ocosingo, near El Real, E of Ocosingo, *Breedlove 56434* (CAS, LL); Mun. of Angel Albino Corzo, just N of Finca Cuxtepec, *Breedlove & Almeda 56982* (CAS, LL, NY); Mun. of Bachajon, 3 km N of Bachajon–Ocosingo rd. at Temo on rd. to Palenque, *Breedlove & Almeda 57190* (CAS, LL, NY); Mun. Villa Corzo, near Colonia Tres Picos, *Breedlove & Almeda 60336* (CAS, MO, NY); 3 mi S of Ishuatan, *P. A. Fryxell & Bates 891* (BH, CAS, CTES, L, NY); Escuintla, *Matuda 90* (GH, MEXU, MO, NY, US); Mun. Yajalón, Chitallik, *Ton 7229* (MEXU, MO, NY); Agua Fría, *Ton 7095* (MEXU, MO, NY); Mun. Tenejapa, Río Cruz Pilal, *Ton 5293* (MEXU, NY). **HIDALGO:** Between Pachuca and Tampico, 18–23 mi S of Huejutla, *Croat & Hannon 65988* (MO, NY). **OAXACA:** Mun. San Gabriel Mixtepec, ca. 1.5 km S of San Gabriel, 45 km N of Puerto Escondido, *Koch, P. A. Fryxell & Wendt 79484* (BR, CAS, CHAPA, ENCB, K, MEXU, MICH, NY, WIS, XAL); Chiltepec, Tuxtepec, *Martínez-Calderón 667* (MEXU, MO, NY, TEX); Mun. San Miguel Chimalapa, Las Guayabitas, *Maya 977* (CHAPA, NY); Sierra Mazateca, Mun. San Bartolomé Ayautla, La Soledad, ca. 3.5 km WNW of Ayautla, *Solheim & Reisfield 1543* (F, NY, WIS). **PUEBLA:** Distr. Huachinango, *G. E. Seler & C. Seler 3767* (GH). **QUERÉTARO:** Mun. Landa, 1.5 km NW of La Mesa de Corozo, *Rubio 2197* (IEB, NY); Mun. Landa, 11 km SE of Agua Zarca, *Rzedowski 42780* (IEB, MO, NY). **SAN LUIS POTOSÍ:** 2 km NE of Xilitla, *King 4414* (F, NY, TEX, US); Mun. Xilitla, Zacatipa, *Rzedowski 10086* (CTES, ENCB). **VERACRUZ:** Valley of Córdoba, *Bourgeau 1513* (CTES, GH, NY, S, US); Barranca de Tenampa, *Purpus 2947* (F, MO, UC, US); Mun. Zongolica, near Zomajapa, *Solheim & Reisfield 1470* (F, NY, WIS); Mun. Tlaltetela, El Limón, *F. Ventura 9339* (CHAPA, CTES, ENCB, MEXU, NY); Mun. Xalapa, Colonia Azteca, *F. Ventura 9870* (CTES, ENCB, MEXU, NY); Mun. Tlapacoyan, Cuauhtojapan, *F. Ventura 15030* (CTES, ENCB, MEXU, NY).

GUATEMALA. Gualan, *Deam 358* (F, GH, MO, NY, S). **ALTA VERAPAZ:** 10–15 km W of San Cristóbal, *L. O. Williams et al. 40578* (F, US); Sepacuité, *Maxon & Hay 3271* (US). **PETÉN:** Lake Petén Itzá, between San José and Remate, *Lundell 17264* (C, LL–2, MO, NY); rd. to Poctún, *Tún Ortíz 1668* (F, US); near Cobán, *von Türckheim 341a* (RB, US). **RETALULEU:** 5 km W of Retaluleu, *Standley 87369* (F). **SANTA ROSA:** Río María Linda, *Heyde & Lux 6070* (GH, K, NY, US); near Oratorio, *Standley 60628* (F, GH).

BELIZE. CAYO: 8.5 km SW of San Ignacio on the rd. to Chaa Creek, *E. Cabrera et al. 7699* (MEXU, TEX); 2.5 mi past Guacamallo bridge, rd. to Millionario, *Dwyer & Liesner 12308* (LL, MO, NY).

HONDURAS. EL PARAÍSO: Quebrada Dantas, *L. O. Williams & Molina 11170* (F, LL, NY, US). **YORO:** Rd. from the city of Yoro to the Buenos Aires Mtn., *C. Nelson & Martínez 1764* (TEFH).

EL SALVADOR. MORAZÁN: Montes de Cacaguatique, *Tucker 704* (K, LL, NY, US).

NICARAGUA. MATAGALPA: Along Río Malacal, 20 km NE of Matagalpa, *L. O. Williams et al. 23998* (F, LL). **NUEVO SEGOVIA:** 3.7 mi S of Las Manos, rd. to Ocotal, *Harriman 14612* (NY, OSH).

COSTA RICA. Cocos Island, Valley of Wafer, *Pittier 16243* (GH, US). **PUNTARENAS:** Río Grande de Térraba, *Herrera et al. 4964* (MO, TEX); El General Valley, *L. O. Williams et al. 28809* (F, NY, US). **SAN JOSÉ:** La Verbena near Alajuelita, *Pittier & Tonduz 9091* (C, NY, US); vic. El General, *Skutch 2926* (GH, K, MO, NY, S), *Skutch 4063* (A, K, MO, NY, S, US).

PANAMA. Without locality, *Seemann 472* (GH, S). **BOCAS DEL TORO:** Changuinola Valley, *W. H. Lewis et al. 824* (GH, K, MO, NY). **CANAL ZONE:** ½ km W of Summit Gardens, *M. Nee 9493* (BISH, LL, MO, POM, US); around Culebra, *Pittier 2140* (NY, US); Farfan beach area, *Tyson et al. 3178* (GH, MO); ca. 6 mi NW of Gamboa, *Wilbur & Teeri 13377* (GH, LL, MO, US). **CHIRIQUÍ:** Vic. of Boquete, *W. H. Lewis et al. 578* (GH, MO, NY). **COCLÉ:** El Valle de Antón, *P. H. Allen 2253* (GH, MO, US), *P. H. Allen 4221* (F, MO). **PANAMÁ:** Ca. 7–8 km N of Gamboa, *Knapp 2260* (LL, MO, NY); Matías Hernández, *Pittier 6879* (US); Barro Colorado Island, *Wetmore & Abbe 150* (GH), *Shattuck 731* (MO). **PEARL ISLANDS:** Saboga Island, *G. S. Miller 1961* (MO, US).

DOMINICAN REPUBLIC. SAMANÁ: El Valle, *Ekman 14987* (S, US); Río Yuna, near Cotuy, *Jiménez 2686* (US); Río San Juan, *Miller 1140* (US).

PUERTO RICO. Bayamon, *Sintenis 1081* (C, GH, S, US).

COLOMBIA. ANTIOQUÍA: Bello, vic. of Medellín, *Díaz & Hodge 325* (GH, US), *Apolinar-María 272* (F); Mun. Medellín, a little above Medellín, *Silverstone-Sopkin & Roldán 2412* (NY). **CAUCA:** Río Calima, La Trojita, *Cuatrecasas 16377* (F, US); San Francisco and La Vibora, near Inzá, *Lehmann 4758* (F, US); Aganche, *Pennell & Killip 6245* (NY, US). **CUNDINAMARCA:** Fusagasugá, *Pennell 2708* (GH, NY, US). **MAGDALENA:** Poponte, Magdalena Valley, *C. Allen 846* (K, MO); ca. 8 km SE of Codazzi, *Haught 3792* (US); Santa Marta, *H. H. Smith 495* (A, CAS, CM, E, F, GH, K, LL, MO, NY, PH, S, U, US). **NORTE DE SANTANDER:** Region of Sarare, Hoya del Río Chitagá, on the Cabuya, *Cuatrecasas et al. 12180* (F, GH, MO, U, US).

VENEZUELA. ARAGUA: Maracay, *Trujillo 2960* (MY, NY); El Limón, Maracay, *Ferrari 162* (F, MY, NY). **CARABOBO:** La Cumaca, near San Diego, *Alston 5909* (NY, US); Guaremales, *Pittier 9145* (GH, NY, US). **DISTRITO FEDERAL:** Guarenas, *Aristeguieta 5465* (MO, NY); near the town of Tovar, *Fendler 90* (GH, MO, PH, US); Lower Cotiza, near Caracas, *Pittier 8284* (GH, US). **LARA:** Dept. Palvecino, S of Barquisimeto, rd. to Caserio Terepaima, *P. A. Fryxell & Burandt 4324* (CTES, MO, NY, PORT); Dept. Jiménez, Yacambú National Park, at the dam site, *Steyermark & Espinoza 108753* (F, MO, NY, U, US). **MÉRIDA:** Near Mérida, *Alston 6747* (NY, U, US); Mérida, below the Mesa de la Virgen, rd. to Tabay, *Dorr & Barnett 7721* (CTES, K, NY, TEX, US). **MIRANDA:** Los Chorros, near Caracas, *Alston 5334* (NY, US); S of the Valle de Sartenejas, *Hoenicka 48* (NY); rd. between Las Canales and El Encanto, *Lasser 659* (US). **MONAGAS:** Río Guarapiche, 2 km SSW of Jusepin, *Pursell et al. 8192* (NY, US). **TRUJILLO:** Near Motatan, *Alston 6398* (NY). **ZULIA:** W of Machiques at base of Sierra Perijá, *Ginés 77* (US).

ECUADOR. AZUAY: N side of León River toward Ona, *Steyermark 53696* (NY). **BOLÍVAR:** Cerro Tiandeagote, La Chorrera, *Játiva & Epling 7* (NY, UC, US). **COTOPAXI:** Tenefuerste, Pilalo River, *Dodson & Embree 13422* (MO, US). **EL ORO:** Above Piedras toward Piñas, *Asplund 18183* (S). **GUAYAS:** Milagro, *Hitchcock 20181* (GH, NY, US), *Hitchcock 20248* (GH, NY, US). **LOS RÍOS:** Hacienda Clementina, *Harling 1670* (GB, NY, S). **TUNGURAHUA:** Agoyán, *Asplund 7641* (CAS, K, NY, S, US); Baños, *Lugo 710* (AAU, F, GB, NY, TEX).

PERU. Andes, *Poeppig 1053* (GH, OXF); Chachapoyas, *Matthews 1244* (GH, OXF). **AMAZONAS:** Santiago River, 1 km below La Poza, *Leveau 293* (MO, NY); Mendoza, *Woytkowski 8097* (GH, MO). **AYACUCHO:** Aina between Huanta and Río Apurimac, *Killip & A. C. Smith 22554* (F, NY, US). **CAJAMARCA:** San Ignacio, 2000 m, *Llatas et al. 9214* (TEX). **CUZCO:** Prov. Convención, Amaybamba, *Marin 1488* (CTES); Prov. Convención, between Potrero and Itama, *C. Vargas 10594* (CTES); between Poterero and Quillabamba, *Núñez & Cárdenas 10327* (MO). **HUÁNUCO:** Huánuco, *Asplund 13470* (NY, S); Tingo María, valley of Huallaga River, *Belshaw 3074* (F, GH, K, LL, MO, NY, UC, US); 69 km NE of Tingo María on rd. to Tocache, *A. H. Gentry et al. 37637* (F, MO, NY, TEX); trail above Chinchao, *Mexia 4151* (GH, MO); Prov. Leoncio Prado, near Aucayacu, *Schunke Vigo 10501* (MO, NY). **JUNÍN:** La Merced, *A. H. Gentry & Prance 16411* (F, MO, NY); Valley of Tulumayo River, 15 km S of San Ramón, *A. H. Gentry et al. 41500* (MO, NY); Agua Dulce to Therma, *Woytkowski 7420* (GH, MO, US). **LORETO:** 1 km S of Yurimaguas, island in Huallaga River, *Straw 2469* (RSA); Iquitos, *Ll. Williams 7933* (F, US); Aguaytia, *Woytkowski 34422* (MO). **PUNO:** San Gabo to Ollachea, *Dillon et al. 1252* (CTES, F, MO, NY, TEX).

BRAZIL. ALAGOAS: Joaquim Gomes, Fazenda Soledade, *Campêlo 1317* (CTES, NY). **CEARÁ:** Baturité, *Fernandes 2788* (CTES). **MATO GROSSO DO SUL:** Ribas do Rio Pardo, Botas River, *Krapovickas & Cristóbal 34363* (CTES). **MINAS GERAIS:** Without locality, *Gardner 4432*

(GH); Mun. Guaxupé, S.J. do Rio Pardo, *Gibbs et al. 1980* (US); Mun. Capinopolis, *Macedo 4950* (S, US). **Pará:** Río Tocantins, *Fróes 27067* (GH); Belém, *Monteiro 1001* (F). **Paraíba:** Areia, Escola de Agronomía do Nordeste, *Vasconcellos 65* (CTES). **Pernambuco:** Tapera, Jaboatão, *Pickel 2634* (GH, NY). **São Paulo:** 30 km S of São José do Rio Preto, rd. to José Bonifacio, *Krapovickas 14388* (CTES); Carioba, *Kuhlmann 751* (CTES, NY).

BOLIVIA. Mapiri, *Bang 1496* (A, C, CM, E, F, GH, MO, NY, PH, S, US); Guanai, *Rusby 1789* (GH, NY, PH, TEX, US). **Beni:** Arroyo just N of Guayaramerin, *W. R. Anderson 12097* (NY). **Cochabamba:** Antahuacana, N of Cochabamba, *Buchtien 80* (C, E, GH, K, MO, NY); Prov. Chapare, Todos Santos, *R. F. Steinbach 390* (BRIT, F, MO, NY, RSA, U, US); Incachaca–S. Antonio, *Werdermann 2094* (MO, S). **La Paz:** Yungas, *Bang 409* (E, F, GH, MO, NY, PH, U, US); Coroico-Yolosa, *S. Beck 7509* (CTES, US); Hacienda Simaco on the rd. to Tipuani, *Buchtien 5459* (GH, S, US); Inquisivi, ca. 2 km N of Licoma Pampa, *M. Lewis 88920* (LPB, MO); 5.4 km S of Yolosa, rd. to Chuspipata, *Luteyn & Dorr 13510* (NY, TEX); Prov. Sud Yungas, 2.9 km NE of Huancané, *Luteyn & Dorr 13749* (CTES, NY, TEX); 5.2 km N of Yolosa, rd. to Coroico, *Solomon et al. 12000* (CTES, F, K, MO, NY).

Local names: Caballo usa amarilla (Peru: Quechua), cadillo altea (Puerto Rico), cadillo anaranjada (Puerto Rico), majagüilla (Dominican Republic; Mexico: Oaxaca), malva del monte (Peru), purma (Peru), tablón (Colombia: Tolima), tamborcito (Colombia: Chocó), tumin chij ak' (Mexico: Chiapas, Tzeltal), tzajal (Mexico: Chiapas, Tzeltal), tzajalzak (Mexico: Chiapas, Tzeltal), yak'il ch'o (Mexico: Chiapas, Tzotzil).

Gürke (1892b: 506) proposed *Pavonia paniculata* forma *hirsuta* Gürke and *P. paniculata* forma *hirta* Gürke to distinguish, respectively, those with long simple hairs on the stems from those that lacked them. Both forms have an understory of stellate and glandular hairs, so the distinction does not seem to merit taxonomic recognition. Also, judging from the brief description, the same type of variation may be the basis for the name *Pavonia paniculata* var. *vitifolia* St.-Hilaire & Naudin (Ann. Sci. Nat. Bot., ser. 2, 18: 43. 1842), based on a collection from Abaité, Minas Gerais (specimen not located).

11. Pavonia patuliloba Hochreutiner ex Chodat & Hassler, Bull. Herb. Boiss., ser. 2, **5:** 297. 1905. *Pavonia subhastata* forma *patuliloba* (Hochreutiner) Hassler, Bull. Herb. Boiss. ser. 2, **7:** 735. 1907. Type. Paraguay. In permanent standing water near Concepción, Sep 1901, *Hassler 7305* (holotype, G; isotypes, B as photo F-9459, CTES [fragment], G, GH, K, MO, NY, S). Fig. 4

Pavonia subhastata subsp. *paludosa* Hassler, Bull. Herb. Boiss.; ser. 2, **7:** 733. 1907. Type. Paraguay. In N Chaco, Santa Elisa lat. 23°10′, Jan 1903, *Hassler 2747* (isotypes, B as photo F-9471, CTES, GH, MO, NY, S).

Pavonia subhastata forma *quinqueloba* Hassler, Bull. Herb. Boiss. ser. 2, **7:** 735. 1907. Type. Paraguay. Gran Chaco, Loma Clavel, lat. S. 23°30′, *Rojas 2469* (isotypes, K, NY, P as photo CTES).

Shrubs 2–4 m tall, the stems stellate-pubescent, the hairs 1 mm long and with dense rows of shorter (sometimes glandular) pubescence. *Leaf* blades usually hastately or palmately 3-lobed (smaller and narrower upward, ovate to subhastate), truncate or subcordate, serrate, acute or acuminate, 4–9 cm long (the central lobe seldom >2 cm wide), the upper surface with erect (but curved) simple hairs ca. 0.5 mm long, the lower surface villous on main veins and with scattered 3–4-armed stellate spicules; petioles 2–22 mm long; stipules 2–8 mm long, subulate or falcate, spreading or reflexed. *Flowers* solitary in the leaf axils, the pedicels 1.5–3 cm long; involucellar bracts 10–12, linear, 11–19 × ca. 1 mm, ciliate; calyx 8–9 mm long, ciliate; corolla yellow, 2–3.5 cm long, minutely pubescent externally, otherwise glabrous; staminal column glabrous, 15–21 mm long, the filaments 3–5 mm long; styles exserted and distinct for distal 4 mm, sometimes exceeding the petals. *Fruits* 6 mm diam., glabrous, the mericarps 5–5.5 mm long, carinate, weakly reticulate (sub)apically acute but scarcely beaked.

Distribution (Fig. 2). Argentina, Bolivia, and Paraguay, mostly at low elevation in riverine habitats. The type locality is at 23°29′S, 57°15′W.

Specimens examined. BOLIVIA. Beni: Prov. Cercado, Puerto Almacén, 8 km SW of the center of the city of Trinidad, *Saldias 584* (MO, NY).

PARAGUAY. Without locality, 1853–1856, *Palmer s.n.* (US); Colonia Risso, near the Apa River, 29 Oct 1893, *Malme s.n.* (S-2). **Central:** Tavarory, *Zardini & Velásquez 27616* (CTES, MO). **Concepción:** Itapucumí, *Rojas 2593* (CTES). **Gran Chaco:** Bahia Negro, marshes of the Paraguay River, *Rojas 13823* (CTES). **Pilar:** At the mouth of the Bermejo River, *Meyer 16114* (CTES).

ARGENTINA. Corrientes: Dept. Esquina, Isla Correntina, *Burkhart 27009* (CTES); Dept. Capital, banks of the Paraná River, *Quarín 2321* (CTES). **Formosa:** Chalma, *Jörgensen 2308* (MO); Dept. Monteagudo, Riacho Inglés, *Krapovickas et al. 13053* (C, CTES, IAC, MBM, NY); Puente Inglés, *Krapovickas et al. 13834* (BAB, CTES, LL, MO, UC).

Pavonia patuliloba is distinguished by its commonly 3-lobed leaves with appressed setae on the upper surface. The flowers are solitary with 10–12 involucellar bracts.

12. Pavonia pseudolaxifolia Fryxell, sp. nov. Type. Brazil. Distrito Federal: Fundação Zoobotânica Capueira, *Murça-Pires, N. T. Silva & Souza 9585* (holotype, CTES; isotype, CTES).

Frutices caulibus pilis stellatis pluri-brachiatis; laminis foliorum ovatis vel hastatis in superficie superioris pilis simplicibus adpressis; inflorescentiis paniculatis; bracteis involucellorum 5–6 ciliatis; corollis 1–1.5 cm longis flavidis; fructibus ca. 6 mm diametro rugulosis.

Shrubs 1.5–2 m tall, the stems stellate-pubescent (the hairs widely scattered, multi-armed, more or less appressed, 1 mm diam. or less), and with narrow longitudinal rows of simple recurved hairs. *Leaf* blades ovate to hastate, to 9 × 7 cm (smaller and narrower upward), cordate to subtruncate in smaller leaves, serrate, acute, palmately 5–9-nerved, somewhat discolorous, the upper surface with abundant appressed simple hairs ca. 0.5 mm long, the lower surface similar but with stellate hairs in addition and rough to the touch; petioles up to 3.5 cm long (reduced upward), densely bearded on adaxial side and with scattered stellate hairs like stem; stipules auriculate or falcate, ca. 4 mm long, reflexed, subglabrous. *Flowers* axillary but developed into a terminal panicle by branching and leaf reduction; pedicels 1–5 cm long, slender, sparsely stellate-pubescent; involucellar bracts 5–6, ligulate, 8–9 × 2 mm, ciliate (hairs 0.5–1 mm long); calyx 6 mm long, ciliate; petals 1–1.5 cm long, yellow (with a red center?), densely bearded at juncture of claws with androecium; staminal column 3 mm long, the filaments 4–5 mm long; styles exceeding the column by 5–8 mm, free for distal 2 mm. *Fruits* ca. 6 mm diam., the mericarps (immature) rugulose, obscurely puberulent.

Distribution (Fig. 2). At present known only from the Distrito Federal, Brazil.

Specimens examined. BRAZIL. DISTRITO FEDERAL: 30 km S of Brasilia, on rd. to Belo Horizonte, 1000 m, *Irwin et al. 18242* (CTES, NY).

Pavonia pseudolaxifolia is similar to *P. laxifolia* but differs in having stems that are stellate pubescent, lacking both long, simple hairs and glandular hairs.

13. Pavonia restiaria Bertoni, Anales Ci. Parag. **9:** 3. 1911. Type. Paraguay. In open areas on the shores of the Alto Paraná River, Puerto Bertoni, *Bertoni 6104* (specimen unknown); neotype, here designated: 40 km E of Iguaçú, *Krapovickas & Cristóbal 40993* (neotype, TEX; isoneotypes, ASU, CTES, F, HAS, K, LIL, NY, SPF, USM, WIS).

Pavonia restiaria forma *sublobata* Urban, Anales Cient. Paraguayos, ser. 1, **9:** 4. 1911. Type. Unknown.

Shrubs 1–2 m tall, the stems with narrow longitudinal lines of recurved simple hairs and with scattered coarse stellate (sometimes simple) hairs (0.5–1.5 mm long), the hairs erect or appressed. *Leaf* blades ovate (sometimes angulate to weakly 3-lobed), up to 16 × 19 cm (usually smaller, and reduced upward), basally cordate to subtruncate, serrate, acute or acuminate, palmately 7–11-nerved, stellate-pubescent (the hairs 3–6-armed, very variable in size), the upper surface also with short (0.3 mm) simple hairs; petioles 2–9(–19) cm long, much shorter than blade to equaling blade, densely and coarsely pubescent; stipules asymmetrically falcate, ca. 10 × 2–3 mm, recurved, sparsely pubescent. *Flowers* usually in a terminal panicle, more or less above the leaves; pedicels 1–2 cm long, slender, densely stellate-pubescent and often distally setose (hairs 1–2 mm long); involucellar bracts 8–9, densely setose, 10–15 mm long; calyx 6–7 mm long, prominently setose on margins and nerves; petals 1.5–2 cm long, yellow; densely bearded on claw; staminal column ca. 3 mm long, declined (?), the filaments 4–6 mm long; styles emergent from the column by ca. 1 cm, declined, free distally for ca. 2 mm. *Fruits* 8–9 mm diam., glabrous; mericarps 5–6 mm tall, narrowed basally, reticulate-angulate.

Distribution (Fig. 2). In S Brazil (Paraná), Paraguay, and NE Argentina (Corrientes, Misiones), in the drainage of the Río Paraná, growing directly on the banks of the Río Paraná or its tributaries (Krapovickas, pers. comm.). The type locality is at 26°20′S, 56°25′W.

Specimens examined. BRAZIL. PARANÁ: Mun. Foz de Iguaçú, Cataracts National Park, *Hatschbach 11217* (CTES); Mun. Guaira, Porto Novo, *Hatschbach 7889* (CTES); Iguaçú National Park, *E. Pereira 5351* (NY, RB).

PARAGUAY. ALTO PARANÁ: 8 km from Hernandarias in the direction of Pto. Presidente Stroessner, *Dlouhy & Fernández-Casas 7335* (MO, NY); Puerto Gibajas, 4 km S of Puerto Iguazú, *Krapovickas et al. 15763* (C, CTES, UC). **CAAGUAZÚ:** Hernandarias, Acaray River, *Gregory et al. 9595* (CTES, NY); Puerto Bertoni, gorge of the Paraná River, *Rojas 7925* (CTES). **ITAPUA:** Kaarendy, *Soria 702* (CTES). **MISIONES:** 10 km N of Ayolas, *Schinini & Vanni 25985* (C, CLEMS, CTES, F, FCQ, G, HAS, LIL, MO, NY).

ARGENTINA. CORRIENTES: Dept. Itazuingo, Rincón Santa María, *Tressens et al. 3432* (CTES, GB, HAS, K, NY, TEX). **MISIONES:** Dept. Candelaria, banks of the Paraná, *Bertoni 2681* (CTES); Dept. Posadas, Garupá, *Bertoni 4211* (CAS, CTES); Dept. Iguazú, cataracts, *Cuezzo & de la Sota 1604* (CTES); Eldorado, *Ferrucci et al. 559* (C, CTES, GH); Dept. San Ignacio, Corpus, Isla Pindoí, *Krapovickas 2389* (CTES, LL, NY), *Krapovickas 2354* (CTES, NY); Dept. Libertador Gral. San Martín, Puerto Rico, *Krapovickas & Cristóbal 13672* (C, CTES, LL, MBM, MO, NY, P, WIS); Dept. Candelaria, San Juan, *Montes 188* (CTES); Dept. Eldorado, Eldorado, banks of the Paraná River, *Schinini & Fernández 6113* (CTES, LIL, LP, MBM, MO, NY, RB); Dept. Montecarlo, Montecarlo, banks of the Paraná River, *Schinini & Fernández 6120* (BA, BAB, CLEMS, CTES, LIL, MO, P, RUC, SI); Dept. Iguazú, Puerto Bemberg, *Schwarz 6948*

(CAS, CTES); Dept. Iguazú, Bosetti, *Schwarz 7079* (CAS, CTES); Dept. Candelaria, San Juan, *Schwindt 31* (CTES); Dept. San Pedro: El Alcázar, *Schwindt 1426* (CAS, CTES).

Local names and uses. Bertoni (1911) proposed the domestication of this species for the production of bast fibers and of cellulose pulp for the paper industry.

Although Bertoni provided no illustration, and the location of the type specimen is unknown, his detailed description indicates a distinct species, closely allied to *Pavonia paniculata* to be sure, but evidently distinguished from it geographically, ecologically, and morphologically. Its involucel is longer than that of *P. paniculata* and its fruits are larger. It also occurs in a distinctive riverine habitat.

14. Pavonia subhastata Triana & Planchon, Ann. Sci. Nat., ser. 4, **17:** 163. 1862. Type. Colombia. near Cali, valley of the Cauca, *Triana s.n. [3115]* (lectotype, here designated, P as photo CTES; isolectotypes, BM as photo CTES, CTES [fragment], NY). Figs. 4, 7

Subshrubs to 2 m tall, the stems more or less setose (hairs 1–2 mm long) and with an understory of minute glandular hairs, often with narrow longitudinal lines of recurved hairs. *Leaf* blades narrowly hastate, basally truncate or subcordate, crenate-serrate, acute, mostly 4–10 × 1–2(–3.5) cm, both surfaces with erect simple hairs (some glandular) 0.2–0.5 mm long; petioles 0.5–4 cm long; stipules falcate, 4–6 mm long, reflexed. *Flowers* solitary in the leaf axils, the pedicels 2–4.5 cm long; involucellar bracts 13–15, linear, 10–14 mm long, ciliate; calyx 8–11 mm long, ciliate; corolla 15–20 mm long, yellow, bearded on margins of claw; staminal column 5 mm long, glabrous, pallid, the filaments 2–3 mm long; styles exserted from the column. *Fruits* 6–7 mm diam., glabrous or obscurely puberulent, the mericarps 4 mm high, rugose.

Distribution (Fig. 8). Endemic to Colombia, where it occurs in savanna and shrubland at elevations of 130–1500 m (Fuertes, 1989). The type locality is at 3°28′N, 76°35′W.

Additional illustration. Fuertes (1989: pl. 14).

Specimens examined. **COLOMBIA.** ICA, Nataima, *Hallman 10* (CTES, US). Cundinamarca: Vic. of La Mesa, *A. Fernández & Pérez-Arbeláez 496* (NY); near La Mesa, Andes of Bogotá, *Triana s.n.* (syntype, B as photo F-9470); Mun. Villeta, toward "Alto de Trigo," *Jaramillo et al. 8266* (COL, NY); between Tocaima and Pubenza, *Killip et al. 38377* (US). Los Llanos: Meta River, Maríz, *Cuatrecasas 3747* (F, US); Matacormena, *Cuatrecasas 4288* (F, US). Meta: Villavicencio, *Cuatrecasas 4722* (F, US); Popayan, *Lehmann 7827* (CAS, F, US). Valle del Cauco: Caliche, Micay River, Pacific coast, *Cuatrecasas 14182* (F); between Gorgona and Puerto Cabuyo, *Cuatrecasas 14517-A* (F, US); Cajambre River, Barco, *Cuatrecasas 17234* (F, US); Mun. Buga, Tiacuante, *Duque & Restrepo 410* (NY); around Cali, *Pittier 616* (NY, US); Mun. Cali, Barrio Olímpico, *Silverstone-Sopkin 2063* (NY).

Pavonia subhastata is characterized by more or less hastate leaves and broadly falcate stipules. It is widely disjunct from other members of *P.* sect. *Pavoniae* and may have its nearest relative in *P. cristobaliae.*

15. Pavonia vitifolia Hochreutiner *ex* Chodat & Hassler, Bull. Herb. Boiss., ser. 2, **5:** 298. 1905. *Pavonia subhastata* forma *vitifolia* (Hochreutiner) Hassler, Bull. Herb. Boiss., ser. 2, **7:** 734. 1907. Type. Paraguay. In the region of the upper Apa River, Nov 1901, *Hassler 7801* (holotype, G; isotypes, B as photo F-9474, CTES, K, MO, NY, S). Fig. 4

Pavonia malachroides Monteiro, Rev. Fac. Ci. Lisboa, ser. 2, **C. 16:** 16. 1969. Type. Brazil. Río de Janeiro: Itaguai, *Monteiro 3207* (holotype, RBR–n.v.).

Shrubs 1–3 m tall, the stems strigillose with minute appressed stellate hairs and narrow lines of dense recurved hairs. *Leaf* blades sometimes ovate, sometimes palmately lobulate, usually hastately 3-lobed, basally truncate to cordate, serrate, acute or acuminate, palmately 5–9-nerved, sparsely pubescent to subglabrous, sometimes with short simple hairs above and stellate hairs beneath, mostly 4–15 cm long; petioles 1.5–12 cm long; stipules 8–12 mm long, falcate, reflexed. *Flowers* solitary or paired in the leaf axils, the pedicels 0.5–2.5 cm long, slender; involucellar bracts (7–)10–12, linear, 9–14 mm long, ciliate; calyx 5–8 mm long, ciliate; corolla 2 cm long, yellow, pubescent on margins of claw; staminal column more or less declined, ca. 1 cm long, pallid, glabrous, the filaments 3–5 mm long; styles exserted from the column, the stigmas subequal to the petals. *Fruits* 8 mm diam., glabrous, the mericarps 4–6 mm high, rugulose, with a small (sub)apical beak.

Distribution (Fig. 8). Argentina, Bolivia, Brazil, and Paraguay, often growing on riverbanks and in swampy habitats, evidently at low elevations. The type locality is at ca. 22°20′S, 57°W; that of *Pavonia malachroides* at 22°53′S, 43°47′W.

Illustrations. Monteiro (1969: pl. 1).

Specimens examined. **BRAZIL.** Amazon River, at the mouth of the Solimões River, *Spruce 1617* (GH). Goiás: 140 km E of Rio Verde, *Gregory et al. 9950* (CTES); BR-452, 96 km E of Rio Verde, by the Corrego Pontal, *Pedersen 12127* (CTES). Mato Grosso do Sul: Campo Grande,

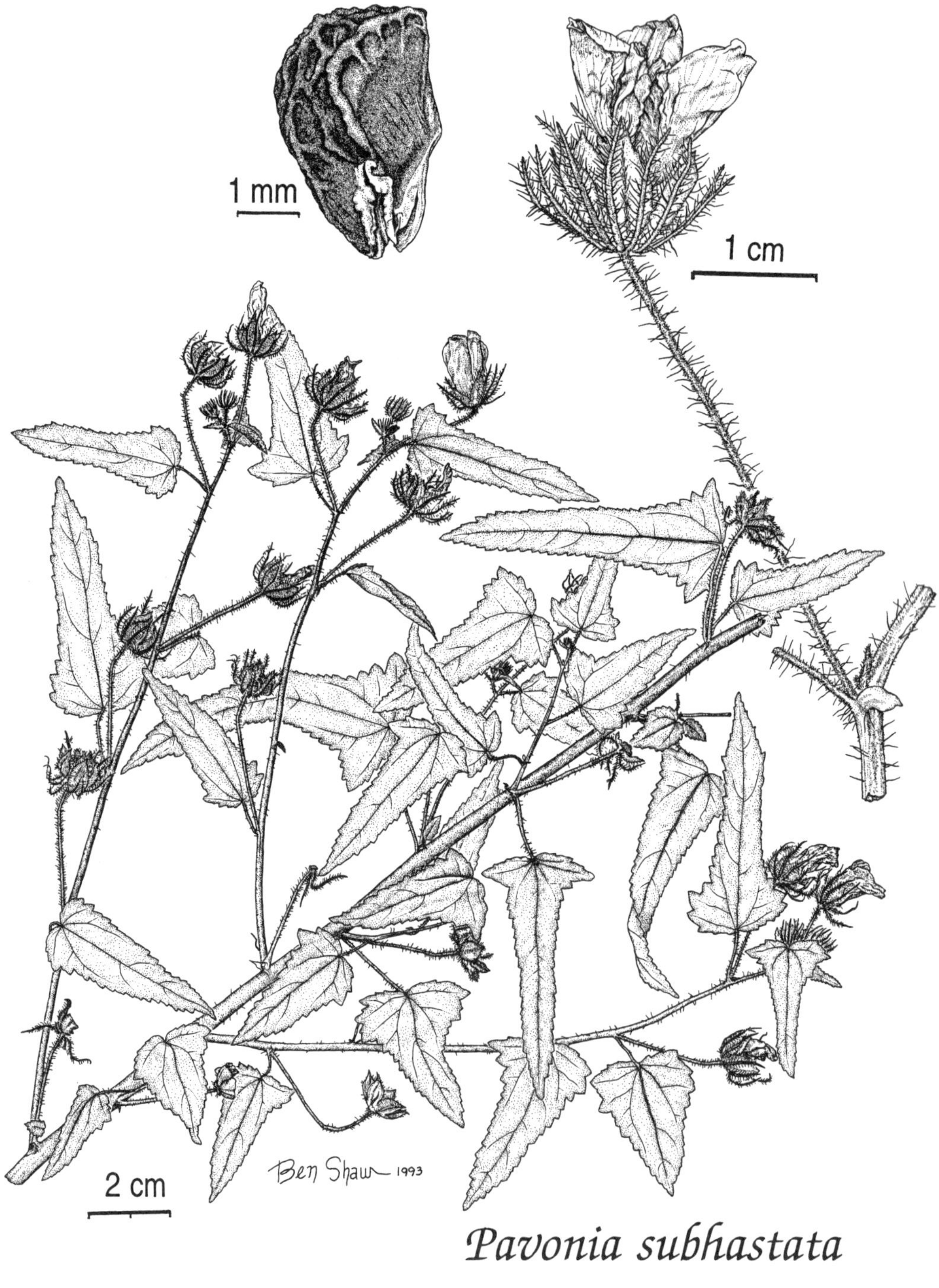

FIG. 7. *Pavonia subhastata*. Flowering and fruiting branch with enlargements of a single flower (top right) and of a mericarp (top left) (*Silverstone2063*).

Campêlo 781 (CTES, F); Mun. Novo Mundo, Rio Paraná, *Hatschbach 45224* (C, CTES); Corumbá, *Krapovickas 29785* (CTES). **Pará:** Mun. Almeirim, Mt. Dourado, *Murça-Pires et al. 703* (NY); Jari, rd. to Munguba, *N. T. Silva 2142* (NY, TEX). **Paraná:** São Mateus River, 23 km E of Cianorte, *Lindeman & de Haas 612* (K, NY).

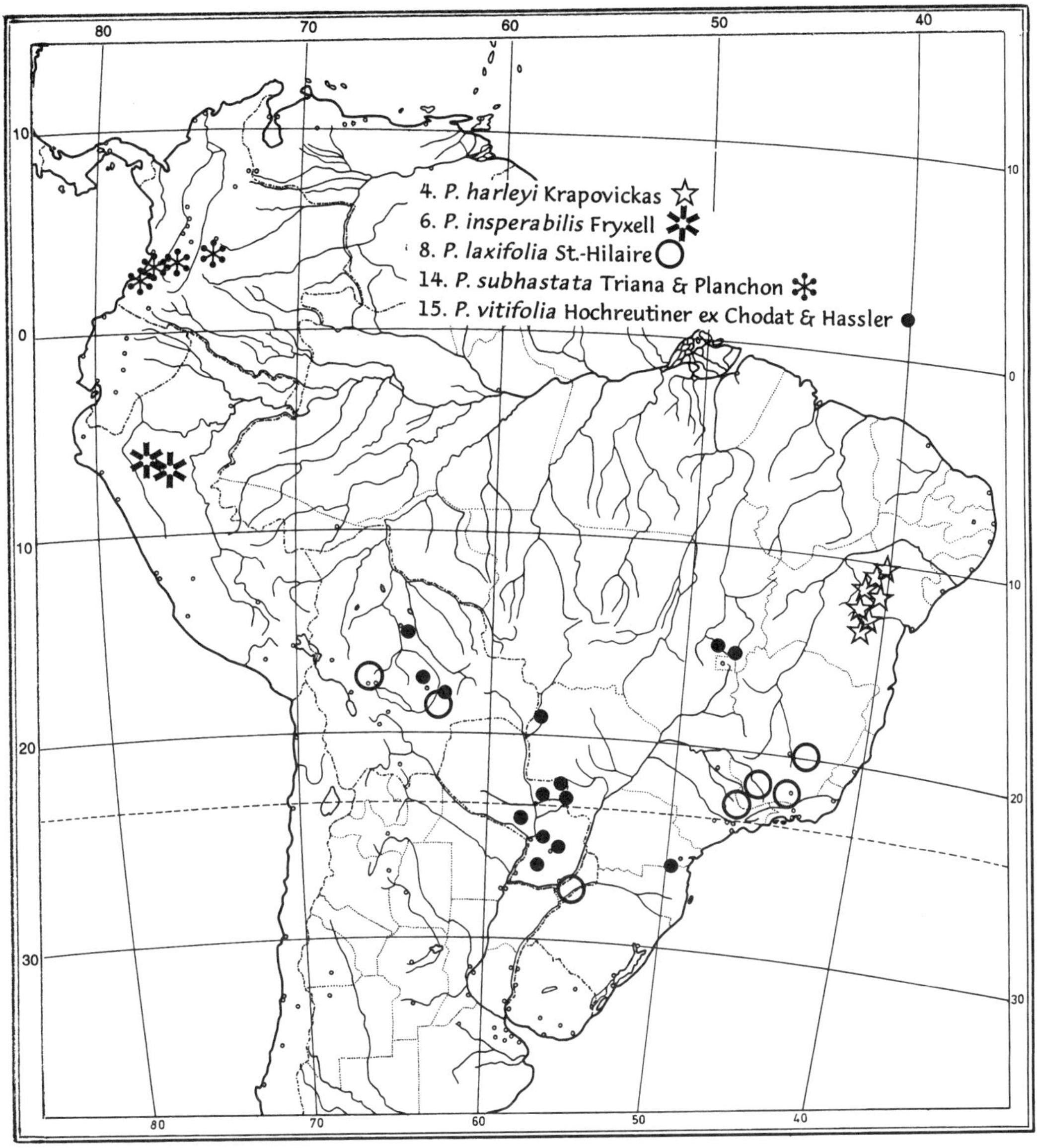

FIG. 8. Distribution map of five South American species of *Pavonia* sect. *Pavonia*.

BOLIVIA. Trinidad, *Cárdenas 22* (K). **Beni:** Arroyo just N of Guayaramerin, *W. R. Anderson & McPherson 12097* (NY, TEX, US). **Santa Cruz:** Near San Antonio-Chapare, *Cárdenas 5033* (CAS); Prov. Santiesteban, E of La Porfía, 17°03′20″S, 63°07′30″W, *M. Nee 45108* (NY, TEX).

PARAGUAY. Alto Paraguay: Chaco, *Fiebrig 1253* (CTES, S); between the Apa River and the Aquidabán River, *Fiebrig 4859* (A, E, GH, K, S); gorge of the Tepui-guazú River, *Rojas 3608* (G). **Amambay:** Toward Bella Vista, vic. of the Aquidabán River, *Mereles & Ramella 588* (CTES); Bella Vista, Apa River, *Schwarz 12214* (CTES). **Caaguazú:** 20 km N of Coronel Oviedo, *Krapovickas et al. 14311* (BAA, CTES, NY, UC). **Guirá:** Cnia. Independencia, *Pedersen 10141* (CTES). **Ñembucú:** Pilar, *Schulz 7908* (CTES, MO, NY). **Paraguarí:** From Ipacaray to Pirayú, *Schinini 4291* (BAA, BAB, CTES, F, G, GH, LIL, SI), *Schinini 4299* (BAA, CTES, LIL, LP, NY); Pte. Hayes, 88 km E from Pozo Colorado, *Schinini & Palacios 25894* (CTES, FCQ, G, MO). **San Pedro:** Cordillera de Villa Rica, *Hassler 8823* (G, GH, K, MO, NY, S); San Estanislao, Río Tapiracuay, *Krapovickas et al. 14306* (BAA, C, CTES, IAC, MBM, NY, UC); Primavera, Alto Paraguay, *Woolston 505* (K, NY), *Woolston 1012* (C, K, NY, S, U).

ARGENTINA. Chaco: Dept. 1° de Mayo, Pto. Antequera, *Schinini et al. 24326* (BAB, C, CTES, G, K, LIL, MBM, NY, UC); Dept. Bermejo, *Schulz 12342* (CTES, NY);

Dept. Bermejo, Isla Guascará, *Tressens & Schinini 1304* (CTES). **CORRIENTES:** Dept. Ituzaingó, 45 km E of Ituzaingó, *Giberti & Ahumada 40* (CTES); Dept. Capital, Facultad de Ciencias Agrarias, *Maidama 182* (CTES); Dept. Gral. Paz, 15 km E of Itá Ibaté, *Mroginski et al. 745* (CTES). **FORMOSA:** Clorinda, *Krapovickas & Schinini 28578* (CTES); Dept. Pilcomayo, Garcete, *Morel 1470* (CAS, CTES, TEX); Clorinda, *Morel 1058* (CTES, US), *Morel 1166* (CAS, CTES, TEX); NE of Puente Angelito, *Morel 7947* (CAS, CTES); 2 km NE of Laguna Primavera, *Morel 9063* (CAS, CTES); 2 km W of Riacho Negro, *Morel 3429* (CAS, CTES); Reventón, *Morel 1998* (CAS); 2 km S of Puesto Porteño, *Morel 2398* (CAS, TEX), *Morel 2105* (CTES); Laguna Oca, *Rojas 12232* (CAS, NY, S). **MISIONES:** Dept. Candelaria, San Javier, *Bertoni 5870* (CTES).

The specimens from Goiás cited above (which evidently represent the northern extreme of the species) are atypical in having flowers that are smaller than is usual in size and number of parts, but having the characteristic (almost diagnostic) scabridulous hairs on the stems.

Pavonia subgen. *Pavonia* sect. *Lopimia*

Pavonia sect. **Lopimia** (Martius) Endlicher, Gen. Pl. 982. 1840. Basionym: *Lopimia* Martius, Nova Acta Phys.-Med. Acad. Caes. Leop.-Carol. Nat. Cur. **11:** 96. 1823. Type. *Sida malacophylla* Link & Otto [≡ *Pavonia malacophylla* (Link & Otto) Garcke].

Shrubs or subshrubs, variously pubescent; leaves usually ovate-cordate, the stipules erect; involucellar bracts 10–24; corolla 1–5.5 cm long, red, rose, yellow, cream-colored, or whitish, sometimes with a dark center; fruits (where known) oblate or retuse, sometimes apiculate, winged or crested.

Type species. *Pavonia malacophylla* (Link & Otto) Garcke.

Key to the species of *Pavonia* sect. *Lopimia*

1. Calyx vestigial, concealed by involucel, 3–6 mm long (except 7–8 mm in *P. setifer*).
    2. Flowers in crowded terminal panicles, the pedicels 1–3 cm long; fruits viscid, blackish.
        3. Involucellar bracts 15–18(–24); corolla 2–3 cm long; androecium manifestly exserted; lower leaves as wide as long. . . . . 30. *P. malacophylla*
        3. Involucellar bracts 11–13; corolla 3–4.5 cm long; androecium slightly exserted; lower leaves slightly longer than wide. . . . . 22. *P. dasypetala*
    2. Flowers solitary in the leaf axils; fruits dry (non-viscid).
        4. Stems stellate-pubescent; plants seldom if at all glandular-pubescent.
            5. Leaf blades 3-lobed, cordate, as wide as long; stipules 3 mm long; involucellar bracts 13–16, 8–10 mm long; corolla 12–15 mm long; Brazil. . . . . 31. *P. piauhyensis*
            5. Leaf blades unlobed, truncate, twice as long as wide, stipules 7–13 mm long; involucellar bracts 11–14, 18–22 mm long; corolla ca. 3 cm long; Venezuela. . . . . 25. *P. imatacensis*
        4. Stems pilose (at least on lower stems); plants often glandular-pubescent.
            6. Involucellar bracts 10–12; petals 1.5–2.5 cm long, red; leaf blades broadly ovate to subrotund; calyx 4 mm long; Brazil. . . . . 34. *P. rosa-campestris*
            6. Involucellar bracts 15–18; petals 2.5–3 cm long, pinkish; leaf blades ovate or deltate to hastately lobed; calyx 7–8 mm long; Mexico. . . . . 36. *P. setifer*
1. Calyx manifest, 6–30 mm long.
    7. Staminal column manifestly exserted.
        8. Leaves elliptic, 2–3.5 times as long as wide.
            9. Corolla 3.5–6.5 cm long, white with red center; leaves 4–11 cm long; calyx 15–25 mm long; along Brazilian-Bolivian border. . . . . 38. *P. windischii*
            9. Corolla 0.7–1 cm long, red; leaves 2–5 cm long; calyx 6–7 mm long; Brazil: Minas Gerais. . . . . 35. *P. schwackei*
        8. Leaves ovate, 1–1.5 times as long as wide.
            10. Calyx 14–18 mm long; staminal column 1.2–1.5 times as long as corolla. . . . . 37. *P. spinistipula*
            10. Calyx 25–35 mm long; staminal column ca. 3 times as long as corolla.
                11. Leaf base deeply cordate, the apex acuminate; petioles 2–8 cm long; stems and petioles usually with simple hairs 3–5 mm long (sometimes lacking) and also short glandular hairs; fruits hirsute, winged. . . . . 28. *P. luetzelburgii*
                11. Leaf base truncate or subcordate, the apex acute or obtuse; petioles 0.5–5 cm long; stems and petioles densely and minutely stellate-pubescent; fruits obscurely scabridulous, not winged. . . . . 29. *P. macrostyla*
    7. Staminal column shorter than corolla.
        12. Involucellar bracts 14–23, linear, exceeding calyx, notably setose.
            13. Leaves 2–3 cm long, triangular to hastately lobed; involucellar bracts 20–23. . . . . 21. *P. blanchetiana*

13. Leaves 2–10 cm long, ovate to subhastate; involucellar bracts 14–21.
    14. Calyx 7–8 mm long; corolla 2.5–3 cm long, pinkish; involucellar bracts 10–12 mm long; Mexico. . . . . 36. *P. setifer*
    14. Calyx 10–15 mm long; corolla 3–4 cm long; involucellar bracts 15–28 mm long; South America.
        15. Involucellar bracts ca. 14; corolla red with yellow venation, the petals narrowly ligulate (<1 cm wide). . . . . 24. *P. hirticalyx*
        15. Involucellar bracts 16–21; corolla yellow with a dark red center, the petals >1 cm wide.
            16. Stems glandular-puberulent, lacking setose hairs; upper leaf surface with scattered glandular hairs. . . . . 18. *P. aschersoniana*
            16. Stems with setose hairs 1–2 mm long in addition to glandular hairs; upper leaf surface with dense stellate-pubescence. . . . . 19. *P. aschersonioides*
12. Involucellar bracts 8–13(–14), shorter than or longer than calyx, setose or not.
    17. Leaves lanceolate or elliptic to narrowly linear, 2–10 times as long as wide; petioles 2–6 mm long; involucellar bracts 7–9. . . . . 17. *P. angustifolia*
    17. Leaves broadly elliptic to ovate (sometimes lobed), 1–2 times as long as wide; petioles longer than 6 mm (except upper leaves subsessile in *P. almasana*, but then the leaves as broad as long); involucellar bracts 10–14.
        18. Leaves short-petiolate to apically subsessile, basally subtruncate, angulate to weakly lobulate; corolla 4–5.5 cm long, rose-pink with a purplish center. . . . . 16. *P. almasana*
        18. Leaves all manifestly petiolate, usually deeply cordate, seldom lobulate; corolla 2–5 cm long, white, pale lavender, pink, or red with yellow venation.
            19. Involucellar bracts prominently setose; stems pilose or setose (hairs 1–5 mm long).
                20. Involucellar bracts 8–9, spatulate (2.5–4 mm wide), red. . . . . 43. *P. decora*
                20. Involucellar bracts 11–13, linear (1 mm wide or less), green.
                    21. Leaves broadly elliptic or subrotund, basally truncate or rounded; corolla white with purple center; plants setose, of reduced stature; Paraguay. . . . . 20. *P. aspera*
                    21. Leaves ovate-cordate; corolla lavender, pink, or red with yellow venation; plants variously pubescent, usually 1 m tall or more, with setose stems (hairs 1–5 mm long); Brazil.
                        22. Leaves more or less obtuse, subentire; calyx 12–13 mm long, deeply divided. . . . . 27. *P. longitricha*
                        22. Leaves acuminate, serrate; calyx 15–25(–30) mm long, ca. half-divided. . . . . 23. *P. grandiflora*
            19. Involucellar bracts stellate- or glandular-pubescent, sometimes also setose; stems sometimes pilose but usually densely short-pubescent; Brazil.
                23. Involucel 1–1.5 cm long (half length of calyx); corolla 4–5 cm long. . . . . . . 26. *P. immitis*
                23. Involucel 1.5–2.8 cm long (⅔–1.5 times length of calyx); corolla 3.5–4.5 cm long.
                    24. Stems non-viscid; Goiás, Dist. Federal. . . . . 32. *P. pohlii*
                    24. Stems viscid; Minas Gerais.
                        25. Leaf blades deeply cordate; involucellar bracts ca. 12, 1.5–2 cm long. . . . . 33. *P. pterocarpa*
                        25. Leaf blades truncate to subcordate; involucellar bracts ca. 14, 2–2.8 cm long. . . . . 24. *P. hirticalyx*

Note: Because its placement is not certain, *Pavonia decora* is keyed out both in *P.* sect. *Lopimia* and in *P.* sect. *Malvaviscoides*.

16. Pavonia almasana Ulbrich, Notizbl. Bot. Gart. Berlin **9:** 43. 1924. Type. Brazil. Bahia: Serra das Almas, *von Lützelburg 100* (isotypes, CTES [fragment], M as photo F-19687 [specimen not found at M]). Fig. 9

Shrubs 1.5–3 m tall, the stems densely stellate-pubescent with intermingled glandular hairs. *Leaf* blades ovate, suborbicular, or weakly 3-lobed, basally truncate to subcordate, serrate, acute, mostly 2–9 × 2–8 cm, glandular-pubescent above and beneath, sometimes also with stellate hairs on lower surface, palmately 7-nerved; petioles 3.5 cm long or less, the uppermost leaves subsessile; stipules 3–4 mm long, subulate, viscid. *Flowers* solitary in the leaf axils, the

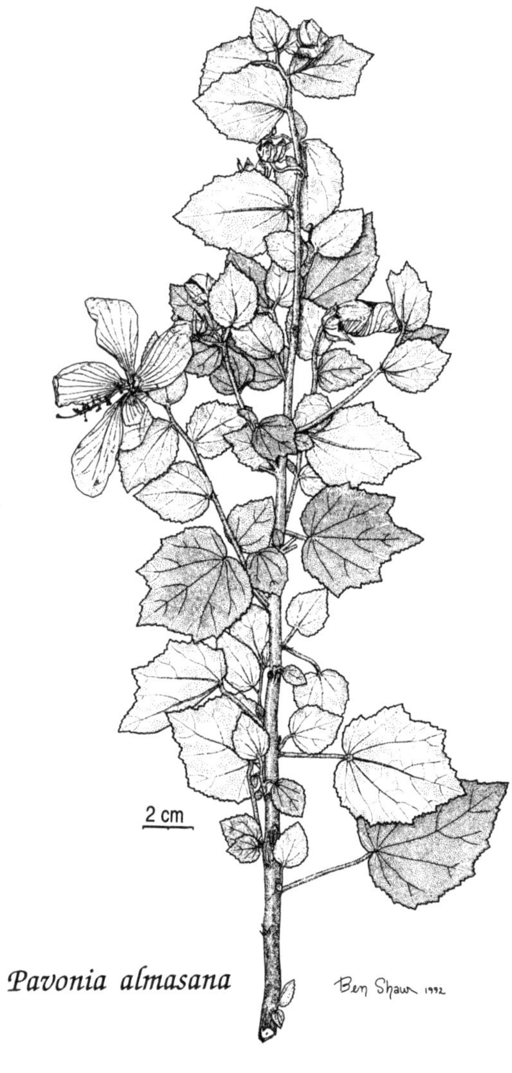

FIG. 9. *Pavonia almasana*. Flowering branch (*Mori et al. 12368*).

pedicels 2–20 mm long; involucellar bracts ca. 10, linear, 7–10 mm long, viscid and ciliate; calyx 9–12 mm long, densely tomentose, marginally ciliate; corolla 4–5.5 cm long, rose-pink with a purplish center, stellate-pubescent externally, pubescent on margins of claw, otherwise glabrous; staminal column 2 cm long, glabrous, pallid, the filaments 3–4 mm long; styles strongly exserted from column, pubescent, united below, free for distal 3–4 mm, the stigmas subequal to corolla. *Fruits* unknown.

Distribution (Fig. 10). Confined to the state of Bahia, Brazil, at elevations of 1000–1400 m. The type locality is at 14°40′S, 42°31′W.

Specimens examined. BRAZIL. Bahia: Rio de Contas, Cachoeira do Fraga, *Esteves 2382* (CTES); Rio de Contas, rd. to Cachoeira do Fraga, 3 km from Mun. do Rio de Contas, *Furlan et al. 1677* (CTES, F, K, NY, SPF); Mun. Rio de Contas, Pico das Almas, *Giulietti et al. 6861* (CTES, NY, SPF); Agua Quente, Pico das Almas, N slope, *Harley et al. 26540* (CEPEC, K, NY, SPF); Rio de Contas, *Hatschbach 46553* (C, CTES, NY, US); 4 km N of Rio de Contas, *Mori et al. 12368* (CTES, NY, RB, US); Mun. Abaíra, rd. from Catolés to Inúbia, 13 km from Catolés, *Stannard et al. 51836* (K).

This very showy species is notable for its large, richly colored flowers and for its suborbicular, short-petiolate leaves that become subsessile upward. It seems to have no close allies.

17. Pavonia angustifolia Bentham, J. Bot. **4:** 119. 1842. *Malache angustifolia* (Bentham) Kuntze, Revis. Gen. Pl. **1:** 70. 1891. Type. Guyana. Piràra, *Schomburgk 731* (holotype, K; isotypes, G as photo F-23693, GH).

Hibiscus rhomboideus Rusby, Mem. New York Bot. Gard. **7:** 301. 1927. Type. Bolivia. Pampas near Lake Rogagua, *Rusby 1383* (holotype, NY).

Few-branched *subshrubs* 1–2 m tall, the stems scabridulous with yellowish stellate hairs. *Leaf* blades elliptic to narrowly linear (rarely 3-lobed: *M. J. Jansen-Jacobs et al. 1335a*, U), mostly 3–10 × 0.3–1.5(–2) cm (2–10 times as long as wide), basally truncate to cuneate, obscurely serrulate, acute to subobtuse, palmately 3–5-nerved (the lateral nerves obscure in the narrowest leaves), harshly stellate-pubescent on both surfaces; petioles 2–6 mm long; stipules obscure (1–2 mm long). *Flowers* solitary in the axils of terminal leaves, the pedicels 3–6(–15) mm long, scabridulous; involucellar bracts 7–9, linear, 5–10 mm long, scabridulous; calyx 10–13 mm long, scabridulous, costate; corolla 3–4.5 cm long, pink to purple with a darker center, apparently glabrous throughout; staminal column 2.5–3 cm long, glabrous, pallid; styles exserted from column, united below, free for distal 2–3 mm, the stigmas subequal to (or slightly exceeding) the corolla. *Fruits* 6 mm diam., obovoid-retuse, glabrous, the mericarps crested and reticulate-veined.

Distribution (Fig. 11). From N South America (Colombia, Venezuela, and Guyana) to Bolivia and Brazil. It occurs principally in savanna vegetation. The type locality is at 3°32′N, 59°50′W. A distributional bimodality is evident in Figure 11, but this is not supported by any morphological differences.

Illustration. Gürke (1892b: pl. 96, fig. 2).

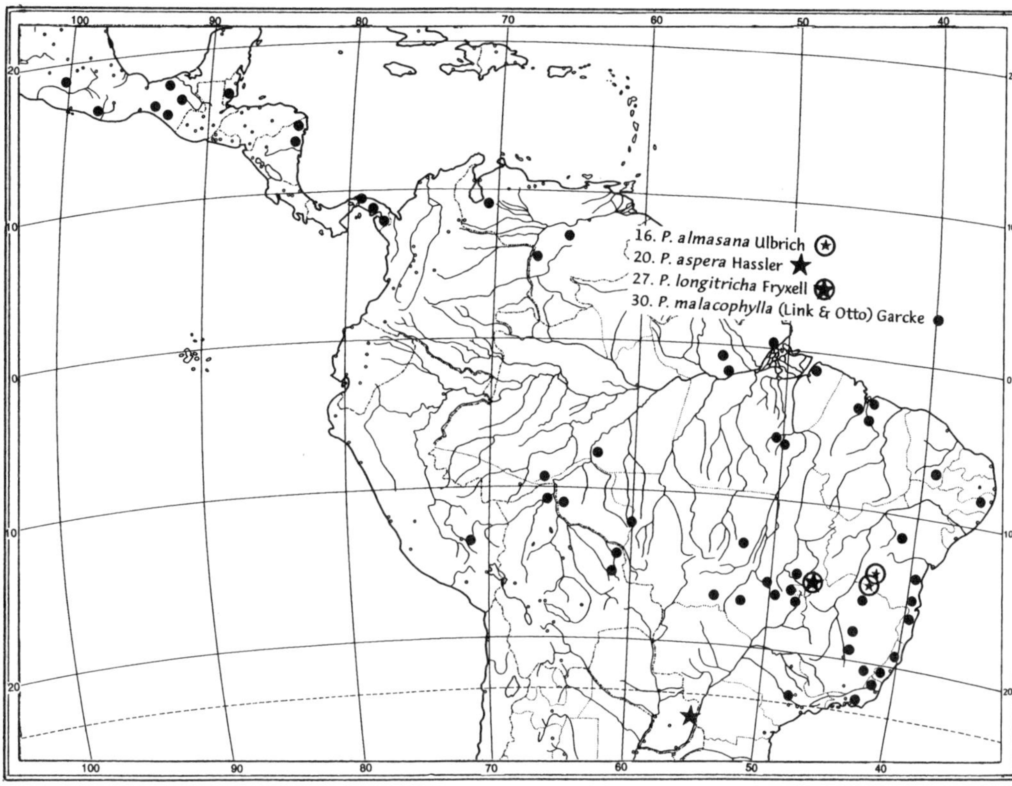

FIG. 10. Distribution map of four neotropical species of *Pavonia* sect. *Lopimia.*

Representative specimens examined. COLOMBIA. LOS LLANOS: Río Casanare, Esmeraldas, *Cuatrecasas 3949* (US).

VENEZUELA. APURE: Along Maporal Caño in Llanos, bridge site near UNELLEZ Módulo, *Phillippe et al. 21398* (ILLS); Dept. Muñoz, between Guaritico and Caicara, *Stergios et al. 2833* (CTES). BOLÍVAR: Dept. Cedeño, S of Caicara del Orinoco, *Aymard & Stergios 3287* (CTES, MO, PORT); Caño Coroso (Río Caripo), between Lake Coroso and the Orinoco River, *Wurdack & Monachino 41190* (NY, TEX).

GUYANA. Upper Takutu District, S Rupununi Savanna, *Hill 27299* (ILLS, TEX); Kanuku Mtns, near Witaru Falls, *Jansen-Jacobs et al. 63* (U, US); Rupununi savanna, Nappi village, *Jansen-Jacobs et al. 1335* (K, MO, NY, U, US); vic. of Karanambo, *Maas et al. 7689* (K, MO, NY, U); Rupununi Basin, near mouth of Charwair Creek, *A. C. Smith 2400* (NY, U, US).

BRAZIL. GOIÁS: 20 km from Ponte Alta do Norte to Porto Nacional, *Rizzo 9511* (CTES); Rio Javaés, Ilha do Bananal, *N. T. Silva 4880* (CTES, NY). MATO GROSSO: Mun. Barra do Garças, 20 km NNE of Xavantina, *G. Eiten & L. T. Eiten 9075* (US); Mun. Corumbá, Rio Paraguai-Porto Manga, *Hatschbach 29486* (C, CTES, NY, US); Mun. Aquidauana, Fazenda Santa Cruz, *Hatschbach & Guimares 21945* (CTES); Paiaguas, 40 km W of the rd. from Cuiabá to Campo Grande, *Krapovickas 29917* (CTES); 50 km SW of Poconó, *Krapovickas 29937* (CTES, F, SI); Mun. Corumbá, Fazenda Marilândia, *Pereira et al. 251* (RB); Mun. Corumbá, Fazenda Acurizal, Nabileque, *Pott et al. 388* (CTES). MINAS GERAIS: Planura, marshland near the Rio Grande, *Gregory et al. 10107* (CTES, F, NY). RORAIMA: Ilha de Maracá, Mun. Alto Alegre, SEMA Estacacão, *Hopkins et al. 537* (K, NY, TEX); Mun. Boa Vista, Reserva Ecológica de Maracá, *G. P. Lewis 1455* (E, K, NY).

BOLIVIA. BENI: Prov. Ballivián, Espíritu in the zone of the Yacuma River, *S. Beck 3298* (CTES, US); Prov. Iturralde, Luisita, *S. Beck 9909* (CTES); Prov. Ballivián, San Borja, *Krapovickas & Schinini 34871* (CTES, NY, SI); Prov. Itenez, S side of Río Guapore, ca. 15 km ESE of Costa Marques, *M. Nee 34607* (NY, TEX). SANTA CRUZ: Velasco, Noel Kempff National Park, 200 m, *Gutiérrez et al. 572* (MO, TEX); Velasco, Camamento El Refugio, 150 m, *Guillén & Coria 1561* (MO, TEX).

Cuatrecasas 3949 from Colombia is an exceptionally broad-leaved form of this species, with lower (juvenile?) leaves 3.5–4 cm wide, and the type of *Hibiscus rhomboideus* Rusby from Bolivia also has relatively broad leaves. More typically, as the name suggests, the very narrow leaves are the hallmark of *Pavonia angustifolia.*

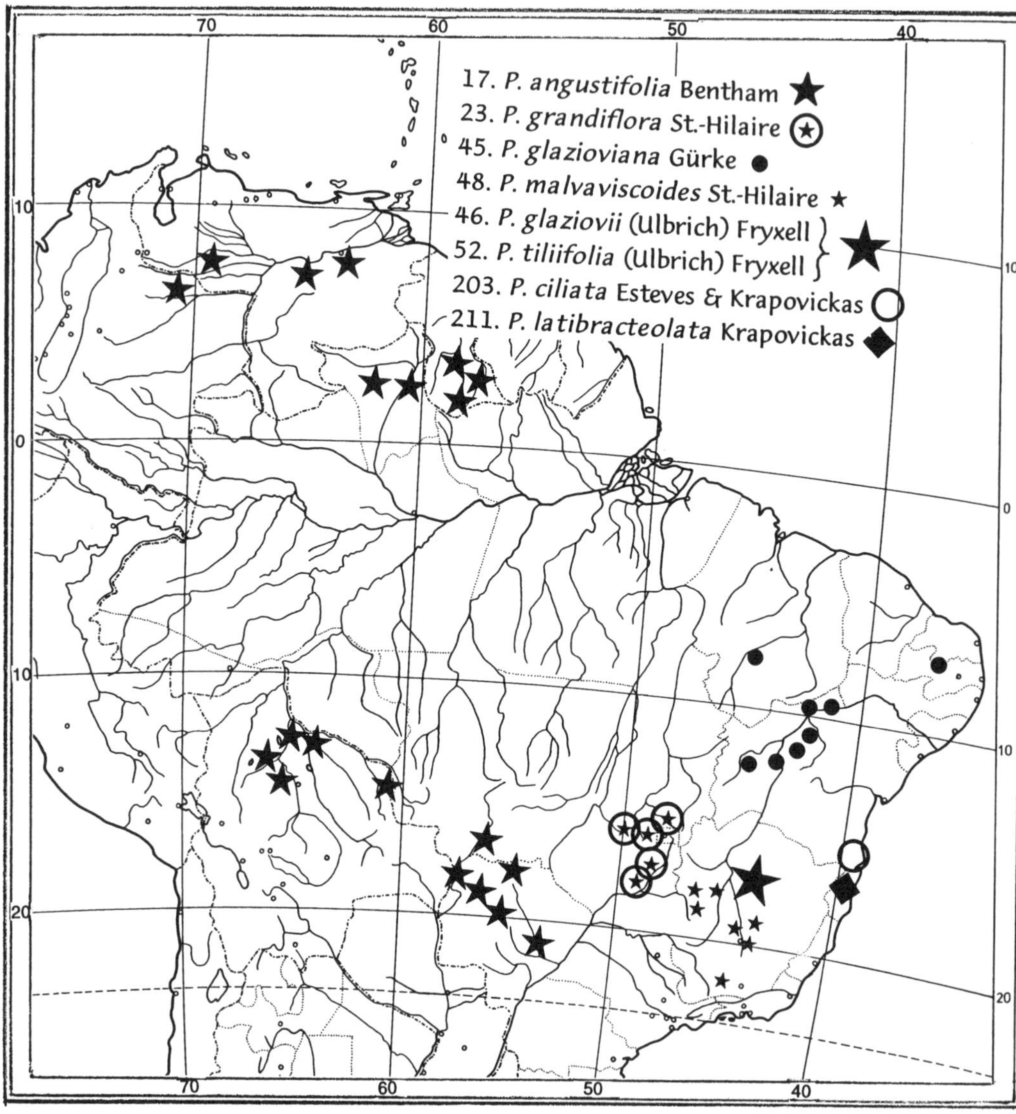

FIG. 11. Distribution map of eight South American species of *Pavonia* sect. *Lopimia*, *P.* sect. *Malvaviscoides*, and *P.* subgen. *Goetheoides*.

18. Pavonia aschersoniana Gürke *in* Martius, Fl. Bras. **12(3):** 518., t. 100., f. 1. 1892. Type. Brazil. Minas Gerais: Fanado, *von Martius s.n.* (lectotype, here designated, M as photo F-19688; isolectotype, CTES [fragment]).

Shrubs 1–2 m tall, the stems finely glandular-puberulent. *Leaf* blades ovate (sometimes weakly lobed), deeply cordate, serrate, acute, to 6 × 5 cm, markedly discolorous, sparsely glandular-puberulent above, densely stellate- and glandular-puberulent beneath, palmately 7-nerved; petioles 2–4 cm long, with pubescence like stem; stipules inconspicuous, early-caducous. *Flowers* solitary in the leaf axils, the pedicels 3–4 cm long, glandular-puberulent; involucellar bracts ca. 18, linear, 20 × 1 mm, copiously ciliate and glandular-puberulent; calyx 10–12 mm long, hirsute; petals 2.5–3 cm long, red with yellow veins, minutely stellate-pubescent externally; androecium subequal to corolla or slightly exserted, purplish; styles exserted early (protogynous?), purplish. *Fruits* unknown.

Distribution (Fig. 14). From central Brazil (Minas Gerais), at elevations of 600–800 m, growing in cerrado vegetation. The type locality (Fanado = Minas Novas) is at 17°15′S, 42°36′W.

Illustration. Gürke (1892b: pl. 100, fig. 1).

Specimens examined. **BRAZIL.** Without precise locality, *Glaziou 13544* (C, K). MINAS GERAIS: Grão Mogol, rd. to Cristália, *Pirani et al. 12985* (CTES); Grão Mogol, hwy. to João Napoles, *Hatschbach & Nicolack 54231* (CTES); without locality, 1841, *Gardner 4433*, (syntypes, CTES [fragment], GH, OXF, P as photo CTES; Alto dos Boys, *von Martius s.n.* (syntype, CTES [fragment]).

Pavonia aschersoniana and *P. aschersonioides* are closely allied, differing in pubescence characters as indicated in the key, as well as in being allopatric (Fig. 14). Both have flowers with a relatively large number (16–21) of long involucellar bracts and yellow corollas with a dark red center.

19. Pavonia aschersonioides Fryxell, sp. nov. Type. Brazil. Goiás: Serra do Caiapó, ca. 30 km (straight line) S of Caiaponia, *W. R. Anderson, Hill, Reis dos Santos & Souza 9403* (holotype, UB–n.v.; isotypes, F, NY, TEX).

Suffrutices caulibus fuscatis glanduloso-pubescentibus et setosis; laminis foliorum ovatis vel subhastatis dense stellato-pubescentibus; floribus solitariis in axillis foliorum; bracteis involucellorum 16–21 copiose ciliatis; calycibus profunde partitis; corollis flavidis centro rubro 4 cm longis; fructibus ignotis.

Shrubs 1–1.5 m tall, the stems more or less darkly pigmented, with minute glandular hairs and scattered simple hairs 1–3 mm long. *Leaf* blades ovate to subhastate, mostly 2–7 × 2–6 cm, deeply cordate, irregularly serrate, acute, discolorous, densely stellate-pubescent above and beneath; petioles ½–1 times length of blade; stipules 3–4 mm long, subulate. *Flowers* solitary in the leaf axils, the pedicels 2–5 cm long (exceeding subtending petiole at anthesis), viscid; involucellar bracts 16–21, narrowly linear, 1.5–2.5 cm long, copiously ciliate (hairs 2–2.5 mm long) and with minute glandular hairs; calyx 12–15 mm long, deeply divided, copiously ciliate; corolla yellow with dark red center, 4 cm long, glabrous externally and on margins of claw; androecium ca. half the length of the corolla, the column glabrous; styles and stigmas equaling corolla. *Fruits* unknown.

Distribution (see Fig. 14). From central Brazil (Goiás, Mato Grosso) at elevations of 600–1000 m, growing in cerrado vegetation. The type locality is at 17°10′S, 51°53′W.

Specimens examined. **BRAZIL**. GOIÁS: Ca. 50 km S of Caiapônia, *Hunt & Ramos 6228* (CTES, K); Serra do Caiapó, 50 km S of Caiapônia, rd. to Jataí, *Irwin et al. 17865* (CTES, F, NY). MATO GROSSO: Alto Araguaia, 1 Jul 1975, *Laroca s.n.* (CTES); Mun. Alto Araguaia, Hwy. MT-299 to Itiquara, 12 km from BR-364, *Windisch & Amorim 6215* (CTES); Mun. Poconé, Fazenda Nova Berlim, *Schessl 91/1-3* (ULM).

Pavonia aschersonioides is closely allied to *P. aschersoniana* (q.v.).

20. Pavonia aspera Hassler, Bull. Herb. Boiss., ser. 2, **7:** 367. 1907. Type. Paraguay. In the region of the Yhú River, *Hassler 9467* (holotype, G; isotypes, CTES, G–2). Fig. 12

Subshrubs 0.1–0.25 m tall, the stems coarsely and densely stellate-setose, the hairs 2–3 mm long, ferruginous, with understory of minute pubescence. *Leaf* blades subrotund to broadly elliptic, mostly 3–7 × 2–4 cm, basally rounded or truncate, serrate, subacute or obtuse, palmately 5-nerved, concolorous, with loose yellow stellate pubescence on both surfaces, the hairs 1–2 mm long; petioles 1–2 cm long; stipules subulate, 4–8 mm long. *Flowers* solitary in the leaf axils, the pedicels mostly 4–10 cm long, with pubescence like stem; involucellar bracts 13, filiform, 13–15 mm long, copiously setose, the hairs 2–3 mm long, yellowish; calyx 1.5–2 cm long, setose and with dense understory of fine stellate hairs; corolla white with purple center, 3 cm long; staminal column ca. 1 cm long, glabrous, the filaments 2–3 mm long. *Fruits* ca. 1 cm diam., glabrous, yellowish, reticulate-veined; mericarps ca. 6 mm high.

Distribution (Fig. 10). Confined to Paraguay, where it is evidently rare. The type locality is at ca. 25°0′S, 55°40′W.

Specimens examined. **PARAGUAY.** Estancia Primera, *Jörgensen 4846* (C, CAS [fragment], MO, NY, S, US).

The placement of *Pavonia aspera* in *P.* subsect. *Lopimia* is tentative, as it seems to have no close allies. The morphology of immature carpels suggests that it fits here better than anywhere else. It is a plant of reduced stature with characteristically elliptic leaves and coarse pubescence.

21. Pavonia blanchetiana Miquel, Linnaea **22:** 801. 1849. Type. Brazil. Without locality, *Blanchet 3662* (holotype, BR as photo CTES; isotypes, C, CTES [fragment], K, P as photo CTES, U). Fig. 13

Shrubs ca. 1 m tall, the stems stellate-scabridulous and with longitudinal lines of dense curled hairs (sometimes also setose, the hairs 2–3 mm long). *Leaf* blades triangular to hastately or palmately 3-lobed, mostly 2–3 × 1–2 cm (the central lobe ca. 1 cm wide),

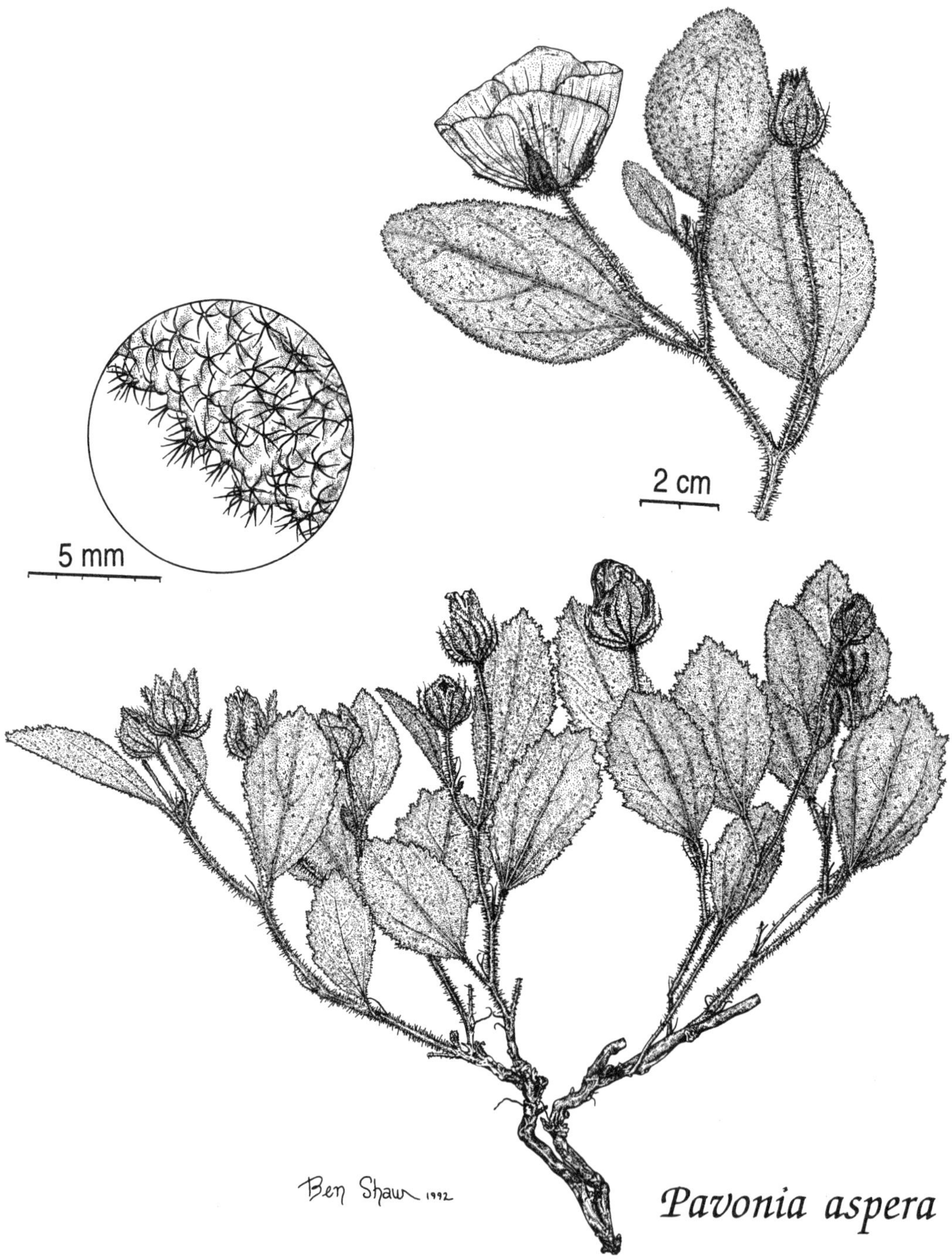

FIG. 12. *Pavonia aspera*. Plant showing reduced stature, with enlargement of leaf margin to show pubescence (*Hassler 9467*).

basally truncate to cordate, serrate, acute, palmately 5-nerved, somewhat discolorous, coarsely pubescent above and beneath, the hairs ca. 1 mm long, stellate beneath, often also with simple and bifurcate hairs above; petioles 1–1.5 cm long; stipules subulate, 3–5 mm long. *Flowers* solitary in the leaf axils, the

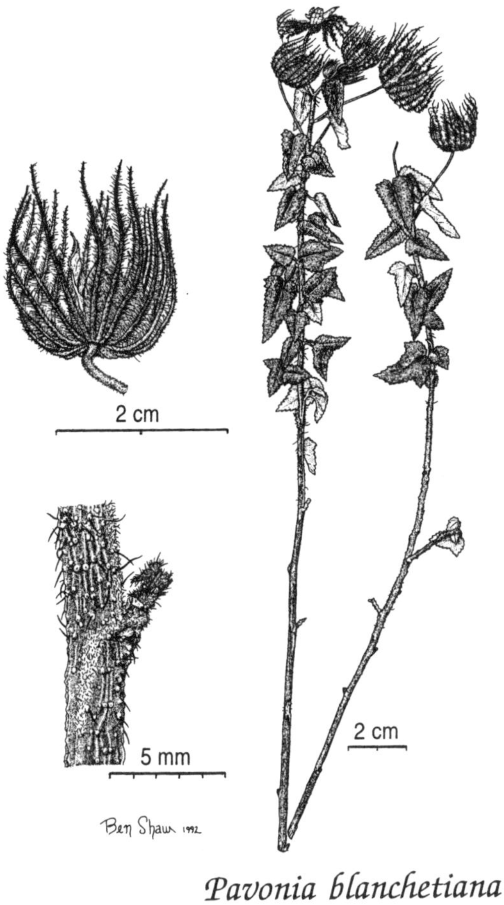

FIG. 13. *Pavonia blanchetiana.* Flowering branch (right), with enlargements of immature fruit (above left) and segment of stem (below left) (*Salgado & Bautista 351*).

pedicels 2–5 cm long; involucellar bracts 20–23, narrowly linear, 15–21 × 0.5 mm , copiously setose, the hairs 2 mm long, ferruginous; calyx 8–10 mm long, ciliate; corolla 3–3.5 cm long, pink with a dark center; staminal column ca. 1 cm long. *Fruits* unknown.

Distribution (see Fig. 14). Known from E-central Brazil (Bahia to Pernambuco), growing in savanna. The type locality is at 11°10′S, 40°28′W.

Additional illustration. Gürke (1892b: pl. 101, fig. 1).

Specimens examined. **BRAZIL. BAHIA:** Mun. Jacobina, rd. from Jacobina to Itaitu, *Amorim et al. 979* (CEPEC, NY); roadside on rd. from Serra da Monta to Itaberaba, *Bastos 317* (CTES); near Jacobina, *Blanchet 313* (G, P as photo US-5974); without locality, *von Martius 10* (CTES [fragment], M as photo CTES); Inhambupe, *Pinto 22/83* (CTES, NY); vic. of Monte Cruziero, *Rose & Russell 20052* (US); Mun. Morro do Chapéu, *Salgado & Bautista 351* (RB). **PERNAMBUCO:** Gravatá, Alto do Morro do Cruzeiro, *Andrade-Lima 62–4061* (IPA, NY); Garanhuns, Monte Sinai, *Andrade-Lima 68–5344* (F, IPA, NY); Buique, Chapada de São José, *Andrade-Lima Lima & de Ataide Sil 74–6527* (CTES).

Local names. Malvarisco (Brazil), malva-rosa (Brazil).

Pavonia blanchetiana is distinctive for the large number of involucellar bracts (perhaps the most numerous in the genus) and for its small leaves, though these are not so small as in *P. martii*. In particularly well-grown specimens of *P. blanchetiana* the leaves become prominently 3-lobed.

22. Pavonia dasypetala Turczaninow, Bull. Soc. Imp. Naturalistes Moscou **31:** 89. 1858. *Lopimia dasypetala* (Turczaninow) Standley, Contr. U.S. Natl. Herb. **18:** 114. 1916. *Malache dasypetala* (Turczaninow) Standley, J. Wash. Acad. Sci. **15:** 459. 1925. Type. Venezuela. Tachira ("Mérida"): San Cristóbal, *Funck & Schlim 1271* (holotype, not located; isotypes, CTES, G as photo F-23695, P–n.v.). Fig. 4

Pavonia velutina St.-Hilaire var. *hookeri* Planchon & Linden *ex* Triana & Planchon, Ann. Sci. Nat. Bot., ser. 4, **17:** 164. 1862. Type. Colombia. "Nouvelle Grenade, prov. de Bogotá," *Linden 852* (lectotype, here designated, BR as photo CTES; isolectotype, CTES [fragment]).

Shrubs or small trees 1–5(–7) m tall, the stems densely stellate-pubescent (hairs 0.5 mm long or less). *Leaf* blades ovate, sometimes weakly angulate or lobulate, mostly 8–25 × 7–19 cm, basally cordate, serrate, acute, palmately 7-nerved, discolorous, minutely stellate-pubescent above and beneath, more densely so on nerves; petioles mostly 8–19 cm long (shorter upward); stipules erect, 1–2.5 cm long, narrowly lanceolate. *Flowers* in crowded terminal nearly leafless panicles; pedicels 1–3 cm long, usually several per axil; involucellar bracts 11–13, valvate in bud, externally stellate-pubescent, internally ciliate-hirsute, narrowly lanceolate, 15–20 × 2–4 mm; calyx 3–4 mm long, wholly concealed by involucel; corolla 3–4.5 cm long, rose-pink (or sometimes white), funnelform, externally stellate-pubescent; androecium subequal to corolla or slightly exserted, the filaments 5–10 mm long; styles slender, purplish. *Fruits* oblate, ca. 1 cm diam., blackish and viscid at maturity; mericarps 6 mm high, carinate and reticulately veined.

Distribution (see Fig. 15). From Central America (as far N as Honduras) and N South America (Colombia and Venezuela), at elevations of 0–1200 m.

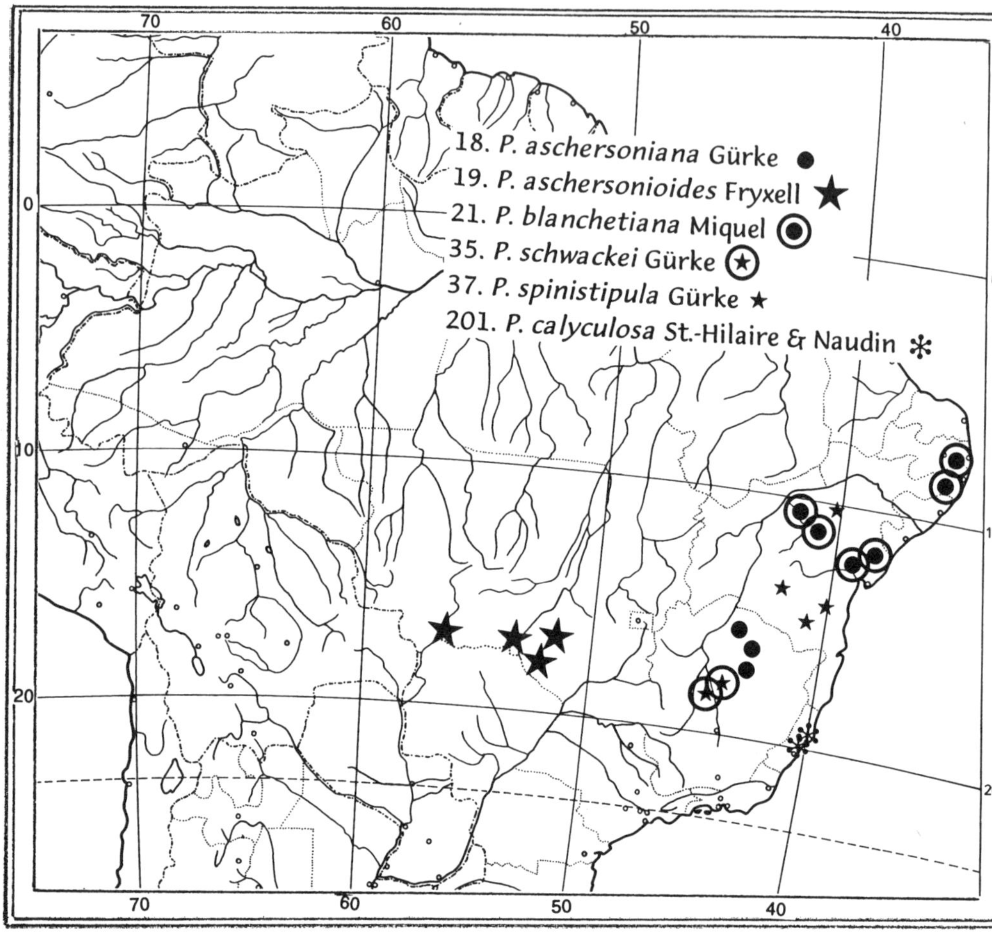

FIG. 14. Distribution map of six South American species of *Pavonia* sect. *Lopimia* and *P.* subgen. *Goetheoides.*

Representative specimens examined. HONDURAS. Cortes: Near Agua Azul, Lake Yojoa, *Molina & L. O. Williams 11302* (F, GH), *Molina 11814* (F, NY, US). **Olancho:** Between Las Marías and La Colonia, *Blackmore & Heath 1663* (BM, MO, NY); tributaries of the Wampú River, *C. Nelson & Clewell 557* (MO). **Yoro:** Bapo Grande, *Edwards P-821* (A, F, NY).

NICARAGUA. Matagalpa: Las Brisas, 15 km W of Waslala, *Moreno 19226* (HNMN, MO, NY); El Comején, *Moreno & Robleto 20555* (HNMN, MO, NY). **Río San Juan:** Bella Vista, ca. 15 km N of Boca de Sábalo, *Grijalva 2350* (HNMN, MO, NY); 1 km NW of the Santa Cruz River, *Moreno 23248* (HNMN, MO, NY). **Zelaya:** 5 km W of Rosita Empalme, *Neill 4116* (MO, NY); ca. 8 km S of Waslala, *Sandino 2437* (HNMN, MO, NY).

COSTA RICA. Alajuela: Llanura de San Carlos, *Molina et al. 17623* (F, GH). **Guanacaste:** N slope of the Orosí Volcano, *Moraga 305* (MO, TEX); lower slopes of Orosí Volcano, ca. 15 km SE of La Cruz, *Wilbur & Stone 9796* (DUKE, LL). **San José:** Shores of the Volcán River, 48 km SE of San Isidro, *A. Jiménez 3754* (C, F); vic. of San Isidro El General, *Skutch 3900* (A, K, MO, NY, S, US). **San Ramón:** Between San Miguel and La Palma, *Brenes 17051* (F).

PANAMA. Canal Zone: Pipeline Rd., 4.7 mi from gate, *Croat 12754* (CTES, MO); Barro Colorado Island, *Fosberg 55631* (NY, POM, US); 5–9 mi N of Gamboa, *A. H. Gentry 2413* (MO, NY, TEX); 8 km NW of Gamboa, *M. Nee 8430* (CTES, E, MO); between Gorgona and Gatún, *Pittier 2273* (NY, US); along Chagres River between jct. with Pequení River and Indio River, *Steyermark & P. H. Allen 16778* (MO, S). **Colón:** Hills just N of Río Guanche, *Davidse & D'Arcy 10082* (MO, NY); 13 km NE of Buenos Aires, *M. Nee 9070* (BISH, LL, MO, POM); Santa Rita Ridge Rd. (9°24′N, 79°39′W), *Sytsma 2030* (MO, NY); Santa Rita Ridge, *Wilbur et al. 15060* (F, MO, NY). **Panamá:** Cerro Azul, *Busey & Croat 253* (F, GH, LL, MO, NY, US); 5 mi SW of Cerro Brewster, *W. H. Lewis et al. 3456* (CTES, MO, UC); 2–3 mi S of Goofy Lake, rd. to

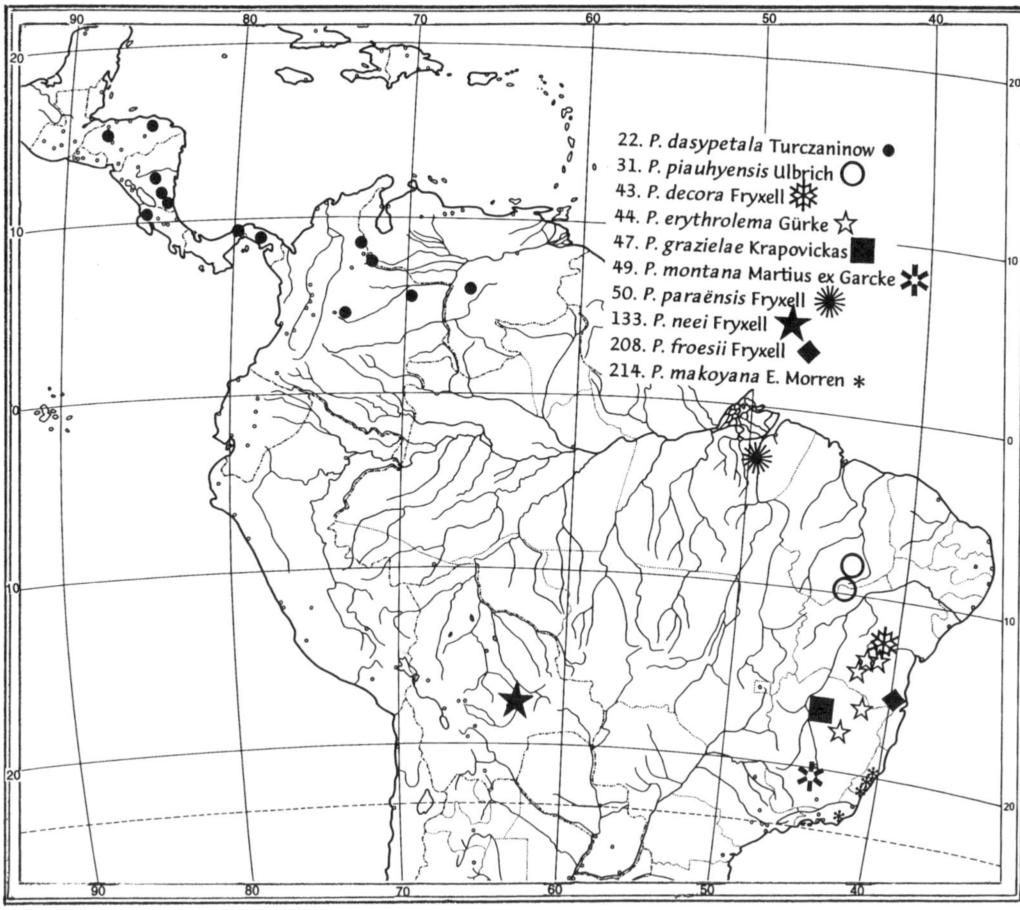

FIG. 15. Distribution map of 10 neotropical species of *Pavonia* sect. *Lopimia*, *P.* sect. *Malvaviscoides*, *P.* subgen. *Asterochlamys*, and *P.* subgen. *Goetheoides*.

Cerro Jefé, *W. H. Lewis et al. 233* (F, GH, K, MO, NY, US); Pipeline Rd., ca. 7–8 km N of Gamboa, *C. E. Smith & H. M. Smith 3296* (F, NY, US); slopes of Cerro Jefé beyond Cerro Azul ca. 5.5 mi on rd. to La Eneida, *Wilbur & Teeri 13602* (F, GH, MO, NY); Cerro Jefé, *Wilbur & Weaver 11343* (DUKE, F, GH, LL, MO, NY, US). SAN BLAS: Trail W of El Llano–Cartí rd., *de Nevers & Herrera 4476* (MO, NY).

COLOMBIA. BOYACÁ: Mun. Tauramena (Casanare), lower part of the Coyoa trail, *Uribe 3941* (US). CUNDINAMARCA: Iconozo, *Pennell 2805* (NY, US); Melgar, *Pennell 2862* (NY). LOS LLANOS: El Meta, *Haught 2461* (A, MO, US). TOLIMA: Icononzo to "Boca de Monte," rd. to Melgar, *García-Barriga 12036* (US).

VENEZUELA. AMAZONAS: Cerro Guanay, Caño Guaviarito, *Maguire et al. 31633* (NY, TEX). TACHIRA: San Cristóbal–Sto. Domingo, *Aristeguieta & Agostini 4118* (F, MO, NY, US).

Local names. Amapola (Honduras), amapolilla (Costa Rica), camilia (Panama), rosalia (Honduras).

Pavonia dasypetala and *P. malacophylla* are closely allied, sharing the character of black, viscid fruits, which character is not found elsewhere in the genus. They are also very similar in their robust growth habit (almost arborescent) and in their flower and inflorescence structure.

Roubik (1982) and McDade and Davidar (1984) describe floral phenology of *Pavonia dasypetala* and conclude that the species is autogamous, although no information is given on the proportions of self- and cross-pollination. Pollination by the Long-tailed Hermit Hummingbird, *Phaethornis superciliosus* (Trochilidae), is described, and Roubik (1982) also discusses nectar-robbing by the bee *Trigona ferricauda*.

23. Pavonia grandiflora St.-Hilaire, Fl. Bras. Merid. **1:** 232. 1827. *Malache grandiflora* (St.-Hilaire) Kuntze, Rev. Gen. Pl. **1:** 70. 1891. Type. Brazil.

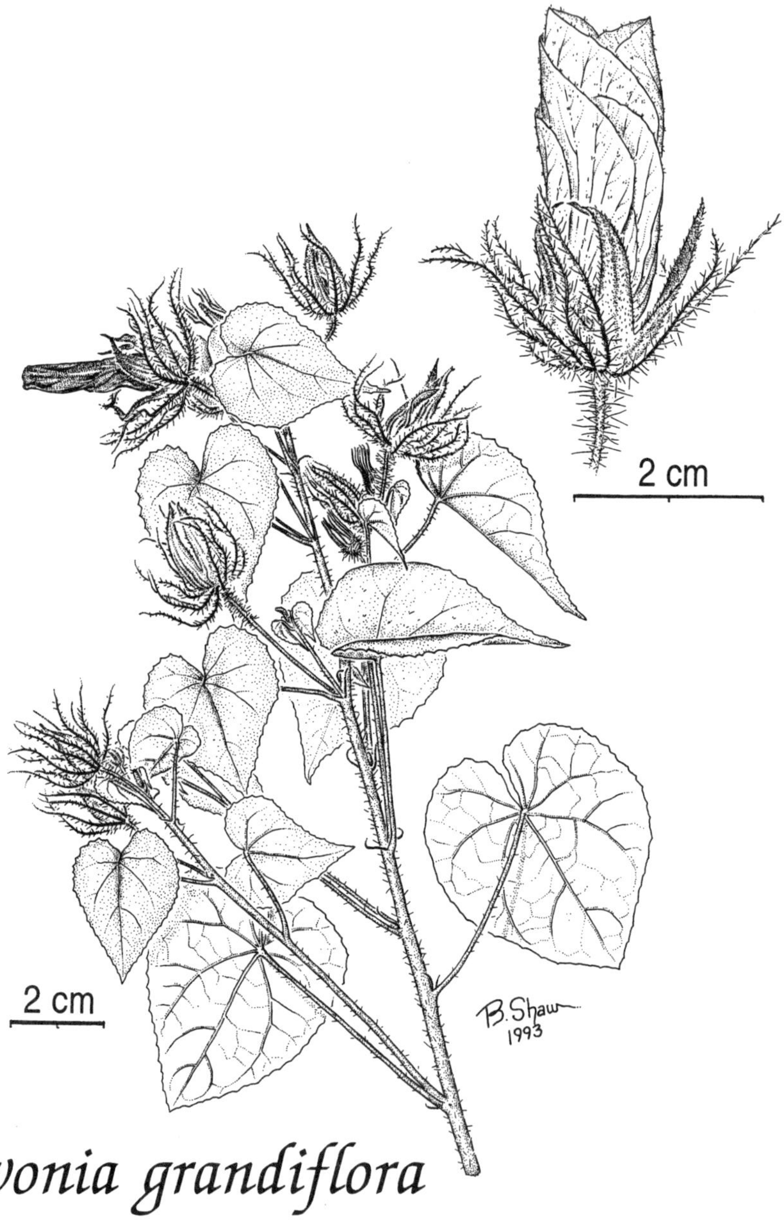

FIG. 16. *Pavonia grandiflora*. Flowering branch with enlargement of single flower (upper right) (*Hatschbach 36281*).

Goiás: Near Corumbá, *St.-Hilaire 680* (holotype, P as photo F-35474; isotype, B as photo F-9446; fragments: CTES, F). Figs. 4, 16

Shrubs 0.5–1 m tall, the stems densely setose (hairs 2–5 mm long) and with understory of scabrid stellate hairs (e.g., *Heringer 15554*) or minute glandular hairs. *Leaf* blades ovate, deeply cordate, serrate, acuminate, slightly discolorous, roughly pubescent on both surfaces, palmately 7-nerved, mostly 3–6 × 2.5–6 cm; petioles 1.5–6 cm long; stipules subulate, 2–6 mm long. *Flowers* solitary in the leaf axils (the flowers sometimes aggregated apically), the pedicels 1.5–5 cm long, with pubescence like stem; involucellar bracts

11–13, linear, 18–26 mm long, viscid and setose; calyx 15–25(–30) mm long, pubescent and ciliate; corolla 3.5–5 cm long, pink with red center; staminal column half length of corolla, glabrous; styles strongly exserted from column, subequal to corolla, united below, free for distal 1–3 mm. *Fruits* 8 mm diam., retuse, glabrous, the mericarps unornamented.

Distribution (Fig. 11). Confined to central Brazil (Goiás and the Distrito Federal), growing in cerrado vegetation at elevations of ca. 1000 m. The type locality is at 15°55′S, 48°48′W.

Representative specimens examined. BRAZIL. DISTRITO FEDERAL: RECOR, near the grounds of the Aeronautica, *Heringer et al. 416* (CTES, NY); km 20 on the rd. to the city of Eclétida, *Heringer et al. 1118* (CTES, NY); basin of the São Bartolomeu River, *Heringer et al. 4396* (IBGE, NY); Reserva das Aguas Emendadas, *Heringer-Salles 92* (IBGE, NY, TEX); ca. 10 km E of Lago Paranoá, *Irwin et al. 26532* (MO, NY, TEX); 1 km N of Barragem Paranoá, *Irwin et al. 26698* (NY, TEX). **GOIÁS:** 12 km S of Luziania, *Arbo et al. 3355* (CTES); between Morrinhos and Caldas Novas, *Gregory et al. 10145* (CTES); Mun. Morrinhos, Chapadão, *Hatschbach & Kummrow 38240* (C, CTES, MO, US, herb. Pedersen); 20 km S Alto Paraíso, *Hatschbach et al. 36281* (C, CTES, MBM, NY); Serra de Caldas Novas, *Heringer 13097* (CTES, US); Pirenópolis, Morro da Caixa dágua, *Heringer 15554* (NY, UB); Chapada dos Veadeiros, Alto de Paraíso (formerly Veadeiros), *Irwin et al. 24792* (NY, TEX, US); rd. connecting Cocalzinho with Barro Alto, *Oliveira et al. 371* (IBGE, TEX).

In the past, the name *Pavonia grandiflora* has been used for plants that are here treated under several different species names, viz. *P. grandiflora*, *P. immitis*, *P. pohlii*, and *P. spinistipula*, species that are clearly related. Some of the variation involved is discussed under *P. immitis*.

24. Pavonia hirticalyx Fryxell, sp. nov. Type. Brazil. Minas Gerais: Mun. Grão Mogol, rd. from Cristália to Cabras, 6 km W of Cristália, 700 m, 21 Jul 1985, *Martinelli et al. 11223* (holotype, RB; isotype, NY). Fig. 17

Frutices caulibus viscidis; laminis foliorum ovatis dense tomentulosis; bracteis involucellorum 14 linearis setosis quam calycis longioribus; calyce dense albo-hirsuto (pilis 2 mm longis); corollis 3.5 cm longis rubris venis flavidis, petalis ligulatis (8 mm latis); fructibus ignotis.

Shrubs ca. 2 m tall, the stems densely glandular-puberulent. *Leaf* blades ovate, 4–10 × 3–8 cm, basally truncate or subcordate, serrulate, acute or acuminate, palmately 5–7(–9)-nerved, discolorous, the upper surface densely and minutely stellate-tomentose with some glandular hairs intermixed, the lower surface densely whitish-tomentose, the epidermis obscured; petioles 1.5–6 cm long, with pubescence like stem; stipules 3 mm long, subulate, puberulent, deciduous. *Flowers* solitary in the upper leaf axils, the pedicels 2–3.5 cm long, with pubescence like stem; involucellar bracts ca. 14, linear, erect or somewhat arcuate, 20–28 × 1 mm or less, ciliate toward base (hairs 1–3 mm long), the abaxial surface densely hirsute; calyx 14–15 mm long, densely white-hirsute (hairs ca. 2 mm long), ca. half-divided, the lobes narrowly triangular; corolla 3.5 cm long, red with yellow venation, externally stellate-pubescent, the petals narrowly ligulate, ca. 8 mm wide; staminal column glabrous, reddish, 2.5 cm long, the filaments 5–6 mm long, the pollen reddish, the androecium subequal to the corolla; styles exserted from the staminal column (and slightly exserted from the corolla), reddish, glabrous. *Fruits* 6–7 mm diam., sparsely pubescent to subglabrous; mericarps 6.5 mm tall, reticulate-veined, twice height of column; seeds 4.5 mm long, minutely stellate-pubescent.

Distribution (Fig. 18). From Minas Gerais, Brazil. The type locality is at 16°43′S, 42°55′W.

Specimen examined. BRAZIL. MINAS GERAIS: Joaquim Felício, Serra do Cabral, rd. to Várzea da Palma, 31 Aug 1985, *Cavalcanti et al. s.n.* [SPF-*39755*] (CTES).

The specific epithet alludes to the distinctive, dense pubescence of the calyx. *Pavonia hirticalyx* is possibly related to *P. pterocarpa*.

25. Pavonia imatacensis Steyermark, Acta Bot. Venez. **3:** 181. 1968. Type. Venezuela. Bolívar: Sierra Imataca, Río Toro (Río Grande) between Río La Reforma and Puerto Rico, N of El Palmar, Dec 1960, *Steyermark 88149* (holotype, VEN; isotypes, F, GH, NY-2, US).

Shrubs 2–4 m tall, "sprawling or subscandent," the stems roughly and minutely stellate-pubescent. *Leaf* blades ovate or subelliptic, 6–15 × 4–9 cm, basally truncate, serrate, acute or acuminate, palmately 5-nerved, discolorous, sparsely and minutely stellate-pubescent; petioles 1–6 cm long; stipules narrowly lanceolate to falcate, 7–13 mm long. *Flowers* solitary in the leaf axils, the pedicels 2–7 cm long; involucellar bracts 11–14, lanceolate and long-acuminate, densely stellate-pubescent externally, ciliate-hirsute internally,

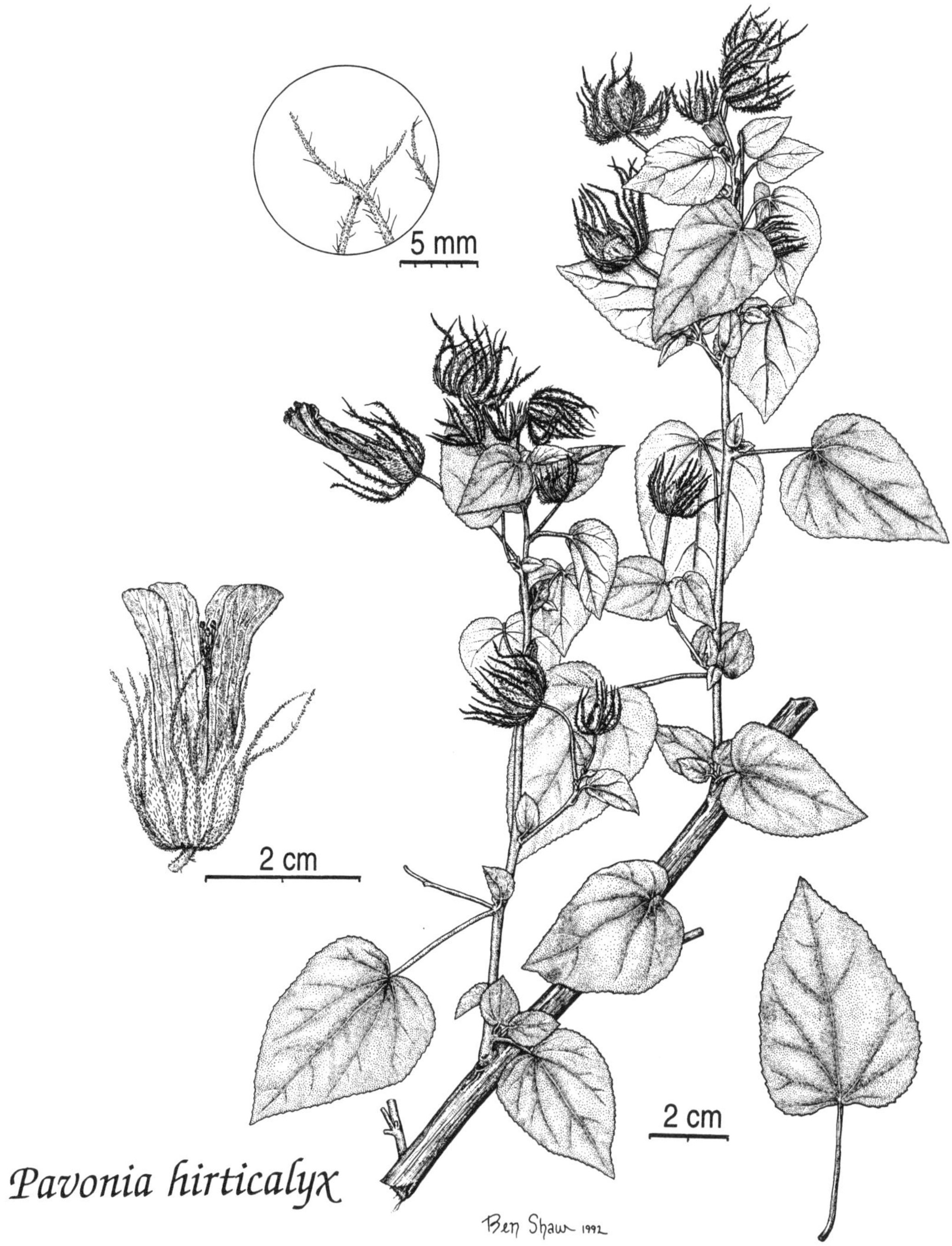

FIG. 17. *Pavonia hirticalyx*. Flowering branch with enlargements of flower (center left) and of bracteole tips (upper left) (*Martinelli et al. 11223*)

18–22 × 1–3 mm, basally connate; calyx 5–6 mm long, densely pubescent; corolla ca. 3 cm long, rose, externally pubescent, tubular; staminal column exserted for ca. 1 cm, glabrous, the filaments 3–5 mm long; styles exceeding the staminal column. *Fruits* ca. 1 cm diam., glabrous, the mericarps 6 mm high, reticulate-veined.

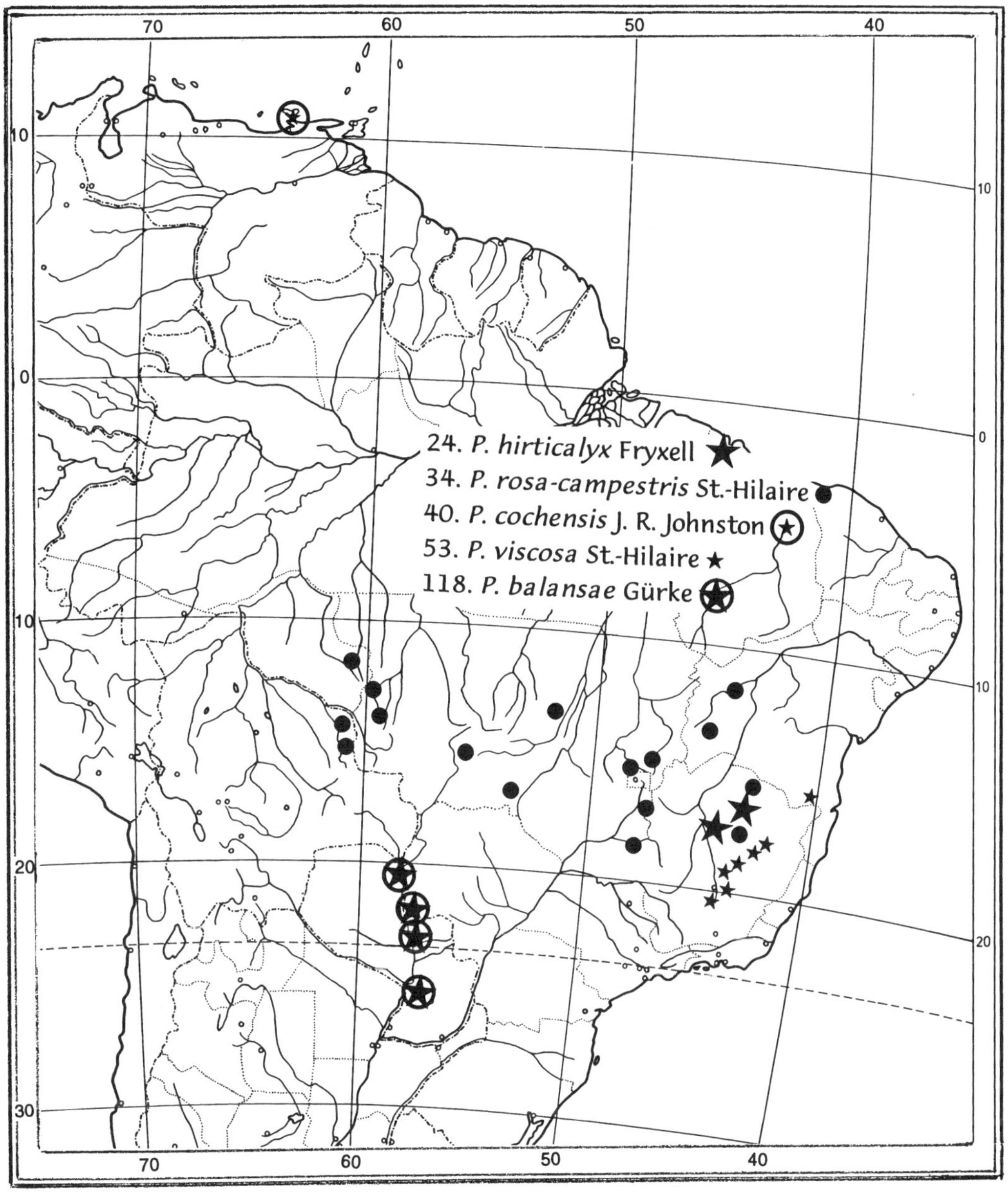

FIG. 18. Distribution map of five South American species of *Pavonia* sect. *Lopimia*, *P.* sect. *Cancellaria*, *P.* sect. *Malvaviscoides*, and *P.* subgen. *Asterochlamys*.

Distribution (see Fig. 20). Endemic to the state of Bolívar in Venezuela, where it occurs in the Sierra Imataca at elevations of 200–300 m. The type locality is at 08°04′N, 61°55′W.

Illustration. Steyermark (1968: fig. 10).

Specimens examined. **VENEZUELA. BOLÍVAR:** Sierra de Imataca, Campamento CVG–La Felicidad, *Colella et al. 1492* (NY, TEX); Gran Sabana, vic. of the Aponguao Falls, 42.5 km NE of the Mission of Santa Teresita de Kavanayén, *Steyermark et al. 117904* (F).

Pavonia imatacensis is possibly related to *P. dasypetala* and *P. malacophylla*, although it differs from these in having solitary, axillary flowers and in lacking the viscid fruits characteristic of them.

26. Pavonia immitis Fryxell, sp. nov. Type. Brazil. Goiás: Ca. 29 km by rd. N of Alto Paraíso, *W. R. Anderson et al. 6741* (holotype, UB; isotype, NY). Figs. 4, 19

Suffrutices caulibus scabrellis; laminis foliorum late ovato-cordatis, aspero-pubescentibus; bracteis involucellorum 10–13 subulatis, quam calyce dimidio longioribus; calycibus scabrellis; corollis 4–5 cm longis roseis; columna staminalis quam corolla dimidio longioribus; stylis stigmatibusque corollam subequantibus, stylis connatis praeter 2–3 mm distales; fructibus obovoideis retusis glabris, leniter reticulatis.

Shrubs 0.5–1 m tall, the stems densely scabridulous, the hairs 0.5 mm long or less, stellate (rarely also setose with simple hairs 2–5 mm long, e.g., *W. R. Anderson et al. 7757*). *Leaf* blades broadly ovate, deeply cordate, serrulate, acute or acuminate, 3–7 × 3–8 cm, palmately 7–9-nerved, discolorous, stellate-pubescent above and beneath, more roughly and less densely so above; petioles 1–3 cm long, with pubescence like that of stems; stipules 3–4 mm long, subulate, pubescent. *Flowers* solitary or paired in the leaf axils, often aggregated distally; pedicels 1–4 cm long, densely pubescent; involucellar bracts 10–13, subulate, 7–9 mm long (ca. half-length of calyx), pubescent, more or less arcuate; calyx 1.5–3 cm long, densely scabridulous, ca. half-divided; corolla 4–5 cm long, pink, stellate-pubescent externally; staminal column ca. half-length of petals, pallid, glabrous, the stamens subsessile or on filaments 1 mm long; styles strongly exserted from staminal column, united proximally, free for distal 2–3 mm, covered with minute white hairs, the stigmas capitate, villous, subequal to petals. *Fruits* 10–12 mm diam., obovoid, retuse, straw-colored, glabrous, weakly reticulate-veined; mericarps 8 mm high; seeds solitary, 4 mm long, basally pubescent.

Distribution (Fig. 20). Confined to the state of Goiás, Brazil, where it grows in cerrado vegetation at elevations of 1000–1800 m. The type locality is at 13°58′S, 47°34′W.

Specimens examined. BRAZIL. Goiás: Chapada dos Veadeiros, ca. 20 km by rd. N of Alto Paraíso, *W. R. Anderson et al. 6226* (NY, TEX); 7 km by rd. N of Alto Paraíso, *W. R. Anderson et al. 6234* (NY, TEX); ca. 25 km by rd. N of Alto Paraíso, *W. R. Anderson et al. 6666* (NY, TEX); Serra Geral do Paraná, 3 km by rd. S of São João da Aliança, *W. R. Anderson et al. 7757* (NY, TEX); Serra Geral do Paraná, ca. 5 km by rd. NE of São João da Aliança, *W. R. Anderson, P. A. Fryxell et al. 9354* (NY, TEX); 5 km N from Alto Paraíso, *Arbo et al. 3609* (CTES, NY); 9 km W de Alto Paraíso, rd. to Niquelandia, *Arbo et al. 3636* (CTES, K); Chapada dos Veadeiros, Paraiso Alto, *Barroso et al. 607* (NY); Mun. Alto Paraiso, 3 km NW Acampamento Chapada dos Veadeiros, 14°S, 47°W, *Filgueras 428* (K, NY); 16 km by rd. N of Alto Paraiso, *Gates & Estabrook 116* (CTES, NY); 15.8 km NE of Alto Paraíso, *Ginzbarg & Cézar 732* (NY, TEX); 7 km N of Alto Paraiso, *Hatschbach & Nicolack 53932* (CTES); Mun. Terezina de Goiás, Hwy. GO-118, Rio das Almas, *Hatschbach & J. M. Silva 54705* (CTES); ca. 15 km W of Veadeiros, *Irwin et al. 12308* (C, CTES, F, NY, US); Chapada dos Veadeiros, 20 km S of Alto Paraíso, *Irwin et al. 24671* (NY TEX); 3 km S of São João da Aliança, *Irwin et al. 31807* (CTES, F, NY, US); 19 km N of Alto Paraíso, *Irwin et al. 32781* (CTES, F, MO, NY); Chapada dos Veadeiros, between São Gabriel de Goiás and São João da Aliança, *Martinelli 3797* (CTES); 13 km S of Alto Paraíso, *Oliviera & W. R. Anderson 478* (CTES, MICH); between Terezina and Alto Paraíso, *Pirani et al. 1849* (NY, SPF); rd. from Alto Paraiso to Campo Belo, km 8, *Shepherd et al. 3694* (NY).

It is tempting to distinguish *Pavonia immitis* and *P. grandiflora* on the absence or presence of long, setose hairs on the stems, because the character is so easily observed, except that some specimens from the Serra Geral do Paraná (*W. R. Anderson et al. 7757, Irwin et al. 31807*) with prominently setose stems clearly conform to *P. immitis* in all other characters. Another collection from the same vicinity (*W. R. Anderson et al. 9354*) lacks this setose pubescence, as is typical of the more abundant material from the Chapada dos Veadeiros. For the present, therefore, *P. immitis* is interpreted as variable for the character of stem pubescence, until further evidence requires the revision of this interpretation.

The specific epithet is from the Latin (*immitis*, harsh or rough) in reference to the harsh pubescence of the herbage.

27. Pavonia longitricha Fryxell, sp. nov. Type. Brazil. Goiás: Mun. Formosa, JK [Juscelino Kubitschek], 9 Jan 1977, *Hatschbach 39368* (holotype, CTES; isotype, C). Fig. 21

Frutices caulibus dense stellato-pubescentibus; laminis foliorum late ovatis plus minusve obtusis insigniter discoloribus; floribus solitariis in axillis foliorum; bracteis involucellorum ca. 12 copiose setosis (pilis 4–5 mm longis) 16–18 mm longis; calycibus profunde partitis, lobis 3-costatis; corollis rubris 4 cm longis; fructibus ignotis.

Shrubs 1 m tall, the stems with setae 5 mm long on lower stem and otherwise densely stellate-pubescent, the hairs <1 mm diam., the arms arching and overlapping. *Leaf* blades broadly ovate, 4–5 × 4–5 cm, deeply cordate, obscurely undulate-crenate to

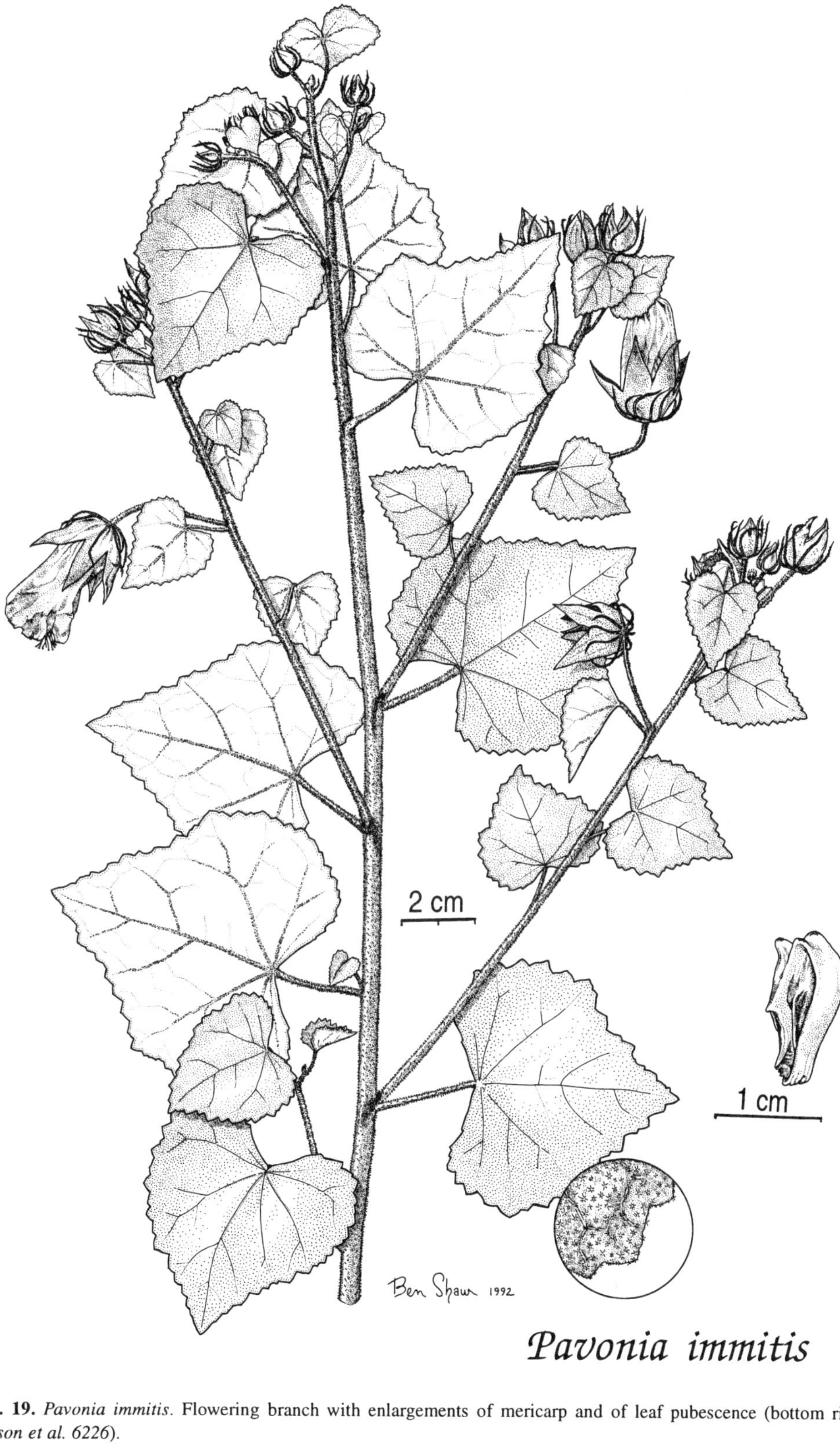

FIG. 19. *Pavonia immitis.* Flowering branch with enlargements of mericarp and of leaf pubescence (bottom right) (*Anderson et al. 6226*).

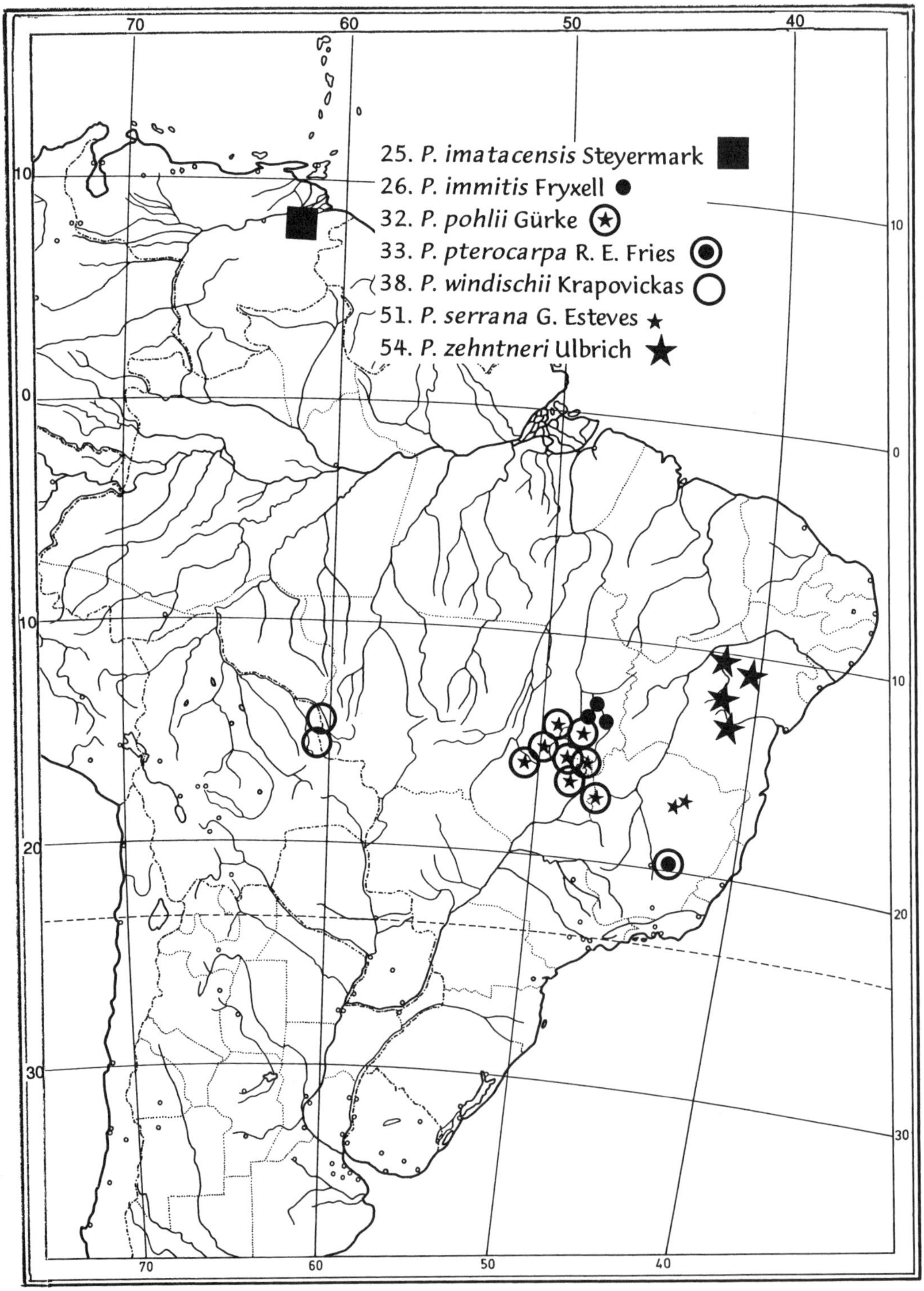

FIG. 20. Distribution map of seven neotropical species of *Pavonia* sect. *Lopimia* and *P.* sect. *Malvaviscoides*.

FIG. 21. *Pavonia longitricha*. Flowering branch with an enlargement of involucel pubescence (*Hatschbach 39368*).

subentire, obtuse or subacute, notably discolorous, palmately 7-nerved, stellate-pubescent above and beneath (the indumentum beneath similar to that of stem, but sparser above); petioles often equaling or exceeding the blade, densely stellate-pubescent; stipules subulate, ca. 3 mm long. *Flowers* solitary in the leaf axils, the pedicels ca. 3 cm long, stellate-pubescent like stem and in distal portion densely setose with ferruginous setae 4–5 mm long; involucellar bracts ca. 12 (?), linear-lanceolate, 16–18 × 1 mm, copiously setose with ferruginous setae 4–5 mm long (radiating outward), with underlying stellate-pubescence; calyx 12–13 mm long, deeply divided, the lobes 3-ribbed, 10 mm long, 4 mm wide, densely stellate-pubescent; petals 4 cm long, red, minutely and obscurely stellate-pubescent externally; genitalia not seen. *Fruits* unknown.

Distribution (Fig. 10). Known only from the type collection from Goiás, Brazil. The type locality is at ca. 15°33′S, 47°20′W (?).

Pavonia longitricha is distinctive for the remarkable long setae on the involucel. Its more or less obtuse leaves are also notable.

28. Pavonia luetzelburgii Ulbrich, Notizbl. Bot. Gart. Berlin-Dahlem **9:** 46. 1924. Type. Brazil. Bahia: Lençois, Carrasco, *von Lützelburg 334* (lectotype, here designated, M, also as photo F-19690; isolectotype, CTES). Fig. 4

Pavonia macrostyla var. *cordata* Gürke *in* Martius, Fl. Bras. **12(3):** 521. 1892. Type. Brazil. Bahia: Near Caiteté, *von Martius 1901* (lectotype, here designated, M as photo F-19691 [pro parte]).

Shrubs 1.5–3 m tall, with variable stem pubescence: with dense setae 3–5 mm long and with understory of minute glandular hairs, or setae lacking and densely stellate-puberulent and sometimes also with glandular hairs. *Leaf* blades broadly ovate, deeply cordate, serrate to subentire, acuminate, discolorous, palmately 7–9-nerved, the upper surface glandular-pubescent, the lower surface stellate-pubescent, mostly 5–12 × 5–12 cm; petioles 2–8 cm long; stipules linear, 5–12 mm long. *Flowers* solitary in the leaf axils, the pedicels 3–12 cm long; involucellar bracts 18–20 (or more), linear, 2–3 cm long, copiously ciliate; calyx 2–3 cm long, puberulent; corolla 2.5–3 cm long, yellow or cream-colored fading reddish, glabrous throughout; staminal column 9–10 cm long (greatly exserted), glabrous, purple, antheriferous in distal 2–3 cm; styles and stigmas slightly exserted from column. *Fruits* 7–9 mm diam., hirsute (hairs 1 mm), the mericarps 7–8 mm high, with broad subapical beak and 2 erect lateral wings. Chromosome number, $2n = 56$ (Fernández, 1981).

Distribution (Fig. 22). Confined to the state of Bahia, Brazil in cerrado and savanna vegetation at elevations of 400–1200 m. The type locality is at 12°33′S, 41°24′W; that of *Pavonia macrostyla* var. *cordata* is at 14°05′S, 42°28′W.

Illustration. Gürke (1892b: pl. 101, fig. 2, as *Pavonia macrostyla*).

Representative specimens examined. BRAZIL. BAHIA: BR-242, 3–8 km W of the rd. to Lençois, *Arbo et al. 5804* (CTES, K, TEX); Mun. Morro de Chapeu, 3–18 km on the rd. to Mundo Novo, *de Carvalho & Saunders 2855* (TEX); km 235 on Hwy. BR-242, Pai Inácio, *de Carvalho & Thomas 3008* (NY); Mun. Jacobina, Serra do Tombador, on the rd. to Morro do Chapéu, *de Carvalho et al. 6144* (CEPEC, NY, TEX); 34 km E de Morro do Chapéu, Chapada da Diamantina, *Davidse et al. 11862* (MO, NY); Mun. Morro do Chapeu, 8–9 km from the city in the direction of Feira de Santana, *Esteves & Lyra-Lemos 2515* (SPF); Morro do Chapéu, *Friere-Fierro et al. 1736* (NY, SPF); 43 km E of Seabra, *P. A. Fryxell 1935* (BH, CAS, NY); 50 km E of Seabra, *P. A. Fryxell 1956* (NY, TEX); Mucugé, *Giulietti et al. 1407* (CTES, F, K–2); Serra do Sincorá, 4 km S of Mucugé, *Harley et al. 16058* (CTES, K, NY, U); Serra da Jacobina, W of Estiva, ca. 12 km N of Senhor do Bonfim, *Harley et al. 16256* (CTES, K); Serra do Brejão ca. 14 km NW of Lençois, *Harley et al. 22346* (CTES, K); Mun. Morro de Chapeu, Morrão, *Hatschbach 39684* (C, CTES); vic. of Morro do Chapéu, *Hatschbach 44245* (RSA, US); Mucugê, 1 km S of Corrego Moreira, *Hatschbach 47517* (CTES); Mucugê, *Hatschbach & Kummrow 47897* (CTES, RB, NY, US); 20 km W of Jacobina, *Krapovickas 12850* (C, CTES, F, MBM, NY, TEX); 6 km E from Pai Inacio, *Krapovickas et al. 38823* (C, CEN, CTES, F, G, UB); Mun. Lençois, Serra Larga W of Lençois, near Caeté-Açu, *Lewis et al. 7164* (F); Mun. Palmeiras, Pai Inácio, ca. 15 km NE of Palmeiras, *Mori 13297* (CEPEC, NY, TEX); 15 km NW of Lençois, *Mori & Boom 14293* (NY); 6 km S of Morro do Chapéu, *Mori & Boom 14456* (NY); Mun. Mucugê, 3 km S of Mucugê, *Mori et al. 12579* (CTES, US); Mucugê, *Pirani et al. 1897* (K, NY, SPF); Mun. Morro de Chapéu, 23 km by rd. from Jacobina to Morro de Chapéu, *Saunders & de Carvalho 3102* (CTES, NY, TEX); Mun. Rui Barbosa, Serra de Rui Barbosa, *N. P. Taylor et al. 1600* (K–2).

Pavonia luetzelburgii and *P. macrostyla* were originally distinguished in varietal rank because of strong similarities in their distinctive flowers. Their fruits, however, are very different (see Fig. 4), as is their foliage. The flowers are notable for the extreme length of the staminal column, which is greatly exserted beyond the corolla.

29. Pavonia macrostyla Gürke *in* Martius, Fl. Bras. **12(3):** 521, t. 101, f. 2. 1892. Type. Brazil. Minas Gerais: Near Domingos, *von Martius 1901* (lectotype,

FIG. 22. Distribution map of eight neotropical species of *Pavonia* sect. *Lopimia*, *P.* subsect. *Exsertae*, and *P.* subsect. *Hastifoliae*.

designated by Krapovickas (1983), M as photo F-19691 [pro parte]). Figs. 4, 23

Pavonia macrostyla var. *ovata* Gürke *in* Martius, Fl. Bras. **12(3):** 522. 1892. Krapovickas (1983) lectotypified *P. macrostyla* such that *P. macrostyla* var. *ovata* = *P. macrostyla* var. *macrostyla*.

Shrubs 1–3 m tall, the stems minutely stellate-pubescent. *Leaf* blades ovate to subelliptic, basally

FIG. 23. *Pavonia macrostyla*. Flowering branch (*Harley et al. 19510, 19947*).

truncate to subcordate, serrate, crenate, or subentire, acute or obtuse, discolorous, palmately 5–9-nerved, stellate-pubescent on both surfaces, the hairs smaller and sparser above than beneath, mostly 3–10 × 2–6(–10) cm; petioles 0.5–5 cm long; stipules 7–10(–27) mm long, lance-linear. *Flowers* solitary in the leaf axils, the pedicels 2–7 cm long; involucellar bracts 18–22, linear, 2–3 cm long, copiously ciliate; calyx 2–3 cm long, with purple glandular hairs; corolla 2.5–3.5 cm long, yellow fading red, apparently glabrous throughout; staminal column (6–)10–12 cm long (greatly exserted), glabrous, purple, with subsessile anthers near apex or with filaments to 5 mm long; styles pubescent, exserted from column, distinct for distal 1–2 mm. *Fruits* sparsely scabridulous, the mericarps 7–8 mm high, with a prominent dorsal rounded crest, the crest retrorsely scabridulous; seeds 4.5 mm long, puberulent.

Distribution (Fig. 22). Nearly confined to the state of Bahia, Brazil (the type from Minas Gerais), in cerrado vegetation, sometimes along rivers, at elevations of 1000–1500 m.

Specimens examined. BRAZIL. Bahia: Mun. Rio de Contas, Mato Grosso, *Esteves & Barros 2370* (CTES); Mun de Piatã, rd. to Capelinha, *Esteves & Lyra-Lemos 2546* (CTES); without locality, *Fróes 12633* (A); Mun. Mucuje, Brejao, *Fróes 12634* (A); Mun. Rio de Contas, 3 km SW of the city, *Giulietti et al. 36377* (CTES); 1 km S of Matio Grosso, *Harley et al. 15181* (CTES, NY); Serra das Almas, lower NE slopes of the Pico das Almas, ca. 25 km WNW of the Vila do Rio de Contas, *Harley et al. 19510* (CTES, K, NY, RB, TEX, U, US); Serra do Rio de Contas, 1 km S of Mato Grosso, on rd. to Vila do Rio de Contas, *Harley et al. 19947* (CTES, K, NY, TEX, U, US); village of Mato Grosso, *Harley et al. 25366* (CEPEC, K, NY, SPF); Mun. Rio de Contas, Pico das Almas, Campo do Queiroz, *Harley et al. 25889* (CEPEC, K, NY, SPF); N of the city, *Harley et al. 26942* (CEPEC, K, NY, SPF); Mun. Barra da Estiva, Morro do Ouro, 9 km S of the city on the rd. to Ituaçú, *Harley et al. 26953* (CEPEC, K, NY, SPF); Mun. Abaíra, Tijuquinho, *Harley et al. 51207* (K); Mun. Rio de Contas, Mato Grosso, *Hatschbach 47404* (CTES); Mun. Rio de Contas, Barra do Brumado, *Hatschbach & Nicolack 53415* (CTES), *Hatschbach et al. 56822* (CTES, K); Mun. Abaíra, Campo de Ouro Fino, *Hind & Queiroz 50073* (K); 6–10 km NW of Rio de Contas, *Mori et al. 12431* (CTES, US).

Pavonia macrostyla is closely similar to *P. luetzelburgii* (q.v.), sharing the remarkably showy flowers but differing in other characters.

30. Pavonia malacophylla (Link & Otto) Garcke, Jahrb. Bot. Gart. Mus. Berlin **1:** 221. 1881. *Sida malacophylla* Link & Otto, Icon. Pl. Select. 67, t. 30. 1822. *Lopimia malacophylla* (Link & Otto) Martius, Nova Acta Phys.-Med. Acad. Caes. Leop.-Carol. Nat. Cur. **11:** 97. 1823. *Malache malacophylla* (Link & Otto) Standley, Contr. U.S. Natl. Herb. **23:** 771. 1923. Type. Berlin Bot. Gard. (specimen unknown). Fig. 4

Brehmia arborescens Schrank, Syll. Pl. Nov. **1:** 85. 1824. Type. unknown.

Pavonia velutina St.-Hilaire, Fl. Bras. Merid. **1:** 233. 1827. *Malache velutina* (St.-Hilaire) Kuntze, Revis. Gen. Pl. **1:** 71. 1891. Type. Brazil. Minas Gerais: Near Itangua, not far from Nossa Sñra. da Penha, *St.-Hilaire s.n.* [*238?*] (holotype, P as photo CTES).

Pavonia plumosa Turczaninow, Bull. Soc. Imp. Naturalistes Moscou **31:** 190. 1858. Type. Brazil. Bahia: *Salzmann s.n.* (lectotype, chosen by Fryxell & Krapovickas, 1990, KW).

Pavonia megalophylla Wright in Grisebach, Cat. Pl. Cub. 27. 1866. Type. Cuba. *C. Wright 2062* (holotype, GOET; isotypes, CTES, GH, K, MA, MO, P as photo CTES).

Shrubs or small trees 2–5 m tall, the stems densely pubescent with stellate and glandular hairs. *Leaf* blades mostly 9–21 cm long, ovate to weakly angulate, cordate, finely serrulate, acute, densely pubescent above and beneath, somewhat discolorous; petioles subequal to the blades on lower leaves, ⅓–½ as long as the blades above; stipules usually 4–7 mm long and subulate but sometimes to 2 cm long and lanceolate. *Flowers* aggregated into dense terminal inflorescences; pedicels of variable length, up to 3 cm long, densely pubescent; involucellar bracts 15–18 (–24), 12–18 mm long, valvate in bud, externally glandular-pubescent, internally long-hirsute, the hairs copious, whitish; calyx 2–3 mm long, wholly concealed by the involucel; petals ca. 3 cm long, rose or red (rarely white), abaxially stellate-pubescent, adaxially glabrous (including margins of claw), erect forming a tubular corolla; staminal column more or less exserted, glabrous, the filaments 4–10 mm long; styles exserted. *Fruits* blackish, viscid, the mericarps twice the height of the column, glabrous.

Distribution (Fig. 10). From Mexico and Cuba to Brazil and Bolivia, but not in all of the intervening countries (i.e., the distribution is sporadic). It occurs principally in forested habitats at elevations of 0–400 m (but to 1100 m in Brazil). The type locality for *Pavonia velutina* is at 21°19′S, 41°45′W.

Illustrations (all as *Lopimia malacophylla*). Fryxell (1988a: frontispiece*; 1992b: fig. 10); Link & Otto (1822: pl. 30); Planchon (1848: pl. 350); Curtis (1848: pl. 4365); Sauget & Liogier (1953: fig. 105).

Representative specimens examined. MEXICO. Chiapas: Between Emiliano Zapata and Palenque, *Cházaro 3316* (CTES, MEXU, MICH, NY); 15 km NW of Boca Lacantum, *E. Martínez 18439* (F, MEXU, RSA, NY). **Guerrero:** Santiago River, Distr. Galeana, *Hinton et al. 11197*

(GH, K, LL, NY, US). **Oaxaca:** Mun. Candelaria Loxicha, 3 km S of Candelaria, *Koch, P. A. Fryxell & Wendt 79520* (CAS, CHAPA, ENCB, MEXU, NY, TEX); Mun. Pluma Hidalgo, 14.5 km N of Pochutla, *Koch, P. A. Fryxell & Wendt 79511* (CHAPA, ENCB, K, MEXU, MO, NY); Palomares, Juchitán, *MacDougall H309* (F, NY, TEX); Distr. Choapam, trail from San Juanito to Trinidad, *Mexia 9297* (GH, K, LL, MO, NY, US); 10 km SW of Santa María (16°52′N, 94°44′W), *Wendt & Rico 4374* (CHAPA, NY). **Tabasco:** Palenque, *Millspaugh s.n.* (F); San Sebastián, *Rovirosa 364* (K, NY, PH).

BELIZE. 17 mi E of Roaring Creek, *Wilbur & Weaver 11528* (DUKE, F, LL, MO).

NICARAGUA. Comarca del Cabo: Near Bilwaskana, *E. B. Nelson 4685* (MO, NY). **Zelaya:** Near Tala Has and Puente Mango, *Stevens 7611a* (MO, NY).

PANAMA. N coast of Colombia and Panama, *Cuming 1143* (E). **Canal Zone:** Ancón Hill, *Piper 5557* (GH, US). **Herrera:** 3–7 km W of Las Minas, rd. to El Toro, *Sytsma & D'Arcy 3182* (MO, NY). **Panamá:** Bique, ca. 12 mi N of Panama City, *Barclay 1045* (BM, F, NY, US); near the City of Panama, *Hayes 482* (E–2); trail from La Esmeralda to La Guinea, *Knapp & Mallet 2965* (MO, NY). **San José Island:** Gulf of Panamá, 55 mi SSE of Balboa, *I. M. Johnston 552* (GH, US), *I. M. Johnston 840* (GH).

CUBA. Isle of Pines: San Pedro and vic., *N. L. Britton & Wilson 14796* (CM, F, GH, MO, NY, US); at Loma Daguilla, *Ekman 12505* (S). **Pinar del Río:** Río Guao, *N. L. Britton et al. 9627* (NY, US); Pinar del Cajálbana, *Ekman 10465* (NY, S); El Guao, Viñales, *Bro. León (J. S. Sauget) & Marie-Victorin 18678* (GH, NY, US).

COLOMBIA. Without locality, *Cuming 1147* (K); Poponte, *C. Allen 911* (K, MO). **Boyacá:** S of Yopal, *Blydenstein 1860* (MO).

VENEZUELA. Amazonas: Hwy. from Puerto Ayacucho to Sanariapo, *Bunting 4273* (F). **Barinas:** Between Santa Barbara and Punta de Piedra, *Steyermark 98274* (US). **Bolívar:** Distr. Cedeño, vic. of Panare, village of Corozal, *Boom & Grillo 6501* (MO, NY).

PERU. Cuzco: Prov. Convención, Abra de Ichiquiato, by way of Palma Real, *C. Vargas 14493* (CTES).

BRAZIL. Acre: Iquiri, 52 km E of Río Branco, *Bockerman 137* (CTES, US). **Amapa:** Km 48 on rd. to Amapa, *Murça-Pires & Cavalcante 51982* (CTES, NY). **Amazonia:** Without locality, *Murça-Pires & M. R. Santos 16144* (CAS, NY); Mun. Humaitá, rd. from Humaitá to Jacarecanga, *Teixeira et al. 1117* (INPA, MO, NY, TEX, US). **Bahia:** Uruçuca on the rd. that links Serra Grande/Itacaré, *Amorim et al. 687A* (NY, TEX, US); rd. between Carnaiba and Itapi curu, 22 Oct 1991, *Menezes et al. s.n.* (NY, SPF); rd. from Andaraí to Mucugê, 4–5 km from Andaraí, *Pirani et al. 2073* (NY, SPF); Itacaré rd. to Barragem, *T. S. Santos 1755* (CTES, NY); Marau, *da Vinha 17* (CTES, NY). **Ceará:** Crato, Chapada do Araripe, *Belo 62* (IPA, NY); Mun. Crato, Araripe National Forest, *Esteves & Barros 2580* (NY, SPF); Gruta de Ubajara, Ubajara, *Fernandes et al. 2895* (CTES); Serra Araripe, *Löfgren 584* (S); Guangeira River, *von Lützelburg 25808* (F). **Distrito Federal:** IBGE Ecological Reserve, *Heringer et al. 489* (CTES, NY); rd. to Planaltina, *Heringer et al. 1535* (NY, US); basin of the S. Bartolomeu River, *Heringer et al. 4564* (CTES, IBGE, K, NY); Paranoá, *Irwin et al. 8494* (CTES, F, GH, MO, NY, TEX, US). **Espírito Santo:** Guarapari, Setiba State Park, *Esteves & Kameyama 2611* (NY, SP); Mun. Linhares, Provoação, *Hatschbach 46753* (C, CTES, F, MO, US, herb. Pedersen); Mun. Ibiraçú, Capoeira, *Hatschbach & J. M. Silva 49432* (C, CM, CTES, NY, S, US); Viana, Valley of the Jucu River, Belo Horizonte–Vitória, *Mello-Silva & Pirani 808* (NY, SPF, TEX); Mun. Serra, rd. from Vitória to Jacaraípe, *Pirani et al. 2783* (NY, SPF, TEX). **Goiás:** Serra Dourada, ca. 17 km (straight line) S of Goiás Velho, *W. R. Anderson et al. 10204* (NY, TEX); without locality, *Gardner 3593* (E, NY, OXF); Hwy. GO-60, 10 km from Goiânia, *Hatschbach 36991* (C, CTES); ca. 30 mi N of Formosa, *Irwin et al. 15585* (CTES, F, GH, MO, NY, TEX, US); Serra do Caiapó, 30 km N of Caiapônia, *Irwin et al. 17984* (CTES, F, NY, TEX). **Maranhão:** Mun. Bom Jardim, ca. 15 km W of Santa Inés, *Balick et al. 1537* (NY, TEX); Fazenda Bacaba, 3 km NW of Lago do Junco, *Daly et al. 451* (NY, TEX); Island of São Luiz, *Snethlage 158* (F, GH, US). **Mato Grosso:** 86 km N of Xavatina, *Irwin et al. 16310* (CTES, F, NY, TEX); Santa Anna da Chapada, *Malme 2002* (S–2); Mun. Alto Garça, Serra da Saudade, *Windisch & Amorim 6297* (CTES). **Minas Gerais:** 5 km by rd. NE of Manso River and Couto de Magalhaes, *W. R. Anderson, P. A. Fryxell et al. 8779* (NY, TEX); 30 km W of Gouveia, *Gregory et al. 10529* (C, CTES); Mun. Dionísio, Rio Doce State Park, *Heringer 16043* (CTES, NY); Serra do Espinhaço, ca. 25 km E of Diamantina, *Irwin et al. 28003* (F, MO, NY, RSA, TEX); Serra de Ventania, near Passos, *Regnell III-178* (F, S–2, US); 10 km N of Belo Horizonte, *L. O. Williams & Assis 6037* (GH, MO). **Pará:** Outskirts of Belém, *Archer 7641* (NY, TEX, US); Ilha de Marajó, Salvaterra, *Kauffmann 12* (F, MO, US); Santarem, rd. from Aramanai to Revolta, *Maciel & Cordeiro 215* (NY, TEX, U); Serra dos Carajás, 6 km SE of AMZA camp N-1, *Sperling et al. 5765* (F, NY, TEX, US); Serra dos Carajás, ferry crossing at Río Itacaiúnas, *Sperling et al. 6100* (NY, TEX); Sete Varas airstrip on Rio Curua, *Strudwick & Sobel 4334* (F, GH, MO, NY, TEX, US). **Paraíba:** Araia, *de Moraes 805* (GH). **Pernambuco:** Vicência, Usina União, *Chiappeta 300* (F); without locality, *Gardner 932* (K, OXF); Tapera, *Pickel 2415* (US). **Rio de Janeiro:** Recife, Curado, *Andrade-Lima 67–5074* (IPA, NY); Mun. Macaé, rd. to Lagoa Comprida, *Araujo 4440* (CTES); near Rio de Janeiro, *Glaziou 10278* (K). **Rondônia:** 17 km N of Vilhena, rd. to Juina, *Krapovickas et al. 40143* (C, CEN, CTES, F, G, K, NY, UC); near Mutumparaná, *Prance et al. 5587* (CTES, F, K, NY, S, U, US). **São Paulo:** Mun. Ibaté, 2.1 km SW of Ibaté, *Eiten et al. 2936* (K, MO–4, US–3); Boa Esperança, vic. of S. Paulo, *Hatschbach & Kummrow 46700* (C, CTES, NY, US); hwy. from Bauru to Marilia km 16, *Leitão et al. 2021* (US).

BOLIVIA. Guanai, *Rusby 1462* (GH, K, NY, PH, TEX, US); Guanai-Tipuani, *Bang 1456* (A, E, F, GH, MO, NY, PH, TEX, US). **Beni:** Prov. Vaca Diez, Guayaramerin, *Krapovickas & Schinini 35104* (CTES). **Pando:** Puerto Candelaria, *M. Nee 31833* (MO, NY, TEX); Madre de Dios, Santo Domingo, *Solomon 6208* (CTES, MO, NY); Prov. Federico Román, Base Naval Abuná, ca. 150 m, *C. Vargas & Foster 841* (F, NY). **Santa Cruz:** Velasco, Campamento Toledo,

180 m, *Guillén & Choré 1849* (MO, TEX); Prov. Velasco, ca. 5 km S of San Simon, *Hopkins et al. 76* (CTES, NY, TEX).

Local names and uses. Amapola (Mexico: Oaxaca), guácima (Brazil: Pará), malva (Brazil: Paraiba); malva rosada (Brazil: Pará), malva veludo (Brazil: Maranhão); manzanilla (Mexico: Tabasco), remela de Cachorro (Brazil: Maranhão). *Pavonia malacophylla* is sometimes used for its bast fibers to produce cordage. A sample of such processed fibers is included as a part of the herbarium specimen of *Kauffmann 12* (US) from Pará, Brazil.

Pavonia malacophylla is closely allied to *P. dasypetala*, and their differences and similarities are discussed under the latter species.

Specimens from San José Island, Panama (*I. M. Johnston 840, 552*) and from the province of Panama (*Knapp & Mallet 2965*) are distinguished by having white corollas (and anthers), but otherwise conform to the species in the remainder of its range.

Gottsberger (1972) observed bee-pollination and postulated hummingbird-pollination in this species.

31. Pavonia piauhyensis Ulbrich, Bot. Jahrb. Syst. **42:** 229. 1908. Type. Brazil. Piauí: Serra Branca, *Ule 7462* (holotype, B as photo F-9460; isotypes:, CTES [fragment], K).

Pavonia philippii Ulbrich, Notizbl. Bot. Gart. Berlin **9:** 47. 1924. Type. Brazil. Bahia: São Bento, 1913, *von Lützelburg 305* (holotype, M as photo F-19692; isotype fragment, CTES).

Erect *shrubs* 0.5–2 m tall, the stems stellate-pubescent (hairs ca. 0.5 mm long). *Leaf* blades triangular to hastately or palmately 3-lobed, cordate, serrulate, acute or acuminate, mostly 3–8 × 3–8 cm, markedly discolorous, densely and softly stellate-pubescent beneath, more sparsely pubescent above with stellate, bifurcate, and simple hairs; petioles 1–5 cm long; stipules 3 mm long, subulate, spreading. *Flowers* solitary in the leaf axils, the pedicels 1–3.5 cm long; involucellar bracts 13–16, narrowly lanceolate, 8–10 mm long, copiously hirsute; calyx 3–5 mm long (mostly concealed by involucel), hirsute; corolla 12–15 mm long, reddish or purplish; androecium included; styles and stigmas slightly exserted. *Fruits* 8–9 mm diam., conical, subglabrous, the mericarps 9 mm tall (including the apical point, which is 3–3.5 mm long), more or less rugulose and verruculate.

Distribution (Fig. 15). Confined to central Brazil (Piauí and Bahia), the habitat probably caatinga (thorn forest/scrub).

Specimens examined. **BRAZIL. PIAUÍ:** Mun. São Raimundo Nonato, Carnaiba, *Del'Arco 492* (CTES, NY); Mun. São Raimundo Nonato, Toca do Vento-Serra Cranca, *Emperaire & Campelo 2757* (CTES); hwy. from São Raimundo Nonato to Canto do Buriti, *Fernandes 6186* (CTES).

Pavonia piauhyensis is notable for its often 3-lobed leaves.

Pavonia philippi was based on scanty material and was acknowledged by Ulbrich to be closely related to *P. piauhyensis*. Since further collections have not come to light that would support the distinction of two species, it seems best to consider the two names synonymous. The type of *P. philippi* is probably a depauperate (drougth-stressed?) specimen.

32. Pavonia pohlii Gürke *in* Martius, Fl. Bras. **12(3):** 517. 1892. Type. Brazil. Goiás: Near S. Cruz, *Pohl 2611* (lectotype, here designated, W as photo F-32651; fragment, F). Fig. 4

Shrubs 1–2 m tall, the stems densely and roughly stellate-pubescent. *Leaf* blades broadly ovate, deeply cordate (the sinus open), crenate-serrate, acute or acuminate, discolorous, mostly 3–5 × 3.5–6 cm, densely stellate-pubescent, the 6–8-armed hairs of the lower surface denser than those of the upper surface; petioles 1.5–3.5 cm long; stipules 4–8 mm long, subulate. *Flowers* solitary in the leaf axils, the pedicels 2–5 cm long; involucellar bracts 12–14, linear, 18–24 mm long, stellate-pubescent; calyx 16–22 mm long, hirsute; corolla 4–4.5 cm long, pink; staminal column 2.5 cm long, glabrous, pallid, the filaments 1 mm long; styles well exserted from column, united below, free for distal 3 mm, the stigmas not quite equaling the petals. *Fruits* 1 cm diam., glabrous, the mericarps 6.5 mm high, basally 2-ridged, otherwise smooth and unornamented.

Distribution (Fig. 20). Confined to central Brazil (Goiás and the Distrito Federal), growing in cerrado vegetation at ca. 1000 m elevation. The type locality is at 17°20′S, 48°28′W.

Illustration. Gürke (1892b: pl. 99, fig. 2).

Specimens examined. **BRAZIL. DISTRITO FEDERAL:** Fazenda Agua Limpa, *Cezar 79* (E, F); Hwy. 15, *Heringer 14453* (CTES, NY, UB); Planaltino, near Colegio Agrícola, *Heringer 15418* (CTES, NY); Zoobotanical Park, *Heringer 7988/250* (CTES, NY, TEX); near lake dam, *Hunt 5493* (CTES); 25 km S of Brasilia, *Irwin et al. 13037* (NY); Brasilia, *E. Pereira 9011* (CTES, NY); 18 km SSW of Brasilia TV tower, *Ratter & da Fonseca 2900* (E, F, MO, NY). **GOIÁS:** Serra dos Cristais, 12 km by rd. N of Cristalina, *W. R. Anderson, P. A. Fryxell et al. 7988* (F, K, NY, TEX, US); 35 km by rd. E of Cristalina, *W. R. Anderson, P. A. Fryxell et al. 8292* (NY, TEX); Serra dos Caipos, *Barbosa 5449*

(CTES); between Planaltino and Formosa, *P. A. Fryxell 1913* (BH, NY, TEX); Campos Natividade, *Gardner 3020* (K); without locality, *Glaziou 20715* (C, K); Niquelandia, *Gregory et al. 10192* (CTES); Mun. São João da Aliança, Fda. Capão da Onça, *Hatschbach 36964* (CTES); Mun. Cristalina, *Hatschbach 43814* (CTES, MO, NY); Mun. Luziania (Fazenda da Pinguela-Rizz.), *Heringer 16934-A* (CTES, NY); 75 km N of Corumbá de Goiás on rd. to Niquelandia, *Irwin et al. 18999* (CTES, NY); Serra dos Caipós 40 km from Amorinópolis on rd. to Rio Verde, *Rizzo 6413* (CTES); Mun. Uruaçu, a 8 km do Rio Maranhão, *Rizzo 7857* (CTES).

Pavonia pohlii is closely related to *P. grandiflora* (q.v.) and *P. immitis* (q.v.) all of which have large, pink flowers.

33. Pavonia pterocarpa R. E. Fries, Bull. Herb. Boiss., ser. 2, **7:** 1000. 1907. Type. Brazil. Minas Gerais: Near the rd. from Sabará to Caeté, Mai 1905, *Damazio 1772* (holotype, G). Fig. 4

Shrubs, the stems densely stellate-pubescent (hairs 0.5 mm long or less) and with glandular hairs intermingled. *Leaf* blades ovate, mostly 3–6.5 × 3–5.5 cm, deeply cordate, serrate, acuminate, palmately 7-nerved, somewhat discolorous, minutely and roughly stellate-pubescent above and beneath, with some glandular hairs present; petioles mostly 2–4.5 cm long; stipules subulate, 4–7 mm long. *Flowers* solitary in the axils, the pedicels 1.5–3.5 cm long, with pubescence like stem and distally with scattered setae 3–5 mm long in addition; involucellar bracts 12–14, linear, 15–20 × 1 mm, stellate- and glandular-pubescent, sometimes setose (like pedicel); calyx 1.5–2 cm long, densely pubescent, costulate; corolla 3.5–4.5 cm long, rose, externally pubescent; staminal column ca. half length of corolla, the styles exserted from the column. *Fruits* 7 mm diam.; mericarps 5–6.5 mm high, glabrous, straw-colored, basally alate, otherwise unornamented.

Distribution (Fig. 20). Known from central Brazil (Minas Gerais). The type locality is at 19°53′S, 43°43′W.

Illustrations. Fries (1908: pl. 4, figs. 7–9).

Specimen examined. **BRAZIL**. Distrito Federal, *A. P. Silva 9* (US).

Pavonia pterocarpa is possibly related to *P. hirticalyx*.

34. Pavonia rosa-campestris St.-Hilaire, Fl. Bras. Merid. **1:** 236. t. 46. 1827. *Malache rosa-campestris* (St.-Hilaire) Kuntze, Revis. Gen. Pl. **1:** 71. 1891. Type. Brazil. Minas Gerais: Minas Novas, near São João, *St.-Hilaire 1259* (holotype, P as photo CTES; isotype fragments, CTES, F). Fig. 4

Pavonia rosa-campestris var. *tomentoso-velutina* Hochreutiner, Annuaire Conserv. Jard. Bot. Genève **10:** 17. 1906. Type. Brazil. Goiás: *Glaziou 20710bis* (holotype, G, as photo F-23707; isotypes, K, P as photo CTES).
Pavonia rosa-campestris var. *tomentella* R. E. Fries, Kongl. Svenska Vetenskapsakad. Handl. **42(12):** 46. 1908. Types. Brazil. Goiás: *Glaziou 20710* (lectotype, here designated, following an annotation by R. E. Fries, S; isolectotypes, C, P as photo CTES).

Subshrubs or decumbent herbs 0.5–1(–2) m tall, with a deep spreading perennial root system, the stems pilose (hairs 2–4 mm long) or rarely glabrous. *Leaf* blades broadly ovate to subrotund (rarely narrower), deeply cordate, crenate-serrate, acute or acuminate, ciliate on margin, with minute stellate hairs to subglabrous on both surfaces or sometimes the upper surface with appressed simple hairs 1–2 mm long; mostly 2.5–4 × 2.5–5 cm; petioles 1.5–3 cm long; stipules subulate, 2 mm long. *Flowers* solitary in the leaf axils, the pedicels 0.5–1.5 cm long, slender; involucellar bracts 10–12, linear-lanceolate, 12–16 × 1.5–2 mm, externally glandular-puberulent, internally hirsute; calyx 4 mm long, glandular-puberulent, mostly concealed by involucel; corolla red, 1.5–2.5 cm long, glabrous throughout, the claw scarcely discernible; staminal column 5 mm long, glabrous, the filaments 2 mm long; styles exserted from column. *Fruits* 1 cm diam., stellate-pubescent, the mericarps 10 mm high, keeled and coarsely rugose, apically pointed.

Distribution (Fig. 18). From Brazil, known from the states of Bahia, Goiás, Mato Grosso, Minas Gerais, Rondônia, and from the Distrito Federal, mostly in cerrado vegetation at 200–1200 m elevation. Two collections are known from Bolivia. The type locality (São João = Itamarandiba) is at 17°52′S, 42°51′W.

Illustrations. Fries (1908: pl. 5, fig. 4); Gottsberger (1972: fig. 19); Gürke (1892b: pl. 98, fig. 2); St.-Hilaire (1827: pl. 46).

Representative specimens examined. **BRAZIL. BAHIA:** Mun. São Deziderio, between Posse and Barrieras, km 50 of Hwy. BR-020, *Coradin et al. 5706* (CTES); Mun. Barrieras, hwy. from Barrieras to Brasilia km 90, *Coradin et al. 7413* (CTES, K). **DISTRITO FEDERAL:** Rio São Bartolomeu, *Heringer 8679/873* (CTES); Paranoá Lake, *Heringer et al. 17415* (CTES, NY); IBGE Ecological Reserve, *Heringer et al. 2129* (CTES, K, MO); Chapada da Contagem, *Irwin et al. 7967* (CTES, NY); Brasilia, *Irwin et al. 8098* (CTES, GH, MO, NY, RB, S); 25 km S of Brasilia, *Irwin & Soderstrom 5594* (C, CTES, F, NY); 38 km NW of Brasilia, *Martinelli et al. 7466* (CTES, RB); Brasilia, *Oldenburger & Mecenas 1501* (CTES, NY, U). **GOIÁS:** Serra dos Cristais, 5 km by rd. E of Cristalina, *W. R. Anderson, P. A. Fryxell et al. 8193* (F, NY, TEX, US); vic. of Campo Alegre de Goiás, *Arbo et al. 3113* (CTES); 20 km N of Formosa, *P. A. Fryxell 1914* (NY); 30

km E of Dianópolis, *Gregory et al. 10423* (CTES); rd. from Alto Paraíso to Terezina, *Heringer et al. 2464* (CTES); 21 km N of São Gabriel de Goiás, *Martinelli et al. 7479* (NY, TEX); Chapada dos Veadeiros, 2 km from Veadeiros, *Prance & N. T. Silva 58188* (CTES, F, GH, K, NY, S, US). **Mato Grosso:** Mun. Camapuã, Capão Redondo, *Hatschbach 32350* (CTES, MO); Serra do Roncador, 210 km N of Xavantina, *Irwin et al. 16270* (CTES, MO, NY, TEX, US); Serra do Roncador, 87 km N of Xavantina, *Irwin et al. 16485* (CTES, GH, MO. NY, S); 1–15 km N of Vilenha, camino a Juina, *Krapovickas et al. 40133* (C, CEN, CTES, F, K, LIL, NY, UC); Mun. Vila Bela, ca. 82 km SE de Vilhena, *Krapovickas et al. 40149* (CEN, CTES, G, NY); 39 km N of Rondonópolis, rd. to Poxoréu, *Krapovickas et al. 40228* (CEN, CTES); beyond Alto Araguaia, rd. to Cuiabá, Brasilia–Acre Hwy., *Maguire et al. 56262* (CTES, NY); Chapada dos Guimarães, just above Cachoeira Furada, Reserva Buriti, *Prance et al. 18861* (CTES, MO, NY, S, US). **Minas Gerais:** Serra de Catiora, *Duarte 3026* (CTES, RB); Grão Mogol, on the rd. to the Ventanis River, *Esteves et al. 13343* (CTES); Mun. Diamantina, Cons. Mata, *Hatschbach 30220* (C, CTES, K, NY, S, US, herb. Pedersen); João Pinheiro, *Heringer 8551/745* (CTES); Morro das Pedras, ca. 30 km NE of Patrocinio, *Irwin et al. 25670* (NY, TEX); between Uberaba and St. Anna, *Riedel 2452* (US); Serra do Espinhaço, between Montes Claros and Riacho dos Machados, *Santos & Castellanos 24273* (CTES, K, NY, TEX, U); without locality, 1841, *Gardner 4430* (syntypes of *P. rosa-campestris* var. *tomentella*, GH, K). **Rondônia:** Mun. Colorado do Oeste, Porto Velho-Cuiabá (12°13′S, 60°61′W), *Cid et al. 4482* (F, NY, TEX, US). Without locality, *Burchell 5547* (syntype of *P. rosa-campestris* var. *tomentella*, P as photo CTES; isosyntypes, CTES [fragment], GH, P as photo CTES).

BOLIVIA. Santa Cruz: Velasco, Noel Kempff National Park, Serranía Huanchaca, 360 m, *Mostacedo et al. 2132* (MO, TEX); Velasco, Noel Kampff National Park, 220 m, *Saldias et al. 2850* (MO, TEX).

Pavonia rosa-campestris is characterized by its red corolla, the petals virtually lacking claws, and its small calyx. Plant pubescence is quite variable.

Gottsberger (1972) describes this species as autogamous. Despite its bright red corolla, in observation for several days it was not observed to be visited by pollinators (neither hummingbirds, butterflies, nor bees), presumably because the overlapping bases of the petals denied access to the floral nectaries. Self-pollination resulted in normal fruit set.

35. Pavonia schwackei Gürke, Notizbl. Bot. Gart. Berlin **1:** 18. 1895. *Lassa schwackei* (Gürke) Kuntze, Deutsch. Bot. Monatsschr. **21:** 173. 1903. Type. Brazil. Minas Gerais: Diamantina, 26 Mar 1892, *Schwacke 8329* (holotype, B destroyed, as photo F-9468).

Shrubs 0.5–1.5 m tall, the stems densely stellate-pubescent, the hairs 0.5–1 mm long. *Leaf* blades elliptic, serrate to subentire, acute or subobtuse, palmately 3-nerved, 2–5 × 1–1.5 cm, 2–3.5 times as long as wide, densely stellate-pubescent on both surfaces, the hairs yellowish, ca. 1 mm diam.; petioles 5–17(–25) mm long; stipules subulate, 5–7 mm long. *Flowers* solitary or paired in the leaf axils, often crowded terminally in dense, narrow inflorescences; pedicels 0.5–2.5 cm long, densely pubescent, the hairs up to 2 mm long; involucellar bracts ca. 12, linear, 7–11 mm long, copiously ciliate, the cilia 1–2 mm long; calyx 6–7 mm long, densely hirsute; corolla 0.7–1 cm long, narrowly tubular, rose or red; staminal column exserted by 5 mm, sometimes purplish, glabrous, the filaments ca. 25, 1–1.5 mm long; styles emergent from the column. *Fruits* ca. 6.5 mm diam., minutely pubescent, the mericarps 4 mm tall, the walls reticulate-veined, fragile; seeds glabrous, 3.5–4 mm long.

Distribution (Fig. 14). Confined to Minas Gerais in the general vicinity of Diamantina. The type locality is at 18°13′S, 43°36′W.

Specimens examined. BRAZIL. Minas Gerais: Mun. Diamantina, 20–26 km WSW of Diamantina, rd. to Conselheiro Mata, MG-220, *Arbo et al. 4349* (CTES); Rio dos Cristais, Diamantina, *Duarte 9652* (CTES); Birybiry, *Glaziou 18885a* (CTES, P–n.v.).

Pavonia schwackei is quite isolated taxonomically, though possibly related to *P. windischii* (q.v.).

36. Pavonia setifer (C. Presl) Fryxell, comb. nov. Basionym: *Hibiscus setifer* Presl, Reliq. Haenk. **2:** 133 (1835). Type. "in America aequatoriale," *Haenke s.n.* (holotype, PR?–n.v.; isotype, W, as photo F-32653).

Pavonia langlassei Hochreutiner, Annuaire Conserv. Jard. Bot. Genève **20:** 151. 1917. Type. Mexico. Guerrero: banks of the Tecpan River, *Langlassé 816* (holotype, G as photo F-23701; isotypes, CTES [fragment], GH, K, MEXU, P as photo CTES, UC, US).

Subshrubs 1.5 m tall, the flowering stems minutely glandular-puberulent, vegetative stems also with simple hairs 5–7 mm long. *Leaf* blades mostly 7–10 × 3–7 cm, sometimes slightly asymmetrical, ovate or deltate to hastately 3-lobed, cordate, finely serrate, acute or (usually) acuminate, discolorous, minutely pubescent with simple and stellate hairs; petioles mostly ⅓–⅔ the length of the blade; stipules 4–5 mm long, subulate, recurved. *Flowers* solitary in the leaf axils, the pedicels 2–5.5 cm long, slender, viscid-puberulent; involucellar bracts 15–18, 10–12 × ca. 0.5 mm, more or less arcuate, densely ciliate and viscid, bearded adaxially; calyx 7–8 mm long, minutely pubescent and ciliate-margined; corolla 2.5–3 cm long, pinkish (?), glabrous (including margins of claw); androecium 1.5 cm long, glabrous, the filaments 1.5–2 mm long; styles glabrous, slightly exceeding the androecium. *Fruits* ca. 7 mm diam., the mericarps more or less reticulate, 4 mm high, slightly ridged at juncture of dorsal and lateral walls, otherwise unornamented.

Distribution (Fig. 24). Endemic to the state of Guerrero, Mexico, where it occurs in pine-oak forest, at 500–700 m elevation.

Specimens examined. MEXICO. GUERRERO: 62.2 mi N of Acapulco, hwy. to Iguala, *Carlson 3105* (F, NY, TEX); Mun. Petatlán, La Guayabera, 26 km N of Coyuquilla, *Soto-Núñez et al. 12193* (MEXU, NY), *Soto-Núñez et al. 12548* (MEXU, MO).

Local name. Malvilla (Mexico: Guerrero).

Pavonia setifer is distinguished from most other members of *P.* sect. *Lopimia* by its Mexican distribution, its pinkish flowers (the petals of which have glabrous claws), and its relatively numerous (15–18) involucellar bracts.

37. Pavonia spinistipula Gürke *in* Martius, Fl. Bras. **12(3):** 513. 1892. Type. Brazil. Bahia: between Cachoeira and Rio das Contas, *von Wied-Neuwied 102* (holotype, B destroyed, as photo F-9469; isotypes, BR, CTES [fragment]). Fig. 25

Shrubs 1.5 m tall, the stems setose (hairs 4–5 mm long) and with sparse underlying stellate and sometimes glandular pubescence, or the setae lacking. *Leaf* blades ovate or somewhat 3-lobed, deeply cordate, serrate, acute or acuminate, somewhat discolorous, with scattered simple and bifurcate hairs (0.5–1 mm long or less) on upper surface, stellate hairs on lower surface, palmately 5-7-nerved, mostly 4–8 × 3–6 cm; petioles 2–5 cm long, with fewer setae than stem except at distal end; stipules 5–6 mm long, subulate. *Flowers* solitary in the leaf axils, the pedicels 2–7 cm long, slender, minutely stellate-pubescent; involucellar bracts 10–12, linear, arcuate, 20–30 × 1 mm or less, sometimes setose, sometimes reddish; calyx 14–18 mm long, stellate-pubescent (sometimes apically pilose), the lobes narrowly triangular; corolla 2–2.5 cm long, red, externally pubescent; androecium more or less exserted. *Fruits* 8–9 mm diam., glabrous or minutely pubescent, the mericarps 7 mm tall, dorsally rounded, subapically pointed, otherwise unornamented.

Distribution (Fig. 14). Confined to the state of Bahia, Brazil. The type locality is at ca. 14°S, 40°W.

Additional illustration. Gürke (1892b: pl. 98, fig. 3).

Specimens examined. BRAZIL. BAHIA: Mun. Caetite, 20 km E of Caetite, rd. to Brumado, *Arbo et al. 5667* (CTES, K, TEX); Mun. Brumado, ca. 15 km on the hwy. from Brumado to Caetité, *de Carvalho et al. 2625* (CTES); from Vitoria da Conquista to Poções, *Duarte 10583* (CTES); Mun. Iraguara, Fazenda Torrinha, entrance 13 km after the entrance to Iraguara, *Esteves & Lyra-Lemos 2533* (CTES); Campo Formoso, Fazenda Sacaiba, *Ferreira 99* (CTES); side rd. ca. 2 km from Estiva, ca. 12 km N of Senhor do Bonfim, *Harley et al. 16511* (CTES); Mun. Poções, 2–4 km W of Poções, *Mori et al. 9492* (CTES, K, NY, RB).

Pavonia spinistipula is distinctive for its 10–12 arcuate, sometimes reddish involucellar bracts and red flowers.

38. Pavonia windischii Krapovickas, Anais XXXVI Congr. Bras. Bot., Curitiba [1985], p. 281. 1990. Type. Brazil. Mato Grosso: Serra Ricardo Franco (15°S, 60°W), campo, alt. ca. 800 m, 22 Mar 1978, *Windisch 1746* (holotype, UNESP–n.v.; isotype, CTES). Figs. 4, 26

Shrubs or trees 2–3 m tall, the stems minutely and densely stellate-pubescent, the hairs 0.1–0.2 mm long. *Leaves* elliptic, basally rounded-truncate to subcordate, shallowly serrate, acute or acuminate, 4–11 × 2.5–5.5 cm, palmately 5-nerved with raised reticulate venation beneath, discolorous, stellate-pubescent above and beneath, more densely and coarsely so beneath; petioles 6–10 mm long, with pubescence like stem; stipules 3–7 mm long, subulate. *Flowers* solitary in the leaf axils, the pedicels 0.5–1 cm long, with pubescence like stem; involucellar bracts 10–14, 12–17 mm long, subulate, more or less condyloid basally, minutely stellate-pubescent; calyx 15–25 mm long, half or more divided, minutely stellate-pubescent, the lobes 3(–5)-nerved; corolla 3.5–6.5 cm long, white with reddish center, externally minutely stellate-pubescent, otherwise glabrous (including claws); androecium 4.5–7 cm long (more or less exserted), slender, glabrous, reddish (fide collectors) or pallid; filaments scattered on exserted portion of tube, especially near apex, 1–3 mm long; styles emergent from the column, free for distal 3–4 mm, slightly pubescent; stigmas capitate, 0.3 mm diam. *Fruits* ca. 7 mm diam. (enclosed by calyx), with minute (0.1–0.2 mm) retrorse pubescence on lower half, with antrorse whitish hairs 1–2 mm long on upper half; mericarps 6 mm high; seeds 4.5–5.5 mm long, glabrous, blackish, not striate.

Distribution (Fig. 20). Known from the Serranía Huanchaca and Serranía de Caparuch in E Bolivia and the Serra Ricardo Franco in W Brazil (Mato Grosso) at 200–850 m elevation. The Bolivian habitat is described as "one of the characteristic granite balds on the sloping parts of this 'lost world' mesa of brazilian shield sandstone and granite" (R. B. Foster, in litt.), where it is evidently not common.

Additional illusration. Krapovickas (1990: fig. 1).

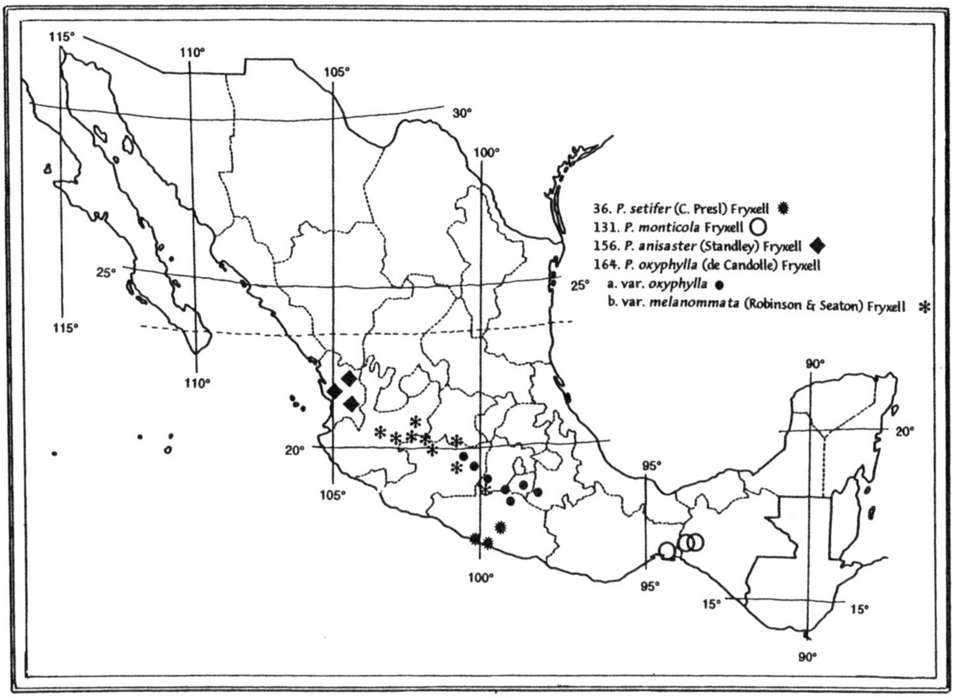

FIG. 24. Distribution map of five Mexican taxa of *Pavonia* sect. *Lopimia*, *P.* sect. *Asterochlamys*, *P.* subsect. *Rubriflorae*, and *P.* sect. *Laminares*.

Specimens examined. **BOLIVIA. Santa Cruz:** Campamento La Torre, 300 m, *Gutiérrez et al. 900* (MO, TEX); Noel Kempff National Park, Serranía de Caparuch, 850 m, *Killeen et al. 5434* (MO, TEX); Prov. Velasco, Campamento Las Torres, banks of the Iténez (Guaporé) River, on the border with Mato Grosso; NE slope of the Serranía Huanchaca, 24 km S of Flor de Oro, ca. 50 km N of the Río Verde, alt. 200–400 m, *Peña & Foster 226* (F, NY, USZ); Serranía Huanchaca, Noel Kempff National Park, some 6 km S of the Iténez or Guaporé River, ca. 15 km SE of Flor de Oro ranch, 500–600 m, Toledo et al. 1 (F, NY, USZ).

Pavonia windischii is apparently allied to *P. schwackei* but differs in a number of characters, as indicated in the key, and also is widely separated from it geographically. Neither species is common (evidently) or well known. A collector's observation describes the flowers of *P. windischii* as having "a beautiful simplicity of pure white and deep red" (R. B. Foster, in litt.).

Pavonia subgen. *Pavonia* sect. *Cancellaria*

Pavonia sect. **Cancellaria** de Candolle, Prodr. **1:** 444. 1824. Type. *Pavonia cancellata* (Linnaeus) Cavanilles. Although *P. zeylanica* (Linnaeus) Willdenow was named lectotype by Mattei (1921), the type is *P. cancellata* according to ICBN, art. 22.5.

Procumbent herbs, the leaves more or less asymmetrical; bracts of the involucel ca. 15, filiform, arcuate; mericarps 3-pointed (see Fig. 4).

Type species. *Pavonia cancellata* (Linnaeus) Cavanilles.

Key to the species of *Pavonia* sect. *Cancellaria*

1. Pedicels 1–1.5 cm long; involucellar bracts ca. 5 mm long; fruits puberulent. . . . . 40. *P. cochensis*
1. Pedicels 2–8 cm long; involucellar bracts 10–15 mm long or longer; fruits (where known) subglabrous.
    2. Leaves markedly asymmetrical, sharply acute; involucellar bracts 1–1.5 cm long; corolla yellow with a dark red center. . . . . 39. *P. cancellata*
    2. Leaves essentially symmetrical, broadly acute to obtuse; involucellar bracts 1–2 cm long; corolla yellow, with or without a dark center.
        3. Leaf blades pubescent, flabelliform, shorter than the petiole, broader than long; stipules divided to the base into 2 or 3 linear elements. . . . . 42. *P. repens*
        3. Leaf blades velutinous, ovate to 5-lobed (aceriform), longer than or equaling the petiole, about as broad as long (see Fig. 27); stipules simple, linear. . . . . 41. *P. humifusa*

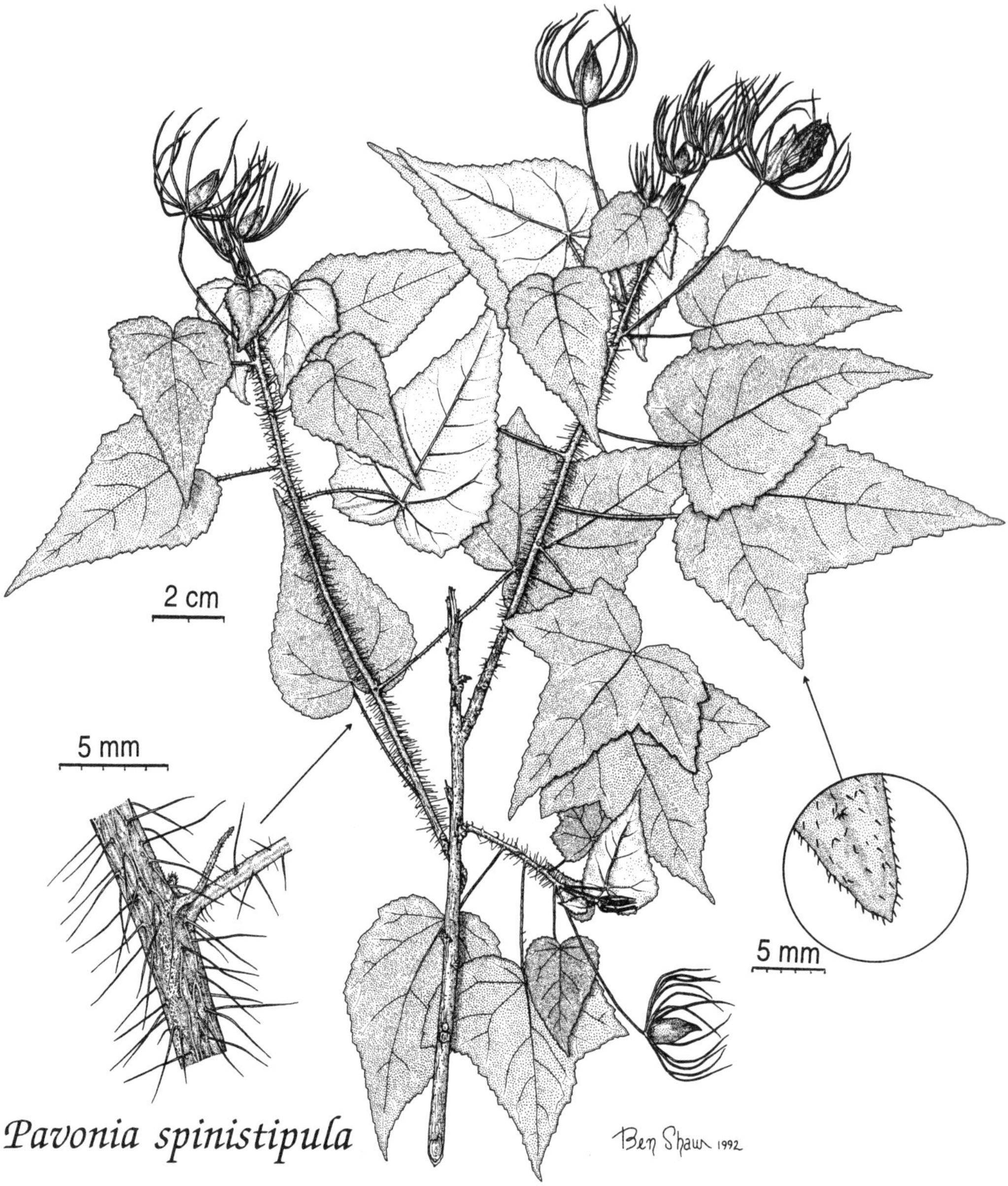

FIG. 25. *Pavonia spinistipula.* Flowering branch with enlargements of stem pubescence (lower left) and leaf tip (lower right) (*Mori et al. 9492*).

39. Pavonia cancellata (Linnaeus) Cavanilles, Diss. **3:** 137. 1787. *Hibiscus cancellatus* Linnaeus, Pl. Surin. 12. 1775. *Malache cancellata* (Linnaeus) Kuntze, Revis. Gen. Pl. **1:** 70. 1891. Type. Surinam. "cancellatus" (holotype, LINN-875.8 as microfiche; isotype, S as microfiche). Fig. 4

Hibiscus anonimus Martius *ex* Colla, Herb. Pedem. **1:** 405. 1833. Type. Brazil. Serra da Cap Frio, 1829, *von Martius s.n.* (holotype, TO).

Pavonia deltoidea Martius, Flora **22 (Beibl.):** 59. 1839. *Pavonia cancellata* var. *deltoidea* (Martius) St.-Hilaire & Naudin, Ann. Sci. Nat. Bot., ser. 2, **18:** 41. 1842. *Malache deltoides* (Martius) Kuntze, Rev. Gen. Pl. **1:** 70. 1891. Type. Brazil. *von Martius 1014* (holotype, BR; isotypes, CTES [fragment], GH, MO, NY, OXF).

Pavonia modesta Martius, Flora **22 (Beibl.):** 59. 1839. *Pavonia cancellata* var. *modesta* (Martius) Garcke, Jahrb. Bot. Gart. Mus. Berlin **1:** 220. 1881. *Malache modesta* (Martius) Kuntze, Revis. Gen. Pl. **1:** 71. 1898.

FIG. 26. *Pavonia windischii*. Vegetative branch (with young bud at tip), showing well-developed foliage and separate flower (*Peña & Foster 226*).

Type. Brazil. Near Ilhéus, *von Martius 403* (holotype, BR as photo CTES; isotypes, ARIZ, CTES [fragment], GH, MO, NY).

Pavonia hirta Klotzsch *ex* Schlechtendal, Linnaea **14:** 301. 1840, non *Pavonia hirta* Schweigger, 1812. Type. Brazil. Bahia: Cruz de Casma, *Luschnath 165* (isotypes, BR–n.v., CTES [fragment], MO).

Pavonia procumbens Casaretto, Nov. Stirp. Bras. Dec. 39. 1842. Type. Brazil. On Itaparica island and in open hills near Badia, *Casaretto s.n.* (holotype, TO?–n.v.).

Pavonia cancellata var. *crassivenosa* Gürke in Martius, Fl. Bras. **12(3):** 516. 1892. Types. Venezuela. Near Tovar, 1854–1855, *Fendler 99* (lectotype, here designated, U; isolectotypes, CTES [fragment], GH, MO, P as photo CTES, PH).

Pavonia cancellata forma *montana* J. Huber, Bull. Herb. Boiss., ser. 2, **1:** 311. 1901. Type. Brazil. Ceará: trailside, near Guaramiranga, *Huber 236* (specimen unknown).

Pavonia cancellata var. *cordata* Hassler, Feddes Repert. Spec. Nov. Regni Veg. **7:** 378. 1909. Type. Paraguay. Caballero-Cué, between the Apa River and the Aquidaban River, Feb 1909, *Fiebrig 4781* (holotype, G as photo CTES; isotypes, CTES [fragment], GH).

Pavonia guanacastensis Standley, Publ. Field Mus. Nat. Hist., Bot. Ser. **18:** 672. 1937. Type. Costa Rica. Guanacaste: NE of Liberia, 13 Jan 1930, *Dodge, Alfaro, & W. S. Thomas 6217* (holotype, F, as photo F-56167; isotype, MO).

Procumbent ("erect" in *Harley 19136*) perennial *herbs*, the stems trailing (to 2 m long), hispid, the hairs 1–2 mm long, with dense lines of shorter hairs. *Leaf* blades mostly 2–3.5 cm long, markedly asymmetrical, deltoid-ovate to angular or hastately lobed, deeply cordate, crenate, acute, the upper surface sometimes with simple hairs 1.5–2 mm long, sometimes with stellate hairs 0.5 mm long, pubescent beneath; petioles ½–1 times the length of the blade; stipules 2–3(–6) mm long, subulate, erect or recurved. *Flowers* solitary or paired in the leaf axils, the pedicels subequal to or exceeding subtending leaves (1–8 cm long); involucellar bracts ca. 15, linear, 10–12(–15) mm long, arcuate, hispid-ciliate, the hairs ca. 2 mm long; calyx 7–8 mm long, hirsute; corolla yellow with dark red center (dark center lacking in *Stephens M2*), 1.5–2.7 cm long; androecium ca. 8–9 mm long, the staminal column glabrous, pallid; styles slightly exserted from column. *Fruits* 7–8 mm diam., obovoid, glabrous, the mericarps 4.5 mm long, with an obscure apical beak. Chromosome number, $2n = 56$ (Krapovickas & Cristóbal, 1965).

Distribution (see Fig. 28). Widespread in Central America (Nicaragua and Costa Rica) and South America (Venezuela, the Guianas, Brazil, Paraguay, and Bolivia), where it is often found on sandy soils growing in open (sometimes arid) habitats, at elevations of 100–1000 m.

Illustrations. Gürke (1892b: pl. 98, fig. 1); Reichenbach (1830, **3:** pl. 215).

Representative specimens examined. MEXICO. VERACRUZ: Santa Lucretia, *Conzatti 3661½* (US).

NICARAGUA. LEÓN: Ca. 1 km SE of jct. of Hwy. 1 & Hwy. 28, *Stevens 11191* (MO, NY, TEX, U). **MANAGUA:** Sierra de Managua, *Garnier 94* (F). **MASAYA:** S slope of the crater of Santiago Volcano, *Grijalva & Vanegas 4090* (HNMN, MO, NY); Masaya Volcano National Park, rim of Santiago Volcano and Jiñocuzba rd., *Stevens 10796* (MO, NY).

COSTA RICA. GUANACASTE: Ca. 3.3 km N of Bagaces on rd. to Aguas Claras, *Almeda et al. 3126* (CAS, NY); Finca Jenny, at S border of Santa Rosa Natl. Park, *Hill et al. 17608* (NY, VT); between Guardia and Liberia, *Jiménez 3347* (F, MO, NY); ca. 10 km S of La Cruz, *Weston 5054* (F, MO, NY, US). **PUNTARENAS:** Rd. from Lagarto to Santa Elena, *Haber & Zuchowski 8903* (CR, MO, NY); Esparta, *Khan et al. 1023* (BM, NY).

VENEZUELA. ISLA MARGARITA: Porlamar, *Stephens M2* (GH, US). **ANZOATEGUI:** 87 km N of El Tigre, *Croat 53960* (MO); 15 km E of Píritu, *Croat 54392* (MO); Mesa Urica, 6 km ESE of La Ceiba, 58 km W of El Tejero, *Davidse et al. 4523* (MO, NY, US); Soledad, *Holt & Gehriger 157* (NY, US). **ARAGUA:** Maracay, El Limon, *Vivos 128* (GH). **BARINAS:** Llanos de Calabozo, between La Encrucijada and Mision de Arriba, *Lasser 119* (US). **BOLÍVAR:** 13 km ESE of Upata along Hwy. 10 to Guasipati, *Davidse et al. 4620* (CTES, MO, NY); Paso de Caruachi, 75 km E of Ciudad Bolívar, *Holst et al. 1964* (MO, NY); Ciudad Bolívar, *López-Palacios 1927* (MER, NY, US); Altaplanicie de Nuria, E of Miano, *Steyermark 88709* (NY, U, US); between Upata and Altagracia, *Steyermark 57695* (F, NY). **CARABOBO:** Vic. of Valencia, *Pittier 8681* (GH, NY, US); El Campo de Carabobo, *Rodríguez 124* (F, MY). **COJEDES:** Dept. El Pao, between La Hacienda Mata Clara and El Baul, *Aymard & Ortega 2325* (PORT, NY). **DISTRITO FEDERAL:** Around Caracas, *Pittier 7439* (GH, US), *Pittier 6084* (US). **GUÁRICO:** Hwy. from Calabozo to San Fernando de Apure, 80 km S of Calabozo, *Agostini et al. 2005* (NY, TEX, VEN); Los Llanos Biological Station, ca. 10 km SSE of Calabozo, *Davidse 3681* (MO, NY); 12 km SE of Calabozo, *Ramírez 2052* (MO, NY); ca. 39 km SSW of Calabozo on Hato Masaquarel, *Rondeau 72* (ARIZ, MO, NY). **MONAGAS:** 2 km SE of Maturín, *Fosberg 45222* (US); 9 km WSW of Jusepín, *Pursell et al. 8019* (NY); 19.5 km SE of Urica, *Steyermark et al. 114280* (MO). **NUEVA ESPARTA:** Guatamare, *Gines 3193* (US); Guayamurí, *Gines 3151* (US). **PORTUGUESA:** Between Ospino and Guanare, *Benítez de Rojas 638* (MY, U); Dept. Guanare, Mesa Alta, *P. A. Fryxell et al. 4351* (CTES, NY). **YARACUY:** Campo Carabobo W of Tocuyito, *Diedrichs 65* (MO).

GUYANA. Karanambo Flats, *Goodland 1026* (NY, US); penal settlement, *Hitchcock 17095* (GH, NY, S, US); Orealla savannah, Corentyne River, Berbice, *Irwin 612* (TEX); 4 mi N of Lethem, Essequibo, *Irwin 1019* (TEX); Rupununi Distr., Warimure, *Jansen-Jacobs et al. 2262* (K, NY, U, US); Manari, *Maas & Westra 3640* (MO, NY, U); Basin of Rupununi River, near mouth of Charwair Creek, *A. C.*

Smith 2387 (GH, K, NY, S, U, US); Roraima and vic., *Tate 122* (NY).

SURINAM. Lower Marowyne River, 25 km N of Albina, *Hekking 1078* (A, U); Marowyne River, *Kappler 1693* (MO, S).

FRENCH GUIANA. 15 km from St. Laurent, rd. to Cayenne, *R. S. Cowan 38940* (NY); Karouany, *Sagot 43* (C, NY, S, U, US).

BRAZIL. **Alagoas:** Mun. Maceió, Pontal da Barra, *Esteves 180* (NY). **Amazonas:** Rd. to Aleixo, ca. 10 km near SUDAM plantation, Manaus, *Hill 13206* (GH, MO, NY, TEX); Ponta Negra, W of Manaus, *Lowe 4245* (E, K); sandy beach at Prainha, *Traill 51* (K–2). **Bahia:** Mun. Una, Reserva Biológica do Mico-leão (IBAMA), *de Carvalho et al. 4531* (CEPEC, NY, TEX); 43 km E of Ibotirama, *P. A. Fryxell 1927* (CTES, NY); 15 km W of Ibotirama, *P. A. Fryxell 1925* (BH, CAS, NY, TEX); 40 km E of Jacobina at Junco, *P. A. Fryxell 1946* (BH, CAS, CTES, NY, TEX); 2 km N of Rio de Contas, *Harley et al. 15619* (CTES, K, MO, NY, U, US); 23 km N of Seabra, rd. to Agua de Rega, *Irwin et al. 30864* (CTES, F, MO, NY, US); Cruz das Almas, *Krapovickas 12789* (C, CTES, F, LL, MBM, NY, SI, UC); 3 km E of Barreiras, *Krapovickas et al. 37872* (CEN, CTES, F, G, UC); Crus de Casma, *Luschnath 5* (BR, CTES [fragment]); vic. of Bahia, *Rose & Russell 19687* (NY). **Ceará:** Mun. Fortaleza, Mucuripe, *Drouet 2170* (F, GH, MO, NY, S, US); Crato, slopes of the Serra do Araripe, *Lima 260* (IPA, NY); Fortaleza, *Lindeman & de Haas 6142* (U); 25 km from Nova Olinda on old Crato rd., *Pickersgill et al. RU72-188* (K); Crateús, *Swallen 4504* (US); Icalaí, Fortaleza, *Tsugaru & Sano B-1183* (NY). **Distrito Federal:** Quadra 713 Norte, *Mendonça & Filgueiras 174* (IBGE, MO, NY); Lagoa Paranoá Norte, *B. A. S. Pereira 231* (CTES, K). **Espírito Santo:** Mun. Linhares, Povoação, *Hatschbach & Cervi 51394* (S). **Goiás:** Mun. Vianópolis, *Walter et al. 170* (CTES). **Maranhão:** Fazenda São Francisco, *Daly et al. D278* (NY); 25 km S of Loreto, *G. Eiten & L. T. Eiten 4120* (NY, US); Island of São Luiz, *Fróes 11889* (A, F, NY, US); Mun. Carolina, 35 km N of Carolina, *E. L. Taylor et al. E1283* (NY, TEX). **Mato Grosso:** Campo Grande, *Krapovickas & Cristóbal 34548* (CTES, F, SI); Mun. Tres Lagoas, *Vália 61* (US). **Minas Gerais:** Rio Pandeiros, ca. 52 km by rd. W of Januária, *W. R. Anderson, P. A. Fryxell et al. 9273* (F, NY, TEX, US); ca. 1 km E of Rio Pandeiros, near rd. to Januária, *W. R. Anderson, P. A. Fryxell et al. 9090* (F, K, NY, TEX); Mun. Buenopolis, between Buenopolis and Joaquim Felicio, *Arbo et al. 4532* (CTES, K, MO); Serra do Espinhaço, ca. 18 km W of Grão Mogol, *Irwin et al. 23654* (C, DS, F, NY, TEX); 60 km SW de Pirapora, *Krapovickas & Cristóbal 42905* (CTES, HAS, K, MO, NY); Pedra Azul, 10 km E of the city, rd. to Almenara, *Harley et al. 25200* (K, NY, SPF). **Pará:** Upper Cupary River, plateau between the Xingu and Tapajos Rivers, *Krukoff 1085* (A, K, NY, S, U); Belem, *Murça-Pires & Black 93* (GH); rd. from Belém to Brasília, *J. Pereira 5052* (CTES). **Pernambuco:** Mun. Exu, BR-122, Rod. Asa Branca, rd. to Sitio Serrinha, *Esteves & Barros 2571* (SPF); Recife, Dois Irmãos, *Krapovickas 8031* (CTES); Serra das Russas, *Lindeman & de Haas 6187* (NY, U); Tapera, *Pickel 107* (CAS, F, GH, PH, US). **Piauí:** Without locality, *Gardner 2049* (GH, OXF); 3 km W of Oeiras, *Krapovickas et al. 38760* (CEN, CTES, F). **Roraima:** Cautá region, 8 km NSE *[sic]* of Boa Vista, *Coradin & Cordeiro 747* (NY, TEX). **Rio de Janeiro:** Environs of Rio de Janeiro, *Glaziou 13548* (C–2, K). **Rio Grande do Norte:** Ponta Negra, Natal, *Tsugaru & Sano B-1227* (NY); S of Natal, ca. 2 mi E of Ponte Velha, *Wurdack B-8* (NY). **Tocantins:** Nub, Paranã, Rio Paranã, *Hatschbach et al. 56052* (C). **Unknown state:** Fazenda Jenipapa before the mouth of the Rio Velha, *Pohl 3064* (syntype of *P. cancellata* var. *crassivenosa*, GH).

BOLIVIA. **Santa Cruz:** Near Ipias, Chiquitos, *Cárdenas 4543* (CAS); Roboré, *Krapovickas 9189* (CTES); Prov. Velasco, 27 km S of San Miguel, *Krapovickas & Schinini 32389* (CTES, F, SI); Prov. Chiquitos, 2 km E of Roboré, near Urasibique River, *Krapovickas & Schinini 36364* (CTES, MICH, NY).

PARAGUAY. **Amambay:** Cerro Corá National Park, *Fernández-Casas & Molero 4084* (MO, NY); Caballero-cué, between the Apa River and the Aquidaban River, *Fiebrig 4781* (E); Aquidaban River, rt. 5 to Bella Vista, *Krapovickas & Schinini 32595* (CTES, SI); Bella Vista, *Mereles & Ramella 586* (CTES, G, MO); ca. 30 km S of P. J. Caballero, *Pedersen 14650* (C, CTES); Cerro Corá, *Schinini & Bordas 20183* (CTES, G).

Local names. Barba-de-bode (Brazil: Piauí), barba-de-veado (Brazil: Piauí), malva chanana (Brazil: Pará).

Pavonia cancellata is a variable species, especially in the character of its pubescence. Gürke (1892b: 516) recognized one extreme form, with densely stellate-pubescent foliage, as *P. cancellata* var. *crassivenosa* Gürke, but this appears to me simply to be an extreme of a continuous range of variation. The asymmetrical leaves and prostrate habit are characteristic.

40. Pavonia cochensis J. R. Johnston, Proc. Amer. Acad. Arts **40:** 690. 1905; Contr. Gray Herb. Harvard Univ. **29:** 690. 1905. Type. Venezuela. Nueva Esparta: Coche Island, San Pedro, 5 Aug 1903, *J. R. Johnston [18]* (holotype, GH; isotype, US). Fig. 4

Perennial *herb*, the stems (presumably trailing) densely stellate-pubescent. *Leaf* blades triangular or pentangular, 1–1.5 × 1.5–2 cm, sometimes slightly asymmetrical, deeply cordate, obscurely crenate, acute, palmately 5-nerved, densely and finely yellowish-tomentose on both surfaces; petioles 1 cm long or less; stipules 2–3 mm long. *Flowers* solitary in the leaf axils, the pedicels 1–1.5 cm long; involucellar bracts ca. 15, ca. 5 mm long (flowers and fruits otherwise not observable).

Distribution (Fig. 18). Known only from the type collection from Coche I., presumably from near sea level. The type locality is at 10°47′N, 63°56′W.

Pavonia cochensis is presumably closely related to the very common *P. cancellata*.

41. Pavonia humifusa St.-Hilaire, Fl. Bras. Merid. **1:** 235. 1827. *Lassa humifusa* (St.-Hilaire) Kuntze, Revis. Gen. Pl. **3(2):** 19. 1893. Type. Brazil. Río de Janeiro: on sandy shores near town called Fasenda de Moribeca, *St.-Hilaire s.n.* [*257*] (holotype, P as photo F-35475; type fragments, CTES, F). Figs. 4, 27

Pavonia prostrata Moricand, Pl. Nouv. Amer. 119. t. 72. 1844. *Malache prostrata* (Moricand) Kuntze, Revis. Gen. Pl. **1:** 71. 1891. Types. Brazil. Near Japira, 1840, *Blanchet 3149* (lectotype, here designated, NY; isolectotypes, CTES [fragment], RB).

Prostrate to ascending perennial *subshrubs* (rarely shrubs to 2.5 m: *Arbo et al. 5443*), the stems variably pubescent (with minute glandular hairs in Pernambuco; stellate-pubescent or glandular in Bahia; setose in Minas Gerais and Bolivia). *Leaf* blades hastately 3-lobed to palmately 5–7-lobed, slightly asymmetrical, deeply cordate, crenate-serrate, acute to obtuse, densely tomentose on both surfaces (sometimes also with scattered 3–4-armed spicules), mostly 2–6 × 3–9 cm; petioles 1.5–4 cm long; stipules subulate or filiform, 3–6 mm long. *Flowers* solitary in the leaf axils, the pedicels 1–4 cm long; involucellar bracts 14–16 (–22 in Pernambuco), 1–2 cm long, arcuate, copiously hirsute; calyx 7–10 mm long, hirsute, usually concealed by involucel; corolla 3 cm long, yellow fading orange. *Fruits* 6–7 mm diam., subglabrous, the mericarps strongly rugose, with 3 erect subapical points 1–2 mm long.

Distribution (Fig. 28). From Brazil (Pernambuco, Bahia, and Minas Gerais) to Bolivia, often growing in sandy soils in open (often arid) habitats. The type locality is near Itabapoana, 21°12′S, 40°39′W; that for *Pavonia prostrata* is at 17°29′S, 39°43′W. There is a gap in the distribution between the Bolivian and the E Brazilian segments of the population (see Fig. 28), but whether this gap is real or an artifact of collecting is not clear. The variation in pubescence (see above) is not correlated well with this distributional hiatus, and no other morphological characters are related to this distributional difference.

Additional illustrations. Gürke (1892b: pl. 99, fig. 1); Moricand (1844: pl. 72, as *P. prostrata*).

Specimens examined. **BRAZIL. Alagoas:** Without locality, 1838, *Gardner 1244* (syntypes of *P. prostrata*, GH–2, NY–2, OXF, US). **Bahia:** Espigão Mestre, ca. 100 km WSW of Barreiras, *W. R. Anderson et al. 36793* (F, K, MO, NY, TEX, US); Mun. Jacobina, 14 km SW of Hwy. BR-324, *Arbo et al. 5443* (CTES, K, NY, SPF, TEX); Caetité, rd. from Caetité to Brumado, ca. 83 km from Caetité, *Esteves & Kamyama 2520* (NY, SP); 50 km E of Ibotirama, *P. A. Fryxell 1958* (BH, NY, TEX); Juazeiro, Escola Agricola, *Krapovickas 12901* (CTES, MICH, NY, P, SI, UC). **Goiás:** Without locality, *Duarte 9528* (CTES). **Espírito Santo:** Itapemerin, near Marataises, *Occhioni 7372, 7946* (CTES). **Minas Gerais:** Pirapora, *Krapovickas & Cristóbal 42863* (CTES, NY). **Pernambuco:** Mun. Petrolina, rd. from Petrolina to Remanso, *Esteves & Lyra-Lemos 2218* (CTES); 40 km E of Remanso, *P. A. Fryxell 1944* (NY).

BOLIVIA. Santa Cruz: Santiago, 650 m, *A. H. Gentry & Foster 75485* (MO, TEX); Prov. Ichilo, 3 km N of Buena Vista, *Krapovickas & Fortunato 43921* (BAB, CTES, LPB); 10 km N of Santa Cruz, *Krapovickas & Schinini 31548* (C, CTES, F, G, MO, NY, SI); Prov. Chiquitos, Roboré, *Krapovickas & Schinini 36511* (CTES, K); Prov. Sara, Buenavista, *J. Steinbach 6303* (A, F, GH, MO, NY, PH, S, U).

Local name. Malva rastrera (Bolivia).

Pavonia humifusa is similar to *P. cancellata* but distinguished by its velutinous pubescence and by its nearly symmetrical leaves.

42. Pavonia repens Fryxell, Brittonia **32:** 266. 1980. Type. Brazil. Lagoa Comprida, Mar 1839, *Gardner 2048* (holotype, K).

Perennial *herbs*, procumbent and rooting at the nodes, the stems pilose, hairs 1.5–3 mm long, and with understory of minute stellate hairs. *Leaf* blades angulately flabelliform, deeply cordate, dentate to subentire, subacute to obtuse, 2–3 × 3.5–5 cm, palmately 7–9-nerved, stellate-pubescent; petioles to 3.5 cm long (usually longer than the blade); stipules 3–5 mm long, usually 2–3-parted. *Flowers* solitary in the leaf axils, the pedicels 6.5–10 cm long (exceeding the leaves); involucellar bracts 13–15, linear, 15–20 mm long, more or less arcuate, ciliate, the hairs 2 mm long; calyx 8–10 mm long, pubescent and ciliate; corolla 2.5–3 cm long, color not known, externally subglabrous; staminal column ca. 8 mm long, the filaments 1–2 mm long, the pollen reddish; styles long-exserted from the column, subequal to the petals, united below, free for distal 1.5–2 mm, the stigmas villous. *Fruits* unknown.

Distribution (Fig. 28). Known only from central Brazil. The type locality is at 7°5′–15′S, 41°25–55′W.

Specimens examined. **BRAZIL. Bahia:** Flamengo, Barrinha River, *Krapovickas 12862* (CTES, G, NY).

Pavonia repens is distinguished from other members of *P.* sect. *Cancellaria* by its repent habit, its distinctive stipules, and its flabelliform, long-petiolate leaves.

Pavonia subgen. *Pavonia* sect. *Malvaviscoides*

Pavonia sect. **Malvaviscoides** St.-Hilaire, Fl. Bras. Merid. **1:** 237. 1827. *Pavonia* subsect. *Malvaviscoides* (St.-Hilaire) Endlicher, Gen. Pl. 982. 1840. *Pavonia* subgen. *Malvaviscoides* (St.-Hilaire) Ulbrich, Bot. Jahrb. Syst. **57:** 65 . 1921. Type. *Pavonia malvaviscoides* St.-Hilaire.

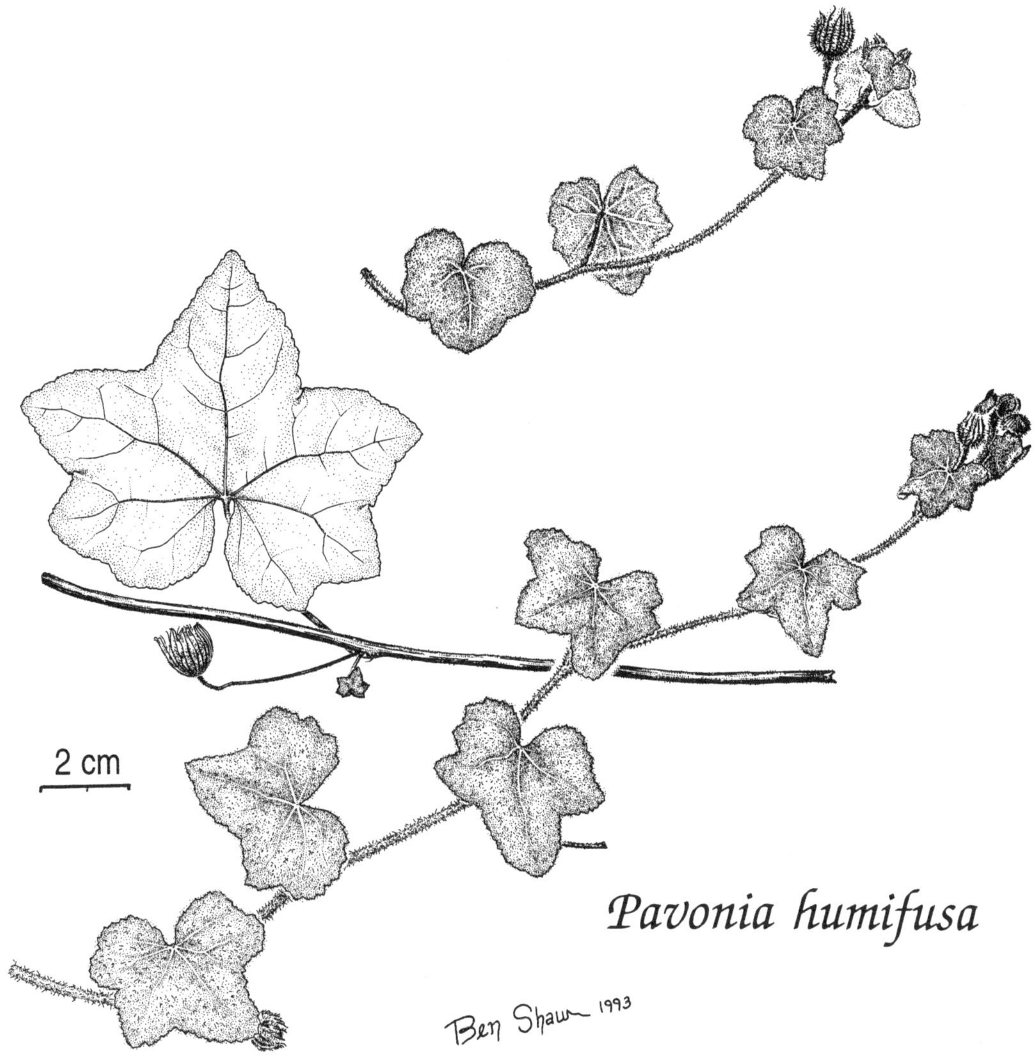

FIG. 27. *Pavonia humifusa.* Procumbent flowering branches (*Fryxell 1944* [large leaf and bud], *Fryxell 1959*).

Shrubs, the leaves elliptic to ovate, unlobed; bracts of the involucel 4–14, sometimes gamophyllous, sometimes reddish or pinkish; calyx tubular, short-lobed, the lobes shorter than the tube.

St.-Hilaire (1827) and Don (1831), included six species in *Pavonia* sect. *Malvaviscoides*, five of which (except the type, *P. malvaviscoides*) pertain to *P.* subgen. *Goetheoides* Gürke. The group was delineated more naturally by Gürke (1892b), whose conception was followed by Ulbrich (1921) and Kearney (1958). Gürke included four species (*P. malvaviscoides, P. erythrolema, P. glazioviana,* and *P. montana*) in the group, to which the more recently described *P. melanostyla, P. grazielae,* and *P. zehntneri* may be added, as well as an earlier species of St.-Hilaire (*P. viscosa*), and two species that were segregated to the genus *Codonochlamys* by Ulbrich but that are better placed in this group as *P. glaziovii* and *P. tiliifolia*. In addition, two new species are added here (*P. decora* and *P. paraënsis*), which brings the total number of species in the section to 12.

The name *Malvaviscoides* means "like the genus *Malvaviscus*" in allusion to the general aspect of the flowers.

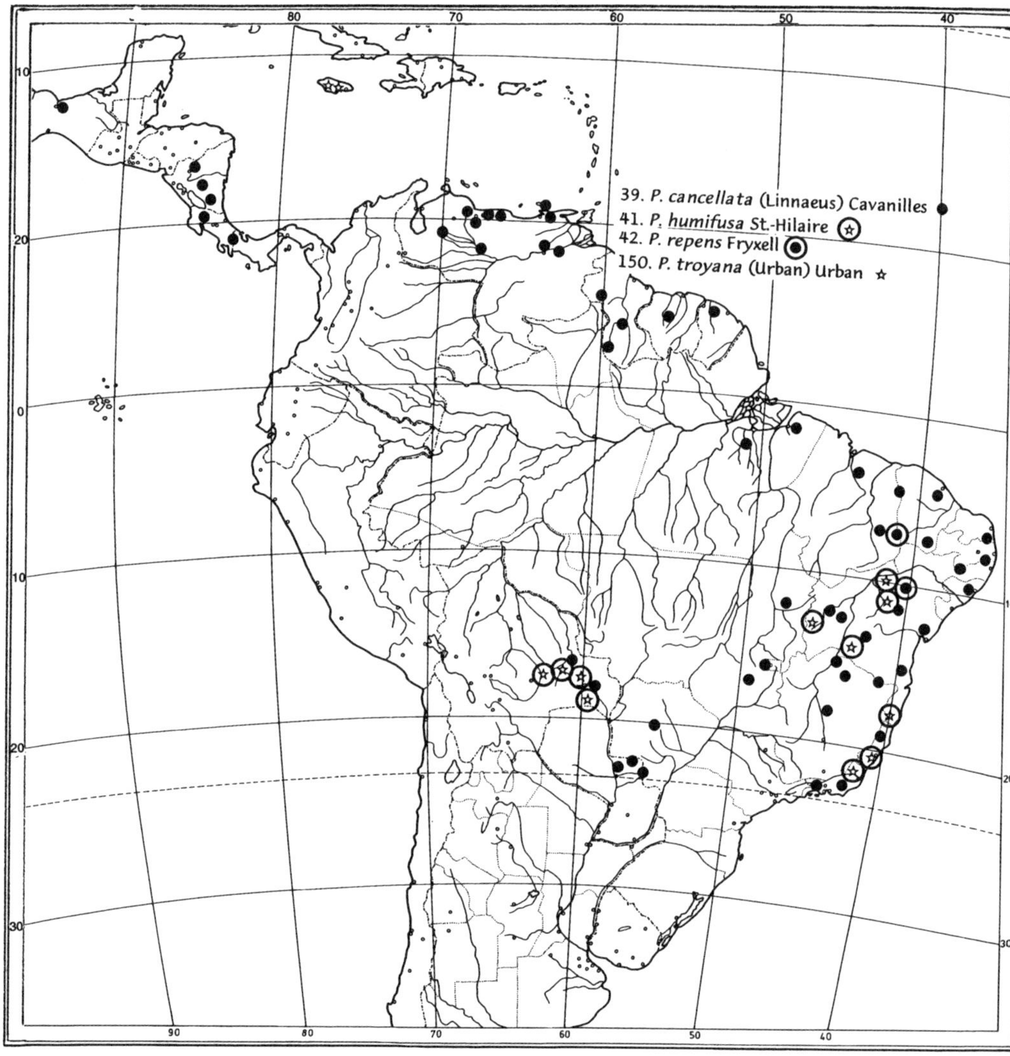

FIG. 28. Distribution map of four neotropical species of *Pavonia* sect. *Cancellaria*, and *P.* sect. *Malache*.

Key to the species of *Pavonia* sect. *Malvaviscoides*

1. Bracts of involucel 4(–6), distinct or gamophyllous, triangular to broadly ovate.
   2. Involucellar bracts distinct, broadly ovate; genitalia subequal to corolla.
      3. Stems viscid; involucellar bracts 10–15 mm wide (rarely narrower), viscid. . . . . 47. *P. grazielae*
      3. Stems tomentose; involucellar bracts 25 mm wide, tomentose. . . . . 51. *P. serrana*
   2. Involucel gamophyllous, forming a tubular or urceolate structure; genitalia exserted.
      4. Leaves tomentose; involucel and calyx velutinous; petals 2.5–3 cm long, erect forming a tubular corolla; pedicels 3–7 mm long. . . . . 52. *P. tiliifolia*
      4. Leaves scabridulous; involucel and calyx scabridulous; petals 3–3.5 cm long, spreading; pedicels to 35 mm long. . . . . 46. *P. glaziovii*
1. Bracts of involucel 8–14, distinct, linear, 0.5–4 mm wide; genitalia exserted or included.
   5. Leaf blades elliptic, palmately 3–5-nerved, 2–5 times as long as wide (rarely wider); petioles 0.3–4 cm long; involucellar bracts 8–10; genitalia exserted. . . . . 44. *P. erythrolema*
   5. Leaf blades ovate-cordate, palmately 7–9-nerved, 1–2 times as long as wide; petioles 1.5–8 cm long; involucellar bracts 8–14; genitalia included or exserted.

6. Involucellar bracts 8–10.
 7. Involucellar bracts 1.5–2 × 0.5–1.5 mm, not reddish; leaves white-velutinous beneath; androecium slightly exserted. . . . . 48. *P. malvaviscoides*
 7. Involucellar bracts 25–30 × 2–4 mm, more or less reddish; leaves coarsely stellate-pubescent beneath; androecium manifestly exserted (in *P. paraënsis*, unknown in *P. decora*).
  8. Stems with appressed stellate hairs; calyx 15–23 mm long; involucellar bracts ligulate. . . . . 50. *P. paraënsis*
  8. Stems with setose hairs 3–5 mm long; calyx 12–13 mm long; involucellar bracts spatulate. . . . . 43. *P. decora*
6. Involucellar bracts 11–14.
 9. Genitalia manifestly exserted; calyx and involucel more or less pink or red distally; herbage non-viscid.
  10. Corolla subequal to calyx; calyx and involucel pink or reddish distally. . . . . 45. *P. glazioviana*
  10. Corolla twice length of calyx; calyx and involucel often yellowish pink (or only yellowish) distally. . . . . 54. *P. zehntneri*
 9. Genitalia included or barely exserted; calyx and involucel not brightly colored; herbage often viscid.
  11. Involucellar bracts filiform, 0.5 mm wide or less; upper leaf surface with appressed simple hairs 0.5–1 mm long. . . . . 49. *P. montana*
  11. Involucellar bracts lance-linear, 1–3 mm wide; upper leaf surface with minute glanduar hairs (seemingly glabrous). . . . . 53. *P. viscosa*

43. Pavonia decora Fryxell, sp. nov. Type. Brazil. Bahia: Mun. Iraquara, Fazenda Torrinha, 12 km S of Iraquara, 640 m, caatinga, 14 Jun 1981, *Mori, Boom & Funch 14431* (holotype, NY). Figs. 29, 30

Frutices caulibus setosis, setis 3–5 mm longis; laminis foliorum ovato-cordatis; pedicellis ubique setosis atque latere abaxiali pilis brevibus recurvatis; bracteis involucellorum 8–9 spathulatis rubris; fructibus laevibus 11–12 mm diametro.

Shrubs 1 m tall, the stems setose, the setae 3–5 mm long, pustular-based, yellowish, spreading or somewhat retrorse, with underlying finer pubescence including scattered glandular hairs and dense longitudinal rows of whitish curled hairs. *Leaf* blades ovate, 4–7 × 4–5.5 cm, basally cordate, crenate, acute, palmately 7-nerved (the principal veins raised above and beneath), concolorous, the upper surface hispid on the veins and with widely scattered simple or stellate hairs (<0.5 mm long) on the blades, the lower surface with scattered 3–4-armed stellate spicules (1 mm long); petioles 2.5–5 cm long, with pubescence like stem, the whitish curled hairs on the adaxial side; stipules subulate, 3 mm long, reflexed (?), deciduous. *Flowers* solitary in the upper leaf axils, the pedicels 4–6 cm long, slender, with pubescence like stem, the whitish curled hairs on the *ab*axial side [!]; involucellar bracts 8–9, distinct, erect, oblanceolate or spatulate, 25 × 2.5–4 mm (gradually narrowed to 1 mm at base), acute, red and showy, minutely pubescent and basally setose, the setae 3–4 mm long, somewhat reddish; calyx 12–13 mm long, minutely stellate- and glandular-pubescent, enclosing the fruit, 5-lobed, the lobes 5 mm long, triangular; flowers otherwise unknown. *Fruits* 11–12 mm diam.; mericarps 6–6.5 mm tall, minutely appressed-pubescent (seemingly glabrous except under high magnification), basally winged, relatively smooth.

Distribution (Fig. 15). Known only from the type collection from 12°23′S, 41°32′W.

This species is included in *Pavonia* sect. *Malvaviscoides*, but it also shows some affinity for species of *P.* sect. *Lopimia*, so it is keyed out there also.

The specific epithet is from the Latin *decorus*, in the sense of ornate, elegant, or beautiful, in reference to the showy involucel. It is distinctive for its 8–10 large, spatulate, reddish involucellar bracts and its exserted androecium.

44. Pavonia erythrolema Gürke *in* Martius, Fl. Bras. **12(3):** 523. 1892. Type. Brazil. Minas Gerais: Fanado (Minas Novas) and Piedade, *von Martius s.n.* (holotype, B as photo F-9441; isotype, M as photo CTES, CTES [fragment]). Fig. 31

Pavonia melanostyla Ulbrich, Bot. Jahrb. Syst. **42:** 228. 1908. Type. Brazil. Bahia: Mtns. near Calderão, *Ule 35* (holotype, B as photo F-9454).

Shrubs 2–3 m tall, the stems roughly and minutely stellate-pubescent. *Leaf* blades lance-elliptic to ovate, mostly 3–8 × 1–5 cm, truncate or subcordate, serrate, acute or acuminate, palmately 3–5(–7)-nerved, discolorous, sparsely stellate-pubescent above, densely to sparsely stellate-pubescent beneath; petioles 0.3–5 cm long; stipules 4–8 mm long, filiform, erect. *Flowers* solitary in the leaf axils, the pedicels up to 6 cm long,

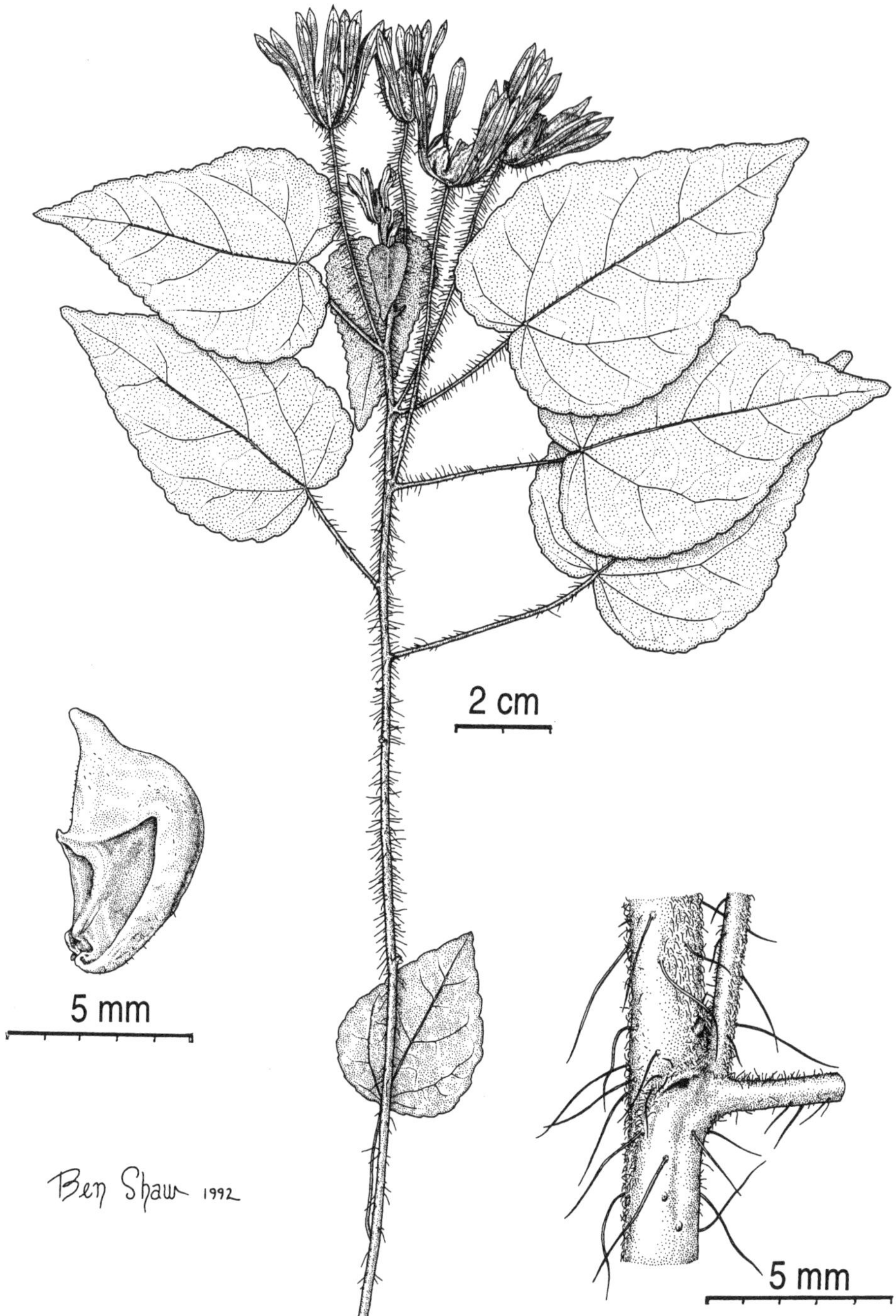

FIG. 29. *Pavonia decora.* Flowering branch, with enlargements of mericarp (lower left) and of stem segment (lower right) (*Mori et al. 14431*).

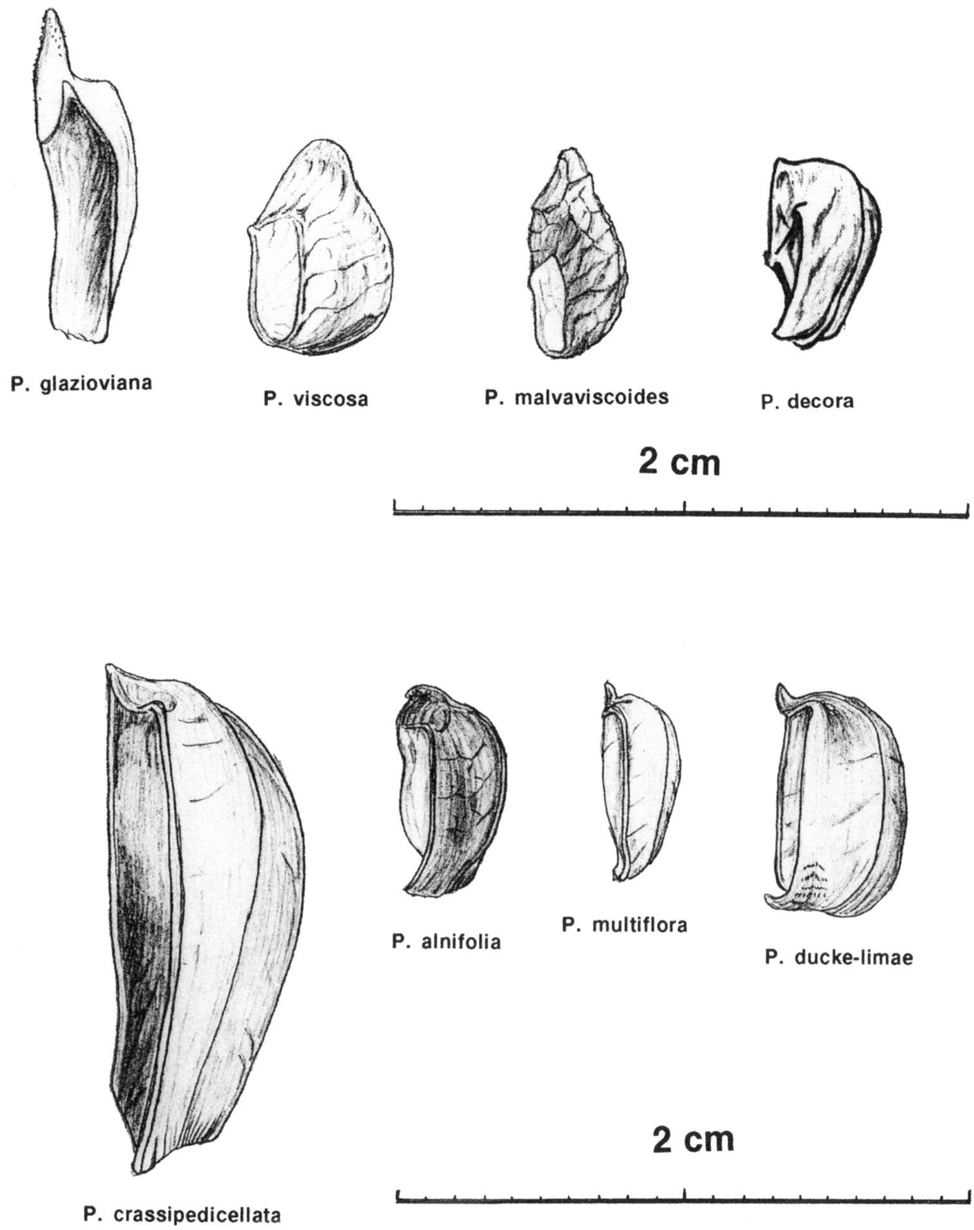

FIG. 30. *Pavonia* mericarps in left lateral view of *P.* sect. *Malvaviscoides* (top row) and *P.* subgen. *Goetheoides* (bottom row). Top row (left to right): *P. glazioviana* (*Blanchet 2843*), *P. viscosa* (*Damazio 933*), *P. malvaviscoides* (*Sazima s.n.*), *P. decora* (*Mori et al. 14431*). Bottom row (left to right): *P. crassipedicellata* (*Silva 374*), *P. alnifolia* (*Duarte 6230*), *P. multiflora* (*de Quieroz 2470*), *P. ducke-limae* (*Krapovickas 12872*).

slender; involucellar bracts 8–10, linear, 20–38 × 1–2 mm, reddish, stellate-pubescent and sometimes also ciliate; calyx 10–19 mm long, densely pubescent, the lobes 4–7 mm long; corolla 2.5–3 cm long; androecium exserted by 1–2 cm, the column purplish, the filaments 2–3 mm long; styles slender, minutely and sparsely pubescent, purplish. *Fruits* unknown (available specimens incomplete).

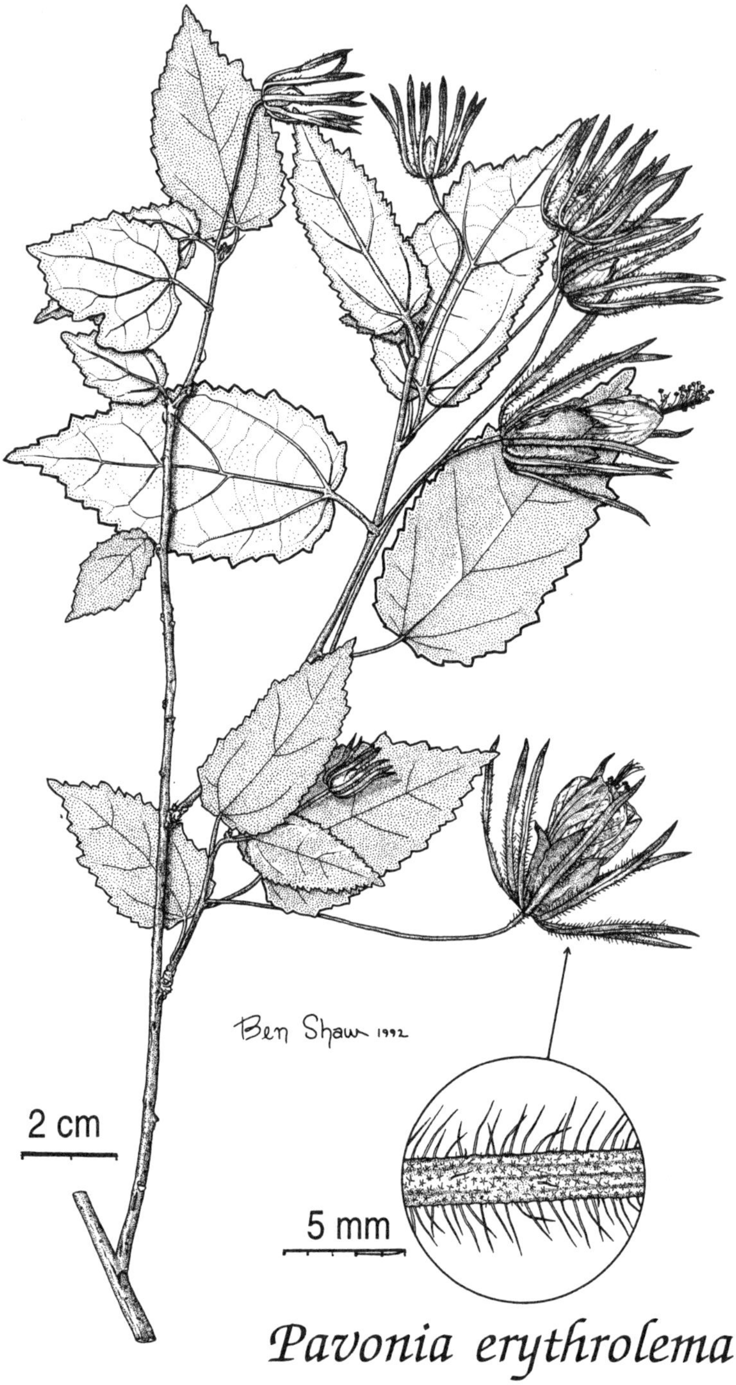

FIG. 31. *Pavonia erythrolema*. Flowering branch, with enlargement of portion of involucellar bract to show pubescence (*Mori et al. 9990*).

Distribution (Fig. 15). From the states of Bahia and Minas Gerais in central Brazil. The type locality is between Minas Novas and Turmalina, 17°16′S, 42°39′W.

Specimens examined. BRAZIL. BAHIA: Mun. Maracás, km 7 on the rd. from Maracás to Contendas do Sincorá, *de Carvalho & Plowman 1567* (CTES, F, NY); Mun. Caitité, Ramal, 29 km on the rd. from Caitité to Brumado, *de Carvalho et al. 3780* (NY, TEX); Maracás, toward Contendas do Sincorá, *Esteves 2384* (CTES, NY, SPF); Mun. Palmeiras, rd. to Guiné, 3 km from the city, *Esteves & Lyra-Lemos 2535* (CTES); between Brejão and Mucugê, *Fróes 19991* (K, NY, US); Contenda do Sincorá, *Hatschbach & J. M. Silva 50073* (C, CM, CTES, US); Mun. Maracás, hwy. from Maracás to Contendas do Sincorá, km 2, *Mattos-Silva et al. 221* (CTES–2, NY); Mun. Maracás, 26 km SW of Maracás, *Mori et al. 9990* (CTES, K, NY, RB); Fazenda da Cova, 18 km from Maracás, toward Contendas and Brumado, *Pabst 8562* (= *E. Pereira 9673*) (NY); Maracás, *Sobral & Silva 5887* (CTES).

Pavonia erythrolema is distinguished by its elliptic leaves and by its exserted androecium with a purplish column.

45. Pavonia glazioviana Gürke *in* Martius, Fl. Bras. **12(3):** 524. 1892. Type. Brazil. Near Certam de Bahia in the Serra Açuruá, *Blanchet 2843* (isotypes, CTES [fragment], G as photo F-23696, MG, MO, NY, P as photo US-5979). Fig. 30

Pavonia andrade-limae Monteiro, Anais Soc. Bot. Brasil, XXIII Congr. Nac. Bot. 131. 1972. Type. Brazil. Pernambuco: between Ibimirim and Arcoverde, *Andrade-Lima 49-253* (holotype, IPA–n.v.).

Shrubs 2–2.5 m tall, the stems densely covered with appressed 6–12-armed stellate hairs. *Leaf* blades subrotund to ovate, sometimes weakly lobed, truncate to cordate, denticulate to subentire, acute to obtuse, more or less discolorous, palmately (sometimes pedately) 7–9-nerved, stellate-pubescent above and beneath, mostly 3–8 × 3–7 cm; petioles 2–4.5 cm long; stipules 3–4 mm long. *Flowers* solitary in the leaf axils, the pedicels 2–6 cm long; involucellar bracts distally pink or reddish, 12–14, linear-lanceolate, 27–35 × 1.5–3 mm, pubescent and ciliate; calyx distally pink or red, 18–25 mm long, tubular, stellate-pubescent; corolla pink, subequal to calyx or slightly exceeding calyx; staminal column exserted 1–2.5 cm, purplish, the filaments 2–4 mm long; styles pubescent, exserted from column. *Fruits* ovoid, 10 mm diam., 13 mm long, beaked, stellate-pubescent; mericarps basally winged.

Distribution (Fig. 11). From the states of Bahia, Ceará, Pernambuco, and Piauí in Brazil, growing in caatinga vegetation at elevations of ca. 1000 m. The type locality of *Pavonia glazioviana* is at 11°15′–45′S, 42°30′W; that of *P. andrade-lima* is at 8°25′S, 37°05–40′W.

Illustrations. Monteiro (1972: figs. 8, 9).

Representative specimens examined. BRAZIL. BAHIA: Mun. Morpará, Carnúba Grande, *Bautista & Salgado 882* (RB); 8.7 km from Boninal, rd. to Seabra, *Esteves & Kameyama 2522* (NY, SP); Mun. Jeremoabo, BR-110, km 24 toward Jeremoabo, *Esteves & Lyra-Lemos 2503* (CTES, NY, SPF); Mun. Seabra, 11 km from the entrance to Boninal, *Esteves & Lyra-Lemos 2544* (CTES); ca. 4 km NE from Gentio do Ouro, rd. to Central, *Harley et al. 18936* (CTES, K, MO, NY, TEX, U); Mun. Seabra, Queimada Nova, *Hatschbach 39528* (C, CTES, NY, herb. Pedersen); Mun. Palmeiras, Rio Lajeadinho, *Hatschbach et al. 56914* (C, CTES, K); Mun. Boninal, Nova Colina, *Hatschbach & J. M. Silva 50154* (C, CTES, NY, US). **CEARÁ:** Serra Bonita, Aiuaba, 15 Jul 1980, *Angélica s.n.* (CTES); Mun. Novo Oriente, Baixa Fria, *de Araujo 46* (IPA). **PERNAMBUCO:** 50 km E of Remanso, *P. A. Fryxell 1943* (BH, CAS, CTES, MICH, NY, TEX); Ibimirim, 2 km from the center of Ibimirim to Petrolina, *Rodal 67* (IPA, NY). **PIAUÍ:** Between São João do Piauí and Boa Esperança, *Castro et al. 1082* (CTES); Mun. Caracol, Serra das Confusões, *Del'Arco & Castro 1113* (CTES); Toca da entrada do Pajaú, Serra da Capivara, São Raimundo Nonato, *Emperaire 2680* (CTES); Cova-Donga, Pio Nono, 21 Feb 1980, *Fernandes & Martins s.n.* (CTES).

Pavonia glazioviana is distinct for its 12–14 reddish involucellar bracts and for its reduced corolla, which is subequal to the reddish calyx.

A specimen from Piauí (Chapa da Serra Branca, São Raimundo Nonato, *Emperaire 2415*, CTES) differs from typical *Pavonia glazioviana* in having pilose hairs 2–4 mm long abundantly present on stems, petioles, pedicels, and both surfaces of the leaf, unlike any of the other specimens cited above. Whether this plant is distinct from *P. glazioviana* is questionable, in part because the specimen is not complete and in part because the locality is the same as that of *Emperaire 2182*, a specimen that clearly represents *P. glazioviana*. Until further material becomes available, the anomalous phenotype will simply be noted for future study.

46. Pavonia glaziovii (Ulbrich) Fryxell, comb. nov. Basionym: *Codonochlamys glaziovii* Ulbrich, Notizbl. Bot. Gart. Mus. Berlin-Dahlem **6:** 331. 1915. Type. Brazil. Minas Gerais: Mendanha, *Glaziou 15842* (holotype, B as photo F-9422; isotypes, C, P as photo CTES).

Subshrubs or shrubs, the stems terete, ferruginous-scabrid, subsequently glabrescent and brownish. *Leaf* blades broadly ovate to suborbicular (sometimes weakly trilobed), basally deeply cordate, irregularly serrate or subdentate, apically acute or subacuminate, stellate-pubescent above and beneath, harshly so (scabrid) beneath, palmately 7-nerved, dark green; petiole straight, ferruginous, scabrid, about half the length of the blade; stipules subulate, ca. 7 mm long, roughly tomentose, caducous. *Flowers* solitary in the leaf axils; pedicels ca. 3.5 cm long, articulated ca. 5 mm below the flower; involucel ca. 15 mm long, gamophyllous, the 5 lobes triangular, 3–4 mm long, striate, densely pubescent with yellowish to ferruginous tomentum; calyx campanulate or somewhat urceolate, slightly exceeding the involucel, the oblong-acuminate lobes 4–5 mm long; corolla red, 3–3.5 cm long, tubular, the petals linear (ca. 5 mm wide), obtuse, spreading following anthesis; staminal column exceeding corolla by 1–1.5 cm, narrowly cylindric, glabrous, reddish, the filaments 1–2 mm long; ovary subglobose, glabrous, 5-parted; styles slender, glabrous. *Fruits* unknown.

Distribution (Fig. 11). Known only from the type collection. The type locality is at 18°05′S, 43°28′W.

Only the isotype (a somewhat incomplete specimen) of this species has been examined, as well as the photographs cited above. The preceding description is consequently derived in part from the original descriptions of Ulbrich, in Latin and in German, together with consideration of the type photos. The photo of the holotype includes an original drawing of a floral dissection, presumably by Ulbrich. Ulbrich described the calyx as bilabiate, but an examination of the isotype at C suggests that this character is an artifact of pressing. *Pavonia glaziovii* is distinctive, along with the similar *P. tiliifolia*, for its gamophyllous involucel.

47. Pavonia grazielae Krapovickas, Revista Bras. Bot. **1:** 89. 1978. Type. Brazil. Minas Gerais: Grão Mogol, 8 Jul 1975, *Barroso s.n.* (holotype, CTES; isotypes, MBM–n.v., P–n.v., RB–n.v.).

Shrubs 2 m tall, the stems densely stellate-puberulent and viscid. *Leaf* blades broadly ovate, 2.5–9 × 2.5–7 cm, cordate, denticulate to subentire, acute or acuminate, discolorous, palmately 7-nerved, the upper surface stellate- and glandular-pubescent, the lower surface predominantly stellate-pubescent, whitish; petioles 1.5–8 cm long; stipules 4–10 mm long, subulate. *Flowers* solitary in the leaf axils, the pedicels 1–6 cm long; involucellar bracts 4(–6), broadly ovate, 20–25 × 10–15 mm (rarely only 5 mm wide, e.g., *Esteves & Kameyama 2503*), viscid; calyx 1.5 cm long, viscid; corolla 2–3 cm long, externally stellate-pubescent, greenish yellow with reddish veins; androecium included, the styles slightly exserted. *Fruits* glabrous, ca. 6 mm diam.; mericarps 10 mm high, with a stout apical spine.

Distribution (Fig. 15). Confined to the state of Minas Gerais, Brazil, growing in cerrado vegetation at elevations of 700–1300 m. The type locality is at 16°35′S, 42°53′W.

Illustration. Krapovickas (1978a: fig. 1).

Specimens examined. **BRAZIL. Minas Gerais:** Fazenda Jambeiro, *Esteves et al. 13315* (CTES); Grão Mogol, 800 m from the bridge over the Itacambiruçu River, on the rd. to Cristália, *Esteves & Kameyama 2503* (NY, SP); Mun. Grão Mogol, ca. 7 km S of the city, *Harley et al. 25062* (K, NY); Mun. Grão Mogol, Itacambiruçu River, *Hatschbach 41298* (C, CTES, NY); 7 km S of Grão Mogol, *Hatschbach 41523* (CTES, US), *Hatschbach 42918* (F); Rio das Mortes, *Hatschbach & Zelma 52015* (TEX, US); Rio das Mortes, *Hensold et al. 30121* (CTES); 4 km from the center of the city Grão Mogol on the rd. to Barrocão, *Mello-Silva et al. 474* (CTES); Córrego Escurinha, *Mello-Silva et al. 43712* (F); Vale do Córrego Escurona, *Pirani et al. 12864* (CTES); banks of the Ribeirão das Mortes, *Simão-Bianchini et al. 12995* (CTES).

Pavonia grazielae is characterized by its few (4–6), broadly ovate, viscid involucellar bracts.

A collection by Pirani et al. (Grão Mogol, Vale do Rio Itacambiruçu, *SPF-39645*, CTES) may represent this species but is anomalous in having usually 6 bracts that are only 3–6 mm wide. However, it comes from within the range of the species and is tentatively included with it.

48. Pavonia malvaviscoides St.-Hilaire, Fl. Bras. Merid. **1:** 237. 1827. Type. Brazil. Minas Gerais: Serra da Candonga, near Tapanhuacanga, *St.-Hilaire s. n.* [*430*] (holotype, P as photo F-35479). Fig. 30

Shrubs or small trees up to 6 m tall, the stems finely puberulent. *Leaf* blades narrowly to broadly ovate, shallowly cordate, serrate to subentire, acute or acuminate, mostly 3–8 × 2.5–8 cm, markedly discolorous, palmately 9-nerved, with minute glandular hairs on upper surface (appearing glabrous), finely and densely canescent on lower surface; petioles 2–6 cm long; stipules 2–3 mm long, subulate. *Flowers* solitary in the leaf axils, the pedicels slender, 3–6 cm long; involucellar bracts 8–10, linear, 1.5–2 × 0.5–1.5 mm, puberulent; calyx 18–30 mm long, finely puberulent, sometimes setose, the lobes 6–10 mm long; corolla 30–38 mm long, red; androecium slightly exserted, the filaments 4–5 mm long; (gynoecium not observed). *Fruits* 6–7 mm diam., ovoid, the mericarps pubescent, 7 mm high (twice height of column), rugose; seeds pubescent.

Distribution (Fig. 11). Confined to the state of Minas Gerais, Brazil, growing in shrubby savanna at ca. 1300 m elevation. The type locality is at 21°53′S, 44°12′W.

Specimens examined. BRAZIL. Minas Gerais: Sierra do Espinhaço, E slopes of Pico de Itambé, *W. R. Anderson et al. 35930* (F, MO, NY, TEX, US); Mun. Diamantina, Serra do Gravião, *Barreto 9918* (F); Serra do Cipó, Santa Luzia, *Barreto & Brade 1291* (F); Serra do Cipó, *Duarte 2587* (RB); Serra do Cipó, km 140, *Duarte 8148* (CTES); Couto Magalhaes, Chapada do Couto, *Furlan et al. 4642* (CTES, NY, SPF); Pico Itambé, *Hatschbach 30131* (C, CTES, K, NY, US); BR-259, 5–10 km from Serro, *Hatschbach & J. M. Silva 50219* (C, CM, CTES, NY, S, US); Serra do Cipó, ca. 153 km N of Belo Horizonte, *Irwin et al. 20317* (CTES, F, GH, MO, NY, US); Mun. Santana do Riacho, km 133 hwy. from Belo Horizonte to Conceição do Mato Dentro, *Pinheiros & Esteves 8571* (CTES); Mun. Jaboticatubas, Serra do Cipó, rd. from Vespaziano to Conceição do Mato Dentro, km 126, 7 Jul 1974, *Sazima s.n.* (NY).

Pavonia malvaviscoides is distinctive for it discolorous foliage, which is prominently white-velutinous beneath.

Sazima (1981) notes that the flowers of this species (voucher: *Sazima s.n.*, cited above) remain open about 3 days and that they are protogynous and odorless. She further describes the Long-billed Hummingbird, *Phaethornis pretrei,* as the most frequent and "best" pollinator.

St.-Hilaire suggested that *Achania cordata* Nees & Martius (Nov. Act. Natur. Cur. 11: 99. 1823) may be the same thing as *Pavonia malvaviscoides*; Gürke transferred it to *Malvaviscus* without having examined it. Gürke's description is unclear (presumably taken from that of Nees & Martius), so the name remains ambiguous. *Index Kewensis* equates it to *Malvaviscus balbisii* de Candolle.

49. Pavonia montana Garcke *ex* Gürke *in* Martius, Fl. Bras. **12(3):** 522. 1892, citing Martius *ex* Garcke, Jahrb. Bot. Gart. Mus. Berlin **1:** 222. 1881, nom. nud. Type. Brazil. 1836, *Sellow* [*1426*] (lectotype, here designated, NY; isolectotypes, B as photo F-9455, BR as photo CTES, CTES [fragment], NY, US).

Pavonia viscosa var. *montana* Martius *ex* St.-Hilaire & Naudin, Ann. Sci. Nat. Bot. ser. 2, **18:** 41. 1842. Type. Brazil. Minas Gerais: *von Martius 1013* (holotype, P as photo CTES; isotypes, BR, CTES [fragment], K, NY).

Shrubs 1.5–2 m tall, the stems minutely glandular-pubescent and also setose, hairs 1–2 mm long. *Leaf* blades ovate, deeply cordate, serrate-crenate, acute or acuminate, discolorous, mostly 4–11 × 3–8 cm, minutely glandular-pubescent and also appressed-villous, the hairs 0.5–1 mm long; petioles 1.5–6 cm long; stipules filiform, 4–10 mm long. *Flowers* solitary in the leaf axils, the pedicels 2–4 cm long; involucellar bracts 12–14, filiform, ca. 10 × 0.5 mm or less, viscid and also setose; calyx 18–21 mm long, hirsute; corolla 3.5 cm long, tubular, orange or red with dark venation, subglabrous; androecium subequal to petals; styles slightly exserted. *Fruits* obovoid, ca. 8 mm diam., glabrous, the mericarps 8–9 mm high, reticulate-veined.

Distribution (Fig. 15). Known from the state of Minas Gerais, Brazil, at an altitude of 1000–1600 m.

Specimens examined. BRAZIL. Without locality, *Riedel s.n.* (OXF); NE Brazil, *Andrade-Lima 72–6988* (IPA, NY). **Minas Gerais:** Without locality, *von Langsdorff s.n.* (K), *Claussen 266* (K); Santa Rita Durao, 28 km N of Mariana, rd. to Santa Barbara, *Arbo et al. 4020* (CTES); Serra do Espinhaço, lower slopes of Pico de Itacolomí, ca. 2 km S of Ouro Preta, *Irwin et al. 29354* (NY, TEX).

Pavonia montana is distinctive for having 12–14 filiform bracts and leaves with their upper surface having appressed setae.

50. Pavonia paraënsis Fryxell, sp. nov. Type. Brazil. Pará: Tucuruí, right bank of the Tocantins River, BR-263, branch to the right at km 16, 29 Jan 1980, *Lisboa, Naciel, M. G. Silva & Bhia 1249* (holotype, NY). Fig. 32

Frutices caulibus pilis stellatis adpressis; laminis foliorum ovato-cordatis, in pagina abaxiali pilis stellatis adpressis; bracteis involucellorum ca. 10 lanceolatis vel loratis calycem excedentibus distaliter rubellis; calyce cylindrico brevi-lobato; corollis 3–4 cm longis rubris; androeciis exsertis purpuratis; fructibus ignotis.

Shrubs 1.5–2 m tall, the stems densely stellate-pubescent, the hairs appressed, ca. 1.5 mm diam., 8–10-armed, the arms arcuate and interlaced. *Leaf* blades ovate, mostly 5–10 × 4–9 cm, basally cordate, serrate, acuminate, palmately 9-nerved, discolorous, stellate-pubescent, the upper surface with hairs 0.5 mm diam., the lower surface with hairs (as on stems) 1–1.5 mm diam., densely mingled with finer pubescence; petioles 2.5–5 cm long, with dense pubescence like stem; stipules 8 mm long, subulate, erect, more sparsely pubescent than adjacent stem or petiole. *Flowers* solitary in the leaf axils, the pedicels 1.5–5 cm long, thickened upward (1.5 mm basally to 2.5 mm distally), with pubescence like stem; involucellar bracts ca. 10, liguliform to lanceolate, acute, 2.5–3 × 2–4(–5) mm (narrowed toward base), stellate-pubescent and hispid-ciliate (the cilia 2–3 mm long), apparently somewhat reddish distally; calyx 15–23 mm long, tubular, densely yellowish-pubescent, shallowly lobed, the lobes 5–7 mm long, triangular; corolla 3–4 cm long, red, the petals erect or flaring distally; staminal column

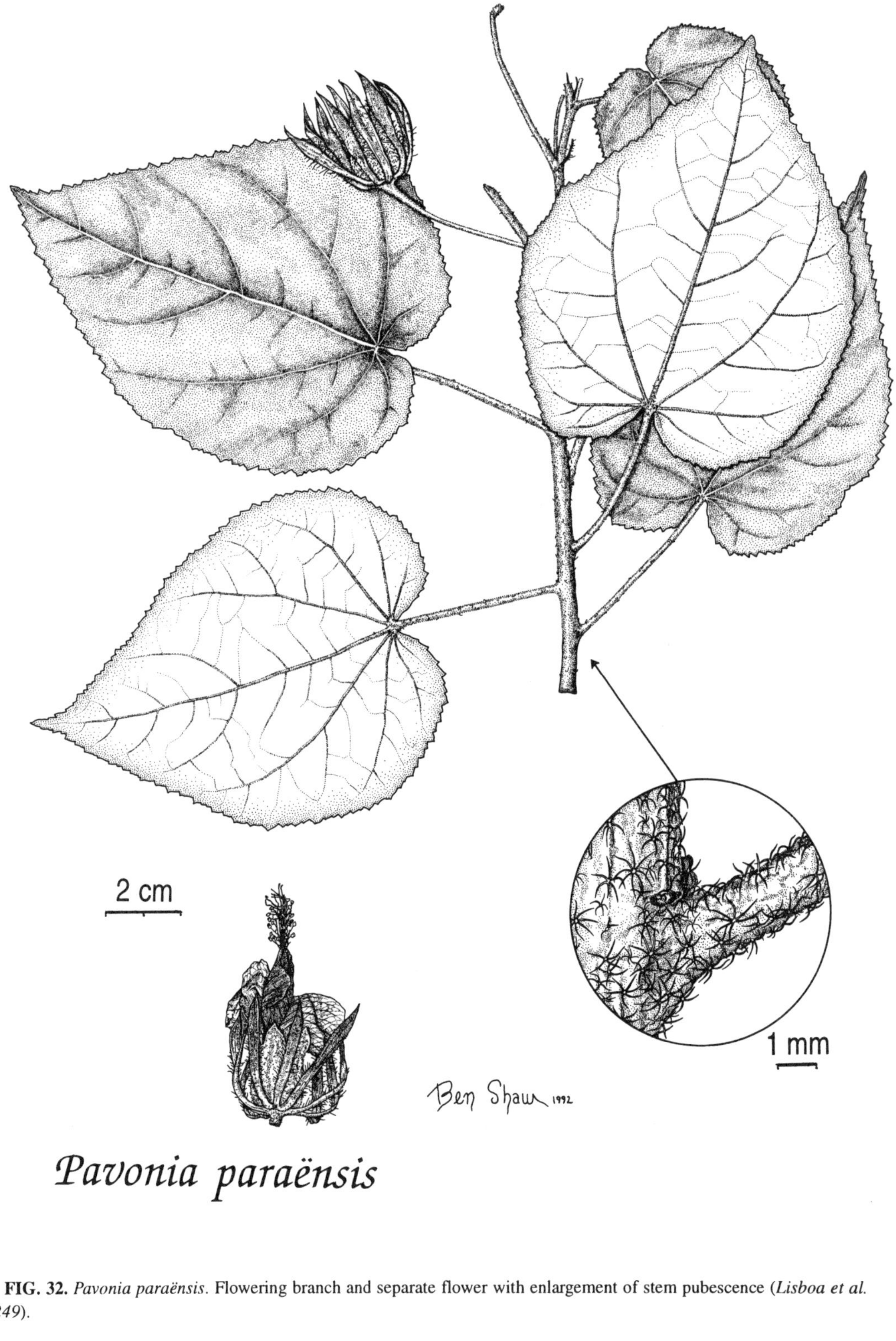

FIG. 32. *Pavonia paraënsis*. Flowering branch and separate flower with enlargement of stem pubescence (*Lisboa et al. 1249*).

exserted, 3.5–4.5 cm long, purple, the filaments ca. 5 mm long, ascending, purple; styles slender, emerging from the column, pubescent or glabrous. *Fruits* unknown.

Distribution (Fig. 15). Known only from the state of Pará, Brazil, along the Rio Tocantins. The type locality is at 3°45′S, 49°42′W.

Specimens examined. **BRAZIL. PARÁ:** 18 km E of Tucuruí and Tocantins River, *Daly et al. 1018* (F, K, NY, TEX, US); right bank of the Tocantins River, 28 May 1980, *M. G. Silva & Rosario 5226* (NY).

The specific epithet signifies "from the state of Pará."

51. Pavonia serrana G. Esteves, Novon **4:** 100. 1994. Type. Brazil. Minas Gerais: Botumirim, exit from the city toward Cachoeira da Noruega, 15 Jun 1991, *Mello-Silva et al. 518* (holotype, SPF–n.v.; isotypes, CTES, SP–n.v.).

Shrubs 2 m tall, the stems densely stellate-tomentose, the hairs 0.5 mm long or less, yellowish. *Leaf* blades broadly ovate, 5–7(–13) cm long, about as wide, deeply cordate, remotely serrate to subentire, acute, strongly discolorous, obscurely pubescent above, the hairs 0.2–0.3 mm long, often bifurcate, densely and softly yellowish-tomentose beneath, palmately 7–9-nerved; petioles 3–5 cm long, with pubescence like stem; stipules subulate, 2–3 mm long, caducous. *Flowers* solitary in the leaf axils, the pedicels 3–5 cm long, densely stellate-tomentose; involucellar bracts 4(–6), broadly cordate-ovate, 2 × 2.5 cm, deeply cordate, densely tomentose; calyx almost as long as involucel, tomentose, short-lobed, largely concealed by involucel; corolla tubular, the petals ligulate, 3–4.5 × 1 cm, exserted from the involucel by 1–2 cm, yellow with red veins (*Mello-Silva et al. 518*) or red with white veins (*Mello-Silva et al. 500*) (perhaps varying with time of day), the androecium reddish, subequal to corolla, the styles slightly exserted. *Fruits* (*ex* descr.) 2.5–3 mm tall, reticulate-veined (otherwise unornamented), apparently glabrous, ca. twice the height of the column.

Distribution (Fig. 20). Known only from the Serra da Canastra in N Minas Gerais, Brazil, in cerrado vegetation. The type locality is at 16°53′S, 43°01′W.

Illustration. Esteves (1994: fig. 1).

Specimen examined. **BRAZIL. MINAS GERAIS:** Botumirim, Serra da Canastra, Tinoco, N extremity of the Serra, near the rd. to Itacambira, *Mello-Silva et al. 500* (CTES).

Pavonia serrana is notable for its 4–6 ovate, involucellar bracts that are 25 mm wide and for its ligulate petals. It is similar to the viscid *P. grazielae*.

52. Pavonia tiliifolia (Ulbrich) Fryxell, comb. nov. Basionym: *Codonochlamys tiliifolia* Ulbrich, Notizbl. Bot. Gart. Mus. Berlin-Dahlem **6:** 330. 1915. Type. Brazil. Minas Gerais: Mendanha, 3 Aug 1881, *Glaziou 13543* (holotype, B as photo F-9423; isotype, P as photo CTES).

Erect *subshrubs*, the young stems yellowish-tomentose. *Leaf* blades broadly ovate or suborbicular (rarely indistinctly 3-lobed), 4–6 × 3.5–5 cm, 7–8-nerved, softly canescent-tomentose beneath, darker and subtomentose above; petioles straight, 2.5–3 cm long (ca. half-length of blade); stipules subulate, 5–7 mm long, cinereous, caducous. *Flowers* solitary in the leaf axils; pedicels 3–7 mm long; involucel campanulate, subequal to calyx, striate, gamophyllous and indistinctly bilabiate, the 3–4 lobes ovate-acuminate, 6–8 mm long; calyx campanulate, 12–14 mm long, yellowish-tomentose, indistinctly striate, more or less bilabiate, with 3 free lobes ca. 3 mm long, 2 connate lobes 1–1.5 mm long; corolla 2.5–3 cm long, narrowly cylindric, the petals red, narrowly linear; staminal column ca. 3.5 cm long, glabrous, very slender, apically 5-dentate and with slender filaments; ovary glabrous, the filiform styles glabrous, ca. 4 cm long (free for distal 5–6 mm), the stigmas 9 (10?), capitate, pilose. *Fruits* more or less included in the calyx; mericarps 5, indehiscent, apically 3-horned, dorsally not or scarcely carinate, yellowish, thin-walled, 8–10 mm long, ca. 5 mm wide; seeds solitary, 6 mm long, dark brown, sparsely pubescent.

Distribution (Fig. 11). Known only from the type collection. The type locality is at 18°05′S, 43°29′W.

No specimen of *Pavonia tiliifolia* has been examined, only the photographs cited above. The preceding description is consequently derived from the original descriptions of Ulbrich, in Latin and in German, together with a consideration of the type photos. The bilabiate nature of the involucel and calyx may be an artifact of pressing, as it was concluded to be in *P. glaziovii* (q.v.). These two species share the characteristic of a gamophyllous involucel.

53. Pavonia viscosa St.-Hilaire, Fl. Bras. Mer. **1:** 236. 1827. Type. Brazil. Minas Gerais: Serra do Caminho da Tapeira, *St.-Hilaire s.n.* (lectotype, here designated, P?–n.v.; fragment, F). Note: St.-Hilaire cites the above specimen and *St.-Hilaire 288* from

Serra do Candonga (CTES [fragment], P as photo CTES) but states that they differ in certain characters and that his description is based on the Serra do Caminho specimen, which is therefore the appropriate choice as lectotype. Fig. 30

Pavonia viscosa var. *velutina* St.-Hilaire & Naudin, Ann. Sci. Nat. Bot., ser. 2, **18:** 41. 1827. Type. Brazil. Minas Gerais: Serra da Caraça, *Lamotte 288* (holotype, P as photo CTES; isotypes, CTES [fragment], G as photo F-23712).
Pavonia montana var. *velutina* Gürke *in* Martius, Fl. Bras. **12(3)**: 523. 1892. Type. Brazil. Near Rio de Janeiro, *Glaziou 14507* (lectotype, here designated, C; isolectotypes, C as photo F-21600, CTES [fragment], K, P as photo CTES).

Shrubs or small trees 1–4 m tall, the stems minutely glandular- and stellate-puberulent (the proportions of the two hair-types varying), rarely also with scattered setae in addition (*Irwin et al. 20693*). *Leaf* blades ovate (rarely lobulate), mostly 4–10 × 2.5–9 cm, subtruncate to deeply cordate, obscurely serrulate to subentire, acute to acuminate, palmately 7–9-nerved, discolorous, the upper surface very minutely glandular-pubescent (usually seemingly glabrous) and sometimes with longer simple hairs on veins, the lower surface minutely velutinous and with glandular hairs; petioles 2–7 cm long; stipules 4–11 mm long, filiform. *Flowers* solitary in the leaf axils, the pedicels 2–7 cm long; involucellar bracts 12–14 (8 in *Mello-Silva et al. 483*), lance-linear, 16–21 × 1–3(–4) mm, glandular-pubescent, sometimes also setose; calyx ca. 17 mm long; corolla 2.5–3.5 cm long, tubular or funnelform, red or orange with dark venation; androecium subequal to corolla or slightly exserted, dark red; styles exserted. *Fruits* 8–10 mm diam., glabrous, prominently reticulately veined, the raised veins dark-pigmented; mericarps 7 mm high, crested; seeds 4.5 mm long, sparsely pubescent.

Distribution (Fig. 18). Bahia and Minas Gerais, Brazil, growing in cerrado vegetation, sedge meadows, and gallery forest, at elevations of 1000–1750 m.

Illustrations. Gürke (1892b: pl. 100, fig. 2); Sazima (1981: figs. 1–5).

Representative specimens examined. BRAZIL. BAHIA: S. Cabeça de Boi, 5 km from Itambé, 13 Jan 1982, *Hensold et al. s.n.* [CFCR-2841] (CTES). **MINAS GERAIS:** Serra do Espinhaço, ca. 18 km SW of Diamantina, rd. to Curvelo, *W. R. Anderson, P. A. Fryxell et al. 8524* (F, K, NY, TEX, US); Mun. Alvadora de Minas, vic. of Itapanhoacan, *Arbo et al. 4311* (C, CTES ["topotipo"], NY); base of the Serra do Caraça, *Badini 19970* (CTES); Mun. Santa Barbara, Serra do Gongo Socco, *Barreto 7946* (F); Mun. Joaquim Felício, Serra do Cabral, *Davis et al. 2406* (E); Serra do Cipó, km 117, *Duarte 11662* (CTES); Caraça, *Emygdio et al. 3507* (CTES, NY); 20 km S of Diamantina, *Gibbs et al. 5251* (NY); Serra de Lenheiró, *Glaziou 20200* (C, G, US); Mun. Gouveia, Serra do Espinhaço, *Hatschbach 27925* (CTES); Mun. Diamantina, km 10 Rod. Guinda-Conc. Mata, *Hatschbach et al. 53024* (C, CTES, S, US); Serra do Cipó, ca. 153 km N of Belo Horizonte, *Irwin et al. 20603* (CTES, F, GH, MO, NY, US); Serra do Espinhaço, at Lapinha, ca. 18 km N of Sêrro on rd. to Diamantina, *Irwin et al. 20693* (CTES, F, MO, NY, TEX, US); Serra do Espinhaço, ca. 7 km N of São João da Chapada, rd. to Campo do Sampaio, *Irwin et al. 28571* (NY, TEX); Serra do Espinhaço, 3 km N of Mariana, rd. to Santa Barbara, *Irwin et al. 29670* (CTES, F, K, NY, US); between Diamantina and Gouveia, *Maguire et al. 49160* (CTES, NY); rd. to Conselheiro Mata, 8 km from Hwy. BR-259, *Mello-Silva et al. 326* (CTES, NY, SPF); Mun. Jaboticatubas, Serra do Cipó, rd. from Vespaziano to Conceição do Mato Dentro, km 114, 8 Jul 1974, *Sazima s.n.* (NY); summit of Belo Horizonte, *L. O. Williams & Assis 7191* (GH, MO, NY, U); Mun. Nova Lima, Serra do Curral, 5 mi SE of Belo Horizonte, *L. O. Williams & Assis 8019* (A, F, MO, NY, U).

Pavonia viscosa is characterized by its broad, lance-linear involucellar bracts and by its viscid herbage.

Sazima (1981) describes the course of floral development in *Pavonia viscosa* (reported as *P. montana*, the voucher: *Sazima s.n.*, cited above) over a period of 3 days, which showed the flowers to be protogynous. Experimental pollinations indicate that fruit-set requires cross-pollination. The most frequent and effective pollinator was observed to be the Long-billed Hummingbird, *Phaethornis pretrei*, photographed visiting a flower (Sazima, 1981: fig. 5).

54. Pavonia zehntneri Ulbrich, Notizbl. Bot. Gart. Berlin **9:** 44. 1924. Type. Brazil. Bahia: Campo Formoso, 1913, *Zehntner 2069* (lectotype, here designated, M as photo F-19693).

Shrubs 1.5–3 m tall, the stems roughly stellate-pubescent, the hairs appressed, 8–10-armed (the arms interlaced), ca. 0.5 mm diam., occasionally stipitate. *Leaf* blades ovate to weakly 3-lobed, up to 10 × 7 cm, basally cordate, coarsely and irregularly serrate, acute or acuminate, palmately 5–7-nerved, discolorous, subglabrous or minutely stellate-pubescent above, densely and softly puberulent beneath; petioles 1–3.5 cm long, with pubescence like stem but less appressed; stipules 3–9 mm long, linear or filiform. *Flowers* solitary in the leaf axils, the pedicels 2–4.5 cm long, exceeding the petioles; involucellar bracts 8–12, 18–22 × 1–2 mm (subequal to calyx or slightly longer), linear, densely stellate-pubescent, sometimes ciliate-margined; calyx tubular to slightly urceolate, 16–19 × 8–9 mm at base, longitudinally 15-ribbed, densely stellate-pubescent, the lobes triangular or somewhat rounded, 2–3 mm long; petals reddish, erect, 30–35 × 3mm; staminal column 4–4.5 cm long, red or purplish, the filaments 2–4 mm long, the anthers "dark violet"; styles exserted from column 5–6 mm. *Fruits* 7–8 mm diam., minutely pubescent; mericarps 8–9 mm long, reticulate-veined.

Distribution (Fig. 20). Confined to the states of Bahia and Minas Gerais, Brazil, at elevations of ca. 1000 m. The type locality is at 10°35′S, 40°19′W.

Specimens examined. BRAZIL. BAHIA: Mun. Campo Formoso, rd. from Alagoinhas to Agua Preta km 3, *Coradin et al. 6053* (CTES); Serra do Curral Feio, 16 km NW of Lagoinha (5.5 km SW of Delfino) on side rd. to Minas do Mimoso, *Harley et al. 17025* (CTES); Mun. Ibiquara, 13 km from Juciapa on the rd. to Morro Branco and Capão da Volta, *Harley et al. 26451* (SPF as photo NY); Serra da Jibóia, Mun. Juciape, *Hatschbach et al. 56833* (C); Mun. Macaúbas, rd. to Canatiba, climbing to Serra Poção near the top, *Hatschbach et al. 65151* (CTES, TEX); 26 km N of Seabra, rd. to Agua de Rega, near Rio Riachão, *Irwin et al. 30793* (CTES, MO, NY). **MINAS GERAIS:** BR-116 (Mun. Aguas Vermelhas), *Hatschbach & J. M. Silva 50020* (CTES, US).

Pavonia zehntneri is related to *P. glazioviana*, from which it differs in its longer corolla and with which it is similar in its colored calyx and corolla.

Pavonia subgen. *Pavonia* sect. *Lebretonia*

Pavonia sect. **Lebretonia** (Schrank) Endlicher, Gen. Pl. 982. 1840. Basionym: *Lebretonia* Schrank, Pl. Rar. Hort. Monac. pl. 90. 1822. Type. *Lebretonia coccinea* Schrank (≡ *Pavonia schrankii* Sprengel).

Herbs or shrubs with elliptic, ovate, hastate, or sagittate leaves; flowers axillary; involucellar bracts 5(–7); mericarps usually carinate, the seeds tufted at each end of the hilum.

Type species. *Pavonia schrankii* Sprengel.

Pavonia sect. *Lebretonia* has been recognized in generic rank by some workers (Britton & Wilson, 1924; Williams & Cheesman, 1929, but for species here excluded from the group) but has more commonly been treated in sectional rank (Ulbrich, 1921; Krapovickas & Cristóbal, 1962; Krapovickas, 1977; Fryxell, 1988a) or more simply submerged in *Pavonia* (Kearney, 1951, 1958; Hutchinson, 1967). The most important treatments are those of Krapovickas & Cristóbal (1962) and Krapovickas (1977). The former treatment recognized five "species groups" but did not formalize them with names or taxonomic rank, although Krapovickas (1977) later described *P.* subsect. *Exsertae*, thereby automatically establishing *P.* subsect. *Lebretonia* for the remainder of the species in the section and thus implicitly removing recognition of the other species groups. Krapovickas and Cristóbal (1962) described two of the "species groups" in detail, but the other three were characterized only by a listing of the included species.

Interestingly, the seed hairs that occur as tufts at either end of the hilum, which characterize *Pavonia* sect. *Lebretonia* and distinguish it from the remainder of *Pavonia*, are also found on seeds of the monotypic genus *Phragmocarpidium*. No other feature seems to tie these two groups together, however.

The subdivision of *Pavonia* sect. *Lebretonia* proposed here closely follows the groupings made by Krapovickas and Cristóbal (1962) and Krapovickas (1977) but with a few modifications.

The name *Lebretonia* commemorates Adrien Lebreton (1662–1736), Jesuit missionary from France in Martinique (McKinley, 1992), and not, as indicated by Fryxell (1988a), Rev. C. Breton (or le Breton) (Stafleu & Cowan, 1979: 780) or Manuel le Breton (Don, 1831; Sweet, 1839).

PAVONIA SUBGEN. PAVONIA SECT. LEBRETONIA SUBSECT. LEBRETONIA

Coarsely pubescent shrubs with lance-ovate or elliptic, unlobed leaves; flowers axillary; involucellar bracts 5, distinct; mericarps carinate, often acute apically.

Type species. *Pavonia schrankii* Sprengel.

Key to the species of *Pavonia* subsect. *Lebretonia*

1. Petals 3–4.5 cm long.
    2. Pedicels 2.5–7 cm long; staminal column basally pubescent; mericarps prominently rugose and distally acute; petals 3–4 times as long as wide. . . . . 60. *P. schrankii*
    2. Pedicels 0.5–1.5(–3) cm long; staminal column glabrous; mericarps slightly rugose and aristate; petals 5 times as long as wide. . . . . 55. *P. angustipetala*
1. Petals 1–2 cm long.
    3. Stipules linear, slightly expanded apically. . . . . 56. *P. latifolia*
    3. Stipules filiform, subulate, or lanceolate.
        4. Pedicels 2–4 cm long; leaf blades bullate; staminal column basally pubescent. . . . . 58. *P. reitzii*
        4. Pedicels 0.2–3 cm long; leaf blades smooth; staminal column glabrous.
            5. Calyx 9–12 mm long, ciliate, the lobes with reddish venation; mericarps apically acute. . . . 57. *P. ramboi*
            5. Calyx 7–9 mm long, densely pubescent, the lobes lacking purplish venation; mericarps apically rounded. . . . . 59. *P. rubra*

55. Pavonia angustipetala Krapovickas & Cristóbal, Lilloa **31:** 51. 1962. Type. Brazil. Rio Grande do Sul: Passo da Guarda, near Bom Jesus, *Rambo 51919* (holotype, LIL–n.v.; isotypes, CTES, S, US). Fig. 33

Pavonia muricata var. *punicea* St.-Hilaire, Fl. Bras. Merid. **1:** 226. 1827. Type. Brazil. Near the city of Itapeva, in the S part of the prov. of São Paulo, *St.-Hilaire s.n.* (holotype, P as photo CTES).

Pavonia schrankii var. *angustifolia* R. E. Fries, Kongl. Svenska Vetenskapsakad. Handl. **42(12):** 51. 1908. Type. Brazil. Rio Grande do Sul: Rincão dos Vallos, Engenho da Sta. Clara, 18 Dec 1898, *Lindman A.1553* (lectotype, here designated, S; isolectotype, S).

Shrubs 1–2 m tall, the stems coarsely stellate-pubescent (hairs 1–1.5 mm long). *Leaf* blades narrowly elliptic, mostly 4–12 × 1–3 cm, basally rounded or subcuneate, crenate-serrate, acute, palmately 3(–5)-nerved, the nerves prominently raised beneath, sunken above, discolorous, densely stellate-pubescent beneath and with scattered bifurcate and simple hairs 1–2 mm long above; petioles 6–22 mm long; stipules subulate, 5–10 mm long. *Flowers*: pedicels 0.5–1.5(–3) cm long; involucellar bracts 5, lance-ovate, 10–14 × 2–4 mm, hirsute; calyx 10–13 mm long, hirsute; corolla 3–4 cm long, red, the petals 3–6(–10) mm wide; staminal column glabrous, nearly equal to corolla or slightly exserted, the filaments 3–4 mm long; styles slender, glabrous or sparsely pubescent. *Fruits* 9–11 mm diam., oblate, blackish, rugulose; mericarps 5–6 mm high, obscurely pubescent with septate hairs.

Distribution (Fig. 34). From S Brazil in the states of Paraná, Rio Grande do Sul, Santa Catarina, and São Paulo, with one collection from Argentina (Misiones), in shrubland. The type locality is at 28°40′S, 50°23′W; that of *Pavonia schrankii* var. *angustifolia* is at 25°07′S, 50°10′W.

Illustrations. Krapovickas & Cristóbal (1962: figs. 2D, 8).

Specimens examined. BRAZIL. PARANÁ: Ponta Grossa, on forest margins (capoeira), *Dusén 2645* (syntypes of *P. schrankii* var. *angustifolia*, CTES, GH); Mun. Guaraniaçu, 9 km E of Guaraniaçu along BR-277 to Curitiba, *Davidse et al. 11281A* (MO); Mun. Palmeira, Fazenda Santa Rita, *Hatschbach 44452* (VT); Fazenda Santa Rita, *Kummrow 2196* (CM). **RIO GRANDE DO SUL:** Neu-Württemberg, *Bornmüller 341* (A, G, GH, U); Canelas, *Emrich 44786* (S); Mun. Bom Jesus, Río Pelotas, rd. from S. Joaquim to Roncinha, *Krapovickas & Schinini 38342* (C, CTES, G, HAS, LIL, MBM, NY, UC). **SANTA CATARINA:** Caçador, Casteanho River, *Krapovickas & Cristóbal 43704* (AAU, C, CLEMS, CTES, MBM, MERL, NY, TEX, U); between Lajes and the bridge over the Pelotas River, on the hwy. from Lajes to Vacaria, *Mattos 16812* (CTES); Mun. Curitibanos, between Ponte Alta and Curitibanos, *L. B. Smith & Klein 8279* (US). **SÃO PAULO:** Mun. Botucatú, *G. Gottsberger 11-14272* (US).

Within *Pavonia* subsect. *Lebretonia*, *P. angustipetala* is characterized by its long, narrow petals and short pedicels.

56. Pavonia latifolia (Nees & Martius) Sprengel, Syst. **3:** 98. 1826. *Lebretonia latifolia* Nees & Martius, Nova Acta Phys.-Med. Akad. Caes. Leop.-Carol. Nat. Cur. **11:** 98. 1823. *Malache latifolia* (Nees & Mart.) Kuntze, Revis. Gen. Pl. **1:** 70 (1891). Type. Unknown.

Low *shrubs*, the stems hirsute, the hairs stellate, grayish. *Leaf* blades ovate-elliptic, 11–14 × 5–7 cm, basally rounded or subcordate, crenate-serrate, shortly acuminate, palmately 5-nerved, pubescent above and beneath; petioles 3–5 cm long, angulate, hirsute; stipules linear or subspatulate, erect, subscabrous. *Flowers* solitary in the leaf axils, the pedicels slightly shorter than the leaves, hirsute; involucellar bracts 5, ovate-acuminate, hirsute; calyx slightly shorter than the involucel, subtomentose, 5-parted; petals reddish, subequal to the involucel; staminal column subequal to the corolla, the anthers yellowish, subglobose; stigmas capitate, purplish. *Mericarps* obovate, more or less reticulate, apically mucronate.

Distribution. Known only from the original description.

This species is virtually unknown, the above description being paraphrased from the original description. It is retained here as distinct following the interpretation of Krapovickas and Cristóbal (1962) and Krapovickas (1977). Krapovickas and Cristóbal (1962) noted that *Pavonia latifolia* was similar to *P. reitzii* in its long pedicels but differed from it in the ornamentation of the mericarps.

57. Pavonia ramboi Krapovickas & Cristóbal, Lilloa **31:** 60. 1962. Type. Brazil. Rio Grande do Sul: Villa Oliva, near Caxias, 3 Dec 1949, *Rambo s.n.* (holotype, LIL–n.v.; isotype, CTES). Fig. 33

Shrubs 1.5 m tall, the stems decumbent to ascending, coarsely stellate-pubescent, the hairs 1–2 mm long. *Leaf* blades lance-ovate to oblong-elliptic, mostly 3–6 × 1–2.5(–3) cm, basally truncate to subcuneate, serrate, acute, palmately 3–5-nerved, the upper surface with appressed simple and bifurcate hairs 2 mm long, the lower surface with stellate hairs 2–3 mm diam.; petioles 6–12 mm long; stipules narrowly lanceolate, 9–14 mm long, hirsute. *Flowers*

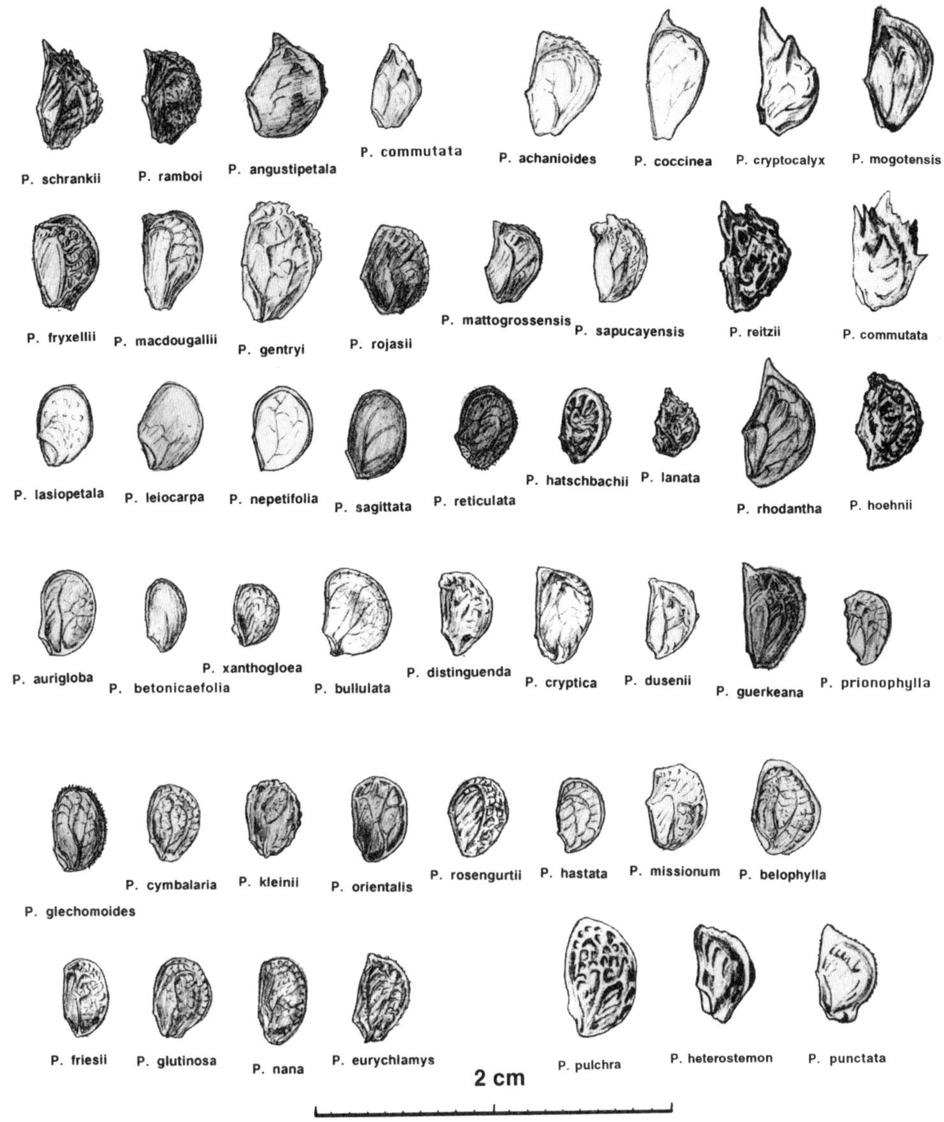

FIG. 33. *Pavonia* sect. *Lebretonia* mericarps in left lateral view. Top row (left to right): *P. schrankii* (*Krapovickas et al. 23136*), *P. ramboi* (*Bueno & Krapovickas 3641*), *P. angustipetala* (*Smith & Klein 8279*), *P. commutata* (*Reitz 5315*), *P. achanioides* (*Palmer & Riley 257*), *P. coccinea* (*Leonard & Leonard 14716*), *P. cryptocalyx* (*Ekman 8974*), *P. mogotensis* (*Ekman 7982*). Second row (left to right): *P. fryxellii* (*Hill 1816*), *P. macdougallii* (*Fryxell & Lott 3383*), *P. gentryi* (*Gentry 7287*), *P. rojasii* (*Rojas 10184*), *P. mattogrossensis* (*Rojas 5632*), *P. sapucayensis* (*Hassler 1930*), *P. reitzii* (*Reitz 3502*), *P. commutata* (*Sellow s.n.*). Third row (left to right): *P. lasiopetala* (*Arsène 6192*), *P. leiocarpa* (*Liogier 11562*), *P. nepetifolia* (*Pringle 13698*), *P. sagittata* (*Regnell III-174*), *P. reticulata* (*Hatschbach 2798*), *P. hatschbachii* (*Hatschbach 15411*), *P. lanata* (*Krapovickas & Cristóbal 33655*), *P. rhodantha* (*Hassler 4385*), *P. hoehnei* (*Brade 6880*). Fourth row (left to right): *P. aurigloba* (*Bartlett 20759*), *P. betonicaefolia* (*Krapovickas et al. 20305*), *P. xanthogloea* (*Krapovickas et al. 17189*), *P. bullulata* (*Krapovickas & Schinini 38380*), *P. distinguenda* (*Rambo 44188*), *P. cryptica* (*Schinini et al. 8348*), *P. dusenii* (*Hatschbach 7802*), *P. guerkeana* (*Smith & Klein 10779*), *P. prionophylla* (*Lindeman et al. 7010*). Fifth row (left to right): *P. glechomoides* (*Venturi 2028*), *P. cymbalaria* (*Grondona 2643*), *P. kleinii* (*Smith & Reitz 10393*), *P. orientalis* (*Wetmore 858*), *P. rosengurttii* (*Krapovickas et al. 22939*), *P. hastata* (*Lindeman 21126*), *P. missionum* (*Schwarz 521*), *P. belophylla* (*Schinini & Bordas 20499*). Sixth row (left to right): *P. friesii* (*Rambo 44159*), *P. glutinosa* (*Krapovickas & Crostóbal 11231*), *P. nana* (*Krapovickas & Cristóbal 11209*), *P. eurychlamys* (*Novara 5162*), *P. pulchra* (*Fiebrig 6083*), *P. heterostemon* (*Wood & Atchison 7436*), *P. punctata* (*Liogier 13274*).

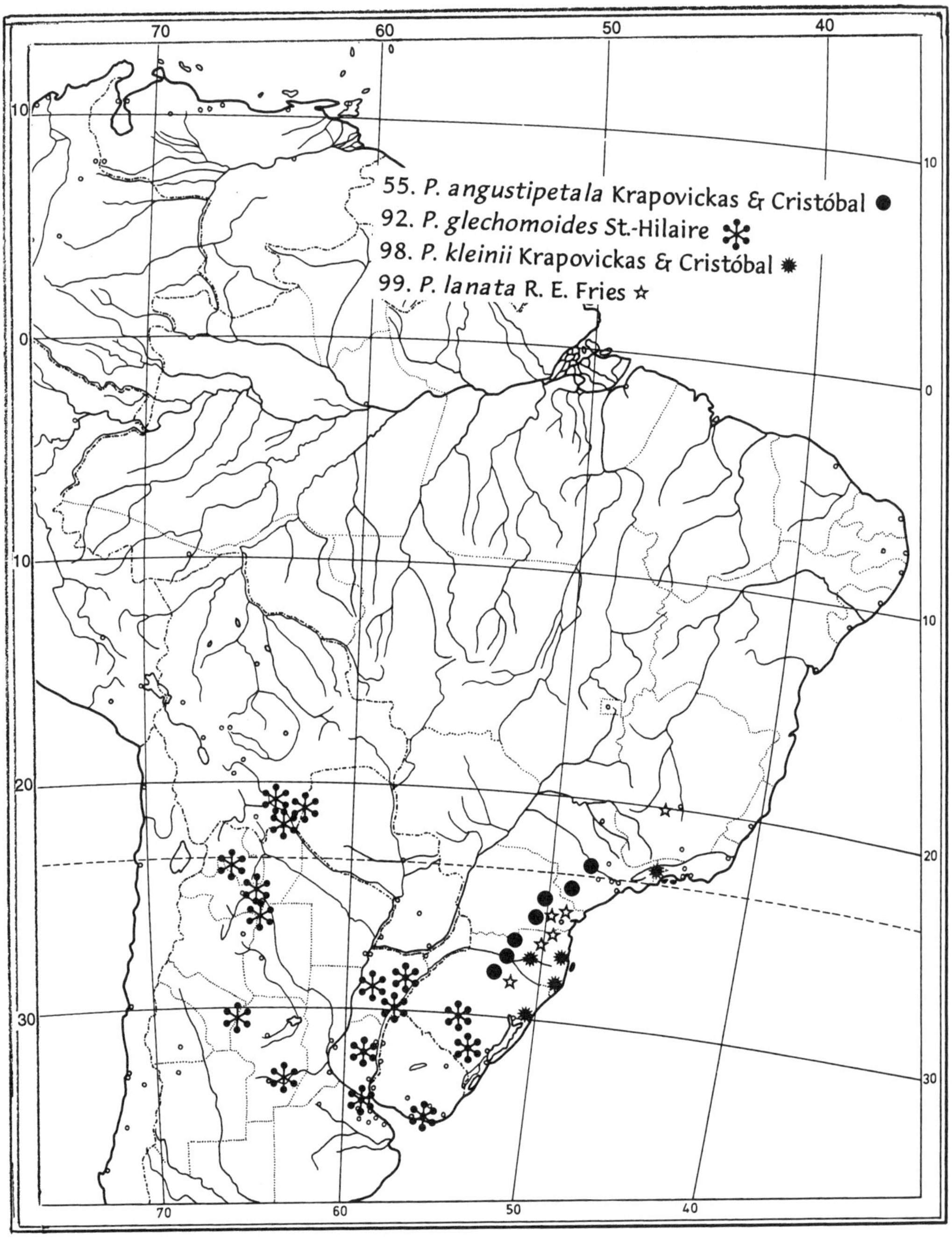

FIG. 34. Distribution map of four South American species of *Pavonia* subsect. *Lebretonia* and *P.* subsect. *Hastifoliae*.

solitary in the leaf axils but often terminally crowded, the pedicels 1 cm long or less; involucellar bracts 5, lanceolate, 9–10 × 1–2 mm, ciliate; calyx 9–12 mm long, ciliate (hairs 1–2 mm long), the lobes with reddish venation; corolla 12–15(–20) mm long, whitish with dark red center and purplish veins; staminal column 5–6 mm long, reddish; styles exserted from column, reddish. *Fruits* ca. 8 mm diam., slightly

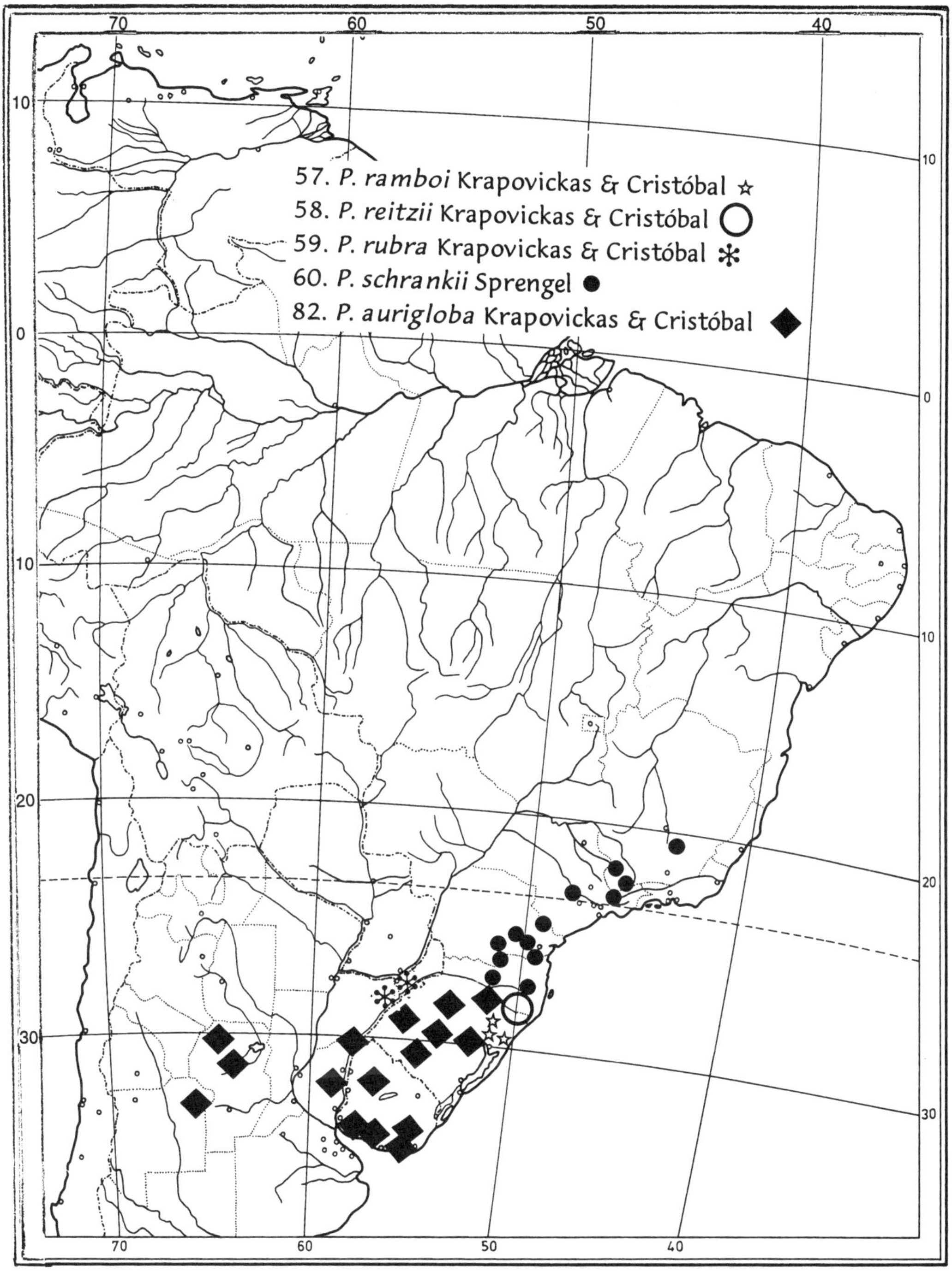

FIG. 35. Distribution map of five South American species of *Pavonia* subsect. *Lebretonia* and *P.* subsect. *Hastifoliae*.

pubescent or subglabrous; mericarps 5 mm long, prominently rugulose, apically acute, with 3 prominent excrescences on each side.

Distribution (Fig. 35). Confined to the state of Rio Grande do Sul, Brazil. The type locality is at 29°13′S, 50°51′W.

Illustrations. Krapovickas & Cristóbal (1962: figs. 2C, 11).

Specimens examined. BRAZIL. RIO GRANDE DO SUL: Caxias do Sul, *Brinker et al. 5509* (US); Mun. Caxias do Sul, Agua Azul, *Bueno 2883* (CTES); Mun. Dos Irmãos, between Dos Irmãos and S. María do Erval, *Bueno & Krapovickas 3641* (C, CTES, NY); 18 km E of São Francisco do Paula, *Krapovickas & Cristóbal 37633* (CTES, F, G, LIL, MEXU, MO, NY, UC, USM); San Pedro near São Salvador, *Leite 2367* (A); Vila Oliva near Caxias, *Rambo 54919* (CTES); Canela, *Richter 7862* (CTES).

The reddish venation of the calyx lobes and the whitish corolla distinguish *Pavonia ramboi* from the other members of *P.* subsect. *Lebretonia.*

58. Pavonia reitzii Krapovickas & Cristóbal, Lilloa **31:** 58. 1962. Type. Brazil. Santa Catarina: Bom Retiro, *Reitz 2810* [= *3502*] (holotype, HBR–n.v.; isotypes, CTES, S). Note: The type is cited as "Reitz 2810 (HBR 3502)" and isotypes as "Reitz 3502 (LIL, R)"; presumably these are duplicates of a single collection and are so treated here. Fig. 33

Shrubs 0.5 m tall, the stems densely and coarsely stellate-pubescent. *Leaf* blades elliptic or lance-oblong, basally truncate or subcuneate, serrate, acute, penninerved, the younger leaves bullate, discolorous, densely stellate-tomentose above and beneath; petioles 0.5–1 cm long; stipules narrowly subulate, 8–12 mm long. *Flowers* solitary in the leaf axils, the pedicels, 2–4 cm long; involùcellar bracts 5 (?), lanceolate, 10–14 × 2–4 mm, hirsute; calyx 8–9 mm long, densely hirsute, the hairs 1.5–2 mm long; petals white (with red veins), 1.5 cm long (or more?). *Fruits* ca. 8 mm diam., the mericarps 5–6 mm long, extremely rugulose, apically acute, with 3–4 excrescences on each side.

Distribution (Fig. 35). Apparently known only from the type collection. The type locality is at 27°48′S, 49°28′W.

Illustrations. Krapovickas & Cristóbal (1962: figs. 2A,B, 10).

Pavonia reitzii is distinctive for its bullate leaves, especially the younger leaves.

59. Pavonia rubra Krapovickas & Cristóbal, Lilloa **31:** 54. 1962. Type. Argentina. Misiones: Dept. San Ignacio, Colonia Ñancanguazú, *Schwarz 5637* (holotype, LIL–n.v.).

Shrubs 0.5–0.75 m tall, the stems coarsely ferruginous-pubescent, the hairs stellate, 1 mm long. *Leaf* blades lance-elliptic, mostly 3–6 (or larger?) × 1–2.5 cm, basally truncate, serrate, acute, palmately 5–7-nerved, with coarse stellate hairs (1.5–2 mm diam.) evenly spaced on both surfaces; petioles 5–9 mm long; stipules subulate 4–6 mm long. *Flowers* solitary in the leaf axils, the pedicels 2–8 mm long; involucellar bracts 5, lanceolate, 8–9 × 2 mm, densely stellate-pubescent; calyx 7–9 mm long, densely stellate-pubescent; corolla rose-pink with red center, ca. 1.5 cm long. *Fruits* subglabrous; mericarps 5 mm long, reticulately rugose.

Distribution (Fig. 35). Known from the Provinces of Corrientes and Misiones, Argentina. The type locality is at 27°18′S, 55°30′W.

Illustrations. Krapovickas & Cristóbal (1962: figs. 2E, 9).

Specimens examined. ARGENTINA. CORRIENTES: Dept. Santo Tomé, Arroyo Chimiray and rt. 40, *Tressens et al. 1848* (C, CTES, F, G, NY, UC).

As the name suggests, *Pavonia rubra* has reddish flowers.

60. Pavonia schrankii Sprengel, Syst. **3:** 98. 1826, nom. nov. *Lebretonia coccinea* Schrank, Pl. Rar. Hort. Monac. t. 90. 1822 (non *Pavonia coccinea* Cavanilles, 1787). *Malache schrankii* (Sprengel) Kuntze, Revis. Gen. Pl. **1:** 71. 1891. Type. Plate 90 of Schrank. Figs. 33, 36

Pavonia muricata St.-Hilaire, Fl. Bras. Merid. **1:** 225. t. 44. 1827. *Malache muricata* (St.-Hilaire) Kuntze, Revis. Gen. Pl. **1:** 71. 1891. Type. Brazil. Minas Gerais: S. María de Baependy, *St.-Hilaire 2441* (holotype, P as photo CTES). Note: In protologue, St.-Hilaire cites "prope vicum Congonhas do campo, in provincia Minas Gerais."

Pavonia hastata forma *longifolia* Gürke *in* Martius, Fl. Bras. **12(3):** 500. 1892. Type. Brazil. Minas Gerais: Caldas, *Regnell II-16* (lectotype, here designated, NY; isolectotypes, C, F, GH, NY, S-3, U).

Pavonia schrankii var. *ovata* Ulbrich, Feddes Repert. Spec. Nov. Regni Veg. **13:** 516. 1915. Type. Brazil. São Paulo: 1836, *Sellow 356* (lectotype, here designated, US; isolectotypes, CTES [fragment], NY, TEX, VT).

Erect *shrubs* 1–2 m tall, the stems densely and roughly stellate-pubescent, more or less ferruginous, the hairs ca. 1 mm long. *Leaf* blades lanceolate-ovate, mostly 5–10 × 2–5 cm, basally truncate or subcordate, serrate, acute, palmately 5-nerved, discolorous, stellate-pubescent on both surfaces, the hairs ca. 1 mm diam., ferruginous on upper surface, whitish or yellowish on lower surface; petioles 1–4 cm long; stipules 7–15 mm long, subulate. *Flowers* solitary in the leaf axils, the pedicels up to 7 cm long (reduced upward to 1 cm); involucellar bracts 5, lanceolate, 10–12 × 2–7 mm, ciliate; calyx ca. 10 mm long, ciliate; corolla rose or red (sometimes white: *Hatschbach 35660, Weir 485*), pubescent externally and on margins of claw,

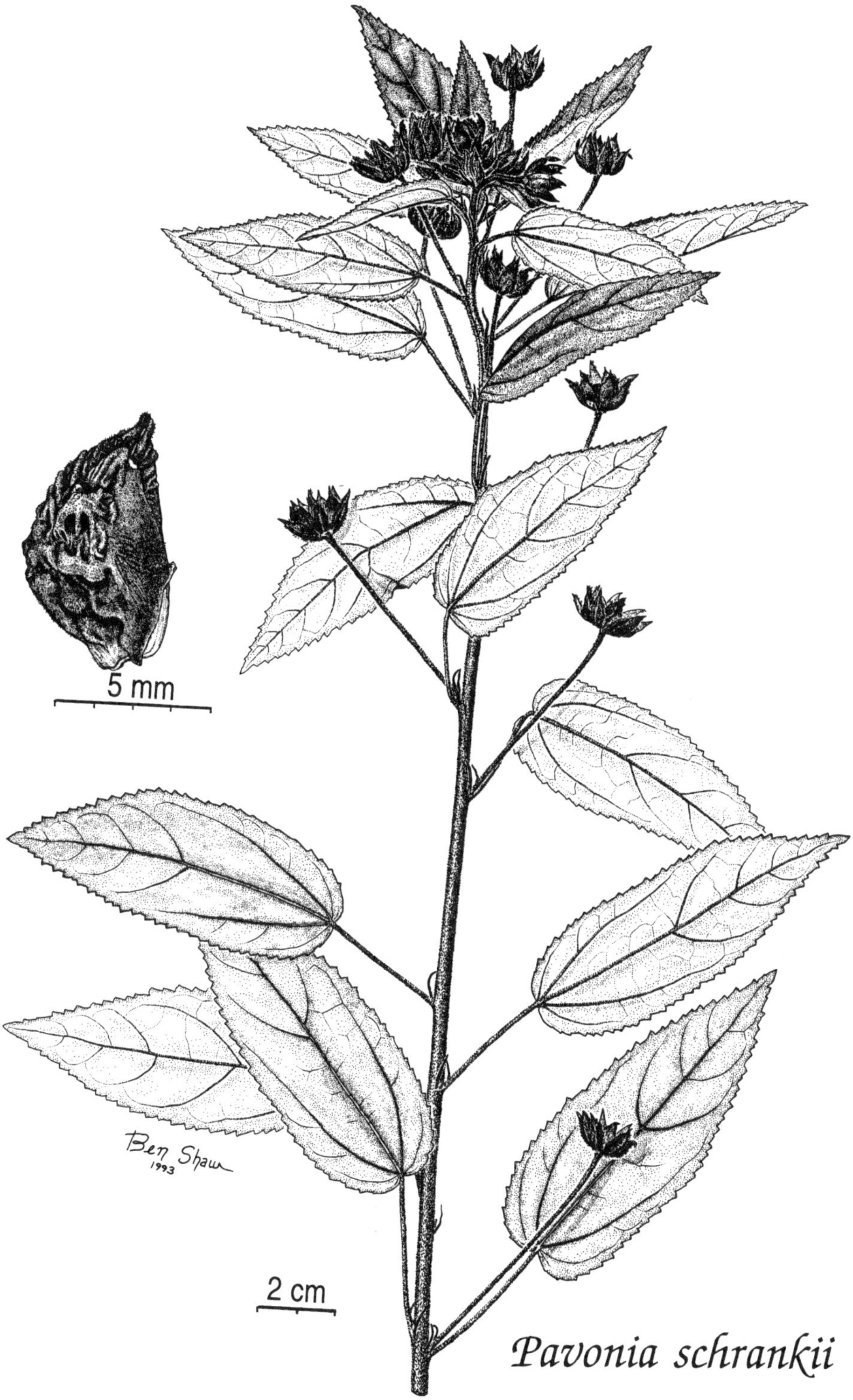

FIG. 36. *Pavonia schrankii*. Fruiting branch with enlargement of a single mericarp (*Krapovickas & Cristóbal 39590*).

3–3.5 cm long; androecium subequal to corolla, the filaments 3–6 mm long, the styles slender, glabrous, pallid, the stigmas reddish. *Fruits* 10–11 mm diam., minutely puberulent; mericarps 6–7 mm high, rugose and carinate, with 3 prominent excrescences on each side and apically acute. Chromosome number, $2n = 56$ (Fernández, 1981).

Distribution (Fig. 35). From Brazil in the states of Minas Gerais, Paraná, Rio de Janeiro, Santa Catarina, and São Paulo. The type locality of *Pavonia muricata* is at 21°58′S, 44°58′W; that of *P. hastata* forma *longifolia* is at 21°55′S, 46°23′W.

Additional illustrations. Bot. Mag. **65:** t. 3692 (1839); Gottsberger (1972: fig. 20); Gürke (1892b: pl. 94); St.-Hilaire (1827: pl. 44, as *P. muricata*).

Representative specimens examined. **BRAZIL.** São Francisco de Paula, Vila Oliva, *Rambo 31102* (CAS). **MINAS GERAIS:** Mun. Ouro Preto, Falcão, *Krapovickas & Cristóbal 33479* (CTES, F, MBM, SI, herb. Pedersen); Mun. Delfim Moreira, rd. to Itajuba, *Kuhlmann & Gehrt SP-40080* (CTES, NY, TEX); Barão de Cocais, serra da Boa Vista, *E. Pereira 2503* (CTES); Serra do Frio, *Vauthier 492* (F). **PARANÁ:** Itaperussú, *Dusén 7163* (GH, S); Jaguariahyva, *Dusén 16308* (C, CAS, F, GH, LL, MO, NY, S); 13 km WSW of Curitiba, *Ferrucci et al. 346* (CTES, G, LIL, MBM, NY); Mun. Curitiba, Barigui River, *Hatschbach 135* (CAS, CTES, DS); Mun. Piraí-Mirim, rd. to Cerne, beginning of the Serra das Furnas, *Hatschbach 507* (CAS, DS); Mun. Pitanga, Borboletinha, *Hatschbach 32858* (C, CTES); Mun. Rio Branco do Sul, Serra do Votuvoru, *Hatschbach 35660* (C, CTES); Mun. Campo Largo, Bateias, *Hatschbach 41073* (CTES, MO); Mun. Curitiba, Col. Orleans, *Hatschbach & L. Smith 16036* (C-2, CTES); Mun. Tijucas do Sul, 46 km S of Curitiba, *Krapovickas & Cristóbal 33656* (CTES, F, SI); Porto Vitoria, *Krapovickas & Cristóbal 39590* (C, CTES, K, MBM, NY); Ponta Grossa, rd. to São Jorge River, *Krapovickas & Cristóbal 40761* (BAA, C, CTES, G, HAS, INPA, LIL, MBM, MO, NY, SPF, TEX, USM); 5 km E of Curitiba, *Krapovickas et al. 23136* (CTES, NY, P, U). **RIO DE JANEIRO:** Without locality, *Weir 374* (F), *Weir 485* (F, K); on the Planalto of the Serra da Bocaina, *Vogel 761* (NY, US). **SANTA CATARINA:** Mun. Ponte Alta, Ponte Alta do Sul, *Krapovickas et al. 23053* (CTES, ENCB, LIL, MBM, MO, NY, P, TEX); 20 km from Ponta Alta on the rd. to Rio do Sul, *E. Pereira 8739* (= *Pabst 8014*) (NY, TEX, U); Pinhal da Companhia, Lauro-Müller-Urussanga, *Reitz & Klein 4103* (CTES, US); Taquara Verde, Caçador, *Reitz & Klein 11761* (CTES); Fazenda Frei Rogério, Pôrto União, *Reitz & Klein 13620* (CTES). **SÃO PAULO:** Villa Emma, *Brade 13014* (RB); Serra do Bocaina, *Epler 70* (RB); 25 km SW of Botucatu, *G. Gottsberger 14-7471* (CTES).

Pavonia schrankii, the commonest species of *P.* subsect. *Lebretonia*, is characterized by long pedicels, 2.5–7 cm long.

Gottsberger (1972) observed *Pavonia schrankii* to be pollinated by hummingbirds.

PAVONIA SUBGEN. PAVONIA SECT. LEBRETONIA SUBSECT. EXSERTAE

Pavonia subsect. **Exsertae** Krapovickas, Trabalh. XXVI Congr. Nac. Bot. [Brasil] 332. 1977. Type. *Pavonia exserta* Krapovickas & Cristóbal (= *P. commutata* Garcke).

Shrubs, the leaves ovate to 3–5-angulate or -lobed, often discolorous; flowers axillary, usually with exserted genitalia; involucellar bracts 5(–7), often more or less gamophyllous basally; mericarps carinate.

Type species. *Pavonia exserta* Krapovickas & Cristóbal.

Three species (*Pavonia lasiopetala*, *P. leiocarpa*, and *P. pabstii*) are included in *P.* subsect. *Exsertae* and not in *P.* subsect. *Hastifoliae* because they conform to the former phytogeographically and in general aspect, even though they lack the exserted androecium that otherwise characterizes members of *P.* subsect. *Exsertae*. The more or less gamophyllous involucel is not constant within *P.* subsect. *Exsertae*, but varies in degree of basal fusion, as is indicated in the following key.

Key to the species of *Pavonia* subsect. *Exsertae*

1. Involucel cupuliform, the bracts basally gamophyllous for 2–5 mm; leaves unlobed.
   2. Calyx 4–6 mm long; petals 1–1.2 cm long; Cuba. . . . . . 66. *P. cryptocalyx*
   2. Calyx 6–11 mm long; petals 1.8–3 cm long; Brazil. . . . . . 65. *P. commutata*
1. Involucellar bracts basally gamophyllous for 1–2 mm or nearly distinct; leaves often lobed, sometimes unlobed; West Indies, North America.
   3. Corolla 1 cm long; calyx 6–8 mm long.
      4. Involucellar bracts 2.5–5 mm wide; stems setose (hairs 1 mm long) and with dense underlying glandular pubescence; Brazil. . . . . . 73. *P. pabstii*
      4. Involucellar bracts 1–1.5 mm wide; stems not setose; Cuba.
         5. Leaves rounded-obtuse, puberulent; involucel subequal to calyx. . . . . . 68. *P. heterostemon*
         5. Leaves acute, puberulent with scattered, large stellate spicules in addition (as on petioles); involucel shorter than calyx. . . . . . 63. *P. calcicola*

3. Corolla 1.5–5 cm long; calyx 6–16 mm long.
   6. Stems glandular-pubescent; calyx 11–16 mm long, usually glandular-pubescent; corolla 3.5–5 cm long, the petals more or less reflexed. . . . . 64. *P. coccinea*
   6. Stems not viscid; calyx 6–13 mm long, glabrous, puberulent, or hispid, but not viscid; corolla 1.5–5 cm long, not reflexed.
      7. Corolla 3–5 cm long; calyx 10–13 mm long.
         8. Stems densely and evenly velutinous; stipules 1–2 mm long; involucel 6–8 mm long, shorter than calyx; corolla reddish; Cuba. . . . . 72. *P. mogotensis*
         8. Stems pilosulose in narrow lines, otherwise subglabrous; stipules 5–8 mm long; involucel 11–14 mm long, subequal to calyx; corolla yellow fading orange; Dominican Republic. . . . . 62. *P. aurantia*
      7. Corolla 1.5–3 cm long; calyx 6–12 mm long.
         9. Genitalia subequal to corolla or shorter; mericarps smooth.
            10. Involucel shorter than calyx, reflexed; calyx 7–11 mm long; Hispaniola. . . . . . 71. *P. leiocarpa*
            10. Involucel subequal to calyx, erect; calyx 9–12 mm long; New Mexico, Texas, NE Mexico. . . . . 70. *P. lasiopetala*
         9. Genitalia exserted; mericarps usually rugulose.
            11. Upper leaf surface with appressed simple hairs 1 mm long; plants usually with glandular hairs (at least on pedicel and involucel); fruits retrorsely hispid. . . . . 61. *P. achanioides*
            11. Upper leaf surface stellate-pubescent or glabrate; plants lacking glandular hairs; fruits stellate-pubescent or puberulent.
                12. Calyx pubescent, prominently ribbed; Hispaniola, Cuba . . . . 74. *P. punctata*
                12. Calyx minutely puberulent or subglabrous, unribbed; Haiti.
                    13. Leaves 2–2.5 times as long as wide, unlobed, acuminate, the principal nerves webbed at base forming domatia; involucel subequal to calyx; filaments 5–8 mm long, the column 3 cm long. . . . . 69. *P. hotteana*
                    13. Leaves 1–1.5 times as long as wide, often lobulate, acute, lacking domatia; involucel ca. ½ length of calyx; filaments ca. 22 mm long, the column 1 cm long. . . . . 67. *P. ekmanii*

61. Pavonia achanioides Grisebach, Cat. Pl. Cub. 27. 1866. *Malache achanioides* (Grisebach) Kuntze, Revis. Gen. Pl. **1:** 70. 1891. Type. Cuba. 1865, *Wright 2066* p.p. (lectotype, GOET-#98; isolectotypes, CTES [fragment], GH, GOET-#99, K, MA, MO-#10245, NY-2, P as photo CTES, S-2, UC, US). The lectotype was chosen by Fryxell & Fuertes (1991: 598). Here excluded is *Wright 2066* p.p. (MO-#10246), which is *Hibiscus phoeniceus* Jacq. Fig. 33

Shrubs with stems densely and minutely pubescent, the hairs 0.2–0.3 mm long. *Leaf* blades ovate-triangular, mostly 4–8 × 2.5–5 cm, basally cordate, serrate-crenate, acute, palmately 7-nerved, concolorous or slightly discolorous, with appressed simple hairs on both surfaces (of variable size, up to 1 mm long); petioles 3–6 cm long; stipules subulate, 2–3 mm long. *Flowers* solitary in the leaf axils, the pedicels 2–3.5 cm long, with some glandular hairs; involucellar bracts 5, subulate, 4–5 × 1–1.5 mm, hispid and glandular-pubescent; calyx 8–9 mm long, densely hispid, the hairs 0.5–1 mm long; corolla 3 cm long, reddish; staminal column ca. 2 cm long, glabrous, the filaments 3–4 mm long; styles exceeding the column, slender, slightly shorter than petals. *Fruits* 8 mm diam., retrorsely hispid, the mericarps strongly carinate, 5–7 mm tall, the crest sometimes prolonged apically and subacute.

Distribution (Fig. 37). A Cuban endemic.

Illustration. Fries (1947: pl. 3, fig. 2).

Specimens examined. **CUBA.** Without locality, *Wright 2067* (syntypes: GH, GOET, K, MA, MO, P as photo CTES). PINAR DEL RÍO: Mtns. near El Guama, *W. Palmer & Riley 257* (NY, US).

Local name. Majagüilla (Cuba).

Pavonia achanioides is distinctive for its relatively large flowers with exserted genitalia, its more or less viscid herbage, and for having appressed simple hairs on the upper leaf surface.

62. Pavonia aurantia Liogier, Mem. New York Bot. Gard. **21:** 134. 1971. Type. Dominican Republic. Cordillera Central: Valley of Bao River, on the way to El Valle, 1–7 Oct 1968, *A. H. Liogier 12747* (holotype, NY; isotypes, GH, P–n.v., US).

Shrubs to 3.5 m tall, the stems subglabrous, with narrow longitudinal lines of simple recurved hairs and (on younger stems) scattered 3–5-armed stellate

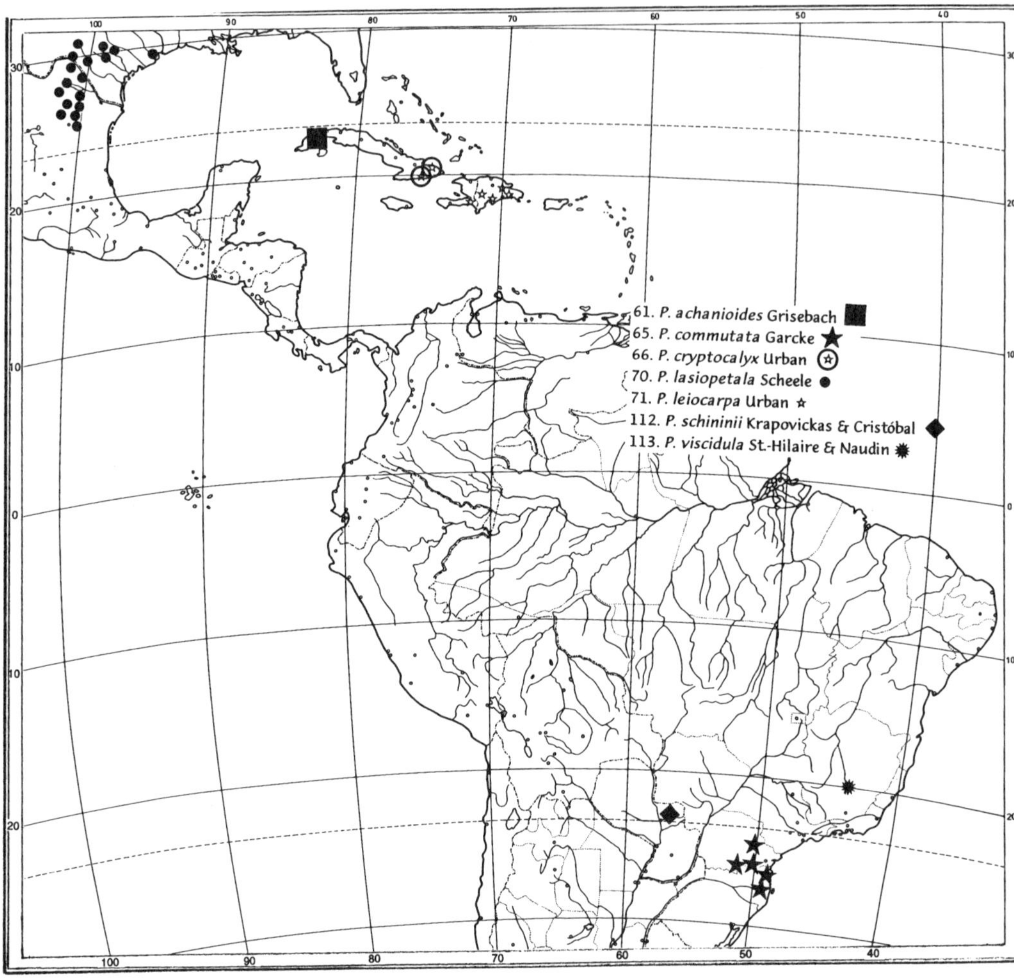

FIG. 37. Distribution map of seven neotropical species of *Pavonia* subsect. *Exsertae* and *P.* subsect. *Hastifoliae.*

spicules. *Leaf* blades ovate, mostly 4–6 × 2.5–3.5 cm, basally truncate to subcordate, serrate, acuminate, palmately 5-nerved, slightly discolorous, the upper surface with appressed simple hairs 1 mm long, the lower surface with appressed simple hairs (especially on veins) and occasional stellate hairs; petioles 1–2 cm long; stipules 5–8 mm long, filiform, ciliate. *Flowers* solitary in the leaf axils, the pedicels 1–2 cm long, arachnoid-pubescent, articulated near base; involucellar bracts 5–7, basally connate for ca. 1 mm, ligulate to lanceolate, 11–14 × 1–2 mm, hispid-ciliate; calyx more or less equaling involucel, hispid-ciliate; corolla 3–3.5 cm long, glabrous (except pubescent on margins of claw), yellow or "yellow turning orange-red"; staminal column ca. 2 cm long, the filaments 2–4 mm long; styles exceeding androecium, shorter than or slightly exceeding corolla. *Fruits* unknown.

Distribution (Fig. 22). Endemic to Dominican Republic, from 1300 to 1400 m elevation. The type locality is at 19°18′N, 70°46′W.

Specimens examined. DOMINICAN REPUBLIC. PROV. DAJABÓN: 10 km from Loma de Cobrera (Pueblo) on the rd. to Capotillo and Alto de la Paloma, *Zanoni & Mejía 17946* (JBSD, TEX).

Pavonia aurantia is distinctive for its yellow corolla (fading orange).

63. Pavonia calcicola (Britton) Ekman, Feddes Repert. Spec. Nov. Regni Veg. **24:** 234. 1928. *Malache calcicola* Britton, Mem. Torrey Bot. Club **16:** 80. 1920. Type. Cuba. Prov. Santa Clara: Río San Juan, 24–25 Mar 1910, *N. L. Britton, Earle & Wilson 5905* (holotype, NY; isotypes, F, GH, US).

Shrubs 2 m tall, the stems densely and very minutely puberulent (including glandular hairs), the younger branches with overlying large stellate spicules. *Leaf* blades ovate to 3-lobed, mostly 2–3 × 1.5–2.5 cm, basally cordate, serrate, acute, palmately 5–7-nerved, discolorous, densely puberulent above, whitish puberulent beneath and with overlying large stellate spicules; petioles 1–1.5 cm long, usually more prominently covered with spicules than stem; stipules subulate, 2–3 mm long. *Flowers* solitary in the leaf axils, the pedicels 10–12 mm long, slender, minutely puberulent (glandular?); involucellar bracts 5, ligulate, 5 × 1 mm, minutely pubescent; calyx 7–8 mm long, minutely pubescent, ecostate; corolla ca. 1 cm long, red, externally pubescent (with large stellate spicules), the petals acuminate; staminal column subequal to corolla to clearly exserted, the filaments 2–3 mm long; styles exserted by 1 cm or more, connate nearly to apex. Fruits ca. 7 mm diam., rugose, glabrous, straw-colored; mericarps prominently crested, apically acute, 6 mm tall; seeds glabrous, obscurely striate.

Distribution (Fig. 22). Endemic to the S coast of Cuba.

Illustration. Sauget & Liogier (1953: fig. 106).

Specimens examined. **CUBA**. SANTA CLARA: Near María Aguilar, *Ekman 18896* (S).

Pavonia calcicola is doubtfully distinct from *P. heterostemon*. The short corolla, barely exceeding the calyx, is distinctive, as are the large spicules found on the outside of the petals.

64. Pavonia coccinea Cavanilles, Diss. **3:** 140, t. 47, f. 1. 1787. Type. Santo Domingo, *Desportes s.n.* (holotype, P-JU no. 12347, as microfiche; possible isotype, MA-476070, cited by Garilleti [1993]). Figs. 33, 38

Shrubs 2 m tall, the stems densely stellate- and minutely glandular-pubescent (sometimes also with large 3–4-armed stellate spicules). *Leaf* blades ovate or somewhat palmately 3-lobed, mostly 6–10 × 3–7 cm, basally cordate, crenate-serrate, acute or acuminate, palmately 5–7-nerved, discolorous, the upper surface with minute glandular hairs, the lower surface densely stellate- and glandular-pubescent, both surfaces with scattered large (2 mm diam.) 3–4-armed stellate spicules, at least on younger leaves; petioles 3–8 cm long; stipules subulate, 5 mm long. *Flowers* solitary in the leaf axils, the pedicels 1.5–7 cm long, articulated 3–8 mm below the flower; involucellar bracts 6, ligulate, 10–12 × 1–1.5 mm, briefly connate basally (essentially distinct), glandular pubescent; calyx 11–16 mm long, glandular-pubescent, the lobes 3-ribbed; corolla 3.5–5 cm long, red, externally pubescent (the petals reflexed according to photograph on *Liogier 16524*, NY); staminal column slightly exserted (or clearly so by reflexion of petals), glabrous, purplish, the filaments 3–5 mm long along upper half of column; styles emergent from the column, pubescent. *Fruits* ca. 1 cm diam., minutely pubescent; mericarps with a dorsal crest or wing prolonged apically into an acute or scimitar-shaped structure, reticulately veined, with 2 or 3 excrescences on each side, the mericarp body 6 mm high (8–9 mm high including the wing); seeds 3.5 mm long.

Distribution (Fig. 39). Relatively common on the island of Hispaniola, on sea cliffs, in shrubland, and in savanna, at elevations of 0–1200 m.

Additional illustrations. Cavanilles (1787: pl. 47, fig. 1); Liogier (1981: fig. 39).

Representative specimens examined. HAITI. Massif des Matheux, Magasin Carries, *Ekman 3055* (S); vic. of Jean Rabel, *E. C. Leonard & G. M. Leonard 13025* (A, E, MICH, US); vic. of Mole St. Nicolas, *E. C. Leonard & G. M. Leonard 13203* (CM, MO, US); vic. of Bassin Bleu, *E. C. Leonard & G. M. Leonard 14716* (GH, US); Presqu'ile Nord-Ouest: Dept. Nord-Ouest, 19 km S of Port-de-Paix (19°49′N, 72°48′W), *Zanoni et al. 34831* (CTES, JBSD, TEX).

DOMINICAN REPUBLIC. ISLA SAONA: *A. H. Liogier & P. Liogier 21851* (JBSD, NY). AZUA: El Número, *Mejía 143* (JBSD, NY); 4 km E of Hatillo de Azua, 18°23′N, 70°31′W, *Mejía et al. 8935* (JBSD, MO, NY). BARAHONA: On the Duvergé River, *Fuertes 373* (A, E, NY); El Firme, Cabral, *Jiménez 3747* (US). DISTRITO NACIONAL: Sierra Prieta de Villa Mella, N of Santo Domingo, *A. H. Gentry & Zanoni 50560* (MO, NY); at town of Sierra Prieta, *Zanoni et al. 15334* (JBSD). Independencia: Bahoruco, vic. of Lake Enriquillo, *Howard 12131* (A, U, US); El Guanito, Puerto Escondido, *A. H. Liogier 14035* (NY, US); El Sabinal, above Puerto Escondido, *A. H. Liogier 14038* (GH, NY, US). LA VEGA: Loma Peguera, Hato Viejo, *A. H. Liogier 17379* (NY); Barrancón, Bonao, *A. H. Liogier 18546* (F, NY); SE of Bonao, Arroyo Hato Viejo, *Zanoni et al. 15700* (JBSD, NY, U). LA ROMANA: NW of Chavón Dam, *Mejía & Zanoni 9138* (JBSD, NY). MONTE CRISTI: Monción, *Ekman 12592* (LL, S, US); near Copey, S of Montecristi, *A. H. Liogier 16385* (GH, NY). PUERTO PLATA: La Isabela, *Jiménez 2317* (CAS, NY, US). SAMANÁ: Cape Samaná, *Abbott 1183* (GH, NY, US); Cabo Samaná, toward El Frontón, *Ekman 15334* (S).

The glandular herbage and reflexed petals are distinctive of *Pavonia coccinea*.

65. Pavonia commutata Garcke, Jahrb. Bot. Gart. Berlin **1:** 212. 1881. Type. Cult. in hort. Berlin, 1856 (holotype, B as photo F-9436). Fig. 33

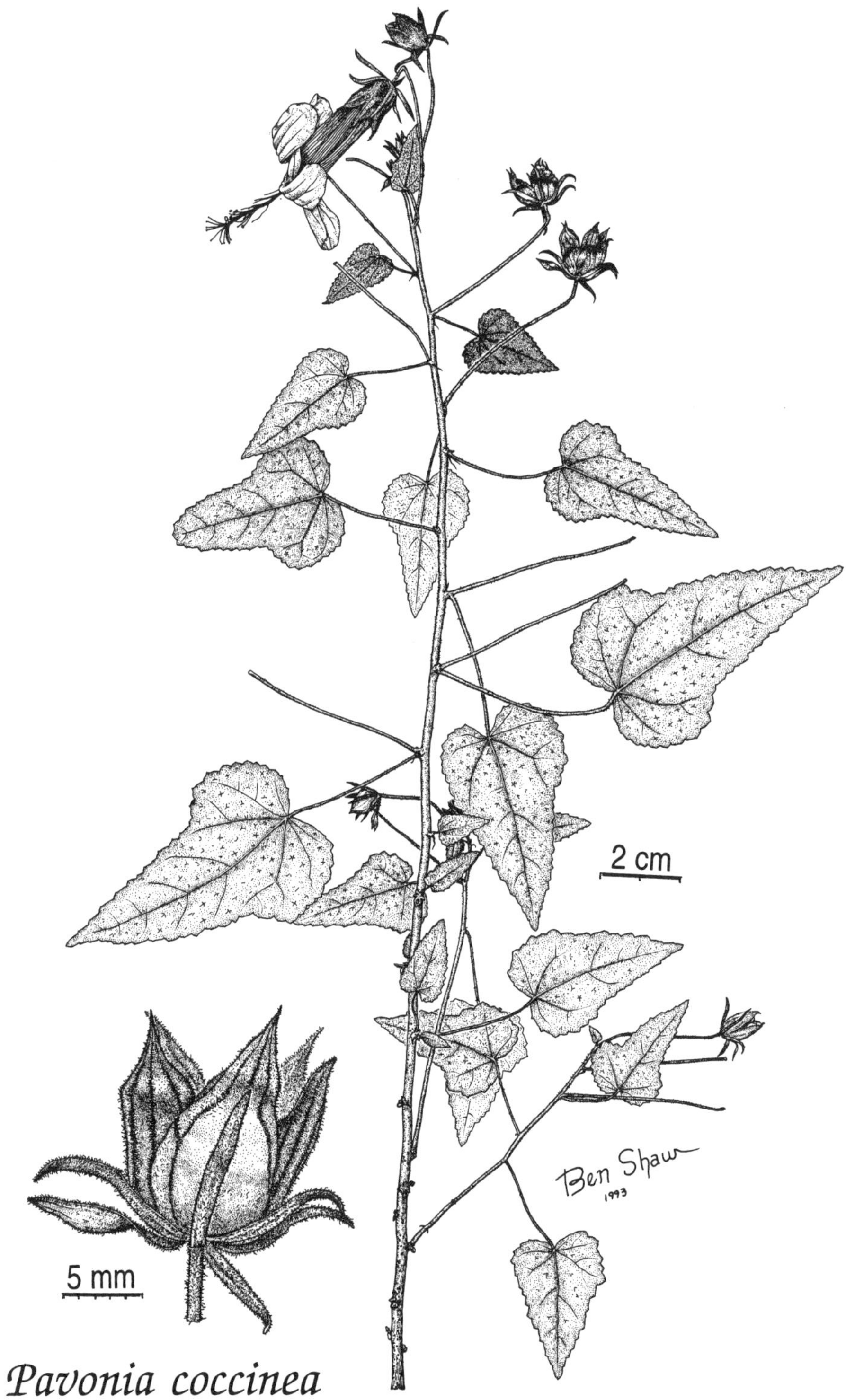

FIG. 38. *Pavonia coccinea*. Flowering branch with enlargement of calyx and involucel (lower left) (*Jiménez2317*).

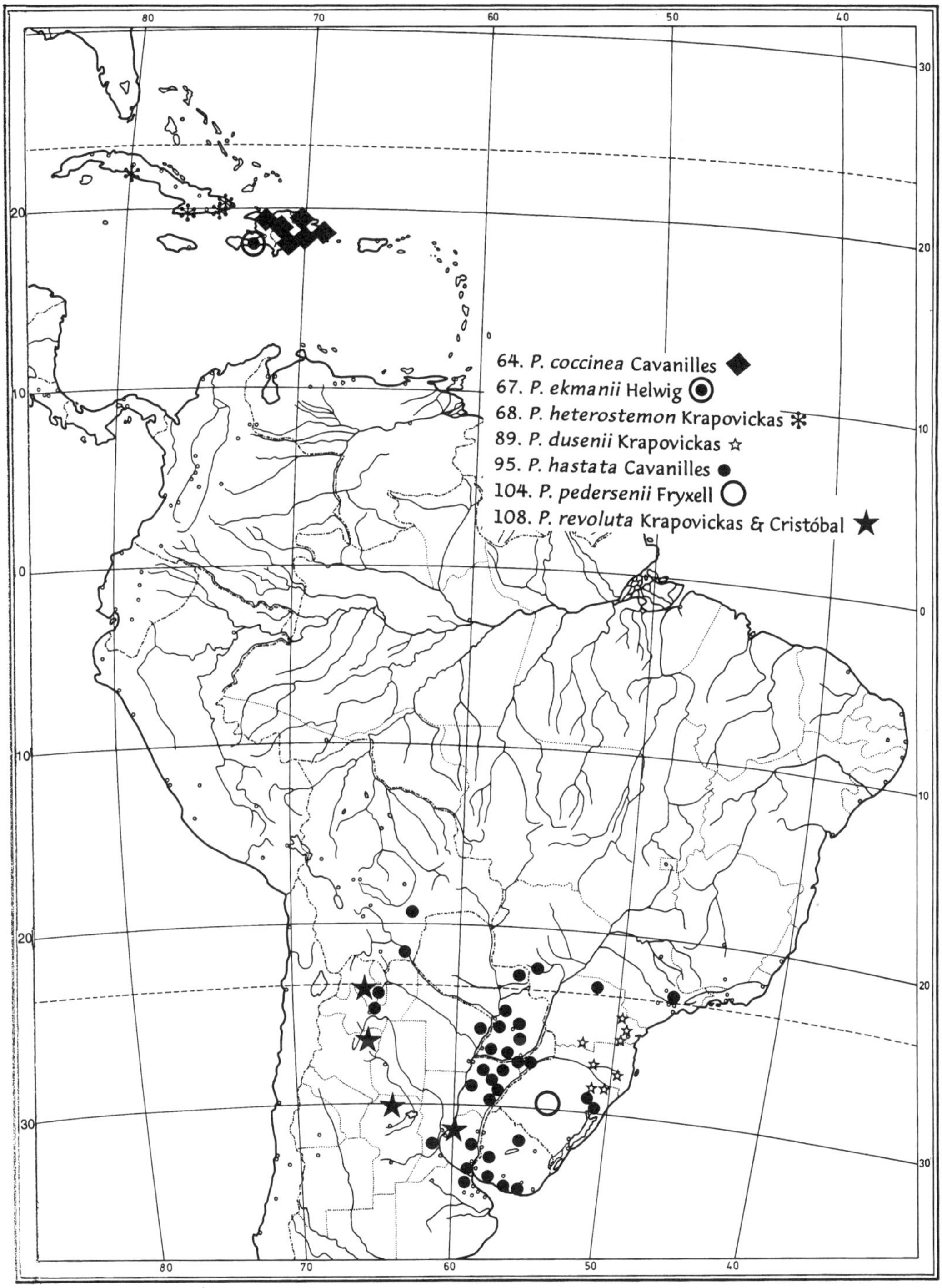

FIG. 39. Distribution map of seven neotropical species of *Pavonia* subsect. *Exsertae* and *P*. subsect. *Hastifoliae*.

Pavonia engleriana Gürke *in* Martius, Fl. Bras. **12(3):** 497. 1892. Type. Brazil. *Sellow 3935* (holotype, B as photo F-9440; isotype, US).

Pavonia exserta Krapovickas & Cristóbal, Lilloa **31:** 63. 1962. Type. Brazil. Santa Catarina: Mun. Mafra, Ruderal, 21 km W of Rio Negrinho, 800–850 m, 11 Mar

1957, *L. B. Smith & Klein 12052* (holotype, LIL–n.v.; isotypes, CTES, HBR–n.v., NY, US).

Pavonia gamophylla Krapovickas & Cristóbal, Lilloa **31:** 66. 1962. Type. Brazil. Santa Catarina: Bom Jardin, S. Joaquim, *Reitz & Klein 7961* (holotype, LIL–n.v.; isotypes, CTES, US).

Shrubs 1–2 m tall, the stems densely and minutely stellate-pubescent. *Leaf* blades ovate or ovate-elliptic, mostly 2–10 × 1–5 cm, basally truncate to cordate, obscurely or prominently crenate-serrate, acute or acuminate, markedly discolorous, sparsely and minutely stellate-pubescent above, densely and finely tomentulose beneath, palmately 5–7-nerved, the nerves sunken above, raised beneath; petioles 0.5–4 cm long; stipules subulate to filiform, 3–7 mm long. *Flowers* solitary in the leaf axils, the pedicels 5–25 mm long; involucel cupuliform (basally gamophyllous), 6–10 mm long, the lobes 5(–6), narrowly triangular, sparsely pubescent (densely so at base); calyx 6–11 mm long, stellate-pubescent, especially on nerves; corolla 1.8–3 cm long, rose or red, tubular, glabrous; androecium exserted, purplish, the filaments 4–8 mm long; styles exceeding the column, free for distal 5 mm. *Fruits* ca. 5 mm diam., minutely puberulent; mericarps 5 mm high, with a dorsal crest or wing (ca. 1 mm wide apically) and with 2 or 3 acute excrescences on each side.

Distribution (Fig. 37). Found in central Brazil (Paraná and Santa Catarina, with one collection from São Paulo without precise locality). The type locality of *Pavonia exserta* is at 26°13′S, 49°37′W; that of *P. gamophylla* is at 28°17′S, 49°43′W.

Illustrations. Gürke (1892b: pl. 93), Krapovickas & Cristóbal (1962: figs. 2F, 12, 13).

Specimens examined. BRAZIL. PARANÁ: Vila Velha Park, Guavirova River, *Hatschbach 8883* (CTES, U); Mun. P. Grossa, Vila Velha Park, Lagoa Dourada, *Hatschbach 12858* (NY, US); Mun. Pôrto Amazonas, Iguazú River, *Hatschbach 16181* (CTES); Mun. Antonio Olinto, Agua Amarela River, *Hatschbach 22264* (CTES, NY); Mun. Paula Freitas, Santa Ana River, *Hatschbach 22727* (CTES, F, MO, NY, S, herb. Pedersen); Mun. Texeira Soares, Rio das Almas, BR-277, *Hatschbach 43531* (F, MO); Mun. Campo do Tenente, Lageado, *Hatschbach & Guimarães 18486* (C–2, CTES, F); Mun. São José dos Pinhais, *Hatschbach & Manosso 51514* (CTES); Mun. Palmeira, Rio Capivra, BR-277, *Ribas & Hatschbach 57* (MBM, TEX); Mun. Palmeira, Rio dos Papagaios, *J. M. Silva & Hatschbach 513* (C, S, US). SANTA CATARINA: Mafra, *E. Pereira 8357* (= *Pabst 7632*) (CTES, NY); Tres Barras, *Krapovickas & Cristóbal 39597* (C, CTES, F, G, K, NY, U, UC); Rio das Pedras, 3 km N of Ponte Alta do Norte, *Krapovickas & Cristóbal 43957* (CTES, K, MO); Mun. São Joaquim, Pelotas River, 3 km W of Bom Jardim da Serra, *Krapovickas & Schinini 38279* (CTES, HAS, UC); Mafra, *Reitz 5315* (BRIT, CAS, CTES, NY, U). SÃO PAULO: Without locality, *Sellow s.n.* (GH).

Krapovickas and Cristóbal (1993) reduce *Pavonia gamophylla* to synonymy with *P. engleriana*. This synonymy is further enlarged by including *P. commutata* and *P. exserta*, because the four are virtually sympatric and essentially impossible to distinguish, in consequence of which they are concluded to represent a single species.

The cupuliform involucel distinguishes *Pavonia commutata* from the other members of *P.* subsect. *Exsertae*, except for *P. cryptocalyx*, which has much smaller flowers and is allopatric.

66. Pavonia cryptocalyx Urban *in* Urban & Helwig, Feddes Repert. Spec. Nov. Regni Veg. **24:** 234. 1928. Type. Cuba. Oriente: Near Santiago de Cuba, 21 Nov 1917, *Ekman 8979* (isotypes, CTES [fragment], F as photo F-56163, NY, S). Fig. 33

Shrubs ca. 1 m tall (?), the stems densely yellowish-puberulent, the hairs 0.1–0.2 mm long. *Leaf* blades ovate to 3(–5)-lobed, 1.5–2 × 1–1.5 cm, cordate, crenate-serrate, acute or subobtuse, discolorous, densely and softly puberulent above and beneath (the hairs very small), prominently reticulate-veined beneath; petioles 1–1.5 cm long; stipules subulate, 1–2 mm long, deciduous. *Flowers* solitary in the leaf axils, the pedicels ca. 1 cm long; involucellar bracts 5, 5–6 × 4 mm, gamophyllous for half of length (simulating calyx), densely puberulent; calyx subequal to involucel and partially obscured by it; corolla evidently rose with red center, 1–1.2 cm long. *Mericarps* 5 mm long, carinate and somewhat rugose.

Distribution (Fig. 37). Endemic to Cuba. The type locality is at 20°5′N, 75°48′W.

Specimens examined. CUBA. GUANTÁNAMO: Meseta de Maisí, Baracoa, *Acuña & Díaz Barreto 17504* (NY); Cabanas Bay, *N. L. Britton & Cowell 12811* (NY); Silla de Gibara, *Bro. León (J. S. Sauget) 20524* (NY).

Pavonia cryptocalyx is notable for its cupuliform involucel coupled with a much reduced calyx and corolla.

67. Pavonia ekmanii Helwig *in* Urban & Helwig, Feddes Repert. Spec. Nov. Regni Veg. **24:** 235. 1928. Type. Haiti. Massif de la Hotte, group Morne Rochelois, Miragoane, limestone cliffs at Carenage, *Ekman 7270* (holotype, S; isotypes, CTES [fragment], K, NY, S, US).

Shrubs 1 m tall, the stems scandent (?), densely covered with hairs <0.1 mm long (the hairs so small the stems seem glabrous) and with a few larger (0.5

mm long) 3–5-armed stellate spicules on younger stems. *Leaf* blades ovate to pentangular, mostly 3–6 × 2–4.5 cm, basally truncate to cordate, serrate, acute, palmately 5–7-nerved, somewhat discolorous, the upper surface very minutely pubescent (seemingly glabrous), the lower surface densely covered with stellate hairs ca. 0.2 mm diam. and with scattered 3–4-armed stellate spicules 0.5–1 mm diam.; petioles 1–2.5 cm long, with pubescence like stem except the large spicules more abundant (especially distally) and sometimes 5-armed; stipules subulate, 3 mm long, often reflexed. *Flowers* solitary in the leaf axils, the pedicels 1–2.5 cm long; involucellar bracts 5–6, ligulate or narrowly triangular, 4–5 × 1–1.5 mm, briefly connate basally, minutely ciliate; calyx 6–9 mm long, subglabrous, virtually ecostate; corolla 2–2.5 cm long, reddish, externally obscurely pubescent; androecium exserted, the column ca. 1 cm long, the filaments 10, ca. 22 mm long, erect; styles slightly exceeding the anthers, connate nearly to the stigmas. *Fruits* ca. 7 mm diam., minutely puberulent; mericarps carinate and coarsely reticulate, ca. 6 mm tall.

Distribution (Fig. 39). Endemic to Haiti, near sea level. The type locality is at 18°28′N, 73°08′W.

Specimens examined. **HAITI.** Anse-à-Veau, *Ekman 5395* (S); Grand Cayemite, Côtes-de-Fer, *Ekman 8952* (F, GH, S, US); Grand Cayemite, Côtes-de-Fer, *Eyerdam 335* (F, GH, NY, US); near Miragoane, *Eyerdam 560* (GH, US); Massif de la Hotte, Dept. Grand'Anse, 4.4 km W of Miragoane, *Zanoni et al. 28746* (JBSD, NY).

Pavonia ekmanii is characterized by a constellation of characters (see key) rather than by any one or two distinctive characters.

68. Pavonia heterostemon Urban *in* Urban & Helwig, Feddes Repert. Spec. Nov. Regni Veg. **24:** 233. 1928. Type. Cuba. Oriente: Sierra Maestra, near Daiquiri, 16 Nov 1916, *Ekman 8370* (lectotype, here designated, S; isotypes, A, CTES [fragment], F as photo F-56162, GH, NY, US). Fig. 33

Shrubs 2 m tall, the stems densely stellate-pubescent. *Leaf* blades ovate, mostly 2–4 cm × 1.5–3.5 cm, basally cordate, coarsely crenate, obtuse or acute, palmately 5–7-nerved, discolorous, minutely stellate-tomentose above and beneath; petioles 8–20 mm long; stipules subulate, 3–4 mm long. *Flowers* solitary in the leaf axils, the pedicels 1–2 cm long, obscurely articulated 2 mm below the flower; involucellar bracts 5, ligulate, 6–8 × 1–1.5 mm, stellate-pubescent; calyx 6–7 mm long, stellate-pubescent; corolla ca. 1 cm long, red; androecium exserted (1.5–2 times length of corolla), the column glabrous, the filaments 5–8 mm long, mostly distal; styles exserted from column by 5–6 mm. *Fruits* ca. 7 mm diam.

Distribution (Fig. 39). Endemic to Cuba, in coastal thickets near sea level. The type locality is at 19°55′N, 75°35′W.

Specimens examined. **CUBA.** **CIENFUEGOS:** Cienfuegos Bay, W bank of San Juan River, on steep dry limestone, *Wood & Atchison 7436* (A). **GRANMA:** Cap Cruz, *Bro. León (J. S. Sauget) 16320* (GH, NY, US); en la Silla de Gibara, *López 1040* (US); near Macambo, Vía Azul, *Bro. Alain (H. Liogier) & López 4211* (GH); Mesa del Chivo, Maisí, *Bro. León (J. S. Sauget) & Marie-Victorin 17134* (NY). **GUANTÁNAMO:** Baracoa Region, base of Loma del Esparto, *Bro. León (J. S. Sauget) 12110* (GH, NY); Loma de Esparto, 10 km E of Cajobabo, on rd. to Jauco, *Morton & Bro. Alain (H. Liogier) 9136* (US); km 7 between Cajobabo and Jauco, *Bro. Alain (H. Liogier) & Morton 5187* (GH).

Pavonia heterostemon and *P. calcicola* (q.v.) are very similar and possibly conspecific.

69. Pavonia hotteana Helwig, Ark. Bot. **22A(17):** 27. 1929. Type. Haiti. Massif de la Hotte, W group Les Roseaux, toward Débarcadère-Plingué, *Ekman 10777* (lectotype, here designated, S; isotypes, CTES, K, NY, S, US).

Scandent *shrubs*, the stems very minutely puberulent. *Leaf* blades narrowly ovate, mostly 4–6 × 1.5–3 cm (2–2.5 times as long as wide), basally truncate or subcordate, coarsely serrate, acuminate, palmately 5 (–7)-nerved (the sinuses of the principal nerves beneath often webbed at base, forming domatia), slightly discolorous, very minutely puberulent above and beneath (rarely with a few 3–4-armed stellate spicules beneath or simple spicules above), seemingly glabrous; petioles 1–3 cm long; stipules subulate, 1–2 mm long. *Flowers* solitary in the leaf axils, the pedicels 1.5–2.5 cm long, articulated 3–4 mm below the flower; involucellar bracts 5, lanceolate, 8–12 × 2–3 mm, sometimes marginally ciliate, otherwise glabrous, subequal to or exceeding the calyx; calyx 8–9 mm long, very minutely puberulent; corolla 2.5 cm long, with scattered pubescence externally; androecium exserted by 1 cm, the column glabrous, purplish, 3 cm long, the filaments 5–8 mm long, distally inserted; styles exserted from column, slender, free for distal 8–10 mm. *Fruits* ca. 6 mm diam.; mericarps carinate and rugulose, minutely puberulent.

Distribution (Fig. 22). Endemic to Haiti, along the coast. The type locality is at 18°35′N, 74°00′W.

Specimens examined. **HAITI.** Massif de la Hotte, W group, Jérémie, Morne des Bordes, *Ekman 10792* (S); Massif de la Hotte, W group, Les Abricots, on the coral reef E of town, *Ekman 10858* (F, GH, S).

The frequent presence of domatia in the axils of the principal nerves on the abaxial surface of the leaves is notable in *Pavonia hotteana*.

70. Pavonia lasiopetala Scheele, Linnaea **21:** 470. 1848. *Malache lasiopetala* (Scheele) Kuntze, Revis. Gen. Pl. **1:** 70. 1891. Type. United States. Texas: Near New Braunfels, 1846, *Lindheimer 352* (lectotype, here designated, US; isotypes, BM, K–3, MO, NY–2, PH–2). Fig. 33

Pavonia wrightii A. Gray, Gen. Amer. Bor. **2:** 76, t. 130. 1849. *Malache wrightii* (A. Gray) Kuntze, Revis. Gen. Pl. **1:** 71. 1891. Type. United States. In cultivation, *ex* seeds from Texas: W Texas, 1848, *Wright s.n.* (holotype, GH; isotypes, K, PH).

Shrubs ca. 1 m tall, the stems densely to sparsely stellate-pubescent, sometimes dark-pigmented. *Leaf* blades mostly 2–5 cm long, slightly longer than wide, ovate, cordate, coarsely dentate, acute, discolorous, stellate-pubescent, more densely and coarsely so beneath; petioles ½–1 times the length of the blade; stipules 2–5 mm long, subulate. *Flowers* solitary in the leaf axils, the pedicels 2–5 cm long; involucellar bracts 5, linear-lanceolate (or narrowly oblanceolate), shorter than to subequal to calyx, 1–2 mm wide, hirsute; calyx 9–12 mm long, hirsute, whitish with green veins; corolla 1.5–2.5 cm long, rose-lavender, ciliate on margins of claw; staminal column 7–13 mm long, glabrous, pallid, the filaments 2–3 mm long; styles exceeding the column, the stigmas usually villous. *Fruits* 8–9 mm diam., oblate, essentially glabrous, the mericarps 3.5–4 mm long, smoothly rounded or obscurely carinate.

Distribution (Fig. 37). *Pavonia lasiopetala* is the northernmost representative of the genus, with the possible exception of adventive plants of *P. hastata* from the state of Georgia (the types of *P. lecontei* and *P. jonesii*). It is in fact wholly extratropical in distribution, occurring only in Coahuila, Nuevo León, New Mexico, and Texas, at 500–1000 m elevation. In Texas it occurs at least as far N as 30°33′N at Marble Falls (Burnet Co.) and near Sheffield (N Terrell Co.). The type locality is at 29°38′N, 98°05′W.

Illustrations. Andrews (1992: 97*); Fryxell (1979: figs. 3F, 5B); Gray (1849: pl. 130, as *P. wrightii*); Meehan (1892: pl. 12, as *P. wrightii*); Powell (1988: fig. 230).

Representative specimens examined. **UNITED STATES. New Mexico:** Without locality, in 1852, *Wright 1333* (BM as photo CTES, NY). **Texas:** "Western Texas," in 1849, *Wright 61* (NY, PH, US–2); Chinati Mtns., *Havard 132* (NY, PH, US); near mouth of the Pecos River near Santa Rosa, 1852–1853, *Bigelow s.n.* (NY); Guadeloupe, 1845, *Lindheimer 503* (MO). **Bexar Co.:** Bracken, near the Cibolo River, *Groth 125* (F, NY); near Selma, *Metz 130* (CM, NY); Classen Ranch, *Schulz 38–43* (F). **Burnet Co.:** Rd. to Max Starcke Dam at bridge over Flatrock Creek, Marble Falls, *Brainard 69* (TEX); banks of the Colorado River, Marble Falls, *Reverchon 1498* (F, MO). **Comal Co.:** Near New Braunfels, 1846, *Lindheimer 242* (MO–2); Comanche Spring, New Braunfels, etc., *Lindheimer 684* (ARIZ, C, F, MEXU, NY, PENN, PH, TEX–2); New Braunfels, 1 mi W of Landis Park, *Warnock 46259* (BISH, BRIT, K, LSU, NY, PENN, PH, TEX). **Galveston Co.:** Stewart Beach Rd. in Galveston, *Mears 7731* (TEX). **Gillespie Co.:** Crab Apple, *Jermy 852* (MO). **Terrell Co.:** 13 mi S of Sheffield, *Webster 193* (TEX). **Travis Co.:** Austin, Bee Creek, *Carsuer & Studhalter 4618* (TEX); Mount Bonnell near Austin, *Tharp et al. 46281* (BRIT, COLO, CTES, F, MEXU, MO, NY, PENN, PH, POM, TEX–2, US); Pedernales River, W of Hamilton Pool, *Webster 510* (COLO, TEX). **Uvalde Co.:** 26 mi N of Uvalde, *Cory 49392* (BRIT); W of Uvalde, *M. E. Jones 28141* (POM); 8 mi N of Uvalde, *S. Jones & G. Jones 6464* (NY); Uvalde, 5–10 Oct 1880, *E. Palmer 89* (MO, NY). **Val Verde Co.:** 8 mi S of Loma Alta, *Cory 41678* (TEX).

MEXICO. Coahuila: Cañón de los Burros, NE part of Serranías del Burro, *Chiang et al. 7515* (TEX); Sierra de Santa Rosa, ca. 10 mi SW of Múzquiz, *P. A. Fryxell 2675* (CTES, NY); ca. 12 air mi E of Boquillas in Sierra del Carmen, *Henrickson 11507* (LL); 11.5 km S of Cuatro Palmas a 8 km S of Mina Aguachile on rd. to Rancho Las Norias, *M. C. Johnston et al. 11940B* (LL); Múzquiz, *Marsh 2133* (BRIT, F, TEX); Sabinas, *E. W. Nelson 6185* (NY, US); Monclova, in 1880, *E. Palmer 90* (NY, PH, U, US); Cañon del Indio Felipe, ca. lat. 28°33′, in Sierra Hechiceros, close to Chihuahua border, *Stewart 45* (LL); El Cedral, Sierra de la Paila, *Villarreal & Carranza 3913* (TEX); Sierra de la Gloria, Cañón El Cono, a side canyon of C. Chilpitín, *Wendt & Riskind 1626* (TEX); Rincón de María, *Wendt et al. 1326* (TEX); Hacienda La Rosita, Mun. Castaños, El Puerto de San Lázaro, *Wynd & Mueller 146* (ARIZ, K, MO, NY). **Nuevo León:** Pastora, near Monterrey, *Arsène 6192* (MO, US); N of Linares, *Clark 6801* (MO, NY); ca. 3 mi S of Bustamante, Grutas, 6 May 1976, *J. E. Fryxell s.n.* (NY); Sierra Anahuac, Distr. Sta. Catarina, *Hinton et al. 16899* (ARIZ, RSA); Mariposa Station (microwave) between Saltillo and Monterrey, *Lane & Leidig 2888* (COLO, TEX); Mun. Villa Santiago, Cañon of Potrero Redondo, *Mueller 2082* (F, MO); near Monterrey, *Pringle 1944* (COLO, E, F, K, NY, PH, TEX, US), *Pringle 2686* (CTES [fragment], E, F, MEXU, MO, NY, TEX, UC), *Pringle 13750* (ARIZ, BRIT, C, COLO, F, NY, TEX, US); Lampazos, *M. Taylor (Edwards) 307* (ARIZ, F, MO, TEX); Horsetail Falls, 30 mi SE of Monterrey, *W. M. Thompson 268* (TEX).

Local names and uses. Rose mallow (Texas) (Nokes, 1986: 269). *Pavonia lasiopetala* is sometimes cultivated in central Texas as an ornamental shrub

(Nokes, 1986) because of its profuse production of pink flowers and its tolerance of relatively dry habitats.

Pavonia lasiopetala stands out for its smooth carpels and flowers with included genitalia. It shares these characters with the closely similar *P. leiocarpa.*

71. Pavonia leiocarpa Urban, Symb. Ant. **7:** 279. 1912. Type. Santo Domingo. Near Constanza, 1200 m, May 1910, *von Türckheim 3189* (isotypes, BM, BR, CTES, E, F, GH, K, MO, NY, P as photo CTES, S, U, US). Fig. 33

Shrubs 2–3 m tall, the stems stellate-pubescent (hairs ca. 0.5 mm long, curled), the pubescence usually in longitudinal rows, with a few larger (1 mm long or more) 3–4-armed spicules scattered on younger growth. *Leaf* blades ovate, mostly 4–9 × 2–5 cm, basally cordate to truncate, crenate, acute or acuminate, palmately 5–7-nerved, discolorous, both surfaces with scattered simple and bifurcate (occasionally stellate) hairs of variable size; petioles 2–4(–6) cm long, stellate–pubescent, densely so on adaxial side, the hairs somewhat longer and denser at distal end; stipules 1–2 mm long, subulate, often reflexed. *Flowers* solitary in the leaf axils, the pedicels 2–5 cm long; involucellar bracts 5, linear or narrowly triangular, 5–6 × 1–1.5 mm, briefly connate at base or essentially distinct, often reflexed, ciliate; calyx 7–11 mm long, sparsely pubescent; corolla 2–2.5 cm long, reddish; androecium subequal to corolla, the filaments 5–7 mm long; styles emergent from the androecium, slender, glabrous. *Fruits* 7–9 mm diam., subglabrous; mericarps 3–4.5 mm long, smoothly rounded, sometimes reticulately veined (but not nerved).

Distribution (Fig. 37). On the island of Hispaniola, growing in shrubland and forest at 1000–1600 m elevation, predominantly in the Dominican Republic but with one collection from Haiti. The type locality is at 18°52′N, 70°45′W.

Specimens examined. HAITI. Massif de la Selle, Morne Brouette, slope toward Rivière Trail, *Ekman 1349* (S).

DOMINICAN REPUBLIC. S of San José de las Matas, *A. H. Liogier 11562* (NY, US); Ciénega de Manabao, Jarabacoa, Los Guanos, *A. H. Liogier 12035* (GH, NY, US); Ciénega de Manabao, along Tablones River, *A. H. Liogier 12078* (GH, NY, US); 6 km W of the Pueblo rural de la Culata, *Zanoni et al. 20281* (JBSD, NY). **Azua:** Sierra de Oca, slope of Loma de los Palos Mojados, *Ekman 11826* (A, C, F, GH, LL, NY, S, US); Prov. de Azua, Valle del Yaque, *Ekman 13705* (S); rd. from Valle Nuevo, *Jiménez 2156* (US). **La Vega:** Between Constanza and Valle Nuevo, *Burch 2530* (MO, NY); SE of Constanza toward Valle Nuevo, *Colella et al. 1376* (NY, TEX); between Constanza and Valle Nuevo, *Jiménez 1572* (CAS, US); El Montazo, from Constanza to Valle Nuevo, *A. H. Liogier 15459* (GH, NY, TEX, US); S of Constanza on rd. to San José de Ocoa, *Mejía & Zanoni 7618* (JBSD, NY). **San Juan:** Piedra del Aguacate to Río del Oro, *R. A. Howard & E. S. Howard 9437* (GH, NY).

Pavonia leiocarpa is similar to *P. lasiopetala* in having smooth carpels and non-exserted genitalia.

72. Pavonia mogotensis (Carabia) Fryxell, comb. nov. Basionym: *Hibiscus mogotensis* Carabia, Mem. Soc. Cubana Hist. Nat. **17:** 15. 1943. Type. Cuba. Pinar del Río: Sumidero, Mogote de la Iglesia, on the hilltop, 1940, *Carabia 3088* (holotype, NY). Fig. 33

Pavonia tuberculata (R. E. Fries) Krapovickas, Trabalh. XXVI Congr. Nac. Bot. Brasil 322. 1977. *Pavonia achanioides* var. *tuberculata* R. E. Fries, Kongl. Svenska Vetenskapsakad. Handl., ser. 3, **24(2):** 26. 1947. Type. Cuba. Pinar del Río: Sierra Guacamayas, Mogote de la Baliza, 10 Aug 1923, *Ekman 17982* (holotype, S; isotypes, CTES [fragment], NY–2).

Shrubs with stems densely and minutely stellate-pubescent, the hairs 0.2–0.3 mm long, yellowish. *Leaf* blades ovate-pentagonal to weakly 5-lobed, becoming narrowly triangular upward, mostly 3–8 × 3–4(–6) cm, basally cordate, serrate-crenate, acute, palmately 7-nerved, somewhat discolorous, minutely pubescent above and beneath, especially on veins, the hairs simple, bifurcate, and stellate, 0.5 mm long or less, and with scattered 3–4-armed stellate spicules beneath; petioles 2–4 cm long; stipules 1–2 mm long, nearly obsolete. *Flowers* solitary in the leaf axils, the pedicels 1–3 cm long; involucellar bracts 5, linear, 6 × 1 mm, hispid-ciliate; calyx 10–11 mm long, costate, densely hispid-ciliate; corolla ca. 3 cm long, reddish, the petals somewhat reflexed distally; staminal column ca. 2.5 cm long, the filaments mostly apical, 3–4 mm long; styles exserted, free for distal 4–5 mm. *Fruits* 8–9 mm diam., minutely scabridulous; mericarps 5.5 mm tall, reticulate-veined and strongly carinate (the ridge apically prolonged and subacute), with weakly developed lateral excrescences.

Distribution (Fig. 22). Endemic to Cuba. The type locality is at 22°28′N, 83°55′W.

Illustrations. Fries (1947: pl. 3, figs. 3–4, as *P. achanioides* var. *tuberculata*).

Specimens examined. **CUBA.** Without locality, *Wright 1575* p.p. (NY, US). (Note: Some other duplicates of *Wright 1575* p.p. are *Hibiscus cryptocarpus* A. Richard.)

Pavonia mogotensis has relatively large, reddish flowers with somewhat reflexed petals and nearly obsolete stipules.

73. Pavonia pabstii Krapovickas & Cristóbal, Bonplandia **7**: 48. 1993. Type. Brazil. Santa Catarina: São Joaquim, banks of the Pelotas River, 23 Oct 1961, *Pabst 6233 & Pereira 6406* (holotype, HB–n.v.; isotypes, CTES, RB–n.v.).

Shrubs 1–1.5 m tall, the stems minutely glandular-pubescent and with scattered whitish setae 1 mm long. *Leaf* blades ovate, basally cordate, coarsely but shallowly crenate, acute, palmately 5–7-nerved, 2.5–4 × 2–3 cm, markedly discolorous, obscurely and minutely pubescent on upper surface, densely whitish-puberulent on lower surface and with scattered setae on principal veins; petioles to 3.5 cm long (subequal to blades), with pubescence like stem but setae primarily at distal end; stipules subulate, 3 mm long. *Flowers* solitary in the leaf axils, the pedicels 1.5–2 cm long (shorter than subtending petiole), with pubescence like stem, the setae denser distally; involucellar bracts 5, lance-ovate, 7–10 × 2.5–6 mm, glandular-pubescent and marginally ciliate; calyx 8–9 mm long, glandular-pubescent and ciliate; petals white, 1 cm long, pubescent on margins of claw; staminal column 6–7 mm tall, glabrous, pallid, distally antheriferous, the filaments 2–3 mm long; styles emergent from the column. *Fruits* unknown.

Distribution (Fig. 22). Known only from the type collection and one other (doubtful) collection from Formosa, Argentina. The type locality is at 28°18′S, 49°55′W (?).

Specimens examined. ARGENTINA. FORMOSA: In the N of Formosa, *Jörgensen 2326* (MO).

Illustration. Krapovickas & Cristóbal (1993: fig. 2).

Krapovickas and Cristóbal (1993) ally *Pavonia pabstii* to *P. lanata* and *P. kleinii*, of *P.* subsect. *Hastifoliae*. I believe it is more likely allied to the other species of *P.* subsect. *Exsertae*, despite its not having exserted genitalia. It is distinctive for its white corollas, broad involucellar bracts, and setose stems. The Jörgensen collection cited above is included only tentatively but may represent this species.

74. Pavonia punctata Urban, Symb. Ant. **5**: 420. 1908. Type. Haiti. Gonaïves, Mare à Collas, Jun 1899, *Buch 182* (isotype, GH). Fig. 33

Shrubs 1–2 m tall, the stems densely and minutely stellate-pubescent and sometimes with scattered 3–4-armed spicules. *Leaf* blades principally 3-lobed (or pentangular), sometimes unlobed, mostly 2.5–5 cm long, about as wide (or narrower if unlobed), basally cordate, serrate, acute, palmately 7(–9)-nerved, discolorous, the upper surface with scattered pubescence (the hairs stellate and bifurcate, the largest 0.5 mm diam.), the lower surface more densely pubescent with minute stellate hairs overlain with large 3–4-armed spicules 1.5–2 mm diam.; petioles 1.5–4 cm long, with minute pubescence like stem and often with a few large 3–4-armed spicules in addition; stipules subulate, 1–3(–5) mm long, often reflexed. *Flowers* solitary in the leaf axils (sometimes aggregated terminally), the pedicels 1–4.5 cm long; involucellar bracts 6–7, nearly distinct, linear, 6–12 × 0.5–2 mm; calyx 6–15 mm long, stellate-pubescent, each lobe prominently 3(–7)-nerved; corolla 2–3.5 cm long, externally with scattered 3-armed spicules, orange-red; staminal column exserted (3.5–4 cm long), glabrous, the filaments 3–4 mm long; styles emergent from the column. *Fruits* ca. 7 mm diam., stellate-pubescent; mericarps carinate with a small apical wing, reticulate-veined, 5.5 mm high.

Distribution (Fig. 40). From Hispaniola and Cuba, although previously thought to be confined to Hispaniola, growing in thickets at 100–1300 m elevation. The type locality is at 19°28′N, 72°44′W.

Specimens examined. CUBA. PINAR DEL RÍO: Region of the mogotes Sierra de la Guazasa, *Ekman 16612* (A, S).

HAITI. Massif de Matheux, Croix des Bouquets, Morne à Cabrites, *Ekman 1002* (LL, NY, S, US); Port-au-Prince, at Blanche River, *Ekman 3042* (S); Massif des Matheux, Cap St. Marc, at the lighthouse, *Ekman 6635* (S); Ile La Gonave, Pte.-à-Raquettes, *Ekman 8810* (S); Thomazeau, Morne-à-Cabrites, *Ekman 9164* (F, GH, S); Morne à Cabrites, *Holdridge 890* (MICH, NY, TEX, US); vic. of St. Marc, *E. C. Leonard 2887* (US).

DOMINICAN REPUBLIC. Loma de Toro, *Phillips 184* (K); E of Santiago Rodríguez on rd. to Moncíón, *A. H. Liogier 13274* (GH, NY, US). **AZUA:** 4 km W of Galindo, *Pimentel et al. 363* (JBSD, NY); Sierra Martín García, 4 km W of Galindo, *Zanoni et al. 32284* (JBSD, NY, TEX). **CORDILLERA CENTRAL:** Prov. Peravia, 2 km from Alto de Los Tramojos on the rd. to El Cercado, *Zanoni et al. 18094* (JBSD, NY). **INDEPENDENCIA:** Sierra de Bahoruco, S of Puerto Escondido, *Zanoni & Pimentel 40398* (JBSD, NY).

The prominently ribbed calyx is the distinctive feature of *Pavonia punctata*.

The spicules that overlie the minute pubescence of the stem are sometimes stipitate, and the stipes are sometimes darkly pigmented. On older stems the spicules are often broken off, leaving only the dark stipes, which could have been the origin of the specific epithet *punctata*, although Urban specifically refers to punctae on the lower leaf surface, which are not evident to me.

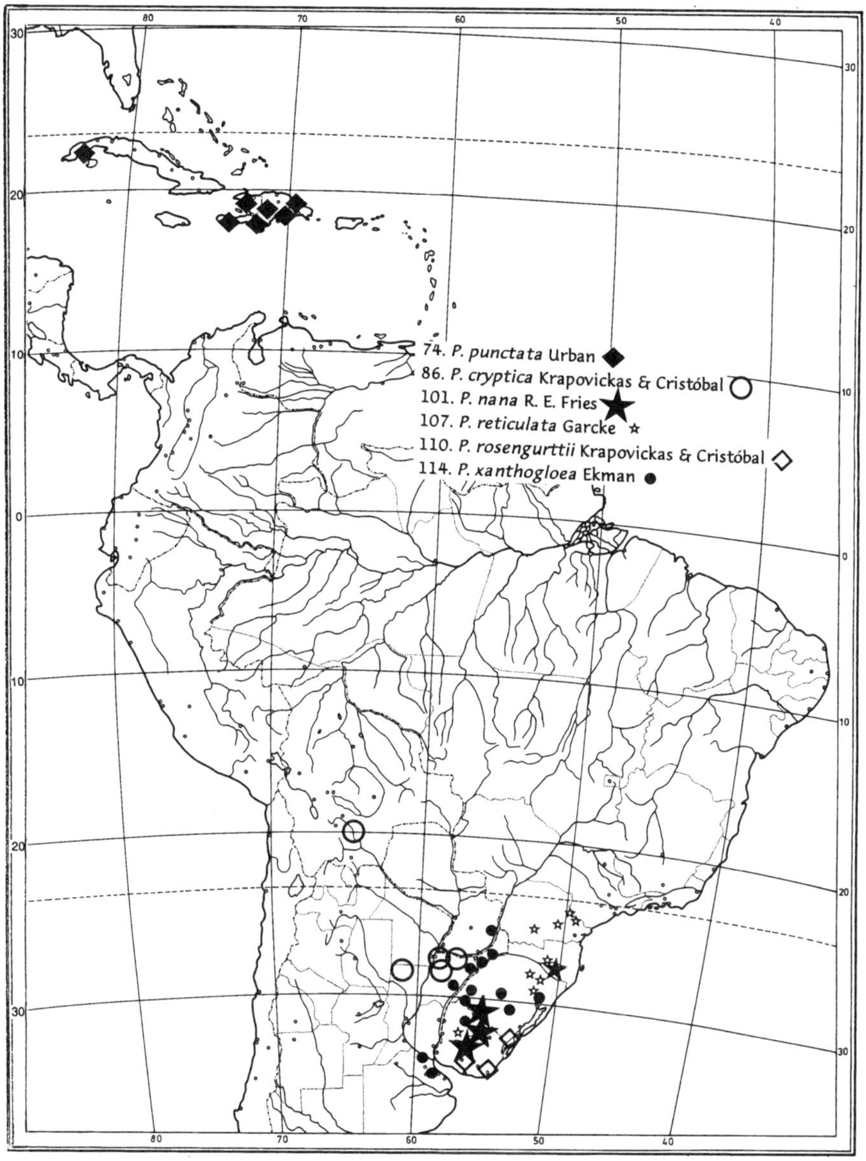

FIG. 40. Distribution map of six neotropical species of *Pavonia* subsect. *Exsertae* and *P*. subsect. *Hastifoliae*.

PAVONIA SUBGEN. PAVONIA SECT. LEBRETONIA SUBSECT. CORDIFOLIAE

Pavonia subsect. **Cordifoliae** Fryxell, subsect. nov. (from the Latin *cordis*, heart, and *folium*, leaf, meaning with heart-shaped leaves).

Frutices, laminis foliorum 4–16 cm longis, late ovato-cordatis, interdum aliquantum 3-lobatis; bracteis involucellorum 5 distinctis (praeter irregulariter connatis in *P. gentryi*); mericarpiis carinatis.

Shrubs, the leaves 4–16 cm long, broadly ovate-cordate, sometimes somewhat 3-lobed; involucellar bracts 5, distinct (except irregularly connate in *P. gentryi*); mericarps carinate.

Type species. *Pavonia fryxellii* Krapovickas.

Key to the species of *Pavonia* subsect. *Cordifoliae*

1. Corolla yellow, 3.5–4 cm long. . . . . 79. *P. rojasii*
1. Corolla white, rose, or purple, 2–5 cm long.
    2. Corolla white, 2.5–5 cm long.
        3. Involucel 8–11 mm long, subequal to calyx; corolla 2.5–3 cm long; Brazil, Argentina, Paraguay, Bolivia.
            4. Calyx 8–10 mm long, ciliate (hairs 1.5–3 mm long). . . . . 78. *P. mattogrossensis*
            4. Calyx 7–8 mm long, stellate-pubescent (hairs 0.2 mm long). . . . . 80. *P. sapucayensis*
        3. Involucel 10–25 mm long, usually exceeding calyx; corolla 2.5–5 cm long; Mexico.
            5. Calyx and involucel densely hirsute, the hairs 0.5–3 mm long; plants not viscid; involucellar bracts 10–20 × 2–4 mm. . . . . 77. *P. macdougallii*
            5. Calyx and involucel glandular-puberulent; plants notably viscid; involucellar bracts 15–25 × 1.5–2.5 mm. . . . . 75. *P. fryxellii*
    2. Corolla rose or purple, 2–3.5 cm long.
        6. Involucellar bracts irregularly divided into 6–8 segments, the segments irregularly connate, 12–15 mm long; corolla 3–3.5 cm long, rich purple with a darker center. . . . . 76. *P. gentryi*
        6. Involucellar bracts 5, distinct, 8–13 mm long; corolla ca. 2 cm long, rose or lavender with a darker center. . . . . 81. *P. venusta*

75. Pavonia fryxellii Krapovickas, Trabalh. XXVI Congr. Nac. Bot. Brasil 316. 1977. Type. Mexico. Oaxaca: 23 km W of Tequisistlán, 9 Sep 1968, *P. A. Fryxell 756* (holotype, CTES; isotypes, CAS, ENCB, MICH, NY, TEX). Figs. 33, 41

Sprawling *shrubs* 1–2 m tall, the stems viscid-puberulent. *Leaf* blades mostly 6–15 cm long (smaller upward), ca. 1.5 times as long as wide, ovate to more or less 3-lobed, cordate, remotely crenate-serrate to subentire, acuminate, stellate-pubescent, markedly discolorous; petioles ¼–¾ the length of the blades; stipules 5–7 mm long, subulate, recurved. *Flowers* solitary in the leaf axils, the pedicels 2–6 cm long, viscid; involucellar bracts 5, 15–25 × 1.5–2.5 mm, densely viscid; calyx 10–14(–17) mm long, densely viscid, deeply divided, whitish with green nerves; corolla 2.5–5 cm long, white (drying yellowish) externally stellate-pubescent; staminal column 2.5–3.5 cm long, pallid, glabrous, the filaments 2–4 mm long; styles exceeding androecium by 6–8 mm, pallid, sometimes pubescent, the stigmas villous. *Fruits* ca. 1 cm diam., sparsely viscid-puberulent, the mericarps 5 mm long, prominently carinate and with raised reticulations.

Distribution (Fig. 42). Endemic to Mexico (Colima, Guerrero, Jalisco, Oaxaca, Puebla, and Veracruz), where it occurs in shrubland and deciduous forest at 500–1200 m elevation (or sometimes lower), flowering from Aug to Jan. The type locality is at 16°25′N, 95°40′W.

Additional illustrations. Cervantes Aceves (1992: fig. 64); Fryxell (1979: fig. 5C); Krapovickas (1977: fig. 2A–C).

Representative specimens examined. **MEXICO. Colima:** Cuyatlán, *Díaz-Luna 3525* (IBUG, MICH); Mun. Manzanillo, 17 km N of Tecomán, *C. P. Cowan et al. 5644* (TEX). **Guerrero:** Mun. Quechultenango, hwy. to Colotlipa and Las Grutas de Justlahuaca, side rd. to Santa Ana, *Koch et al. 87242* (CHAPA, CTES, MEXU, MICH, NY, TEX). **Jalisco:** Chamela, *M. G. Ayala & R. Ayala 209* (F, MEXU, NY, TEX); Chamela, *Bullock 986* (CAS, MEXU, MO); near Los Pozos, *Lott 509* (MEXU, MO); Chamela, rd. to Nacastillo, *Magallanes 454* (ARIZ, MEXU); Chamela, *Pérez 1361* (MEXU, TEX). **Oaxaca:** Ca. 60 km SE of Oaxaca, *P. A. Fryxell 1145* (CTES, GH, MEXU, MICH, NY, US); Mun. San Bartolo Yautepec, 23 mi W of Tequisistlán, *P. A. Fryxell & Bates 913* (BH, NY); Mun. Asunción Tlacolulita, rd. to Sta. María Ecatepec, *Koch & P. A. Fryxell 78338* (BM, CANB, CAS, CHAPA, ENCB, MICH, MO, NY, TEX); Mun. San Carlos Yautepec, 8 km NE of Hondo River (or 87 km NE of Tehuantepec), *Koch & P. A. Fryxell 78361* (CHAPA, ENCB, K, NY, TEX, WIS); Barranca del Río Chico, E Cuicatlán, *Miranda 4610* (MEXU, NY); Pochutla, *Reko 3581* (NY, US); 4 km NW of Totolapan, hwy. to Oaxaca, *Rzedowski 35423* (CTES, ENCB, NY); Mun. Salina Cruz, 11.4 km SW of Morro Mazatán, hwy. to Pochutla, *Torres & Cortes 9266* (MEXU, MO, NY). **Puebla:** Tecuatitlán San Martín, W of Acatlán between Tecomatlán and Axutla, *P. A. Fryxell 760* (CAS, CHAPA, CTES, ENCB, MICH, NY). **Veracruz:** Near Amamaloya, beyond Chinameca, after the side rd. to Pajón toward Soleapan, *Cházaro & Robles 3320* (NY, XAL).

Pavonia fryxellii is distinctive for its large white flowers and its exceedingly viscid herbage.

76. Pavonia gentryi Fryxell, Brittonia **25:** 81. 1973. Type. Mexico. Sinaloa: Sierra Surutato above La Jolla, 17–24 Mar 1945, *H. S. Gentry 7287* (holotype, UC; isotypes, ARIZ, CAS, DS, F as photo F-64923, NY, PH, US). Fig. 33

Shrubs to 2 m tall, the stems with minute stellate hairs, long simple hairs (1–2 mm), and a few glandular

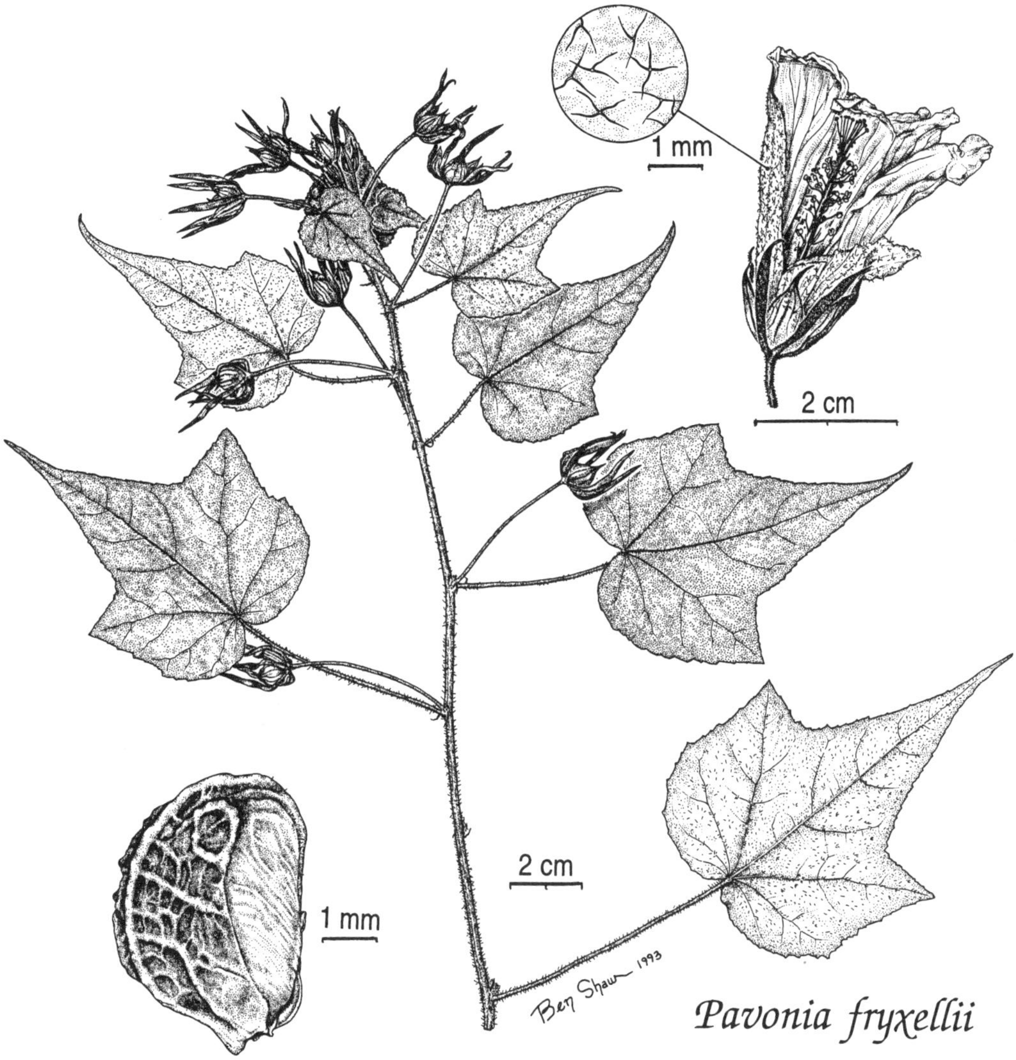

FIG. 41. *Pavonia fryxellii.* Fruiting branch with enlargements of a flower (upper right) and of a mericarp (lower left) (*Hill 1816* [flower and mericarp], *Cházaro & Robles 3320* [branch]).

hairs. *Leaf* blades up to 13 cm long, almost as wide as long, broadly ovate or shallowly 3-lobed, cordate, irregularly serrate, acuminate, somewhat discolorous, with appressed simple hairs 0.5–1 mm long above and beneath, somewhat more densely pubescent beneath; petioles ½–⅔ the length of the blade; stipules 4–5 mm long, subulate. *Flowers* solitary in the leaf axils, the pedicels 3–6 cm long, viscid and with long simple hairs, articulated 5–10 mm below the flowers; involucellar bracts 6–8, irregularly connate below, viscid and ciliate, 12–15 mm long; calyx 15–20 mm long, viscid and ciliate; corolla 3–3.5 cm long, rich purple with a darker center, pubescent on claw, otherwise glabrous; staminal column 12–15 mm long, glabrous, pallid, the filaments 2–4 mm long, purple, the anthers yellowish, sometimes paired; styles emergent from the androecium, almost equaling the petals, glabrous and distally purplish. *Fruits* 8–9 mm diam., minutely puberulent, subglobose, retuse; mericarps ca. 6 mm long, strongly carinate and reticulate-veined.

Distribution (Fig. 42). Known only from Sinaloa, in oak-pine forest at about 1500 m elevation, flowering in March. Gentry (1946) provides a

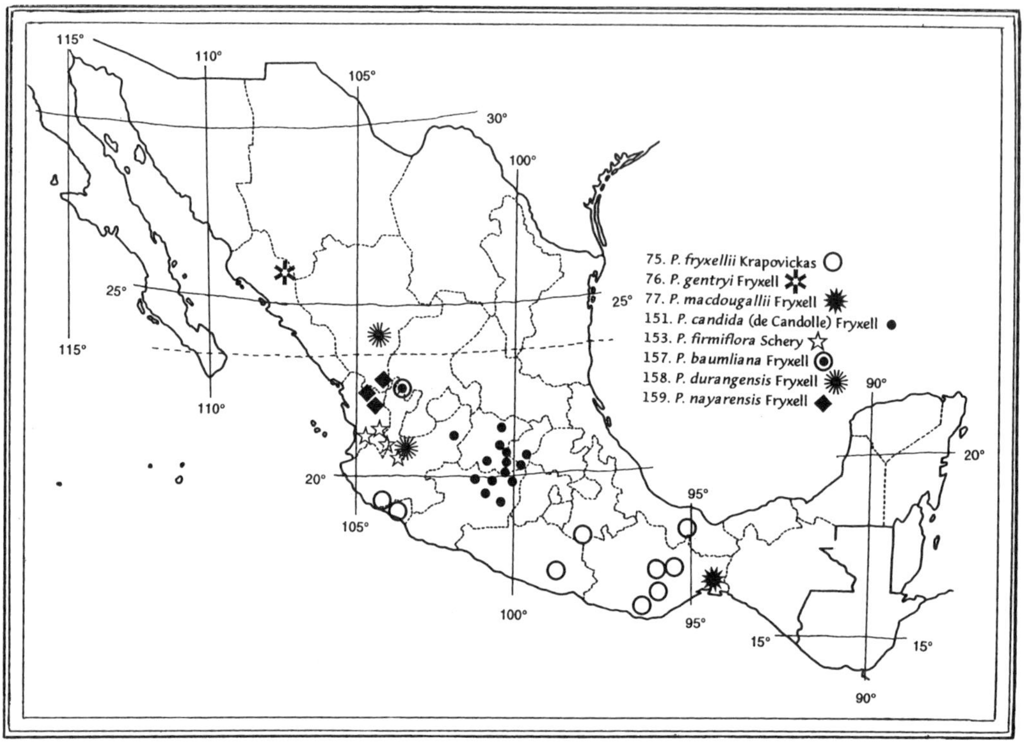

FIG. 42. Distribution map of eight Mexican species of *Pavonia* subsect. *Cordifoliae* and *P.* sect. *Collicolae.*

description of the habitat and a precise localization of the type locality.

Illustration. Fryxell (1973: fig. 3).

Specimens examined. MEXICO. SINALOA: Sierra Surutato, above La Jolla, *H. S. Gentry 7279* (ARIZ, DS).

Pavonia gentryi is distinctive for its irregular involucel and for its richly colored corolla. Its large, showy flowers merit horticultural consideration, if living material could be obtained.

77. Pavonia macdougallii Fryxell, Phytologia **37:** 296. 1977. Type. Mexico. Oaxaca: Tapesco, S of Tres Cruces, *MacDougall H54* (holotype, NY; isotype, TEX). Fig. 33

Shrubs 1–2 m tall, the stems stellate-pubescent, the hairs 0.5–2(–3) mm long. *Leaf* blades mostly 6–12 × 9 cm, ovate or weakly trilobulate, basally cordate, crenate-serrate to subentire, the lateral lobes acutish, the apex long-acuminate, palmately 7–9-nerved, discolorous, softly stellate-pubescent, densely so beneath; petioles ca. half the length of the blade or less; stipules 2–5 mm long, hirsute, subulate. *Flowers* solitary in the leaf axils, the pedicels 1–5 cm long, prominently hirsute; involucellar bracts 5, lanceolate or lance-elliptic, 10–20 × 2–4 mm, densely pubescent and marginally ciliate (cilia 1–2 mm long); calyx 8–12 mm long, densely pubescent and ciliate; corolla 3–4 cm long, white, ciliate on claw, densely pubescent externally; staminal column ca. 2 cm long, white, glabrous, declined, the filaments 3–5 mm long, the pollen white; styles exceeding the column, sparsely pubescent, white, the stigmas villous. *Fruits* 8–9 mm diam., minutely pubescent, the mericarps 5 mm long, prominently carinate and reticulate-veined, often with 3 or 4 excrescences on each side.

Distribution (Fig. 42). Endemic to the Tehuantepec Peninsula in Oaxaca, southern Mexico, occurring in seasonally dry deciduous forest at 400–800 m elevation.

Illustrations. Fryxell (1977: fig. 9; 1979: fig. 5A).

Specimens examined. MEXICO. OAXACA: Cerro Guiengola, *P. A. Fryxell & Lott 3383* (BH, BM, CAS, CHAPA, CTES, ENCB, F, MEXU, MO, NY, TEX, WIS); Tehuantepec, Potrero Villalobos, 15 Nov 1970, *MacDougall s.n.* (NY); Cerro San Pedro, Tehuantepec, *MacDougall 498.S* (F, NY); Cerro Guiengola, *MacDougall 617.S* (F, NY); Cerro Guiengola, *Torres 1992* (MEXU, MO); Cerro Guiengola, 11 km NW of Tehuantepec, *Torres & Campos 14056* (MEXU, MO).

Pavonia macdougallii is distinguished by its large white flowers and by its dense stellate (non-viscid) pubescence.

78. Pavonia mattogrossensis R. E. Fries, Kongl. Svenska Vetenskapsakad. Handl. **42(12):** 49. 1908. Type. Brazil. Mato Grosso: Corumbá, *Malme II-3021* (holotype, S; isotypes, S–2). Fig. 33

Pavonia mattogrossensis var. *lobata* Hassler, Feddes Repert. Spec. Nov. Regni Veg. **7:** 377. 1909. Type. Paraguay. Estancia San Luis, between the Apa River and the Aquidaban River, Dec 1908, *Fiebrig 4418* (holotype, G as photo F-23702; isotypes, E–2, GH, K).

Shrubs 1–2.5 m tall, the stems densely and minutely glandular-pubescent. *Leaf* blades broadly ovate, sometimes 3-lobulate, mostly 4–8 × 3.5–7 cm, cordate to subtruncate, crenate-dentate, acuminate, palmately 7-nerved, sometimes discolorous, with minute simple and bifurcate hairs above and beneath, the hairs 0.2–0.3 mm long, and with a few large 2–3-armed spicules scattered on lower surface, at least on younger leaves; petioles 2–6 cm long; stipules subulate, 3 mm long. *Flowers* solitary in the leaf axils, the pedicels 3–8 cm long, slender; involucellar bracts 5, lance-linear, 8–11 × 2–4 mm, minutely glandular-pubescent; calyx 8–10 mm long, ciliate (hairs 1.5–3 mm long); corolla 2.5–3 cm long, white; staminal column subequal to petals, glabrous, the filaments 3–4 mm long; styles exserted, often exceeding the corolla. *Fruits* subglabrous, ca. 1 cm diam.; mericarps 4.5–5 mm high, reticulate-rugose, the crest prolonged and acute apically. Chromosome number, $2n = 28$ (Krapovickas & Cristóbal, 1962).

Distribution (Fig. 44). Known from Brazil, Paraguay, and Bolivia. The type locality of *P. mattogrossensis* is at 19°00′S, 57°38′W; that of *P. mattogrossensis* var. *lobata* is at 22°13′S, 57°14′W.

Illustrations. Fries (1908: pl. 5, figs. 8–10).

Specimens examined. **BRAZIL. MATO GROSSO:** Mun. Corumbá, Bodoquena, *Hatschbach 38649* (C, CTES).

BOLIVIA. SANTA CRUZ: Prov. Obispo Santisteban, La Fortuna (ca. 15 km from Montero), *Krapovickas 9119* (CTES, MO, NY, SI).

PARAGUAY. DEPT. CONCEPCIÓN: Near Estancia San Luis, SW of Fuerte San Carlos, Concepción, *Bernardi 19408* (NY, U); San Lázaro, *Rojas 5527* (CTES, G, NY).

Pavonia mattogrossensis is closely allied to *P. sapucayensis*, from which it differs in calyx size and pubescence (see key).

79. Pavonia rojasii Hassler, Feddes Repert. Nov. Spec. Regni Veg. **7:** 73. 1909. Type. Paraguay. Sierra de Amambay, near Estrella, Feb 1908, *Rojas* (for Hassler) *10184* (holotype, G; isotypes, B as photo F-9466, CTES, G–3, K, MVM–n.v., NY). Figs. 33, 43

Shrubs 1 m tall, the stems densely and finely stellate-puberulent and with scattered (longer) glandular hairs. *Leaf* blades broadly ovate, up to 6 × 4–6 cm, deeply cordate at base, crenate-serrate, acute or somewhat acuminate, palmately 7-nerved, somewhat discolorous, densely and softly stellate-tomentose on both surfaces; petioles 3–5.5 cm long; stipules subulate, 2–4 mm long. *Flowers* solitary in the leaf axils, the pedicels erect in flower, geniculate in fruit, up to 3.5 cm long; involucellar bracts 5, ligulate, 14–18 × 1.5–3 mm, densely stellate-tomentose and with scattered glandular hairs; calyx 14–18 mm long, whitish at base, deeply divided, with pubescence like involucel, the lobes many-veined; corolla yellow, 3.5–4 cm long. *Fruits* 11–12 mm diam., sparsely puberulent; mericarps 5 mm high, carinate, with prominently raised rugosities.

Distribution (Fig. 44). Confined to Paraguay. The type locality is at 22°04′S, 57°21′W.

Specimens examined. **PARAGUAY. AMAMBAY:** Pedro Juan Caballero, *Schwarz 11811* (CTES, NY).

Pavonia rojasii is notable for its large yellow corollas, the only example of this in *P.* subsect. *Cordifoliae*.

80. Pavonia sapucayensis Hassler, Feddes Repert. Nov. Spec. Regni Veg. **8:** 42. 1910. Type. Paraguay. Forest margins near Sapucay, *Hassler 1930* (holotype, G as photo F-23714). Fig. 33

Pavonia mattogrossensis var. *minor* Hassler, Feddes Repert. Nov. Spec. Regni Veg. **12:** 267. 1913. Type. Paraguay: in the valley of the Upper Apa River, May 1913, *Hassler 11223* (isotypes, A, C, MO, NY, RB, S, UC as photo CTES, US).

Subshrubs 0.5 m tall, the stems minutely glandular-pubescent. *Leaf* blades ovate, sometimes 3-lobulate, mostly 4–8 × 3–4(–8) cm, shallowly to deeply cordate, crenate-serrate, acuminate, palmately 5–7-nerved, discolorous, the upper surface with minute (0.2–0.3 mm) simple and bifurcate hairs and minute glandular hairs, the lower surface with minute stellate hairs and minute glandular hairs, and with scattered larger (1.5 mm diam.) 2–3-armed stellate spicules; petioles 3–6 cm long; stipules 3–4 mm long, subulate. *Flowers* solitary in the leaf axils, the pedicels 3.5–6.5 cm long; involucellar bracts 5, lanceolate, 9–11 × 2.5–4 mm, minutely stellate-pubescent (hairs 0.2 mm long); calyx 7–8 mm

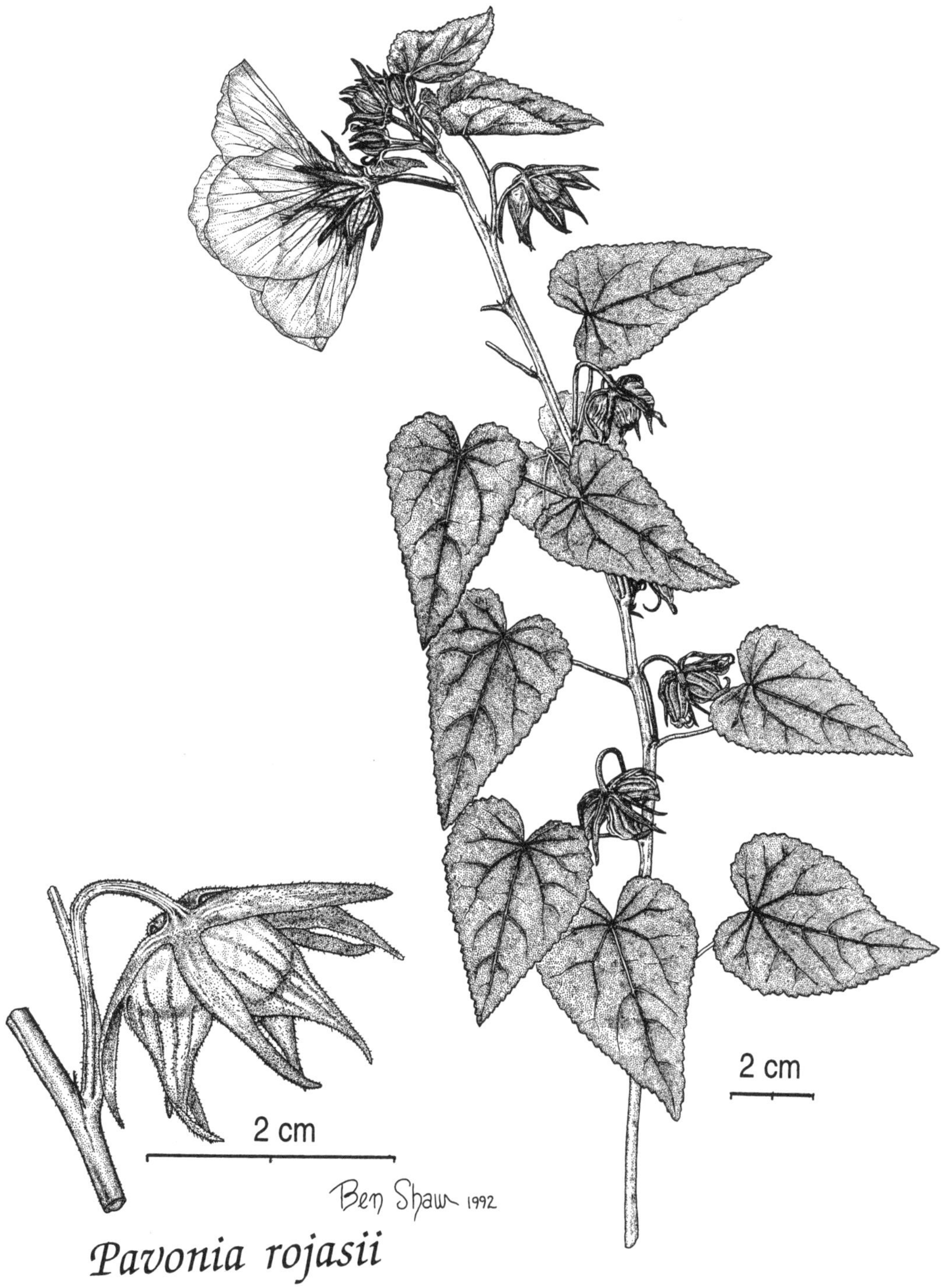

FIG. 43. *Pavonia rojasii*. Flowering branch (right) with enlargement of developing fruit (lower left) (*Hassler 10184*).

long, minutely stellate-pubescent; corolla white, 2–3 cm long, glabrous on claw; staminal column more or less declined, 2.5 cm long, glabrous, the filaments 2–3 mm long; styles exserted. *Fruits* 8–9 mm diam., glabrous; mericarps 4–5 mm high, reticulate-veined, with erose apical excrescences.

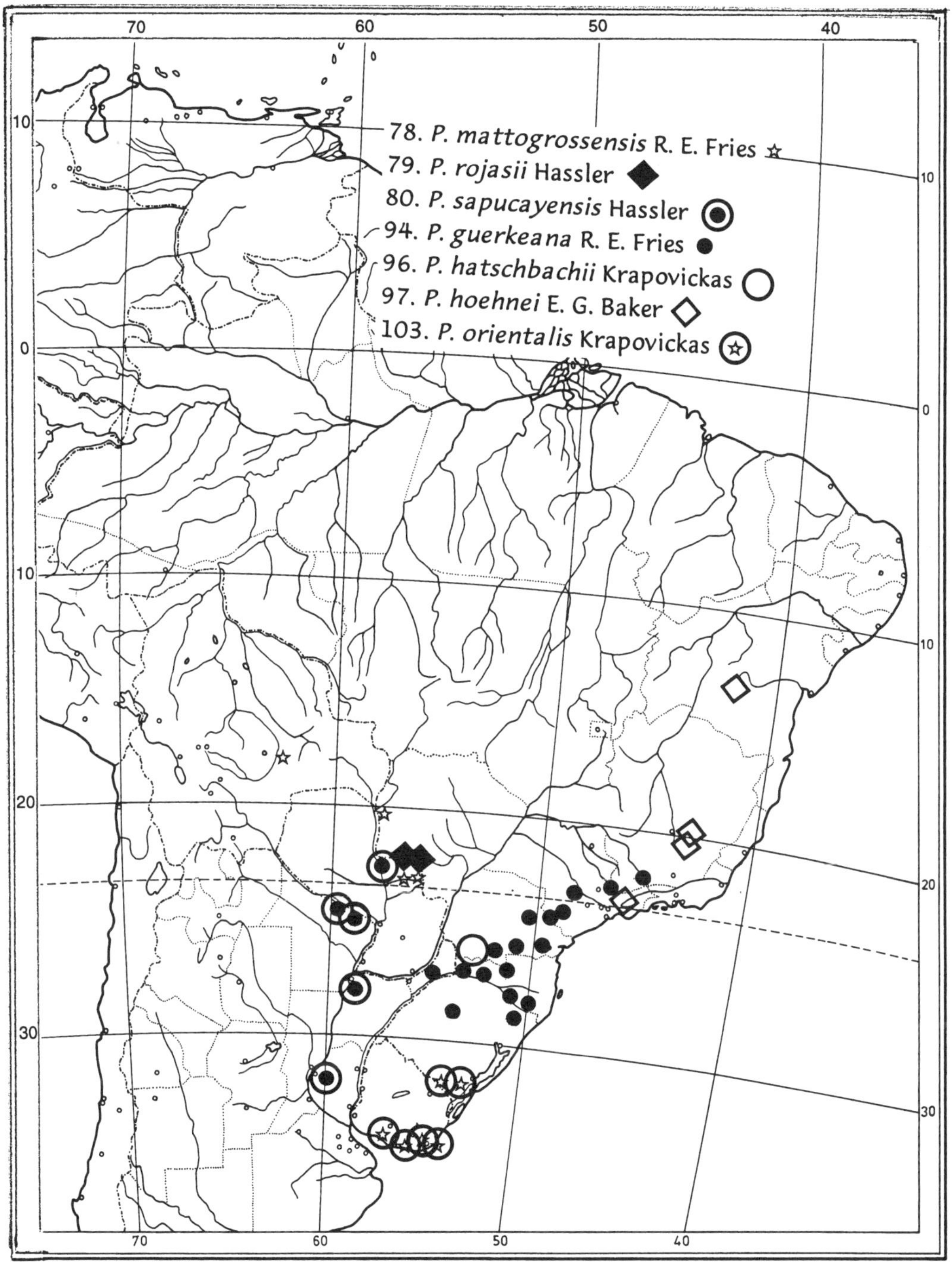

FIG. 44. Distribution map of seven South American species of *Pavonia* sect. *Lebretonia*.

Distribution (Fig. 44). From Paraguay and adjacent Argentina. The type locality of *Pavonia sapucayensis* is at 25°38′S, 56°57′W; that of *P. mattogrossensis* var. *minor* is at ca. 22°10′S, 56°40′W.

Specimens examined. PARAGUAY. Upper Apa River, *Hassler 12070* (A). **Guairá:** Tobatinguá, *Montes 16625* (CTES).

ARGENTINA. Chaco: 10 km SE of Puerto Velaz, *Krapovickas 13108* (BAA, CTES, P, UC); Dept. Bermejo,

Arroyo Cangüi Chico, *Krapovickas & Cristóbal 13171* (CTES, MO); Dept. 1° de Mayo, Colonia Benítez, *Schulz 26* (CTES); Colonia Benítez, *Schulz 543* (CTES, S); Colonia Benítez, *Schulz 563* (CTES, S); Dept. San Fernando, Antequera, *Schulz 2079* (C, CTES, NY); Colonia Benítez, *Schulz 2080* (CTES, MO, NY); Dept. 1° de Mayo, Arroyo Guaycurú, *Schulz 2174* (CTES, NY); Arroyo Guaycurú, *Schulz 2190* (CTES); La Loma, *Schulz 6851* (CTES); Colonia Benítez, *Schulz 8044* (C, CTES); Dept. 1° de Mayo, La Loma, *Schulz 8808* (CTES). CORRIENTES: Corrientes [cult. *ex* Chaco], *Krapovickas 44410* (CTES, TEX). ENTRE RÍOS: Dept. Paraná, Hernandarias, *Muñoz 3099* (CTES).

Pavonia sapucayensis is closely allied to *P. mattogrossensis*, from which it differs in calyx size and pubescence (see key).

81. Pavonia venusta Fryxell, Syst. Bot. **12:** 279. 1987, nom. nov., based on *Pavonia discolor* Fryxell, Syst. Bot. **9:** 415, t. 1. 1984 (non *Pavonia discolor* Ulbrich, 1921). Type. Mexico. Nayarit: SW running canyon on W slope of Cerro San Juan, 23 Jan 1970, *Boutin & Brandt 2909* (holotype, MEXU; isotypes, CAS, HNT).

Shrubs ca. 3 m tall, the stems minutely and densely stellate-pubescent, the younger stems also viscid. *Leaf* blades mostly 9–16 cm long, about as wide, broadly ovate or weakly lobulate, cordate, remotely and shallowly denticulate, acuminate, puberulent; petioles ⅔ the length of the blades to slightly exceeding the blades; stipules 2–5 mm long, filiform. *Flowers* solitary or paired in the leaf axils but more or less aggregated apically, the pedicels up to 6 cm long, slender, viscid; involucellar bracts 5, lance-ovate, 8–13 mm long, viscid; calyx equaling involucel or slightly shorter, viscid; corolla ca. 2 cm long, rose or lavender with a diffuse purplish center, glabrous throughout; staminal column 14–15 mm long, glabrous, pallid, the filaments 2–4 mm long; styles slightly exceeding the androecium. *Fruits* (immature) sparsely glandular-puberulent, the mericarps dorsally ridged and with some reticulate venation.

Distribution. Known only from the type collection, in wet forest at ca. 1500 m elevation. The precise location of Cerro San Juan has not been ascertained.

Illustrations. Fryxell (1984b: fig. 1, as *P. discolor*; 1988a: fig. 86).

Pavonia venusta is poorly known but is distinguished by its rose-colored corollas and its more or less viscid herbage.

PAVONIA SUBGEN. PAVONIA SECT. LEBRETONIA SUBSECT. HASTIFOLIAE

Pavonia subsect. **Hastifoliae** Fryxell, subsect. nov. (from the Latin *hasta*, spear, and *folium*, leaf, meaning spear-shaped leaf).

Herbae perennes procumbentes seu suffrutices vel frutices; laminis foliorum ovato-triangularibus, vulgo 2–3-plo longioribus quam latioribus, saepe hastiformibus vel sagittiformibus; mericarpiis carinatis.

Perennial, procumbent herbs to erect subshrubs or shrubs, the leaves ovate-triangular, often 2–3 times as long as wide, often hastately or sagittately lobed at base; mericarps carinate.

Type species. *Pavonia hastata* Cavanilles.

A word of clarification is needed explaining the usage of the terms *hastate* and *sagittate*. "Sagittate" is used to describe leaves whose basal lobes extend straight back from the main part of the blade in such a way that the margin of the blade is continuous. Among other species, *Pavonia reticulata* exhibits this leaf form, as is illustrated in Fig. 52. "Hastate" is used to describe leaves whose small basal lobes diverge at an angle from the central portion of the blade. Species such as *P. hastata* exhibit this leaf form most clearly.

Key to the species of *Pavonia* subsect. *Hastifoliae*

1. Corolla yellow.
   2. Leaves ovate-triangular, somewhat discolorous, slightly longer than wide, 1.5–3 cm long, coarsely and prominently crenate; calyx 9–10 mm long; Mexico. . . . . 102. *P. nepetifolia*
   2. Leaves narrowly oblong-ovate to sagittate, strongly discolorous, about twice as long as wide, 2–5 cm long, coarsely crenate to subentire; calyx 4–9 mm long; Brazil. . . . . 98. *P. kleinii*
1. Corolla lavender, rose, or red, sometimes almost white.
   3. Involucellar bracts suborbicular or broadly spatulate; leaves often sagittate, discolorous.
      4. Stems hirsute, the hairs 1–2.5 mm long; leaves prominently sagittate.. . . . . 111. *P. sagittata*
      4. Stems stellate-pubescent, the hairs 0.5 mm long or less; leaves subsagittate.
         5. Involucellar bracts 4 mm wide; calyx lobes 7–9 mm wide. . . . . 83. *P. belophylla*
         5. Involucellar bracts 5–9 mm wide; calyx lobes 4–5 mm wide. . . . . 97. *P. hoehnei*
   3. Involucellar bracts ovate or lanceolate to linear; leaves variously shaped, if sagittate then involucellar bracts clearly longer than wide.
      6. Petals (2–)2.5–4 cm long; calyx lobes 4–8(–12) mm wide.

7. Calyx lobes 6–8(–12) mm wide, cordate, plicate. . . . . 106. *P. pulchra*
7. Calyx lobes 4–5 mm wide, ovate, not plicate.
8. Mericarps acute or apiculate at apex. . . . . 94. *P. guerkeana*
8. Mericarps rounded at apex. . . . . 88. *P. distinguenda*
6. Petals 1–2(–2.5) cm long (only 0.2 cm in *P. schininii*); calyx lobes seldom >4 mm wide (except 5–6 mm wide in *P. bullulata*).
9. Corolla red; calyx 8–15 mm long.
10. Leaves sagittate, the upper surface stellate-pubescent (hairs 1–2 mm long); plants ferruginous-hirsute; calyx hirsute, without pigmentation on veins. . . . . 109. *P. rhodantha*
10. Leaves ovate or ovate-triangular, sometimes weakly lobed, the upper surface with hairs <1 mm long; plants pubescent or puberulent, often with glandular hairs, not ferruginous; calyx lobes often with red- or purplish-pigmented veins.
11. Upper leaf surface with mostly simple appressed hairs 0.5–1 mm long; involucellar bracts 1.5–2.5 mm wide; fruits 8–10 mm diam., obscurely pubescent. . . . . 100. *P. missionum*
11. Upper leaf surface minutely puberulent; involucellar bracts1–1.5 mm wide; fruits ca. 7 mm diam., nearly glabrous.. . . . . 93. *P. glutinosa*
9. Corolla rose-lavender or paler; calyx 5–12 mm long.
12. Leaves markedly sagittate, the basal lobes subequal to the petioles, the blades reflexed and parallel to stem (see Fig. 52); involucellar bracts ligulate or ovate, equaling or exceeding the calyx, 10–13 mm long, 3–7 mm wide.
13. Calyx and involucel ciliate and with minute glandular hairs; leaves subglabrous except prominently ciliate on margins.. . . . . 107. *P. reticulata*
13. Calyx and involucel hirsute, lacking glandular hairs; leaves softly tomentose.
14. Corolla 1 cm long; pedicels 1 cm long; plants decumbent, the stems purplish with scattered stellate hairs. . . . . 99. *P. lanata*
14. Corolla 1.5–2.5 cm long; pedicels 1.5–2.5 cm long; plants erect, the stems densely stellate-pubescent. . . . . 96. *P. hatschbachii*
12. Leaves hastate, oblong, ovate, or suborbicular (or if sagittate then the petiole longer than the basal leaf lobes); involucellar bracts various.
15. Veins of the calyx lobes red- or purplish-pigmented (see Fig. 50).
16. Plants decumbent-ascending; involucel 6–8 mm long; calyx 6–7 mm long. . . . . 101. *P. nana*
16. Plants erect; involucel 8–11 mm long; calyx 6–12 mm long.
17. Petals 7 mm long; staminal column 4.5 mm long; leaves sagittate. . . . . 110. *P. rosengurttii*
17. Petals 15–25 mm long; staminal column 6–8 mm long; leaves hastate-triangular or somewhat palmately lobed.
18. Upper leaf surface with simple hairs 0.5–1 mm long; stipules 2–3 mm long. . . . . 90. *P. eurychlamys*
18. Upper leaf surface with shorter stellate (or simple) hairs or subglabrous; stipules 8–11 mm long. 105. *P. prionophylla*
15. Veins of the calyx not dark-pigmented (or sometimes so in *P. betonicaefolia*).
19. Plants procumbent; leaf blades 1–3 cm long.
20. Stems and upper leaf surface very minutely stellate-pubescent (almost lepidote). . . . . 87. *P. cymbalaria*
20. Stems and upper leaf surface coarsely pubescent, the hairs ca. 1 mm long.
21. Involucellar bracts 5–6 mm long, 2.5–4 mm wide. . . . . 112. *P. schininii*
21. Involucellar bracts 7–11 mm long, 1–2 mm wide. . . . . 92. *P. glechomoides*
19. Plants erect; leaf blades up to 7 cm long.
22. Involucellar bracts usually 2 mm wide or less.
23. Stems coarsely stellate-pubescent, the hairs 0.5–0.9 mm long; petioles evenly stellate-pubescent; leaves ovate-oblong. . . . . 84. *P. betonicaefolia*
23. Stems stellate-pubescent, the hairs 0.2–0.4 mm long; petioles pilose distally (hairs 1 mm) near juncture with blade; leaves more or less hastately lobed.. . . . . 103. *P. orientalis*
22. Involucellar bracts usually >2 mm wide.

24. Leaves with coarse stellate pubescence, the hairs ca. 1 mm diam., 3–4-armed on upper surface, ca. 8-armed on lower surface. . . . . 85. *P. bullulata*
24. Leaves with fine stellate pubescence, the hairs 0.5 mm diam. or less.
    25. Stem pubescence coarse, the hairs 0.5–1 mm long; involucellar bracts 3–4 mm wide. . . . . 89. *P. dusenii*
    25. Stem pubescence fine, the hairs 0.3 mm long or less; involu-cellar bracts 2–4 mm wide.
        26. Calyx and involucel hirsute (hairs 1–2 mm long), especially basally. . . . . 91. *P. friesii*
        26. Calyx and involucel minutely pubescent (or sometimes pilose at base in *P. revoluta,* but then bracts spreading).
            27. Calyx and involucel glandular-pubescent.
                28. Base of blade and distal end of petiole setose, the hairs 1 mm long. . . . . 114. *P. xanthogloea*
                28. Blade and petiole pubescent (hairs 0.5 mm long) but not setose. 113. *P. viscidula*
            27. Calyx and involucel stellate-pubescent, lacking glandular hairs.
                29. Stellate hairs on stems sublepidote, 0.1–0.2 mm diam.; involucel often longer than calyx. . . . . 82. *P. aurigloba*
                29. Stellate hairs on stems fine (0.3 mm long or less) but not sublepidote; involucel mostly shorter than calyx.
                    30. Leaves usually >4 cm long, crenate-serrate. . . . . 95. *P. hastata*
                    30. Leaves mostly <4 cm long, often coarsely serrate.
                        31. Leaves with mostly simple hairs 0.5 mm long and some stellate hairs; petals 1.5–2 cm long; involucellar bracts spreading or somewhat reflexed in bud.. . . . . 108. *P. revoluta*
                        31. Leaves minutely stellate-pubescent; petals 1–1.5(–2) cm long; involucellar bracts erect in bud.
                            32. Involucellar bracts constricted at base, 5–8 mm long, shorter than the calyx. . . . . 86. *P. cryptica*
                            32. Involucellar bracts not basally constricted, 10–11 mm long, exceeding the calyx. . . . . 104. *P. pedersenii*

82. Pavonia aurigloba Krapovickas & Cristóbal, Lilloa **31:** 29. 1962. Type. Argentina. Córdoba: Dept. Punilla, Mallín, *Gutiérrez 312* (holotype, LIL–n.v.). Fig. 33

Shrubs ca. 1 m tall, the stems covered with minute (sublepidote) multi-armed stellate hairs. *Leaf* blades narrowly hastate-triangular, or subsagittate, mostly 2–3(–7) × 0.5–1.5(–3) cm, serrate, acute or subobtuse, palmately 3–5-nerved, strongly discolorous, minutely stellate-pubescent above, densely tomentulose beneath; petioles 0.5–1 cm long; stipules 3–5 mm long, subulate. *Flowers* solitary in the leaf axils, the pedicels 1–4.5 cm long, slender; involucellar bracts 5–6, lance-ovate, 6–10 × 3–4 mm, minutely pubescent; calyx 7 mm long, densely yellowish pubescent; corolla 1–2 cm long, lavender with a dark red center and dark veins, externally pubescent. *Fruits* oblate, 8–9 mm diam.; mericarps 4.5 mm long, with minute septate hairs, dorsally carinate, with raised reticulations. Chromosome number, $2n = 56$ (Krapovickas & Cristóbal, 1962; Fernández, 1974).

Distribution (Fig. 35). From S Brazil (Rio Grande do Sul), Uruguay, and N Argentina (Buenos Aires,

Córdoba, Entre Ríos, and San Luis), at elevations of 800–2000 m. The type locality is at 31°10′S, 64°47′W.

Illustrations. Krapovickas & Cristóbal (1962: figs. 1G, 5).

Representative specimens examined. **BRAZIL.** **RIO GRANDE DO SUL:** Fiusa River, *Bornmüller 462* (A, GH); Sta. Barbara do Sul, *Krapovickas & Cristóbal 33723* (CTES, F, MBM); 28 km SE of Puerto Xavier, *Krapovickas & Cristóbal 33732* (CTES, MBM, SI); 91 km SE of San Borja, *Krapovickas & Cristóbal 38388* (CTES, HAS, LIL); 4 km S of the access to Torres, BR-101, *Krapovickas & Cristóbal 44726* (CTES, TEX); BR-285, km 61, 17 km E of Lagoa Vermelha, *Krapovickas & Schinini 38227* (CTES, F, HAS, UC); Mun. Rosario do Sul, 8 km from the Rio Grande do Sul toward São Gabriel, *Krapovickas et al. 22830* (CTES, MBM, NY, P); Mun. Itaquí, between São Caruto & Sobradinho, *Pedersen 11665* (A, C, CTES, K, MO); Fiao near São Leopoldo, *Rambo 44403* (CTES, MO, NY).

ARGENTINA. **BUENOS AIRES:** Without locality, *L. Née s.n.* (F); Cerro Vacuá, *Rossi-Bachmann 424* (CAS). **CÓRDOBA:** La Cumbre, Sierra de Córdoba, *Bartlett 20111* (US); Sierra Chica, Estancia La Reducción, *Burkhart 7311* (F); Dept. Colón, Sierra Chica, Quebrada of the Ceballos River, *Caro 3434* (MO); Dept. Punilla, La Cumbre, *Cuezzo & Balegno 2122* (K); Dept. Colón, Santa Catalina, *G. Dawson 644* (NY); Copina, *Krapovickas 6627* (CTES); Dept. Calamuchita, Villa General Belgrano, *Krapovickas & Cristóbal 14758* (C, CTES, MO, P, UC, WIS); Dique, Dec 1891, *Kuntze s.n.* (F, NY); Dept. San Alberto, Mina Clavero, *Meyer 13906* (CAS); Alta Gracia, *O'Donell & Rodríguez 515* (U); Cofrina, *O'Donell & Rodríguez 992* (F, U); Dept. Santa María, Cerro La Cocha, *Pierotti 5048* (CAS, CTES, LL); Cosquin, *Rodrigo 130* (NY); Alta Gracia La Isla, *de la Sota 1479* (CAS); Dept. Punilla, Villa del Lago, *de la Sota 1577* (TEX); Dept. Santa María, La Serranita, *de la Sota 1781* (CTES). **CORRIENTES:** Dept. Monte Caseros, km 173, Campo General Avalos, Arroyo Curupí, *Schinini et al. 17599* (CTES). **ENTRE RÍOS:** Without locality, 26–29 Feb 1880, *Lorentz s.n.* (PH); Dept. Uruguay, Concepción del Uruguay, *Lorentz 115* (VT). **SAN LUIS:** Dept. Junin, Merlo, Piedra Blanca, *Digilio-Grassi 2032* (CAS, CTES); Dept. Pringles, between Río Grande and Los Tapiales, *Ferraro et al. 4136* (CTES); Dept. Capital, El Trapiche, *Roighez 19028* (CTES).

URUGUAY. Dept. Colonia, Arroyo de Pintos, Artilleros, near Puerto Platero, *Bartlett 20759* (US); Dept. de Rivera, near Rivera, *Bartlett 21094* (NY); Montevideo, Apr 1836, *Gaudichaud s.n.* (F); Dept. Colonia, Colonia, *Krapovickas 6574* (MO); Dept. Maldonado, Piriápolis, Cerro San Antonio, *Krapovickas & Cristóbal 11126* (CTES); Tacuarembó, Rincón del Bonete, *Krapovickas & Cristóbal 11204* (CTES); Dept. San José, Sierra Mahoma, *Krapovickas & Cristóbal 16292* (CTES); Montevideo, *L. Née s.n.* (F); Dept. Minas, Verdun, *Osten 5189* (G); Dept. Rivera, Cuñapirú, *Pedersen 11638* (C, CTES, MO); Cerro Largo, Cerro de Las Cuentas, *Rosengurtt B2524* (US).

Pavonia aurigloba is readily recognized by the globular calyx (in bud) with its even, yellowish pubescence, unlike any of the other members of *P.* subsect. *Hastifoliae*.

83. Pavonia belophylla Hochreutiner, Annuaire Conserv. Jard. Bot. Genève **6:** 43. 1902. *Pavonia hastata* forma *belophylla* (Hochreutiner) Hassler, Bull. Herb. Boiss. ser. 2, **7:** 740. 1907. Type. Paraguay. Sierra de Maracayú, in the region of the Curuguaty River, Sep 1900, *Hassler 4602* (lectotype, G; isotypes, B as photo F-9432, CTES [fragment], G as photo F-23694, GH, K, NY). Fig. 33

Subshrubs to 0.5 m tall, erect to decumbent, the stems densely stellate-pubescent. *Leaf* blades hastately or sagittately oblong-triangular, mostly 2–5 × 1–2 cm, basally sagittately lobed, the sinus narrowly to broadly acute, serrate, acute to obtuse, palmately 3–5-veined, strongly discolorous, with scattered stellate hairs above, canescent to tomentulose beneath; petioles 1–2.5 cm long (longer than lobes of blade); stipules subulate, 1–3 mm long. *Flowers* solitary in the leaf axils, the pedicels 0.5–2.5 cm long; involucellar bracts 5, broadly ovate or suborbicular, 4 × 4 mm, reflexed; calyx 8–9 mm long, pubescent, the lobes broadly cordate, 7–9 mm wide, plicate-pentangular in bud; corolla 1.5–2 cm long, lavender with dark red center, pubescent on margins of claw; staminal column 5–7 mm long, glabrous. *Fruits* 8–9 mm diam., oblate, with minute septate hairs; mericarps ca. 5 mm high, dorsally carinate, with raised reticulations, enclosed by calyx.

Distribution (Fig. 22). From S Brazil (Paraná, Mato Grosso, and Rio Grande do Sul) and Paraguay. The type locality is at 24°10′S, 55°36′W.

Specimens examined. **BRAZIL.** **MATO GROSSO:** 70 km S of Campo Grande, *Gregory et al. 9808* (CTES); Rio Amambay, rd. to Ponta Porá-Amambay, *Krapovickas et al. 14152* (CTES); Mato Grosso do Sul, 9 km W of Caarapó, *Krapovickas & Cristóbal 34325* (CTES, NY, SI). **PARANÁ:** Laranjeiras do Sul, *Hatschbach 15544* (CTES, F, US), *Hatschbach 20618* (CTES). **RIO GRANDE DO SUL:** Cruz Mtns., 20 Jan 1902, *Malme s.n.* (S).

PARAGUAY. Pastoreo, rd. to Jhu (Caaguazú), *Gregory et al. 9611* (CTES, NY). **AMAMBAY:** 4 km W of Pedro Juan Caballero, *Krapovickas et al. 14076* (BAA, CTES); Colonia Yvypyté, *Schinini & Bordas 20499* (CTES, NY).

Hassler published *Pavonia hastata* var. *grandiflora* Hassler (Bull. Herb. Boiss., ser. 2, **7:** 739. 1907) with a description, but with no citation of specimens. Under this variety he included forma *belophylla* (Hochreutiner) Hassler and forma *pulchra* (Hochreutiner) Hassler (≡ *P. pulchra*, q.v.), in each case citing the types of the basionyms. Because he made no provision for a forma *grandiflora*, the var. *grandiflora* cannot be treated as validly published, since it cannot be typified.

Pavonia belophylla is characterized by its plicate calyx with broad (7–9 mm), cordate lobes and by its suborbicular involucellar bracts. It is apparently allied to *P. pulchra* and possibly also to *P. hoehnii*.

84. Pavonia betonicaefolia C. Presl, Reliq. Haenk. **2:** 128. 1835. Type. "in Peruvia," *Haenke s.n.* (holotype, PR?–n.v.). Fig. 33

Pavonia consobrina N. E. Brown, Trans. Proc. Bot. Soc. Edinb. **20(1)**: 47. 1894. *Pavonia hastata* var. *consobrina* (N. E. Brown) Hassler, Ostenia 341. 1933. Types. Uruguay. Montevideo, *Gibert 351* (lectotype, here designated, K).

Shrubs 1–1.5 m tall, the stems coarsely stellate-pubescent, the hairs 0.5–0.9 mm long. *Leaf* blades lance-oblong, mostly 2–4(–8) × 1–1.5(–2.5) cm, basally rounded-truncate or subcordate, crenate-serrate, acute or subobtuse, slightly discolorous, stellate-pubescent above and beneath; petioles 5–12 mm long; stipules filiform, 3 mm long. *Flowers* solitary in the leaf axils or aggregated apically, the pedicels 2–26 mm long; involucellar bracts 5, oblanceolate or spatulate, 6–9 × 1.5–3 mm (slightly exceeding calyx), stellate-pubescent; calyx 5–7 mm long, costate, stellate-pubescent, sometimes with reddish coloration; corolla ca. 1 cm long, white (drying reddish) with red veins, the margins with stellate hairs. *Fruits* 6–7 mm diam., minutely puberulent; mericarps 3.5 mm high, obscurely carinate on dorsum, with two small lateral wings (or ridges) basally, otherwise smooth. Chromosome number, $2n = 56$ (Krapovickas & Cristóbal, 1962).

Distribution (Fig. 45). From southernmost Brazil, Uruguay, Paraguay, and N Argentina (Buenos Aires, Corrientes, Chaco, Entre Ríos, Formosa, Misiones, and Santa Fé). The type locality of *Pavonia consobrina* is at 34°50′S, 56°15′W.

Illustrations. Fries (1908: pl. 6, figs. 19–20); Krapovickas (1965: fig. 53F-H).

Representative specimens examined. BRAZIL. Rio Grande do Sul: Uruguayana, Arroio Imbaá, *Sobral et al. 3451* (CTES); Mun. Santana do Livramento, BR-293 at the bridge over Arroio Sarandi II, *Bueno 484* (CTES).

PARAGUAY. Dept. Pte. Hayes, Chaco-í, *Schinini 26696* (CTES, NY).

ARGENTINA. Buenos Aires: Dept. Partido de la Plata, Punta Lara, *Krapovickas 4802* (LIL, LL); Campana, *Krapovickas 4808* (BRIT, LIL); Partido de Zárate, Las Palmas, *Hunziker 1746* (K). **Chaco:** Dept. Bermejo, rt. 11, km 1084, *Bacigalupo et al. 9684* (BAA, NY); Colonia Benítez, *Schulz 17664* (C, F, MO); Dept. Fontana, Enrique Urín, *Schulz 17770* (CTES); Río Pilcomayo, near Fortín Page, *Kerr 66* (syntype of *P. consobrina*, CTES [fragment], K). **Corrientes:** Dept. Esquina, 47 km W of Sauce, *Ahumada et al. 1404* (BAA, BAB, CTES, MICH); Dept. General Paz, 6 km SW of Caa-Catí, rt. 13, *Ahumada 1945* (CTES); Dept. Monte Caseros, Mocoretá, *Ahumada & Schinini 3926* (CTES, MEXU); Dept. Sauce, Río Guayquiraró, 18 km S of Sauce, *Cristóbal et al. 1603* (CTES, NY); Dept. Paso de los Libres, Arroyo Capiquice, *Krapovickas & Cristóbal 11299* (CTES); Dept. Santo Tomé, Gobernador Virasoro, rt. 14, *Krapovickas et al. 16720* (BAA, BAB, CTES, MBM, NY); Dept. San Martín, 11 km N of Carlos Pellegrini, Arrocera Drews, *Krapovickas et al. 20305* (BAA, BAB, CTES, F, GH, NY); Dept. Mburucuyá, Estancia "Santa Teresa," *Pedersen 4791* (C, CTES, K, MO, NY, S, U, US, herb. Pedersen); Dept. Mercedes, Miriñay River, rt. from Mercedes to Paso de los Libres, *Pedersen 11364* (C, CTES, MO, NY); Dept. Curuzú Cuatiá, Estancia Araguá, *Pedersen 12542* (C, CTES, NY); Dept. San Miguel, Carandaytí, *Pedersen 13407* (C, CTES, MO); Dept. San Roque, route 23, 3 km W of Santa Lucía River, *Schinini & Ahumada 15987* (CTES, MEXU, SI). **Entre Ríos:** Dept. Paraná, 90 km S of La Paz, *Bacigalupo et al. 611* (CTES); Dept. Concordia, Balneario N of Concordia, *Bacigalupo et al. 992* (CTES); Delta del Paraná, Ceibo River, *Cabrera 2523* (NY, S–2, US); Dept. Feliciano, 26 km W of Feliciano, San Victor, *Cristóbal et al. 1553* (CTES); Dept. La Paz, 1 km N of La Paz, *Krapovickas & Cristóbal 43436* (CTES, NY); Dept. Federación, Estancia "Buena Esperanza," *Pedersen 7172* (C, herb. Pedersen). **Formosa:** Dept. Pirané, El Colorado, *Insfran 445* (CTES); Guayenlec, *Jörgensen 3163* (MO, US). **Misiones:** Dept. Leandro N. Alem, Güemes, San José and Co. Azul, *Cabrera et al. 28624* (CTES); Dept. Capital, rt. 12, 1 km W of Arroyo Garupá, *Schinini 5430* (CTES, ENCB, G, R); Dept. Candelaria, rt. 12, 3 km W of Arroyo San Juan, *Schinini 5540* (CTES). **Santa Fé:** Dept. Gral. Obligado, Malabrigo, *Blanchoud 2369* (CTES); Dept. Gral. Obligado, El Sombrerito, *Blanchoud 2115* (CTES); Dept. San Justo, 15 km S from Vera and Pintado, *Krapovickas 14323* (CTES, GB, HB).

URUGUAY. Montevideo, *Gibert 126* (syntype of *P. consobrina*, E? K? MVM?); Dept. Cerro Largo, marshes of Medina, *Rosengurtt B-2503* (NY). **Canelones:** Marshes of Carrasco, *Krapovickas 4398* (CAS); ruta Interbalnearia km 27.5, *Lema 6743* (CTES). **Paysandú:** Meseta de Artigas, *Del Puerto 5409* (CTES, NY). **San José:** Santiago Vásquez, *Del Puerto 2257* (K); Barra de Santa Lucía, *Krapovickas & Cristóbal 11225* (CTES, S). **Tacuarembó:** Rincón del Bonete, *Krapovickas & Cristóbal 11202* (CTES, P, UC); Valle Edén, *Rosengurtt et al. 70724* (CTES).

From the description, *Pavonia betonicaefolia* belongs to *P.* sect. *Lebretonia* subsect. *Hastifoliae* and is evidently the same as *P. consobrina*. Members of *P.* sect. *Lebretonia* are not known from Peru, the locality given by Presl, and the collection, in fact, is probably from elsewhere, given the unreliability of locality data often associated with Haenke collections. The range of *P. consobrina* (= *P. betonicaefolia*) includes S Uruguay and the vicinity of Buenos Aires (see Fig.

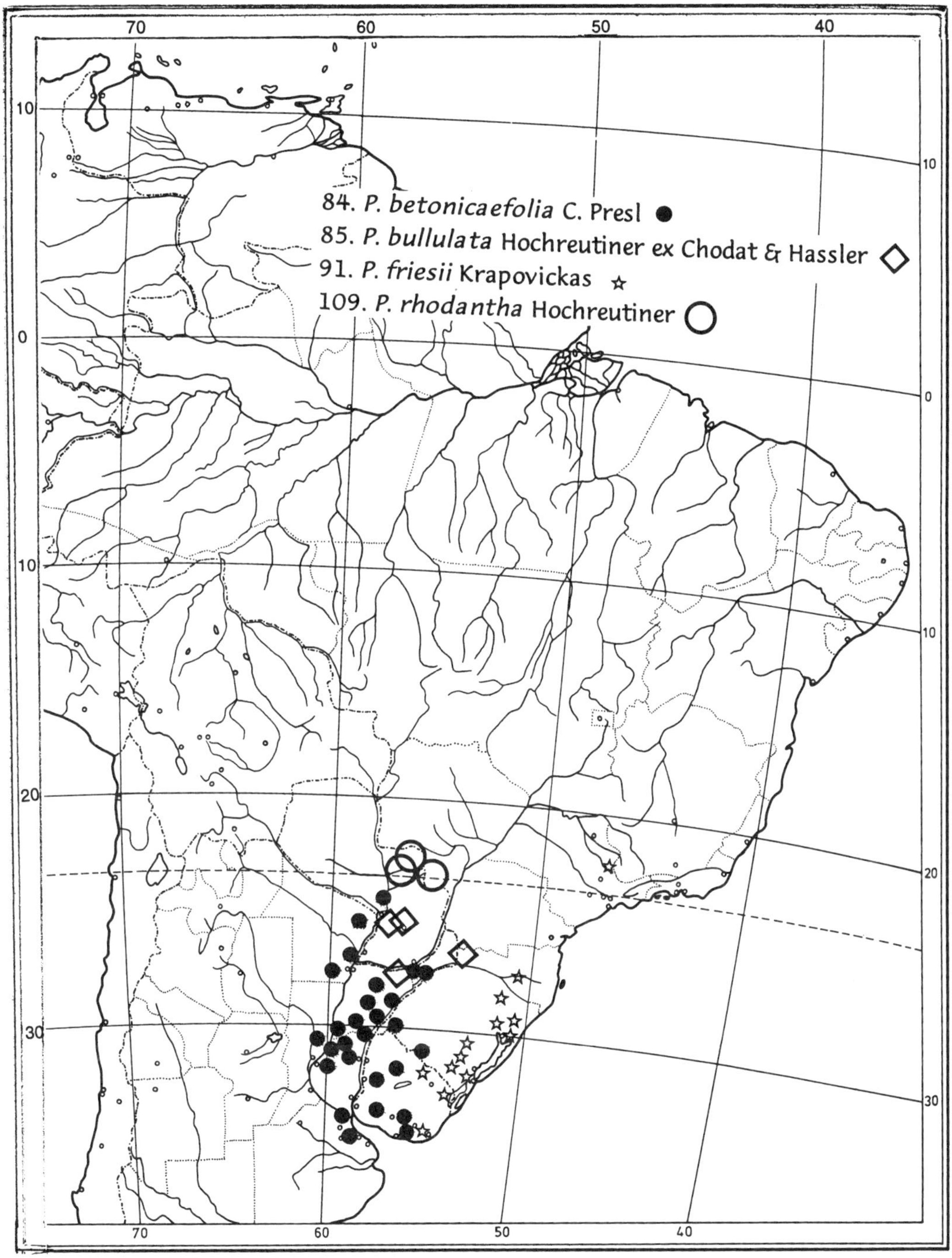

FIG. 45. Distribution map of four South American species of *Pavonia* subsect. *Hastifoliae*.

45); thus Haenke presumably collected the type of *P. betonicaefolia* during his overland travels between Buenos Aires and N Chile.

Pavonia betonicaefolia is characterized by relatively narrow involucellar bracts, coupled with coarsely pubescent stems and evenly pubescent petioles.

85. Pavonia bullulata Hochreutiner *ex* Chodat & Hassler, Bull. Herb. Boiss. **5:** 299. 1905. *Pavonia hastata* forma *bullulata* (Hochreutiner) Hassler, Bull. Herb. Boiss. **7:** 739. 1907. Type. Paraguay. In the region of the upper Yacá River, near Valenzuela, Jan 1900, *Hassler 6987* (holotype, G; isotypes, B destroyed, as photo F-9433, CTES [fragment], G, NY, S). Fig. 33

? *Pavonia hastata* subforma *piliosuscula* Hassler, Bull. Herb. Boiss. ser. 2, **7:** 739. 1907. Type. Paraguay. Near Villa Rica, *Hassler 8757* (specimen unknown).

Subshrubs 0.5–1 m tall, the stems with coarse stellate hairs ca. 1 mm long. *Leaf* blades lance-oblong, mostly 2–4.5 × 1–1.5 cm, basally truncate or subhastate, serrate, acute, palmately 3(–5)-nerved or seemingly penninerved, with scattered stellate hairs 0.5–1 mm diam. above and beneath, very sparse above; petioles 8–18 mm long; stipules filiform, 3–4 mm long. *Flowers* solitary in the leaf axils, the pedicels 0.5–2.5 cm long; involucellar bracts 5, lanceolate, 5–6 × 2–2.5 mm, stellate-pubescent; calyx 8–9 mm long, stellate-hirsute, plicate-pentangular in bud, the lobes cordate, 5–6 mm wide; corolla lavender with a dark center, 1.5–2 cm long, glabrous throughout; staminal column 6–7 mm long, glabrous, reddish, the filaments 2 mm long. *Fruits* ca. 7 mm diam., minutely pubescent; mericarps 4 mm high, dorsally carinate, with raised reticulations, enclosed by calyx.

Distribution (Fig. 45). From S Brazil (Santa Catarina), Paraguay, and NE Argentina (Corrientes). The type locality is at 25°35′S, 56°50′W.

Specimens examined. BRAZIL. SANTA CATARINA: Mun. da Tepeja, S. Joaquim, *Reitz 3512* (CAS).

PARAGUAY. CAAGUAZÚ: Pastoreo, rd. to Yhú, *Gregory et al. 9612* (CTES [fragment *ex* LIL]).

ARGENTINA. CORRIENTES: Dept. Santo Tomé, Gdor. Virasoro, Ea. Vuelta del Ombús, *Krapovickas & Schinini 38380* (CTES, G, NY).

Hassler published *Pavonia hastata* var. *subconcolor* Hassler (Bull. Herb. Boiss., ser. 2, **7:** 739. 1907) with a description but with no citation of specimens. He included two formae in his variety, forma *bullulata* (Hochreutiner) Hassler, as a new combination, and forma *tomentosa* (based on *Rojas 384*, n.v.) as a new forma. It seems to me that the best way to handle this situation nomenclaturally is to regard forma *tomentosa* as an incorrect name for forma *subconcolor*, with the latter name based on *Rojas 384*. The placement of this taxon remains dubious until the type can be located and examined.

Pavonia bullulata is characterized by its coarsely pubescent leaf blades.

86. Pavonia cryptica Krapovickas & Cristóbal, Lilloa **31:** 41. 1962. Type. Argentina. Corrientes: Dept. Ituzaingó, Las Casuarinas Creek, *Pierotti 6193* (holotype, LIL–n.v.). Fig. 33

Shrubs 0.5 m tall, the stems minutely stellate-pubescent and with some glandular hairs. *Leaf* blades ovate-oblong, mostly 1–2.5(–5) × 8–15(–20) mm, basally hastate-cordate, coarsely serrate, acute to obtuse, discolorous, basally 3-nerved, minutely stellate-pubescent above, canescent beneath; petioles 5–14 mm long; stipules subulate, 2–3 mm long. *Flowers* solitary in the upper leaf axils, the pedicels 1–2 cm long, slender; involucellar bracts 5, ovate, 5–8 × 2–4 mm (constricted at base), sometimes spreading or reflexed, the margins revolute; calyx 8–9 mm long, densely pubescent, sometimes with glandular hairs; corolla 1–2 cm long, pale lavender (to almost white) with a purplish center and veins. *Fruits* 7–8 mm diam., minutely pubescent; mericarps 4–5 mm tall, carinate, reticulate-rugose or with 3 or 4 prominent excrescences on each side.

Distribution (Fig. 40). From Uruguay, Argentina (Córdoba, Corrientes), and Bolivia. The type locality is at 27°38′S, 56°40′W.

Illustrations. Krapovickas & Cristóbal (1962: figs. 1C, 7).

Specimens examined. **BOLIVIA.** Rosal, 6000 ft, *Brooke 5956* (U).

ARGENTINA. CÓRDOBA: Sierra Chica, *Lossen 210* (MO, PH). **CORRIENTES:** Dept. Concepción, 6 km E from Santa Rosa, *Arbo et al. 933* (CTES); Estancia "Santa Teresa," *Krapovickas & Cristóbal 11414* (BH, CORD, CTES, G, K, P, SI, UC, US); 21 km W of Ituzaingó, *Krapovickas & Schinini 38179* (CTES, ENCB, F, G, HAS, IPA, LIL, MICH, NY, UC); 12 km NE of San Miguel, *Krapovickas et al. 24805* (BH, CTES); Dept. Ituzaingó, Ituzaingó, *Meyer 6197* (U); Concepción, Estancia "Santa Rosalia," *Pedersen 12002* (C–2, CTES, MO); Dept. Mburucuyá, Estancia "Santa Teresa," *Pedersen 1020* (C, MO, S, US, herb. Pedersen); 21 km S of Loreto, *Schinini et al. 8239* (CTES); 12 km N of Loreto, *Schinini et al. 8348* (CTES, NY); Itazuingó, *Schwarz 444* (NY); Dept. Empedrado, Manuel Derqui, *Schwarz 10183* (CAS); Dept. Gral. Paz, 12 km NE from San Miguel, *Tressens et al. 2060* (CTES, HAS); Dept. San Miguel, Loreto, *Ybarrola 4115* (CAS).

URUGUAY. COLONIA: Riachuelo, *A. L. Cabrera 4031* (F).

87. Pavonia cymbalaria St.-Hilaire & Naudin, Ann. Sci. Nat. Bot., ser. 2, **18:** 42. 1842. *Pavonia hastata* forma *brevifolia* Gürke in Martius, Fl. Bras. **12(3):** 500. 1892 (based on *P. cymbalaria* St.-Hilaire & Naudin). *Pavonia hastata* forma *cymbalaria* (St.-Hilaire & Naudin) Hassler, Ostenia 340. 1933, nom. superfl. Type. Uruguay. Near Montevideo, *Gay s.n.*

(holotype, P as photo F-35470; isotype fragments, CTES, F). Fig. 33

Subshrubs, procumbent to ascending, to 0.5 m tall, the stems covered with minute appressed stellate (sublepidote) hairs. *Leaf* blades hastate-triangular, mostly 10–20 × 8–15 mm, basally hastate-cordate, crenate, acute to obtuse, palmately 3–5-nerved, discolorous, minutely stellate-pubescent above, densely canescent beneath; petioles 3–14 mm long; stipules subulate, 2(–4) mm long. *Flowers* solitary in the leaf axils, the pedicels 1–2 cm long; involucellar bracts 5–6, lance-ovate, 4–7 × 2.5–3 mm; calyx 6–7 mm long, densely puberulent; corolla 1–1.5 cm long, lavender with a purplish center; staminal column 5 mm long, glabrous, purplish, the filaments 1.5–2 mm long; styles purplish, the stigmas pallid and villous. *Fruits* ca. 7 mm diam., minutely pubescent; mericarps 4 mm high, carinate, reticulate-veined. Chromsome number, $2n$ = 56 (Krapovickas & Cristóbal, 1962).

Distribution (Fig. 47). From Uruguay and Argentina. The type locality is at 34°50′S, 56°15′W.

Representative specimens examined. ARGENTINA. Buenos Aires: Inst. Nac. Tecnol. Agropec., *Dalby H22* (MSC, NY); Pdo. Balcarce, Estación Experimental, Co. de Tambo, *Grondona 2643* (BAA, NY); Est. San Simón, Co. de la Virgen, *Grondona & G. Dawson 6367* (BAA, NY); Pdo. Tandil, Canteras San Luis, *Huidobro 1732* (NY, S); Pdo. Tornquist, Sierra de la Ventana, *Krapovickas 2174* (CTES, MO, NY, S, US); Tandil, Cerro Leones, *Krapovickas 2956* (CTES, NY); Pdo. Olavarría, Cerro Dos Hermanas, *Krapovickas 3389* (CTES); Tandil, Nov 1892, *Kuntze s.n.* (NY, US); near Balcarce, *Pedersen 1059* (C, US, herb. Pedersen); Pdo. Saavedra, Abra del Hinojo, *Pedersen 13172* (C, CTES); Pdo. Cnel. Suarez, Sierra de las Tunas, *Rossi-Bachmann 430* (CTES); Pdo. Tornquist, Arroyo San Bernardo, Las Piletas, *Rossi-Bachmann 440* (CAS, CTES, G, TEX); Pdo. Tornquist, Sierra de Curamulal, *Spegazzini 5556* (BAB, NY); Parque Tornquist, *Villamil 3926* (NY).

URUGUAY. Maldonado: Solís, *Krapovickas & Cristóbal 11156* (CTES, F, S); Piriápolis, foot of the Cerro del Toro, *Krapovickas & Cristóbal 11167* (CTES, SI, UC); Pan de Azucar Creek, *Krapovickas & Cristóbal 11173* (CTES). **Minas:** In rocky hills, *Osten 16983* (NY). **Treinta y Tres:** Carajá del Olinar 2 leagues from Sta. Clara, *Rosengurtt 4841* (CTES).

Pavonia cymbalaria is a procumbent plant with discolorous leaves and small flowers (corolla 1–1.5 cm long). It is similar in growth habit to *P. glechomoides* and *P. schininii*, both of which have coarse stem hairs in contrast to the very minute stem hairs of *P. cymbalaria.*

88. Pavonia distinguenda St.-Hilaire & Naudin, Ann. Sci. Nat., ser. 2, **18:** 42. 1842. Type. Brazil. Rio Grande do Sul: 1833, *Gaudichaud s.n.* (holotype, P as photo F-35484; isotype fragment, F). Figs. 33, 46

Pavonia urbaniana Gürke *in* Martius, Fl. Bras. **12(3):** 501. 1892. Type. Argentina. Entre Ríos: Dept. Uruguay, Concepción del Uruguay, banks of Cupalén Creek, *Lorentz 1775* (lectotype, here designated, B as photo F-9473; isolectotypes, BAF, CTES [fragment], UC). Syntypes, *Arechavaleta 991*; *Lorentz 649, 1775* ; *Niederlein 40*; *Sellow 1622*, US. Note: Gürke (1892b: 501) described *P. urbaniana* var. *tomentoso-velutina* and *P. urbaniana* var. *villosa* under *P. urbaniana*. Krapovickas and Cristóbal (1962) designated the former as the lectotype of the species, i.e., var. *tomentoso-velutina* = var. *urbaniana* (ICBN, art. 26), but did not designate a lectotype specimen, although two of the above syntypes were cited by them: *Lorentz 649* (UC) and *Lorentz 1775* (BAF, UC), both from the same locality.

Pavonia malmeana R. E. Fries, Kongl. Svenska Vetenskapsakad. Handl. **42(12):** 61. 1908. Type. Brazil. Rio Grande do Sul: Cachoeira, *Malme II-1059* (holotype, S; isotypes, CTES [fragment], S).

Pavonia urbaniana var. *macrantha* R. E. Fries, Kongl. Svenska Vetenskapsakad. Handl. **42(12):** 60. 1908. Type. Brazil. Rio Grande do Sul: Cruz Alta, *Malme II-1132* (holotype, S).

Subshrubs 0.5–1 m tall, the stems densely stellate-pubescent, the hairs 0.5 mm long. *Leaf* blades oblong-lanceolate, mostly 4–6(–9) × 1–2(–4) cm, basally truncate to sagittate (with an acute sinus), serrate, acute, palmately 3–5-nerved, discolorous, the upper surface with simple, bifurcate, and stellate hairs, the lower surface densely stellate-tomentulose; petioles 1–2 cm long; stipules subulate, 2–4 mm long. *Flowers* solitary in the leaf axils, the pedicels 0.5–3 cm long; involucellar bracts 5, lanceolate, 7–9(–18) × 2–4(–5) mm, hirsute; calyx 7–10 mm long, hirsute; corolla 2.5–3 cm long, white or pale lavender with reddish veins and a purplish center, densely pubescent on margins of claw; staminal column 6–10 mm long, glabrous, purplish, the filaments 1.5–3 mm long; styles purplish, the stigmas pallid, villous. *Fruits* 7 mm diam., minutely puberulent; mericarps 5 mm high, prominently rugulose. Chromosome number, $2n$ = 56 (Krapovickas & Cristóbal, 1962).

Distribution (Fig. 47). From Brazil (Paraná, Rio Grande do Sul, and Santa Catarina), Uruguay, and Argentina (Misiones and Corrientes). The type locality for *Pavonia urbaniana* is at 32°30′S, 58°15′W; that for *P. malmeana* is at 30°0′S, 52°55′W.

Additional illustrations. Fries (1908: pl. 6, figs. 21–23 and pl. 7, fig. 4, as *Pavonia malmeana*; pl. 6, figs. 24–27, pl. 7, fig. 5, as *P. urbaniana*); Gürke (1892b: pl. 95, fig. 3, as *P. urbaniana*).

FIG. 46. *Pavonia distinguenda*. Flowering branch with enlargement of calyx and involucel (lower right) (*Krapovickas 14493*).

Representative specimens examined. BRAZIL. PARANÁ: Jaguariahyva, 23 Apr 1911, *Dusén s.n.* (GH); Rio Capinzal, 1908–1912, *Dusén s.n.* (S). **RIO GRANDE DO SUL:** Mun. Santa María, Ibicuí-Mirim Biological Reserve, *Bueno 5431* (CTES, HAS); 20 km W of São Francisco de Paula, *Krapovickas & Cristóbal 37624* (CTES, HAS); BR-290, 92 km W of Porto Alegre, *Krapovickas & Cristóbal 38392* (C, CTES, F, G, HAS, LIL, UC); 17 km E of Lagoa Vermelha, *Krapovickas & Schinini 39349* (CTES, NY); 10 km E of Vacaria, rd. to Bom Jesus, *Krapovickas & Vanni*

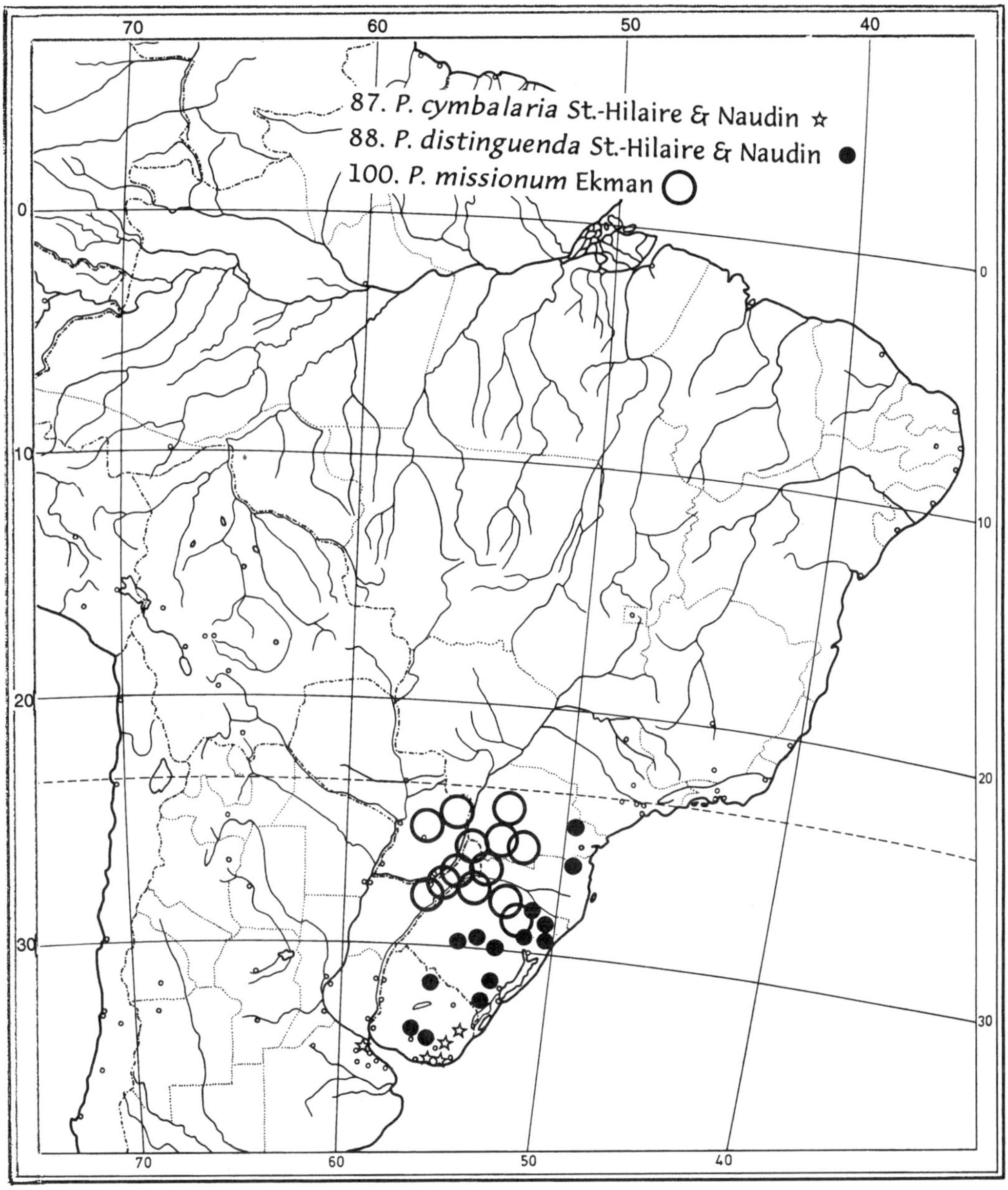

FIG. 47. Distribution map of three South American species of *Pavonia* subsect. *Hastifoliae*.

37013 (CTES, WIS); Pelotas, IPEAS, near Botanic Garden, *Krapovickas et al. 22938* (BAB, CTES, ENCB, MBM, MO, NY); 4 km N of Nova Hamburgo, *Krapovickas et al. 22986* (CTES, MBM, NY, U); 13 km E of São Pedro do Sul, *Lindeman et al. 21167* (CTES, U); Morro Sapucaia, *Palacios-Cuezzo 447* (CAS, F); Santa Cruz do Sul, *Palacios-Cuezzo 1184* (CAS, CTES); vic. of São Leopoldo, *Rambo 44135* (CTES, MO); Viamao, *Rambo 44188* (CAS, CTES, NY); Vila Oliva near Caxias, *Rambo 44603* (CTES, US); Lagoa dos Quadros, near Torres, *Rambo 45972* (CAS, CTES); Santa Rita, near Farroupilha, *Rambo 56748* (CTES, NY, TEX); Porto Alegre, near São João, *Reineck & Czermak 266* (E, S). SANTA CATARINA: Mafra, *Reitz 5238* (CAS).

ARGENTINA. CORRIENTES: Dept. Paso de los Libres, Paso de los Libres, *Ahumada & Schinini 3942* (CTES, F, ICN, LIL); Dept. Capital, Corrientes [cult.], *Krapovickas 14493* (CTES, NY); Dept. Santo Tomé, Estancia Garruchos, *Krapovickas et al. 21599* (BAA, BAB, CTES, F, GH, LIL,

NY, SI, WIS). MISIONES: Dept. Concepción, Arroyo Persiguero, *Schwarz 3631* (AS, CAS, LIL, TEX).

URUGUAY. CANELONES: Arroyo Solís Grande, *Krapovickas & Cristóbal 11151* (CTES); Canelones, *Rosengurtt B-2864* (NY). RIVERA: Rt. 5 near the rd. to Tranqueras, *Del Puerto 3215* (CTES, NY); Trojas Viejos, Salinas, *Rosengurtt 10866* (CTES). SAN JOSÉ: Barra de Santa Lucía, *Krapovickas 4850* (C, NY), *Krapovickas & Cristóbal 11229* (CTES), *Osten 21816b* (F, S).

89. Pavonia dusenii Krapovickas, Trabalh. XXVI Congr. Nac. Bot. Brasil 313. 1977. Type. Brazil. Santa Catarina: Mun. Lajes, 20 km S of Lajes, 30 Jan 1973, *Krapovickas, Cristóbal & Maruñak 23036* (holotype, CTES; isotypes, K, MBM, MO, NY, P, S). Fig. 33

Subshrubs 0.5–1 m tall, the stems coarsely stellate-pubescent, the hairs 0.5–1 mm long. *Leaf* blades sagittate-oblong or sometimes ovate, mostly 1.5–6 × 0.5–2 cm, basally sagittate or cordate, coarsely crenate-serrate, acute, palmately 3–5-nerved, discolorous, stellate-pubescent above and beneath, the hairs ca. 0.5 mm long; petioles 6–18 mm long; stipules subulate, 3–4 mm long. *Flowers* solitary in the leaf axils, the pedicels 0.5–2.5 cm long, stellate-hirsute, the hairs longer (to 1 mm) distally; involucellar bracts 5–6, lance-ovate, 7–8 × 3–4(–5) mm, hirsute; calyx 8–9 mm long, stellate-hirsute (more coarsely so than involucel); corolla 1.5–2 cm long, pale lavender with purplish center and venation. *Fruits* 7–9 mm diam., minutely pubescent; mericarps carinate, rugulose, 4.5–5 mm high, basally winged.

Distribution (Fig. 39). Brazil (Paraná, Rio Grande do Sul, and Santa Catarina). The type locality is at 27°56′S, 50°23′W.

Illustrations. Krapovickas (1977: fig. 2A-C).

Representative specimens examined. BRAZIL. STATE UNKNOWN: Serra São Luis, 50 km out of Timatuba, *Ll. Williams 18729* (NY). PARANÁ: Mun. Curitiba, S. Lourenço, *Hatschbach 1194* (CAS); Mun. Rio Negro, rd. from Lageado to Rio Negrinho, *Hatschbach 3996* (US); Mun. Rio Branco do Sul, Itapiruçú, *Hatschbach 7802* (CTES, U); rd. from Palmas to Mangueirinha, *Hatschbach 15439* (CTES, F, MO, NY, US); 43 km from Horizonte, rd. to Palmas, *Krapovickas & Cristóbal 39722* (C, CTES, F, G, K, LIL, MBM, MICH, MO, NY, UC). RIO GRANDE DO SUL: Caxias do Sul, Ana Rech, Faxinal, *Brinker et al. 5510* (US); Mun. Bom Jesus, Pelotas River, rd. from S. Joaquim to Rocinha, *Krapovickas & Schinini 38335* (CTES, F, HAS, UC); 10 km E from Vacaria, rd. to Bom Jesus, *Krapovickas & Vanni 37012* (ASU, CTES); Aparados da Serra (Bom Jesus), Fazenda Bernardo Velho, *Rambo 34567* (S); Caí, *Rambo 41241* (CAS, CTES); Passo do Socorro, near Vacaria, *Rambo 51628* (S, US); Canela, *Sobral & Silva 4942* (CTES). SANTA CATARINA: Mun. Ponte Serrada, Campo de Palmas, *Hatschbach et al. 28283* (C, CTES); Mun. São Joaquim, Río Lavatudo, rd. from São Joaquim to Lajes, *Krapovickas & Cristóbal 37807* (BAB, CTES, NY); Mun. Lajes, access rd. to Painel, *Krapovickas & Cristóbal 41956* (CLEMS, CTES, HAS, LIL, MO, NY); 23 km E of Campos Novos BR-470, *Krapovickas & Cristóbal 43572* (CTES); 33 km S of Urubichi, SC-438, *Krapovickas & Cristóbal 43721* (CTES, NY); 23 km S of S. Joaquim, rd. to Bom Jardim da Serra, *Krapovickas & Schinini 38263* (ASU, CTES); Pelotas River, 3 km W of Bom Jardim da Serra, *Krapovickas & Schinini 38283* (CTES, F, LIL, WIS); 14 km E of Lajes, rd. to São Joaquim, *Krapovickas & Vanni 36886* (CTES, G, K, MBM, MICH, MO); Mun. Ponte Alta, Ponte Alta do Sul, *Krapovickas et al. 23052* (CTES, G, LIL, NY, RB, SI); Mun. Agua Doce, near Morro de Capão Doce, 8 km S of border of Paraná, *Pedersen 13724* (C–2, CTES, NY); Canoinhas, *Reitz & Klein 11538* (CTES); Abelardo Luz, *Reitz & Klein 16570* (CTES).

As noted by Krapovickas in his original description, *Pavonia dusenii* is sufficiently similar to *P. guerkeana* that the two species are difficult to distinguish. The coarsely hirsute stems and calyces of *P. dusenii*, however, are notable.

90. Pavonia eurychlamys Ulbrich, Feddes Repert. Nov. Spec. Regni Veg. **13:** 514. 1915. Type. Argentina. Sierra de Catamarca, Singuil, 1888, *Schunck 199* (holotype, B as photo F-9442). Fig. 33

Shrubs 1.5–2 m tall, the stems pubescent with minute appressed stellate hairs 0.5 mm long or less. *Leaf* blades triangular-ovate, sometimes weakly lobed, mostly 1–5 × 1–3.5 cm, basally truncate to cordate, crenate-serrate, acute, palmately 5-nerved, somewhat discolorous, with scattered hairs 0.5–1 mm long, mostly simple above, mostly stellate beneath; petioles 7–20 mm long, with pubescence like stem and with setae 0.5–1.5 mm long in addition; stipules subulate, 2–3 mm long. *Flowers* solitary in the leaf axils, the pedicels 1.5–4.5 cm long, slender, distally pilose; involucellar bracts 5, broadly spatulate or ovate, acute, 8–10 × 3–7 mm, narrowed and yellowish at base, sparsely and minutely pubescent; calyx 8–11 mm long, sparsely pubescent to densely hirsute, the lobes usually with purplish venation; corolla 15–18 mm long, rose with dark red veins. *Fruits* ca. 7 mm diam., minutely pubescent; mericarps carinate, apically acute, reticulately rugose, with 2 or 3 excrescences on each side, 4–4.5 mm high.

Distribution (Fig. 48). From NW Argentina (Jujuy, Salta, and Tucumán) and Bolivia, at elevations of 1500–1700 m. The type locality is at 27°50′S, 65°54′W.

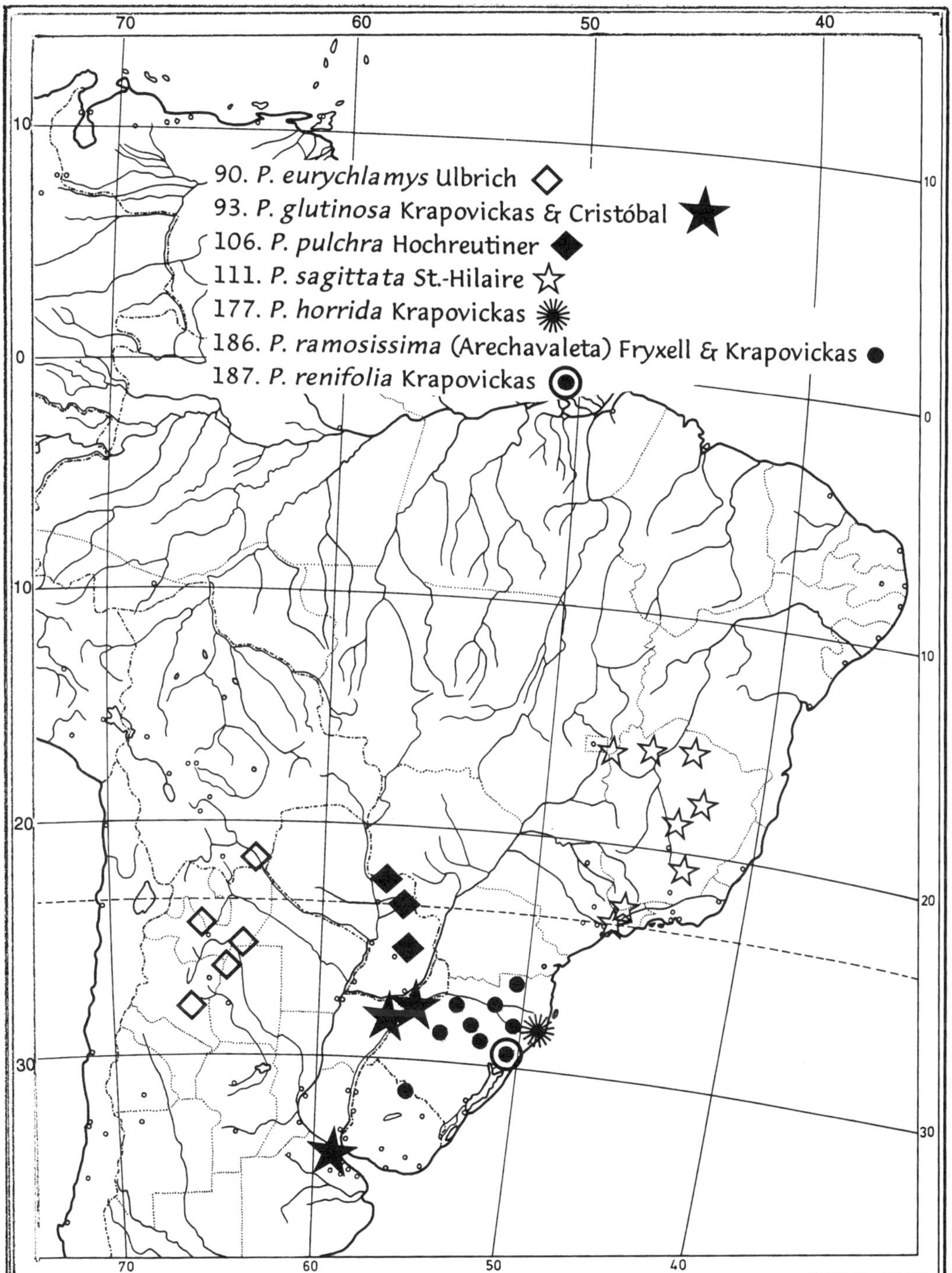

FIG. 48. Distribution map of seven South American species of *Pavonia* subsect. *Hastifoliae* and *P.* sect. *Urenoideae*.

Specimens examined. BOLIVIA. TARIJA: Prov. O'Connor, Narváez, rd. from Tarija to Entre Ríos, *Krapovickas & Schinini 38835* (CTES, NY).

ARGENTINA. JUJUY: Dept. Capital, Cerro Zapla, *Ahumada 5280* (CTES). SALTA: Dept. Anta, Co. Chañar, El Rey National Park, *Brown 1626* (CTES, NY); Dept. Santa

Victoria, between Los Toldos and El Arrozay, Novarra 5162 (G). TUCUMÁN: Dept. Río Chico, Escoba (El Chorro), *Monetti 1759* (U); Dept. Burruyacu, Sierra Nogalito, W slope, *de la Sota 2676* (CTES).

Pavonia eurychlamys is one of the species of *P.* subsect. *Hastifoliae* having red-veined calyx lobes.

91. Pavonia friesii Krapovickas, Trabalh. XXVI Congr. Nac. Bot. Brasil 318. 1977. Type. Brazil. Rio Grande do Sul: Near Viamão, *Rambo 44192* (holotype, LIL–n.v.; isotypes, CTES, MO, NY, S, US). Fig. 33

Shrubs to 2 m tall, the stems covered by minute multi-armed stellate hairs. *Leaf* blades hastate-triangular or sagittate-oblong, mostly 2–5 × 1–2 cm, basally cordate-hastate or -sagittate (the sinus acute), serrate, acute or subobtuse, discolorous, palmately 5–7-nerved, minutely stellate-pubescent, densely so beneath; petioles 1–2 cm long; stipules subulate, 4 mm long. *Flowers* solitary in the leaf axils, the pedicels 1–3 cm long; involucellar bracts 5, lanceolate or spatulate, 7–8 × 2.5 mm, stellate-puberulent and hirsute with hairs 1–2 mm long (especially basally); calyx 7–9 mm long, with pubescence like involucel; corolla 1.5–2 cm long, rose with dark red center, bearded on margins of claws; staminal column 5 mm long, reddish, glabrous, the filaments 3 mm long. *Fruits* ca. 7 mm diam.; mericarps minutely pubescent, 4–5 mm long, carinate, reticulately rugose.

Distribution (Fig. 45). Known from S Brazil (Minas Gerais, Rio Grande do Sul, and Santa Catarina), Uruguay, and with one geographically anomalous collection from Bolivia. The type locality is at 30°06′S, 51°02′W.

Illustrations. Krapovickas (1977: fig. 3D-F).

Representative specimens examined. **BRAZIL.** MINAS GERAIS: Poços de Caldas, *Barreto 11122* (CAS); without locality, *Widgren 513* (S–2), 1845, *Widgren s.n.* (S–2). RIO GRANDE DO SUL: Caxias do Sul, Ana Rech, Faxinal, *Brinker et al. 5508* (US); Mun. Gravatai, Morro Morungava toward Sítio dos Ferrari, *Bueno 1251* (CTES); Mun. Viamão, Morro da Pedreira, Itapoã, *Bueno 1793* (CTES); Mun. Pelotas, Botanic Garden, *Bueno 2823* (CTES); São Vicente do Sul, Fazenda Santa Fé, *Bueno 3394* (CTES); Mun. Caçapava do Sul, Minas do Camaquá, *Krapovickas & Cristóbal 34196* (CTES, F, SI); Campo Bonito, ca. 8 km SW of Torres, *Krapovickas & Cristóbal 38493* (AK, CTES, HAS, NY, UC); 47 km SW from Caçapava do Sul, rd. to Lavras, *Krapovickas & Cristóbal 44659* (CTES, TEX); Lavras do Sul, *Krapovickas & Cristóbal 44665* (CTES, TEX, US); 17 km E from Lagoa Vermelha, *Krapovickas & Schinini 38223* (CTES, HAS); 39 km S of Cruz Alta, Ivaí River, *Krapovickas & Vanni 37086* (CTES, ICN, NY); Mun. Bagé, 70 km NE of Bagé Vassoural, *Lindeman et al. 7025* (CTES, U); Mun. Arroio Grande, around Arroio Grande, *Pedersen 13842* (CTES, NY); Rio dos Sinos, *Rambo 38699* (GH); Steinkopf, near São Leopoldo, *Rambo 39010* (CAS, CTES); Vila Manresa, near Porto Alegre, *Rambo 44159* (CTES, NY, TEX); Furna do Tigre, near Itapoana, *Rambo 44434* (CTES, NY, US). SANTA CATARINA: Mun. Campos Novos, 19–20 km S of Campos Novos, *L. B. Smith & Klein 11972* (US).

BOLIVIA. COCHABAMBA: Rosal, *Brooke 5756* (LL, NY).

URUGUAY. MONTEVIDEO: Cerro (Lit. Platens), *Herter 253a* (US). CERRO LARGO: Ca. 45 km E of Melo on rd. to Río Branco, *Pedersen 13861* (CTES). TACUAREMBÓ: Rincón del Bonete, *Krapovickas & Cristóbal 11189* (C, CTES, HAS, MICH).

Pavonia friesii is notable for its coarsely hirsute calyx and involucel coupled with very fine pubescence on the stems.

92. Pavonia glechomoides St.-Hilaire, Fl. Bras. Merid. **1:** 227. t. 45. 1827. *Malache glechomoides* (St.-Hilaire) Kuntze, Revis. Gen. Pl. **1:** 70. 1891. *Lassa glechomoides* (St.-Hilaire) Kuntze, Revis. Gen. Pl. **3(2)**: 19. 1898. Type. Uruguay. Near Montevideo, *St.-Hilaire s.n.* (holotype, P?–n.v.). Note: *St.-Hilaire 2309*, cited by Gürke (1892b) and distributed as a photograph F-23698, is not the type, as it was collected in Brazil in 1828, a year after the publication of the name. Fig. 33

Pavonia hirta Sprengel, Syst. Veg. **3:** 99. 1826 (non *Pavonia hirta* Schweigger, 1812). *Malache hirta* (Sprengel) Kuntze, Revis. Gen. Pl. **1:** 70. 1891. Type. Uruguay. Montevideo, *Sellow s.n.* (specimen unknown).

Procumbent perennial *herb*, the stems loosely stellate-pubescent. *Leaf* blades subovate to subrotund (sometimes triangular to pentagular), mostly 1–2.5 cm long, about as wide, basally cordate, coarsely crenate-dentate, acute to obtuse, palmately 5-veined, with scattered stellate hairs beneath, with simple, bifurcate, and stellate hairs above, the hairs ca. 1 mm long; petioles 1–1.5 cm long; stipules subulate, 2–3 mm long. *Flowers* solitary in the leaf axils, the pedicels 2–4 cm long; involucellar bracts 5, ligulate to spatulate, 7–11 × 1–2 mm, ciliate; calyx 7–10 mm long, sparsely ciliate; corolla 1.5–2.5 cm long, rose with purplish center and veins, pubescent on margins of claw; staminal column 7–8 mm long, glabrous, the filaments 1–2 mm long; styles emergent from the column. *Fruits* ca. 7 mm diam., minutely pubescent; mericarps 4 mm high, carinate, reticulately veined but relatively smooth.

Distribution (Fig. 34). From S Brazil, Uruguay, Argentina, and Bolivia, from low elevation to 2600 m. The type locality is at 34°50′S, 56°15′W.

Illustrations. Krapovickas (1965: fig. 53A–E); St.-Hilaire (1827: pl. 45).

Representative specimens examined. **BRAZIL. RIO GRANDE DO SUL:** 30 km NE of Bajé, on BR-153, *Arbo & Schinini 2481* (CTES); 49 km E of São Gabriel, *Krapovickas & Cristóbal 34173* (CTES); 40 km S of Bajé, 10 Nov 1976, *Pedersen s.n.* (herb. Pedersen).

BOLIVIA. SANTA CRUZ: Near Pampa Grande, Aug 1991, *Acevedo s.n.* (NY). **TARIJA:** Prov. Méndez, Sama, *Ehrich 113* (CTES); Padcaya, *Fiebrig 2535* (E, GH, S–2, U, US); from Tarija to Narváez, *Kiesling et al. 3770* (CTES); Prov. O'Connor, Narváez, rd. from Tarija to Entre Ríos, *Krapovickas & Schinini 38838* (CTES); rt. from Tarija to Villa Montes, Entre Ríos, *Krapovickas et al. 19069* (CTES, NY); Prov. Arce, 39.9 km S of jct. of rd. to Entre Ríos on rd. to Padcaya, *Solomon 10201* (MO); E of Tarija on Villa Montes rd., between km 80 and 65, *West 8272* (GH, MO).

ARGENTINA. BUENOS AIRES: Pdo. Tandil, Cerro Nocetti, *G. Dawson 3013* (BAB, NY); La Cascada, W of Tandil, *Hjerting & Rahn 3054* (C); Sierra de la Ventana, *Huidobro 1348* (NY, S); Pdo. Tornquist, Abra Pantanoso Viejo, rd. to Abra de la Ventana, *Pertusi 147* (CTES). **CATAMARCA:** From La Merced to Balcozna, between La Huigera and Balcozna, *Hunziker 18775* (CTES). **CORRIENTES:** Dept. Paso de los Libres, 13 km SW of the access to P. de los Libres, rt. 126, *Cristóbal et al. 1947* (CTES); Dept. Santo Tomé, Gobernador Virasoro, *Krapovickas et al. 16681* (CTES); Dept. Mercedes, 22 km SE of Mercedes, *Tressens et al. 2400* (CTES, MICH). **CÓRDOBA:** Sierra Grande, *Hunziker 7149* (NY); Dept. Río Cuarto, 10 km E of Achiras, *Saenz et al. 685* (CTES). **ENTRE RÍOS:** Dept. Federación, Charají to S. Jaime, *Bacigalupo et al. 750* (CTES); Dept. Uruguay, Cerro del Uruguay, *Troncoso & Bacigalupo 2951* (CTES). **JUJUY:** Near El Potrerillo, San Antonio, *Balls 5926* (F, US); Dept. Santa Barbara, El Fuerte, *Cabrera et al. 22246* (CTES); Dept. Capital, Juan Galán, *Cabrera et al. 27329* (CTES); Dept. Tumbaya, Abra del Volcán, *Schinini & Vanni 22494* (CTES, USM); Dept. Humiahuaca, Sierra de Feuta, *Venturi 8370* (US). **SALTA:** Dept. Caldera, *Cabrera et al. 19092* (CTES, NY); rt. 9 toward Jujuy, *G. Dawson & Celastremé 2045* (BAB, NY); La Angostura, 30 km N of Salta, *Krapovickas & Cristóbal 40296* (C, CTES, G, NY, USM); Dept. Molinos, Seclantos, *Romero s.n.* (CAS); rd. to Cornisa between Salta and Jujuy, *Schulz 6616* (CTES, NY); La Calderilla, *Sparre 1240* (LL); Dept. San Carlos, Cerro de Cachi, *Venturi 6913* (US); Dept. Rosario de Lerma, Campo Quijano, *Venturi 8028* (US). Capital, N of Caseros, *Wall & Sparre 1180* (S). **TUCUMÁN:** Dept. Tafi, *Olea 159* (NY, S).

URUGUAY. CANELONES: Los Bañados de Carrasco, *Bartlett 20796* (NY, TEX, US). **COLONIA:** Estancia "Cerros de San Juan," 21 Jan 1977, *Pedersen s.n.* (herb. Pedersen). **FLORIDA:** Estancia Rincón de Santa Elena, Picada Castro, *Rosengurtt 5817* (BRIT, CAS, MO); A° Mansavillagia, *Rosengurtt B711* (NY). **LAVELLEJA:** 47 km SSW of Treinta y Tres, rd. to Montevideo, *Arbo & Schinini 2498* (CTES). **MALDONADO:** Lago del Cisne, prosc. a Laguna del Sauce, *Del Puerto 3892* (CTES); Piriápolis, *Krapovickas & Cristóbal 11127* (CTES); Solís, *Krapovickas & Cristóbal 11157* (CTES); La Barra, El Placer, *Krapovickas & Cristóbal 11187* (CTES). **MONTEVIDEO:** Montevideo, 30 Nov 1876, *Fruchard s.n.* (F); Montevideo, *Gibert 954* (K); Atahualpa, *Herter 711* (F, MO, NY, U, US); Sayago, *Herter 11456* (S). **SAN JOSÉ:** Kiyú, *Izaguirre et al. 9741* (CTES). **TACUAREMBÓ:** Rincón de Bonete, *Krapovickas & Cristóbal 11207* (CTES, NY).

Pavonia glechmoides is a procumbent plant that commonly has small, suborbicular leaves. It is distinguished from the similarly procumbent *P. cymbalaria* by its coarser pubescence and from *P. schininii* by its longer, narrower involucellar bracts

93. Pavonia glutinosa Krapovickas & Cristóbal, Lilloa **31:** 26. 1962. Type. Argentina. Buenos Aires: Ptdo. La Plata, Punta Lara, *Krapovickas 4804* (holotype, LIL–n.v.; isotypes, CORD, CTES, GH, K, MO, NY, SI). Fig. 33

Subshrubs 1 m tall, the stems densely and finely puberulent with stellate hairs and glandular hairs intermingled, 0.1–0.2 mm long. *Leaf* blades triangular-ovate, sometimes weakly lobed, 1.5–3 × 1–2 cm, basally cordate, crenate-dentate, acute, palmately 5–7-nerved, minutely stellate-puberulent above and beneath, with some intermingled glandular hairs (at least above); petioles 1.5–3 cm long; stipules subulate, 4–5 mm long. *Flowers* solitary in the leaf axils, the pedicels 1–4.5 cm long, slender; involucellar bracts 5, ligulate, 7–8 × 1–1.5 mm, glandular-puberulent; calyx 8–9 mm long, glandular-puberulent and ciliate, the lobes with purplish venation; corolla red, 10–15 mm long, externally pubescent; staminal column 6–7 mm long, glabrous, the filaments 1 mm long. *Fruits* ca. 7 mm diam., nearly glabrous; mericarps carinate, reticulate-rugose, 4 mm high.

Distribution (Fig. 48). From Uruguay and adjacent Argentina. The type locality is at 34°55′S, 58°00′W.

Illustrations. Krapovickas & Cristóbal (1962: figs. 1E, 6).

Specimens examined. **ARGENTINA. BUENOS AIRES:** Pdo. La Plata, Punta Lara, *Krapovickas 4811* (CTES). **CORRIENTES:** Dept. Paso de los Libres, 21 km E of Bonpland, Uruguay River, *Schinini et al. 17365* (CTES, SI). **MISIONES:** Dept. Candelaria, Santa Ana, *Bertoni 98416* (U).

URUGUAY. SAN JOSÉ: Barra de Santa Lucía, *Krapovickas & Cristóbal 11231* (BH, CTES, G, K, NY, P, UC, US).

Pavonia glutinosa is distinguished by having red corollas coupled with purplish-veined calyx lobes. The specific epithet is derived from the glandular hairs that are found on the calyx and involucel (and elsewhere), a character it shares with *P. viscidula* and *P. xanthogloea*, both of which have paler corollas.

94. Pavonia guerkeana R. E. Fries, Kongl. Svenska Vetenskapsakad. Handl. **42(12):** 57, t. 7, f. 1. 1908. Type. Brazil. Minas Gerais: Caldas, 21 May 1873, *Regnell II-15b* (lectotype, here designated, S; isolectotypes, B as photo F-9447, C, K, P as photo). Fig. 33

Pavonia sagittata var. *conjugens* R. E. Fries, Kongl. Svenska Vetenskapsakad. Handl. **42(12):** 51. 1908. Type. Brazil. Minas Gerais: Caldas, 13 Dec 1877, *Regnell II-15* (syntypes?, CTES, F–2, U, US).

Shrubs 1–2 m tall, the stems stellate-pubescent, the hairs of variable size. *Leaf* blades ovate-triangular to lance-oblong, mostly 2–8(–10) × 1–2.5 cm, basally cordate to subsagittate, coarsely crenate-dentate, acute, palmately 5-nerved, discolorous (whitish beneath), stellate-pubescent on both surfaces; petioles 1–2.5 cm long; stipules subulate, 2 mm long. *Flowers* solitary in the leaf axils, the pedicels 1–4 cm long; involucellar bracts 5, ovate, 5–8 × 2–4 mm, stellate-pubescent; calyx 6–9 mm long, ciliate; corolla (sometimes cleistogamous) 2–3 cm long, rose-pink with dark red center, minutely pubescent on margins of claw; staminal column 5–9 mm long, the filaments 2–3 mm long; styles emergent from the column, purplish, free for distal 3–4 mm. *Fruits* 7–9 mm diam., minutely pubescent; mericarps 5 mm high, carinate (the ridge apically prolonged and acute), rugose, usually with acute excrescences on each side.

Distribution (Fig. 44). From Brazil (Minas Gerais, Paraná, Rio Grande do Sul, Santa Catarina, São Paulo) and Argentina (Misiones). The type locality is at 21°56′S, 46°22′W; that of *Pavonia sagittata* var. *conjugens* is the same.

Illustrations. Fries (1908: pl. 6, figs. 11, 12; pl. 7, figs. 1, 2); Krapovickas & Cristóbal (1962: fig. 1B).

Representative specimens examined. **BRAZIL. MINAS GERAIS:** Poços de Caldas, Véu das Noivas, *Krapovickas & Cristóbal 35355* (CTES, MBM, MO, NY). **PARANÁ:** Pinhaes, *Dusén 8922* (F, GH, NY, S); Curitiba, *Dusén 14322* (F, GH, MO, NY, S); 39 km E of Guarapuava, *Ferrucci et al. 208* (A, C, CTES, MBM, UC); Mun. Tibagi, hwy. to Café, *Hatschbach 13218* (CTES); Mun. Palmas, Estr. Palmas–Mangueirinha, *Hatschbach 15456* (CTES, F, US); Mun. Lapa, Rio Passo Dois, *Hatschbach 15984* (CTES); Mun. Laranjeiras do Sul, Laranjeiras do Sul, *Hatschbach 21148* (CTES, MO); Mun. Curitiba, Barigui, *Hatschbach 26227* (CTES, NY); Mun. Mandirituba, Rio Mauricio, *Hatschbach 41432* (CTES, MO); Mun. Curitiba, Lago Azul, *Hatschbach 47575* (CM, CTES); Mun. Jaguariaiva, Jaguariaiva, *Hatschbach & Haas 13946* (CTES); Rio Branco do Sul, rd. to Cerro Azul, *Landrum 4063* (CAS, CTES, NY, TEX); 7 km E de Contendas, *Krapovickas & Cristóbal 39629* (CTES, NY); 17 km S of Curitiba, *Krapovickas et al. 23085* (CTES, ENCB); 27 km N of Bocaiuva do Sul, *Krapovickas et al. 23275* (CONC, CTES, HUS, NY, P, TEX); Mun. Colombo, Ressaca, *Kummrow 891* (CTES); Mun. Curitiba, Rio Iguaçú Park, *Kummrow 1285* (CTES, MU, NY); Mun. Campina Grande do Sul, Mandasaia, *Paciornik & Oliveira 335* (CTES); Mun. Prudentópolis, BR-277, where the rd. to Prudentópolis turns off, *Pedersen 15951* (TEX); Curitiba, rd. to the airport, *E. Pereira 5142* (NY); Rio Cavernoso, Guarapuava, *Reitz & Klein 17709* (CTES, NY, US). **RIO GRANDE DO SUL:** Mun. Passo Fundo, Punto Boa Vista, 19 km E of Passo Fundo, *Krapovickas & Cristóbal 37575* (CTES, MBM); BR-285, access to Panambí, *Krapovickas & Schinini 38185* (CTES, HAS, UC); BR-286, km 61, 17 km E from Lagoa Vermelha, *Krapovickas & Schinini 38226* (CTES, HAS); 53 km W de Vacaría, Santa Rita River, *Krapovickas & Vanni 37036* (CTES, ICN); Campestre BR-116, Mun. Vacaria, *Krapovickas et al. 22998* (BAA, BAB, CTES, ENCB, MO, NY, P, R, RB, TEX, UC, WIS). **SANTA CATARINA:** Mun. San Miguel Oeste, Canela Gaúcha, *Castellanos 24770* (CTES); 10 km E of Ponte Serrada, *Krapovickas & Cristóbal 33703* (CTES); 16 km from BR-116, rd. to Campo Belo do Sul, *Krapovickas & Cristóbal 41978* (CTES, HAS, NY); Fraiburgo, *Krapovickas & Cristóbal 43980* (C, CTES, MBM); Pelotas River, *Krapovickas & Vanni 36872* (CTES, ICN, MBM, NY); 14 km E of Lajes, rd. to São Joaquim, *Krapovickas & Vanni 36887* (CTES, G, ICN, NY); Mun. Ponte Alta, Ponte Alta do Sul, *Krapovickas et al. 23050* (CONC, CTES, NY, P, TEX); Mun. Pôrto União, E of Valões Irineópolis) on rd. to Canoihas, *L. B. Smith & Klein 10720* (CTES, NY, US); Mun. Pôrto União, SE of Pôrto União, *L. B. Smith & Klein 10779* (US); Mun. Caçador, 8 km N of Caçador, *L. B. Smith & Klein 10952* (US). **SÃO PAULO:** Guapyra, *Brade 5972* (S–2); 9 km S of Atibaia, *Davidse & Ramamoorthy 10520A* (MO, NY); Mun. Botucatu, 10 km SE of the town Botucatu, *G. Gottsberger 12-25171* (CTES); Mun. Amparo, Monte Alegre, *Kuhlmann 130* (CTES, NY).

ARGENTINA. MISIONES: 7 km from Bernardo de Irigoyen, rd. to San Pedro, *Krapovickas et al. 23384* (BAA, BAB, CTES, LIL); Dept. Gral. M. Belgrano, Bernardo de Irigoyen, *Maruñak 93* (BAA, BAB, CTES, LL, NY, WIS). **CORRIENTES:** Dept. Capital, Corrientes [cult. *ex* Brazil], *Krapovickas 14494* (CTES, NY).

Pavonia guerkeana R. E. Fries was described with two varieties, var. *major* and var. *minor*, based on differences in leaf size. The former was based on *Regnell II-16* from Caldas in Minas Gerais, *Mosén 819* (Río Capivary, P as photo CTES), and *Dusén s.n.* (= *Dusén 9358*, S?) (Prov. Paraná) and the latter on *Widgren s.n.*, *Widgren 513* (both without locality), and *Regnell II-15b* (from Caldas). This presents a tangled question of typification, since *Regnell II-16* already typified *P. hastata* var. *longifolia* (= *P. schrankii*) and Regnell 15b typified *P. guerkeana*, thus rendering var. *minor* = var. *guerkeana*. Moreover, the Widgren collections represent yet another species. The problem of typification is complicated by the fact that numerous Regnell collections may be found with the collection number *15* or *15b* bearing a variety of collection dates 1845, 1861, 1864, 1873, and 1877 (see Krapovickas & Cristóbal, 1962: 40). The matter is further complicated by the fact that Fries cited *Regnell 15* and *Regnell 15a* (also from Caldas) as the basis for *P. sagittata* var. *conjugens* R. E. Fries. These names can be lectotypified by specimens in the Regnellian herbarium (S), but the matter of identifying isotypes will require careful study.

95. Pavonia hastata Cavanilles, Diss. **3:** 138, t. 47, f. 2. 1787. *Malache hastata* (Cavanilles) Kuntze, Revis. Gen. Pl. **1:** 70. 1891. *Lassa hastata* (Cavanilles) Kuntze, Revis. Gen. Pl. **3(2):** 19. 1898. Type. Uruguay. *Commerson s.n.* (holotype, P-JU no. 12340 as microfiche; isotypes, P-LA as microfiche, P as photo CTES; a specimen at NY labeled "Buenos Ayres et Monte Video, *Commerson 246*" may be type material; another specimen at C labeled "Commerson e monte video, ded. D. Thouin" may also be type material). Fig. 33

Sida bonariensis Willdenow *ex* Sprengel, Syst. Veg. **3:** 120. 1826. Type. Argentina. Buenos Aires, *Commerson s.n.* (holotype, B–W no. 12719 as microfiche).

Malva lecontei Buckley, Amer. J. Sci. **45:** 176. 1843. *Pavonia lecontei* (Buckley) Torrey & Gray, Mem. Amer. Acad. Arts n.s. **4** (Pl. Fendl.): 16. 1849. Type. United States. Georgia: *LeConte s.n.* (holotype, NY?–n.v.; isotypes, GH, PH–2).

Greevesia cleisocalyx F. Mueller, Trans. Proc. Victorian Inst. Adv. Sci. **1854/1855:** 115. 1855. Type. From the Melbourne Botanic Garden from seeds collected by Bunce on the Leichhardt expedition (lectotype, MEL, as photocopy).

Pavonia jonesii Feay *ex* Wood, Class-Book Bot. 269. 1862. Type. United States. Georgia: Liberty Co., *Jones 4468* (lectotype, MO *ex* herb. Redfield).

Pavonia hastata var. *pubescens* Gürke *in* Martius, Fl. Bras. **12(3):** 500. 1892. Treated by Hassler (1907: 737) as equivalent to "var. genuina" (i.e., var. *hastata*), in effect a lectotypification.

Pavonia hastata forma *longifolia* Gürke *in* Martius, Fl. Bras. **12(3):** 500. 1892. From the brief description and the context, this name was evidently intended by Gürke to represent the "typical" form, i.e., forma *hastata*, and is so interpreted here. The isonym published by Hassler (*Pavonia hastata* forma *longifolia* Hassler, Bull. Herb. Boiss., ser. 2, **7:** 738. 1907), was basically a new combination in quadrinomial form, i.e., *P. hastata* var. *pubescens* f. *longifolia* Gürke = *P. hastata* var. *genuina* f. *longifolia* (Gürke) Hassler.

Pavonia hastata subforma *pubescens* Hassler, Bull. Herb. Boiss., ser. 2, **7:** 738. 1907. Type. Paraguay. Near Villarica, Jan 1905, *Hassler 8760* (isotypes, K, MO, NY, S).

Pavonia hastata subforma *tomentosula* Hassler, Bull. Herb. Boiss., ser. 2, **7:** 738. 1907. Type. Paraguay. Sierra de Maracayú, Jejuiguazú River, *Hassler 5684* (lectotype, here designated, NY; isolectotypes, GH, K, MO, S).

Subshrubs ca. 0.5 m tall, the stems with minute stellate pubescence, the hairs 0.1 mm long. *Leaf* blades up to 7 cm long (usually smaller), ovate-triangular to hastate-oblong, basally cordate or hastate with an open acute sinus, coarsely crenate, acute or subobtuse, discolorous, minutely stellate-pubescent, more densely so beneath; petioles ⅓ the length of the blade or less; stipules ca. 2 mm long, subulate. *Flowers* sometimes cleistogamic, the pedicels solitary in the leaf axils, 17–23(–40) mm long; involucellar bracts 5, ovate, 4–6 × 2–3 mm; calyx 6–8 mm long, stellate-pubescent; corolla 15–22 mm long, deep rose with a dark center. *Fruits* ca. 6 mm diam., oblate, minutely pubescent, the mericarps ca. 4 mm long, dorsally rounded and keeled, reticulate-costate. Chromosome number, $2n = 56$ (Krapovickas & Cristóbal, 1962).

Distribution (Fig. 39). From Argentina, Bolivia, Brazil, Paraguay, Uruguay, and as an adventive or introduction elsewhere (Australia, Hawaii [where cultivated and escaped], Mexico, United States).

Illustrations. Cavanilles (1787: pl. 47, fig. 2); Fries (1908: pl. 6, figs. 7–10; pl. 7, fig. 3); Fryxell (1979: fig. 3E, as *P. nepetifolia*); Gürke (1892b: pl. 95, fig. 1); Krapovickas (1953: fig. 107A, C); Reichenbach (1830, **3:** pl. 227).

Representative specimens examined. **UNITED STATES. Florida:** *Buckley s.n.* (C, MO, NY). **In cultivation:** Cambridge Botanic Garden, Aug 1862, *Gray s.n.* (NY); Missouri Botanical Garden, St. Louis, 25 Aug 1911, *Kellogg s.n.* (MO).

MEXICO. Socorro Island: S of Academy Bay, *Moran 5868* (RSA). **Tamaulipas:** Mun. Aldama, region of Rancho Las Yucas, ca. 40 km NNW of Aldama, Sierra La Borada (Sierra de San Juan) ca. highest point in Sierra, W of Las Yucas, *Dressler 1988* (MO).

BRAZIL. Mato Grosso do Sul: 1 km from Amambay River, *Gregory et al. 9660* (C, CTES); 9 km W of Caarapó, *Krapovickas & Cristóbal 34323* (CTES, F, SI). **Paraná:** Mun. Londrina, Leroville-Campo das Pedras, *Hatschbach 22910* (US); Mun. Campo Mourão, Barragem rio Mourão, *Hatschbach 44843* (CTES). **Rio Grande do Sul:** Mun. Arroio do Tigre, Barragem de Itaúba, *Bueno et al. 667* (CTES); Mun. Rio Pardo, 10 km W of Pantano Grande, Arroio do Marcos, BR-290 km 134, *Krapovickas & Cristóbal 38391* (AK, CTES, HAS, LIL, UC); roadside 56 km from Santiago, 94 km from S. Borja, *Lindeman et al. 21126* (CTES, U); Porto Alegre, *Rambo 203* (LL–2, MO, NY, TEX, US); Caí, *Rambo 32848* (CTES); Vasconcellos Jardim, Canoas, *Rambo 41389* (CTES, NY). **São Paulo:** Indaiatuba [S of Campinas], *Hambleton 9* (CTES); Sete Barras, Carlos Botelho Biological Reserve, *Sakane 575* (NY, SP); Cunha, *Viégas 5805* (CAS).

BOLIVIA. Santa Cruz: Comarapa, *Cárdenas 4623* (CAS); Samaipata, *Krapovickas 8968* (CTES). **Tarija:**

Prov. Arce, rd. between El Mezón and La Mamora, *Legname et al. 4843* (CTES); Prov. Arce, close to La Capilla, *Meyer et al. 21747* (CTES).

PARAGUAY. Central Paraguay, near Caballero, *Morong 428* (NY); N of Tobatí, *Fiebrig 812* (A, E–3, F, K). **CAAGUAZÚ:** 15 km N of Caaguazú, *Hahn et al. 887* (CTES, MO); 46 km N of Coronel Oviedo, *Krapovickas & Cristóbal 44875* (CTES, MO, TEX); 2 km E of Caaguazú, *Krapovickas et al. 12520* (BH, CTES, G, P); Guyraunguá, *Sparre & Vervoorst 2149* (NY). **CAAZAPA:** Tavai, *Soria 2545* (CTES, MO). **CANENDIYÚ:** 3 km N of Itambú, *Hahn et al. 998* (CTES, MO, NY). **GUAIRÁ:** Itapúa, Trinidad, *Krapovickas & Cristóbal 44513* (CTES, TEX); Costa Jhú, *Montes 16532* (CTES); Borja, *Montes 16332* (CTES); Colonia Independencia, *Rojas 4702* (AS, CTES, NY). **LA CORDILLERA:** 10 km E of Ayala, *Krapovickas & Cristóbal 13257* (CTES, P); 4 km from Itacurubí, *Krapovickas & Cristóbal 13299* (CTES). **MISIONES:** 3 km S of Santiago, Estancia La Soledad, *Pedersen 3214* (C, NY, US, herb. Pedersen). **ÑEEMBUCÚ:** Between Dist. Naranjito and Est. Espinillo, *Schulz 7797* (CTES). **PARAGUARÍ:** Villarica, *Jörgensen 3808* (A, F, K, MO, NY, PH, S, US); spine shrubland on the banks of Caañave Creek, *Rojas 3284* (AS, CTES, NY). **SAN PEDRO:** Primavera, *Woolston 782* (C, NY, S).

ARGENTINA. BUENOS AIRES: Pdo. de La Plata, Punta Lara, *Cabrera 6348* (F, NY, TEX, US); Pdo. General Uribaru, Atucha, *Krapovickas 3268* (CTES); Colonia del Sacramento in Buenos Aires, *L. Née s.n.* (C). **CHACO:** Dept. Resistencia, Margarita Belán, *Aguilar 657* (CTES); Puerto Bermejo, *Rojas 11862* (NY); Dept. 1° de Mayo, Colonia Benítez, *Schulz 547* (CTES, F, NY); Potrero Mendiburu, *Schulz 4030* (CTES, MO, NY). **CORRIENTES:** Dept. San Cosme, Costa Toledo, *Krapovickas & Cristóbal 11338* (CTES, F, MEXU); Dept. Empedrado, Empedrado, *Krapovickas & Cristóbal 11413* (CTES, F); Dept. Paso de los Libres, Laguna Mansa, *Krapovickas & Cristóbal 21661* (BAA, CTES, NCU, NY); Dept. Santo Tomé, Estancia San Francisco (23 km NW of Gobernador Virasoro), *Krapovickas et al. 16865* (BAA, C, CTES, IAC, LL, MBM, MO, NY, WIS); Estancia Garruchos, Arroyo Chimiray, *Krapovickas et al. 21113* (BA, BAA, CTES, F, GH); Dept. Itazuingó, Isla Apipé Grande, Puerto San Antonio, *Krapovickas et al. 24061* (BH, CONC, CTES, ENCB, G, LIL, MBM, RB); Dept. Gral. Paz, 12 km E of Itá Ibaté, banks of Paraná River, *Mroginski et al. 597* (CTES, F, NY, TEX); Dept. Concepción, Estancia "Fortin del Iberá," *Pedersen 8753* (C, CTES, MO, NY, herb. Pedersen); Dept. Itatí, rt. 12, rd. to Itatí 8 km, *Schinini 4579* (BAA, CTES, F, LIL, LP, NY); Ea. Santa Rita, *Tressens et al. 3286* (CTES, NY); Arroyo Santa Isabela and rt. 12, *Vanni & Radovancich 992* (CLEMS, CTES, HAS, NY). **ENTRE RÍOS:** Delta del Paraná, Arroyo Martínez, *Boelcke 861* (US); Concepción del Uruguay, *Lorentz 1262* (PH); Dept. Uruguay, Estancia "La Selmira," low land by the Uruguay River, *Pedersen 7225* (C, K, MO, herb. Pedersen); Dept. La Paz, Piedras Blancas, *Pedersen 8268* (C, CTES, K, herb. Pedersen); Colón, *Pedersen 8999* (C, CTES, herb. Pedersen). **FORMOSA:** Dept. Capital, Salado River, rt. 11, *Krapovickas & Cristóbal 13218* (CTES, F); Dept. Pilcomayo, 4 km NW of Siete Palmas, *Morel 8608* (CAS, CTES). **JUJUY:** Rd. to Las Capillas, *Legname 3237* (CTES). **MISIONES:** Dept. San Juan, Posadas, Bonpland, *Ekman 168* (S); Leandro N. Alem, rt. from Cerro Azul to San José, at 100 m from Liso Creek, *Maruñak 37* (CTES, NY, WIS); Santa Ana, *Montes 1552* (C, COLO, NY); Loreto, *Montes 14729* (CTES, F, NY, US); Dept. San Ignacio, Teyuenaré, *Montes 14797* (NY, TEX); Dept. San Pedro, Monte Carlo, *Montes 27710* (NY, TEX); Dept. Candelaria, between Loreto and San Ignacio, *Quarín et al. 160* (CTES, NY); Dept. San Javier, Puerto Rosario, *Schwarz 3880* (CAS, TEX); Dept. Cainguas, Campo Grande, *Vanni & Cáceres 688* (CTES, NY). **SALTA:** 8 km N of Salta, *Krapovickas & Schinini 30389* (CTES, LIL, MO). **SANTA FÉ:** Dept. San Jerónimo, Isla Campo Rico, *Franceschi 282* (CTES); Rosario, *Schulz 9101* (C, CTES).

URUGUAY. COLONIA: Puerto Colonia, *Archer 4972* (US); Colonia, *Krapovickas 6576* (CTES); Carmelo, *Rosengurtt et al. 8579* (CTES). **MALDONADO:** Piriápolis, Cerro San Antonio, *Krapovickas & Cristóbal 11133* (CTES); Solís, *Krapovickas & Cristóbal 11165* (CTES, HAS); Cerro Pan de Azúcar, *Lavia et al. 20* (CTES, TEX). **MONTEVIDEO:** Montevideo, Apr 1836, *Gaudichaud s.n.* (NY); Miguelete, *Herter 253* (F, MO, NY, S, U); between Sayago and Colón, *Herter 2502* (S); Montevideo, *L. Née s.n.* (MA as photo F-29773); Barra Sta. Lucia, *Osten 5297* (US). **PAYSANDÚ:** Meseta Artigos, *Rosengurtt 9309* (CTES). **SALTO:** Rt. 31, by the bridge on Río Itapebi, *Pedersen 15643* (C–2, CTES, MO, NY). **SAN JOSÉ:** Playa Fomento, *Arrillaga et al. 2689* (CTES, NY); Kiyú, *Del Puerto 4* (CTES). Arazatí, *Rosengurtt 8133* (CTES). **TACUAREMBÓ:** Rincón del Bonete, *Krapovickas & Cristóbal 11195* (C, CTES, F, G, K, MBM, MEXU, MO, NY).

Local names and uses. Escotadura rosella (Argentina: Misiones), rosellina (Argentina: Misiones). Mitchell (1982) reports economic aspects of *Pavonia hastata* in Australia as an ornamental and as a weed.

96. Pavonia hatschbachii Krapovickas, Trabalh. XXVI Congr. Nac. Bot. Brasil 315. 1977. Type. Brazil. Paraná: Mun. Palmas, Horizonte, *Hatschbach 28137* (holotype, CTES; isotypes, C, US). Fig. 33

Shrubs 1.5 m tall, the stems densely stellate-pubescent, the hairs 1–1.5 mm long. *Leaf* blades sagittate-oblong, mostly 2.5–6 × 1.5–2 cm, basally deeply sagittate, the lobes parallel to and often equaling the petiole, coarsely crenate to subentire, acute to subobtuse, discolorous, softly tomentose on both surfaces (hairs 1 mm long), palmately 5–7-nerved; petioles 0.5–1.5 cm long; stipules filiform, 4 mm long. *Flowers* solitary in the leaf axils, the pedicels 1.5–2.5 cm long; involucellar bracts 5, broadly ligulate, 7–12 × 2.5–6 mm, densely pubescent, often with prominent nerves; calyx 8–12 mm long, pubescent; corolla 1.5–2.5 cm long, rose with darker veins; genitalia included. *Fruits* 8–9 mm diam., minutely pubescent; mericarps 4–4.5 mm high, carinate, prominently rugose with 3–4 excrescences on each side.

Distribution (Fig. 44). From Brazil (Paraná) and Uruguay. The type locality is approximately at. 26°30′S, 52°0′W.

Illustrations. Krapovickas (1977: fig. 2F-H).

Specimens examined. BRAZIL. PARANÁ: Mun. Palmas, S. Agostino, *Hatschbach 15411* (C, CTES, F, NY, US); Morro da Baliza, *Hatschbach 30728* (CTES).

URUGUAY. MONTEVIDEO: Carrasco, *Legrand 2844* (CAS).

97. Pavonia hoehnei E. G. Baker, J. Bot. **63:** 239. 1925. Type. Brazil. Minas Gerais: Caeté, *Hoehne 6134* (holotype, BM as photo CTES, as photo NY-317). Fig. 33

Subshrubs 0.6 m tall, the stems densely stellate-pubescent, the hairs 0.2–0.5 mm long, yellowish. *Leaf* blades mostly 4–8 × 2–3 cm (7–10 × 4.5–5.5 cm in type), oblong to lance-ovate, basally cordate or subcordate (scarcely sagittate), crenate, acute or rounded-acute, discolorous, basally palmately 5(–7)-nerved, both surfaces evenly stellate-pubescent (densely so beneath), the hairs 0.5 mm long, the lower surface with raised reticulate venation; petioles 1–1.5 cm long, evenly pubescent like stem; stipules filiform, 3–4 mm long, densely pubescent, deciduous. *Flowers* solitary in the leaf axils, the pedicels 1–3 cm long (exceeding the petioles), densely pubescent like stem; involucellar bracts 5, broadly ovate or suborbicular, 5–10 × 5–9 mm, basally narrowed to a stipe and 3–5-ribbed, stellate-pubescent; calyx 7–9 mm long, the lobes 4–5 mm wide, acute, stellate-pubescent; corolla 1.5–2 cm long, rose (?). *Fruits* ca. 7 mm diam., minutely pubescent; mericarps 4.5–5 mm high, prominently rugulose.

Distribution (Fig. 44). From central Brazil (Bahia, Minas Gerais, and São Paulo) and Uruguay. The type locality is at 19°52′S, 43°40′W.

Specimens examined. BRAZIL. BAHIA: Iracema, region of the Serra de Sincorá, *Fróes 20184* (NY). **MINAS GERAIS:** Belo Horizonte, *Gehrt [SP-3208]* (CTES); Serra do Curral, S of Belo Horizonte, *L. O. Williams & Assis 6370* (GH). **SÃO PAULO:** S. Caetano, *Brade 7364 [SP-6880]* (CTES, NY).

URUGUAY. Bords du St. Lucia (Banda oriental), *Gibert 691* (K).

Pavonia hoehnii is notable for its suborbicular, stipitate involucellar bracts that are 5–9 mm wide and also its relatively broad calyx lobes. This species may be related to *P. belophylla* and *P. pulchra*.

98. Pavonia kleinii Krapovickas & Cristóbal, Lilloa **31:** 70. 1962. Type. Brazil. Santa Catarina: Mun. Lajes, 34 km NE of Bocaino do Sul, 11 Feb 1957, *L. B. Smith & Klein 11265* (holotype, LIL–n.v.; isotypes, NY, US). Fig. 33

Subshrubs 0.5 m tall, the stems stellate-pubescent. *Leaf* blades ovate-oblong, mostly 2–5 × 1–2.5 cm, basally cordate, obscurely crenate or undulate, acute or subobtuse, palmately 5-nerved, discolorous, densely canescent beneath, sparsely pubescent above with simple hairs 0.5–1 mm long and some stellate hairs; petioles 1–3 cm long; stipules subulate, 2–3 mm long. *Flowers* solitary in the leaf axils, the pedicels 1–2 cm long; involucellar bracts 5, lance-ovate, 5–6 × 2–3 mm, stellate-pubescent; calyx 4–5 mm long, densely hirsute; corolla yellow, ca. 1 cm long. *Fruits* nodding and hidden by leaves, ca. 6 mm diam., minutely pubescent; mericarps 4 mm high, more or less carinate and prominently rugose.

Distribution (Fig. 34). From central and S Brazil (Minas Gerais, São Paulo, Rio Grande do Sul, Santa Catarina). The type locality is at 27°35′S, 49°43′W.

Illustrations. Krapovickas & Cristóbal (1962: figs. 1A, 15).

Specimens examined. BRAZIL. MINAS GERAIS: Mun. Cananducaia, Monte Verde, *Leitão et al. 1834* (NY). **RIO GRANDE DO SUL:** Mun. Cambará do Sul, Itaimbezinho [29°3′S, 50°6′W], *Krapovickas & Cristóbal 38408* (CTES, G, HAS, NY). **SANTA CATARINA:** Serra da Boa Vista, São José, *Reitz & Klein 10215* (CTES); Mun. Bom Retiro, Campo dos Padres, *L. B. Smith & Klein 7808* (NY, US), *L. B. Smith & Reitz 10393* (US). **SÃO PAULO:** Serra do Bocaina, *Brade 21186* (CTES, NY).

Pavonia kleinii is distinguished for its yellow flowers, a character shared in *P.* subsect. *Hastifoliae* only with the Mexican *P. nepetifolia*. Elsewhere in *P.* sect. *Lebretonia*, yellow flowers are also found in *P. aurantia* and *P. rojasii*.

99. Pavonia lanata R. E. Fries, Kongl. Svenska Vetenskapsakad. Handl. **42(12):** 52. 1908. Type. Brazil. *Sellow s.n.* (holotype, B, destroyed, as photo F-9451). Fig. 33

Decumbent *subshrubs*, the stems purplish, with scattered stellate hairs ca. 1 mm long. *Leaf* blades sagittate, mostly 3–7 × 1.5–3 cm, deeply cordate-sagittate basally (the lobes sometimes equaling the petiole), clearly serrate to obscurely undulate-serrate or subentire, acute, palmately 7-nerved, notably discolorous, evenly stellate-pubescent on upper surface (hairs 0.5 mm long), densely matted-canescent beneath; petioles 8–15 mm long; stipules filiform, 3–4 mm long. *Flowers* solitary in the leaf axils, the pedicels ca. 1 cm long, slender; involucellar bracts 5, lance-ovate, 8–9 × 3–4 mm, softly tomentose; calyx subequal to involucel, softly tomentose; corolla ca. 1 cm long, rose with darker veins. *Fruits* 6–7 mm diam., minutely pubescent; mericarps rugulose, 4 mm high, apiculate at apex. Chromosome number, $2n$ = 56 (Fernández, 1981).

Distribution (Fig. 34). From S Brazil (Minas Gerais, Paraná, Rio Grande do Sul, and Santa Catarina).

Illustrations. Fries (1908: pl. 6, figs. 5, 6).

Specimens examined. **BRAZIL. Minas Gerais:** Mun. Belo Horizonte, near Lagoa Pampulha, *L. O. Williams & Assis 6059* (GH). **Paraná:** Mun. Colombo, Bacaitava, *Hatschbach 7316* (CTES); Mun. São José dos Pinhaes, Hwy. Gob. Lupion, Pequeño River, *Hatschbach 8550* (CTES); Pequeño River (S. José dos Pinhais), *Hatschbach 22831* (C, CTES, LL, NY, S); Pequeño River, *Hatschbach & Fontella 20813* (CTES); Mun. Tijucas do Sul, Campina, 46 km S of Curitiba, *Krapovickas & Cristóbal 33655* (CTES, F, SI, herb. Pedersen); Mun. Colombo, Campestre, *Nicolack & Ribas 107* (CTES). **Rio Grande do Sul:** Aparados da Recinha, near Bom Jesus, *Rambo 45491* (CAS, CTES, NY, S). **Santa Catarina:** Campo Alegre, *Reitz & Klein 5150* (NY, US); Morro do Iquererim, Campo Alegre, *Reitz & Klein 6466* (CTES, US).

Pavonia lanata is notable for having strikingly discolorous, sagittate leaves. In *P.* subsect. *Hastifoliae* only five other species have prominently sagittate leaves, viz., *P. hatschbachii*, *P. reticulata*, *P. rhodantha*, *P. rosengurttii*, and *P. sagittata*, although a few others have a weak tendency in this direction (e.g., *P. belophylla* and *P. hoehnii*). Of these, *P. hatschbachii* is distinguished from *P. lanata* in the key to the species of subsect. *Hastifoliae* (couplet 14); *P. reticulata* has concolorous, subglabrous leaves; *P. rhodantha* has red corollas; *P. rosengurttii* has red-veined calyx lobes; and *P. sagittata* has coarsely hirsute stems (hairs ferruginous, 1–2.5 mm long), all in contrast to the character states in *P. lanata*.

100. Pavonia missionum Ekman, Ark. Bot. **9(4):** 33. t. 7. 1909. Type. Argentina. Misiones: Posadas, Bonpland, *Ekman 173* (lectotype, here designated, S; isotypes, MO, NY, S–2). Fig. 33

Shrubs 1–2 m tall, the stems minutely glandular-pubescent and sometimes with occasional setae 1 mm long. *Leaf* blades ovate to ovate-triangular, rarely obscurely 3–5-lobed, mostly 2–6(–9) × 1.5–3.5(–8) cm, basally more or less cordate, crenate-serrate, acute, discolorous, the upper surface with simple appressed hairs 0.5–1 mm long, the lower surface sparsely to moderately pubescent with stellate and simple hairs (especially on veins); petioles 1–4(–7) cm long; stipules subulate, 3–4 mm long. *Flowers* solitary in the leaf axils, the pedicels 1–5 cm long; involucellar bracts 5, lanceolate, 6–7 × 1.5–2.5 mm, glandular-pubescent; calyx 9–11 mm long (to 15 mm in fruit), deeply divided, glandular-pubescent and ciliate, sometimes with reddish-pigmented veins; corolla 1.5–2 cm long, red, glabrous on claw and throughout; staminal column 7–10 mm long, glabrous, the filaments 0.5–1 mm long. *Fruits* 8–10 mm diam., obscurely pubescent; mericarps 4–5 mm high, carinate and prominently reticulate-rugose. Chromosome number, $2n$ = 56 (Krapovickas & Cristóbal, 1962).

Distribution (Fig. 47). From Brazil (Paraná, Rio Grande do Sul, Santa Catarina), Argentina (Corrientes, Misiones), and Paraguay. The type locality is approximately at 27°30′S, 55°32′W.

Illustrations. Ekman (1909: fig. 7); Krapovickas & Cristóbal (1962: fig. 1I).

Representative specimens examined. **BRAZIL. Paraná:** BR-277, km 417, *Ferrucci et al. 359* (CTES, MBM, NY); Mun. Laranjeiras do Sul, Campo Novo, *Hatschbach & Guimarães 20569* (C, CTES, F, MO, NY); Mun. Campo Mourão, *Hatschbach 8870* (CTES, U); Mun. Guarapuava, Candoi, *Hatschbach & Ribas 52562* (C, CAS, MO, MU, US); Capinzal River, *Dusén 17804* (S); Cavernoso River, *Hatschbach et al. 13376* (CTES, NY, US), *Reitz & Klein 17703* (CTES, F, US). **Rio Grande do Sul:** 53 km W of Vacaría, Santa Rita River, *Krapovickas & Vanni 37034* (CTES, ICN, NY, WIS); Mun. Giruá, 11 km Giruá-Santa Rosa, *Bueno et al. 3982* (CTES); 17 km E of Lagoa Vermelha, *Krapovickas & Schinini 38225* (CTES, HAS); Iraí, in front of Balneario Oswaldo Cruz, *Krapovickas & Cristóbal 44006* (CTES, MBM); 13 km SE of Santa Rosa, *Lindeman et al. 8975* (CTES, U); 18 km E from Santa Barbara del Sul, *Krapovickas & Vanni 36782* (CTES, ENCB, ICN); Cerro Largo, near S. Luis, *Rambo 53173* (LL, S). **Santa Catarina:** Mun. São Miguel Oeste, Rio das Flores, *Castellanos 3383* (CTES); São Luis, Guaraciaba, *Reitz & Klein 16952* (CTES).

ARGENTINA. Corrientes: Garruchos, *Huidobro 5350* (CAS, LL, RSA, S); Dept. Santo Tomé, 29 km E of rt. 14, rd. to Colonia Garabí, *Krapovickas et al. 16997* (CTES, NY). **Misiones:** Dept. Apostoles, San Juan, *Huidobro 4956* (MO, NY); Cerro San Juan, *Krapovickas & Cristóbal 12051* (BH, CTES, K, P, US); Dept. Candelaria, Santa Ana, *Krapovickas et al. 14835* (C, CTES, LL, MBM, MO, P, UC, WIS, herb. Pedersen); Dept. San Pedro, Macao Creek, 4–5 km before San Pedro, *Maruñak 26* (CTES, LL, NY, UC, WIS); Dept. Gral. Belgrano, 5 km S of Bernardo de Irigoyen, *Maruñak 90* (C, CTES, HUT, L, MBM, P, UC, WIS); Dept. Iguazú, 20 km E of Wanda, *Schinini & Fernández 6023* (CTES, G, RB); Dept. San Ignacio, Puerto Cajador, *Schwarz 1985* (CTES, MO); Dept. Guaraní, rt. 14, km 304, near Chafariz Creek, *Schwindt 3186* (BRIT, C, CAS, K, TEX); Dept. Cainguas, 3 km S from Campo Grande, rd. to A. Posse, *Vanni & Cáceres 680* (CTES, NY); Dept. Oberá, rt. 14 and Yazá Creek, *Vanni et al. 572* (CTES); Dept. Montecarlo between San Pedro and Pro. Piray, *Zuloaga et al. 2079* (NY, SI).

PARAGUAY. ALTO PARANÁ: Tati Yupi Biological Reserve, *Marmori 244* (CTES); Puerto Kurupa'yty, *Marmori 1056* (CTES); Encarnación, *Rojas 10943* (CTES, NY). **GUAIRÁ:** Borja, *Montes 16219* (CTES).

Pavonia missionum is distinguished by its red corollas and red-veined calyx lobes, together with characteristic leaves, which are ovate-triangular to weakly lobed and relatively sparsely pubescent with appressed simple hairs on the upper surface.

101. Pavonia nana R. E. Fries, Bull. Herb. Boiss., ser. 2, **7:** 999. 1907. *Pavonia hastata* subforma *nana* (R. E. Fries) Hassler, Ostenia 341. 1933. Type. Uruguay. Durazno: Cuesta de Cuadras, *Osten 4299* (holotype, G; isotypes, CTES [fragment], MVM–n.v.). The type was cited by Fries in the protologue as "4229." Fig. 33

Decumbent-ascending *herb* or subshrub, the stems stellate-pubescent, the hairs 0.5–1 mm long. *Leaf* blades triangular to oblong, mostly 1–3.5 × 0.6–1.5 cm, basally hastate or sagittate, crenate-dentate, acute, coarsely stellate-pubescent on both surfaces, the hairs ca. 1 mm long; petioles 0.5–1.5 cm long; stipules subulate, 2–4 mm long. *Flowers* solitary in the leaf axils, the pedicels 1–2.5 cm long; involucellar bracts 5, broadly lanceolate, 6–8 × 1.5–2.5 mm, hirsute; calyx 6–7 mm long, more or less hirsute (hairs 0.5–1 mm long), the nerves with reddish pigmentation; corolla 12–15 mm long, rose with dark red center; staminal column 3–4 mm long, reddish, the filaments ca. 1 mm long; styles emergent from the column, purplish. *Fruits* oblate, ca. 6 mm diam., minutely pubescent; mericarps carinate, rugulose, 4–5 mm high.

Distribution (Fig. 40). From S Brazil and Uruguay. The type locality is at 33°30'S, 56°35'W.

Illustrations. Fries (1908: pl. 4, figs. 3–6).

Specimens examined. BRAZIL. RIO GRANDE DO SUL: Mun. Alegrete, rd. from Chão to Livramento, *Bueno 3260* (CTES). **SANTA CATARINA:** Cambajuva-S. Joaquim, *Reitz 3271* (CAS, S); Passo das Contas, *Reitz 3320* (CAS).

URUGUAY. TACUAREMBÓ: Rincón del Bonete, *Krapovickas & Cristóbal 11209* (CTES, F, NY, SI); Villaboas al Sur, on the rd. from Durazno to Paso de los Toros, *Rosengurtt 4715* (CTES).

Pavonia nana is a decumbent plant of low stature, having red-veined calyx lobes and small coarsely stellate-pubescent leaves.

102. Pavonia nepetifolia (Standley) Standley, Publ. Field Columbian Mus., Bot. Ser. **4:** 231. 1929. *Malache nepetaefolia* Standley, Contr. U.S. Natl. Herb. **23:** 772. 1923. Type. Mexico. Coahuila: General Cepeda, 7 Oct 1905, *Pringle 13698* (holotype, US; isotypes, CTES [fragment], GH, MICH, US, VT–2). Fig. 33

Subshrubs perhaps 30 cm tall, the stems densely stellate-pubescent, the hairs 6–12-armed, 0.2–0.5 mm long. *Leaf* blades 1.5–3 × 1–2(–3) cm, ovate or oblong to triangular-hastate, cordate, coarsely crenate-dentate, obtuse to acute, discolorous, minutely pubescent with mostly simple hairs above, mostly stellate hairs beneath; petioles subequal to blades or somewhat shorter; stipules 2–4 mm long, filiform. *Flowers* solitary in the leaf axils, the pedicels 1–1.5 cm long, pubescent (including a few glandular hairs); involucellar bracts 5, elliptic, 5–9 × 2–4 mm, densely pubescent, the pubescence sometimes including glandular hairs; calyx 9–10 mm long, minutely and densely stellate-pubescent, the lobes cordate-plicate, 5–6 mm wide, sometimes purplish; corolla 9–12 mm long, yellow, bearded on margins of claw; staminal column ca. 4 mm long, glabrous, with ca. 10 yellowish anthers, the filaments 1–2 mm long; styles emergent from the column, sparsely hairy, the stigmas reddish. *Fruits* 5–6 mm diam., puberulent, the mericarps ca. 4 mm long, rounded, reticulate-veined, carinate.

Distribution (Fig. 49). Found in NE Mexico in the states of Coahuila, Guanajuato, Nuevo León, San Luis Potosí, and Tamaulipas. Its distribution is principally extra-tropical, with two tropical records (*E. Ventura & E. López 6734* and *E. Palmer 579*). The type locality is at 25°28'N, 101°32'W.

Illustration. Krapovickas (1977: fig. 2D).

Specimens examined. MEXICO. COAHUILA: Mun. Villa Acuna, Serranías del Burro, Rancho El Bonito, *Riskind 2363* (TEX); Mun. Múzquiz, Rancho Agua Dulce, Sierra de San Manuel, *Wynd & Mueller 336* (ARIZ, CTES [fragment], MO, NY, US). **GUANAJUATO:** Mun. Xichú, Cerro Gordo, 12 km N of Molinitos, *E. Ventura & E. López 6734* (IEB, NY). **NUEVO LEÓN:** 3 km E of Pablillo, *Hinton et al. 18881* (TEX); Zaragoza, Cerro Viejo, *Hinton et al. 23405* (TEX); Dulces Nombres, just E of Tamaulipas border, *Meyer & Rogers 2713* (CTES, MO); Hacienda Pablillo, Galeana, *Taylor 120* (CTES [fragment], F). **SAN LUIS POTOSÍ:** Alvarez, 1905, *E. Palmer 579* (US). **TAMAULIPAS:** Mun. Bustamante, 3.2 mi N of La Presita and 1.1 mi S of rd. to Bustamante, *Dorr et al. 2652* (NY, TEX); Ejido Ricardo García or La Presita, km 66 on hwy. from Victoria to Tula, *M. Martínez 377* (NY).

Pavonia nepetifolia is distinctive for its yellow corollas, a character it shares with *P. kleinii* from

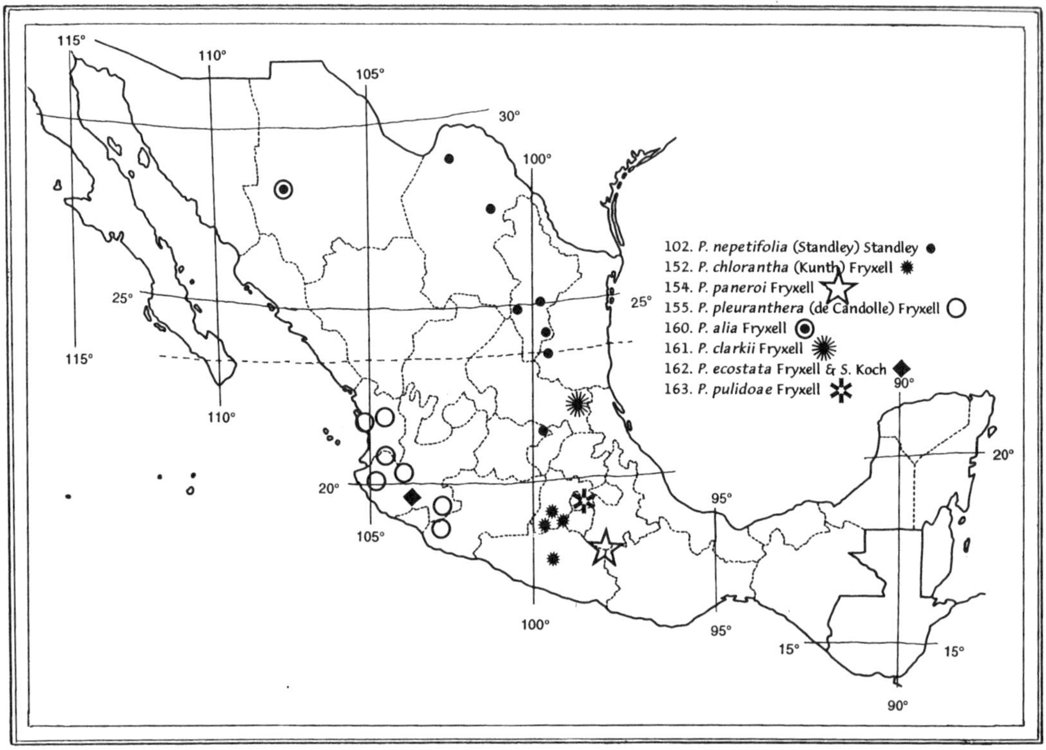

FIG. 49. Distribution map of eight Mexican species of *Pavonia* subsect. *Hastifoliae* and *P.* sect. *Collicolae*.

Brazil, the only other member of *P.* subsect. *Hastifoliae* to have this flower color. It also has relatively low stature and relatively small leaves.

According to Pringle's diary (Davis, 1936: 227), on 7 Oct 1905 Pringle was collecting "at the base of mountains about a dozen miles south" of General Cepeda. On that day he mentions collection numbers from 10039 to 10088 but no number as high as that of the type of *Pavonia nepetifolia* (no. 13698). Collections in the days preceding and following 7 Oct are also in the ten-thousand series. How this discrepancy arose is not clear.

103. Pavonia orientalis Krapovickas, Trabalh. XXVI Congr. Nac. Bot. Brasil 320. 1977. Type. Uruguay. Maldonado: Cerro Pan de Azúcar, 2 Jan 1964, *Krapovickas & Cristóbal 11178* (holotype, CTES; isotypes, CTES, GB, HB, K, LIL, MBM, MO, NY, S, SI, TEX, UC, US, WIS). Fig. 33

Shrubs or subshrubs 0.5–1 m tall, the stems stellate-pubescent (almost scabridulous), the hairs 0.1–0.2 mm long (the hairs to 0.4 mm long and denser at the stem tips). *Leaf* blades ovate-triangular to hastately lobed, mostly 1.5–3 × 1–1.5 cm, basally cordate (or hastately lobed), dentate, acute, palmately 5-nerved, discolorous, the upper surface evenly stellate-pubescent, the hairs 1 mm diam., the lower surface densely stellate-pubescent, the hairs of variable size; petioles 1–2 cm long, often pilose distally; stipules subulate to filiform, 4–7 mm long. *Flowers* solitary in the leaf axils, the pedicels 1–3 cm long; involucellar bracts 5, lanceolate or spatulate, 5–6 × 1–2.5 mm, stellate-pubescent (sometimes also setose-hirsute); calyx 5–7 mm long, stellate-pubescent (sometimes also setose-hirsute); corolla 1–1.5 cm long (flowers sometimes cleistogamous). *Fruits* 6 mm diam., minutely pubescent; mericarps 3.5–4.5 mm high, carinate and rugulose but not prominently so.

Distribution (Fig. 44). From Uruguay and southernmost Brazil. The type locality is at 34°45′S, 55°15′W.

Illustrations. Krapovickas (1977: fig. 3G-H).

Specimens examined. **BRAZIL.** Without locality, *Sellow s.n.* (VT). **Rio Grande do Sul:** Pelotas, *da Costa Sacco 471* (F), *da Costa Sacco 1015* (F); 5 km from Pinheiro Machado, on hwy. to Bagé, *Mattos 18765*, *18771* (CTES); near Pelotas, BR-2, km 185, *Trinta 1142* (CTES).

URUGUAY. La Valleja: 15 km S of Minas, *Krapovickas 4832* (CTES); 5 km S of Minas, *Krapovickas*

4833 (CTES, GB); Pororó, *Krapovickas 4856* (CTES, LIL, WIS); La Valleja, Paradero Pororó (Aiguá), *Krapovickas 4861* (CTES, MO); La Valleja, *Krapovickas 16171* (BAB, CTES, G, NY), *Krapovickas & Cristóbal 16164* (CTES); Pororó, *Rosengurtt 5309* (CTES). **MALDONADO:** Piriápolis, Cerro del Toro, *Krapovickas & Cristóbal 11144* (CTES, NY); La Barra, "El Placer" Park, *Krapovickas & Cristóbal 11183* (BAA, CTES, G, NY), *Krapovickas & Cristóbal 11184* (CTES, RB, NY), *Krapovickas & Cristóbal 11185* (CTES, S); Pan de Azúcar, *Krapovickas & Cristóbal 11170* (CTES, NY), *Krapovickas & Cristóbal 11176* (CTES, S), *Krapovickas & Cristóbal 11177* (CTES, G, NY). **MINAS:** Mtn. opposite Jungfrau Mtns., *Herter 16983* (US). **ROCHA:** San Vicente, *Wetmore 858* (US). **SAN JOSÉ:** Santa Lucía River, *Del Puerto 2254* (CTES); desembocadura or. Cufré, *Lema 6319* (CTES).

Pavonia orientalis is characterized by its relatively narrow involucellar bracts, coupled with minutely pubescent stems and petioles that are often distally pilose. The name indicates its relatively easterly distribution.

104. Pavonia pedersenii Fryxell, sp. nov. Type. Brazil. Rio Grande do Sul: Mun. Restinga Seca, near Restinga Seca, in dried-out swamp, 21 Jan 1994, *Pedersen 15902* (holotype, TEX; isotypes, C–n.v., CTES–n.v., herb. Pedersen–n.v.).

Frutices caulibus dense stellato-pubescentibus; laminis foliorum oblongis 2–3 cm longis dense et minute stellato-pubescentibus; floribus solitariis; bracteis involucellorum 5, 10–11 mm longis; calycibus 8–9 mm longis; corollis subroseis; fructibus stramineis valde rugulosis.

Shrubs 1 m tall, the stems densely stellate-pubescent, the hairs 0.3 mm long. *Leaf* blades oblong or more or less hastately lobed, 2–3 × 0.7–1.5 cm, palmately 3–5-nerved, densely and minutely stellate-pubescent, more or less discolorous, basally cordate (the sinus 2–6 mm deep), crenate, subacute to subobtuse; petioles 6–11 mm long, with pubescence like stem; stipules 2–3 mm long, subulate, pubescent. *Flowers* solitary in the leaf axils, the pedicels 5–15 mm long, densely pubescent like stem; involucellar bracts 5, distinct, connivent in bud concealing calyx, liguliform to slightly elliptic, 10–11 × 3 mm, acute, erect at anthesis, externally densely stellate-pubescent, internally densely pubescent in distal half, glabrous basally; calyx 8–9 mm long, ca. half-divided, the lobes 3–4 mm wide, densely stellate-pubescent; petals ca. 16 mm long, "pinkish or nearly white with darker bases and veins," externally glabrous; genitalia included, ca. half length of corolla. *Fruits* yellowish at maturity; mericarps 4.5 mm tall, prominently rugulose with 2 or 3 lateral excrescences.

Distribution (Fig. 39). At present known only from the type collection from Rio Grande do Sul, Brazil, at ca. 29°55′S, 53°30′W.

The new species is named to honor Troels Myndel Pedersen (b. 1916), knowledgeable botanist and collector of the type specimen and assiduous collector of high-quality specimens of South America, especially of Argentina and Paraguay.

105. Pavonia prionophylla R. E. Fries, Kongl. Svenska Vetenskapsakad. Handl. **42(12):** 56. 1908. Type. Brazil. Rio de Janeiro and Minas Gerais: *Glaziou 15836* (lectotype, here designated, K; isotypes, CTES [fragment], B as photo F-9461). Figs. 33, 50

Pavonia psilophylla Ekman, Ark. Bot. **9(4):** 30, t. 6. 1909. Type. Argentina. Bonpland, Colonia Finlandesa, *Ekman 172* (holotype, S; isotypes, NY, S).

Pavonia hastata var. *glabriuscula* Gürke *in* Martius, Fl. Bras. **12(3)**: 501. 1892. Type. Brazil, *Sellow s.n.* (holotype, B destroyed, as photo F-9445).

Shrubs to 1.5 m tall, the stems sparsely pubescent with minute glandular and stellate hairs intermingled. *Leaf* blades triangular to lance-oblong, mostly 2.5–7 × 1–3 cm, basally cordate-sagittate, coarsely dentate, acute, palmately 5-nerved, with widely scattered simple and stellate hairs above and beneath, becoming subglabrous, except prominently hirsute at base of blade beneath; petioles 1–3 cm long, prominently hirsute distally (hairs 1 mm long) especially on abaxial side; stipules filiform, 8–11 mm long. *Flowers* solitary in the leaf axils, the pedicels 1.5–4 cm long, articulated ca. 5 mm below the flower, often hirsute above articulation especially at juncture with involucel; involucellar bracts 5, ovate (subtrullate, basally constricted), acuminate, 8–11 × 3–5 mm, sparsely stellate-pubescent; calyx 8–12 mm long, deeply divided, the lobes cordate, acuminate, sparsely pubescent to nearly glabrous, the veins with purplish pigmentation, the intercostal areas whitish; corolla 2–2.5 cm long, pale rose with purplish center; staminal column 6–7 mm long, purplish, the filaments 1–3 mm long; styles emergent from the column, purplish. *Fruits* ca. 8 mm diam., minutely pubescent; mericarps 4–5 mm high, carinate, rugulose with 3 excrescences on each side.

Distribution (Fig. 51). From Argentina (Misiones) and Brazil (Rio Grande do Sul, Santa Catarina). The type locality for *Pavonia psilophylla* is at 27°30′S, 55°32′W.

Illustrations. Ekman (1909: fig. 6, as *P. psilophylla*); Fries (1908: pl. 6, figs. 13–15).

Representative specimens examined. BRAZIL. RIO GRANDE DO SUL: Mun. Arroio do Tigre, Barragem de Itaúba, *Bueno et al. 664* (CTES); Mun. Santa María, Ibicuí-Mirim Biological Reserve, *Bueno 5234* (CTES, HAS); Mun.

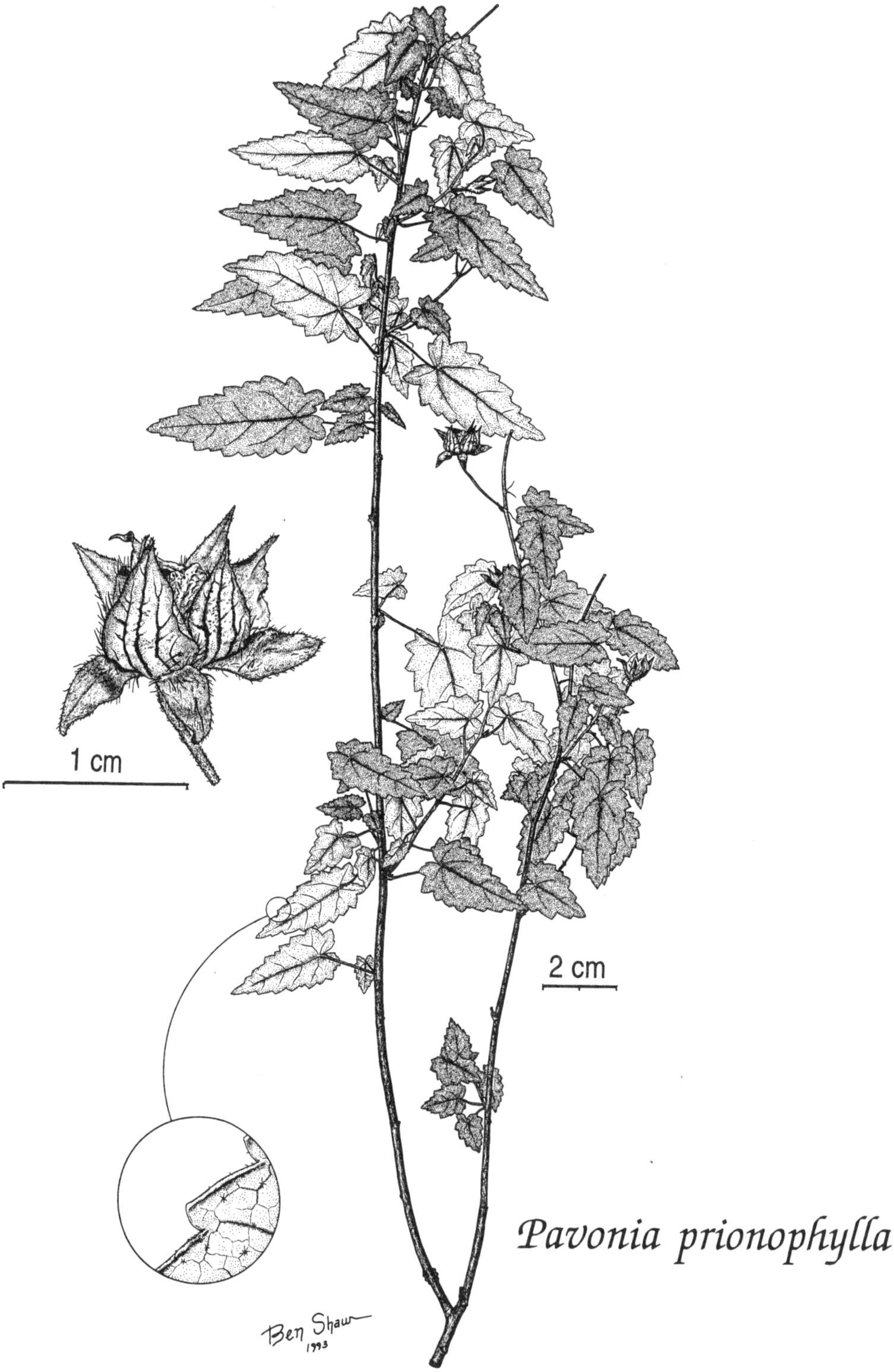

FIG. 50. *Pavonia prionophylla.* Flowering and fruiting branch, with enlargement of fruiting calyx showing the purple venation of the calyx lobes, and an enlargement of the leaf margin (*Vanni et al.569*).

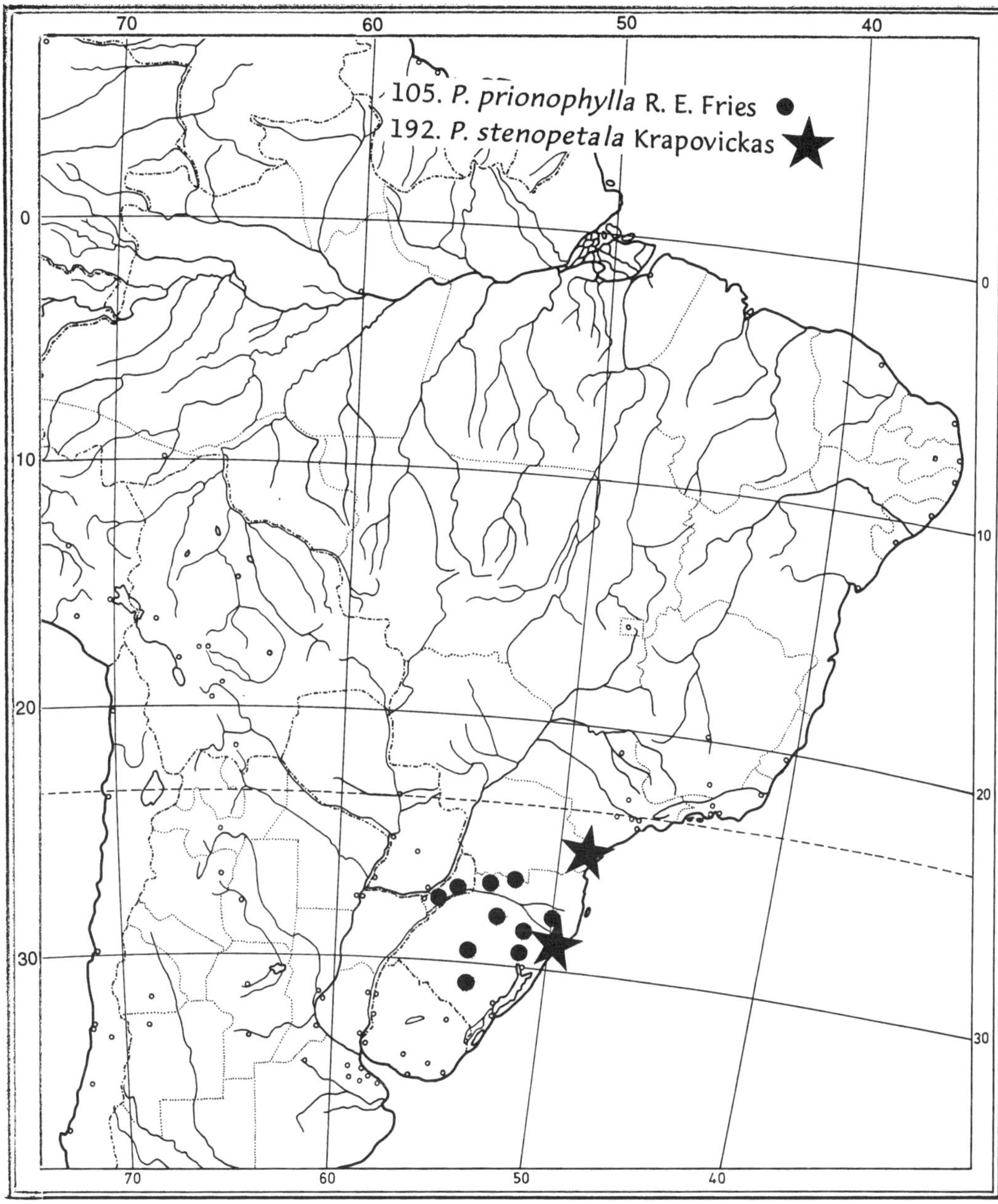

FIG. 51. Distribution map of two South American species of *Pavonia* subsect. *Hastifoliae* and *P.* sect. *Urenoideae.*

Passo Fundo, Punto Boa Vista, 19 km SE of Passo Fundo, *Krapovickas & Cristóbal 37576* (CTES, MBM); 17 km E of Lagoa Vermelha, *Krapovickas & Schinini 38228* (CTES, G, HAS, NY, UC); 2 km N of Caxias do Sul, *Krapovickas et al. 22992* (BAB, CTES, LIL, NY, P, RB); Campestre BR-116, Mun. Vacaria, *Krapovickas et al. 22999* (BAA, CTES, ENCB, LIL, MBM, MO, NY, P, RB, WIS); Mun. Bagé, Passo dos Enforcados, 73 km NE of Bagé, bank of Camaquã River, *Lindeman et al. 7010* (CTES, U); Mun. Caxias do Sul, 10 km from Caxias do Sul, grounds of Hotel Samuara, *Pedersen 15921* (TEX); Pinhal Queimado, *E. Pereira 8559* (= *Pabst 7834*) (NY, TEX); Aparados da Serra (Bom Jesus), Fazenda Bernardo Velho, *Rambo 34560* (S); Cerro Largo, *Sehnem 3566* (CAS, CTES). **SANTA CATARINA:** Mun. São Miguel do Oeste, Flores River, *Klein 5074* (CTES); Mun. Lajes, BR-116, 9 km N of the Pelotas River, *Krapovickas & Schinini 38254* (C, CTES, F, HAS); Mun. São Joaquim, Pelotas River, 10 km S of Encrucilhada, *Krapovickas &*

Schinini 38329 (CTES, HAS, UC); Mun. Chapecó, 24 km W of Campo Eré, *Smith et al. 9385* (US).

ARGENTINA. Misiones: Dept. San Javier, 11 km NE of San Javier, *Krapovickas & Cristóbal 28881* (CTES, G); Posadas, Bonpland, *Lilliesköld s.n.* (S); Dept. San Pedro, *Maruñak 27* (CTES, NY); Dept. Obera, 3 km S of Guaraní, *Murça-Pires & Mroginski 301* (CTES, NY); San Javier, *Schulz 7004* (CTES); rt. 14 and Yazá Creek, *Vanni et al. 569* (CTES, HAS, NY).

Pavonia prionophylla is distinguished by purplish-veined calyx lobes and herbage that is sparsely pubescent to subglabrous. It shows similarity to *P. missionum*, which differs in pubescence characters and corolla color.

106. Pavonia pulchra Hochreutiner, Annuaire Conserv. Jard. Bot. Genève **6:** 42. 1902. *Pavonia hastata* forma *pulchra* (Hochreutiner) Hassler, Bull. Herb. Boiss., ser. 2, **7:** 740. 1907. Type. Paraguay. Sierra de Maracayú, in the region of the Capibary River, *Hassler 4384* (holotype, G; isotypes, A, B destroyed, as photo F-9462, CTES [fragment], G, GH, K, NY, S). Fig. 33

Subshrubs 0.5–1.5 m tall, the stems minutely puberulent, the hairs 0.1 mm long. *Leaf* blades ovate to subhastate or subsagittate, mostly 2–3.5 × 2–3 cm, basally deeply cordate or sagittately lobed (lobes 8–12 mm long), shallowly crenate-serrate, acute, palmately 3–5-nerved, markedly discolorous, both surfaces minutely stellate-pubescent (hairs 0.2–0.3 mm diam.), the lower surface more densely so, the blades more or less geniculate with the petioles and parallel to the stem; petioles 2–3 cm long, with pubescence like stem but distally with longer hairs (to 0.5 mm); stipules subulate to filiform, 2–6 mm long. *Flowers* solitary in the leaf axils, the pedicels 5–20 mm long; involucellar bracts 5–7, narrowly lanceolate, 10 × 1–2(–5) mm, minutely puberulent (some of the hairs glandular); calyx 10–16 mm long, minutely puberulent and viscid, the lobes 6–8(–12) mm wide; corolla 2.5–4 cm long, white or rose with a red center, pubescent on margins of claw; staminal column 7–8 mm long, glabrous, reddish, the filaments 2 mm long. *Fruits* 8–9 mm diam.; mericarps minutely pubescent, strongly carinate and reticulate, 7 mm high.

Distribution (Fig. 48). Endemic to Paraguay. The type locality is at 25°40′S, 55°40′W.

Illustration. Krapovickas & Cristóbal (1962: fig. 1D).

Specimens examined. PARAGUAY. In the region of the Alto Paraná River, *Fiebrig 6083* (CTES [fragment], E, G as photo F-23699, GH, S); 5 km E of Caaguazú, *Gregory et al. 9601* (CTES, K, NY); in vic. of Caaguazú, *Hassler 9091* (A, CTES [fragment], K, MO, NY, S, TEX); Sierra de Amambay, *Hassler 9894* (CTES [fragment], MO, NY, S), *Hassler 10073* (CTES [fragment], K, NY), *Hassler 11392* (A, CTES [fragment]); in the upper Apa River, *Hassler 11953* (A, CTES [fragment]); Ihú, *Jörgensen 4850* (A, CTES, F, MO, NY, S); Dept. Amambay, Cerro Torin, *Rojas 4129* (CTES).

Several collections from the Sierra de Amambay (*Hassler 9894*, *10073*, *11392*, and *Fiebrig 6083*) conform to *Pàvonia pulchra* but are anomalous in having broadly ovate involucellar bracts (4–5 mm wide) and calyx lobes up to 12 mm wide (the calyx thus plicate-pentangular). In these particulars they resemble the condition in *P. belophylla*, but in all other respects they conform to *P. pulchra*. Although the two collections recombine characters of *P. pulchra* and *P. belophylla*, they do not appear to represent hybridization between the two species since they are not characterized by intermediacy. They are thus retained in *P. pulchra* but noted as anomalous. Although both species occur in the Sierra de Amambay, it is not known to what extent they may in fact be sympatric. Annotations on some of these specimens by A. Krapovickas (dated 1974) agree with the above interpretation, except for one duplicate of *Hassler 9894* (S), which is annotated as *P. belophylla*.

Hassler's publication of *Pavonia hastata* var. *grandiflora*, which included *P. hastata* forma *pulchra*, is discussed under *P. belophylla* (q.v.). I conclude that the name var. *grandiflora* is not validly published.

Pavonia pulchra is notable for its plicate calyx with broad, cordate lobes that are 6–8(–12) mm wide and for its relatively large corollas. It is allied to *P. belophylla*, perhaps also to *P. hoehnii*.

107. Pavonia reticulata Garcke, Jahrb. Bot. Gart. Berlin **1:** 212. 1881. Type. Brazil. *Sellow 3919* (holotype, B, destroyed, as photo F-9463). Figs. 33, 52

Decumbent to ascending perennial *herbs* or subshrubs, the stems minutely stellate-puberulent and with glandular hairs on younger stems. *Leaf* blades sagittate (except juvenile leaves subrotund), up to 4.5 (excluding the lobes) × 1–2 cm, basally sagittately lobed (the lobes 8–20 mm long), crenate-dentate to subentire, acute, the blade geniculate with the petiole and thus more or less parallel to the stem, the margin prominently ciliate (hairs 1 mm long) or rarely eciliate (*Hatschbach 43047*), both surfaces sparsely pubescent (with some glandular hairs) to subglabrous; petioles 6–10 mm long; stipules subulate, 1–5 mm long. *Flowers* solitary in the leaf axils, often apically crowded, the pedicels 0.5–1 cm long, distally pilose; involucellar bracts 5, lance-ovate, 11–15 × 3–7 mm,

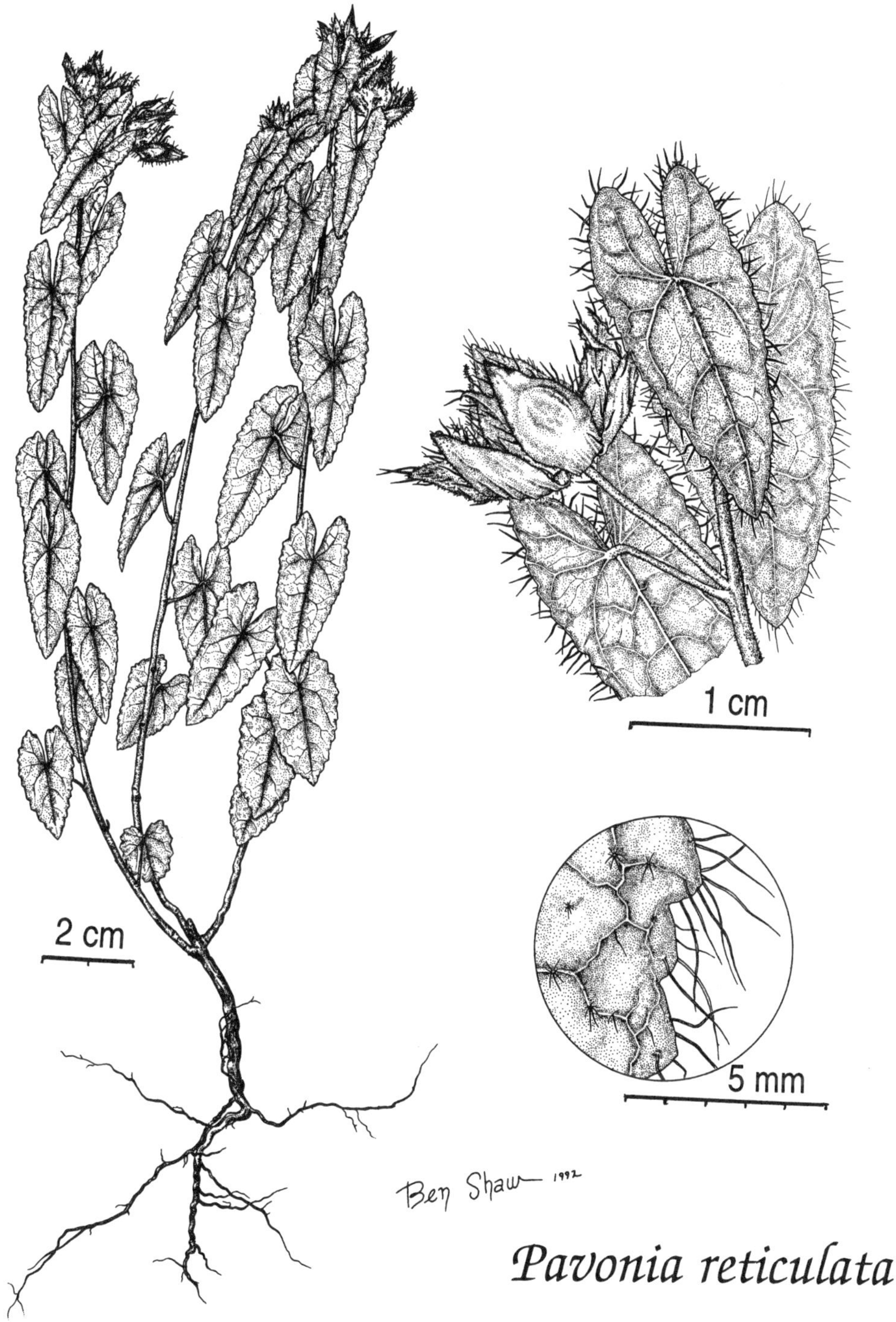

FIG. 52. *Pavonia reticulata*. Plant habit (left) with enlarged detail of flower bud (upper right) and of leaf margin (lower right) (*Krapovickas & Cristóbal 41980, Krapovickas & Vanni 37011*).

ciliate-pilose (hairs 1–2 mm long); calyx 8–9 mm long, ciliate and with minute glandular hairs; corolla 1.5 cm long, rose with red center, pubescent on margins of claw; staminal column 2–3 mm long, the filaments 1–1.5 mm long; styles emergent from the column, purple. *Fruits* 7–8 mm diam., minutely puberulent; mericarps 4 mm high, carinate, reticulately veined.

Distribution (Fig. 40). From S Brazil (Paraná, Rio Grande do Sul, and Santa Catarina), with one record from Uruguay.

Additional illustrations. Fries (1908: pl. 6, figs. 16–18).

Representative specimens examined. **BRAZIL. PARANÁ:** Jaguariahyva, *Dusén 9587* (F, GH, NY, S); Capão Bonito, *Dusén 16895* (NY, S); Porto Amazonas, *Gurgel 49* (RB); Mun. Tibagi, Fazenda Monte-Alegre, Harmonia, *Hatschbach 2798* (NY, US); Mun. Arapoti, Rio das Cinzas, Barra do Perdizes, *Hatschbach 8788* (CTES); Mun. Palmas, rd. from Palmas to Mangueirinha, *Hatschbach 15444* (CTES); Mun. Guarapuava, Canta Galo, *Hatschbach 19870* (C, CTES); Rod. do Café, Capivari River, *Hatschbach 22937* (C, CTES, MO, NY, S, US); Quatigua, *Hatschbach 43047* (MO, NY); Mun. Jaguariaiva, hwy. from Jaguaraiva to Arapoti, *Hatschbach et al. 12206* (CTES). **RIO GRANDE DO SUL:** Mun. Caxias do Sul, Agua Azul, 1–1.5 km from the Barragem Piai, *Bueno 2893* (CTES); Mun. Esmeralda, 10 km antes da sede da cidade, *Bueno 4726* (CTES); BR-285, 17 km E of Lagoa Vermelha, *Krapovickas & Schinini 38224* (C, CTES, F, G, HAS, MICH, NY); 10 km E of Vacaria, rd. to Bom Jesus, *Krapovickas & Vanni 37011* (CTES, G, HAS, ICN, NY); BR-285, Cruz Altinha, *Krapovickas & Vanni 37060* (CTES); 25 km S from Vacaría, *Krapovickas et al. 23002* (CTES); Passo do Socorro, near Vacaria, *Rambo 51558* (CAS, S, US); Farroupilha, *Rambo 61753* (S). **SANTA CATARINA:** Lajes, Vacas Gordas, *Klein 4471* (CTES); 16 km from BR-116, rd. to Campo Belo do Sul, *Krapovickas & Cristóbal 41980* (CLEMS, CTES, HAS, NY); Rio das Pedras, 3 km N of Ponte Alta do Norte, BR-116, *Krapovickas & Cristóbal 43956* (C, CTES, LIL, MBM); Mun. Urubici, 6 km N of Pericó, *Krapovickas & Schinini 38305* (C, CTES, HAS, UC); 5 km E of Jardín da Serra, *Krapovickas & Vanni 36910* (CONC, CTES, ENCB, ICN, MBM, NY, WIS).

URUGUAY. TACUAREMBÓ: Grutas de los Cuervos, *Berro 4779* (CTES).

Pavonia reticulata is distinguished by its sagittate leaves that are held more or less parallel to the stem (Fig. 52), these leaves being subglabrous except prominently ciliate on the margins. The flowers (and fruits) are crowded apically on erect pedicels and are presented more or less above the leaves.

108. Pavonia revoluta Krapovickas & Cristóbal, Lilloa **31:** 26. 1962. Type. Argentina. Córdoba: Dept. Colón, Jesús María, La Toma, *Meyer 13033* (holotype, LIL–n.v.).

Shrubs 1.5 m tall, the stems minutely pubescent with stellate subarachnoid hairs. *Leaf* blades triangular, mostly 2–3 × 1–1.5 cm, basally cordate, very coarsely crenate-dentate, acute, palmately 5-nerved, evenly pubescent above and beneath with mostly simple hairs ca. 0.5 mm long and some stellate hairs; petioles 1–1.5 cm long; stipules subulate, 1–2 mm long. *Flowers* solitary in the leaf axils, the pedicels 1–2 cm long, slender, distally pilose; involucellar bracts 5, lance-ovate, 6–7 × 3–3.5 mm, basally pilose, spreading or somewhat reflexed in bud; calyx 6–8 mm long, minutely stellate-pubescent; corolla 1.5–2 cm long, white with red center and red veins; staminal column 4–5 mm long, the filaments 1 mm long; styles emergent from the column, purplish. *Fruits* ca. 7 mm diam., minutely puberulent; mericarps 4 mm high, carinate, the crest prolonged and acute apically, reticulate-veined and with small excrescences on each side toward apex. Chromosome number, $2n = 56$ (Krapovickas & Cristóbal, 1962).

Distribution (Fig. 39). From Argentina (Córdoba, Entre Ríos, San Luis, Tucumán) and Bolivia. The type locality is approximately at 31°S, 64°W.

Illustrations. Krapovickas & Cristóbal (1962: figs. 1H, 4).

Specimens examined. BOLIVIA. TARIJA: Prov. O'Connor, 32.2 km E of Junacas (ca. 5.2 km above Narváez) on rd. to Entre Ríos, *Solomon 10956* (MO, NY); 1.5 km E of (below) Narváez on rd. to Entre Ríos, *Solomon 11007* (CTES, MO, NY, TEX).

ARGENTINA. CÓRDOBA: Dept. Colón, Salsipuedes, *G. Dawson 47* (S); Sierra Grande, Pampa de Achala, 10 km below Hotel El Condor, *Goodspeed 17241* (K); Dept. de las Minas, Cacapiche, *Hieronymus 839* (NY); Dept. Punilla, Lago San Roque, *Hunziker 5999* (S); Dept. San Justo, between Miramar and the mouth of the Plujunta River, *Hunziker 13311* (MO); Dept. Río Seco, San Miguel Creek, *Hunziker et al. 21751* (CTES); Torquin [sp.?], hill above "La Toma," *Rodrigo 333* (NY); Villa Warcalde, *Sayago 23* (CTES). **ENTRE RÍOS:** Paraná, *Geoffrey 35* (CTES); Dept. La Paz, Piedras Blancas, 6 Apr 1967, *Pedersen s.n.* (herb. Pedersen); Paraná, *Schulz 21* (CTES); without locality, *Schulz 6311* (CTES); Puesto del Paraiso, *Stuckert 1247* (S). **SAN LUIS:** Dept. Gral. Pedernera, Mercedes, *Zöllner 885* (CTES). **TUCUMÁN:** Without locality, 1837, *Tweedie s.n.* (K).

The Stuckert specimen (*Stuckert 1247*) from Puesto del Paraíso (Córdoba, Argentina) was given a tentative manuscript name by R. E. Fries, together with an appended description (in Swedish), but this name was evidently never published. Fries regarded *Pavonia hastata* forma *brevifolia* Gürke (see *P. cymbalaria*) as a synonym of his proposed new species. *Pavonia revoluta*, as the name suggests, is characterized by the possession of spreading to reflexed involucellar bracts.

109. Pavonia rhodantha Hochreutiner, Annuaire Conserv. Jard. Bot. Genève **6:** 44. 1902. Type. Paraguay. Sierra de Maracayú, in the region of the Capibari River, *Hassler 4385* (holotype, G; isotypes, B as photo F-9464, CTES, G, GH, K, NY, S). Fig. 33

Pavonia rhodantha var. *discolor* Hassler, Feddes Repert. Spec. Nov. Regni Veg. **7:** 377. 1909. Type. Paraguay. Sierra de Amambay, in dry areas near Estrella, *Rojas* [for Hassler] *10585* (isotype, CTES).
Pavonia rhodantha var. *pusilla* Hassler, Feddes Repert. Spec. Nov. Regni Veg. **7:** 378. 1909. Type. Paraguay. Sierra de Amambay, Esperanza, *Rojas* [for Hassler] *10585a* (lectotype, G as photo CTES).

Subshrubs 0.2–0.5 m tall, the stems densely stellate-hirsute, the hairs ferruginous, 2–3 mm long. *Leaf* blades sagittate, mostly 3–5 × 1.5–3.5 cm, basally sagittate (the lobes 1–1.5 cm long), coarsely serrate, acute, palmately 5–7-nerved, discolorous, coarsely stellate-pubescent on both surfaces (the hairs 1–2 mm long), the blades geniculate with the petioles and more or less parallel to stem; petioles 1.5–2 cm long; stipules subulate, 3–4 mm long (obscured by pubescence of stem and petiole). *Flowers* solitary in the upper axils, the pedicels 1–1.8 cm long, ferruginous-hirsute; involucellar bracts 5, spatulate, 7–9 × 4 mm, coarsely hirsute; calyx 12–14 mm long, with stellate ferruginous hairs 1–2 mm long; corolla 15–21 mm long, red, pubescent on margins of claw, otherwise essentially glabrous; staminal column 8–9 mm long, glabrous or with pubescence at base, reddish, the filaments 2–5 mm long; styles emergent from the column, glabrous. *Fruits* 8–9 mm diam., minutely puberulent; mericarps 6–7 mm high, acute apically, carinate and with 3 or 4 raised excrescences on each side.

Distribution (Fig. 45). From Paraguay and adjacent Brazil. The type locality is at 25°40′S, 55°40′W; that of *Pavonia rhodantha* var. *discolor* is at 22°20′S, 55°55′W.

Illustration. Krapovickas (1977: fig. 2E).

Specimens examined. **BRAZIL. MATO GROSSO:** 10 km W of Amambay, rd. to Capitan Bado, *Gregory et al. 9671* (CTES).
PARAGUAY. CANENDIYU: Río Corrientes, *Krapovickas & Schinini 32566* (CTES).

The type of *Pavonia rhodantha* var. *discolor* differs from that of *P. rhodantha* var. *rhodantha* in having a whitish underlying tomentum on the lower leaf surface (which gives rise to the name) and more prominently ciliate margins of the leaves, but they differ in little else and are here considered conspecific. The species is distinctive for its red corollas, sagittate leaves, and densely and coarsely hirsute herbage.

110. Pavonia rosengurttii Krapovickas & Cristóbal, Lilloa **31:** 68. 1962. Type. Uruguay. San José: Barra de Santa Lucia, 8 Dec 1948, *Krapovickas 4848* (holotype, LIL; isotypes, CTES, F as photo F-64929, K, NY, P, SI, UC, US). Fig. 33

Shrubs 1.5 m tall, the stems densely stellate-tomentose. *Leaf* blades sagittate-oblong, mostly 2–6 (excluding the lobes) × 1.5–3 cm, basally deeply cordate-sagittate (the lobes 6–15 mm long), obscurely crenate, acute to subobtuse, palmately 5–7-nerved, discolorous, evenly stellate-pubescent beneath, the upper surface with simple and bifurcate (and some stellate) hairs; petioles 1–4 cm long; stipules subulate, 3–4 mm long. *Flowers* solitary in the leaf axils, the pedicels 1–3 cm long; involucellar bracts 5, lance-ovate, 8–10 × 3–4 mm, puberulent and ciliate; calyx 9–11 mm long, ciliate, the lobes with red veins; corolla (not seen) 7 mm long. *Fruits* 9 mm diam., sparsely and minutely puberulent; mericarps 4 mm high, carinate and reticulately rugose. Chromosome number, $2n = 56$ (Krapovickas & Cristóbal, 1962).

Distribution (Fig. 40). From S Brazil and Uruguay. The type locality is at 34°28′S, 56°28′W.

Illustrations. Krapovickas & Cristóbal (1962: figs. 1F, 14).

Specimens examined. **BRAZIL. RIO GRANDE DO SUL:** Pelotas, IPEAS, *Krapovickas et al. 22939* (CTES, LIL, MO, NY, P).
URUGUAY. CERRO LARGO: Río Negro, Estancia Palleros, *Gallinal et al. B-2352* (NY). **ROCHA:** Camino de los Indios, km 303, *Del Puerto & Marchesi 6147* (CTES).

Pavonia rosengurttii is characterized by having sagittate leaves, red-veined calyx lobes, and small corollas.

111. Pavonia sagittata St.-Hilaire, Fl. Bras. Merid. **1:** 229. 1827. *Malache sagittata* (St.-Hilaire) Kuntze, Revis. Gen. Pl. **1:** 71. 1891. *Lassa sagittata* (St.-Hilaire) Kuntze, Revis. Gen. Pl. **3(2):** 20. 1893. Type. Brazil. Minas Gerais: Near the town of S. João del Rey, *St.-Hilaire 3749* (holotype, P as photo CTES; isotype fragments, CTES, F). Fig. 33

Pavonia sagittata var. *sessiliflora* St.-Hilaire, Fl. Bras. Merid. **1:** 230. 1827. Type. Brazil. Minas Gerais: Córrego de Mathias, *St.-Hilaire s.n.* (holotype, P as photo F-35481; isotype, G as photo F-23708 [p.p.]; fragment: F).
Pavonia affinis St.-Hilaire, Fl. Bras. Merid. **1:** 229. 1827. Type. Brazil. Minas Gerais: Minas Novas, near Itacaramby, not far from Nossa Sñra. da Penha, *St.-Hilaire s.n.* (holotype, P as photo F-35482).

Erect *shrubs* 1–1.5 m tall, the stems coarsely ferruginous-hirsute, the hairs stellate, 1–2.5 mm long. *Leaf* blades sagittate-oblong to sagittate-elliptic, mostly 3–8(–10) (excluding lobes) × 1–5.5 cm, basally cordate-sagittate (the lobes 5–15 mm long), shallowly dentate, acute to obtuse, palmately 5–7-nerved, discolorous, the upper surface with erect (i.e., tufted) stellate hairs ca. 1 mm long, the lower surface whitish, with denser and finer stellate pubescence; petioles 1–2 cm long; stipules 1–3 mm long, subulate, often hidden by density of pubescence. *Flowers* solitary (or paired) in the leaf axils, the pedicels 1–4 cm long; involucellar bracts 5, broadly ovate or suborbicular, 5–8 × 3–7 mm, distally revolute, densely hirsute, discolorous; calyx 8–10 mm long, densely ferruginous-hirsute; corolla pink with red center and red venation, 1.5–3.5 cm long. *Fruits* 7 mm diam., minutely pubescent; mericarps 4–5 mm high, carinate, smooth. Chromosome number, $2n = 56$ (Krapovickas & Cristóbal, 1962; Fernández, 1981).

Distribution (Fig. 48). From central Brazil. The type locality is at 21°10′S, 44°13′W; that of *Pavonia affinis* is at 15°5′S, 44°7′W.

Illustrations. Fries (1908: pl. 6, figs. 3–4); Gürke (1892b: pl. 95, fig. 2).

Representative specimens examined. **BRAZIL**. **DISTRITO FEDERAL:** Basin of the S. Bartolomeu River, *Heringer 3689* (CTES). **GOIÁS:** Serra do Rio Preto, *Irwin et al. 10437* (F, GH, MO, NY, S, TEX, US). **MATO GROSSO:** Mun. Bataguaçú, BR-267, *Hatschbach 24253* (C, CTES, LL, MO, NY, S, US, herb. Pedersen); Bataguaçú, 42 km W of Porto 15 de Novembro, *Krapovickas 14398* (CTES, LP, MBM, MO, NY); 80 km W of Bataguaçú, Fazenda S. Jorge, *Gregory et al. 9787* (CTES, NY). **MINAS GERAIS:** 5 km by rd. NE of Rio Manso and Couto de Magalhães, *W. R. Anderson, P. A. Fryxell et al. 8746* (NY, TEX); Cardoso, *Barreto 2953* (A); Diamantina, BR-259, toward Gouveia, *Esteves 2316* (CTES, NY, SPF); 30 km W of Gouveia, *Gregory et al. 10527* (CTES, F); Mun. Jaboticatuba, Serra do Cipó, *Hatschbach 29923* (CTES); São Tomé das Letras, *Hatschbach & Ahumada 31316* (CTES); Bacia de Três Marías, Paraoeba River, *Heringer 6993* (CTES); Fazenda das Pindaibas, *Heringer 7405* (CTES); Serra do Espinhaço, ca. 48 km W of Montes Claros, *Irwin et al. 23958* (F, K, MO, NY, RB, TEX, US); Congonhas, *Krapovickas & Cristóbal 33550* (CTES, F, MBM, SI, herb. Pedersen); Poços de Caldas, Véu das Noivas, *Krapovickas & Cristóbal 35349* (C, CLEMS, CTES, F, K, MBM, MO, NY, TEX); São Gonçalo do Abaeté, *Krapovickas & Cristóbal 42821* (CTES, K, MBM, NY); 53 km E of Araxá, Quebra Anzol River, *Krapovickas et al. 33366* (CTES); Contendas, 1892, *Kuntze s.n.* (NY, US); Allegro, *Pohl 2928* (F); Selvíria, Fazenda de Ensino e Pesquisa da UNESP-"Campus" de Ilha Solteira, *Tiritan 555* (SPF); Tabara, *Vauthier 491* (F, GH). **PARANÁ:** Morungava, *Dusén 16641* (F, GH, NY, S-2). **PIAUHY:** Without locality, *Warming 1350* (C). **SÃO PAULO:** Capital, *Hoehne [SPF-10546]* (CTES); Cascata, *Kiehl 4064* (CAS).

Pavonia sagittata is characterized by the possession of sagittate leaves and broadly ovate to suborbicular involucellar bracts. It is also distinctive for its ferruginous-hirsute herbage.

112. Pavonia schininii Krapovickas & Cristóbal, Bonplandia **7:** 45. 1993. Type. Paraguay. Alto Paraná: Estancia Santa Elena, Pira Pyta, 11 Oct 1990, *Schinini & Caballero Marmori 27162* (holotype, CTES; isotypes, FCQ, G, TEX).

Decumbent perennial (?) *herb* or *subshrub*, the stems coarsely stellate-pubescent, the hairs 1 mm long. *Leaf* blades ovate, basally cordate, dentate, acute to subobtuse, palmately 5–7-nerved, 1.5–3 × 1.5–2 cm, prominently reticulate-veined beneath, strongly discolorous, the upper surface with appressed simple and bifurcate hairs 1 mm long, the lower surface densely and softly stellate-tomentose; petioles 9–17 mm long; stipules subulate, 3 mm long. *Flowers* solitary in the leaf axils, the pedicels 1.5–3.5 cm long, exceeding the subtending petioles; involucellar bracts 5, lance-ovate, 5–6 × 2.5–4 mm, stellate-pubescent and marginally ciliate; calyx subequal to involucel, stellate-pubescent; corolla not seen ("al marchitarse, rojisa"), the flowers apparently cleistogamous and the petals much reduced (2 mm long); staminal column 2 mm long, the sessile anthers few (5?); styles strongly recurved. *Fruits* 7–8 mm diam., minutely pubescent; mericarps 5 mm tall, carinate, reticulate-veined.

Distribution (Fig. 37). Known only from the type collection. The type locality is at 25°17′S, 54°35′W.

Illustration. Krapovickas & Cristóbal (1993: fig. 1).

Krapovickas & Cristóbal (1993) relate this species to *Pavonia rhodantha*, for its similar indumentum, and to *P. glechomoides* and *P. cymbalaria*, for its procumbent habit. The latter affinity seems the more likely.

113. Pavonia viscidula St.-Hilaire & Naudin, Ann. Sci. Nat. Bot., ser. 2, **18:** 42. 1842. Type. Brazil. Minas Gerais: Near Lagoa Santa, *Vauthier s.n.* (holotype, P as photo F-35485).

Shrubs ca. 1 m tall, the stems coarsely stellate-pubescent, the hairs 0.5 mm long or more, and with a few glandular hairs. *Leaf* blades lance-oblong, 5–7 × 1–3 cm, basally sagittate (the lobes shorter than the petiole), serrate, acute, essentially penninerved, apparently reflexed and parallel to stem, somewhat discolorous, subglabrous above (except on veins), densely pubescent beneath, prominently reticulate-veined; petioles 1–1.5 cm long; stipules subulate, 5–7 mm long. *Flowers* solitary in the leaf axils, the pedicels 1–3 cm long; involucellar bracts 5, broadly ovate, acute, 7–8 × 5 mm, hirsute; calyx subequal to involucel, more densely hirsute, the lobes 3-ribbed; corolla ca. 1.5 cm long. *Fruits* 7–8 mm diam., minutely and obscurely pubescent, the mericarps 4.5–5 mm long, strongly carinate, reticulate-veined.

Distribution (Fig. 37). Known from Minas Gerais, Brazil. The type locality is at 19°37′S, 43°52′W.

Specimens examined. BRAZIL. MINAS GERAIS: Herculano Pena, *Duarte 607* (CTES).

Pavonia viscidula was treated as a synonym of *P. sagittata* by Gürke (1892b), who was followed by Kearney (1958), but this placement is clearly incorrect. Krapovickas (1977) equated *P. viscidula* and *P. prionophylla*, using the former as the older name. However, in the detailed original description of *P. prionophylla*, Fries did not mention viscid pubescence, so this placement does not seem correct. In concordance with Krapovickas (pers. comm.), the species is here recognized as distinct.

Pavonia viscidula has broadly ovate involucellar bracts and ribbed calyx lobes. It has glandular-pubescent calyces, similar to *P. xanthogloea*, from which it differs in petiole pubescence, and also similar to the red-flowered *P. glutinosa*.

114. Pavonia xanthogloea Ekman, Arkiv Bot. **9(4):** 27. t. 5. 1909. Type. Argentina. Misiones: Posadas, Bonpland, *Ekman 170* (lectotype here designated, S; isolectotype, NY); Posadas, Bonpland, *Ekman 171* (syntypes, MO, S).

Shrubs 1.5–2 m tall, the stems minutely puberulent, the hairs stellate with intermingled glandular hairs, 0.1–0.2 mm long. *Leaf* blades triangular-ovate to oblong-hastate, mostly 2–6(–8) × 1–2.5 cm, basally truncate to subcordate, sometimes subhastately lobed, serrate, acute, palmately 3–5-nerved, discolorous, both surfaces with minute stellate hairs and simple glandular hairs intermingled, the pubescence of lower surface whitish, with setae 1 mm long at base of the blade beneath and often also on the distal end of the petiole; petioles 1–3 cm long, minutely puberulent and often setose distally on the abaxial side; stipules subulate, 3–5 mm long, puberulent. *Flowers* solitary in the leaf axils, the pedicels 1–3 cm long, slender; involucellar bracts 5, broadly spatulate or ovate, 5–7 × 2–3 mm, glandular-puberulent; calyx 5–7 mm long, glandular-puberulent; corolla 1.5–2 cm long, rose-pink with dark red center and dark veins, densely pubescent on margins of claw, otherwise glabrous; staminal column 5–7 mm long, dark red, the filaments 1–1.5 mm long; styles emergent from the column, reddish. *Fruits* 6–7 mm diam., minutely puberulent; mericarps 3.5 mm high, rugulose, obscurely carinate. Chromosome number, $2n = 56$ (Krapovickas & Cristóbal, 1962).

Distribution (Fig. 40). From S Brazil, Paraguay, Argentina (Buenos Aires, Corrientes, and Misiones), and Uruguay. The type locality is at 27°30′S, 55°32′W.

Illustrations. Ekman (1909: fig. 5); Krapovickas (1953: fig. 108D).

Representative specimens examined. BRAZIL. RIO GRANDE DO SUL: Mun. Arroio del Tigre, Barragem de Itaúba, *Bueno et al. 663* (CTES); Mun. Alegrete, bridge over Inhanduí River, *Bueno 3258* (CTES); Mun. Santa María, Ibicuí-Mirim Biological Reserve, *Bueno 5381* (CTES, HAS); 5 km S of Caçapava do Sul, *Krapovickas & Cristóbal 34190* (CTES, F, SI); 48 km SW of Caçapava do Sul, rd. to Lavras, *Krapovickas & Cristóbal 44661* (CTES, TEX); 18 km E of Santa Barbara do Sul, *Krapovickas & Vanni 36781* (CONC, CTES, ENCB, ICN, LIL, NY); Mun. Quaraí, Passo da Guarda, *Pedersen 11660* (A, C, CTES); Espírito Santo near P. Alegre, *Rambo 39162* (CTES).

PARAGUAY. In the region of the Alto Paraná River, *Fiebrig 5913* (GH, S).

ARGENTINA. BUENOS AIRES: Zárate, Estancia "Las Palmas," *Boelke et al. 1586* (CTES); Ptdo. de La Plata, Punta Lara, *Krapovickas 2892* (CAS, NY, S), *Cabrera 2093* (F, NY, S-2, US). **CORRIENTES:** Dept. Santo Tomé, 26 km SE of Colonia Garabí, *Ahumada & Schinini 4093* (CTES, ICN, MEXU); Dept. San Martín, Yapeyú, *Krapovickas & Cristóbal 28972* (CTES); Dept. Santo Tomé, Estancia San Francisco, 23 km NW of Gobernador Virasoro, *Krapovickas et al. 17189* (C, CTES, IAC, TEX); 15 km from Gdor. Virasoro, rd. to Garruchos, *Krapovickas et al. 21039* (CTES, GH, LIL, NY); Dept. Santo Tomé, Garruchos, Estancia San Juan Bautista, *Krapovickas et al. 25861* (CTES); Dept. Santo Tomé, Chimiray Creek and rt. 40, *Schinini 6813* (BAF, CTES, LIL, NY, USM). **MISIONES:** Dept. Apostoles, San José, *Huidobro 4954* (CAS, CTES, E, MO, TEX); Dept. Concepción, Santa María, *Krapovickas & Cristóbal 28833* (CTES, G, NY, SI); Dept. San Javier, San Javier, Krapovickas & Cristóbal 28897 (CTES, G, NY, SI); Olegario V. Andrade, *Krapovickas et al. 14823* (C, CTES, MBM, MO, NY); Panambí (Oberá), Uruguay River, *Sandeman 4796* (K); Dept. Leandro N. Alem, from L. N. Alem to Cerro Azul, *Schinini 5590* (CTES, ENCB, G, RB); Puerto Nuevo, *Schwarz 2209* (CAS, CTES, LIL, TEX); Puerto Viejo, *Schwarz 2258* (CAS, CTES); Dept. San Ignacio, Gobernador Roca, *Schwartz 5175* (CAS, LIL, TEX); Dept. Cainguás, Capiovysiño, *Schwarz 7691* (CAS, CTES, TEX); Puerto Cazador, *Schwarz 8011* (CTES, US).

URUGUAY. TACUAREMBÓ: Valle Edén, *Rosengurtt et al. 10132* (CTES).

Pavonia xanthogloea has glandular-puberulent calyces, similar to *P. viscidula*, from which it is distinguished by having hairs 1 mm long at the distal end of the petiole, and similar to *P. glandulosa*, which is distinguished by its red corollas.

PAVONIA SUBGEN. ASTEROCHLAMYS

Pavonia subgen. **Asterochlamys** Fryxell, nom. nov. et stat. nov. Replaced basionym: *Asterochlaena* Garcke, Bot. Zeitung (Berlin) **8:** 666. 1850 (non *Asterochlaena* Corda, 1845). *Pavonia* sect. *Asterochlaena* (Garcke) Ulbrich, Bot. Jahrb. Syst. **57:** 64. 1921. Type. *Asterochlaena cuspidata* Garcke (= *Pavonia mutisii* Humboldt, Bonpland & Kunth).

Pseudopavonia Hassler, Feddes Repert. Spec. Nov. Regni Veg. **7:** 74. 1909. *Asterochlaena* sect. *Pseudopavonia* Hassler, Feddes Repert. Spec. Nov. Regni Veg. **7:** 76. 1909, nom. provis. *Pavonia* subsect. *Pseudopavonia* Hassler *ex* Ulbrich, Bot. Jahrb. Syst. **57:** 64. 1921, nom. nud. Type. *Pseudopavonia tenax* Hassler. (= *Pavonia balansae* Gürke).

Perennial procumbent herbs to erect shrubs, often soft-pubescent, sometimes with glandular hairs, the leaves usually ovate-cordate, sometimes weakly lobulate; involucellar bracts 4–12; mericarps chartaceous, unornamented, usually more or less loculicidally dehiscent, subequal to column; seeds usually striate.

Type species. *Pavonia mutisii* Humboldt, Bonpland & Kunth.

Synopsis of *Pavonia* subgen. *Asterochlamys*

1. **Pavonia** sect. **Asterochlamys** (29 species).
2. **Pavonia** sect. **Albae** Fryxell (3 species).

Pavonia sect. *Asterochlamys* was originally described in generic rank as *Asterochlaena* (Garcke, 1850) and was retained there by some workers (e.g., Hassler, 1910). The same name was also published for a fossil fern in 1845 (*Asterochlaena* Corda) and for a modern member of the Convolvulaceae in 1893 (*Asterochlaena* Hallier f.). In consequence, Hochreutiner (1931) raised the question of conservation of the name, should the malvaceous plant be recognized in generic rank, but he concluded that it should not be so recognized. Subsequently, Hassler proposed the genus *Pseudopavonia* Hassler, but a year later he transferred the single species to *Asterochlaena* as *A. tenax* (Hassler) Hassler (Hassler, 1910). However, most recent workers either have submerged *Asterochlaena* in *Pavonia* (Hutchinson, 1967), have treated it as an undifferentiated part of "sect. *Eupavonia*" (Gürke, 1892b; Kearney, 1951, 1958), or have treated it in sectional rank (Ulbrich, 1921; Fryxell, 1988a). It is here elevated to subgeneric rank in recognition of its distinctive fruit morphology (see above).

The subgenus is divisible into two groups on the basis of differences in seed morphology and more subtle differences in "aspect." *Pavonia* sect. *Asterochlamys*, the larger of the two sections, could be easily divided into a yellow-flowered group and a rose-flowered group, and this difference is used for convenience in the key to species, but it is not recognized with formal taxonomic rank. A few species alliances can be discerned. Within the rose-flowered group, for example, the following species form a series that seems to represent a single lineage: *P. garckeana*, *P. biflora*, *P. krapovickasii*, *P. hexaphylla*, and *P. cracens*. A second group that similarly represents another lineage includes *P. fonsecana*, *P. mollis*, *P. mutisii*, *P. purpusii*, and *P. spuria*. In the same fashion, among the yellow-flowered species, there is clearly a group centering around *P. sidifolia*, but it is not simple to delimit this group, or to place a boundary around it, or to interrelate the species within it in a clear-cut way. Other species stand isolated from all of the these groups, notably *P. intermixta*. Consequently, I believe it is premature to propose any further subdivisions at this time.

The original description of *Pavonia falconensis* noted a whorl of staminodes at the base of the staminal column. Further study reveals that such staminodes are characteristic of the entire subgenus, at least as far as material has been available for examination. These staminodes appear to be "filaments without anthers," although they are often thicker than the filaments of the fertile stamens and have the same pigmentation as the column. They are more readily observable in fresh flowers than in dried specimens but are commonly demonstrable in the latter.

The name *Asterochlaena* is derived from the Greek *astero* (star) and *chlaina* or *chlamys* (cloak), presumably in allusion to the involucel.

Because Garcke's name is a later homonym, a new name is required (see above). The new name, *Asterochlamys*, is chosen to retain the meaning of the replaced name and to make only minimal change in the orthography.

Pavonia subgen. *Asterochlamys* sect. *Asterochlamys*

Soft-pubescent shrubs or subshrubs, sometimes decumbent; corolla yellow, rose-pink, or white; seeds striate.

Key to the species of *Pavonia* sect. *Asterochlamys*

1. Corolla yellow.
   2. Corolla with a reddish or purplish center.
      3. Plants procumbent; leaves subrotund; fruits 8–10 mm diam., glabrous or minutely scabridulous.
         4. Leaves as wide as long or wider, more or less obtuse; calyx 7–10 mm long; fruits 9–10 mm diam., glabrous; leaves softly puberulent; petal spot always present. . . . . 140. *P. subrotunda*
         4. Leaves as wide as long or longer, acute; calyx 6–7 mm long; fruits ca. 8 mm diam., scabridulous; leaves very densely and softly tomentose; petal spot sometimes absent.. . . . . 141. *P. vannii*
      3. Plants erect or subscandent; leaves ovate, longer than wide, acute or acuminate; fruits 5–8 mm diam., glabrous or pubescent.
         5. Involucellar bracts 6–7, more or less rhombic, 1–1.5 times as long as calyx, 2.5–4 mm wide; stems densely stellate-pubescent and with a few glandular hairs. . . . . 138. *P. sidifolia*
         5. Involucellar bracts 6–9, linear or lanceolate, 1–2 times as long as calyx, 0.5–3 mm wide; stems setose or glandular-pubescent or both.
            6. Claws of petals glabrous; fruits subglabrous; involucellar bracts 8–9. . . . . 133. *P. neei*
            6. Claws of petals pubescent; fruits more or less pubescent; involucellar bracts 6–8.
               7. Upper leaf surface with appressed simple and bifurcate hairs 0.5–2 mm long; stems setose; corolla 1–1.5 cm long; involucel 1–1.5 times as long as calyx. . . . . 116. *P. argentina*
               7. Upper leaf surface puberulent; stems glandular-pubescent; corolla 1.5–2.5 cm long; involucel twice length of calyx or longer. . . . . 126. *P. hieronymi*
   2. Corolla lacking a dark center (or sometimes present in *P. vannii* and *P. apiculata*?).
      8. Leaves lance-elliptic, <1 cm wide, 6–10 times as long as wide (except in juvenile leaves); seeds smooth; Cuba. . . . . 128. *P. intermixta*
      8. Leaves ovate, >1 cm wide, 1–3 times as long as wide; seeds striate; Mexico and South America.
         9. Involucellar bracts 8–10.
            10. Plants reclining or decumbent, the herbage very densely and softly tomentose; petal spot sometimes present. . . . . 141. *P. vannii*
            10. Plants more or less erect, the herbage "normally" pubescent; petal spot always absent.
                11. Leaf blades wider than long; stems glandular-pubescent and with scattered setae 2–3 mm long; stipules 4–6 mm long. . . . . 118. *P. balansae*
                11. Leaf blades longer than wide; stems stellate-pubescent (hairs 0.2–0.4 mm long); stipules 8–9 mm long. . . . . 137. *P. rupestris*
         9. Involucellar bracts 6–8.
            12. Involucellar bracts (and herbage generally) prominently viscid, lacking setae; stipules 5–12 mm long; Mexico. . . . . 131. *P. monticola*
            12. Involucellar bracts puberulent, tomentose or setose (sometimes also with glandular hairs); stipules 1.5–7 mm long; Paraguay, Bolivia, Brazil.
                13. Involucellar bracts broadly spatulate or subrotund, acute or obtuse, 4–6.5 mm wide.. . . . . 134. *P. platyloba*
                13. Involucellar bracts linear or lancelate, 0.5–2 mm wide.
                    14. Mericarps laterally winged and apically apiculate; petals 1.5–3.5 cm long.
                        15. Petals 1.5–2.5 cm long; stems minutely stellate-pubescent and with scattered setae.. . . . . 115. *P. apiculata*
                        15. Petals 3.5 cm long; stems finely puberulent. . . . . 143. *P. yatarendana*
                    14. Mericarps neither winged nor apiculate; petals 1–1.5 cm long. . . . . . . 127. *P. immaculata*
1. Corolla rose-pink or white (pale pink drying yellowish in *P. geminiflora*).
   16. Bracts of involucel 4, cordate-ovate, 7–17 mm wide.
       17. Involucellar bracts prominently cordate basally, green throughout, 10–17 mm wide; petals 3–5 cm long; androecium shorter than corolla.. . . . . 122. *P. garckeana*
       17. Involucellar bracts not cordate, 7–9 mm wide, the involucel cupuliform, basally yellowish; petals 2.5 cm long; androecium subequal to corolla. . . . . 136. *P. quadrifida*
   16. Bracts of involucel (4–)5–14 (if only 4 then narrowed to a cuneate base), lanceolate to linear, 0.5–7 mm wide.
       18. Corolla 3–5.5 cm long; bracts of involucel (4–)5–6, lanceolate to ovate, 5–7 mm wide.. . . . . . 119. *P. biflora*

18. Corolla 0.5–4 cm long; bracts of involucel 5–10, (ovate)-lanceolate to linear, 0.5–4(–7) mm wide.
   19. Genitalia manifestly exserted; corolla white. . . . . 124. *P. hassleriana*
   19. Genitalia included or subequal to corolla; corolla white or rose-pink (drying yellowish in *P. geminiflora*).
      20. Corolla white, the claws of the petals glabrous (except pubescent in *P. fonsecana*).
         21. Stems densely viscid and sparsely pilose (hairs 2–2.5 mm long); fruits obscurely scabridulous on suture, otherwise usually glabrous; involucellar bracts 6–7; Colombia, Ecuador. . . . . 132. *P. mutisii*
         21. Stems viscid, lacking long, simple hairs; fruit pubescence otherwise; involucellar bracts 7–9; Mexico, El Salvador.
            22. Fruits glabrous; calyx 6–7 mm long with sparse cilia 1 mm long; stems hairs glandular, 0.2–0.3 mm long; involucel viscid; El Salvador. . . . . 121. *P. fonsecana*
            22. Fruits pubescent; calyx 7–9 mm long with dense cilia 1.5 mm long; stems hairs glandular, 0.5 mm long; involucel viscid and pilose (hairs 1–2 mm long), especially at base; Mexico. . . . . 139. *P. spuria*
      20. Petals rose-pink (except drying yellowish in *P. geminiflora* and apparently white in *P. fonsecana*), the claws pubescent.
         23. Stems pilose with simple hairs 2–4 mm long in addition to short glandular hairs; fruits usually pubescent.
            24. Fruits glabrous, the mericarps apically apiculate; flowers densely bearded at juncture of petals and staminal column. . . . . 123. *P. geminiflora*
            24. Fruits pubescent, the mericarps not apiculate; flowers sparsely ciliate on margins of claw.
               25. Stipules subulate-triangular, 2–4 mm long; seed glabrous; Panama, Venezuela, Colombia, Ecuador, Peru. . . . . 130. *P. mollis*
               25. Stipules linear, 4–9 mm long; seeds minutely pubescent; Mexico, Guatemala, Honduras. . . . . 135. *P. purpusii*
         23. Stems not pilose or rarely with a few long hairs; fruits glabrous.
            26. Involucellar bracts 5–7 × 0.5–0.7 mm; corolla 1.5 cm long; leaves <2 cm long. . . . . 120. *P. cracens*
            26. Involucellar bracts 7–11 × 1–4(–7) mm; corolla 1–4 cm long; leaves usually >3 cm long.
               27. Flowers often geminate; stems often purpish; involucellar bracts 5–7.
                  28. Lower leaf surface pubescent (especially on veins) but epidermis visible through the hairs; involucellar bracts 1(–2) mm wide. . . . . 125. *P. hexaphylla*
                  28. Lower leaf surface covered by dense whitish tomentum, the epidermis obscured; involucellar bracts 1.5–4 mm wide.
                     29. Calyx and involucel glandular-pubescent; involucellar bracts 5. . . . . 129. *P. krapovickasii*
                     29. Calyx and involucel glandular-pubescent and also ciliate or hirsute; involucellar bracts 7 (6–8). . . . . 142. *P. vinosa*
               27. Flowers solitary; stems not purplish; involucellar bracts 7–8.
                  30. Corolla 1.2 cm long; plants viscid; mericarps obscurely apiculate; El Salvador. . . . . 121. *P. fonsecana*
                  30. Corolla 4 cm long; plants puberulent; mericarps prominently apiculate; Paraguay. . . . . 117. *P. aurelii*

Note: *Pavonia fonsecana* is keyed out under both leads of couplet 20, because corollas are as yet poorly known in this species.

115. Pavonia apiculata R. E. Fries, Kongl. Svenska Vetenskapsakad. Handl. **42(12):** 45. 1908. *Asterochlaena hieronymi* var. *apiculata* (R. E. Fries) Hassler, Feddes Repert. Spec. Nov. Regni Veg. **8:** 117. 1910. Type. Paraguay. Estancia Tagatiya, *Anisits 2639* (lectotype, here designated, S; isotype, S). Fig. 53

Subshrubs 0.5–1.5 m tall, the stems minutely stellate- or glandular-pubescent and with scattered setae 1.5–2 mm long. *Leaf* blades ovate, 3–5(–8) × 2–4(–6) cm, basally cordate, serrate, acute, palmately 5-nerved, discolorous, the upper surface minutely pubescent with

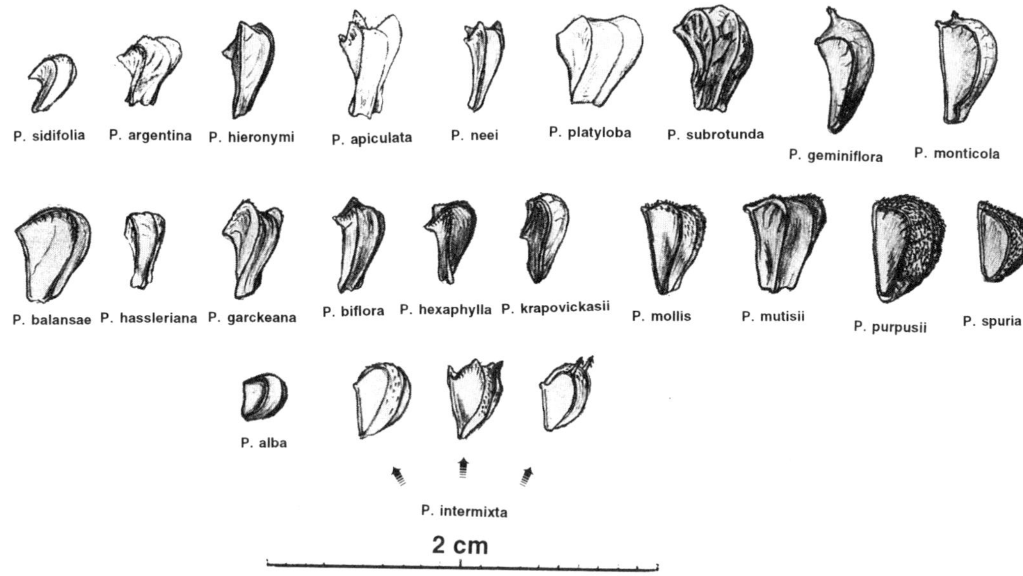

FIG. 53. *Pavonia* subgen. *Asterochlamys* mericarps in left lateral view (drawn as if indehiscent). Top row (left to right): *P. sidifolia* (*Lindeman & Haas 4414*), *P. argentina* (*Mandon 811*), *P. hieronymi* (*Rojas 5416*), *P. apiculata* (*Anisits 2639*), *P. neei* (*Rojas 5741*), *P. platyloba* (*Anisits 2893*), *P. subrotunda* (*Cabral et al. 183*), *P. geminiflora* (*Blanchet 944*), *P. monticola* (*Breedlove & Raven 13716*). Center row (left to right): *P. balansae* (*Rojas 13858*), *P. hassleriana* (*Hassler 1922*), *P. garckeana* (*Gottsberger 11–16271*), *P. biflora* (*Cross s.n.*), *P. hexaphylla* (*Hopkins et al. 115*), *P. krapovickasii* (*Krapovickas et al. 40229*), *P. mollis* (*Fosberg 21998*), *P. mutisii* (*Dodson et al. 11041*), *P. purpusii* (*Jones et al. 5396*), *P. spuria* (*Koch et al. 78258*). Bottom row (left to right): *P. alba* (*Wilbur et al. 12927*), *P. intermixta* (left to right: *León 15082*, *Killip 45751*, *Palmer & Riley 941*).

stellate and bifurcate hairs, the lower surface softly tomentose, also with scattered 2–3-armed stellate spicules or these sometimes absent; petioles 0.5–2 cm long; stipules filiform, 2–3 mm long. *Flowers* solitary or paired in the leaf axils, the pedicels 3–6 cm long; involucellar bracts 6–7, lanceolate, 11–13 × 1–3 mm, minutely stellate-pubescent and also ciliate, especially toward the base, the hairs 1–2 mm long; calyx 6–8 mm long, minutely stellate-pubescent and marginally ciliate, the hairs 1–2 mm long; corolla 1.5–2.5 cm long, pubescent on margins of claw, yellow, with small red center. *Fruits* 6–7 mm diam., minutely and retrorsely scabridulous; mericarp body ca. 5 mm tall, with 2 lateral wings and a subapical apiculum 1–2 mm long; seeds 3–4 mm long, glabrous, striate.

Distribution (Fig. 54). Reported from S Brazil, Paraguay, Bolivia, and Argentina.

Illustrations. Fries (1908: pl. 5, figs. 11–14).

Specimens examined. **BRAZIL**. **Mato Grosso do Sul:** Mun. Corumbá, Aterro Murundum-Alegre, *Bueno & Pott 33* (CTES); Serra das Araras, Vale da Camarinha S of Bauxi, Dubs 2125 (E, K, TEX, UFMT, Z); Mun. Aquidauana, Serra de Maracaju, Estação Repetidora de Paxixi, *Hatschbach et al. 58951* (K, MBM); Mun. Cuiabá, Coxipó da Ponte, *Krapovickas & Cristóbal 43043* (CTES); 28 km S of BR-070, rd. to Poconé, *Krapovickas & Cristóbal 43056* (CTES); Paraguai River, *Schaller 98* (NY).

BOLIVIA. **Santa Cruz:** Lomerio, *Jardim 1892* (MO, TEX, USZ); Ñuflo de Chávez, Lomerio, *Mamani & Saucedo 646* (MO, TEX, USZ).

ARGENTINA. **Formosa:** Dept. Pilagá, Estancia Jacaré, *Pierotti 20* (NY).

116. Pavonia argentina Gürke *in* Martius, Fl. Bras. **12(3):** 508. 1892. *Lassa argentina* (Gürke) Kuntze, Revis. Gen. Pl. **3(2):** 19. 1898. *Asterochlaena argentina* (Gürke) Stuckert *ex* Seckt, Fl. Cordob. 347. 1929. Type. Bolivia. Prov. Larecaja: Vic. of Sorata, Mar 1861, *Mandon 811* (lectotype, here designated, NY; isolectotypes, CTES, F, GH, MO, NY, S). Fig. 53

Pavonia campestris R. E. Fries, Kongl. Svenska Vetenskapsakad. Handl. **42(12):** 43. 1908. *Asterochlaena sidifolia* var. *campestris* (R. E. Fries) Hassler, Feddes Repert. Spec. Nov. Regni Veg. **8:** 114. 1910. Type. Paraguay. Garai-Cué, Campo Abente, 18 Dec 1898, *Anisits 2751* (lectotype, here designated, S).

Pavonia bangii Ulbrich, Notizbl. Bot. Gart. Mus. Berlin-Dahlem **11:** 543. 1932. Type. Bolivia. Vic. of Cochabamba, 1891, *Bang 1053* (isotypes, E, F as photo F-56161, G, GH, MO, NY, TEX, PH, US as photo

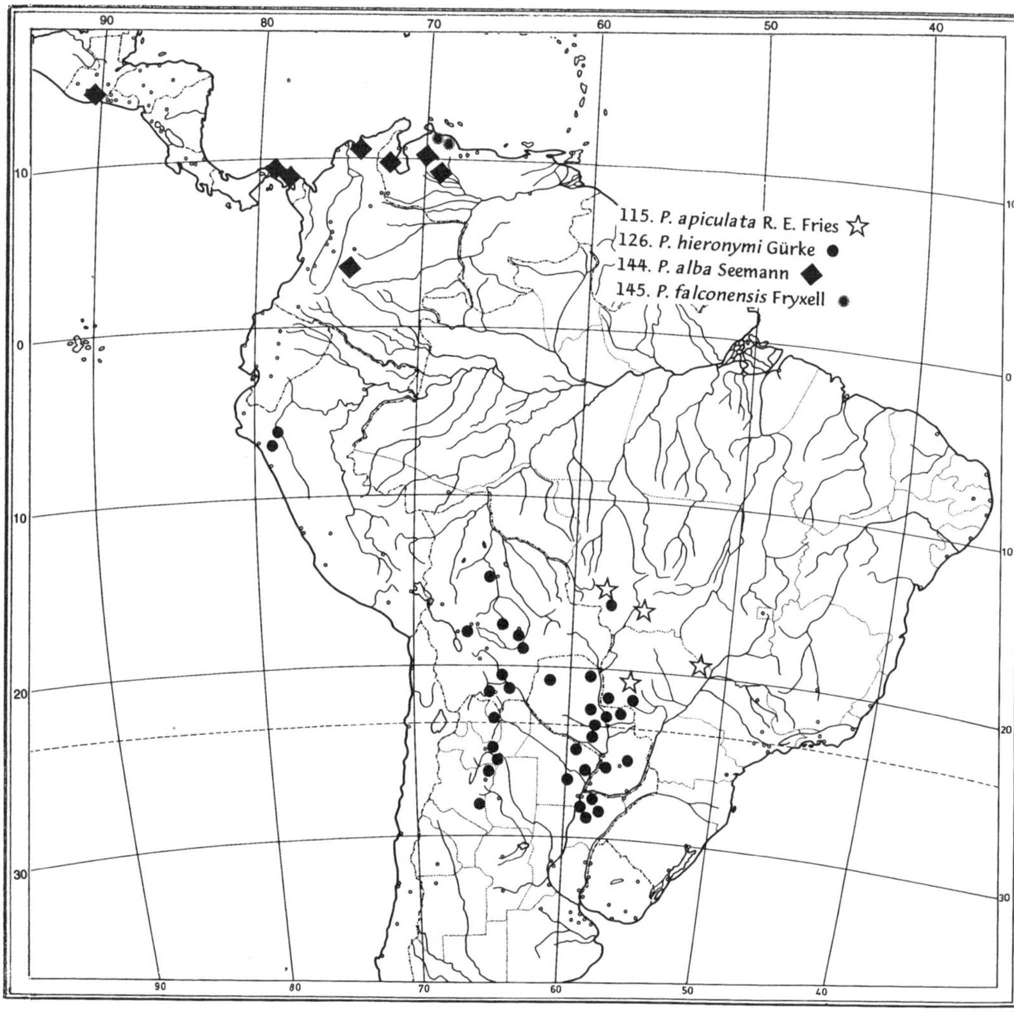

FIG. 54. Distribution map of four neotropical species of *Pavonia* subgen. *Asterochlamys*.

US-3010). The type was cited incorrectly as "1853" by Ulbrich.

Shrubs 1–2 m tall, the stems often setose (hairs 1.5–3 mm long, or these lacking) and with an understory of short stellate and septate glandular hairs. *Leaf* blades ovate, mostly 2–8 × 1.5–7 cm, cordate, crenate-serrate, acuminate, usually unlobed (or weakly lobulate), with simple and stellate hairs (and scattered 3-armed spicules) beneath and with appressed simple and bifurcate hairs above 0.5–2 mm long; petioles 1.5–7 cm long; stipules subulate, 2–3 mm long. *Flowers* solitary in the leaf axils, the pedicels 2.5–6.5(–11) cm long, slender; involucellar bracts 6–8, broadly lanceolate, 6–12 (–15) × 2–3 mm, ciliate (especially toward base); calyx 4–7 mm long, ciliate (sometimes densely so); corolla 1–1.5 cm long, yellow with red center, sparsely pubescent externally, bearded on margins of claw, otherwise glabrous; staminal column 5 mm long, glabrous, the filaments 0.5–1 mm long; styles emergent from the column, glabrous. *Fruits* oblate, 5–7 mm diam., pubescent, each mericarp with a minute triangular subapical protrusion.

Distribution (Fig. 55). From Bolivia, Paraguay, and NW Argentina, at elevations up to 2500 m. The type locality is at 15°47′S, 68°38′W; that of *Pavonia campestris* is at 22°42′S, 57°22′W; that of *P. bangii* is at 17°20′S, 66°18′W.

Illustration. Gürke (1892b: pl. 97, fig. 2).

Representative specimens examined. **BOLIVIA.** Camataquí, *Fiebrig 3057* (F, GH, MO, NY, S). **CHUQUISACA:** Pcia. L. Calvo, 3 km S of Isirenda, *Saravia et al. 11603* (CTES, TEX). **COCHABAMBA:** Pampas near Lake Rogagua, *Cárdenas 1341* (PH), *Cárdenas 1347* (US). **SANTA CRUZ:** Santa

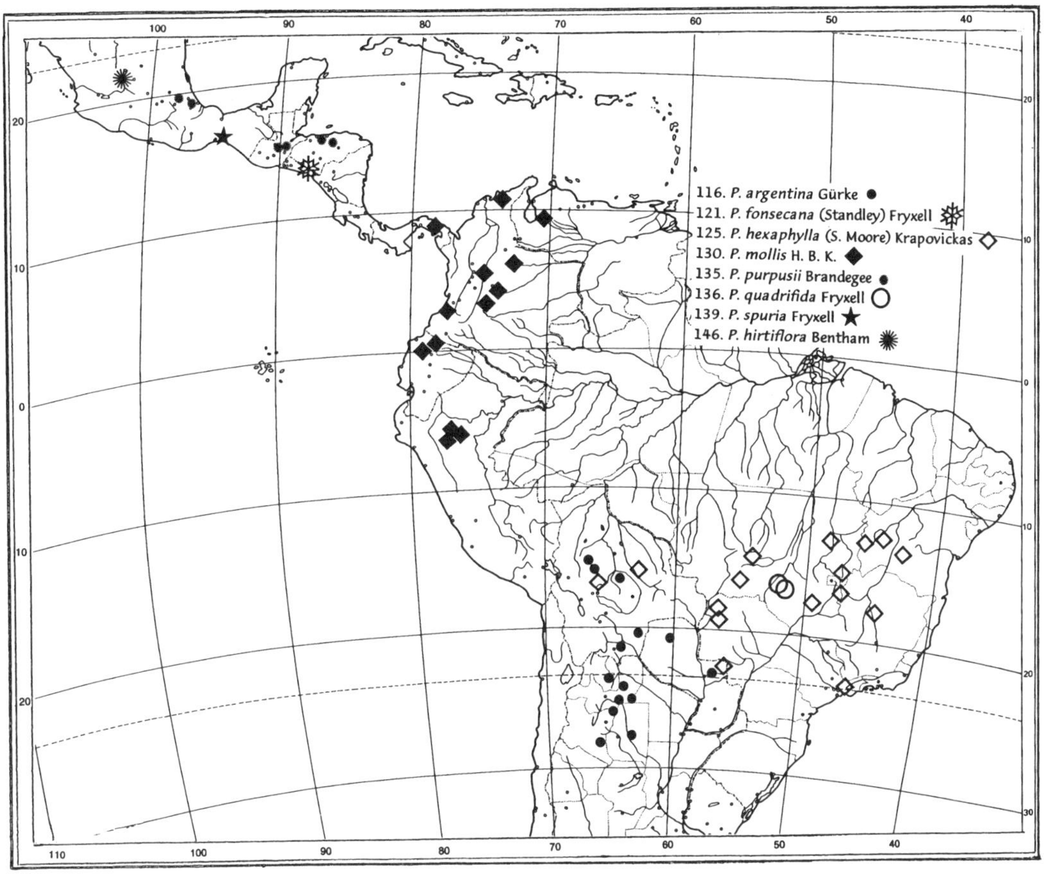

FIG. 55. Distribution map of eight neotropical species of *Pavonia* sect. *Asterochlamys*.

Cruz, *Badcock 412* (K); Prov. Cordillera, Camiri, *Cárdenas 4767* (CAS); Prov. Sara, Bañado del Piray, *J. Steinbach 5258* (GH). TARIJA: Prov. Gran Chaco, 15 km N of Yacuiba, *S. Beck 11462* (CTES, US); Prov. Gran Chaco, 7 km N of Villa Montes, rd. to Camiri, *Krapovickas & Schinini 39217* (CTES, G, NY, UC).

PARAGUAY. CHACO: Cerro León, *Schinini & Bordas 17854* (BAB, C, CTES, MO, NY); Garai-Cué, Campo Abente, *Anisits 2747* (syntype, S).

ARGENTINA. CATAMARCA: Dept. Capital, rd. to "La Quebrada," *Krapovickas 1796* (CAS); Catamarca, *Schunck 7908* (G). CÓRDOBA: Valley near "La Reducción," *Hieronymus 662* (syntype, B as photo F-9476); Dept. Río Seco, Serranías between Villa de María and San Miguel, *Hunziker 10688* (CTES, MO, NY); Dept. Colón, Sierra Chica, Quebrada del Río Primero, El Diquecito, *Hunziker 15404* (CTES); Alta Gracia, Estancia Los Molles, 14 Feb 1944, *Pierotti s.n.* (CAS, U, US); Dept. Colón, El Payador, *de la Sota 2268* (CAS, CTES). FORMOSA: La Rinconada, 2.5 km S of the Pilcomayo River, *Molina & Cabral 345* (CTES). JUJUY: Near Yuto, *Cabrera & Fabris 16194* (C, CTES); Dept. Santa Barbara, rd. to Palma Sola, *Cabrera & Fabris 19918* (CTES, NY); Dept. San Pedro, 7 km from the Río Grande, rd. to Santa Clara, *Krapovickas & Schinini 39296* (CTES, F, G, UC). LA RIOJA: Sierra de los Llanos, Quebrada de Polco, 6 km SE of Chamical, *Hunziker & Di Fulvio 16927* (CTES); Dept. Gral. Belgrano, vic. of Dique de Olta, *Hunziker et al. 13950* (CTES, NY). SALTA: Urundel, Colorado River, *Krapovickas 1567* (CTES, NY, S); Dept. Anta, near Estancia "San Jorge," *Pedersen 10824* (C, CTES, MO); Dept. San Martín, Gral. Ballivián, *Pedersen 12868* (C, CTES, NY); Dept. Orán, km 10 Vespucio to Hickman, *Schreiter 11463* (CAS, CTES, F, U). SAN JUAN: Dep. Valle Fertil de Sa. from Rivero to Astica, *Kiesling & Meglioli 6661* (CTES). SANTIAGO DEL ESTERO: Dept. Choya, near Villa la Punta, *Pedersen 12806* (C, CTES, NY); Dept. C. Pellegrini, Estancia El Remate, *Venturi 5852* (CAS, US). TUCUMÁN: Dept. Burruyaco, Cerro del Campo, *Bailleti 274* (G); Dept. Leales, San Francisco, *Venturi 742* (US); Dept. Cruz Alta, Corre río el Timbó, *Venturi 2375* (US).

Among the yellow-flowered species of the *Pavonia sidifolia* complex, *P. argentina* is distinguished by the dense covering of appressed simple hairs on the upper leaf surface.

117. Pavonia aurelii Fryxell, sp. nov. Type. Paraguay. Amambay: Cerro Corá, Cerro Yryvuy, 17 Aug 1980,

Schinini & Bordas 20153 (holotype, CTES; isotype, TEX). Fig. 56

Suffrutices dense tomentosi; laminis foliorum ovatis discoloribus; floribus solitariis; bracteis involucellorum 7–8 quam calyces subaequantibus; corollis 4 cm longis roseis; fructibus transverse costatis oblatis-pentagularibus, mericarpiis apiculatis plus minusve dehiscentibus.

Shrubs with stems densely tomentose, the hairs 0.3–0.4 mm long. *Leaf* blades ovate, 6–10 × 4.5–8 cm, basally cordate to subtruncate, serrate, acute or acuminate, palmately 7–9-nerved, discolorous, softly tomentose on both surfaces; petioles 3.5–8.5 cm long (subequal to blade to half length of blade), with pubescence like stem; stipules filiform, 12–13 mm long, tomentose. *Flowers* solitary in the leaf axils, the pedicels 1–5 cm long, tomentose like stem; involucellar bracts 7–8, ligulate, 10–11 mm long, softly tomentose (hairs 0.3–0.4 mm long); calyx 9–10 cm long (subequal to involucel or a little shorter), softly tomentose, half-divided, the lobes 3-ribbed; petals 4 cm long, rose, externally with scattered simple and stellate hairs; staminal column ca. 2 cm long, slender, glabrous, the filaments 1–1.5 mm long; styles exserted from column for ca. 9 mm, free for distal 4 mm, minutely glandular-pubescent. *Fruits* ca. 8 mm diam., transversely ribbed, oblate-pentangular; mericarps more or less dehiscent, glabrous except pubescent on triangular subapical apiculum ca. 1 mm long.

Distribution (Fig. 57). Known only from the type collection. The type locality is at 22°20′S, 56°2′W.

The specific epithet is chosen to honor the collector of the type specimen, Aurelio Schinini (b. 1943), discriminating collector of South American plants.

118. Pavonia balansae Gürke *in* Martius, Fl. Bras. **12(3):** 507. 1892. *Asterochlaena balansae* (Gürke) Hassler, Feddes Repert. Spec. Nov. Regni Veg. **7:** 382. 1909. Type. Paraguay. Near Paraguari, Cerro Hué, 8 Apr 1883, *Balansa 4385* (lectotype, here designated, F; isolectotypes, B as photo F-9431, F, K, S [fragment]). Fig. 53

Pseudopavonia tenax Hassler, Feddes Repert. Spec. Nov. Regni Veg. **7:** 74 (1909). *Asterochlaena balansae* subsp. *tenax* (Hassler) Hassler, Feddes Repert. Spec. Nov. Regni Veg. **7:** 382 (1909). Type. Paraguay. Gran Chaco: In hills near Fuerte Olimpo, Dec. 1907, *Fiebrig 1466* (holotype, G; isotypes, CTES, K). Note: The holotype is marked "unicum," evidently incorrectly.

Asterochlaena balansae forma *longipetiolata* Hassler, Feddes Repert. Spec. Nov. Regni Veg. **7:** 383 (1909). Type. Paraguay. Dept. Olimpo, Centurion, between the Apa River and the Aquidaban River, *Fiebrig 4197* (isotypes, CTES [fragment], E-2, GH, M as photo CTES).

Asterochlaena sidifolia forma *crenata* Hassler, Feddes Repert. Spec. Nov. Regni Veg. **8:** 114 (1910). Type. Paraguay. Between the Apa River and the Aquidaban River, sandy rocky hill with sparse woody vegetation, Dec 1908, *Fiebrig 4434* (isotypes, CTES, GH, M as photo CTES).

Shrubs 1–2.5 m tall, the stems densely glandular-puberulent and with scattered setae 2–3 mm long, also with narrow lines of dense recurved hairs. *Leaf* blades broadly cordate-ovate (suborbicular), sometimes angulate or weakly lobulate, crenate, subobtuse to acute or acuminate, mostly 4–9 × 4.5–10.5 cm, palmately 9-nerved, discolorous, softly puberulent; petioles 3–11 cm long, with pubescence like that of stem but the setae few or absent; stipules subulate, 4–6 mm long. *Flowers* solitary or paired in the leaf axils or terminally aggregated, the pedicels 1–4 cm long, articulated 4–5 mm below the flower; involucellar bracts 9–10, viscid and setose, 9–13 × 1 mm or less, linear; calyx 6–8 mm long, pubescent; corolla yellow, 1.5–2.5 cm long, densely pubescent on margins of claw; staminal column 11–14 mm long, glabrous, pallid, the filaments 2–3 mm long, the styles emergent from the column. *Fruits* 7–8 mm diam., 4–5 mm high, oblate, subglabrous, nearly smooth, straw-colored; seeds 4 mm long, glabrous, strongly striate.

Distribution (Fig. 18). From Brazil and Paraguay. The type locality is at 25°40′S, 57°10′W; that of *Pseudopavonia tenax* is at 21°00′S, 57°50′W; that of *Asterochlaena sidifolia* forma *crenata* is approximately at 22°30′S, 57°30′W.

Illustration. Gürke (1892b: pl. 97, fig. 1)

Specimens examined. **BRAZIL. MATO GROSSO DO SUL:** 15–20 km W of Caracol, *Hatschbach & J. M. Silva 49204* (C, CTES, US).

PARAGUAY. Dept. Olimpo, Fuerte Olimpo, *Rojas 13858* (C, CAS, CTES, K, NY, TEX, US); Ipacarai, *Rojas 14183* (CAS, CTES, F, NY, TEX, US); Cerro Peran, near Paraguari, *Balansa 1612* (syntype, CTES).

119. Pavonia biflora Fryxell, Phytologia **37:** 295. 1977. Type. Brazil. Distrito Federal: Cachoeira Piripiripau, ca. 15 km S of Planaltina, 20 Feb 1970, *Irwin, da Fonsêca, Souza, Reis dos Santos & Ramos 26416* (holotype, UB; isotypes, CTES [fragment], F as photo F-56160, MO, NY, TEX, US). Fig. 53

Shrubs 2–2.5 m tall, the stems densely pubescent with glandular hairs, sometimes also stellate hairs, and occasionally with long setae. *Leaf* blades ovate-cordate, crenate-serrate, acuminate, markedly discolorous, softly tomentulose beneath, with predominantly glandular hairs above, mostly 5–11 × 2.5–7 cm, palmately 7–9-nerved; petioles 1–4 cm long, with pubescence like that of stem; stipules subulate, 5–6

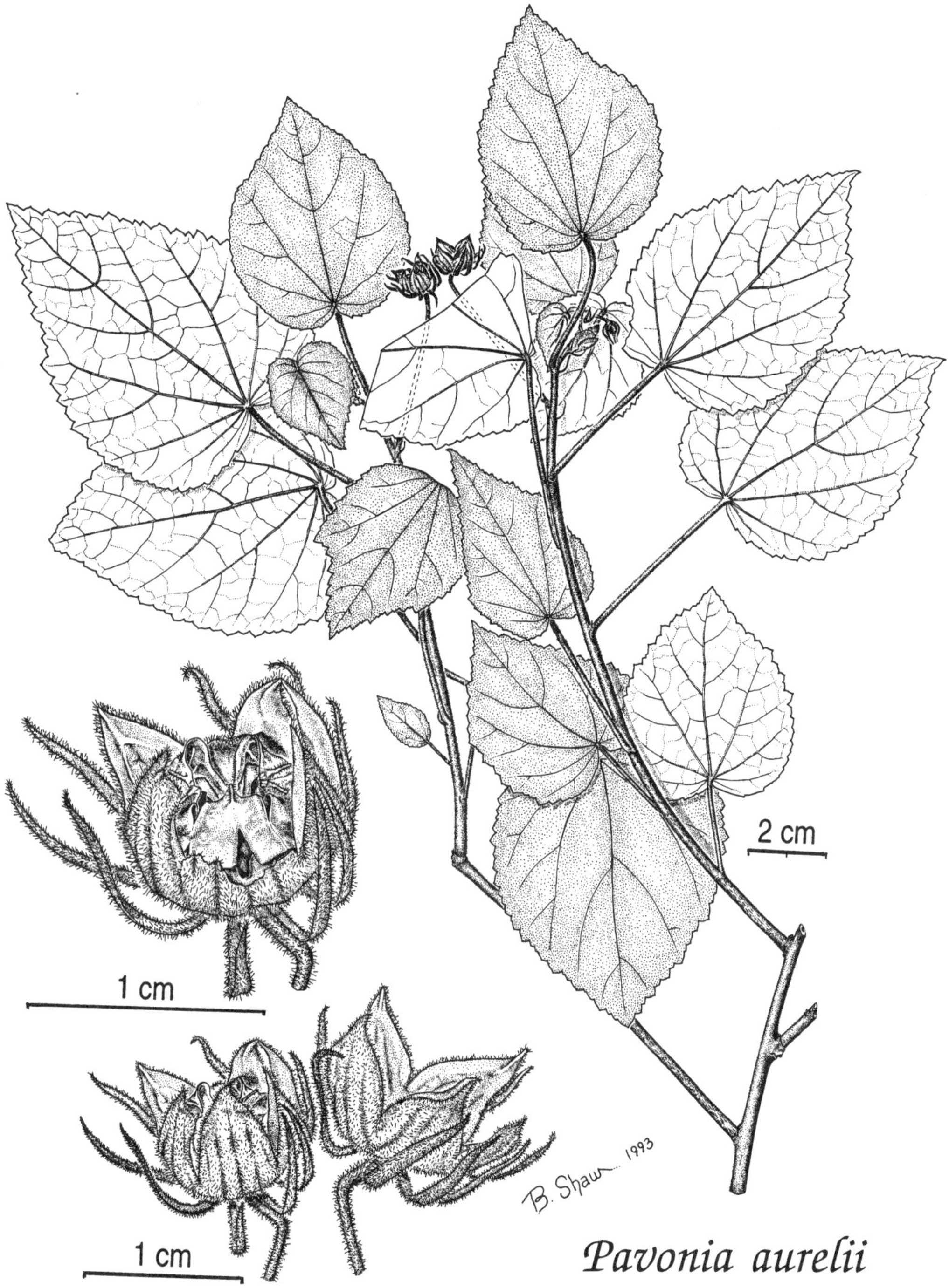

FIG. 56. *Pavonia aurelii*. Flowering branch with enlargements of fruits (*Schinini & Bordas 20153*).

mm long. *Flowers* solitary in the leaf axils, the pedicels 4–8 cm long, with pubescence like that of stem, commonly bearing two flowers (or flowers solitary); involucellar bracts (4–)5–6, ovate to lanceolate, 10–14 × 5–7 mm, puberulent; calyx subequal to involucel, densely ciliate; corolla 3–5.5 cm long, pink with

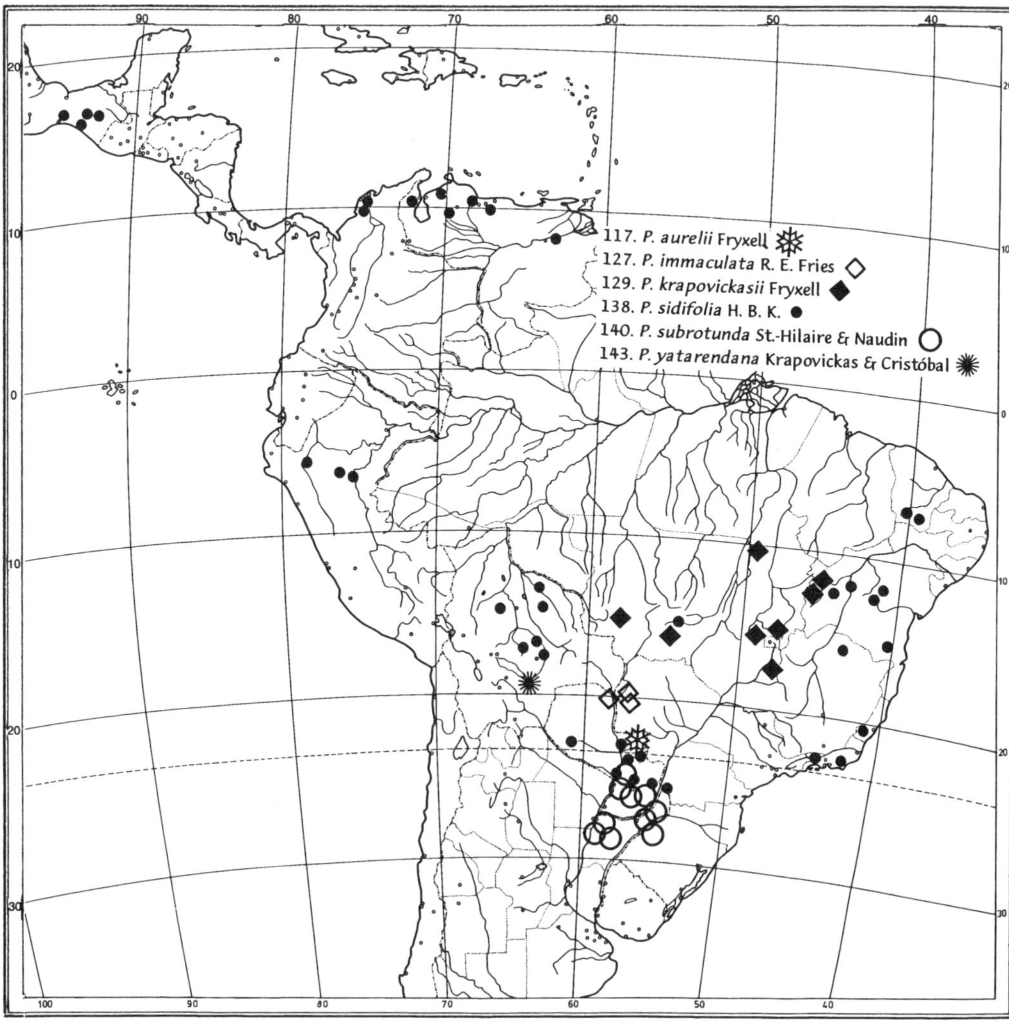

FIG. 57. Distribution map of six neotropical species of *Pavonia* sect. *Asterochlamys*.

red center, glabrous on margins of claw; staminal column 2 cm long, glabrous, the filaments 1 mm long; styles emergent from column, with capitate, villous stigmas. *Fruits* oblate, 7 mm diam., 4–5 mm high, nearly glabrous, each mericarp with a triangular subapical protrusion.

Distribution (Fig. 58). Central Brazil (Distrito Federal, Mato Grosso, Minas Gerais, Paraná, and São Paulo) with one collection from Bolivia, growing in gallery forest and adjacent cerrado. The type locality is at 15°31′S, 47°36′W.

Illustration. Fryxell (1977: fig. 8).

Specimens examined. BRAZIL. Distrito Federal: Sobradinho, Piripiripau, *Heringer 14315* (CTES); basin of the São Bartolomeu River, vic. of Cachoeira do Piripiripau, *Heringer et al. 6641* (CTES, K, MO, US); 20 km S of Frutal, *Krapovickas et al. 33037* (CTES, F, MBM, herb. Pedersen). **Mato Grosso:** Rd. to Xavantina, 5–7 km N of Balsa Río Pardo, *Gregory et al. 9760* (CTES); Mun. Bataguaçu, Corrego Feio, *Hatschbach 23559* (CTES); Grego bridge, Aquidauna River, 48 km NE of Terenos, *Krapovickas & Cristóbal 34494* (CTES, MO). **Minas Gerais:** Jequitinhonha River, 31 km SE of Diamantina, *Arbo et al. 5169* (CTES); rd. from Grão Mogol to Virgem da Lapa, banks of Ventania River, *Cordeiro et al. 974* (NY, SPF); 6 km W of Alexandrita, 6 Feb 1984, *Cross s.n.* (NY). **Paraná:** Mun. Guaira, Sete Quedas, *Buttura 426* (CTES). **São Paulo:** Mun. Botucatú, 18 km N of Botucatú, *I. S. Gottsberger & G. Gottsberger 1004* (TEX, ULM).

BOLIVIA. Santa Cruz: Prov. Chiquitos, 1 km NE of Roboré, *Krapovickas & Schinini 36433* (CTES).

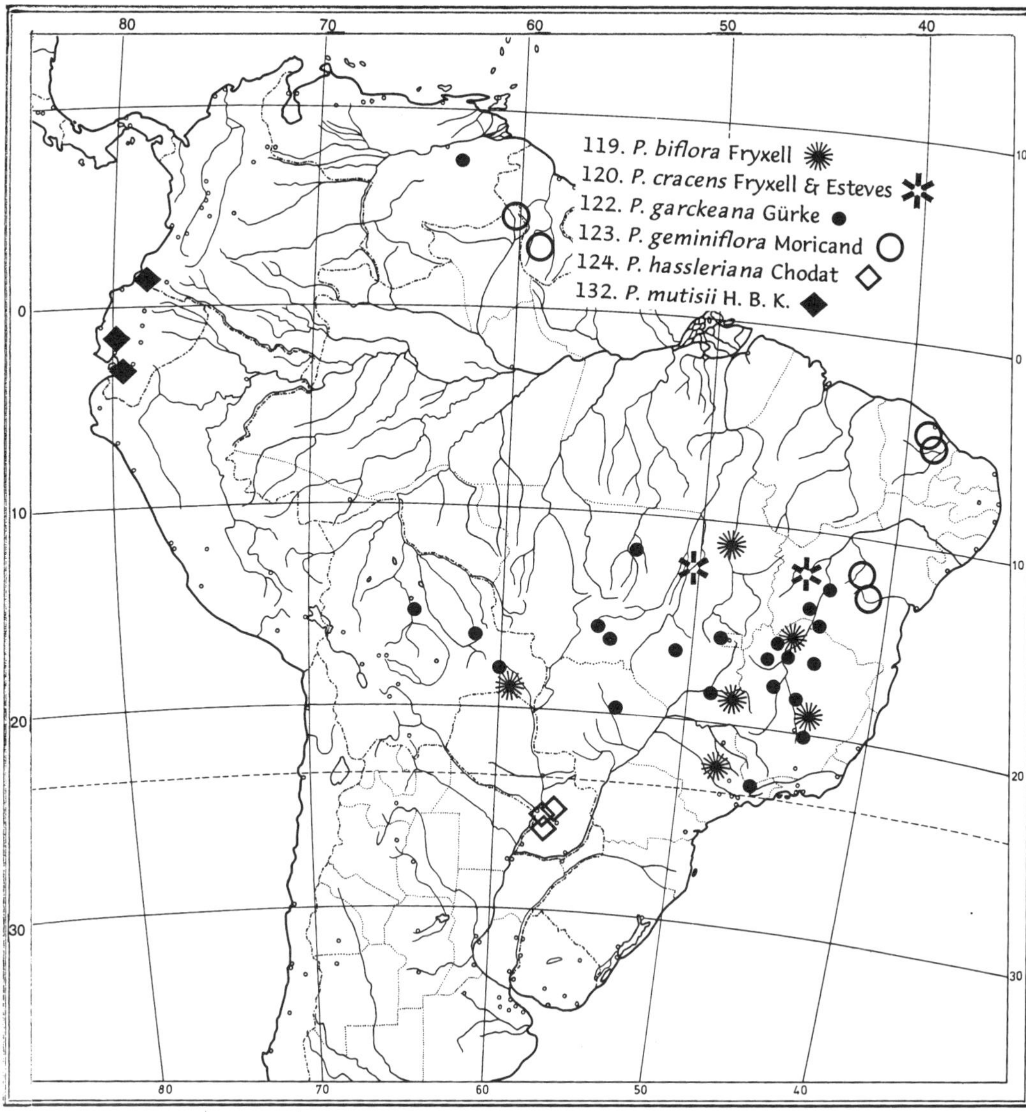

FIG. 58. Distribution map of six South American species of *Pavonia* sect. *Asterochlamys*.

Pavonia biflora is similar to *P. garckeana* and undoubtedly closely related to it, but with narrower, more numerous involucellar bracts.

120. Pavonia cracens Fryxell & Esteves, Bol. Bot. Univ. São Paulo **17:** 4. 1998. Type. Brazil. Bahia: Mun. Barreiras, Serra do Saco, 10 Oct 1981, *Hatschbach 44113* (holotype, MBM; isotypes, C, CEPEC, CTES, NY, RB, US, herb. Pedersen). Fig. 59

Frutex gracilis viscidus; laminis foliorum parvis anguste ovatis dense pubescentibus; pedicellis 1.5–2 cm longis axillaribus; bracteolis involucellorum 6 linearis viscidis; calyce quam involucello breviore; corollis 1.5 cm longis subroseis; columnis staminalis inclusis, antheris subsessilibus; fructibus ignotis.

Slender-stemmed *shrub* (scandent?), the stems densely glandular-puberulent (hairs 0.5 mm long). *Leaf* blades ovate, mostly 12–18 × 5–8 mm, basally truncate, serrulate, acute, palmately 3–5-nerved, discolorous, the upper surface densely pubescent with stellate and glandular hairs, the lower surface velutinous (the epidermis hidden), the stellate hairs of variable size, including scattered 3–4-armed spicules ca. 1 mm diam.; petioles 2–5 mm long, glandular-puberulent; stipules 3 mm long, subulate,

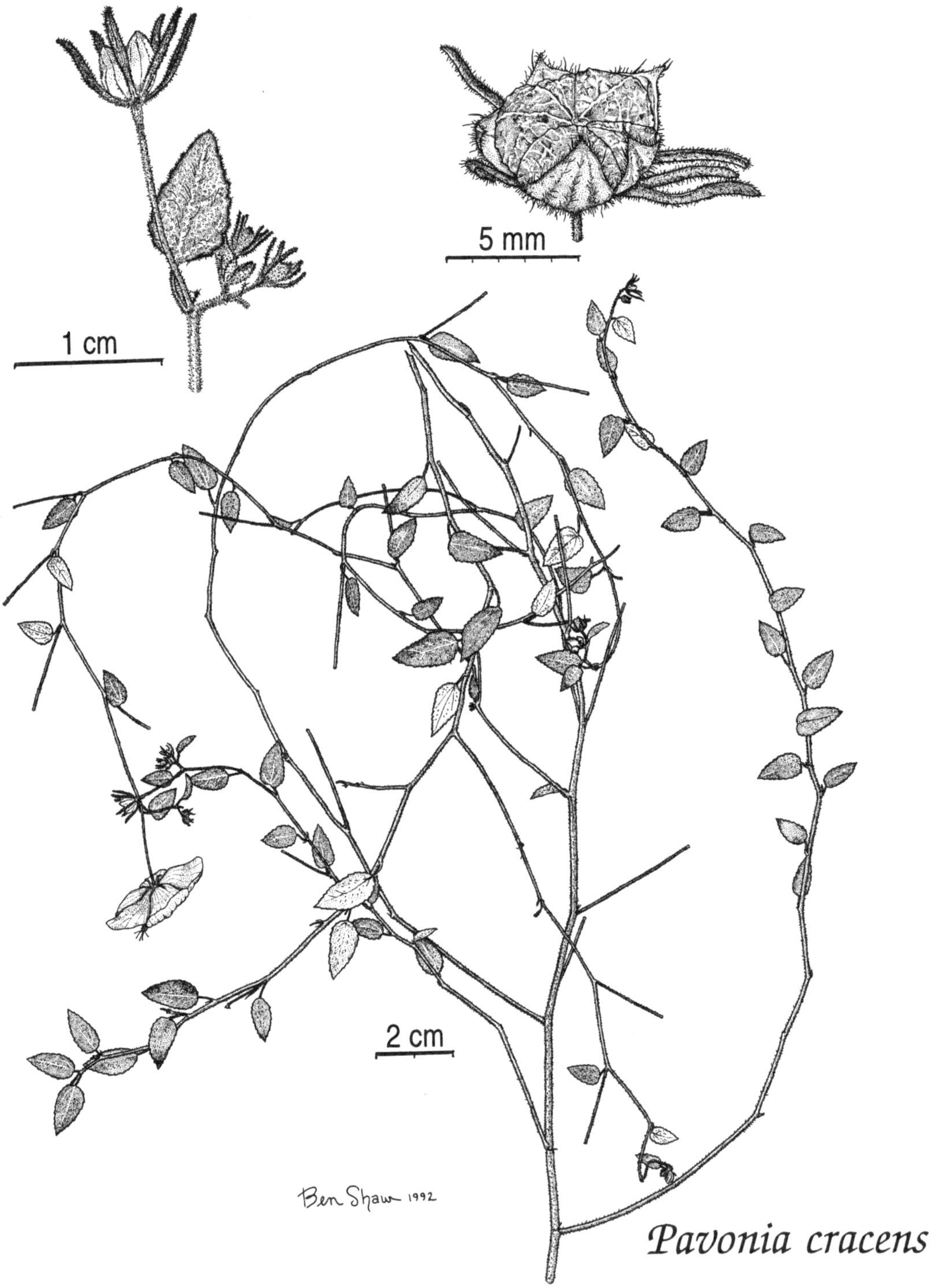

FIG. 59. *Pavonia cracens*. Plant habit (presumably decumbent), with enlargements of young buds (upper left) and mature fruit (upper right) (*Hatschbach 44113*).

glandular-puberulent. *Flowers* solitary in the leaf axils, the pedicels 1.5–2 cm long, glandular-puberulent; involucellar bracts 6, linear, 5–7 × 0.5–0.7 mm, glandular-puberulent; calyx 5 mm long, minutely pubescent; corolla 1.5 cm long, pink, externally stellate-pubescent, densely bearded on margins of claw; staminal column 8–9 mm long, the anthers subsessile; styles exserted from column by 3 mm, the stigmas pinkish, subequal to corolla. *Fruits* oblate, 6 mm diam.; mericarps 3.5 mm high, glabrous, verruculose.

Distribution (Fig. 58). From central Brazil (Bahia and Mato Grosso). The type locality is approximately at 12°10′S, 45°0′W.

Specimens examined. BRAZIL. MATO GROSSO: Serra do Roncador, 260 km NNE of Xavatina, *G. Eiten & L. Eiten 8975* (US).

The Mato Grosso specimen (*Eiten & Eiten 8975*) is assigned to this species only provisionally, as the specimen is an incomplete one.

The specific epithet is from the Latin (*cracens*; graceful, slender) in reference to the growth habit of the plant. *Pavonia cracens* is the smallest-flowered member of the species group that includes *P. biflora*, *P. garckeana*, *P. hexaphylla*, *P. krapovickasii*, and *P. vinosa* and probably also has the smallest stature of these species.

121. Pavonia fonsecana (Standley) Fryxell *in* Dorr & Berendsohn, Cuscatlania **1(10)**: 173–8. 1997, comb. nov. Basionym: *Malache fonsecana* Standley, J. Wash. Acad. Sci. **14:** 239. 1924. Type. El Salvador. La Uni6n: Vic. of La Uni6n on Gulf of Fonseca, 13–21 Feb 1922, *Standley 20677* (holotype, US; isotypes, GH, NY).

Shrubs 1–2 m tall, the stems densely glandular-puberulent ("very viscid, strong guava odor" fide collector), the hairs 0.2–0.3 mm long. *Leaf* blades ovate, mostly 3–4.5 × 2–2.5 cm, basally cordate, serrate, acute or acuminate, palmately 5–7-nerved, discolorous, densely glandular-puberulent above, densely whitish stellate-pubescent beneath; petioles 1–2 cm long; stipules subulate, 3 mm long, viscid. *Flowers* solitary in the leaf axils, the pedicels 3–5.5 cm long, viscid; involucellar bracts 7–8, linear, 10 × 1 mm, viscid-puberulent; calyx 6–7 mm long, hispid-ciliate (hairs ca. 1 mm), pallid with purplish veins (3 per lobe); corolla externally pubescent, 12 mm long, apparently white, pubescent on margins of claw. *Fruits* 7 mm diam., oblate, glabrous; mericarps unornamented, smooth, 5 mm tall, with a minute subapical beak.

Distribution (Fig. 55). Known only from the type collection. The type locality is at 13°21′N, 87°52′W.

The citation of the 1997 combination given above is a very nonconventional method of pagination, a method sure to give problems to librarians and bibliographers, as well as to botanists. The page citation "173-8" refers to family no. 173 (i.e., the Malvaceae), the eighth page of that family, but *not* the eighth page of issue no. 10. The sequence of pagination of vol. 1, no. 10 is as follows: i–ii (introductory material), 170-1, 172-1 to 172-3, 173-1 to 173-11, 174-1 to 174-2, 175-1 to 175-3, and iii–vi (bibliography). There are thus 26 pages in the issue, devoted to the Malvales, but it is difficult to cite a reference unequivocally. One hopes that this method of pagination will not be emulated widely.

Pavonia fonsecana is related to *P. purpusii* and *P. spuria*, the three being very similar morphologically. In contrast to the other two species, *P. fonsecana* has glabrous fruits.

122. Pavonia garckeana Gürke *in* Martius, Fl. Bras. **12(3):** 496. 1892. *Goethea garckeana* (Gürke) Garcke, Bot. Jahrb. Syst. **21:** 395. 1896. Type. Brazil. Without locality, *Glaziou 13545* (lectotype, here designated, C; isolectotypes, BR, CTES [fragment], F [fragment]). Fig. 53

Shrubs 1–2 m tall, the stems invested with short glandular hairs and with narrow lines of whitish stellate and recurved simple hairs, rarely also with occasional long setose hairs. *Leaf* blades ovate (rarely lobulate), cordate, crenate-serrate, acuminate, markedly discolorous, softly stellate-tomentulose, the hairs on the upper surface smaller than those on the lower surface and often bifurcate, mostly 4–10 × 3–9 cm, palmately 7-nerved; petioles 1–4 cm long; stipules subulate, 3–7 mm long. *Flowers* solitary in the leaf axils (rarely paired), the pedicels 2–8 cm long, with pubescence like stem, usually 1-flowered (sometimes 2-flowered, the pair crowded); involucellar bracts 4, broadly ovate-cordate, 9–17 × 10–17 mm, puberulent; calyx usually shorter than the involucel, ciliate; corolla 3–5 cm long, pink with red center, the margins of the claws glabrous; staminal column 2–3 cm long, glabrous, pallid, the filaments 1–2 mm long, disposed in several whorls; styles pallid, the stigmas capitellate. *Fruits* oblate, 7–8 mm diam., nearly glabrous; mericarps 3.5–5 mm high, lacking a subapical projection but with minute basal wings; seeds 3.5 mm long, striate. Chromosome number, $2n = 56$ (Fernández, 1981; Krapovickas & Cristóbal, 1965).

Distribution (Fig. 58). From Venezuela, Brazil (Bahia, Distrito Federal, Goiás, Mato Grosso, Minas Gerais, and São Paulo), and Bolivia, in cerrado vegetation and savanna, at 300–1000 m elevation.

Illustration. Gottsberger (1972: fig. 21, as *P.* cf. *lindmannii*); Gürke (1892b: pl. 92).

Representative specimens examined. **VENEZUELA.** **Bolívar:** Cerro Altamira, *Maguire et al. 35911* (NY, TEX); Chinese slide on N lower slope of Cerro Bolívar, *Maguire et al. 35963* (NY, TEX).

BRAZIL. Bahia: 60 km E de Barreiras, *Krapovickas et al. 37881* (CEN, CTES, G). **Distrito Federal:** Paranuá dam, *Heringer 8894/1088* (CTES); Brasilia, basin of São Bartolomeu River, *Heringer et al. 5471* (K, NY). **Goiás:** Mun. Cristalina, Lageado, *Hatschbach 43833* (C, CTES, MO, NY, US); Mun. Rio Verde, BR-060, *Hatschbach & Kummrow 34957* (CTES). **Mato Grosso:** Km 8 on rd. from Vila Bela to Pontes and Lacerda, *Carreira et al. 822* (NY); Mun. Rondonopolis, Serra da Petroliva, *Hatschbach 34729* (CTES); Mun. Sto. Antonio do Leverger, 4 km on BR-364, rd. to Barão de Melgaço, *Krapovickas & Cristóbal 43128* (C, CTES, HAS, K, MBM, NY); Mun. Cáceres, 18 km W of Cáceres, *Krapovickas et al. 40111* (C, CEN, CTES, F, K, NY). **Minas Gerais:** Serra do Espinhaço, by Jequití River, 2 km from Mendanha, *W. R. Anderson, P. A. Fryxell et al. 8834* (F, K, NY, RB, TEX, US); 6 km S of Rio São Francisco, rd. from Montes Claros to Januária, *W. R. Anderson, P. A. Fryxell et al. 9017* (F, MO, NY, TEX); 1 km E of Rio Paneiros, near rd. to Januária, *W. R. Anderson, P. A. Fryxell et al. 9102* (F, NY, TEX, U); Mun. Belo Horizonte, Freitas, *Barreto 7965* (F); Diamantina, toward Mendanha, *Esteves 2331A* (CTES, NY, SPF); Francisco Sá, *Hatschbach 42831* (CTES, NY); Mun. Grão Mogol, Itacambiruçú River, *Hatschbach 41310* (CTES, MO, NY, herb. Pedersen); Mun. Jequitié, BR-367, between km 70 and 72, *Hatschbach & Ribas 55240* (CTES, TEX); Mun. Cristalia, Bem Querer, *Hatschbach et al. 55000* (C, CTES, MO); Serra do Espinhaço, ca. 8 km W of Grão Mogol, *Irwin et al. 23333* (F, K, MO, NY, TEX, US); 43 km E of Ituiutaba, intersection of BR-153 & BR-365, *Krapovickas & Cristóbal 42812* (CTES, MBM, NY); 40 km NE of Patos de Minas, Côrrego Segredo, *Krapovickas & Cristóbal 42843* (CTES, K, MBM, NY); Grão Mogol, near Corrego da Bonita, *Zappi et al. 12086* (NY, SPF). **São Paulo:** Mun. Botucatú, along São Marcel–Piracicaba hwy., *G. Gottsberger 11-16271* (US); Rancharia, *Gregory et al. 9737* (CTES, NY). **Without locality:** *Pohl 3344* (syntype, not seen); 1883, *Claussen s.n.* (syntype, not seen).

BOLIVIA. Santa Cruz: Prov. Velasco, 10 km E of San Ignacio, *Krapovickas & Schinini 32265* (C, CTES, F, G, MBM, MICH, MO, NY, SI); Prov. Chiquitos, Roboré, *Krapovickas & Schinini 36456* (CLEMS, CTES, K); Prov. Chiquitos, 10 km S of San José, *Krapovickas & Schinini 36657* (CLEMS, CTES, MICH, MO, NY); Velasco, Jul 1892, *Kuntze s.n.* (B as photo F-9444, CTES [fragment], NY).

Pavonia garckeana is characterized by having only four ovate involucellar bracts, a characteristic it shares with *P. quadrifida*, whose bracts, however, are not basally cordate as they are in *P. garckeana*. The nearest relative of *P. garckeana* is evidently *P. biflora*, which has 5–6 lanceolate or narrowly ovate bracts.

Gottsberger (1972) observed this species (as *P.* cf. *lindmannii*) to be bee-pollinated.

123. Pavonia geminiflora Moricand, Pl. Nouv. Amer. 120. t. 73. 1844. Type. Brazil. Bahia: *Blanchet 944, 1334* (lectotype, here designated, G; isotype, G). The lectotype is a single specimen bearing two numbers "944" and "1334"; the isotype bears the number "944." The lectotype was clearly the subject for Moricand's plate no. 73. Fig. 53

Herbs or subshrubs 0.5–2 m tall, the stems setose (hairs 2–3 mm long) and with dense understory of minute glandular hairs. *Leaf* blades ovate to shallowly 3-lobulate, mostly 4–8 × 2–5 cm, cordate, serrate, acute to long-acuminate, palmately 5–7-nerved, both surfaces with appressed setae (especially on veins) and obscure but abundant glandular hairs; petioles 2–5 cm long; stipules filiform, 4–5 mm long, somewhat reflexed. *Flowers* solitary (occasionally paired) in the leaf axils, the pedicels 2–7.5 cm long (subequal to or exceeding subtending petioles, or the upper ones exceeding the leaves), slender; involucellar bracts 8–10, linear, viscid and setose, 6–11 mm long; calyx 5–7 mm long, ciliate; corolla 12–22 mm long, pale pink (drying yellowish), glabrous except white-bearded at juncture of claw and staminal column; staminal column 9–15 mm long, glabrous, pallid, declined and somewhat curved, the filaments secund, 2–3 mm long; styles united below, free for distal 3–4 mm, emergent from the column and nearly equaling the petals. *Fruits* 7–8 mm diam., glabrous, reticulate-veined; mericarps 5–6 mm high, subapically apiculate, minutely winged basally; seeds obscurely striate.

Distribution (Fig. 58). Occurs in Guyana and in the states of Bahia and Ceará in Brazil, in scrub forest at 500–1000 m elevation.

Illustrations. Gürke (1892b: pl. 96, fig. 3); Moricand (1844: pl. 73).

Specimens examined. **GUYANA**. Sakuth, *Appun 2311* (K); Rupununi District, Dadanawa, Rupununi River, Bush Island, *Jansen-Jacobs et al. 2753* (MO, NY, TEX, U); near Roraima, *Schomburgk 468* (OXF).

BRAZIL. Bahia: 34 km E of Moro do Chapéu, Chapada da Diamantina, *Davidse et al. 11863* (MO, NY); Cruz das Almas, *Pinto 351* (NY). **Ceará:** Rd. from Barra de Ceará to Soure, 24 Jun 1955, *Black & Fernandes 1280* (CTES); Mun. Moranguapa, S. Antonio do Berraco, *Drouet 2591* (CAS, F, GH, MO, NY, S, US); Maracanahú, *Fernandes 29* (F); Soledade, Caucaia, 11 Jun 1974, *Fernandes 2428* (CTES).

Pavonia geminiflora is similar in general aspect, especially in their horizontal presentation of flowers, to *P. harleyi*, although the two species are not closely related; see comparative comment under the latter species.

124. Pavonia hassleriana Chodat, Bull. Herb. Boiss., ser. 2, **1:** 402. 1901. Type. Paraguay. on woodland margin near Piribebuy, *Hassler 1922* (holotype, G; isotypes, K, NY). Fig. 53

Subshrubs 0.5–1.5 m tall, the stems with glandular hairs and sometimes a few stellate hairs. *Leaf* blades ovate (sometimes 3–5-lobulate), cordate, crenate, acute or acuminate, palmately 7–9-nerved, discolorous, mostly 3–7 × 2–5 cm, softly puberulent above and beneath; petioles 1–5.5 cm long; stipules subulate, 3–4 mm long. *Flowers* solitary in the leaf axils (sometimes geminate, rarely umbellate), the pedicels 2–5 cm long; involucellar bracts 5(–6), lanceolate, 5–9 × 1–2 mm, densely glandular-pubescent and sometimes with a few large stellate hairs; calyx subequal to involucel and with similar pubescence; corolla white, 1.5–2.5 cm long, externally covered with coarse 3–4-armed stellate hairs, the petals bearded on margins of claw; staminal column glabrous, exserted at anthesis, the filaments 2–3 mm long; styles emergent from the column, glabrous, with capitate stigmas. *Fruits* 5–7 mm diam., subglabrous, the mericarps verruculose, dehiscent, basally minutely winged; seeds 3 mm long, glabrous, striate.

Distribution (Fig. 58). Endemic to Paraguay. The type locality is at 25°27′S, 57°05′W.

Illustrations. Fries (1908: pl. 6, figs. 1–2).

Specimens examined. **PARAGUAY. CORDILLERA CENTRAL:** Near Chololo, in valley of the Yacá River, *Hassler 6785* (C, G as photo F-23700, GH, K, MO, NY, S). **PARAGUARÍ:** Near Pirareta Falls, *Arbo et al. 1715* (CTES); Cerros de Tobatí, *Fiebrig 692* (A, E–3, F, GH, M as photo CTES); near Pirareta Falls, *Krapovickas & Cristóbal 44588* (CTES, TEX); Paraguarí, *Lindman 3847* (B as photo F-9448, S–2); Pirareta, *Schinini 4230* (CTES, G); near Pirareta Falls, *Schinini 4369* (CTES, G, K, NY).

Pavonia hassleriana is distinguished by its manifestly exserted genitalia from a white corolla. It may be allied to *P. quadrifida*.

125. Pavonia hexaphylla (S. Moore) Krapovickas, Bonplandia **5:** 269. 1983. *Pavonia mutisii* var. *hexaphylla* S. Moore, Trans. Linn. Soc., ser. 2, **4(3):** 314. 1895. Type. Paraguay. Near Barra de San Lorenzo, *Moore 10* (holotype, BM as photo CTES; isotype, CTES [fragment]). Fig. 53

Pavonia kunthii var. *pohlii* Gürke in Martius, Fl. Bras. **12(3):** 512. 1892. Type. Brazil. *Burchell 6992* (lectotype, here designated, K; isolectotypes, BR as photo CTES, CTES [fragment], GH).

Spreading or subscandent *subshrubs* 1–2.5 m tall, the stems purplish, slender, covered with reddish glandular hairs and with narrow lines of recurved whitish hairs. *Leaf* blades ovate (sometimes weakly lobed, e.g., *Hatschbach 34048*), mostly 2.5–8.5(–16) × 1.5–5(–8) cm, crenate, acuminate, softly stellate-tomentulose, more sparsely so above than beneath, with some glandular hairs intermingled; petioles 0.5–5 cm long; stipules subulate, 2–3 mm long. *Flowers* solitary in the leaf axils, the pedicels 2–9 cm long, slender, usually 1-flowered, sometimes 2-flowered (rarely 3-flowered and subumbellate); involucellar bracts 6, linear-lanceolate, glandular-pubescent, 7–10 × 1(–2) mm; calyx 5–7 mm long, ciliate; corolla 1.5–3 cm long, pink with dark red center, pubescent on margins of claw; staminal column glabrous, shorter than petals, the filaments ca. 1 mm long. *Fruits* oblate, 6 mm diam., verruculate, subglabrous; seeds 3 mm long, striate. Chromosome number, $2n = 56$ (Krapovickas & Cristóbal, 1965, as *P. kunthii* var. *pohlii*).

Distribution (Fig. 55). From Bolivia, Brazil (Bahia, Goiás, Mato Grosso, Minas Gerais, and São Paulo), and Paraguay, in cerrado and forest vegetation at 400–800 m elevation.

Representative specimens examined. BRAZIL. BAHIA: Espigão Mestre, 10 km N of Roda Velha River, ca. 100 km WSW of Barrieras, *W. R. Anderson et al. 36883* (F, NY, TEX, US); Mun. Barreiras, Roda Velha, *Hatschbach 39448* (CTES); Espigão Mestre, 32 km W of Barrieras, *Irwin et al. 31528* (NY). **GOIÁS:** 20 km N of São João da Aliança, Chapada dos Veadeiros, *E. Y. Dawson 14425* (CAS, RSA); Mun. Jaraguá, Rio das Almas, *Hatschbach 38024* (C, CTES, MO, NY, herb. Pedersen); 12 km W of Caldas Novas, *Heringer & Eiten 14184* (MO, US); Serra do Morcêgo, Côrrego Estrema, 35 km NW of Formosa, *Irwin et al. 14974* (C, CTES, F, NY, US); 20 km N of Itumbiara, *Krapovickas et al. 33107* (CTES, SI, herb. Pedersen); near the banks of the Araguaia River, *Lima 58-3030* (K, NY); Mun. Mozarlandia, Cavalo Branco, *Pereira 397* (CTES); 51 km N of Aragarças, *Philcox & Fereira 4044* (CTES, K). **MATO GROSSO:** Ca. 20 km (straight line) ENE of Barra do Garças, *W. R. Anderson et al. 9854* (C, F, K, MO, NY, TEX, US); Urucum, Corumbá, *Dorrien Smith 86* (K); Mun. Itaquire, Rio Corrente, *Hatschbach 34048* (CTES); ca. 30 km S of Xavatina, *Irwin et al. 16982* (NY); Mun. Rondopolis, Serra da Petroliva, *Hatschbach 34727* (C, CTES, MU, S, US); Terenos, *Hatschbach 23694* (C, CTES, K, NY); behind Acurizal, Paraguai River, *Prance et al. 26096* (GH, NY). **MINAS GERAIS:** 10 km W of Iturama, 6 Feb 1984, *Cross s.n.* (NY); Mendanha, near the Jequitionha River, *Esteves & Kameyama 2477* (NY, SP); Serra do Espinhaço, 25 km E of Diamantina, near Jequití River, *Irwin et al.*

27556 (CTES, MO, NY, TEX); Itapagipe, MG-255, *Krapovickas & Cristóbal 42755* (CTES, K, MO, NY). **São Paulo:** Matão, *Gomes 463* (RB); Jaboticabal, rd. to Araraquara, *Gregory et al. 9972* (CTES, NY); Mun. Nova Granada, BR-153, km 47, *Pedersen 12049* (CTES).

BOLIVIA. Cochabamba: Cochabamba, *Cárdenas 3778* (CAS). **Santa Cruz:** Prov. Ñuflo de Chaves, km 11 on rd. from San Ignacio to Concepción, *Hopkins et al. 115* (CTES, NY, TEX).

PARAGUAY. Amambay: Estrella, *Soria 3733* (MO, NY). **Cordillera:** Tobatí, *Gregory et al. 9617* (MO).

Syntypes of *Pavonia kunthii* var. *pohlii* include: *Pohl 1160, 1303, 2667* (F) and *Regnell III-177* (BR as photo CTES, CTES [fragment], S–2).

Pavonia hexaphylla is related to *P. krapovickasii* and *P. vinosa*, from which it is distinguished by linear involucellar bracts and different leaf pubescence. More distantly related (with greater stature and larger flowers) are *P. biflora* and *P. garckeana* and (with smaller stature and smaller flowers) *P. cracens*.

126. Pavonia hieronymi Gürke *in* Martius, Fl. Bras. **12(3):** 509. 1892. *Asterochlaena hieronymi* (Gürke) Hassler, Feddes Repert. Spec. Nov. Regni Veg. **8:** 116. 1910. *Asterochlaena hieronymi* subsp. *brevipila* Hassler (loc. cit. = subsp. *hieronymi*). Type. Argentina. Salta: Near San José, *Hieronymus & Lorentz 237* (holotype, B, destroyed, as photo F-9449; isotypes, CTES, US). Fig. 53

Pavonia morongii S. Moore, Ann. New York Acad. Sci. **7:** 58. 1892. *Asterochlaena morongii* (S. Moore) Hassler, Feddes Repert. Spec. Nov. Regni Veg. **7:** 382. 1909. *Pavonia sidifolia* forma *morongii* (S. Moore) Hassler, Trab. Mus. Farm. Fac. Med. Buenos Aires **21:** 84. 1909. Type. Paraguay. Pilcomayo River, 12 Jan 1890, *Morong 872* (isolectotypes, CAS-fragment, GH, MO, NY–2, PH, US).

Pavonia geminiflora var. *grandiflora* Chodat & Hassler, Bull. Herb. Boiss., ser. 2, **5:** 300. 1905. *Asterochlaena hieronymi* var. *grandiflora* (Chodat & Hassler) Hassler, Feddes Repert. Spec. Nov. Regni Veg. **8:** 117. 1910. Type. Paraguay. In the region of the Apa River, *Hassler 7568* (lectotype, here designated, NY; isolectotypes, CTES, G, P as photo CTES, NY, S, UC as photo CTES). Note: In the 1910 publication, Hassler describes the stems of var. *grandiflora* as "laxe hirsuto, pilis ad 5 mm longis," but an examination of the type shows the stems to be densely stellate-pubescent, the hairs approximately 0.5 mm long, which is more in agreement with the protologue "caulibus aequaliter cinereo pubescentibus."

Asterochlaena hieronymi var. *montana* Hassler, Feddes Repert. Spec. Nov. Regni Veg. **8:** 117. 1910. Type. Paraguay. Centurion, on mtn. slope, *Fiebrig 4056* (isotype, CTES).

Asterochlaena hieronymi forma *angustiphylla* Hassler, Feddes Repert. Spec. Nov. Regni Veg. **8:** 117. 1910; *A. morongii* subforma *viscosissima* Hassler, Feddes Repert. Spec. Nov. Regni Veg. **8:** 118. 1910. Type. Paraguay. Centurion, between the Apa River and the Aquidaban River, *Fiebrig 4021* (holotype, G; isotype, K).

Asterochlaena hieronymi var. *pseudoapiculata* Hassler, Feddes Repert. Spec. Nov. Regni Veg. **8:** 117. 1910. Type. Paraguay. Gran Chaco: Puerto Talavera, on dry banks of the Paraguay River, *Fiebrig 1434* (holotype, G–n.v.; isotypes, CTES, S).

Pavonia hieronymi subsp. *tomentella* R. E. Fries, Kongl. Svenska Vetenskapsakad. Handl. **24(2):** 29. 1947. Type. Argentina. Chaco: Colonia Benítez, *Schulz 894* (holotype, S; isotype, CTES).

Suberect to trailing or subscandent *subshrubs* to 1–2 m tall, the stems glandular-pubescent, sometimes also sparsely setose. *Leaf* blades ovate (to triangular in distal leaves, rarely subhastate; *Woolston 1150*), mostly 2–7 × 1.5–6 cm, cordate, crenate-serrate, acute or acuminate, discolorous, sparsely to densely soft-puberulent, with scattered 3-armed spicules beneath or these absent; petioles 1–4.5 cm long; stipules subulate, 2–3 mm long. *Flowers* solitary in the leaf axils, the pedicels 1–5 cm long, slender; involucellar bracts 6–8, lanceolate, 6–10 × 0.5–2 mm, puberulent, sometimes viscid; calyx 4–5 mm long, pubescent; corolla 1.5–2.5 cm long, yellow with a dark red center, densely wooly on margins of claw; staminal column ca. 5 mm long, purplish, the filaments 1 mm long. *Fruits* oblate, 6 mm diam., 4.7 mm high, often retrorsely scabridulous on central ridge, otherwise obscurely pubescent to subglabrous, more or less reticulate-veined, with vestigial subapical protruberance, or this suppressed.

Distribution (Fig. 54). From Peru, Brazil, Bolivia, Paraguay, and Argentina, at 100–2800 m elevation, growing in shrublands and forest margins, sometimes in flooded areas. The type locality is approximately at 25°S, 66°W; that of *Asterchlaena hieronymi* var. *montana* is at 22°17′S, 57°33′W; that of *A. hieronymi* var. *pseudoapiculata* is at 20°45′S, 57°55′W; that of *Pavonia hieronymi* subsp. *tomentella* is at 27°20′S, 58°57′W.

Illustration. Gürke (1892b: pl. 97, fig. 3).

Representative specimens examined. PERU. Amazonas: Prov. Bongará, 18–19 km below Caclic, Utcubamba River, *Hutchison & Wright 5858* (BH, F, NY, UC, US, USM); Prov. Bongará, rd. S from Pte. Ingenio to Chachapoyas, *Plowman 5562* (CTES, F, GH). **Cajamarca:** Pucará, *Hutchison & Wright 3597* (BH, F, MO, NY, UC, US, USM); Distr. Ichocán, *Sánchez et al. 2456* (CTES).

BRAZIL. Mato Grosso: Fazenda Salina, Pantanal do Rio Negro, *Dubs 172* (NY); Mun. Bela Vista, Apa River, MS-384, *Hatschbach et al. 58799* (CTES, MBM, TEX, Z); Mun. Pto. Murtinho, Porto Murtinho, *Krapovickas 14427* (CTES, NY); Mun. Cáceres, 75 km E of Cáceres, *Krapovickas et al. 40203* (CEN, CTES, NY).

BOLIVIA. Beni: Trinidad, *Krapovickas & Schinini 34708* (C, CTES, F, G, MBM, UC). **Cochabamba:** 5 km

SE of Cochabamba, *Eyerdam 24908* (GH). **Potosí:** Prov. Oropeza, 1 km N of Chuqui on rd. to Aiquile, *Spooner & García-Fernández 6625* (NY, WIS). **Santa Cruz:** Prov. Florida, Tembladeras around Rancho Chavez, 17 km from Pampa Grande, *Acevedo Rdgz. et al. 4507* (NY, TEX, US); Prov. Cordillera, 30 km S of Charagua, *Killeen et al. 4260* (LPB, MO, TEX); Andres Ibañez, 5.8 km E of La Bélgica, *Solomon 18621* (LPB, MO). **Tarija:** Camiri–Villa Montes rd., *Badcock 236* (K); Chaco region, near Charagua, *Cárdenas 2578* (US); Prov. O'Connor, entrance to the Cuesta San Simón, *Krapovickas & Schinini 39047* (CTES); 28 km E of Villa Montes, *Krapovickas et al. 19303* (BAA, CTES, LL, SI, UC); Yacuiba, *J. Steinbach 1642* (F).

PARAGUAY. Alto Paraguay: Palmas Chicas, *Krapovickas & Schinini 32753* (CTES); Trinidad-Asunción, on the banks of Paraguay River, *Rojas 5416* (CTES, G); Chaco Paraguayo, Fuerte Olimpo, *Rojas 13696* (CAS, E, LIL, TEX). **Boquerón:** Puerto Casado and vic., near Guajhó, *Pedersen 4076* (C, herb. Pedersen). **Central:** Tavarory, 1 km E of Paraguay River, *Zardini 23594* (CTES, MO). **Cordillera:** San Lorenzo, Quinta Guaraní, *Archer & Rojas 4759* (F, G, K, NY, US); 14 km W of Arroyos y Esteros, *Zardini & Velásquez 17090* (CTES, MO). **San Pedro:** Estero Patiño, km 164 of Trans-Chaco Hwy., *Schinini & Palacios 25906* (CTES, FCQ, G, MO, NY); Primavera, Alto Paraguay, *Woolston 1150* (C, K, NY, S, U). **Villa Hayes:** Trans-Chaco rt., km 162, *Krapovickas et al. 37462* (BAB, CTES, F, G, ICA, IPA, LIL, MBM, NY); Villa Hayes, 4 km from the Paraguay River, *Schinini 19600* (C, CTES, F, G, MBM, NY, UC). **Pte. Hayes:** 70 km E of Pozo Colorado, *Schinini & Palacios 25890* (CTES, NY).

ARGENTINA. Catamarca: Dept. Paclin, La Merced, *Krapovickas & Cristóbal 11261* (C, CTES, G). **Chaco:** Dept. 1° de Mayo, Colonia Benítez, *Schulz 2097* (CTES); La Amalia, Resistencia, *Schulz 894a* (S). **Corrientes:** Dept. Empedrado, Estancia "El Peata," *Pedersen 12013* (C, CTES, MO); Dept. San Luis del Palmar, 15 km SE of San Luis del Palmar, *Cristóbal et al. 1339* (CTES); Dept. Capital, 12 km E of Corrientes, *Cristóbal 1425* (CTES); Dept. Itatí, Isleta de Chaco, *Krapovickas & Cristóbal 11393* (CTES, F, MBM, UC); Molina Punta, *Krapovickas & Cristóbal 11759* (CHR, CTES, EACS, F, GB, HAS, K, LG, MICH, MO, NY, SI, TEX, TIAC, U); Dept. Mburucuyá, Estancia "Santa Teresa," *Pedersen 3118* (CTES, K, MO, NY, S, U, US, herb. Pedersen); Dept. San Cosme, Paso de la Patria, along the Paraná River, *Schinini 4856* (BA, CLEMS, CTES, IPA, MBM, NY, PAL, SI, WIS). **Formosa:** Dept. Formosa, 2 km S de ruta 81, camino a Mariano Boedo, *Krapovickas & Cristóbal 46534* (CTES, TEX); Dept. Pirané, Loto 4 Riacho Pilaga, *Morel 742* (TEX, US); Dept. Pilcomayo, rt. 11 at km 101, *Morel 4276* (CAS); Dept. Boca del Bermejo, Estancia "Herradura," near Arroyo Saldillo, *Pedersen 1279* (C, herb. Pedersen). **Jujuy:** Dept. Santa Bárbara, between El Piquete and Palma Sola, *Burkhart et al. 30530* (CTES); Cuesta de las Lajitas, *Cabrera et al. 13778* (CTES, US); Dept. El Carmen, Perico, *Krapovickas et al. 28503* (CTES, NY); Dept. San Pedro, San Pedro, *Venturi 5184* (US). **La Rioja:** Near Cantadero, *Hunziker 5300* (CTES). **Salta:** Dept. Orán, Caraparí, *Bridarolli 3225* (F); Dept. Gral. Güemes, Unchimé, *Cuezzo & Meyer 5369* (CTES); Dept. La Viña, 20 km E of Colonel Moldes, *Krapovickas & Cristóbal 24531* (C, CTES); Dept. Capital, 20 km E of Salta, *Krapovickas & Schinini 30429* (CTES, F, NY, SI); Dept. Metán, Metán Viejo, Metán River ("topotipo"), *Krapovickas et al. 27924* (BAA, CTES, HAS, NY); Dept. Chicoana, Chicoana, *Krapovickas et al. 28370* (CTES, MBM); Pala-Pala, slope of Cerro Colorado, *Villa 2326 & Legname 3235* (CAS). **Tucumán:** Dept. Trancas, San Pedro de Colalao, *Krapovickas & Cristóbal 15337* (C, CTES, UC); Dept. Graneros, between Rumipunco and El Túnel, *Meyer et al. 22388* (CTES, F); Cerro del Duraznillo, *Monetti 15812* (G); Dept. Burruyacu, Valley of the Nió River, *Venturi 10176* (US).

Some would distinguish *Pavonia morongii* (with stellate stem pubescence) from *P. hieronymi* (with glandular stem pubescence), but the distinction does not appear to be absolute, and there is considerable geographic overlap in the two character expressions. I have been unable to find any reliable supporting characters, so I do not maintain the distinction.

127. Pavonia immaculata R. E. Fries, Kongl. Svenska Vetenskapsakad. Handl., ser. 3, **24(2):** 28. 1947. Type. Brazil. Mato Grosso: Corumbá, 11 Apr 1903, *Malme s.n.* (lectotype, S; isotype, S).

Decumbent *subshrubs* to 0.5 m tall, the stems often reddish, minutely glandular-pubescent and sometimes also with scattered setae. *Leaf* blades ovate-triangular, mostly 1.5–3.5 × 10–22 mm, basally cordate, irregularly serrate, acute or acuminate, essentially concolorous, palmately 5-nerved, the upper surface with minute glandular hairs to glabrate, the margin minutely glandular-ciliate, the lower surface with minute glandular hairs and widely scattered 3-armed spicules 1.5 mm diam. (the spicules sometimes lost in age); petioles 1–2.5 cm long; stipules subulate, 3 mm long. *Flowers* solitary in the leaf axils, the pedicels 2.5–6 cm long; involucellar bracts 6–8, linear, 8–14 × 0.5–1 mm, viscid and also setose; calyx 5–8 mm long, setose, with white intercostal areas; corolla 1–1.5 cm long, yellow, glabrous on margins of claw; staminal column 6 mm long, pallid, the filaments 1 mm long; styles yellowish, exceeding the staminal column by 2–3 mm. *Fruits* oblate, 6–7 mm diam., minutely punctulate and scabridulous along suture, otherwise glabrous, more or less reticulate-veined; seeds striate, glabrous, 3 mm long.

Distribution (Fig. 57). From S Brazil and Paraguay. The type locality is at 19°00′S, 57°39′W.

Illustration. Fries (1947: pl. 7, fig. 1).

Specimens examined. **BRAZIL. Mato Grosso:** Corumbá, *Krapovickas 29789* (C, CTES, MBM, NY, UC); Corumbá, rd. to Tamarindeiro, *Pott et al. 2796* (CTES).

PARAGUAY. Chaco Paraguayo: F. Olimpo, *Rojas 13693* (CAS, CTES, MO).

128. Pavonia intermixta A. Richard, Hist. Phys. Cuba, Pl. Vasc. 128. 1845 (1841); Hist. Fis. Cuba, Bot. 45. 1845 (1841). Type. Cuba. 1836, *de la Sagra s.n.* (holotype, P [*ex* herb. Richard] as photo CTES) Fig. 53

Pavonia linearis A. Richard, Hist. Phys. Cuba, Pl. Vasc. 127, t. 13. 1845 (1841); Hist. Fis. Cuba, Bot. 44, t. 13. 1845 (1841). Type. Cuba. *de la Sagra s.n.* (holotype, P [*ex* herbarium Richard] as photo CTES).

Pavonia lagunarum Hadač, Folia Geobot. & Phytotax. Praha **5:** 431. 1970. Type. Cuba. Pinar del Río: lake near La Fé, *Hadač 225* (holotype, PR).

Delicate *subshrubs* or perennial herbs from a woody rootstock, probably ascending or scandent, the stems slender, minutely and obscurely stellate-pubescent and with narrow longitudinal lines of dense curled hairs. *Leaf* blades narrowly lance-elliptic or lance-oblong, mostly 25–50 × 3–7 mm, mostly 6–10 times as long as wide (except juvenile leaves shorter and wider, broadly elliptic or sometimes suborbicular), basally truncate or subcordate, shallowly serrulate to subentire, acute, palmately 3-nerved or penninerved, somewhat discolorous, minutely stellate-pubescent on both surfaces; petioles 2–4 cm long (longer on juvenile leaves); stipules subulate, 2–3 mm long. *Flowers* solitary in the leaf axils, the pedicels 3–8 mm long; involucellar bracts ca. 10, narrowly oblanceolate, 2–5 × 0.5 mm or less, ciliate; calyx 2.5–4(–6) mm long, stellate-pubescent and ciliate; corolla 8–11 mm long, yellow, glabrous (including margins of claw); staminal column slightly shorter than corolla, glabrous, apparently asymmetrically declined, the anthers subsessile; styles and stigmas subequal to corolla or slightly exserted. *Fruits* oblate, 4.5–6 mm diam., 3.5–4 mm high, with appressed simple and bifurcate hairs (rarely glabrous), the apex rounded, the two lateral spines usually suppressed (e.g., *Shafer 10483*) or up to 1.5 mm long with a few retrorse hairs; seeds glabrous, not striate.

Distribution (Fig. 60). Endemic to W Cuba, where it grows near sea level in savannas, wet coastal thickets, palmetto swamps, and stream banks, often in sand.

Illustration. Richard (1845: pl. 13).

Specimens examined. CUBA. Isle of Pines: Vic. Siguanea, *N. L. Britton et al. 14350* (CM, F, GH, NY, US); near Nueva Gerona, *Curtiss 290* (CM, CTES, E, F, GH, MO, NY, P as photo CTES, TEX, US); Santa Barbara, *Ekman 12090* (F, S); on rd. to Santa Isabel, near SE corner of Cerro Daguilla, *Killip 43946a* (US); ca. 4 km S of Santa Barbara, *Killip 44122* (CAS, US); between San Francisco de las Piedras and Cerro La Cañada, *Killip 44626* (US); rd. to San Francisco de las Piedras, *Killip 44879* (CAS, US); along rd. from Nueva Gerona to Santa Bárbara, *Killip 45026* (US); Los Indios, *Killip 45145* (US); Playa Roja, Bahia de Siguanea, *Killip 45751* (US); near Nueva Gerona, *W. Palmer & Riley 941* (NY, US). **Pinar del Río:** 13 km S of Pinar del Río, *Bro. Alain (H. Liogier) 2605* (GH); Herradura, *C. F. Baker 3466* (POM); Herradura, *N. L. Britton et al. 6517* (NY); vic. of Pinar del Río, *N. L. Britton et al. 6973* (NY, US); vic. of Pilar del Río, *N. L. Britton et al. 6976* (F); between Pinar del Río and Coloma, *N. L. Britton et al. 10060* (K, NY, US); Herradura, *Earle 711* (NY, US); Herradura, *Ekman 10807* (S); near Los Palacios, *Ekman 10877* (S); Sabalo, *Ekman 11437* (NY, S); near La Fé, *Hadač 446* (PR); on Lake Jovero, *Hadač 1914* (PR); E of Los Palacios, *Bro. León (J. S. Sauget) 7391* (NY); between Pinar del Río and Coloma, *Bro. León (J. S. Sauget) 12865* (GH); Bacunagua, near San Cristóbal, *Bro. León (J. S. Sauget) 15082* (GH, NY, US); Las Martinas, 28 Jun 1940, *Roig & Acuña s.n.* (CAS); Herradura, *Shafer 521* (NY); palm barrens W of Guane, *Shafer 10483* (A, MO, NY, TEX, US); Herradura to Paso Real, *Shafer 11740* (NY); El Salado (locality stated on S duplicate), *Wright 2059* (GH, K, MA, MO, NY, S, US).

Local name. Majagüilla (Cuba).

Sauget and Liogier (1953: 248) suggest that *Pavonia linearis* and *P. intermixta* are conspecific. Although the descriptions of Richard seem to indicate clear-cut differences between *P. intermixta* and *P. linearis*, Richard acknowledges that they are very similar and that the specimens in the la Sagra herbarium were completely intermingled. No specimens have come to hand that conform to the description of *P. linearis*, and an examination of a photograph of the type of *P. linearis* (cited above, ex herb. Richard) confirms that only a single species is involved. I therefore conclude that Sauget & Liogier are correct in equating the two names.

The placement of the species in *Pavonia* sect. *Asterochlamys* is more problematical, however. It seems to fit here as well as anywhere else, largely on the basis of mericarp morphology, despite the fact that it has smooth rather than striate seeds, a characteristic that is essentially diagnostic for the section. The mericarp morphology is rather variable (see Fig. 53), which is notable because the morphology of the entire plant (other than the mericarps) is very uniform. *Pavonia intermixta* appears to have no close relatives and is easily recognized by its small flowers and small, linear leaves.

129. Pavonia krapovickasii Fryxell, sp. nov. Type. Brazil. Mato Grosso: 39 km N of Rondonopolis, rd. to Paxoréu, *Krapovickas, Valls, Simpson & G. Silva 40229* (holotype, CTES; isotypes, CEN, NY). Figs. 53, 61

Frutices caulibus viscidis; laminis foliorum ovato-cordatis puberulentis valde discoloribus; bracteis involucellorum 5, late lanceolatis viscidis calycem excedentibus; calycibus

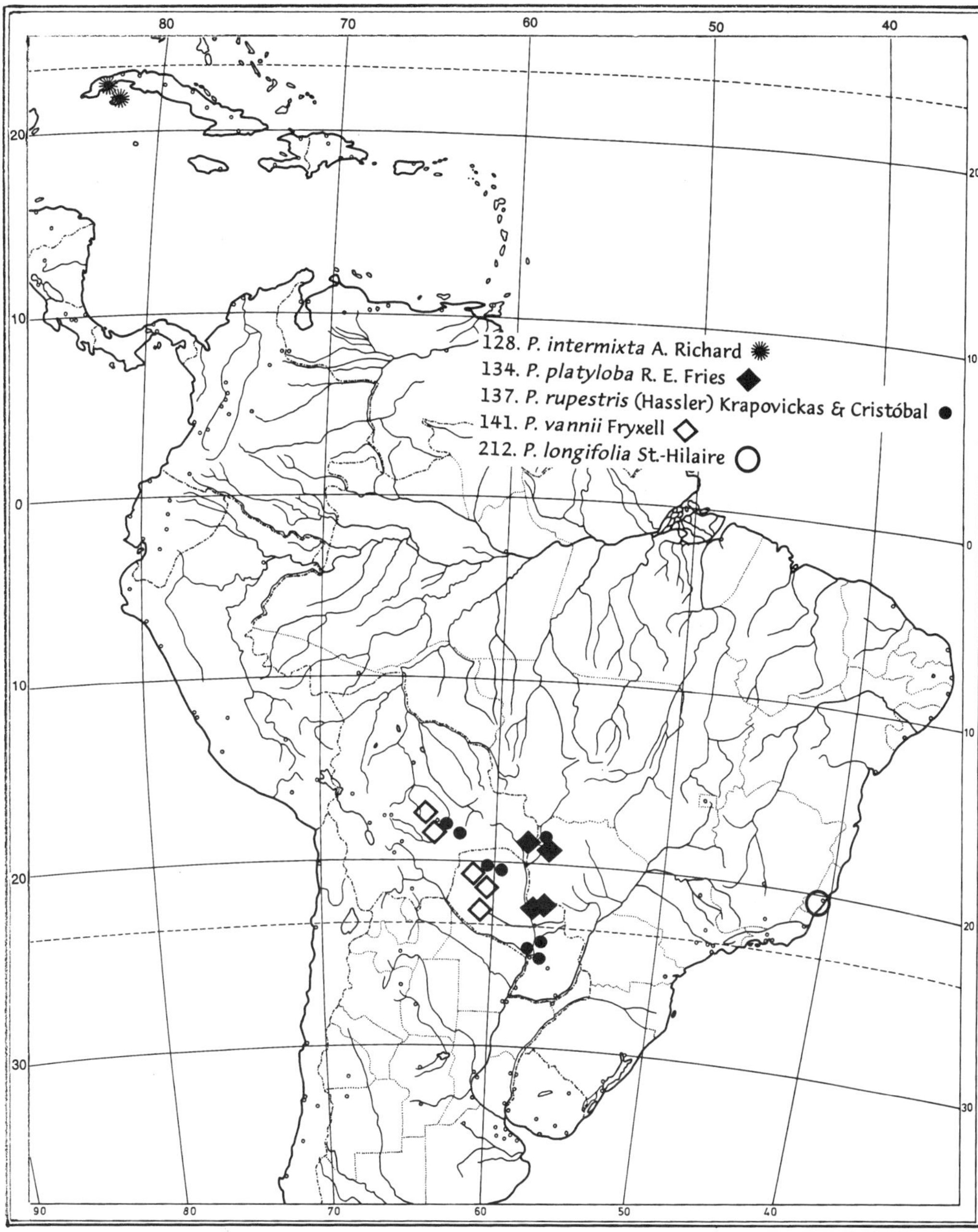

FIG. 60. Distribution map of five neotropical species of *Pavonia* sect. *Asterochlamys* and *P.* subgen. *Goetheoides*.

ciliatis; corollis 3–3.5 cm longis roseis; columna staminalis quam corollis brevioribus, filamentis brevissimus verticillis pluralibus; stylis pubescentibus; stigmatibus subroseis; fructibus glabris verruculosis.

Shrubs 1.5–2 m tall, the stems densely covered with purplish glandular hairs 0.5 mm long or less, also with narrow lines of recurved white hairs, sometimes also with scattered simple hairs 1.5–2 mm long. *Leaf* blades ovate (sometimes slightly lobed), cordate, crenate or serrate, acuminate, mostly 4–13 × 1.5–8 cm, palmately 7–9-nerved, strongly discolorous, minutely puberulent, the lower surface densely and softly puberulent with

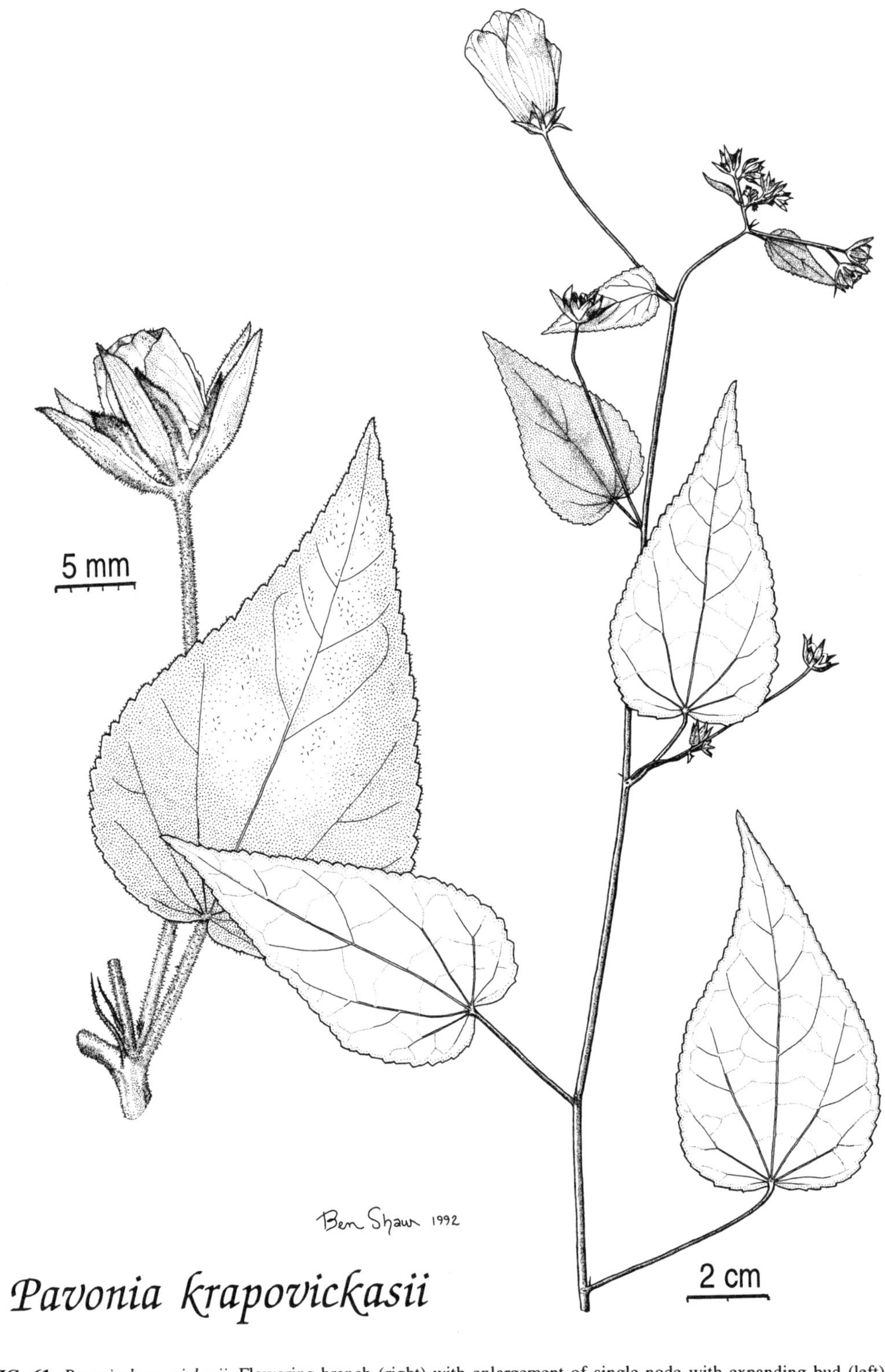

FIG. 61. *Pavonia krapovickasii.* Flowering branch (right) with enlargement of single node with expanding bud (left) (*Anderson et al. 6888*).

whitish stellate hairs, the upper surface with simple glandular hairs and scattered stellate hairs; petioles 1–6.5(–8) cm long; stipules 2–4 mm long, subulate. *Flowers* solitary in the leaf axils (occasionally geminate), sometimes aggregated terminally; pedicels 2–6 cm long, slender; involucellar bracts 5, lanceolate, 7–9 × 1.5–4 mm (sometimes narrower), acute, usually exceeding calyx, densely pubescent with glandular and stellate hairs; calyx 5–8 mm long, ciliate; corolla (2–)3–3.5 cm long, campanulate, pink or rose with a dark center, externally with a few scattered hairs; staminal column 18–20 mm long, glabrous, dark red at base, gradually paler distally, apically 5-dentate, the filaments short (1 mm or less), arranged in ca. 5 whorls, the anthers and pollen yellow; styles exceeding the staminal column, covered with minute white hairs, the capitate stigmas pink, villous. *Fruits* minutely verruculose, glabrous, 8–9 mm diam., the mericarps 4.5 mm high.

Distribution (Fig. 57). Confined to central Brazil (Bahia, Goiás, Mato Grosso, Tocantins), in forest and cerrado vegetation at elevations of 600–800 m. The type locality is at 16°06′S, 54°32′W.

Specimens examined. BRAZIL. BAHIA: 5 km S of Rio Roda Velha, ca. 150 km SW of Barrieras, *Irwin et al. 14897* (CTES, F, NY); Espigão Mestre, 32 km W of Barreiras, *Irwin et al. 31529* (CTES, NY). **GOIÁS:** 24 km by rd. SW of Monte Alegre de Goiás, *W. R. Anderson et al. 6779* (C, F, MO, NY, US); Monte Alegre de Goiás, Serra da Atalaia, ca. 25 km by rd. SW of Monte Alegre de Goiás, *W. R. Anderson et al. 6888* (F, K, NY, TEX); Formosa, Rio Tiquiri-Formosa, *Heringer 11465* (CTES); Pirenópolis, *Heringer 15827* (US). **MATO GROSSO:** Buriti, NE of Cuyabá, *Dorrien Smith 297* (K, NY). **TOCANTINS:** 32 km N of Paraíso do Norte de Goiás, BR-153, km 782, Côrrego São Jose, *Krapovickas et al. 37837* (C, CEN, CTES, G, NY, UC).

The specific epithet is chosen to honor Antonio Krapovickas (b. 1921) of Corrientes, Argentina, co-collector of the type specimen and knowledgeable and productive specialist in the Malvaceae. *Pavonia krapovickasii* is related to *P. hexaphylla* and *P. vinosa*. It is distinguished from the former by the discolorous leaves and from the latter by the fewer involucellar bracts and eciliate calyx.

130. Pavonia mollis Humboldt, Bonpland & Kunth, Nov. Gen. Sp. **5:** 282 (folio ed., p. 220). 1822. *Pavonia mutisii* var. *mollis* (Humboldt, Bonpland & Kunth) Triana & Planchon, Prodr. Fl. Novo-Granat. 163. 1862. *Malache mollis* (Humboldt, Bonpland & Kunth) Kuntze, Revis. Gen. Pl. **1:** 70. 1891. *Pavonia kunthii* var. *mollis* (Humboldt, Bonpland & Kunth) Gürke in Martius, Fl. Bras. **12(3):** 511. 1892. Type. "Novo-Granat., S. Ana," *Humboldt & Bonpland s.n.* (lectotype, here designated, P-HBK as microfiche; isolectotype, B as photo F-9450). Fig. 53

Pavonia mutisii var. *intermedia* Triana & Planchon, Prodr. Fl. Novo-Granat. 163 (1862). *Pavonia kunthii* var. *intermedia* (Triana & Planchon) Gürke *in* Martius, Fl. Bras. **12(3):** 511. 1892. Type. Colombia. near Fusagasugá, *Triana s.n.* (lectotype, here designated, P as photo CTES; isolectotype, W as photo F-32650).

Pavonia mutisii var. *involucrosa* Triana & Planchon, Prodr. Fl. Novo-Granat. 163. 1862. Type. Colombia. Near La Mesa, prov. of Bogotá, 1200 m, *Triana s.n.* (holotype, P as photo CTES).

Subshrubs 0.5–3 m tall, the stems with short glandular hairs and setose (hairs 3–4 mm long). *Leaf* blades mostly 4–10 × 3–7 cm, ovate (sometimes with a weak tendency to lobing), cordate, coarsely crenate, acute or acuminate, palmately 5–7-nerved, softly pubescent (the veins sometimes also pilose), somewhat discolorous; petioles 1–7 cm long; stipules 2–4 mm long, subulate. *Flowers* solitary in the leaf axils, the pedicels 4–8 cm long; involucellar bracts 6–8, lanceolate, 8–11 × 1–1.5 mm, ciliate; calyx 5–9 mm long, pilose; corolla 1.5–3.5 cm long, pink, sparsely ciliate on claw; staminal column 1–1.5 cm long, glabrous, the filaments 1–1.5 mm long, in 3 or 4 whorls; styles emergent from the column, the stigmas villous. *Fruits* oblate 8–9 mm diam., short-pubescent; seeds 3 mm long, obscurely striate, sparsely and minutely pubescent.

Distribution (Fig. 55). From Panama, Venezuela, Colombia, Ecuador, and Peru, growing in forest and savanna at 300–1600 m elevation.

Illustration. Fuertes (1989: pl. 12, as *P. mutisii*).

Representative specimens examined. PANAMA. SAN BLAS: Mulatuppu (Río Ibedi), *Duke 8465* (MO).

COLOMBIA. STATE UNKNOWN: La Mesa, prov. Bogotá, *Hartweg 914* (OXF); near Honda, *Humboldt & Bonpland s.n.* (syntype of *P. mollis*: B, destroyed, as photo F-9450); San Antonio, *Langlassé 33* (K); Inza, *Lehmann 4765* (F). **ANTIOQUÍA:** Hwy. from Chocó de Cocorná to Cocorná, *Castañeda 1469* (F); rd. from Medellín to Rionegro, *Killip et al. 39915* (US); El Cairo below Santa Barbara, *Pennell 10871* (US); Segovia, *Sandeman 5565* (K); Armería, near Medellin, *Toro 772* (NY). **CALDAS:** Near La Dorada, *Haught 2127* (S, US). **CUNDINAMARCA:** Fusagasugá, *André 1492* (NY); S of Silvania, rd. to Fusagasugá, *Barclay et al. 3459* (US); below Fusagasugá, *Dugand 3533* (US); between Fusagasugá and Pandi, *Linden 859* (syntypes, BR as photo CTES, CTES [fragment], OXF, P as photo CTES); La Mesa to San Javier, *Duque-Jaramillo 3674* (CM); between Pacho and Río Negro, *García-Barriga 10749* (CAS). **HUILA:** 21 km E of Neiva, valley of Río Fortalecillas, *Fosberg 19808* (CAS, US); Río Saldaña to Natagaima, *Rusby & Pennell 241* (NY, US); Cordillera Oriental, E of Neiva, *Rusby & Pennell 494* (NY, US); Santa Marta, *H. H. Smith 492* (A, CAS, CM, E, F, GH, K, LL, MO, NY, PH, S, U, US). **SANTANDER:** Between Pie de Cuesta and Las Vegas, *Killip & A. C. Smith 15470* (A, GH, NY, US). TOLIMA: Rd. to Melgar, *García-Barriga 12017* (CM, NY, US). VALLE: Cali, *Fosberg 21998* (CAS, US); Mun. Yumbo, Corregimiento de Mulaló, 4 km above the town by the pipeline hwy., *J. E. Ramos & L. H. Ramos 2663* (NY).

VENEZUELA. Ocaña, *Schlim 49* (BR as photo CTES, CTES [fragment]); Ocaña, *Kalbreyer 251* (K). TÁCHIRA: Along rd. between San Cristóbal and Choro, *Croat 38426* (MO); Quebrada La Cordera, near entrance to town of Cordero, *Luteyn et al. 5876* (NY).

ECUADOR. CARCHÍ: SE of Chical, *Barfod 41413* (AAU, MO, NY, QCA, TEX); between Ibarra and San Lorenzo, near La Carolina, *Eriksen 59252* (AAU); Peñas Blancas, 20 km below Maldonado on the San Juan River, *Madison et al. 4640* (F, QCA, SEL).

PERU. SAN MARTÍN: Prov. Lamas, near Lamas, *Ferreyra 17288* (CTES); 6 km S of Tarapoto on rd. to Juanjui, *A. H. Gentry et al. 37696* (F, MO, NY, TEX); Zepalacio near Moyobamba, *Klug 3538* (F, GH, MO, NY, S, US); Juan Jui, upper Huallaga River, *Klug 4221* (A, F, K, MO, NY, S, U, US).

Local name. Obeja-quivan (Peru).

Garcke (1881) synonymized *Pavonia mollis* and *P. mutisii* and used the name *P. mutisii* for the combined species. Some more recent authors have followed suit (e.g., Fuertes, 1989; Krapovickas, 1983; Macbride, 1956); Kearney (1958) also combined them but incorrectly used the name *P. mollis* for the combined species; incorrect because Garcke's choice should have been followed. Gürke (1892b) distinguished them in varietal rank, but I believe they are sufficiently distinct (see key) to merit recognition in specific rank, following Humboldt, Bonpland and Kunth, who provided the correct names. There may also be an elevational difference in their distributions.

131. Pavonia monticola Fryxell, Phytologia **37:** 300. 1977. Type. Mexico. Chiapas: 11 mi E of Tapanatepec, 20 Oct 1965, *Breedlove & Raven 13716* (holotype, DS; isotypes, F, MEXU, MO, NY). Fig. 53

Shrubs to 2 m tall, the stems densely glandular-pubescent. *Leaf* blades mostly 8–12 cm long, ca. half as wide, narrowly ovate, cordate, crenate-serrate, acuminate, discolorous, stellate-pubescent, with denser coarser pubescence beneath and with intermixture of glandular hairs above; petioles ½–¾ the length of the blades; stipules 5–12 mm long, linear, glandular-pubescent. *Flowers* solitary or paired in the leaf axils, the pedicels to 5.5 cm long, viscid, articulate 7–13 mm below the flower; involucellar bracts 8, linear, 8–15 mm long, viscid; calyx 6–8 mm long, ciliate, whitish (except for green veins); corolla 10–12 mm long, yellow, densely bearded on margins of claw, forming a cushion at juncture with staminal column; staminal column 6–10 mm long, glabrous, the filaments 1–3 mm long; styles glabrous, slightly exserted. *Fruits* oblate or somewhat obovate, 6 mm diam., obscurely scabridulous on margin, otherwise glabrous, smooth or weakly reticulate, the mericarps 5 mm high, narrowly winged laterally, with a small subapical cusp; seeds glabrous, striate.

Distribution (Fig. 24). Known only from S Mexico in a small area along the Chiapas-Oaxaca state boundary at 700–1000 m elevation, growing in forests and along rivers. The type locality is at 16°20′N, 94°00′W.

Illustrations. Fryxell (1977: figs. 11, 13E; 1979: fig. 7F).

Specimens examined. **MEXICO**. CHIAPAS: Mun. Cintalapa, 12 km S of Hwy. 190 near Rizo de Oro, *Breedlove & Thorne 20685* (CAS, MO, NY); Cinco Cerros, 11 mi NE of Oaxaca boundary on Hwy. 190, *P. A. Fryxell & Lott 3233* (BM, CAS, CHAPA, CTES, ENCB, MEXU, MO, NY, WIS).

132. Pavonia mutisii Humboldt, Bonpland & Kunth, Nov. Gen. Sp. **5:** 283 (folio ed., p. 220). 1822, nom. nov. *Hibiscus cordifolius* Linnaeus fil., Suppl. 309. 1781 (non *Hibiscus cordifolius* Miller, 1768). *Malache cordifolia* (Linnaeus fil.) Kuntze, Revis. Gen. Pl. **1:** 70. 1891. *Pavonia kunthii* Gürke *in* Martius, Fl. Bras. **12(3):** 510. 1892, nom. superfl. *Pavonia kunthii* var. *cordifolia* (Linnaeus fil.) Gürke *in* Martius, Fl. Bras. **12(3):** 511 (1892). *Pavonia cordifolia* (Linnaeus fil.) Hieronymus, Bot. Jahrb. Syst. **20 (Beibl. 49):** 45. 1895 (non *Pavonia cordifolia* Wright in Grisebach, 1866). Type. Colombia. *Mutis 5186* (holotype, MA; isotype, US). Fig. 53

Asterochlaena cuspidata Garcke, Bot. Zeitung (Berlin) **8:** 668. 1850. *Pavonia cuspidata* (Garcke) Garcke, Bot. Jahrb. Syst. **21:** 392. 1896. Type. Ecuador. Guayaquil, *Ruíz 1800* (holotype, B destroyed, as photo F-9438).

Pavonia albiflora R. E. Fries, Kongl. Svenska Vetenskapsakad. Handl., ser. 3, **24(2):** 29. 1947. Type. Ecuador. Guayas: Chongón, *Asplund 5882* (holotype, S; isotypes, CAS, CTES, K, S, US).

Subshrubs 1 m tall, the stems densely viscid and sparsely pilose, with scattered simple hairs 2–2.5 mm long, or these lacking. *Leaf* blades mostly 4–8(–13) × 2–5(–9) cm, ovate, cordate, prominently to obscurely serrate, acuminate, palmately 5–7-nerved, minutely and softly stellate-tomentose (the hairs 0.2–0.5 mm long beneath, smaller above), also pilose on the veins, discolorous; petioles 1–2.5(–7) cm long; stipules 2–5 mm long, subulate. *Flowers* solitary in the leaf axils or forming a terminal panicle, the pedicels 2.5–10 cm long, equaling or exceeding the subtending leaves; involucellar bracts 6–7, 10–15 x ca. 1 mm, viscid, pilose only toward the base; calyx 6–7 mm long, ciliate, often whitish in intercostal areas; corolla 2.5–3 cm long, white, glabrous on margins of claw; staminal column 15–17 mm long, pallid, glabrous, the filaments 0.5–1 mm long, often in 3 or 4 whorls; styles exceeding the column, the stigmas villous. *Fruits* oblate, 7–9 mm diam., 4.5 mm high, obscurely scabridulous on suture, otherwise glabrous.

Distribution (Fig. 58). From Colombia and Ecuador.

Illustrations. Fries (1947: pl. 8, as *P. albiflora*); Fuertes (1989: pl. 11, as *P.* aff. *cuspidata*).

Specimens examined. COLOMBIA. Mun. La Argentina, Quebrada del Pueblo, *Lozano et al. 4121* (F). **Norte de Santander:** Between Chinacota and La Esmeralda, *Killip & A. C. Smith 20872* (A, GH, NY, US). **Valle:** Yotoco, *Llanos & Velosa 17* (U).

ECUADOR. Guayas: Nobol, *Asplund 16675* (C, K, LL, NY, S, US); Capeira, km 21 Guayaquil to Daule, *C. H. Dodson & P. M. Dodson 11486* (MO); 9 km N of Guayaquil, *C. H. Dodson & Thein 1312* (MO); Capera, 23.5 km N of Guayaquil on rd. to Daule, *C. H. Dodson et al. 11041* (F, MO, US); rd. from Guayaquil to Nobol, km 24, *Harling & Andersson 21077* (GB); 19 km SW of Guayaquil, *Worth et al. 8947* (UC, US). **El Oro**: Vic. of Portovelo, *J. N. Rose & G. Rose 23375* (GH, NY, US); between Portovelo and Río Cabra, *Steyermark 54081* (NY).

Pavonia mutisii is related to *P. mollis*. See the latter for further discussion.

133. Pavonia neei Fryxell, sp. nov. Type. Bolivia. Santa Cruz: Prov. Andrés Ibáñez, 12 km E of Santa Cruz, *M. Nee 35916* (holotype, TEX; isotypes, CAS, CM, CTES, NY, TEX). Fig. 53

Asterochlaena morongii Hassler var. *intermedia* Hassler, Feddes Repert. Spec. Nov. Regni Veg. **8:** 118. 1910. Type. Paraguay. Gran Chaco: Santa Elisa lat. 23°10', Dec 1903, *Rojas* (for Hassler) *2693* (isotypes, CTES, NY, P as photo CTES).

Suffrutices caulibus glanduloso-pubescentibus et sparse setosis; laminis foliorum ovato-cordatis minute stellato-pubescentibus; floribus solitariis; bracteis involucellorum 8–9 quam calyce longioribus; corollis 2–2.5 cm longis flavidis centro rubro, unguibus glabris; fructibus 7 mm diametro fere glabris laevibus.

Spreading *subshrubs* 0.5 m tall, the stems glandular-pubescent and sparsely setose with hairs 3–5 mm long. *Leaf* blades ovate, cordate, serrate, acute, mostly 3–4 × 1.5–2.5 cm, minutely stellate-pubescent, somewhat discolorous; petioles 1–2 cm long; stipules subulate, 3 mm long. *Flowers* solitary in the leaf axils, the pedicels 2–7 cm long, slender; involucellar bracts 8–9, linear, 11–13 × 1 mm, ciliate; calyx 6–7 mm long, ciliate; corolla 2–2.5 cm long, yellow with red center, the margins of the claw glabrous; staminal column glabrous, ca. 7 mm long, purplish (at least basally), the filaments 1 mm long; styles emergent from the column, pallid, the stigmas capitate, villous. *Fruits* oblate, 7 mm diam., 4 mm high, nearly glabrous, smooth.

Distribution (Fig. 15). From Bolivia, Paraguay, and N Argentina, at elevations up to 400 m. The type locality is at 17°46'S, 63°04'W.

Specimens examined. PARAGUAY. Chaco Paraguayo, Pilcomayo River, *Morong 947* (E, F); Puerto Casado, *Rojas 5741* (G).

ARGENTINA. Formosa: Dept. Pilcomayo, Salvación, *Morel 4174* (CAS).

134. Pavonia platyloba R. E. Fries, Kongl. Svenska Vetenskapsakad. Handl. **42(12):** 45. 1908. *Asterochlaena platyloba* (R. E. Fries) Hassler, Feddes Repert. Spec. Nov. Regni Veg. **7:** 383. 1909. Type. Paraguay. Estancia Santa María, *Anisits 2893* (lectotype, here designated, S; isolectotype, S). Fig. 53

Asterochlaena platyloba var. *grandiflora* Hassler, Feddes Repert. Spec. Nov. Regni Veg. **12:** 268. 1913. Type. Paraguay. Margarita, *Hassler 11070* (isotypes, CTES, K).

Asterochlaena platyloba var. *hirsutula* Hassler, Feddes Repert. Spec. Nov. Regni Veg. **12:** 267. 1913. Type. Paraguay. In the region of "Cerro Margarita," *Hassler 10998* (isotype, CTES).

Shrubs 1–1.5 m tall, the stems densely tomentose, the hairs stellate, 0.5–1.5 mm long. *Leaf* blades ovate, mostly 4–9 × 3–8 cm, basally cordate, crenate-dentate (or subentire in juvenile leaves), acute to subobtuse, palmately 7–9-nerved, discolorous, the upper surface stellate-tomentose, the lower surface more densely so and with larger 3–4-armed spicules scattered among the softer hairs; petioles 2.5–5 cm long; stipules subulate, 4 mm long. *Flowers* solitary in the leaf axils, the pedicels 2–5.5 cm long, slender; involucellar bracts 6 (rarely 7), broadly spatulate to subrotund, 6–10 × 4–6.5 mm, obtuse or acute, softly tomentose, more densely so toward base; calyx 5–7 mm long, wooly-pubescent; corolla 1.5–2(–3) cm long, externally stellate-pubescent, yellow (without red center, or with weakly developed red center in *Hassler 11070*); androecium subequal to corolla or shorter. *Fruits* 7–8 mm diam., oblate, 10-ribbed, subglabrous but obscurely scabridulous along ridges; mericarps 3.5 mm high, dehiscent.

Distribution (Fig. 60). From S Brazil, Bolivia, and Paraguay. The type locality is at 22°44'S, 57°27'W.

Illustrations. Fries (1908: pl. 5, figs. 5–7).

Specimens examined. BRAZIL. Mato Grosso do Sul: Mun. Corumbá, rd. to Gaturama, *Hatschbach 29494* (CTES); Mun. Antonio João, campestre, *Hatschbach & Zelma 49102* (CTES, US); Antonio João, km 7 of the hwy. to Caracol, *Hatschbach et al. 58784* (K, MBM, TEX);

Corumbá, *Pott & Eiten 194* (CTES); 7 km S of Corumbá, Pott et al. 7794 (CPAP, CTES, TEX).

BOLIVIA. Puerto Suarez, hwy. to Corumbá, *Hatschbach et al. 52517* (C, CTES, MO, US).

PARAGUAY. Estancia Santa María, *Anisits 2335* (syntype of *P. platyloba*: S), *Anisits 2511* (syntype of *P. platyloba*: S). CHACO PARAGUAYO: Puerto Casado, *Rojas 2600* (CTES, G).

Pavonia platyloba is notable for the broadly spatulate or subrotund involucellar bracts. Stem pubescence is evidently variable in this species. It may be short (0.5 mm) and soft (as in *Hatschbach et al. 52517* and *Rojas 2600*) or long (1.5 mm) and harsh (as in *Hatschbach & Zelma 49102*). The plants are otherwise the same and are here considered conspecific.

135. Pavonia purpusii Brandegee, Zoe **5:** 250. 1908. *Malache purpusii* (Brandegee) Standley, Contr. U.S. Natl. Herb. **23:** 772. 1923. Type. Mexico. Veracruz: Zacuapán, Jan 1907, *Purpus 1946* (holotype, UC; isotypes, BM, CTES, E, F as photo F-56164, GH, MO, NY, US). Fig. 53

Pavonia liebmannii Ulbrich, Feddes Repert. Spec. Nov. Regni Veg. **13:** 516. 1915. Type. Mexico. 1841–1843, *Liebmann s.n.* (holotype, B destroyed, as photo F-9453; probable isotype, C [numbered 418, labeled "Mirador," and dated Apr 1842]).

Shrubs 1–2 m tall, the stems densely covered with short (0.3–0.5 mm) glandular hairs and sometimes also with long (2–3 mm) simple hairs. *Leaf* blades ovate, mostly 4–8(–12) × 2–5 cm (smaller upward), basally cordate, serrate, acuminate, palmately 7-nerved, somewhat discolorous, the upper leaf surface with mostly simple hairs (0.3–0.4 mm long, arched) interspersed with stellate and glandular hairs, the lower surface with soft stellate pubescence; petioles 3–7 cm long, stipules 4–9 mm long, linear. *Flower* solitary in the leaf axils, the pedicels 4–8 cm long; involucellar bracts 7–10, linear or narrowly ligulate, 9–13 × 0.5–1 mm, viscid and ciliate; calyx 6–10 mm long, ciliate, whitish basally with green veins distally; corolla 1.5–2.5 cm long, ciliate on margins of claws, otherwise glabrous, pink or lavender with darker center; staminal column 15–17 mm long, pallid, glabrous, the filaments ca. 1 mm long or subsessile, with inconspicuous staminodes (sterile filaments) at base of column; styles emergent from the column, pink. *Fruits* 8–9 mm diam., densely pubescent; mericarps 4–5.5 mm high, dorsally rounded.

Distribution (Fig. 55). From S Mexico (Veracruz, Oaxaca, Chiapas), Guatemala, and Honduras, growing in shrubland and forest at elevations of 100–900 m. It apparently flowers throughout the year.

Specimens examined. MEXICO. CHIAPAS: Mun. Villa Corzo, 58 km S of Hwy. 190 on rd. to Nueva Concordia, *Breedlove 37625* (CAS). VERACRUZ: Near Huatusco, *S. Jones & G. Jones 5396* (F, MEXU, NY, TEX); Zacuapan, *Purpus 4311* (F, GH, UC, US); Barranca de Tenampa, *Purpus 8379* (GH, MO, NY, UC, US); near Huatusco, *von Rozynski 767* (F, NY); ca. 4.5 mi W of Palmilla, rd. to Huatusco, *Stevens 1407* (DS, MO, MSC, NY, TEX); Mun. Xico, El Paponal, *F. Ventura 888* (ARIZ, CTES, ENCB, LL); Mun. Totula, Encinal, *F. Ventura 7884* (ARIZ, CTES, ENCB, LL, NY).

GUATEMALA. ALTA VERAPAZ: Between Cajval and Cahabon, *Pittier 211* (GH, US); vic. of caves SW of Lanquín, *Steyermark 44045* (F). IZABAL: NE of Quiriguá, *P. A. Fryxell 1187* (BH, CAS, ENCB, NY); Los Amates, *Kellerman 5091* (US); between Los Amates and Izabal, *Kellerman 7290* (F, NY).

HONDURAS. ATLANTIDA: Congrejal River, between Yaruca and La Ceiba, rd. from Olanchito, *Blackmore & Chorley 4113* (BM, NY). YORO: Vic. of Coyoles, Aguan river valley, *Yuncker et al. 8635* (F, NY).

Pavonia purpusii is related to *P. fonsecana* and *P. spuria*. It is distinguished from these by its pink or lavender corolla, which is 1.5–2.5 cm long, in contrast to white corollas barely over 1 cm long.

136. Pavonia quadrifida Fryxell, sp. nov. Type. Brazil. Goiás: Drainage of the upper Araguaia River, 87 km SE of Aragarças, 700 m, 21 Jun 1966, *Irwin, Souza, Grear & Reis dos Santos 17552* (holotype, NY; isotypes, CTES, F, NY, US). Fig. 62

Frutices ubique plus minusve glanduloso-pubescentes; laminis foliorum cordato-ovatis vel leniter 3-lobatis molliter tomentosis longi-petiolatis; floribus axillaribus; involucellis cupuliformibus basaliter gamophyllis, bracteis involucellorum 4 basin versus luteolis apicem versus virentibus 10–12 mm longis, calycem occultandibus; corollis ut videtur tubularis 2.5 cm longis; androecio corollam subaequanti vel parum excedenti; fructibus 8 mm diametro puberulis reticulato-venosis.

Shrubs ca. 1 m tall, the stems minutely glandular-puberulent. *Leaf* blades ovate to shallowly 3-lobed, mostly 5–15 × 3.5–14 cm, deeply cordate, irregularly serrulate, acuminate, palmately 7–9-nerved, discolorous, densely and softly tomentose on both surfaces (and with minute glandular hairs intermingled); petioles 2.5–12 cm long, minutely glandular-puberulent; stipules subulate, 6 mm long, glandular-puberulent, spreading or recurved, deciduous. *Flowers* solitary (or paired) in the leaf axils, these sometimes aggregated apically, the pedicels 3.5–7 cm long; involucellar bracts 4, basally gamophyllous and cupuliform, ovate, 10–12 × 7–9 mm, 5–7-veined, yellowish basally, greenish distally, softly tomentulose (with glandular hairs intermingled) and basally also with a few long

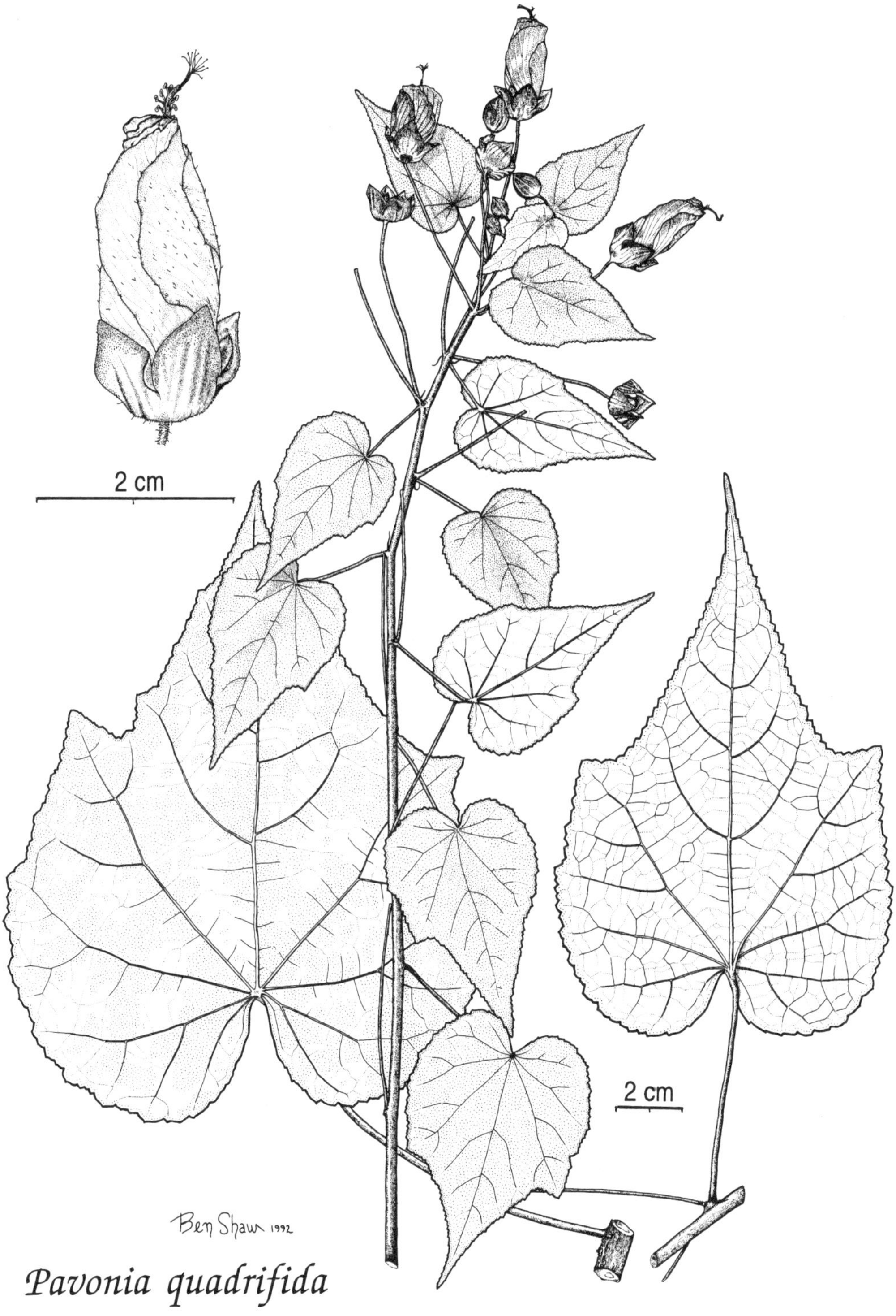

FIG. 62. *Pavonia quadrifida.* Flowering branch (center), enlargement of flower (above left), and maximally developed leaves (bottom) (*Irwin et al. 17552*).

simple hairs 1–1.5 mm long; calyx 6–7 mm long (largely concealed by enclosing involucel), tomentulose, ciliate-margined; corolla 2.5 cm long, "red-orange," tubular, the petals dark-margined in bud, externally stellate-pubescent, the hairs 3–5-armed, ca. 1 mm diam.; androecium subequal to corolla or slightly exserted; styles exserted from column by 5–8 mm, slender, glabrous, the stigmas villous. *Fruits* ca. 8 mm diam., yellowish, verruculate and reticulate-veined, minutely scabridulous.

Distribution (Fig. 55). Endemic to Goiás at 700 m elevation. The type locality is approximately at 16°35′S, 52°20′W.

Specimen examined. **BRAZIL.** GOIÁS: Caiaponia-Aragarças rd., ca. 80 km from Aragarças, *Hunt 6131* (CTES, K).

The specific epithet is chosen in reference to the 4-parted, gamophyllous involucel, an unusual arrangement in *Pavonia*, although this arrangement is superficially similar to that found in *P. tiliifolia*.

137. Pavonia rupestris (Hassler) Krapovickas & Cristóbal, Bonplandia **7:** 55. 1993. Basionym: *Asterochlaena sidifolia* var. *rupestris* Hassler, Feddes Repert. Spec. Nov. Regni Veg. **8:** 115. 1910. Type. Paraguay. Cerros de Tobatí, Sep 1900, *Hassler 6143* (lectotype, here designated, G–n.v.; isolectotypes, CTES, GH, MO, NY, S, UC as photo CTES).

Shrubs 2–3 m tall, the stems densely stellate-pubescent (hairs 0.2–0.6 mm long), sparser on older stems. *Leaf* blades ovate or ovate-oblong (juvenile leaves sometimes suborbicular), 4–7 × 3–6 cm, basally cordate, shallowly crenate-serrate, acute, palmately 7–9-nerved, discolorous, densely and softly pubescent on both surfaces; petioles (1–)2–6 cm long, with pubescence like that of stem; stipules filiform, 8–9 mm long. *Flowers* solitary (or paired) in the leaf axils, the pedicels 2–7 cm long; involucellar bracts usually 9 (less often 8 or 10), 1–1.5 × 1.5–3 mm, softly pubescent, sometimes with glandular hairs intermingled and with a few setae ca. 1 mm long; calyx shorter than the involucel, 8–9 mm long, densely stellate-pubescent and with marginal setae 1 mm long; petals 1.5–2 cm long, coarsely stellate-pubescent externally, yellow with or without a dark center, pubescent on margins of claw; staminal column 7–8 mm long, the filaments 1–2 mm long; styles and stigmas exserted from staminal column but not equaling the corolla. *Fruits* 7–8 mm diam., verruculate, glabrous; mericarps 4.5–5.5 mm long, dehiscent; seeds 3.5 mm long, striate, glabrous, dark brown.

Distribution (Fig. 60). Found in Paraguay and adjacent parts of Bolivia and Brazil, apparently occurring in Paraguay in two separated localities, according to the following citations. The Brazilian collections cited are included only tentatively. The type locality is at 25°15′S, 57°05′W.

Specimens examined. **BRAZIL.** Corumbá, Fazenda Carandá, *Conceição 132* (CTES); Mun. Nioaque, BR-060, *Hatschbach et al. 52430* (C, CTES, US); Corumbá, *Malme 3063* (S-2).

BOLIVIA. SANTA CRUZ: Prov. Andrés Ibáñez, 5.5 km S of Basilio on rd. from Santa Cruz to Camiri, *M. Nee 44587* (NY, TEX); Prov. Andrés Ibáñez, 8 km NW of Río Grande Planta de Gas, *M. Nee 44668* (NY, TEX).

PARAGUAY. CENTRAL: Trinidad, Asunción, *E. Pérez 504* (MO, US); Ñemby, 100 m from the church, *Pérez de Molas et al. 16* (MO, US). CHACO: Defensores del Chaco National Park, Cerro León, *Hahn 1569* (CTES, MO); Cerro León, *Schinini & Bordas 17853* (CTES). CORDILLERA: Caacupé, Barrio Kennedy, *Arbo et al. 6079* (CTES, MO, TEX); Tobaty, *Hassler 3978* (NY); 7 km from Caacupé, Ytú Creek, *Krapovickas & Cristóbal 11473* (CTES, NY); Tobatí, "Ybitú Silla" mesa, *Zardini 4305* (CTES, MO); Cerro Tobatí, *Zardini & Degen 3803* (CTES, MO). PTE. HAYES: 80 km from Asunción, por ruta Trans-Chaco, *Duré 419* (CTES); Asunción, *Gibert 1038* (K). Cursis superioris fluminis Y-acá, Dec 1900, *Hassler 6872* (syntypes, GH, K, MO, NY, S).

138. Pavonia sidifolia Humboldt, Bonpland & Kunth, Nov. Gen. Sp. **5:** 283 (folio ed., p. 220). 1822. *Malache sidifolia* (Humboldt, Bonpland & Kunth) Kuntze, Revis. Gen. Pl. **1:** 71. 1891. *Lassa sidifolia* (Humboldt, Bonpland & Kunth) Kuntze, Revis. Gen. Pl. **3**(2): 20. 1898. *Asterochlaena sidifolia* (Humboldt, Bonpland & Kunth) Hassler, Feddes Repert. Spec. Nov. Regni Veg. **8:** 114. 1910. Type. Between Angostura and Trapiche de D. Felix Farreras (Orinoco), *Humboldt & Bonpland 1077* (holotype, P-HBK as microfiche; isotype, B-W no. 12694 as photo F-9796). Fig. 53

Pavonia diuretica St.-Hilaire, Fl. Bras. Merid. **1:** 234. 1827. *Malache diuretica* (St.-Hilaire) Kuntze, Revis. Gen. Pl. **1:** 70. 1891. *Pavonia sidifolia* subsp. *diuretica* (St.-Hilaire) Gürke in Martius, Fl. Bras. **12(3):** 509. 1892. *Asterochlaena sidifolia* subsp. *diuretica* (St.-Hilaire) Hassler, Feddes Repert. Spec. Nov. Regni Veg. **8:** 114. 1910. Type. Brazil. Minas Gerais: Near Mangahy, *St.-Hilaire s.n.* (holotype, P as photo F-35471; isotype fragments, CTES, F).

Asterochlaena sidifolia var. *paraguariensis* Hassler, Feddes Repert. Spec. Nov. Regni Veg. **8:** 114. 1910. Lectotype (here designated). Paraguay. Cordillera de Altos, *Hassler 3154* (isolectotypes, CTES, GH, NY, S, UC as photo CTES).

Malache chiapensis Standley, Contr. U.S. Natl. Herb. **23:** 772. 1923. *Pavonia chiapensis* (Standley) Standley, Publ. Field Columbian Mus., Bot. Ser. **4:** 231. 1929. Type. Mexico. Chiapas: Jiquipilas, 31 May 1904, *Goldman 1037* (holotype, US).

Widely branching *shrubs* or subshrubs 1.5 m tall, the stems sometimes trailing, stellate-pubescent and with some glandular hairs especially on older stems, sometimes also (e.g., *M. Nee 38211*, *Pedersen 9410*) with setae to 3 mm long. *Leaf* blades mostly 4–8 cm long, 1.1–1.6(–2) times as long as wide, ovate (in well-grown plants sometimes weakly lobulate), cordate, serrate, acute or acuminate, discolorous, soft-pubescent above and beneath; petioles ⅓–⅔ the length of the blades in vegetative branches but shorter in flowering branches, the leaves sometimes subsessile; stipules 2–4 mm long, filiform, recurved. *Flowers* solitary or paired in the leaf axils, sometimes forming a terminal panicle, the pedicels 2–6.5(–9.5) cm long; involucellar bracts 6–7(–9 in *Hatschbach 23721*!), exceeding the calyx, broadly lanceolate or subrhombic (2.5–4 mm wide, rarely narrower); calyx 6–7 mm long, hirsute; corolla (1–)2–2.5 cm long, bright yellow with purple center; staminal column 5–7 mm long, maroon, the filaments 1–2 mm long; styles emergent from the column, glabrous, the stigmas pallid, villous. *Fruits* 6–8 mm diam., oblate, with appressed barbs along medial costa and lateral margins of mericarps; mericarps 2.5 mm high, smooth or verruculate, essentially glabrous. Chromosome number, $2n = 56$ (Bates, 1976, as *P. chiapensis*).

Distribution (Fig. 57). Widespread in South America (Colombia, Venezuela, Peru, Brazil, Bolivia, Paraguay) and disjunctly in S Mexico. The type locality for *Asterochhlaena sidifolia* var. *paraguariensis* is at 25°32′S, ca. 57°W.

Illustrations. Fryxell (1979: fig. 7E); St.-Hilaire (1824: pl. 53).

Representative specimens examined. MEXICO. CHIAPAS: Mun. Ocozocuautla, 1 km N of Ocozocuautla, *Breedlove 21598* (CAS, NY); 5 km W of Rizo de Oro, Mun. Cintalapa, *Breedlove 36689* (CAS, NY); near Las Cruces junction, 25 mi N of Arriaga, *P. A. Fryxell 1165* (BH, CAS, CTES, GH, MICH, NA, NY, UC); Hacienda Monserrate, *Purpus 9224* (F, GH, MO, NY, US). **OAXACA:** Mun. San Miguel Chimalapa, 12–15 km from the hwy. from Juchitán to Tuxtla Gutiérrez, *Koch & P. A. Fryxell 78356* (CHAPA, CTES, ENCB, F, NY); Mun. Santo Domingo Petapa, 1 km N of Santo Domingo, *Koch, P. A. Fryxell & C. P. Cowan 78255* (BR, CAS, CHAPA, CTES, ENCB, K, MO, NY, TEX, WIS).

PANAMA. Los Santos, Monagre beach, *Dwyer 4154* (MO).

COLOMBIA. Barranquilla and vic., *Elias 651* (US). **ATLÁNTICO:** Chiriguana, Magdalena Valley, *C. Allen 122* (MO); Palmar de Varela, along the Magdalena, *Dugand 3129* (US). **BOLÍVAR:** N of Arjona, *Killip & A. C. Smith 14491* (GH, NY, US). **MAGDALENA:** La Jagua, *C. Allen 534* (MO); Becerril, 60 km NE of Chiriguaná, *Haught 2294* (F, S, US); La Jagua, *Haught 3649* (GH).

VENEZUELA. Near Tovar, *Fendler 1932* (GH). **ARAGUA:** Potrero de Las Delicias, *Lasser 834* (US). **BOLÍVAR:** Distr. Páez, Laguna Larga, 8 km SE of Upata, *Delascio & Liesner 6855* (CTES, MO, NY); Paso de Caruachi, 75 km al E de Ciudad Bolívar, *Holst et al. 2274* (MO). **CARABOBO:** Near Guacara, *Pittier 8181* (GH, US). **FALCÓN:** Hwy. from Falcón to Zulia, between the Borojó and Maticora Rivers, *Wingfield 6063* (CORO, CTES, NY). **PORTUGUESA:** Guanare, Mesa Alta, *P. A. Fryxell et al. 4356* (NY). **TRUJILLO:** Valera, *Pittier 10715* (GH, NY). **ZULIA:** Dept. Bolívar, Palito Blanco, *P. A. Fryxell et al. 4406* (CTES, MO, NY); vic. of Mene Grande, *Pittier 10643* (GH, NY, US).

PERU. CAJAMARCA: Jaén, crossing to Bellavista, *Llatas-Quiroz 2203* (F, HUP, NY); Mandanguia, *Woytkowski 6805* (GH, MO, US). **SAN MARTÍN:** 1–4 km NE of Tarapoto, *Belshaw 3357* (F, GH, LL, MO, NY, RSA, U, UC, US); Prov. Tarapoto, along Río Cumbaza, *Croat 51122* (MO); Prov. Lamas, near Lamas, *Ferreyra 17306* (CTES); 20–30 km NE of Juanjui, *A. H. Gentry & D. N. Smith 44956* (MO, NY).

BRAZIL. Falls of Madeira, *Rusby 1459* (GH, NY–2, PH, TEX–2, US). **BAHIA:** 15 km W of Ibotirama, *P. A. Fryxell 1926* (BH, CAS, NY); 70 km W of Seabra, *P. A. Fryxell 1932* (CTES, TEX); 25 km E of Jacobina, *P. A. Fryxell 1948* (BH, CAS, NY, TEX); 3 km E of Barreiras, *Krapovickas et al. 37875* (C, CEN, CTES, F); Mun. Lençois, *Pinto 67/86* (RB). **CEARÁ:** Near Crato, *Gardner 1457* (GH, K, NY, OXF, TEX, US); 8 km W of Crato, *A. H. Gentry et al. 50176* (MO, NY); 29 km W of Crato, *A. H. Gentry et al. 50103* (CM, GH, MO, NY). **ESPÍRITO SANTO:** Vic. of Guarapari, *Hoehne 5513* (CTES); near Vitória, *Trinta 1065* (CTES). **GOIÁZ:** Without locality, *Gardner 3591* (E, K, NY, OXF), *Gardner 3599* (K). **MATO GROSSO:** Terenos, *Hatschbach 23721* (C, CTES, MBM, MO, NY); Mun. Miranda, Miranda River, *Hatschbach 38639* (C, NY); 41 km E of Cuiabá, *Krapovickas et al. 40222* (CTES); Mun. Antonio João, a few km W of Antonio João, *Pedersen 16022* (TEX). **MINAS GERAIS:** 13 km W of Januária, *W. R. Anderson, P. A. Fryxell et al. 9132* (C, F, MO, NY, TEX, US); 15 km N of Montalvânia, *W. R. Anderson et al. 37162* (F, K, MO, NY, TEX, US); Mun. Janauba, Ribeirão Poções, *Hatschbach & Guimarães 55105* (CTES); Pirapora, *Krapovickas & Cristóbal 42866* (CTES, MBM). **PARANÁ:** Iguazú Falls, International Bridge, *Hatschbach 9759* (CTES); Guaira, 7 Quedas, *Lindeman & Haas 4414* (NY, U). **RIO DE JANEIRO:** Cabo Frio, *Campêlo & Campêlo 572* (CTES); near Rio de Janeiro, *Glaziou 12448* (C, K); between Campos and Morro do Côco, *Trinta 1043* (= *Fromm 2119*) (CTES, NY). **SÃO PAULO:** SP-425, 25 km S of Tieté River, *Krapovickas et al. 32997* (CTES, MBM).

BOLIVIA. BENI: Prov. Itenez, Magdalena, *Krapovickas & Schinini 34802* (CTES); Prov. Ballivián, San Borja, *Krapovickas & Schinini 34884* (CTES, LPB, NY); Prov. Moxos, San Ignacio de Moxo, *Krapovickas & Schinini 34929* (C, CTES, LPB); Trinidad-Missiones Guarayos, *Werdermann 2361* (K, MO, S). **SANTA CRUZ:** Prov. Gutiérrez, 8 km E of Portachuelo, *Krapovickas & Schinini 31564* (CTES, MO); Avenida Pirai, W side of Santa Cruz, *M. Nee 38211* (CM, CTES, MO, NY, TEX); Prov. Velasco, 400 m,

Seidel 157 (US); Prov. Sara, Buenavista, *J. Steinbach 1435* (GH, NY); Prov. Cercado, *J. Steinbach 2718* (GH, NY); Angostura, *R. F. Steinbach 330* (BRIT, F, GH, MO, NY, RSA, S, U, US); Prov. Ichilo, Buena Vista, *R. F. Steinbach 788* (BRIT, F, GH, MO, RSA, S, U, US).

PARAGUAY. Central Paraguay, without locality, *Morong 702* (E, GH, K, MO, NY, PH, TEX, US); Cerros de Tobatí, Cerro Penitente, *Fiebrig 758* (E, F). **AMAMBAY:** Bella Vista, *Krapovickas & Cristóbal 44982* (CTES, TEX); 45 km E of Yby Yaú, *Krapovickas et al. 45881* (CTES, TEX). **BOQUERÓN:** 4 km N of 25 Laguas, rt. 9, *Calvert 115* (NY); Chaco Paraguayo, F. Olimpo, *Rojas 13882* (CAS, CTES, TEX). **CENTRAL:** 2 km SW of the Salado River, rd. from Limpio to Emboscada, *Arbo et al. 1597* (CTES, F, K); Asunción, *Hahn 764* (CTES, MO, NY, TEX); in the region of Lake Ypacaray, *Hassler 11459* (A, C, E, F, GH, MO, NY, S, US); Tarumandy, *Schinini 6172* (CTES, G, LIL, NY). **CONCEPCIÓN:** Horquetas, *Gregory et al. 10594* (S); 27 km W of Horqueta, rd. to Loreto, *Schinini et al. 29343* (CTES, TEX). **LA CORDILLERA:** 5 km SE of Emboscada, *Zardini & M. Velázquez 20767* (CTES, MO); 1 km E of Nueva Colombia, *Zardini & R. Velázquez 20878* (CTES, MO). **PARAGUARI:** Arroyo Yuquity, 7 km E of Nueva Italia, *Zardini & U. Velázquez 21534* (MO). **PRESIDENTE HAYES:** Puente Remansito, on the Paraguay River, *Krapovickas et al. 45526* (CTES, MO, TEX). **SAN PEDRO:** Villa Elisa, *Pedersen 3122* (C, NY, S, US, herb. Pedersen); Villarica, *Jörgensen 3812* (A, C, CTES, GH, MO, NY, PH, S).

ARGENTINA. CHACO: Colonia Benítez (cult.), *Schulz 9205* (CTES, NY). **CORRIENTES:** Dept. Ituzaingó, Ituzaingó, gorge of the Paraná River, *Quarín et al. 3379* (CTES, MO); Dept. San Miguel, Ea. Santa Ana Ñú, bridge over the Paraná River, *Tressens et al. 3448* (CTES, MO).

Local name. Malva grande (Bolivia).

There are many collections in CTES from Corrientes and Misiones with setae 3–4 mm long on the stems. These plants may represent something other than *Pavonia sidifolia*. Setose plants from Brazil and Bolivia may be still different; the setae are neither as long nor as abundant, although they may be very abundant in some collections [e.g. *Krapovickas & Schinini 31564* (Bolivia. Santa Cruz: Prov. Gutierrez, 8 km E de Portachuelo)] and 2–3 mm long. This problem needs further study.

139. Pavonia spuria Fryxell, sp. nov. Type. Mexico. Oaxaca: Mun. Santo Domingo Petapa, 1 km N of Santo Domingo, 12 km SW of Matias Romero, 13 Nov 1978, *Koch, P. A. Fryxell & C. P. Cowan 78258* (holotype, CHAPA; isotypes, BR, CTES, ENCB, K, MO, NY–2, WIS). Figs. 53, 63

Suffrutices caulibus viscidis; laminis foliorum ovato-cordatis pilis adpressis simplicibus atque aliquot glandulosis; bracteis involucellorum 7–10 lineari-lanceolatis, quam calyce parum longioribus; calycibus valde ciliatis; corollis albis (erubescentibus) 12 mm longis interne glabris (unguibus includentibus); columna staminalis et stylis et stigmatibus glabris inclusis; fructibus oblatis minute pubescentibus.

Subshrubs 1.5 m tall, well branched, the stems viscid, the glandular hairs 0.5 mm long. *Leaf* blades ovate, cordate, serrate, acuminate, mostly 3–6 × 1.5–4.5 cm, palmately 5–7-nerved, discolorous, with appressed, predominantly simple hairs (0.3–0.4 mm long) on both surfaces, with glandular hairs at least on margin; petioles 1–3.5 cm long, densely glandular-pubescent; stipules subulate, 2 mm long, pubescent. *Flowers* solitary in the leaf axils, the pedicels 2.5–5.5 cm long, slender, glandular-pubescent; involucellar bracts 7–10, linear-lanceolate, 8–10 × 1 mm, glandular pubescent and with basal cilia 1–2 mm long; calyx 7–9 mm long, prominently ciliate on margins and nerves (hairs 1.5 mm); corolla 12 mm long, white (fading pink), externally pubescent, internally glabrous (including margins of claws); staminal column 6–8 mm long, glabrous, pallid, the filaments 1–1.5 mm long; styles emergent from the staminal column, glabrous; stigmas capitate, glabrous. *Fruits* oblate, 7 mm diam., minutely pubescent; mericarps dehiscent, 4 mm high; seeds 3 mm long, glabrous, striate.

Distribution (Fig. 55). Confined to the state of Oaxaca in S Mexico, found at ca. 350 m elevation.

Specimen examined: **MEXICO. OAXACA:** Vic. of San Juan Guichicovi, *E. W. Nelson 2716* (US).

The specific epithet, from the Latin (*spurius*, false), refers to the close similarity to *Pavonia purpusii* and the easy confusion with the latter species, as well as with the similar *P. fonsecana*. However, *P. spuria* is distinguished from those two species by having claws of the petals that are entirely glabrous.

140. Pavonia subrotunda St.-Hilaire & Naudin, Ann. Sci. Nat. Bot., ser. 2, **18:** 42. 1842. Type. Brazil. Rio Grande do Sul: 1835, *Isabelle s.n.* (holotype, G-DEL as photo F-23711). Fig. 53

Pavonia orbicularis Ulbrich, Bot. Jahrb. Syst. **42:** 123. 1908. *Asterochlaena orbicularis* (Ulbrich) Hassler, Feddes Repert. Spec. Nov. Regni Veg. **8:** 115. 1910. Type. Paraguay. Atyrá: Loma, Dec 1902, *Fiebrig 513* (lectotype, here designated, B destroyed, as photo F-9457; isolectotypes, A, GH, K), (syntype. *Bettfreund 96*).

Asterochlaena orbicularis var. *decumbens* Hassler, Feddes Repert. Spec. Nov. Regni Veg. **8:** 115. 1910. Type. Paraguay. In sand on the Capibary River, *Hassler 5906* (holotype, G–n.v.).

Asterochlaena orbicularis var. *ekmanii* Hassler, Feddes Repert. Spec. Nov. Regni Veg. **8:** 116. 1910. Type. Argentina. Misiones: San Ignacio, *Ekman 176* (lectotype, here designated, S; isotype, S).

Procumbent perennial *subshrubs*, the stems trailing from a woody rootstock, hirsute with hairs 1–2 mm long and with an understory of fine stellate hairs, or

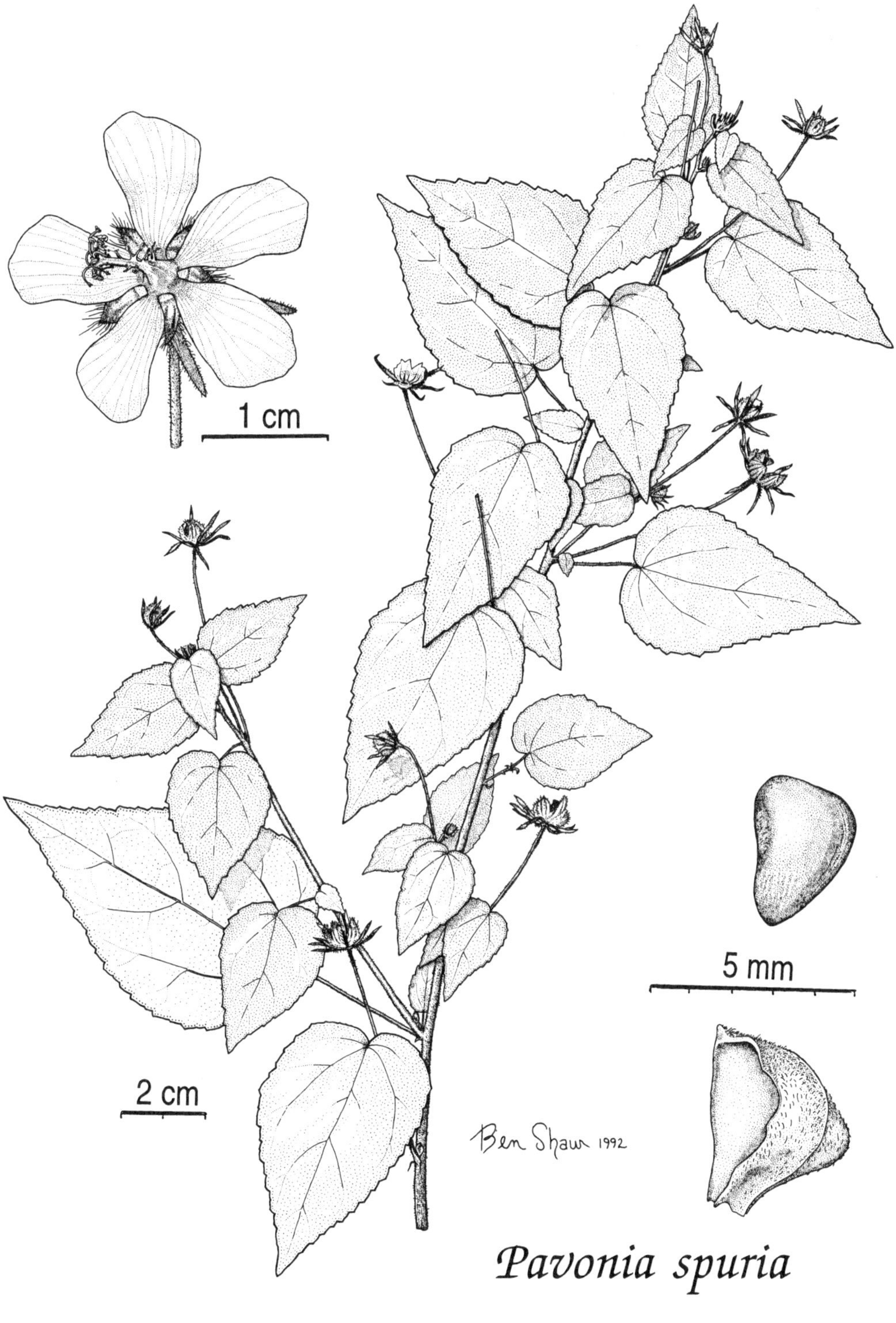

FIG. 63. *Pavonia spuria*. Flowering and fruiting branch (center) with enlargements of flower (upper left) and mericarp and seed (lower right) (*Koch et al. 78258*).

sometimes with the long hairs absent, rarely glabrate. *Leaf* blades orbicular, deeply cordate, crenate-dentate, obtuse or subacute, 1.5–5 × 2–6.5 cm, palmately 7-nerved, softly stellate-puberulent, somewhat discolorous; petioles 1–6 cm long; stipules subulate, 4–8 mm long. *Flowers* solitary in the leaf axils, the pedicels 2–9 cm long; involucellar bracts 9–11, linear-lanceolate, 6–10 × 1 mm (shorter than to longer than the calyx), puberulent or ciliate, sometimes viscid; calyx 7–10 mm long, pubescent; corolla 1.5–2.5 cm long, yellow (drying reddish) with a sharply defined red center, pubescent on margins of claw; staminal column 7–9 mm long, reddish (at least basally), the filaments 1–2 mm long, the anthers purplish; styles emergent from the column, united except for distal 1 mm, nearly equaling the petals. *Fruits* oblate, 9–10 mm diam., glabrous, the mericarps 5–6 mm high, more or less ridged and reticulate, sometimes the ridge projected subapically into a triangular protruberance; seeds 5 mm long, striate.

Distribution (Fig. 57). From southernmost Brazil, Paraguay, and Argentina (Corrientes and Misiones). The type locality for *Pavonia orbicularis* is at 25°17′S, 57°10′W; that for *Asterochlaena orbicularis* var. *decumbens* is at 25°40′S, 55°40′W; that for *A. orbicularis* var. *ekmanii* is at 27°15′S, 55°34′W.

Illustration. Ekman (1909: fig. 8, as *P. orbicularis*).

Representative specimens examined. BRAZIL. Rio Grande do Sul: 70 km SE of São Borja ("topotipo"), *Krapovickas & Cristóbal 38385* (C, CTES, F, G, HAS, LIL, MICH, NY, UC, WIS).

PARAGUAY. Caazapá: Distr. Yuty, 20 km E of San Miguel, *Arbo et al. 2879* (CTES); vic. of San Pedro del Paraná, *Schinini et al. 27757* (CTES, TEX). **Central:** Tavarory, *Zardini & U. Velázquez 24277* (MO). **Itapuá:** Capitán Miranda, 4.2 km N of the Hotel Tirol, behind the Barrio CONAVI, *Krapovickas & Cristóbal 44455* (CTES, TEX). **San Pedro:** Rd. between Caacupé and Tobaty, *Archer & Rojas 4862* (A, US); 8 km E of San Estanislao, Tapiracuay River, *Schinini 8844* (CTES);

ARGENTINA. Corrientes: Dept. Concepción, 6 km E of Santa Rosa, *Arbo et al. 828* (CTES, NY); Dept. San Roque, Estancia Añá Cuá, *Carnevali 1668* (CTES); Tabay, *Krapovickas & Cristóbal 11661* (BH, CTES, G, K, P, S, SI, US); Mburucuyá, Loma Alta, *Pedersen 3004* (C, CTES, K, MO, NY, S, U, US, herb. Pedersen); Médano El Carmen, *Pedersen 5325* (C, K, MO, NY, herb. Pedersen); Dept. San Roque, Estancia Caaguazú, *Pedersen 5859* (C, CTES, herb. Pedersen); Dept. Saladas, Estancia Bovril, *Schwarz 9137* (CAS, CTES); Dept. Concepción, 11 km NW of Santa Rosa, *Tressens et al. 2019* (CTES, ICN, MEXU, MICH, NY). **Misiones:** Teyú Cuaré, 8 km SW of San Ignacio, *Cabral et al. 183* (CTES, NY); Dept. San Ignacio, Teyú Cuaré, 3 km S of Gendarmería, *Vanni & Radovancich 1037* (C, CLEMS, CTES, F, FCQ, HAS, MICH, MO, NY, SPF, UC).

141. Pavonia vannii Fryxell, sp. nov. Type. Paraguay. Boquerón: Col. Menno, Loma Plata, enclosure of the Loma Plata Reserve, *Vanni, Radovancich & Schinini 1948* (holotype, CTES). Fig. 64

Suffrutices decumbentes vel ascendentes caulibus dense pubescentibus et pilosis; laminis foliorum ovatis vel suborbicularis molliter et perdense pubescentibus; floribus solitariis; bracteis involucellorum 8–9 quam calyce longioribus; corollis 15–18 mm longis flavidis cum vel sine centro rubro; fructibus oblatis ca. 8 mm diametro minute scabridulosis.

Decumbent or ascending *subshrubs*, often supported on other vegetation, the stems densely stellate-pubescent (hairs 0.5 mm long or less) and pilose with simple hairs 1.5–2 mm long. *Leaf* blades ovate to suborbicular, 3–4 × 2.5–3.5 cm, basally more or less cordate, serrate, acute, palmately 5–7-nerved, very densely and softly tomentose on both surfaces; petioles 1–3.5 cm long, half length of blade to subequal to blade, with pubescence like stem but more densely pilose distally; stipules filiform, 3–4 mm long, caducous. *Flowers* solitary in the leaf axils, the pedicels 3–6.5 cm long, minutely stellate-pubescent and pilose, more densely so distally; involucellar bracts 8–9, lanceolate, 9–15 × 2–3 mm, densely stellate-tomentose, sometimes also pilose; calyx 6–7 mm long, stellate-pubescent, ca. half-divided, the lobes 3-ribbed; petals 15–18 mm long, yellow with or without a red center, externally with simple and 3-armed stellate hairs, internally glabrous, pubescent on margins of claw; staminal column ca. 8 mm long, glabrous, the filaments 1–1.5 mm long; styles emergent from the column, with capitate, villous stigmas. *Fruits* ca. 8 mm diam., oblate, minutely scabridulous; mericarps ca. 5 mm high.

Distribution (Fig. 60). Paraguay and Bolivia, at low elevation. The type locality is at 22°21′S, 59°50′W.

Specimens examined. BOLIVIA. Santa Cruz: Prov. Ibañez, Santa Cruz ca. 22 km S toward Abapó, *S. Beck 6506* (CTES); Prov. Andrés Ibáñez, along Quebrada Peji, on rd. from Santa Cruz to Comiri, *M. Nee 45835* (NY, TEX); 30 km S of Santa Cruz, *Saravia-Toledo 2683* (CTES).

PARAGUAY. Boquerón: 4 km N of 25 Leguas, *Calvert 115* (CTES, NY); 25 km S of Filadelfia, *Krapovickas & Cristóbal 44212* (CTES, TEX); 9 km S of Loma Plata, *Vanni et al. 2198* (CTES); Parque "Valle Nacional," 12 km S of Filadelfia, *Vanni et al. 2470* (CTES). **Chaco:** Mayor Pedro Lagerenza (20°S, 60°45′W), *Schinini & Bordas 15102* (CTES). **Cordillera:** Caacupé, Instituto Agronómico Nacional (cult.), *Krapovickas 41872* (CTES).

The Bolivian specimens cited have more densely pilose involucels than the Paraguayan specimens, but they otherwise appear to be the same and are here treated as conspecific.

FIG. 64. *Pavonia vannii.* Flowering branch (*Krapovickas & Cristóbal 44212*).

The specific epithet is chosen to honor the collector of the type, Ricardo Oscar Vanni (b. 1954) of Corrientes, Argentina.

142. Pavonia vinosa Esteves, Bol. Inst. Bot. São Paulo [in press, 1999]. Type. Brazil. Ceará: Serra do Araripe, 18 Apr 1910, *Löfgren 595* (holotype, R–n.v.).

Shrubs 1–2.5 m tall, the stems with purplish glandular hairs and also dense longitudinal lines of simple twisted hairs and occasional long spreading simple hairs. *Leaf* blades ovate, basally cordate, 4–7.5 × 2–4.5 cm, crenate-serrate, acute, palmately 7–9-nerved, strongly discolorous, the upper surface with scattered stellate and glandular hairs to glabrescent, the lower surface densely canescent-tomentose with stellate and glandular hairs, the veins with appressed simple hairs; petioles 1–2.5 cm long; stipules filiform, 2.5–4 mm long. *Pedicels* 2.5–7.5 cm long, articulated 2.5–5 mm below the flower; involucellar bracts 7 (6–8), elliptical, 8–11 × 2–3.5 mm, subequal to calyx, ciliate and with glandular and stellate hairs, basally hirsute; calyx 8–10 mm long, stellate- and glandular-pubescent, the lobes ciliate; petals ca. 3 cm long, rose or whitish; staminal column 1.5–1.7 cm long, the filaments 1–2 mm long. *Mericarps* 3.5–4 mm long, trigonal, umbonate, glabrous, verrucose, the dorsal surface with reticulate venation, laterally winged; seeds 3–3.5 mm long obovoid, striate.

Distribution. From the Brazilian states of Ceará and Maranhão.

Additional specimens examined. BRAZIL. CEARÁ: Crato, Chapada do Araripe, *Martins 8070* (EAC–n.v., SPF–n.v.), *Bezerra 446* (RBR–n.v., CTES–n.v.). **MARANHÃO:** Passagem Franca, Canto do Brejo, *Nunes & Martins 5824* (CTES–n.v., EAC–n.v., SPF–n.v.).

This species, material of which I have not examined, appears to be very similar to *Pavonia krapovickasii*. It differs from *P. krapovickasii* in having more involucellar bracts and somewhat different pubescence characters (especially the ciliate calyx and involucel), and it also has an allopatric distribution. On the basis of these characters, it is retained as distinct.

143. Pavonia yatarendana Krapovickas & Cristóbal, Bonplandia **7:** 51. 1993. Type. Bolivia. Santa Cruz: Prov. Cordillera, 5 km N of Yatarenda, 17 Apr 1977, *Krapovickas & Schinini 31482* (holotype, CTES; isotypes, K, LPB, MO, TEX).

Shrubs 1–1.5 m tall, the stems finely puberulent, the hairs 0.1 mm long. *Leaf* blades ovate, basally cordate, crenate-serrate, acute or acuminate, 6–8 × 3–6.5 cm (1.5–2 times as long as wide), palmately 7-nerved, discolorous, minutely puberulent above, more densely and coarsely so beneath, also with scattered 3-armed spicules 1 mm diam.; petioles 2–7 cm long, minutely puberulent; stipules subulate, 5 mm long. *Flowers* solitary in the leaf axils, the pedicels 6–8 cm long, minutely puberulent; involucellar bracts 8, narrowly lanceolate, 15–18 × 1.5–2 mm, minutely puberulent, sometimes also with a very few setae 1–2 mm long at base; calyx 10–12 mm long, minutely puberulent and setose along the midrib of the lobes, with a few 2–3-armed spicules at tips of lobes; corolla yellow without red center, 3.5 cm long, bearded on margins of claw; androecium purplish, the staminal column 9 mm tall, the filaments 2 mm long; styles exserted from column by 8 mm, free for distal 2 mm. *Fruits* 8–9 mm diam., minutely pubescent, 5-winged; mericarps 6 mm tall (excluding wings), dehiscent, with lateral wings and a prominent apical point 3 mm long; seeds 4 mm long, striate.

Distribution (Fig. 57). Known only from the type collection. The type locality is at 19°12′S, 63°32′W.

Illustration. Krapovickas & Cristóbal (1993: fig. 3).

Krapovickas and Cristóbal (1993: 54) note that this species is unusual for having a recurved stylar column, as is shown in their illustration.

Pavonia subgen. *Asterochlamys* sect. *Albae*

Pavonia sect. **Albae** Fryxell, sect. nov. (the name taken from that of the type species, *P. alba*, in reference to its corolla color).

Suffrutices vel frutices, caulibus pilis recurvatis minutis atque pilis longis simplicibus; laminis foliorum ovatis; floribus solitariis axillaribus; corollis albis interdum centro atropurpureo; columna staminalis verticillum staminodiorum inconspicuorum in base columnae; fructibus oblatis fragilis; mericarpiis dehiscentibus; seminibus non-striatis minute pubescentibus.

Subshrubs or shrubs, the stems with minute recurved hairs and also long simple hairs. Leaves ovate, unlobed. Flowers solitary in the leaf axils; corolla white, with or without a dark center; staminal column with a whorl of inconspicuous staminodes at base of column. Fruits oblate, fragile; mericarps dehiscent; seeds non-striate, minutely pubescent.

Type species. *Pavonia alba* Seemann.

Key to the species of *Pavonia* sect. *Albae*

1. Involucellar bracts ca. 13; pedicels 0.5–2.5 cm long; corolla 6–8 mm long, white with maroon center. . . . . 145. *P. falconensis*
1. Involucellar bracts 6–8; pedicels 2–5.5 cm long.
   2. Corolla 22–24 mm long, white with red center; Mexico. . . . . 146. *P. hirtiflora*
   2. Corolla 4–7 mm long, white, lacking red center; Guatemala to Venezuela. . . . . 144. *P. alba*

144. Pavonia alba Seemann, Bot. Voy. Herald 81. 1853. *Malache alba* (Seemann) Kuntze, Revis. Gen. Pl. **1:** 70. 1891. Type. Panama. *Seemann 90* (holotype, BM as photo MO; isotypes, CTES, GH, K). Fig. 53

Malache panamensis Standley, Contr. U.S. Natl. Herb. **18:** 116. 1916. *Pavonia panamensis* (Standley) Standley, J. Wash. Acad. Sci. **17:** 168. 1927. Type. Panama. Canal Zone: Sabana of Panama, 24 Jan 1911, *Pittier 2548* (holotype, US; isotypes, CTES [fragment], GH, NY).

Subshrubs 1 m tall, the stems evenly covered with minute recurved hairs, sometimes also with scattered setae 1.5–3 mm long. *Leaf* blades lance-ovate, mostly 4–8 × 1.5–3 cm, basally rounded or subcordate, coarsely serrate, acute or acuminate, sparsely and obscurely pubescent; petioles 1–5 cm long; stipules subulate or filiform, 3–5 mm long. *Flowers* solitary or paired in the leaf axils, sometimes in a terminal panicle, the pedicels 2–4.5 cm long, slender, articulated 4–10 mm below the flower, minutely pubescent, sometimes setose distally; involucellar bracts 6–8, linear-lanceolate, 7–9 × 1 mm or less, ciliate; calyx 4–5(–7) mm long, ciliate; corolla 4–7 mm long, white. *Fruits* 5–6 mm diam., oblate, glabrous, smooth, the mericarps 3 mm high; seeds 2 mm long, not striate, minutely pubescent, the hairs slightly curled.

Distribution (Fig. 54). *Pavonia alba* has previously been treated as confined to Panama and Colombia, but collections are now also known from Venezuela and (disjunctly) Guatemala. It grows in forest and savanna, at 0–800 m elevation. The type locality for *Malache panamensis* is at 09°00′N, 79°35′W.

Specimens examined. **GUATEMALA. ESCUINTLA:** Without locality, *Aguilar 1735* (F).

PANAMA. CANAL ZONE: 1 mi SW of Cocoli, *Wilbur et al. 12927* (F). **PANAMÁ:** Hills between Capira and Potrero, *Dodge & Hunter 8634* (C, CTES [fragment], MICH, MO, UC); Portrero River, ca. 2 mi from Arraijan, *Duke 4677* (MO, NY); along Las Cruces trail, *Hunter & P. H. Allen 757* (MO); SE slope of Cerro Campana, *W. H. Lewis et al. 3123* (MO); Arraiján (hwy. of the Congo River), Las Lajitas, *Marianela 38* (MO); Chepo, *Pittier 4448* (GH, NY); Matías Hernández, *Pittier 6907* (US); along the Corozal Rd. near Panamá, *Standley 26846* (GH, US); Nuevo Emperador, *Troetsch 14* (MO).

COLOMBIA. BOLÍVAR: Mun. San Jacinto, Cerro Maco, 16 km NW of San Jacinto, *Zarucchi & Cuadros 4024* (MO, NY, TEX). **MAGDALENA:** La Jagua, Magdalena Valley, *C. Allen 554* (MO); Poponte, Magdalena Valley, *C. Allen 819* (MO). **META:** Uribe, Duda River, *Fosberg 19477* (CAS, NY, US); vic. of Turbaco, *Bro. Heriberto 443* (NY, US).

VENEZUELA. 3.5 km from Sirapta on rd. to Machiques, *Jeffrey & Trujillo 2268* (F). **PORTUGUESA:** 32 km NW of Guanare, *Aymard & Ortega 4939* (CTES, NY); 30 km W of Guanare by air, along Tucupido River, *Liesner et al. 12646* (CTES, MO); 38 km al N de Guanare on hwy. from Guanare to Ospino, *Steyermark et al. 127138* (CTES, MO), *Steyermark et al. 127187* (CTES, MO, NY, VEN).

145. Pavonia falconensis Fryxell, Contr. Univ. Michigan Herb. **17:** 165. 1990. Type. Venezuela. Falcón: Cerro Píritu, 61 km E of Coro, 24 Feb 1984, *Wingfield 12402* (holotype, MO; isotypes, CORO, NY, TEX).

Shrubs 1–2.3 m tall, the stems pubescent with a mixture of minute recurved hairs, some glandular hairs, and scattered setae ca. 1 mm long. *Leaf* blades ovate, up to 8 × 5.5 cm (smaller upward), cordate, crenate-serrate, acuminate, palmately 7–9-nerved, glabrate above, sparsely pubescent beneath; petioles up to 5(–8) cm long; stipules 4–7 mm long, filiform. *Flowers* solitary in the leaf axils (often crowded apically), the pedicels 0.5–2.5 cm long, densely hirsute; involucellar bracts ca. 13, lance-linear, 9–12 × 0.5 mm, prominently ciliate (hairs 2–3 mm long) and viscid; calyx 6 mm long, ciliate; corolla 6–8 mm long, whitish (or "salmon") with maroon center, glabrous throughout (including margins of claw); staminal column ca. 4 mm long, glabrous, pallid, the filaments 1 mm long, with a whorl of staminodes ca. 1 mm long at base of column; styles glabrous, slightly exceeding the column. *Fruits* oblate, 5–6 mm diam., subglabrous; mericarps 3 mm tall, loculicidally dehiscent; seeds 2.8 mm long, not striate, minutely and sparsely pubescent, the hairs slightly curled.

Distribution (Fig. 54). Confined to the state of Falcón in N Venezuela, to the E and SE of Coro, growing at elevations of 100–360 m. The type locality is at 11°22′N, 69°10′W.

Illustration. Fryxell (1990: fig. 2).

Specimens examined. **VENEZUELA. FALCÓN:** S of Dos Bocas, at dam site, *Steyermark & A. González 113575* (CTES, F, MO, VEN); Cerro El Caballo, 53 km E of Coro, *Wingfield 11970* (CORO, MICH, NY); Fila Barigua, 22 km E of Coro, *Wingfield 12157* (CORO, NY, US); Cerro Mampostal, 44 km E of Coro, *Wingfield 12255* (CORO, NY); Cerro Agua María, 75 km SE of Coro, *Wingfield 12423* (CORO, NY).

The staminodes at the base of the staminal column were first noted in *Pavonia falconensis* and were later found to occur commonly in various other species of *P.* subgen. *Asterochlamys*, such as *P. purpusii.*

146. Pavonia hirtiflora Bentham, Pl. Hartw. 7. 1839. *Malache hirtiflora* (Bentham) Kuntze, Revis. Gen. Pl. **1:** 70. 1891. Type. Mexico. Aguascalientes: *Hartweg 23* (holotype, K as photo MICH-604; isotypes, BM, CTES [fragment], E, G as photo F-23697, K, NY, P as photo F-35472).

Erect *subshrubs*, the stems minutely stellate-pubescent. *Leaf* blades to 8 × 4 cm, ovate, somewhat cordate, coarsely dentate, acuminate, gradually reduced upward, minutely stellate-pubescent above and beneath, discolorous; petioles half length of blade or more; stipules 4–5 mm long, filiform. *Flowers* solitary in the leaf axils, the pedicels up to 5.5 cm long, exceeding the subtending petioles, minutely stellate-pubescent, articulated 5–7 mm below the flower; involucellar bracts 6, linear, 10–11 × 0.5–1 mm, minutely pubescent; calyx 7–10 mm long, puberulent; corolla 22–24 mm long, white (?) with dark center; staminal column ca. 1 cm long, bearded at base, the filaments 2–3 mm long. *Fruits* oblate, ca. 6 mm diam., the mericarps 4 mm long, glabrous, weakly reticulate-veined; seeds unknown.

Distribution (Fig. 55). Known only from the type collection, the plant not having been collected again in more than 150 years. The type locality is approximately at 22°N, 102°W.

PAVONIA SUBGEN. MALACHE

Pavonia subgen. **Malache** (Vogel) Fryxell, stat. nov. Basionym: *Malache* Vogel *in* Trew, Pl. Select. **9:** 50. 1772, nom. rejic. *Pavonia* sect. *Malache* (Vogel) de Candolle, Prodr. **1:** 443. 1824. *Pavonia* subsect. *Malache* (Vogel) Endlicher, Gen. Pl. 982. 1840. Type. *Malache scabra* Vogel (≡ *Pavonia paludicola* Nicolson)

Erect or scandent shrubs or small trees; flowers solitary or in terminal leafless racemes; involucellar bracts 6–12; fruits (see Fig. 65) indurate or woody, sometimes winged or ribbed but otherwise smooth and unornamented, glabrous or pubescent.

Synopsis of *Pavonia* subgen. *Malache*

1. **Pavonia** sect. **Malache**. Type species: *P. paludicola* Nicolson (4 species).
2. **Pavonia** sect. **Collicolae** Fryxell.
   - i. **Pavonia** subsect. **Collicolae.** Type species: *P. pleuranthera* (de Candolle) Fryxell (5 species).
   - ii. **Pavonia** subsect. **Rubriflorae** Fryxell. Type species: *P. anisaster* (Standley) Fryxell (4 species).
   - iii. **Pavonia** subsect. **Conditiflorae** Fryxell. Type species: *P. ecostata* Fryxell & S. Koch (4 species).
3. **Pavonia** sect. **Laminares** Fryxell. Type species: *P. oxyphylla* (de Candolle) Fryxell (1 species with 2 varieties).

The group of species included in *Pavonia* subgen. *Malache* has not previously been treated coherently, except indirectly by Fryxell (1988a: 312). A few workers (e.g., Britton & Millspaugh, 1920; Britton & Wilson, 1924) have treated the group in generic rank (at least in a narrow sense), but others (e.g., Grisebach, 1859; Kearney, 1951; Adams, 1972) have simply included it in *Pavonia* without infrageneric subdivision. De Candolle's conception of *P.* sect. *Malache* was a highly artificial assemblage of 11 species but was based on Vogel's genus and included the type species, *M. scabra* Vogel (≡ *Pavonia paludicola* Nicolson). Fryxell (1988a) recognized the alliance of a group of western-Mexican species of upland distribution with the predominantly mangrove-distributed species that Vogel had initially dealt with. One species (*P. nayarensis*) is described as new herein, and *P. oxyphylla* is included as well. The latter species accords well with this group in fruit morphology but not as well in other characters, and it thus stands rather isolated. This enlarged group of 17 species is united most clearly by characters of fruit morphology (see Fig. 65). It is treated here in subgeneric rank, to recognize its distinctiveness from the rest of the genus and to permit the further subdivision of the group (see "Synopsis", above), to more sensitively reflect taxonomic relationships among the species and to provide a parallel classification to treatments of other coordinate groups within *Pavonia.*

The name *Malache* means mallow, probably derived from the Greek *malache* or *malakos* (soft or gentle).

Pavonia subgen. *Malache* sect. *Malache*

Scandent shrubs, mostly in mangrove habitats; leaves unlobed, subentire; involucellar bracts usually 6–8; corolla greenish yellow; mericarps woody, sometimes winged.

Type species. *Malache scabra* Vogel ≡ *Pavonia paludicola* Nicolson.

Key to the species of *Pavonia* sect. *Malache*

1. Scandent shrubs of mangrove habitats; leaves obscurely crenate or denticulate to subentire; corolla 1–2 cm long, greenish yellow; involucellar bracts 1–3 mm wide; Florida (U.S.A.), West Indies, Central America, N South America.
   2. Leaves narrowly lanceolate, palmately 3-nerved, 3–4 times as long as wide; pedicels 0.5–1.5 cm long; fruits beaked, retrorsely scabridulous; Pacific coasts of Panama, Colombia, and Ecuador. . . . . . 149. *P. rhizophorae*
   2. Leaves ovate-cordate, palmately 5–7-nerved, ca. twice as long as wide; pedicels mostly >2 cm long; fruits winged but not beaked, glabrous; of more or less circum-Caribbean distribution.
      3. Genitalia exserted; flowers and fruits axillary, among the leaves; involucellar bracts 1.5–3 mm wide; Bahama Archipelago. . . . . . 147. *P. bahamensis*
      3. Genitalia subequal to corolla; flowers and fruits in terminal leafless racemes; involucellar bracts 1–2 mm wide; S Florida (U.S.A.), Antilles, and Central America to N South America. . . . . . 148. *P. paludicola*
1. Similar to preceding species but growing in upland habitats; involucellar bracts 2–5 mm wide, oblong; carpels broadly winged, the wings sometimes retrorsely scabridulous; Jamaica. . . . . . 150. *P. troyana*

147. Pavonia bahamensis Hitchcock, Rep. Missouri Bot. Gard. **4:** 63, t. 11. 1893. *Malache bahamensis* (Hitchcock) Britton & Millspaugh, Bahama Fl. 271. 1920. Type. Bahamas. Fortune Island: Nov 1890, *Hitchcock s.n.* (holotype, MO as photo CTES, NY; isotype, F as photo F-56238). Fig. 65

Malvaviscus cokeri Britton ex Coker in Shattuck, Veget. Bahama Isl. 259. 1905. Type. Bahamas. 1903, *Coker s.n.* (holotype, NY).

Shrubs 2–4 m tall, the stems minutely and very sparsely stellate-pubescent. *Leaf* blades ovate, mostly 4–6 × 2.5–5 cm, basally cordate to subtruncate, entire, acute, palmately 5–7-nerved, minutely and sparsely stellate-pubescent on both surfaces, densely bearded (the hairs 0.5–0.8 mm long, whitish) in axils of principal nerves beneath; petioles 1–3 cm long; stipules linear, 5–6 mm long. *Flowers* solitary in the leaf axils, the pedicels 1–3 cm long, slender; involucellar bracts 5–6 ("as many as 10" fide Correll & Correll [1982], 5–9 per their illustration), ligulate or lanceolate, 10–12 × 1.5–3 mm, briefly connate and minutely stellate-pubescent basally, subglabrous distally; calyx subequal to involucel, sparsely stellate-pubescent to glabrate externally, densely puberulent internally; corolla tubular, greenish yellow ("green veined with red" fide *N. L. Britton & Millspaugh 6145bis*), externally pubescent, 1.5–2 cm long; androecium exserted, 3 cm long. *Fruits* 10–11 mm diam., glabrous, subglobose and 5-winged; mericarps 6–7 mm tall, with broad lateral wings.

Distribution (Fig. 66). *Pavonia bahamensis* is endemic to the Bahama Archipelago and occurs principally in the S (tropical) parts of it, at or near sea level, growing on rocky coasts or in mangrove associations (Rathcke, 1999).

Illustrations. Hitchcock (1893: pl. 11, cited in error as plate 1).

Specimens examined. **BAHAMA ARCHIPELAGO. ACKLIN'S ISLAND:** Spring Point, *Brace 4334* (F, NY); Atwood Harbor, *Gillis 13165* (A, CM). **ANDROS:** Nicols Town, *J. I. Northrup & A. R. Northrup 736* (NY). **CAICOS GROUP:** North Caicos, W side of Whitby, *Correll 43262* (F, LL, NY); North Caicos, beach near Seven Keys, *Gillis 11840* (A); *Gillis 12300* (A); Pine Cay, *C. F. Millspaugh & C. M. Millspaugh 9215* (F, NY); Middle Caicos, vic. of Conch Bar, *Proctor & Gillis 34052* (A). **CROOKED ISLAND:** Marine View Rd., *Brace 4622* (F, NY); Jingo Hill, Brace 4763 (F, NY, US); W of Church Grove along rd. to Turtle Sound, *Correll 44322* (NY); Crooked Island, *Correll & Proctor 48693* (NY, US); rd. from Church Grove to landing on sound, *Gillis 10614* (A). **LONG CAY:** Rd. to S side, *Brace 4048* (F, NY, PH). **RUM CAY:** E side of Port Nelson Salt Lake, *Correll & Wasshausen 46838* (F, GH, MO, NY, US). **SAN SALVADOR:** On peninsula near Coast Guard Station, *Correll 43863* (F, FTG, NY, TEX). **WATLINGS ISLAND:** *N. L. Britton & Millspaugh 6187* (F, NY), *N. L. Britton & Millspaugh 6145* (F, NY, TEX); Cockburn Town, *Wilson 7206* (GH, NY), *Wilson 7230* (F, GH, K, NY), *Wilson 7268* (F, GH, K, NY).

Local name. Bahama swamp bush (Bahamas).

Pavonia bahamensis is distinguished from the other members of *P.* sect. *Malache* by its axillary flowers (among the leaves rather than above them), exserted genitalia, shorter mericarps, and its relatively broad involucellar bracts (although they are also broad in *P. troyana*).

The species is bird-pollinated, and the floral presentation and the exserted genitalia are doubtless a part of this relationship. Bananaquits (*Coereba*

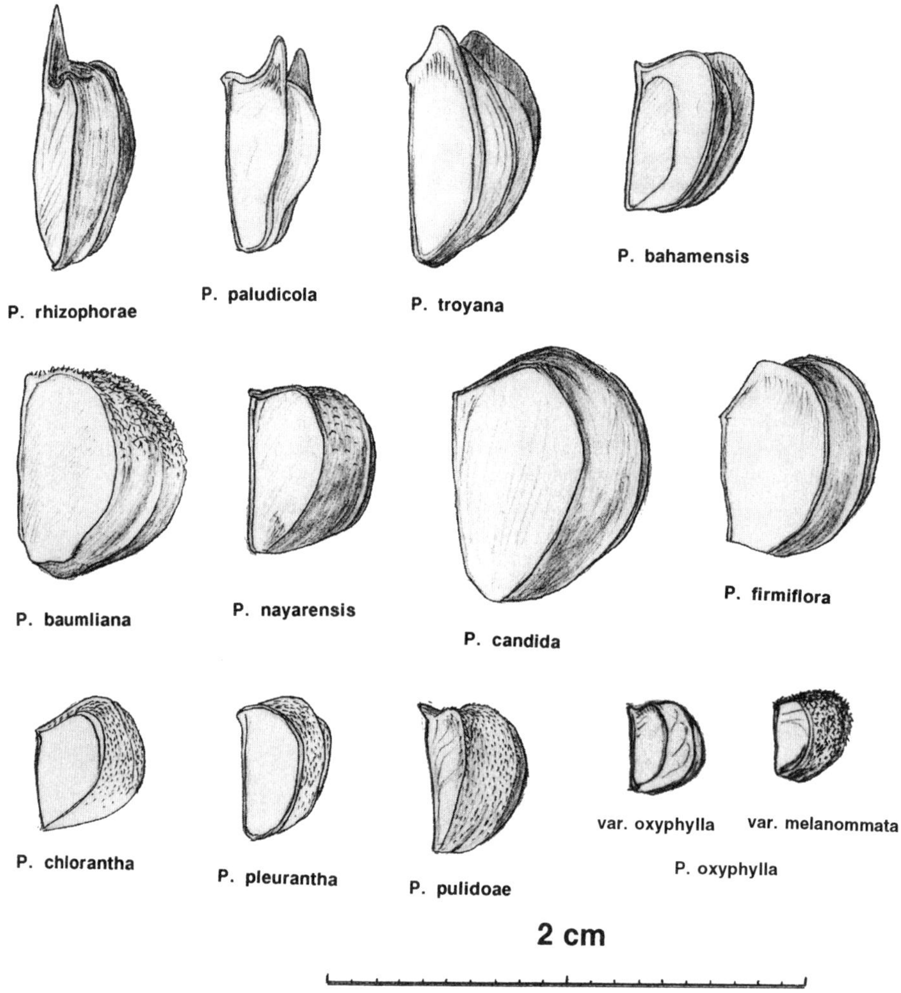

FIG. 65. *Pavonia* subgen. *Malache* mericarps in left lateral view. Top row, *P.* sect. *Malache* (left to right): *P. rhizophorae* (*Killip & Cuatrecasas 38834*), *P. paludicola* (*Gentle 3816*), *P. troyana* (*Harris 9457*), *P. bahamensis* (*Correll & Wasshausen 46838*). Center row, *P.* sect. *Collicolae* (left to right): *P. baumliana* (*Bauml 2020*), *P. nayarensis* (*Flores et al. 1661*), *P. candida* (*Koch & Fryxell 78392*), *P. firmiflora* (*Gentry et al. 19502*). Bottom row, *P.* sect. *Collicolae* and *P.* sect. *Laminares* (left to right): *P. chlorantha* (*Hinton et al. 7218*), *P. pleuranthera* (*Nelson 4061*), *P. pulidoae* (*Pulido 23*), *P. oxyphylla* var. *melanommata* (*Pringle 4343*), *P. oxyphylla* var. *oxyphylla* (*Pringle 8718*).

flaveola: Coerebidae) have been observed to be the principal pollinators and Bahama Woodstars (*Calliphlox evelynae*: Trochilidae) are also active (Rathcke, 1996, 1999). The plants are self-incompatible and thus obligate outcrossers but nevertheless fruit set is high (82%), indicating effective pollination (Rathcke, 1996) by these birds. However, hurricane-induced reduction in the bird population in October 1996 resulted in a marked reduction in fruit set (to 11%) in subsequent months (Rathcke, 1999), further indicating the importance of pollinators to this species. Flowers remain open for three days and are copious producers of nectar; further detail on floral phenology is given by Rathcke (1996).

148. Pavonia paludicola Nicolson *in* Howard, Fl. Lesser Ant. **5:** 241. 1989, nom. nov. *Malache scabra* Vogel *in* Trew, Pl. Select. **9:** 50. t. 90. 1772. *Pavonia*

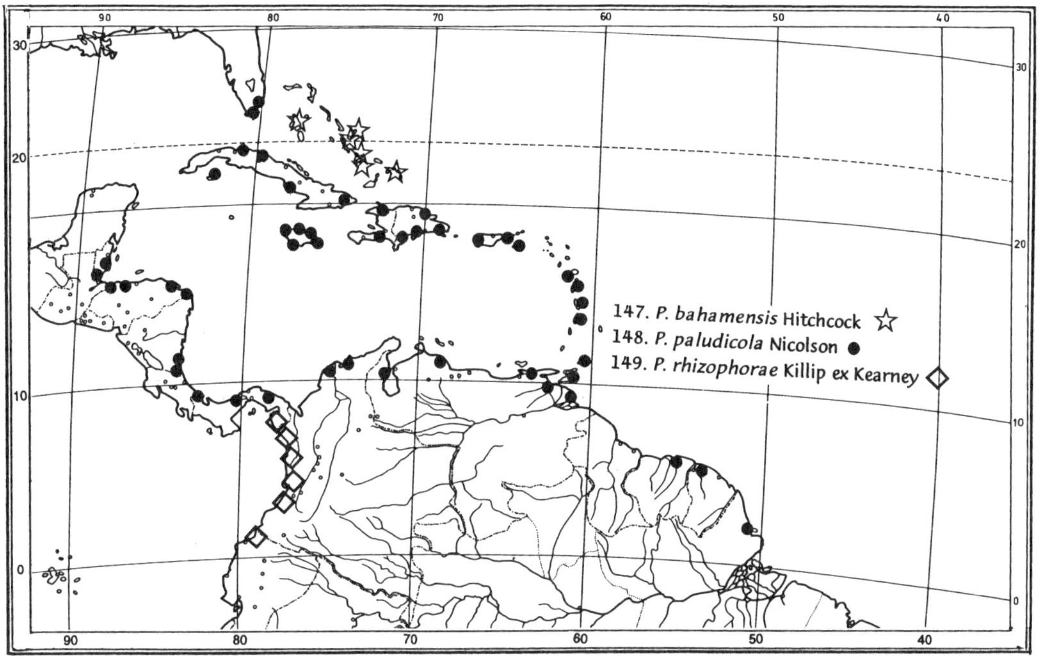

FIG. 66. Distribution map of three species of *Pavonia* sect. *Malache*.

spicata Cavanilles, Diss. **3:** 136. 1787. *Althaea racemosa* Swartz, Prodr. 102. 1788. *Malache spicata* (Cavanilles) Kuntze, Revis. Gen. Pl. **1:** 70. 1891. *Pavonia racemosa* (Swartz) Swartz, Fl. Ind. Occid. **2:** 1215. 1800. *Pavonia scabra* (Vogel) Ciferri, Atti Ist. Bot. Univ. Pavia, ser. 4, **8:** 321. 1936. (Non *Pavonia scabra* C. Presl, 1835.) Type. Trew, Pl. Select. (1772) pl. 90. Fig. 65

Semi-scandent *shrubs* 1–4 m tall, the stems minutely stellate-pubescent. *Leaf* blades 6–18 × 4–10 cm, ovate, subcordate, obscurely dentate or subentire, acuminate, palmately nerved, sparsely and minutely stellate-pubescent; petioles 2–8 cm long (shorter upward). *Inflorescence* an elongated terminal leafless raceme; pedicels 1–4 cm long; involucellar bracts ca. 8, lanceolate, shorter than to equaling the calyx; calyx 8–11 mm long, minutely stellate-pubescent; corolla 12–18 mm long, pale yellow or yellow-green, externally pubescent; androecium subequal to petals, the stigmas slightly exserted. *Fruits* 10–13 mm diam., glabrous, usually apically beaked or depressed and 5-pointed, the mericarps 7–9 mm tall, ligneous, each apically 3-pointed.

Illustrations. Cavanilles (1787: pl. 46, fig. 1, as *P. spicata*); Correll & Correll (1982: fig. 397, as *P. spicata*); Fryxell (1979: fig. 7B, as *P. spicata*); Fryxell (1989: fig. 94); Robyns (1965: fig. 6C, as *P. spicata*); Vogel *in* Trew (1772, pt. 10: pl. 90).

Distribution (Fig. 66). From mangrove habitats and river mouths around the periphery of the Caribbean, in S Florida (United States), Central America, N South America, and the Greater and Lesser Antilles, growing at or near sea level.

Representative specimens examined. UNITED STATES. Florida: Dade Co.: Coral Gables, *Churchill 7542560* (RSA); Arch Creek, *Curtiss 386** (CM, F, NY, PH, TEX); Miami, Jul 1877, *Garber s.n.* (NY, PH, TEX); near Fairchild Tropical Garden, *Gillis 9839* (MO, RSA); Addison Hammock, *Small 7483* (NY); Arch Creek, head of Bay Biscayne, *J. K. Small & G. K. Small 4598* (C, K, MO, NY, PENN, POM, TEX).

GUATEMALA: Izabál: Chacón River, *Johnson 1235* (F, US); vic. of Puerto Barrios, *Standley 72173* (A, NY, F); Mun. Livingston, Laguna Escondida, 15°45′N, 88°48′W, *Tenorio et al. 14905* (MEXU, MO).

BELIZE. Distr. Toledo, Punta Gorda, *Arvigo et al. 612* (NY, TEX); 6–7 mi NW of Belize, *Croat 23288* (GH, MO, NY, POM, TEX, US); at mouth of Temash River, *Dwyer 12895* (MO, NY); North River, *Gentle 1338* (A, F, NY); Monkey River in mangrove swamp, *Gentle 3816* (A, F, MO–2, NY, US); Sittee River, *Schipp 635* (A, F, GH, MO, NY, S); Belize City, *Worthington 21260* (MO, TEX, UTEP).

HONDURAS. Atlantida: Vic. of Tela, *Standley 53568* (A, F, US); Bradley's Plantation, Pto. Sierra, *Wilson 394* (MO, NY). **Gracias a Dios:** Cauquira, *Barkley & C. Nelson*

40665 (GH); near mouth of Río Plátano, *P. A. Fryxell 2805* (BM, BRIT, CAS, CTES, ENCB, F, LL, NY, TEX); tributary stream of Río Plátano, village of Ras, *A. H. Gentry et al. 7559* (MO, NY, US). **Roatán Island:** Sea shore, *Molina 20628* (F, NY, S, US).

NICARAGUA. Zelaya: Vic. of Bluefields, *Bunting & Licht 1296* (F, NY, US); Bahia de Bluefields, Schooner Cay, *Molina 2035* (F, GH); Monkey Point, Caño El Pato, *Moreno 12432* (MO, NY); El Bluff, N of El Muelle, edge of El Puerto, 10–15 m, *Sandino 2222* (MO, MU).

COSTA RICA. Punta Mina, Atlantic coast, *Pittier 12687* (US). **Limón:** Parque Tortuguero, Estación Cuatro Esquinas, *Chavarría 136* (CR, K, MO, NY); between Punta Cocles and Quebrada Ernesto, *Grayum & Sleeper 3660* (MO, NY); 4 mi N of Limón, *Stork 3200* (F, NY).

PANAMA. Bocas del Toro: Water Valley, *von Wedel 738* (GH, MO); Old Bank Island, vic. of Chiriquí Lagoon, *von Wedel 1997* (GH, MO, US). **Colón:** Miguel de la Borda, *Croat 9827* (CTES, F, MO, NY); Coclé del Norte, *Hammel 4582* (MO, NY); along Culebra River above Santa Isabel, *Pittier 4168* (NY, US). **San Blas:** Near Mandinga, on the mainland, *Duke 8887* (MO); small coral island NW of Ailigandi, *Hammel & D'Arcy 5060* (MO); lower Cangandi River, *de Nevers 4635* (MO, NY).

CUBA. Marshes on the river Mata, *Wright 2058* (GH, K, MA). **Camaguey:** Maximo River, *Shafer 594* (F, NY, US). **La Habana:** Marianao beach, *N. L. Britton & Wilson 4534* (F, NY); near Guanímar, *Bro. León (J. S. Sauget) 5053* (GH, NY). **Isle of Pines:** Bibijagua, *Ekman 12530* (S); Las Casas River, *Killip 43553* (CAS); near Nueva Gerona, *E. Palmer & Riley 1031* (US). **Matanzas:** Cienega de Zapata, Mar 1923, *Acuña s.n.* (CAS); Cayo Delicias, Ciénega de Zapata, *Bro. León (J. S. Sauget) et al. 19576* (GH); vic. of Matanzas, *N. L. Britton et al. 198* (CM, NY); near Tomary River, *Rugel 91* (COLO, K, MO, NY, S). **Santiago de Cuba:** Lower valley of Miel River, *Shafer 4336* (F, NY, US); Moa Bay, E of Río Moa, *Shafer 8342* (A, NY, U, US). **Pinar del Río:** Galafre, *N. L. Britton & Colwell 9836* (NY, U). **Santa Clara:** Abreus, Damuji River, *R. Combs 222* (F, GH, MO, NY).

JAMAICA. Near Port Antonio, *Fredholm 3151* (NY, US); Ferry River, *Harris 8629* (NY); vic. of Hunts Bay, *Maxon & Killip 324* (A, F, GH, NY, US). **Clarendon:** Salt River, *Proctor 6111* (NY). **St. Ann:** W of Dry Harbor, *Powell 971* (MO). **St. James:** Near mouth of Great River, *Proctor 32656* (F, LL, U). **St. Catherine:** Jun 1847, *McNab s.n.* (CM). **St. Marys:** 1843, *McNab s.n.* (E). **St. Thomas:** Ca. 1 mi W of Morant Point lighthouse, *Webster et al. 4841* (A, US); Point Morant, *Yuncker 18180* (F, NY, S). **Trelawney:** Montego Bay to Falmouth, *E. G. Britton 2920* (NY, US); Florida Beach, E of Falmouth, *West & Arnold 797* (GH). **Westmoreland:** Negril and vic., *N. L. Britton & Hollick 2015* (NY); vic. of Kingston, Jan 1849, *McNab s.n.* (C, CM); New Broughton, *Proctor 33471* (LL, RSA, U).

HAITI. Massif de la Selle, Port-au-Prince, *Ekman 9474* (NY, S); Massif de la Hotte, Dame Marie, Bariodèle, *Ekman 10451* (K, LL, S, US); Mariani, vic. of Port-au-Prince, *E. C. Leonard 10090* (NY, US); Tortue Island, vic. of Basse Terre, *E. C. Leonard & G. M. Leonard 13900* (A, GH, NY, US); Bayeux to Borgue, *Nash 403* (F, NY).

DOMINICAN REPUBLIC. Barahona, *Fuertes 16* (F, GH, K); Prov. María Trinidad Sánchez, El Diamante beach, *Zanoni et al. 29962* (NY). **El Seibo:** E of Jovero, *Abbott 2852* (C, GH, US); Naranjo Arriba, Samaná Bay, *Mejía & Pimentel 23950* (JBSD, NY, TEX); on Ozama River, *von Türckheim 2621* (DS, E, F, GH, K, MO, NY, S, U, US). **La Romana:** Shores of Nagua River, *Jiménez 1288* (CAS, US); 2 km up the Chavón River, *Zanoni et al. 14882* (JBSD, NY, TEX, U). **Samaná:** Playa del Valle, *Mejía & Zanoni 6673* (JBSD, NY); El Valle, *G. L. Smith 10457* (NY); 0.5–1 km N of the mouth of the Barracote River, Samaná Bay, *Zanoni et al. 34261* (JBSD, NY). **San Pedro de Macorís:** Cumayasa River, CEA ranch, *Pimentel & García 607* (JBSD, MO, NY).

PUERTO RICO. Coco Beach, *Boom 8002* (NY, TEX); Santurce, *A. A. Heller & E. G. Heller 824* (C, F, K, NY, US); along railroad N of Mayaguez, *A. A. Heller 4570* (A, E, F, GH, MO, NY, PH, POM, US); Cabo Rojo to Joyuda, *Sintenis 798* (C, CTES, GH, S).

VIRGIN ISLANDS. St. Croix: *Isert s.n.* (C); Salt River, Oct 1871, *von Eggers s.n.* (C–2). **St. Thomas:** *Deam 6051* (GH, US).

WINDWARD ISLANDS. Guadeloupe: *Duss 2786* (NY, US); Pointe-à-Pitre, *H. Stehlé 302* (S, US); Marie Galante, at mouth of St. Louis River, *Proctor 20230* (A, US). **Dominica:** St. John Parish, Estuary of Indian River, Portsmouth, *W. H. Hodge & B. T. Hodge 3744* (GH, US). **Martinique:** *Duss 841* (A, F, GH, MO, NY, PH, US). **St. Lucia:** Anse Galette, *Proctor 18245* (A).

COLOMBIA. Antioquía: Barranquilla, *Elias 1301* (F, S); Turbo, 1 km W of Turbo, *Feddema 2030* (MICH, NY, US); Santa Marta, *H. H. Smith 2019* (A, CAS, CM, E, F, GH, K, LL, MO, NY, PH, S, TEX, U, US).

VENEZUELA. T.F. Delta Amacuro: Dept. Pedernales, SE of Pedernales, *Steyermark et al. 114292* (CTES, MO); Pedernales–Orinoco delta, *Curran & Haman 1305* (CAS, GH, US); Dept. Antonio Díaz, along Caño Atoiba, *Steyermark et al. 114700* (CTES, MO). **Falcón:** S of Chichiriviche, *Steyermark & Manara 110449* (MO). **Monagas:** Forest Reserve of Guarapiche (Caño Colorado), *Aristeguieta et al. 7238* (F, NY, US); Caño Paijano–Isla de San Carlos, *Curran & Haman 788* (A, GH, NY, US). **Sucre:** 53 km NE of Maturin, San Juan River, *Breteler 4672* (MO, NY, S, U, US); La Bodega Lagoon, E of Santa Fé, *Steyermark et al. 108414* (MO, NY, U, US). **Zulia:** In the mangroves of Lake Maracaibo, *Aristeguieta & Montoya 2045* (CAS); Dept. Paéz, vic. of Lake Sinamaica, *Aristeguieta et al. 12501* (CTES); Mun. Santa Bárbara, Caño Concha, *Marcano-Berti et al. 164-981* (MER, NY).

TRINIDAD. Banks of the Caroni River, 5 Mar 1933, *Broadway s.n.* (A, K, MO, U), *Curran & Haman 1340* (GH). **Tobago:** Richmond Estate near the sea, *Broadway 4572* (CAS, F, GH, NY, S, U).

GUYANA. upper Rupununi River, near Dadanawa, *de la Cruz 1432* (CM, GH, MO, NY, PH, US); Pomeroon River, *de la Cruz 3005* (GH, MO, NY, PH, U); Kamakusa, upper Mazurumi River, *de la Cruz 4094* (CM, GH, NY, PH, US); Hacket Creek, lower Waini River, *Fanshawe 5060* (K, NY, U); Vreed-en-Hoop, opposite Georgetown, *Hitchcock 16707* (GH, NY, S, US).

SURINAM. Lower Surinam River, plantation of Purmerend, *Jonker-Verhoef & Jonker 510* (K, U); lower part of Huntley Creek, *Lanjouw & Lindeman 3109* (NY); N of Paramaribo, *Stahel 37* (NY, U).

FRENCH GUIANA. Base of Montagnes des Cheveaux, *Oldeman B-850* (U); Mana, *Sagot 1086* (CTES, S, U, US).

BRAZIL. PARÁ: Amapá, *Huber 1195* (G).

Local names. Cadillo de ciénega (Puerto Rico), caye moho (Belize), cotton (Belize), mahout (Dominican Republic), majagua (Venezuela); majagüilla (Cuba, Dominican Republic), mangrove mallow (Uhited States: Florida), saltwater moho (Belize), sea cotton (Guyana), smaller mahoe (Jamaica), swamp bush (Bahamas, Puerto Rico), tjawassie (Surinam), wild cotton (Belize), yahuballi (Guyana).

Pavonia paludicola is the most widely distributed species of *P.* sect. *Malache* and is distinguished by the combination of terminal racemes, ovate-cordate leaf blades, and a mangrove distribution.

149. Pavonia rhizophorae Killip *ex* Kearney, Leafl. W. Bot. **7:** 118. 1954. Type. Colombia. El Valle: El Forge, near Buenaventura, 5 June 1944, *Killip & Cuatrecasas 38834* (holotype, US; isotype, CAS). Figs. 65, 67

Shrubs 1–3 m (or more) tall, the stems stellate-pubescent. *Leaf* blades mostly 6–14 × 2–5 cm, lance-oblong, truncate, remotely and obscurely denticulate to subentire, acuminate, palmately 3-nerved, sparsely and minutely stellate-pubescent (especially on veins) to glabrate; petioles 1–4 cm long; stipules inconspicuous, caducous. *Inflorescence* an elongated terminal leafless raceme; pedicels 0.5–1.5 cm long; involucellar bracts 6–8, linear-lanceolate, 6–10 × 1–2 mm; calyx 11–13 mm long, minutely stellate-pubescent; corolla 1.5–2 cm long, yellowish green; externally pubescent; staminal column ca. 15 mm long, the filaments 1–2 mm long. *Fruits* ca. 1 cm diam., subequal to calyx, obscurely 10-ribbed, beaked, 10–11 mm long (including beak), retrorsely scabridulous to glabrate, the mericarps ligneous.

Distribution (Fig. 66). *Pavonia rhizophorae* occurs on the Pacific coast of Colombia, extending also to N coastal Ecuador and to adjacent Panama, growing in mangrove habitats at or near sea level. The type locality is at 3°55′N, 77°6′W.

Additional illustration. Robyns (1965: fig. 6D).

Specimens examined. PANAMA. DARIEN: Piñas River, *Duke 10552* (MO); Puerto St. Dorotea, on bank of creek near ocean, *Dwyer 2286* (MO–2).

COLOMBIA. Isla del Gallos, *Dryander 2597* (US); between Punta Las Barcas and Turbo, *Duke 9751* (MO). **CHOCÓ:** Solano Bay, along El Valle River, *Espina et al. 2715* (CHOCO, MO, NY); mouth of San Juan River, below Docordó, *Forero et al. 4409* (MO); area of Baudó, ca. 9 km upstream of estuary, *Fuchs & Zanella 22068* (MO, US); near mouth of El Valle River, *A. H. Gentry & Forero 7380* (F, MO, NY); Nuquí-Pangui, banks of Chicui River, *A. Gómez et al. 419* (LL, MO); banks of Quebrada Togoromá, *Killip & Cuatrecasas 39131* (F as photo F-56154, US). **EL VALLE:** Buenaventura, *Alston 8646* (BM, US); Pacífico coast, Yurumanguí River, *Cuatrecasas 15917* (F, NY, US); Buenaventura Bay, Quebrada de Aguadulce, *Cuatrecasas 19722* (F, US); Mun. Buenaventura, Corrigimiento Puerto Merizaldo, Isla Ají, Caño El Zancudo, Río Naya, Devia et al. 2929 (MO, TEX); Malaga Bay, mouth of Quebrada Sierpe, *A. H. Gentry et al. 40414* (MO, NY, TEX); Malaga Bay, edge of Mora swamp, *A. H. Gentry et al. 53412* (MO, NY); Mun. Buenaventura, between Naya River and Yurumanguí River, *Juncosa 2090* (MO, NY); Punta Arenas, N shore of Buenaventura Bay, *Killip & Cuatrecasas 38636* (US), *Killip 33007* (US), *Killip 5237* (GH, NY, PH, US), *Killip 11692* (GH, NY, PH, US); Buenaventura Bay, *Killip 34964* (S, US).

ECUADOR. ESMERALDAS: San Lorenzo, Río San Antonio, Finca El Horizonte, *Sparre 18341* (S).

Pavonia rhizophorae is distinguished from other members of *P.* sect. *Malache* by its narrowly elliptical leaves and apically pointed mericarps, as well as by its Pacific coastal distribution.

150. Pavonia troyana (Urban) Urban, Feddes Rep. Spec. Nov. Regni Veg. **13:** 459. 1914. *Pavonia racemosa* var. *troyana* Urban, Symb. Ant. **5:** 530 (1908). *Malache troyana* (Urban) Britton, Bull. Torrey Bot. Club **35:** 345. 1908. *Pavonia spicata* var. *troyana* (Urban) Fawcett & Rendle, Fl. Jamaica **5:** 132. 1926. Type. Jamaica. Near Troy, 14 Sep 1906, *Harris 9457* (isotypes, BM as photo CTES, F, NY, US). Fig. 65

Trees or shrubs 3–7 m tall, the young stems minutely stellate-pubescent, glabrate in age. *Leaf* blades broadly ovate, mostly 6–15 × 4–10 cm, deeply cordate, obscurely crenate to entire, acuminate, palmately 7-nerved, more or less discolorous, the upper surface essentially glabrous and more or less vernicose, the lower surface hirsute in the axils of the principal nerves, slightly pubescent on principal nerves, otherwise glabrous; petioles up to 6 cm long, progressively shorter upward; stipules subulate, 4–5 mm long, deciduous, becoming connate in the inflorescence to form persistent bifid bracts subtending the pedicels. *Inflorescence* a terminal leafless raceme; pedicels 1–1.5(–2.5) cm long; involucellar bracts 6–7, ligulate to ovate, basally connate, 6–9 × 2–5 mm, sparsely stellate-pubescent; calyx 9–10 mm long, sparsely stellate-pubescent externally, densely whitish-puberulent internally; corolla 13–16 mm long, greenish yellow, externally pubescent, tubular; androecium

FIG. 67. *Pavonia rhizophorae*. Flowering branch with enlargement of inflorescence tip (*Cuatrecasas 15917*).

subequal to corolla or slightly exserted. *Fruits* 10–12 mm diam., crowned with 5 broad alate ridges (the ridges sometimes scabridulous or retrorsely barbed), otherwise glabrous; mericarps ligneous, reticulate-veined, 7–10 mm high (including wings).

Distribution (Fig. 28). Endemic to Jamaica, where it forms a part of the interior forests at 400–800 m elevation. The type locality is at 18°15′N, 77°38′W.

Specimens examined. **JAMAICA. CLARENDON:** Near Aenon Town, ca. 2300 ft, *Crosby & W. R. Anderson 1236* (LL–2, MICH); Peckham Woods, *Harris 10868* (F, NY, US), *Harris 10964* (F, NY, US), *Harris 12775* (F, GH, K, MO, NY, US); 4–5 mi NW of Kellits, *Pruski 1532* (MO, NY). **ST. ANN:** 3 mi NW of Kellits, *Webster 13574* GH). **ST. ELIZABETH:** Ipswich, *Harris 12364* (F, MO, NY, US); Island View Hill, Wilson Valley Distr., *Proctor 21415* (GH, NY); between Mulgrave and Elderslie, *Proctor 22565* (NY, U). **ST. JAMES:** 1 mi S of Sweet Water, *Proctor 7964* (NY). **TRELAWNEY:** Cockpit Country, several miles NW of Crown Lands, 1700 ft, *W. R. Anderson & Sternberg 3228* (GH, LL, MICH, MO, US); Troy, *N. L. Britton 515* (NY, TEX); 5 mi W of Troy, *Hespenheide et al. 1143* (GH, LL, NY, US); Troy, *Howard 14112* (A); vic. of Ramgoat Cave, *Howard & Proctor 14416* (A);Troy, *Perkins 1377* (GH); vic. of Ramgoat Cave, *Proctor 10615* (A, US); Wilson Valley District, Island View Hill, *Proctor 21415* (NY); Wilson Valley District, "Miss Laura's Hill," *Proctor 24457* (LL); N of Burnt Hill, *Proctor 38596* (NY); vic. of Burnt Hill, *Thorne & Proctor 48095* (NY, RSA).

Local name. Congo mahoe (Jamaica).

Pavonia troyana stands out from the other members of *P.* sect. *Malache* by its upland (non-mangrove) occurrence and by its broad (3–5 mm wide) involucellar bracts.

Pavonia subgen. Malache sect. Collicolae Fryxell, sect. nov.

Frutices vel arbores parvae; floribus solitariis vel in inflorescentiis aggregatis, corollis albis luteolis roseis vel atrorubris; fructibus ligneis vel subligneis.

Shrubs or small trees, the flowers solitary or aggregated into inflorescences, the corolla white, yellowish, rose, or dark red, the fruits woody or nearly so.

Type species. *Pavonia pleuranthera* (de Candolle) Fryxell. All of the species of this section are endemic to Mexico.

PAVONIA SUBGEN. MALACHE SECT. COLLICOLAE SUBSECT. COLLICOLAE

Pavonia subsect. **Collicolae** Fryxell, subsect. nov. (from the Latin *collis*, hill, and *cola*, dweller, meaning from the uplands).

Frutices vel arbores parvae; laminis foliorum ovatis vel 3–5-angulatis vel -lobatis; bracteis involucellorum 6–12; floribus aggregatis apicem versus, floribus paucis, corollis albis vel luteolis, genitaliis inclusis vel exsertis; fructibus ligneis.

Shrubs or small trees with leaves ovate to 3–5-angled or -lobed; involucellar bracts 6–12; flowers in few-flowered terminal aggregations, the corolla white or yellowish, the genitalia included or exserted; fruits woody.

Type species. *Pavonia pleuranthera* (de Candolle) Fryxell.

Key to the species of *Pavonia* subsect. *Collicolae*

1. Corolla 6–8 cm long, white; genitalia notably exserted; fruits glabrous, 16–20 mm diam. . . . . 151. *P. candida*
1. Corolla 1–3.5 cm long, usually yellowish; genitalia included (or if exserted, then corolla <2 cm long); fruits glabrous or pubescent.
    2. Flowers solitary in the leaf axils; corolla yellowish, 3–3.5 cm long, the genitalia included; fruits 16–19 mm diam., retrorsely scabrid. . . . . 153. *P. firmiflora*
    2. Flowers in terminal, leafless racemes; corolla of various colors; genitalia included or exserted.
        3. Fruits 16–19 mm diam., retrorsely scabrid; corolla 3–4 cm long, the genitalia included. . . . 153. *P. firmiflora*
        3. Fruits 8–13 mm diam., glabrous, puberulent, or scabridulous; corolla 1–3 cm long, the genitalia exserted or subequal to corolla.
            4. Corolla 2.5–3 cm long; genitalia subequal to corolla; fruits 12–13 mm diam.; involucel often hirsute at base; leaves ovate, usually unlobed.
                5. Leaf blades basally truncate; fruits glabrous; stems scabridulous; lowermost pedicels up to 2 cm long. . . . . 154. *P. paneroi*
                5. Leaf blades basally cordate; fruits scabridulous; stems softly pubescent; lowermost pedicels up to 4 cm long. . . . . 152. *P. chlorantha*
            4. Corolla 1–1.5 cm long; genitalia exserted; fruits 8–9 mm diam., retrorsely pubescent; involucel stellate-pubescent at base; leaves often lobed. . . . . 155. *P. pieuranthera*

151. Pavonia candida (de Candolle) Fryxell, Bol. Soc. Bot. Méx. **38:** 17. 1979. *Malvaviscus candidus* de Candolle, Prodr. **1:** 445. 1824. *Hibiscus pedunculatus* Sessé & Mociño, Pl. Nov. Hisp. 113. 1889 (non *Hibiscus pedunculatus* Linnaeus fil., 1781). Type. *Icones Florae Mexicanae 409* (Torner Collection acc. no. 6331.760, Hunt Institute). A copy of this drawing from the Calque des Dessins (G) is available as a Field Museum photo F-30508. Figs. 65, 68

Malvaviscus pleurogonus de Candolle, Prodr. **1:** 446. 1824. Type. *Icones Florae Mexicanae s.n.* (Torner Collection acc. no. 6331.980, Hunt Institute).

Malvaviscus pringlei E. G. Baker *ex* Robinson & Greenman, Amer. J. Sci., ser. 3, **50:** 175. 1895. Type. Mexico. Michoacán: Near Lake Cuitzeo, 20 July, 26 October 1892, *Pringle 4132* (holotype, K; isotypes, BM, COLO, CTES, F, GH, MA, MASS, MEXU, MIN, MO, MU, NY, PH, TEX, UC, US, VT, WIS).

Shrubs or small trees 2–6 m tall, the stems minutely stellate-tomentose. *Leaf* blades mostly 6–13 cm long, about as wide as long, palmately 3–5-angled or -lobed, crenate-serrate, more or less acuminate, softly stellate-tomentose, slightly discolorous; petioles ⅓–1 times the length of the blades, shorter near inflorescence; stipules 5–6 mm long, subulate. *Flowers* sometimes solitary in the leaf axils but more commonly aggregated terminally in a condensed raceme of 2–6 flowers; pedicels 2–6(–8.5) cm long, stout; involucellar bracts 10–13, shorter than calyx, linear, stellate-pubescent; calyx 17–25 mm long, sparsely stellate-pubescent; corolla tubular, 6–8 cm long, white; androecium prominently exserted, 9.5–12 cm long, nearly glabrous, pallid, the filaments 6–15 mm long; styles exceeding androecium, minutely glandular-puberulent, the stigmas capitate, <1 mm diam. *Fruits* 16–20 mm diam., oblate, blackish at maturity; mericarps 8–10 mm high, glabrous, indehiscent, unornamented, woody or bony, the walls 1–2 mm thick; seeds ca. 6 mm long.

Distribution (Fig. 42). Central Mexico, probably confined in nature to the states of Querétaro, Guanajuato, Michoacán, and Jalisco, at elevations of 1800–2800 m. *Pavonia candida* is sometimes cultivated as an ornamental shrub for its showy flowers, flowering from Jul to Oct. Some of the early collections of this species are from cultivation, but the labels are often incomplete or uninformative. The natural range of the species, therefore, is probably not as extensive as the following specimen citations suggest. A specimen from Europe (Wiener Bot. Garten, 1859, *Halàcz s.n.*, NY) indicates it was cultivated in Vienna more than 100 years ago.

Additional illustrations: Bull. Soc. Hort. Prat. Rhône 152* (1846, as *Malvaviscus pleurogonus*); Cervantes Aceves (1992: fig. 63); Fryxell (1979: figs. 6B, 7A; 1993a: 110); Schery (1942: 207, as *Malvaviscus candidus*).

Specimens examined. **MEXICO. COAHUILA:** Saltillo, 28 Jul 1905, *E. Palmer 686* (CM, F, MO, NY). **COLIMA:** Manzanillo, *Matuda 18310* (MEXU). **GUANAJUATO:** San Miguel de Allende, *Argüelles 1927* (ARIZ, MEXU); 6 km S of Acambaro, *Iltis & Cochrane 267* (CTES, ENCB, WIS); San Miguel de Allende, *Kishler 802* (MEXU); 5 km S of Valle de Santiago, hwy. to Morelia, *Rzedowski 38450* (ENCB, IEB, NY); Mun. Salvatierra, Eménguaro, *Rzedowski 40363* (IEB, MO, NY, TEX, WIS); Mun. San Luis de la Paz, Cerro Las Antenas de los Pozos, *E. Ventura & E. López 9645* (IEB, NY). **JALISCO:** Rancho "La Granchina" 1 km W of Valle de Guadalupe, *S. González 128* (IBUG, NY); Río Blanco, *E. Palmer 48* (PH, US); Monte Escobedo to Colotlán, *Rose 2670* (GH, MO, US); Guadalajara, *Rose & Hay 6279* (US). **MÉXICO:** Molino [cult.], *MacDaniels 526* (F); Mun. Acolman, Tepexpan, 10 km N of Texcoco [cult.], 16 Aug 1981, *Ortíz s.n.* (CHAPA, NY); Mun. Tezoyuca, *A. Ventura 3734* (CTES, ENCB, NY). **MICHOACÁN:** Edge of the village of Acuitzio del Canje, *Díaz-Barriga 2477* (IEB); Mun. Pátzcuaro, near Sanubria, *Díaz-Barriga 3479* (ENCB, IEB); Mun. Pátzcuaro, Colonia Nueva, *Escobedo 1211* (IEB, NY); Aguatibia, Paruándiro, *Hernández & Vázquez de Hernández 748* (MEXU, MO, NY); Mun. Zinapécuaro, Lago Cuitzeo, 2 km N of San Bernardino, rd. to Coro, *Koch & P. A. Fryxell 77392* (CHAPA, CTES, ENCB, F, LL, MEXU, NY, OSH, TEX); Mun. Cuitzeo, edge of Cuitzeo, rd. to Morelia, *Martínez 1681* (ENCB, IEB); Queréndaro, *Pringle 27* (MEXU). **QUERÉTARO:** Rd. to Huimilpan, through San Rafael, *Argüelles 1504* (MEXU, NY); Los Cues, Mun. Pedro Escobedo, *Argüelles 1812* (MEXU); without locality, *Arsène 10621* (US).

Local names and uses. Acauita, lirio (Sinaloa), monacillo (Jalisco). Sometimes cultivated as an ornamental shrub.

The type plate (Sessé & Mociño #409) shows a petal with a prominent basal auricle such as is characteristic of the genus *Malvaviscus*, which may have been the basis (together with the prominently exserted genitalia) for placing this species in *Malvaviscus*. However, the examination of corollas of living specimens (progeny of *Koch & P. A. Fryxell 77392*) shows that the petals lack any basal auricle, and that the Sessé & Mociño drawing is in error in this particular.

Pavonia candida is notable for its large white corollas (6–8 cm long) and for its very long staminal column (9.5–12 cm long). The only other species of *Pavonia* to have staminal columns of comparable dimensions are the Brazilian *P. luetzelburgii* and *P. macrostyla*, of *P.* sect. *Lopimia*, which are thus not closely related to *P. candida* and, to a lesser extent, *P.*

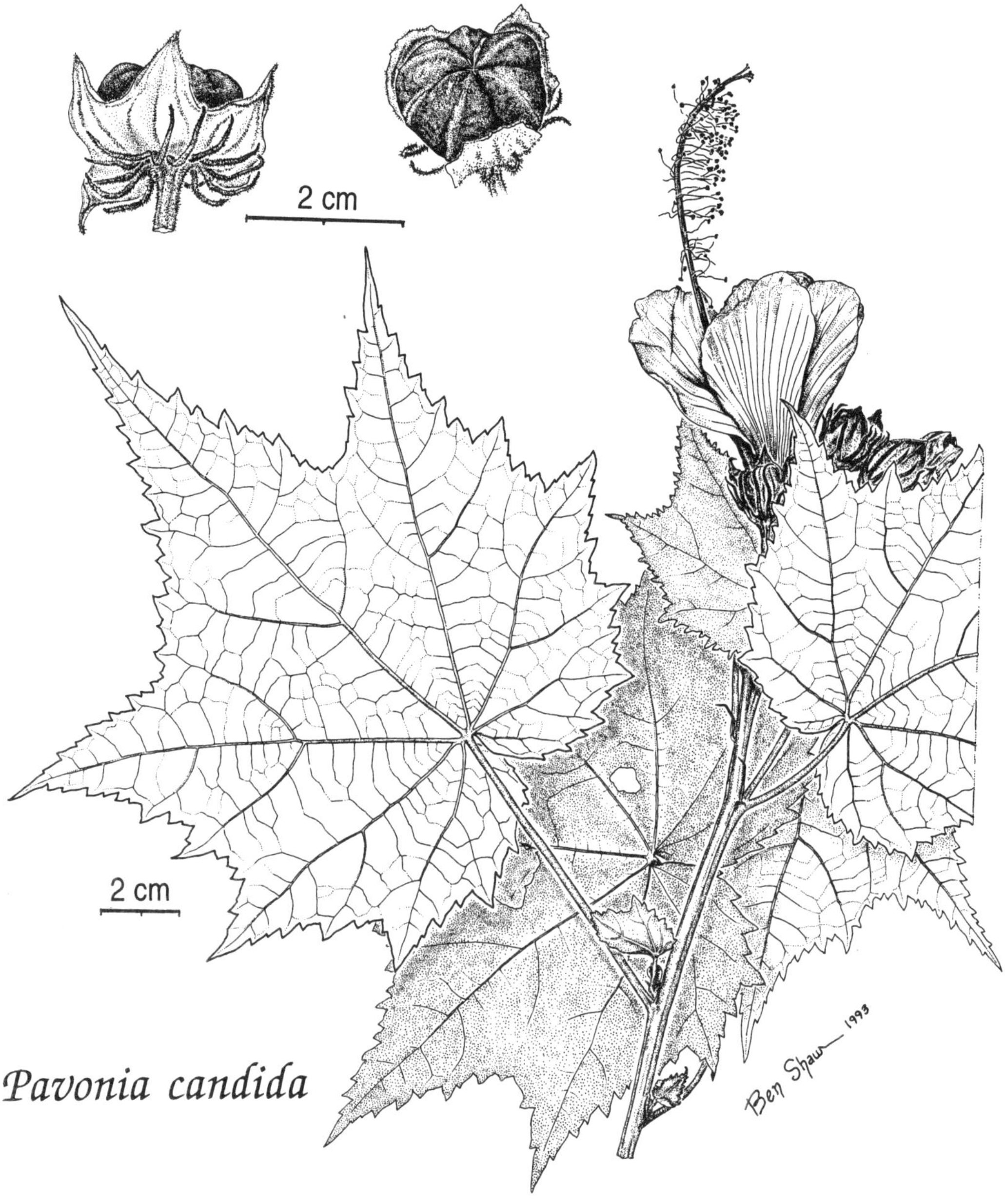

FIG. 68. *Pavonia candida.* Flowering branch (right) with fruit enlargements (top left) (*Rzedowski 40363* [branch], *Ventura & López 9645*, *Hernández 697* [fruits]).

ducke-limae (of *P.* subgen. *Goetheoides*): lesser in that, although the staminal column is equally exserted, it is somewhat smaller in absolute length. In all three cases, the corollas are much smaller than they are in *P. candida.*

152. Pavonia chlorantha (Kunth) Fryxell, Bol. Soc. Bot. Méx. **38:** 18. 1979. *Anotea chlorantha* Kunth, Ind. Sem. Hort. Berol. **1846:** 13. 1846; Ann. Sci. Nat. Bot., ser. 3, **7:** 188. 1847. Type. Cultivated, Bot. Gard. Berlin, Nov 1845, *Kunth 1627* (holotype, B as photo F-9421). Fig. 65

Hibiscus racemosus Sessé & Mociño, Fl. Mex., ed. 2, 161. 1894 (non *Hibiscus racemosus* Lindley, 1825). Type. Mexico. *Sessé & Mociño 3548* (holotype, MA; isotypes, BM *ex* herb. Pavón as photo CTES, F as photo F-46747).

Shrubs 1–7 m tall, the stems variably pubescent, with minute stellate hairs and with longer (1 mm) somewhat curly simple hairs. *Leaf* blades mostly 6–13 × 8 cm (reduced upward), ovate (sometimes weakly 3-lobed), cordate, obscurely to prominently crenate-serrate, acute to acuminate, palmately 7–9-nerved, sparsely stellate-pubescent; petioles ⅓–½ the length of the blades, reduced upward with the leaves subsessile immediately below the inflorescence; stipules 5–8 mm long, subulate. *Inflorescence* a terminal, leafless raceme; pedicels up to 4 cm long proximally, to 0.5 cm long distally; involucellar bracts 6–7, connate and forming a basal cup, free for distal 1 cm, linear to subspatulate (1–3 mm wide), slightly exceeding calyx, sparingly hirsute, sometimes retrorsely bearded at base; calyx 10–14 mm long, minutely puberulent to subglabrate, sometimes hispid; corolla 2.5–3 cm long, purplish (drying yellowish), externally minutely pubescent; androecium subequal to petals; stigmas slightly exserted. *Fruits* 12–13 mm diam., oblate, woody, minutely puberulent, 10-ribbed; mericarps 5–6 mm high, indehiscent.

Distribution (Fig. 49). Confined to W-central Mexico in the states of Guerrero and Mexico. The species grows in mixed forest and shrubland, at elevations of 1400–2300 m.

Illustration. Fryxell (1979: fig. 6D).

Specimens examined. MEXICO. GUERRERO: Above Amojileca SW of Omiltemi, *Breedlove 61882* (NY, TEX); along Milpillas–Atoyac rd., 3 mi E of Xochipala, *Reveal et al. 4174* (NY). **MÉXICO:** Tejupilco, *Hinton et al. 1923* (NY, RSA, US); La Labor, *Hinton et al. 3680* (NY, RSA); Ypericomes, Distr. Temascaltepec, *Hinton et al. 3892* (RSA); Nanchititla, *Hinton et al. 5326* (K, NY, RSA); Distr. Temascaltepec, San Lucas, *Hinton et al. 7091* (K, US); Cajones, *Hinton et al. 7218* (K, NY, US); Cierrita, *Hinton et al. 8351* (K, US); Zacualpan, Cerro de Corona, *Matuda et al. 38687* (LL–2); near Amatepec, *Matuda 30027* (MEXU, NY); Mun. Sultepec, 4 km S of Sultepec, near La Cienega, *Moreno 40* (ENCB, NY); Mun. Tejupilco, 5 km N of Ocotepec, *Rzedowski 25313* (ARIZ, CTES, ENCB, LL, NY).

In some specimens (e.g., *Hinton et al. 7218*, *Matuda 30027*) the pubescence is relatively sparse, but this is simply an extreme of a range of expression. The distinguishing feature of *Pavonia chlorantha* is the more or less hirsute calyx that is often retrorsely bearded at the base.

153. Pavonia firmiflora Schery, Ann. Missouri Bot. Gard. **29:** 227. 1942. Type. Mexico. Jalisco: Tequila, 6 Jul 1893, *Pringle 5447* (holotype, MO; isotypes p.p., BRIT, CAS, ENCB, F as photo F-56168, G, GH, K, LL, MEXU, MICH, NMC, NY, OSH, TEX, UC, US). *Note:* The holotype includes only a flowering branch. All of the isotypes include flowering material (collected 6 Jul 1893) and fruiting material (collected 22 Sep 1893). Thus, only the flowering portions (and not the fruiting portions) can be considered to be type material Fig. 65

Shrubs ca. 1.5 m tall, the stems densely and roughly stellate-pubescent. *Leaf* blades mostly 7–10 cm long, broadly ovate and often 3-lobulate, subtruncate to deeply cordate, crenate to subentire, rounded-acute, minutely stellate-pubescent above, densely so beneath, discolorous; petioles ¼–¾ the length of the blades; stipules 4–5(–10) mm long, subulate. *Flowers* solitary in the leaf axils or in a terminal raceme of 2–6 flowers; pedicels 2.5–4.5 cm long; involucellar bracts 8–10, narrowly triangular, 6–12 mm long, stellate-pubescent; calyx 1.5–1.8 cm long, stellate-pubescent; corolla tubular, 3–4 cm long, yellowish, externally sparsely pubescent; androecium ca. 2 cm long, the filaments 2–3 mm long; styles and stigmas exceeding the androecium, shorter than the petals. *Fruits* 16–19 mm diam., oblate-depressed, 5- or 10-ribbed, retrorsely scabrid (hairs 0.5 mm long), the mericarps 8–10 mm high, woody, dorsally ridged.

Distribution (Fig. 42). From the states of Jalisco and Nayarit in W Mexico, at elevations of 900–1200 m. The type locality is at 20°54′N, 103°49′W.

Illustrations. Fryxell (1988a: fig. 83); Schery (1942: 228).

Specimens examined. MEXICO. JALISCO: 2 mi NW of Tequila, *McVaugh 18618* (MICH, NY, US); 5 km NW of Tequila, *Rzedowski 14271* (CTES, ENCB). **NAYARIT:** 10 mi SE of Ahuacatlán, *Feddema 303* (DS, ENCB, MICH, TEX), *Feddema 337* (ENCB, MICH, TEX); ca. 18 mi SE of Tepic, *H. S. Gentry et al. 19502* (ARIZ, US).

Pavonia firmiflora is distinguished by its relatively large woody fruits (16–19 mm diam.) coupled with relatively small corollas (3–3.5 cm long) with included genitalia.

154. Pavonia paneroi Fryxell, Lundellia **1:** 75. 1998. Type. Mexico. Puebla: Izúcar de Matamoros, hwy. from Izúcar to Huahuapan de León, side rd. to Microondas Raboso, deciduous woodland; shrub 1.5–2 m tall; flowers dark orange; 1300 m, 30 Oct 1994, *Panero, Manrique & Calzada 5311* (holotype, TEX; isotypes, MEXU + 4 duplicates to be distributed). Fig. 69

Shrubs to 2 m tall, the stems scabridulous, with stiff stellate hairs 0.5–1 mm diam. and with an understory of very minute stellate hairs. *Leaf* blades ovate-elliptic,

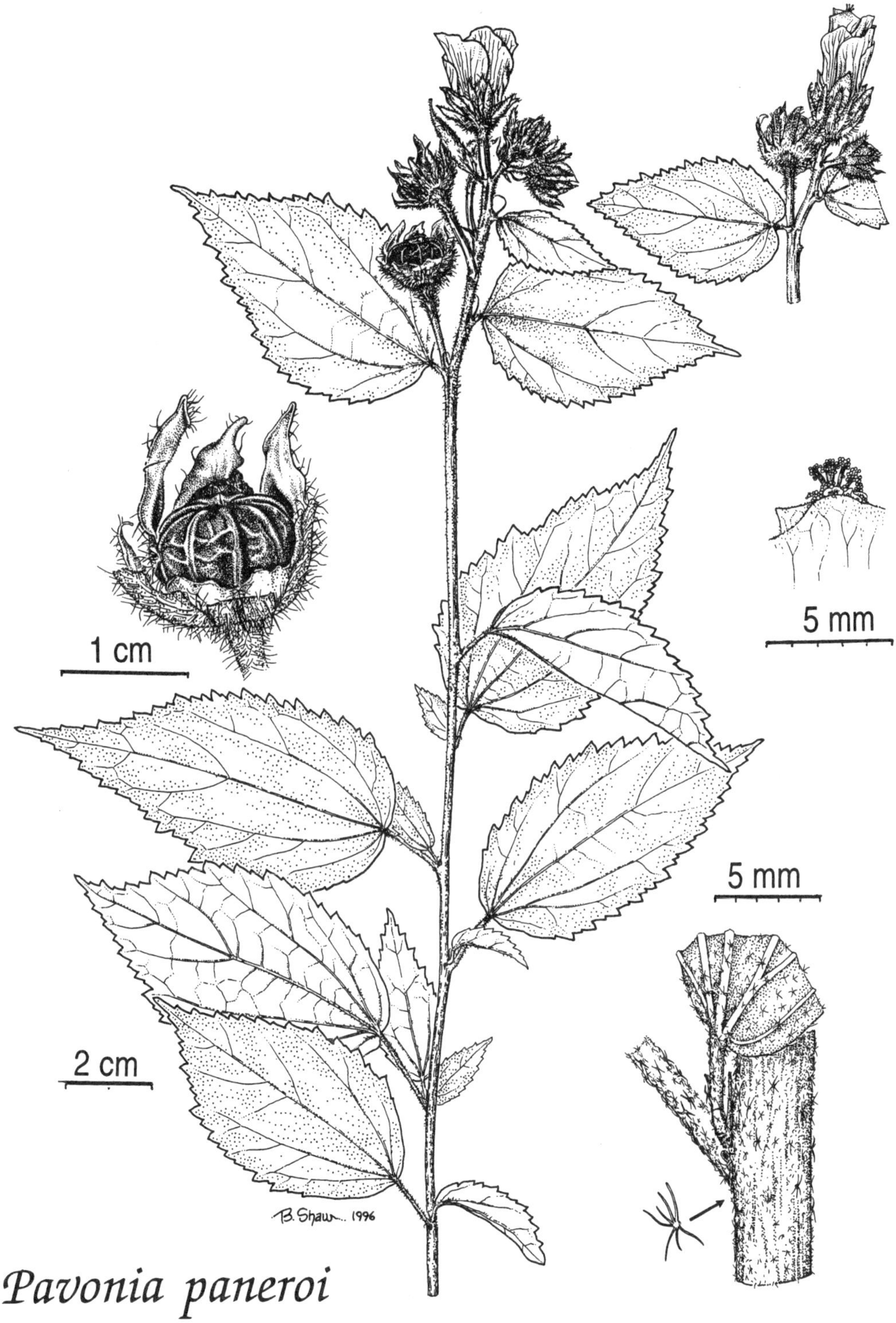

FIG. 69. *Pavonia paneroi*. Flowering branch (center) with enlargement of fruit (center left), of flower tip showing barely exserted stigmas (center right), and of node showing pubescence (bottom right) (*Panero et al. 5311*).

to 12 × 6 cm, gradually reduced upward, basally truncate, serrate, gradually acuminate, roughly stellate-pubescent above and beneath (subscabrid), the hairs up to 1 mm diam. (commonly smaller), palmately or subpedately 5-nerved from the base; petioles up to 2 cm long (shorter upwards) with pubescence like that of stem; stipules 5–8 mm long, subulate or linear. *Flowers* axillary or forming a terminal raceme by reduction of the terminal leaves, the lowermost pedicels 1.5–2 cm long, these shorter (to subsessile) upwards; involucel of ca. 8 ligulate bracts, these 10–13 × 2 mm, acute, marginally ciliate distally, densely hirsute at base, the hairs 1–2 mm long, connate below forming a cup; calyx ca. 15 mm long, about half-divided, externally hirsute like involucel, internally whitish-puberulent; corolla dark orange (yellowish when dried), 2.5 cm long, minutely glandular-puberulent externally, the petals erect forming a tubular corolla; genitalia slightly shorter than to equaling the corolla; anthers and pollen yellow, the stigmas slightly exserted from the staminal column. *Fruits* woody, oblate, 12 mm diam., glabrous, carinate and more or less rugulose.

Distribution (Fig. 49). Presently known only from the type locality from Puebla, Mexico, at 18°31′33.4″N, 98°25′5.6″W, growing in deciduous woodland at an elevation of 1300 m. It is known from this locality by only a few plants (J. L. Panero, pers. comm.).

Pavonia paneroi is distinguished from other members of *P.* subsect. *Collicolae* by its glabrous fruits and by its truncate-based leaf blades. Its corolla color also differs from that of the other species.

The new species is named to honor José L. Panero (b. 1959), the collector of the type specimen, knowledgeable specialist in the Asteraceae, and assiduous collector of neotropical plants.

155. Pavonia pleuranthera (de Candolle) Fryxell, Bol. Soc. Bot. Méx. **38:** 18. 1979. *Malvaviscus pleurantherus* de Candolle Prodr. **1:** 446. 1824. Type. *Icones Florae Mexicanae s.n.* (Torner Collection acc. no. 6331.823, Hunt Institute). Fig. 65

Pavonia racemifera W. J. Hooker & Arnott, Bot. Beechey Voy. 277. 1838. *Malache racemifera* (W. J. Hooker & Arnott) Kuntze, Revis. Gen. Pl. **1:** 71. 1891, as "racemiflora." Type. Mexico. Nayarit: probably from Tepic, *Beechey s.n.* (holotype, K as photo MICH-605; isotype, E).

Malvaviscus palmeri E. G. Baker *ex* Rose, Contr. U.S. Natl. Herb. **3:** 313. 1895. *Malvaviscus cinereus* E. G. Baker, J. Bot. **37:** 347. 1899, nom. superfl. *Pavonia palmeri* (E. G. Baker) Schery, Ann Missouri Bot. Gard. **29:** 230. 1942. Type. Mexico. Nayarit: Tepic, 1892, *E. Palmer 1990* (lectotype, US-49232; isotypes, BM, C, F as photo F-56663, GH, NY, US).

Malache ortegiana Standley, Proc. Biol. Soc. Wash. **37:** 46. 1924. *Pavonia ortegiana* (Standley) Standley, Publ. Field Columbian Mus., Bot. Ser. **4:** 231. 1929. Type. Mexico. Nayarit: San Blas, Rancho de los Aguacates, 1923, *González-Ortega 68* (holotype, US; isotypes, CTES [fragment], K, PH).

Pavonia amplifolia Standley, Publ. Field Columbian Mus., Bot. Ser. **4:** 230. 1929. Type. Mexico. Jalisco: San Sebastián, Arroyo de las Castillas, 14 Jan 1927, *Mexia 1480* (holotype, F as photo F-56169; isotypes, A, BM, CAS, CTES [fragment], MICH, MO–2, NY, UC, US).

Shrubs or trees 3–5(–6) m tall, the stems stellate-pubescent or glabrate. *Leaf* blades mostly 6–17(–22) cm long, almost as wide or slightly wider, 3–5(–7)-lobed, deeply cordate, crenate-serrate, usually acute, minutely and sparsely stellate-pubescent, more densely and coarsely so beneath, occasionally glabrate above; petioles subequal to (or exceeding) the blades (up to 30 cm), progressively reduced upward to ¼ the length of the blades; stipules inconspicuous. *Inflorescence* a crowded terminal leafless raceme; pedicels 5–10(–25) mm long, stellate-pubescent; involucellar bracts 7–8, narrowly linear, 10–15 mm long, stellate-pubescent; calyx 7–8 mm long, stellate-pubescent (especially on nerves); corolla tubular, 10–15 mm long, pale yellow, externally glandular-puberulent; staminal column 18–22 mm long, exserted, pallid, glabrous, the filaments 2 mm long; styles exceeding the androecium, glandular-pubescent, the stigmas minutely villous. *Fruits* 8–9 mm diam., retrorsely pubescent, 10-costate; mericarps 5.5–6 mm high, indurate.

Distribution (Fig. 49). Confined to W-central Mexico in the states of Jalisco, Michoacán, and Nayarit, at 400–1800 m elevation in shrubland, deciduous forest, and pine-oak forest. The species flowers from Dec to Apr.

Illustrations. Cervantes Aceves (1992: fig. 66); Fryxell (1979: fig. 7C; 1988a: fig. 83); Schery (1942: 229, as *P. palmeri*).

Specimens examined. **MEXICO. Jalisco:** 27.8 mi SE of Puerto Vallarta, *Almeda 2541* (CAS, NY); 5.6 km W of Puerto Los Mazos, Sierra Manantlán, *Cochrane et al. 10845b* (IBUG, NY, WIS, ZEA); Mascota, 18 km SW of San Sebastián, *A. Flores & Cervantes Aceves 208* (IBUG, NY); Mun. Talpa, rd. to Desmoronada, *Magallanes 3457* (MEXU, NY); 15–18 mi by rd. SW of Autlán, *McVaugh 11947* (MEXU, MICH, US); 2.5–4 mi N of La Cuesta, rd. to Talpa de Allende, *McVaugh 21254* (MICH); Puente San Pedro, 8 km SW of Tecalitlán, *McVaugh 22938* (ENCB, MICH, NY); 20–22 km S of Talpa de Allende, *McVaugh 23322* (ENCB, MICH, NY); Mascota, *E. W. Nelson 4061* (US); 17–18 km NE of Cuautitlán, *Santana 4255* (WIS); Mun. Talpa, rd. to Desmoronada, *J. A. Vázquez 1169* (ENCB, IBUG); 20 km NE of Tecalitlán, *Villa & Chávez 537* (CHAPA, NY). **Michoacán:** Coalcomán, Pto. Zarzamora, *Hinton et al. 12717* (F, MEXU, MICH, MO, NY, RSA, TEX, UC, US). **Nayarit:** 3 mi W of Jalcocotán, *Ayers 115* (NY, TEX); between km 15–14, rd. to Jalcocotán, *Croat 45199* (MEXU, MO, NY); 9 km NW of Mesillas, *G. Flores et al. 897* (MEXU, MO, NY); Mun. Compostela, 7 km W of Compostela, *P. A. Fryxell 4943* (CHAPA, CTES, MEXU, MICH, NY); 10 mi E of Jalcocotán, *McVaugh 12099* (MEXU-2, MICH, US); Tepic, in 1892, *E. Palmer 1835* (F); 500 m on the dirt rd. at km 6 of the Hwy. Compostela–Las Varas, *Téllez 9489* (MEXU); Mun. Tepic, 6 km NE of Camichin de Jauja, *Téllez 10136* (MEXU, MO, NY); La Escondida, 15 km N of Tepic, *Téllez & G. Flores 11673* (MEXU, MO, NY); 10 km W of Compostela, *Téllez & G. Flores 11842* (MEXU, MO, NY).

Pavonia pleuranthera differs from other members of *P.* subsect. *Collicolae* in its small corolla (1–1.5 cm long) with exserted genitalia and by its small fruits (8–9 mm diam.). Its often palmately lobed leaf blades are also notable.

Pavonia subgen. Malache sect. Collicolae subsect. Rubriflorae

Pavonia subsect. **Rubriflorae** Fryxell, subsect. nov. (from the Latin *ruber*, red, and *floris*, flower, in reference to the dark red corolla).

Frutices laminis foliorum ovatis vel ellipticis stellato-pubescentibus; bracteis involucellorum 8–12; floribus in racemis terminalibus aphyllis dispositis, corollis atrorubris tubularis, genitaliis inclusis vel parum exsertis; stylis fere ad apicem connatis, stigmatibus subsessilibus; fructibus ligneis plus minusve costatis.

Shrubs with ovate or elliptic leaves, stellate-pubescent; involucellar bracts 8–12; flowers in terminal leafless racemes, the corolla dark red, tubular, the genitalia included or slightly exserted; styles connate nearly to the apex, the stigmas subsessile; fruits more or less ribbed, woody.

Type species. *Pavonia anisaster* (Standley) Fryxell.

Key to the species of *Pavonia* subsect. *Rubriflorae*

1. Fruits 18 mm diam., densely pubescent; pedicels 4–14 mm long, stout (3 mm diam.); calyx ca. 15 mm long; involucellar bracts ca. 12, ovate-lanceolate with revolute margins, ciliate (hairs 1–2 mm long) . . . . 157. *P. baumliana*
1. Fruits 10–15 mm diam., glabrous; pedicels 10–40 mm long, slender (ca. 1 mm diam.); calyx 6–12 mm long; involucellar bracts 8–10, subulate or linear-lanceolate with straight margins, minutely puberulent.
    2. Leaves narrowly lanceolate (2.5–5 times as long as wide), basally truncate; fruits 10 mm diam., more or less verruculate; calyx (at anthesis) 6–7 mm long . . . . 159. *P. nayarensis*
    2. Leaves ovate, sometimes trilobed (0.9–2 times as long as wide), basally cordate; fruits 10–15 mm diam., ribbed but not verruculate; calyx (at anthesis) 7–12 mm long.
        3. Calyx 7–8 mm long; involucel (at anthesis) exceeding calyx; fruits ca. 10 mm diam.; leaf blades 1.5–2 times as long as wide; petioles more densely and coarsely pubescent adaxially than abaxially . . . . 158. *P. durangensis*
        3. Calyx 10–12 mm long; involucel (at anthesis) ca. ½ length of calyx; fruits ca. 15 mm diam.; leaf blades 0.9–1.2(–1.5) times as long as wide; petioles uniformly pubescent . . . . 156. *P. anisaster*

156. Pavonia anisaster (Standley) Fryxell, Sida **7:** 223. 1977. *Hibiscus anisaster* Standley, Publ. Field Columbian Mus., Bot. Ser. **4:** 230. 1929. Type. Mexico. Nayarit: Between Tepic and Santiago, 15 Sep 1926, *Mexia 634* (holotype, US; isotypes, CAS, F, GH, UC).

Shrubs 0.5–2 m tall, the stems minutely and roughly stellate-pubescent. *Leaf* blades mostly 5–9 cm long, ovate or weakly 3-lobed, cordate or truncate, crenate or serrate, acute, stellate-pubescent, slightly discolorous; petioles ⅓–⅔ the length of the blades, reduced upwards to almost sessile; stipules 4–5 mm long, linear. *Inflorescence* generally a terminal leafless raceme; pedicels 1–2.5 cm long; involucellar bracts 8–10, shorter than calyx; calyx 10–12 mm long at anthesis, accrescent to 15 mm in fruit; corolla tubular, 1.5–2 cm long, dark red, externally pubescent; staminal column glabrous, pallid, shorter than petals, the filaments 2–3 mm long; stigmas subglabrous, slightly exceeding the stamens. *Fruits* 1.5 cm diam., oblate, glabrous, prominently 10-ribbed; mericarps glabrous, unornamented.

Distribution (Fig. 24). Endemic to the state of Nayarit in W-central Mexico, in the vicinity of Tepic, in oak forest and savanna at 400–1100 m elevation.

Illustrations. Fryxell (1979: fig. 6C; 1988a: fig. 80).

Specimens examined. **MEXICO. Nayarit:** NE of Santa María del Oro, *Feddema 678* (MICH, NY); Mirador del Aguila, ca. 14 mi N of Tepic, *Feddema 820* (MICH, NY); 10 mi SE of Tepic (2 mi SE of El Refugio), *McVaugh 13375* (MICH, NY), *McVaugh 16595* (MICH); ca. 12 mi W of Tepic, 6.5 mi E of Jalcocotán, *McVaugh 18836* (MICH,

NY); between Pochititán and Francisco Madero, *Miller & Téllez 3172* (MO, NY, UCR); between Pochititán and Francisco Madero, *Téllez & Miller 10459* (MEXU, MO, NY).

Pavonia anisaster is distinguished from the other red-flowered species of *P.* subsect. *Rubriflorae* by a combination of characters, as given in the key.

157. Pavonia baumliana Fryxell, Syst. Bot. Monogr. **25:** 317. 1988. Type. Mexico. Jalisco: Mun. Mezquitic, just E of Las Guayabas, *Bauml & Voss 2020* (holotype, RSA; isotype, NY). Fig. 65

Shrubs to almost 3 m tall, the stems densely stellate-pubescent, the hairs more or less appressed. *Leaf* blades 8–10 × 7–9 cm, progressively reduced upward, broadly ovate, cordate, irregularly crenate-serrate, acute, palmately 7–9-nerved, densely and minutely stellate-pubescent; petioles up to 5.5 cm long (progressively shorter upward); stipules 3–5 mm long, subulate. *Inflorescence* a terminal leafless raceme; pedicels 4–14 mm long, ca. 3 mm diam., stellate-pubescent; involucellar bracts 11–12, ca. half length of the calyx, ovate-lanceolate, the margins revolute and undulate, prominently ciliate (hairs 1–2 mm long); calyx 1.5 cm long, stellate-puberulent; corolla tubular, 1.5–2.5 cm long, dark red, externally pubescent, internally glabrous (including margins of claw); staminal column ca. 2 cm long, glabrous, pallid, the filaments 2 mm long. *Fruits* 18 mm diam., oblate, weakly 10-ribbed, densely stellate-pubescent, the hairs more or less retrorse; mericarps 8–10 mm high, ligneous.

Distribution (Fig. 42). Known only from the Municipio de Mezquitic in Jalisco, in oak woodland and deciduous forest at 1000–1800 m elevation, flowering in Oct and Nov.

Illustration. Cervantes Aceves (1992: fig. 62); Fryxell (1988a: fig. 80).

Specimens examined. **MEXICO**. **Jalisco:** Mun. Mezquitic, vic. of Las Guayabas, toward the river, *Chávez 8.9.9* (IBUG–n.v.); rd. to the Las Guayabas ranch, to the S of San Andrés Cohamiata, *Reynoso & Ramírez 335* (IBUG–n.v.).

Pavonia baumliana is distinguished by its revolute, ciliate involucellar bracts, its stout pedicels, and its large fruits (18 mm diam.).

158. Pavonia durangensis Fryxell, Syst. Bot. Monogr. **25:** 321. 1988. Type. Mexico. Durango: 10 km SW of Mezquital, *Tenorio, Ignacio & Romero de Tenorio 9596* (holotype, MEXU; isotype, NY).

Shrubs to 2 m tall, the stems stellate-pubescent, the hairs 0.5 mm long. *Leaf* blades 6–8.5 × 3–5 cm (progressively reduced upward), narrowly ovate or somewhat 3-lobed, cordate to subtruncate, coarsely crenate-serrate, acute, palmately 7-nerved, discolorous, sparsely stellate-pubescent above, densely so beneath; petioles 2–3.5 cm long (½ length of blade or less), stellate-pubescent. *Inflorescence* a leafless terminal raceme; pedicels 1–2.5 cm long, sparsely stellate-pubescent; involucellar bracts 8–9, linear-lanceolate, 9–12 mm long, minutely puberulent, basally connate and forming a cup; calyx 7–8 mm long, minutely puberulent; corolla 16–18 mm long, tubular, dark red, externally pubescent, glabrous on margins of claw; staminal column subequal to the petals, glabrous; stigmas capitate, villous, maroon. *Fruits* ca. 1 cm diam., oblate, glabrous.

Distribution (Fig. 42). From the state of Durango in N Mexico and in adjacent Jalisco, growing in deciduous forest at 1400–1800 m elevation. The type locality is at 22°45′N, 104°20′W.

Illustration. Fryxell (1988a: fig. 80).

Specimens examined. **MEXICO. Durango:** 3 km from Temohaya by the rd. to Mezquital, *M. González & Rzedowski 1596* (NY, TEX); 4 km E of El Mezquital, rd. to Guadalajara, *S. González & Rzedowski 1848* (ENCB, NY). **Jalisco:** Mun. Tonalá, Barranca de Huentitan, microcuenca del Cerro del Diablo, 4 km NE of Tonalá, *A. Flores 2486* (TEX, WIS).

Pavonia durangensis is distinguished from the other red-flowered members of *P.* subsect. *Rubriflorae* by a combination of characters, as given in the key.

159. Pavonia nayarensis Fryxell, sp. nov. Type. Mexico. Nayarit: Mun. Nayar, 5 km SW of the village of Cerro Cangrejo, in a ravine; elev. 1400 m; ecotone of oak woodland-selva baja caducifolia; shrub to 2 m, scarce; fl. dark red, fr. green, 20 Oct 1989, *G. Flores, Téllez, Tenorio & Cadena 1661* (holotype, MEXU; isotypes, MO, NY; other isotypes to be distributed by MEXU). Figs. 65, 70

Frutices 1–3 m alti, caulibus heterotrichis pilis stellatis adpressis; laminis foliorum anguste lanceolatis; flores plerumque in racemis terminalibus aphyllis; bracteolis involucellorum 8–10, quam calycis brevioribus; calycibus floriferis 6–7 mm longis fructiferis 10 mm usque; corollis 1.2–2 cm longis atrorubris erectis tubum formantibus; androeciis corollam subaequantibus; fructibus oblatis 1 cm diametro glabris plus minusve verruculosis leniter costatis.

Shrubs 1–3 m tall, the stems with appressed stellate hairs of different size classes, the largest ca. 1 mm diam. *Leaf* blades narrowly lanceolate, 3–11 × 1.5–4.5 cm (2.5–5 times as long as wide), basally truncate or subcordate, crenate-serrate, acuminate, palmately 5-nerved, discolorous, stellate-pubescent

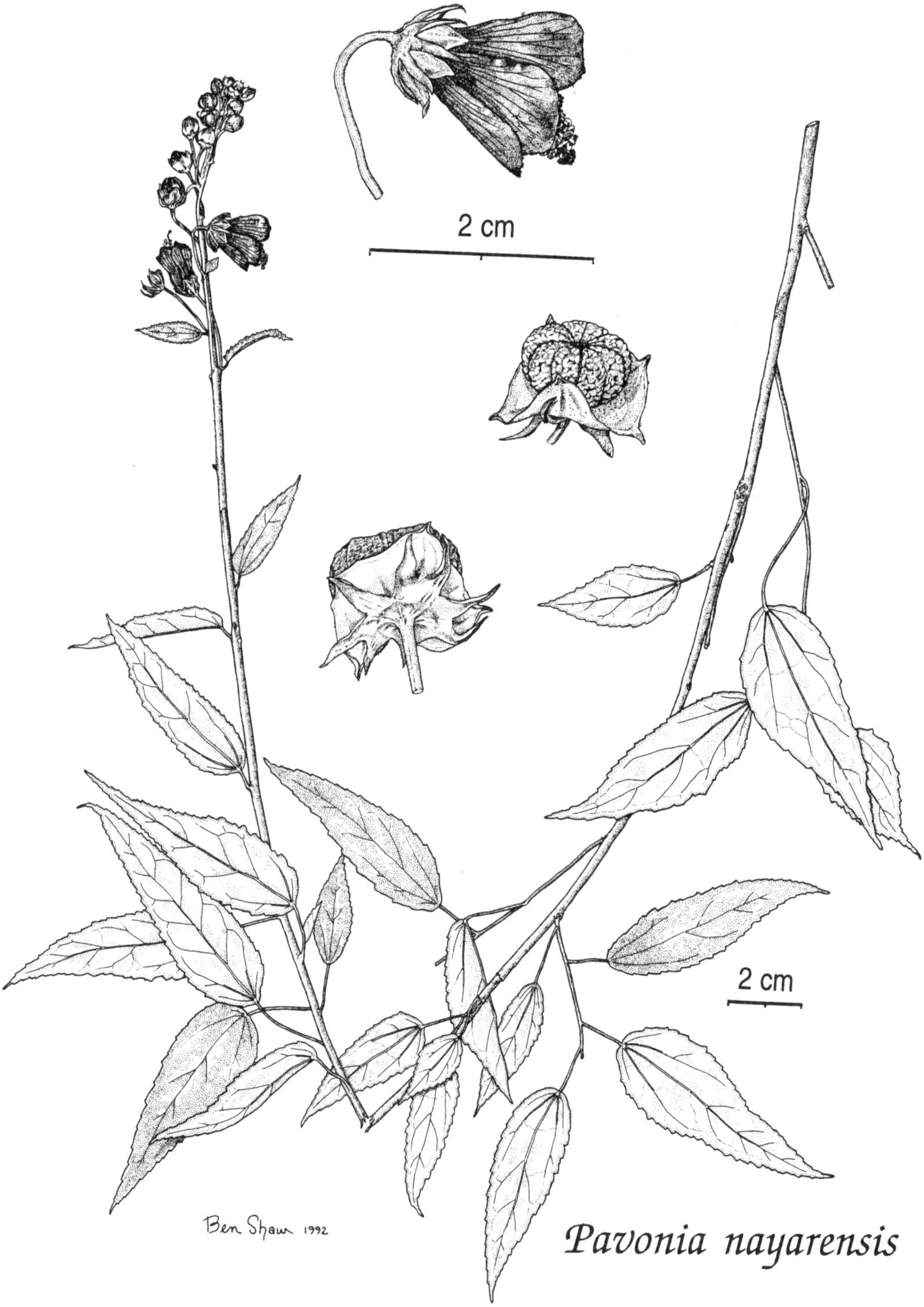

FIG. 70. *Pavonia nayarensis*. Flowering branch with enlargements of flower (above) and (at center) upper and lower views of mature fruit (*Tenorio & Flores 16192* [branch]; *Flores et al. 1661* [fruits]).

on both surfaces; petioles 0.5–4 cm long (progressively shorter upward), densely and evenly pubescent like stem; stipules subulate, 2–3 mm long, caducous. The lowermost *flowers* occasionally axillary but mostly borne in terminal leafless racemes; pedicels 3–4 cm long basally to 0.5 cm long distally, with minute stellate pubescence; involucellar bracts 8–10, basally connate (forming a broad disc in fruit), linear-lanceolate, ca. 1 mm wide, shorter than calyx, stellate-pubescent; calyx 6–7 mm long in flower to 10 mm long in fruit, costate, stellate-pubescent; corolla 1.2–2 cm long, dark red, tubular (the petals erect), externally pubescent, internally glabrous; staminal column glabrous, subequal to corolla or slightly exserted, the filaments 1–2 mm long, the pollen yellow-orange; styles slightly exserted, dark red or purplish, connate nearly to apex, the stigmas subsessile. *Fruits* oblate, 1 cm diam., 7 mm high, glabrous, weakly ribbed or unribbed, more or less verruculose.

Distribution (Fig. 42). *Pavonia nayarensis* is nearly confined to the Municipio de Nayar (hence the choice of specific epithet) in the state of Nayarit, Mexico, except for a single collection from farther east in the state of Jalisco. Most of the collection sites are within the geographic bounds of 22°11–16′N, 104°33–45′W, with the exception noted from ca. 103°55′W. It occurs in the ecotone between oak forest and deciduous forest, at elevations of 970–1960 m. The type locality is at 22°15′N, 104°40′W.

Specimens examined. MEXICO. Jalisco: Mun. Huejuquilla, Rancho San Nicolás, 25 km S of Huejuquilla, rd. to Tenzompa, 1600 m, *A. Flores 2075* (TEX). **Nayarit:** 5–8 km NW of the Mesa de Nayar, from La Mesa del Nayar to Villa de Guadalupe (Cerro Cangrejo), 1300 m, *G. Flores et al. 1131* (MEXU, NY); 5 km NW of the village of Cerro Cangrejo, in a ravine, 1300 m, *G. Flores et al. 1595* (MEXU, NY); Mun. Nayar, 2.8 km NW of Jesús María, rd. from Jesús María to Mesa del Nayar, 970 m, *G. Flores & Tenorio 1302* (MEXU, MO, NY); 6.4 km NW of La Mesa del Nayar, rd. to Santa Teresa, 1960 m, *G. Flores & Tenorio 1445* (MEXU, MO, NY); 3–5 km N of the Mesa del Nayar, rd. to Cerro Cangrejo, *Téllez et al. 12144* (MEXU, MO, NY); ravine SE of the village of Villa de Guadalupe, rd. from Herradura to El Cañaveral and/or to Jesús María, 1200–1400 m, *Tenorio & G. Flores 16080* (MEXU, MO, NY); ravine NE of the village of Villa de Guadalupe, 1200–1400 m, *Tenorio & G. Flores 16192* (MEXU, MO, NY); 2.8 km NW of Jesús María, *Tenorio & G. Flores 16396* (MEXU, MO); 6 km NW of La Mesa de Nayar, rd. to Santa Teresa, 1900 m, *Tenorio & G. Flores 16512* (MEXU, MO).

The specimens cited were predominantly flowering in Sept and beginning to mature fruits in Oct.

Pavonia nayarensis is notable for its narrowly lanceolate leaf blades with truncate bases and for its verruculate fruits.

Pavonia subgen. Malache sect. Collicolae subsect. Conditiflorae

Pavonia subsect. **Conditiflorae** Fryxell, subsect. nov. (from the Latin, *conditus*, hidden, and *floris*, flower, in reference to the relatively obscure flowers).

Frutices laminis foliorum ovatis stellato-pubescentibus; bracteis involucellorum 8–9; floribus axillaribus solitariis, corollis rubellis vel salmoneis anguste tubularis, genitaliis exsertis; fructibus subligneis.

Shrubs with ovate leaves, stellate-pubescent; involucellar bracts 8–9; flowers axillary, solitary, the corolla reddish or salmon-colored, narrowly tubular, the genitalia exserted; fruits semi-woody.

Type species. *Pavonia ecostata* Fryxell & S. Koch.

Key to the species of *Pavonia* subsect. *Conditiflorae*

1. Stem pubescence sparse, in narrow longitudinal lines; stipules 2–3 mm long; leaves sparsely pubescent; Mexico: Chihuahua. . . . . 160. *P. alia*
1. Stem pubescence dense, uniform; stipules 2–7 mm long; leaves densely and softly pubescent.
   2. Calyx 10–12 mm long; anthers yellowish; fruits 8–9 mm diam., scabridulous, ribbed; leaves 2–5 cm long.
      3. Involucellar bracts 1.5–3 mm wide distally, exceeding calyx; leaves 2–3 cm long; stipules 3–6 mm long, erect; Mexico: Mexico.. . . . . 163. *P. pulidoae*
      3. Involucellar bracts 1–1.2 mm wide distally, subequal to calyx; leaves 3–5 cm long; stipules 2 mm long, reflexed; Mexico: San Luis Potosí. . . . . 161. *P. clarkii*
   2. Calyx 5–7 mm long; anthers purplish; fruits 11–12 mm diam., glabrous, unribbed; leaves 4–8 cm long; Mexico: Jalisco. . . . . 162. *P. ecostata*

160. Pavonia alia Fryxell, Syst. Bot. Monogr. **25:** 315. 1988. Type. Mexico. Chihuahua: Mun. Urique, 4 km S of Quirare, rd. to La Bufa, *Tenorio, Romero de Tenorio, Ignacio & Dávila 9961* (holotype, MEXU; isotypes, MEXU, NY).

Shrubs to 1.5 m tall, the stems sparsely pubescent in narrow lines of short curled hairs. *Leaf* blades mostly 3–5.5 × 2–3 cm, ovate, truncate to subcordate, coarsely crenate-serrate, acute, sparsely pubescent above, with scattered stellate hairs beneath; petioles

1.5–2.5 cm long, adaxially pubescent; stipules 2–3 mm long, subulate. *Flowers* solitary in the upper leaf axils, the pedicels 0.5–1.5 cm long, sparsely pubescent to subglabrous; involucellar bracts 8–9, linear-lanceolate, 8–9 × 1 mm or less, subglabrous; calyx 6–7 mm long, subglabrous; corolla 2.5–3 cm long, narrowly tubular, salmon or reddish; staminal column exserted, glabrous, pallid, the filaments 1–2.5 mm long; styles exceeding the column, with capitate reddish stigmas. *Fruits* unknown.

Distribution (Fig. 49). Known only from the type locality, in oak forest at ca. 1900 m elevation.

Illustration. Fryxell (1988a: fig. 79).

Specimen examined: **MEXICO. CHIHUAHUA:** 4 km SW of Quirare, rd. to Batopilas, *Torres & Tenorio 3733* (MEXU, NY).

Pavonia alia is distinguished from the other species of *P.* subsect. *Conditiflorae* by its relatively sparse pubescence.

161. Pavonia clarkii Fryxell, sp. nov. Type. Mexico. San Luis Potosí: S of Valles, 16 Jun 1935, *Clark 6863* (holotype, MO).

Frutices graciles sparse pubescentes; laminis foliorum late ovatis; stipulis parvis reflexis; floribus solitariis axillaribus; bracteis involucellorum 8 linearibus vel leniter spatuliformibus calycem subaequantibus; corollis rubellis, androecio exserto ut videtur; fructibus ignotis.

Slender *shrubs*, the stems sparsely and minutely stellate-pubescent and with denser lines of recurved simple hairs running longitudinally. *Leaf* blades broadly ovate with a weak tendency to be 3-lobed, 3–5 × 2–4 cm, basally cordate, coarsely crenate (2–3 teeth per cm), acute, palmately 5-nerved, slightly discolorous, sparsely stellate-pubescent, the hairs ca. 0.5 mm diam.; petioles ca. ⅓ length of blade, with pubescence like stem; stipules 2 mm long, subulate, reflexed. *Flowers* solitary in the leaf axils, the pedicel 1.5 cm long in flower, more densely pubescent than stem, with stellate and recurved hairs; involucel of 8 linear or weakly spatulate, distinct bracts, each 12–13 × 1–1.2 mm at anthesis, sparsely pubescent; calyx ca. 11 mm long at anthesis, essentially ecostate, sparsely stellate-pubescent, less than half-divided, the lobes triangular, 5 × 3.5 mm; petals 3 cm long, apparently rose-pink; staminal column slender, incomplete on specimen but reaching as far as the distal margin of the petals without filaments emerging, therefore presumably exserted, the styles and stigmas unknown. *Fruits* unknown.

Distribution (Fig. 49). Known only from the type collection, "in jungles" S of Ciudad Valles, San Luis Potosí, Mexico.

Fryxell (1988a) cited *Clark 6863* as pertaining to *Pavonia pulidoae*, but reexamination of the specimen indicates that it is allied to, but distinct from, *P. pulidoae*. Unfortunately, fruits are lacking on the specimen, the single flower is incomplete, and no further collections from this area have come to light. The principal distinctions between the two species are given in the key. The new species is named in honor of the collector of the type, Ora Melvin Clark (1889–1952).

162. Pavonia ecostata Fryxell & S. Koch, Brittonia **43:** 24. 1991. Type. Mexico. Jalisco: Mun. Autlán, Sierra Cacoma, 8 km N of Autlán, thence 15 km W by dirt rd. to the microwave tower, 19 Nov 1989, *Koch & P. A. Fryxell 89198* (holotype, CHAPA; isotypes, CTES, ENCB, F, IBUG, K, MEXU, MICH, MO, NY, TEX, US, WIS).

Shrubs to 3 m tall, the young stems densely stellate-pubescent. *Leaf* blades broadly ovate, mostly 4–8 × 3–5.5 cm, subtruncate to shallowly cordate, crenate-dentate, acute, palmately 5–7-nerved, densely and softly stellate-tomentose; petioles 7–30(–40) mm long, evenly tomentose; stipules 4–7 mm long, subulate. *Flowers* solitary, the pedicels 2–10 mm long at anthesis, to 25 mm in fruit; involucellar bracts 8, lanceolate to spatulate, 7–9 (accrescent to 15) × 2–3 mm, tomentose, basally connate; calyx 5–7 mm long; corolla 21–23 mm long, narrowly tubular, red, externally pubescent and marginally ciliate; staminal column exserted, glabrous, the filaments 0.5–1 mm long, the anthers purplish; styles purplish, barely exserted from column. *Fruits* oblate, 11–12 mm diam., glabrous, smooth, the mericarps 7–9 mm high, ligneous. Chromosome number, $2n = 56$ (Fryxell & Stelly, 1993).

Distribution (Fig. 49). Endemic to central Jalisco, where known in highly disturbed vegetation at 800–1700 m elevation.

Illustration. Fryxell & Koch (1991: fig. 1).

Specimen examined: **MEXICO. JALISCO:** Mun. Autlán, rd. to the San Francisco mine, 4.9 km NW of hwy. from Autlán to El Grullo, *Lott & Dávila 2734* (MEXU, NY).

Pavonia ecostata is distinguished by its small calyx, purplish anthers, and smooth (unribbed) fruits.

163. Pavonia pulidoae Fryxell, Phytologia **46:** 392. 1980. Type. Mexico. México: Mun. Texcoco, Cerro Tetzcutzingo, 7 km E of Texcoco, 19 Jun 1979, *Pulido 23* (holotype, CHAPA; isotypes, BH, CAS, CTES, ENCB, F as photo F-64926, GB, IBUG, K, MEXU, MICH, MO, NY, TEX, US). Fig. 65

Shrubs to 1.5 m tall, the young stems densely stellate-pubescent. *Leaf* blades 2–3 × 1.2–2 cm, ovate, truncate to cordate, remotely dentate or crenate, rounded-acute, somewhat discolorous, soft-pubescent; petioles ca. ¼ the length of the blade or less; stipules 3–6 mm long, subulate, erect. *Flowers* solitary in the leaf axils, the pedicels ca. 1 cm long in flower to ca. 2 cm long in fruit; involucellar bracts 8–9, linear-lanceolate or weakly spatulate, 12–15 × 1.5–3 mm (distally), stellate-pubescent; calyx 10–12 mm long, sparsely pubescent; corolla 2–3 cm long, narrowly tubular, salmon-colored (drying reddish); staminal column ultimately exserted, pallid, glabrous, the filaments 1 mm long; styles pallid with reddish stigmas. *Fruits* 8–9 mm diam., oblate, minutely appressed-scabridulous, more or less 10-ribbed; mericarps relatively woody, 6–7 mm high.

Distribution (Fig. 49). Known from two collections in central Mexico in the state of Mexico.

Illustration. Fryxell (1980: fig. 2).

Specimen examined. MEXICO. MÉXICO: Mun. Texcoco, La Purificación, 10 km from Texcoco, *Pulido 635* (CHAPA, TEX).

Pavonia pulidoae is distinguished from *P. clarkii* by the characters given in the key.

Pavonia subgen. *Malache* sect. *Laminares*

Pavonia sect. **Laminares** Fryxell, sect. nov. (from the Latin *lamina*, blade, in reference to the distinctive laminar staminodes).

Frutices molliter pubescentes, pilis glandularis; laminis foliorum ovatis; corollis quasi-rotatis staminodeis laminaris purpuratis in base columnae staminalis, columna plus minusve declinata, floribus quamobrem aliquantum zygomorphis; mericarpiis columnam aequantibus indehiscentibus ligneis.

Softly pubescent shrubs, glandular-pubescent, with ovate leaves; corolla rotate with purple laminar staminodes at base of staminal column, the column more or less declined, the flower thus somewhat zygomorphic; mericarps equaling the column, woody, indehiscent.

Type species. *Pavonia oxyphylla* (de Candolle) Fryxell. The only species of this section is endemic to Mexico.

The following species (with two varieties) is placed in subgenus *Malache* principally on the basis of its mericarp morphology. Within this group it appears to be isolated and is therefore assigned its own section. The epithet "*Laminares*" refers to the distinctive laminar staminodes, which seem to have no counterpart elsewhere in the genus.

Key to the varieties of *Pavonia oxyphylla*

1. Leaves with sparse, coarse pubescence beneath, the surface green; petioles of midstem leaves usually >½ the length of the blades; involucellar bracts 10–11 × 1.5–2 mm; staminal appendages (staminodes) 2–4 mm wide; fruits puberulent.................. 164a. *P. oxyphylla* var. *oxyphylla*
1. Leaves with dense white tomentum beneath; petioles of midstem leaves usually <½ the length of the blades; involucellar bracts 8–10 × 1 mm; staminal appendages (staminodes) 1–1.5 mm wide; fruits glabrous................ 164b. *P. oxyphylla* var. *melanommata*

164a. Pavonia oxyphylla (de Candolle) Fryxell, Sida **7:** 223. 1977 var. **oxyphylla**. *Hibiscus oxyphyllus* de Candolle, Prodr. **1:** 455. 1824. Type. Mexico. "in montibus Xochipici," *Icones Florae Mexicanae 392* (Torner Collection, acc. no. 6331.8, Hunt Institute). [cf. photo F-30506 of copy at G]. Figs. 65, 71

Pavonia glandulosa C. Presl, Reliq. Haenk. **2:** 129. 1835. *Malache glandulosa* (C. Presl) Kuntze, Revis. Gen. Pl. **1:** 70. 1891. Type. Mexico. *Haenke s.n.* (lectotype, here designated, PR-197888A; isotypes, BM, MO, NY [fragment], PR, W as photo F-32648). The lectotype was indicated as "holotype" by Fryxell (1988a).

Pavonia melanommata var. *pringleana* R. E. Fries, Kongl. Svenska Vetenskapsakad. Handl. **42(12):** 43. 1908. Type. Mexico. Morelos: S of Yautepec, 7 Nov 1902, *Pringle 8718* (lectotype, S?–n.v.; isotypes, BM, C, CM, CTES, E, F as photo F-64924, K, MEXU, MICH, MIN, MO, NY, P as photo CTES, PH, POM, TEX, UC, US, VT).

Shrubs 1–2 m tall, the stems minutely glandular-puberulent. *Leaf* blades mostly 6–15 cm long (smaller upward), 1.5–2 times as long as wide, ovate, cordate, coarsely serrate, acuminate, sparsely stellate-pubescent, the hairs often simple above, stellate beneath (the largest hairs 3–4-armed), nearly concolorous; petioles (of midstem leaves) usually >½ the length of the blades; stipules 4–6 mm long, subulate, often recurved. *Flowers* principally in a terminal panicle with reduced leaves; pedicels 1.5–8.5 cm long, slender, glandular-puberulent; involucellar bracts 5, linear, 10–11 × 1.5–2 mm, minutely pubescent; calyx 5–8 mm long, minutely pubescent; corolla 1.5–2 cm long, pale pink to white with a yellowish center, the center often edged with red or purple, ciliate on claw; staminal column 1.5 cm long, dark purple, glabrous, with distinctive purplish spatulate appendages at base (staminodes), each 2–4 mm wide and distally erose, the filaments 1–3 mm long. *Fruits* oblate, 7–8 mm diam., puberulent, the mericarps 3.5–4 mm high, unornamented.

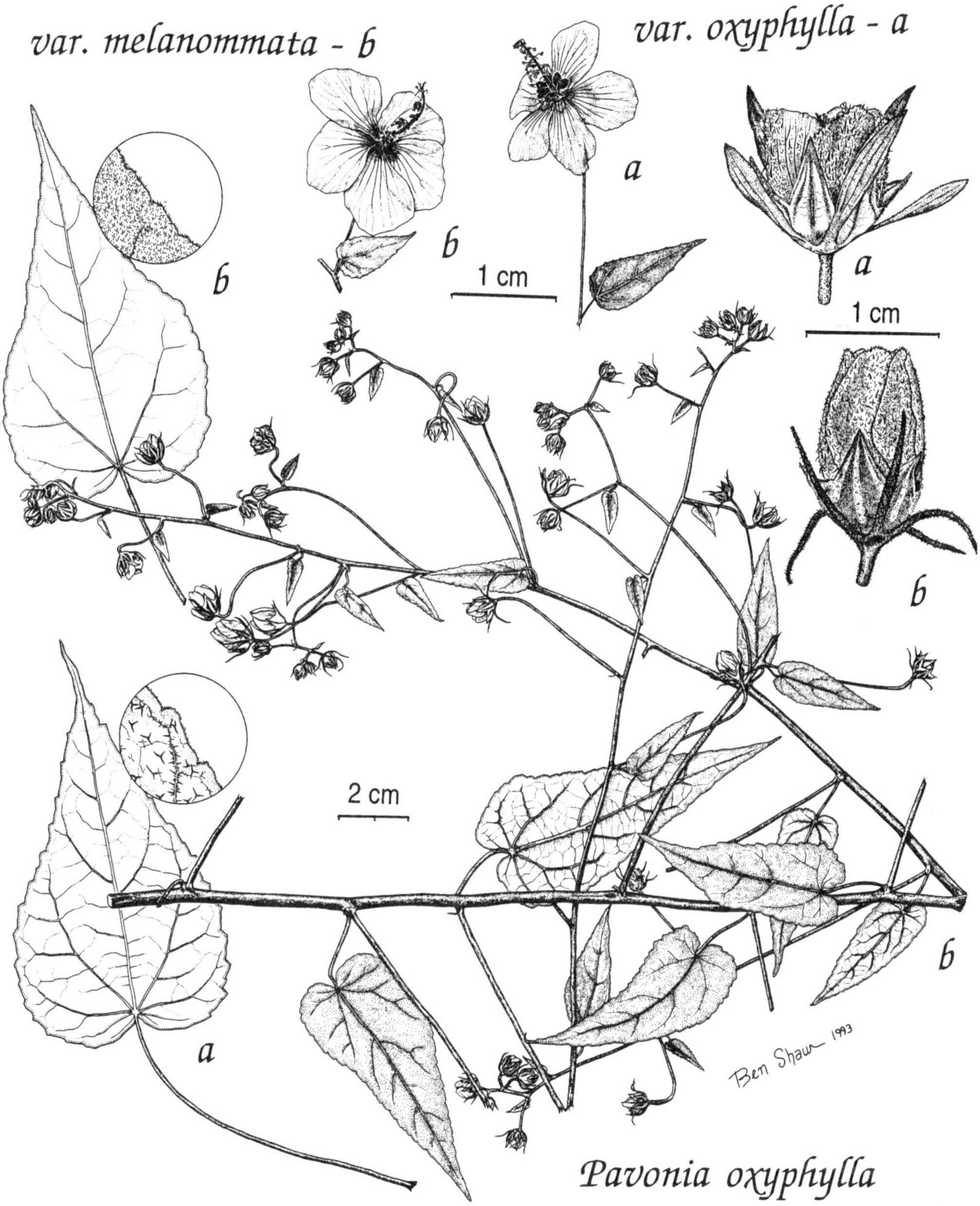

FIG. 71. *Pavonia oxyphylla.* (**A**) Bud, flower, and leaf of var. *oxyphylla*; (**B**) flowering branch, bud, flower, and leaf of var. *melanommata*. (var. *oxyphylla*: *Fryxell et al. 1651*; var. *melanommata*: *Koch & Fryxell 8322* [branch and bud], *Fryxell et al. 1649* [leaf], *Fryxell 566* [flower]).

Distribution (Fig. 24). Endemic to central Mexico in the states of Puebla, Guerrero, México, Michoacán, and Morelos, at elevations of 700–1700 m. It grows in deciduous forest and shrubland, flowering principally in Oct and Nov.

Additional illustration. Fryxell (1979: figs. 2A, 7D).

Specimens examined. MEXICO. WITHOUT LOCALITY: *Ghiesbreght 214* (K); *Sessé & Mociño 3564* (F, MA as photo F-46880). **GUERRERO:** Ca. 25 mi W of Iguala, *P. A. Fryxell*

607 (BH, MEXU, NY, WIS); between El Mogote and Azizintla (N of Taxco), *P. A. Fryxell & Bates 2163* (BH, ENCB, MO, MSC, NY); near Huitzuco, E of Iguala, *P. A. Fryxell & Bates 2172* (BH, NY); Iguala, *Holway 49* (US); ca. 2 km SE of Iguala, *Koch, P. A. Fryxell & Wendt 7986* (BR, CHAPA, ENCB, F, K, MEXU, MICH, MO, NY, TEX, WIS); ca. 2 mi S of Cacahuamilpa, *Webster & Breckon 16171* (DAV, NY). **México:** Temascaltepec, *Hinton et al. 8579* (ARIZ, CTES [fragment], F, K, LL, NY, PH, POM, US). **Michoacán:** 5 mi N of Tuzantla, *P. A. Fryxell et al. 1651* (BH, CAS, NY); Distr. Hidalgo, between Tuxpam and Tlaxmalac, *G. E. Seler & C. Seler 4227* (US). **Puebla:** Agua Escondida, *Miranda 2296* (MEXU).

164b. Pavonia oxyphylla var. **melanommata** (B. L. Robinson & Seaton) Fryxell, Sida **7:** 223. 1977. *Pavonia melanommata* Robinson & Seaton, Proc. Amer. Acad. Arts **28:** 104. 1893. *Malache melanommata* (Robinson & Seaton) Standley, Contr. U.S. Natl. Herb. **23:** 772. 1923. Type. Mexico. Michoacán: [Mpio. de Yurécuaro,] Monte León, 12 Nov 1892, *Pringle 4343* (holotype, GH; isotypes, ARIZ, BM, BR, COLO, CTES [fragment], E, F as photo F-56165, GH, GOET, K, MA, MEXU, MIN, MO, MU, NY, PH, TEX, UC, US, VT). Figs. 65, 71

Subshrubs or shrubs 2–3 m tall, the stems densely viscid-puberulent. *Leaf* blades mostly 5–11 cm long (smaller upward), 1.5–2 times as long as wide, ovate, cordate, shallowly crenate, acuminate, densely and softly pubescent, the hairs simple above, stellate-tomentose beneath, discolorous; petioles (of midstem leaves) usually <½ the length of blades; stipules 3–8 mm long, often recurved. *Flowers* principally in a terminal panicle with much-reduced leaves; pedicels 1–5.5 cm long, slender, glandular-puberulent; involucellar bracts 5, linear, 8–10 × 1 mm, minutely pubescent; calyx 6–8 mm long, minutely pubescent; corolla 2–2.5 cm long, pink with a dark purple center, the margins of the claws ciliate; staminal column 1.5 cm long, dark purple, glabrous, with distinctive purplish appendages (staminodes) at base, each 1–1.5 mm wide, distally erose, the filaments 1–3 mm long. *Fruits* oblate, 6–7 mm diam., glabrous, the mericarps 3.5–4 mm high, unornamented; seeds 2.5–3 mm long. Chromosome number, 2*n*= 56 (Bates, 1976).

Distribution (Fig. 24). Endemic to central Mexico in the states of Guanajuato, Jalisco, México, and Michoacán, at elevations of 1000–2000 m. It grows in deciduous forest and shrubland and flowers Sept to Apr, but principally in Oct and Nov. The two varieties of *Pavonia oxyphylla* are allopatric (see Fig. 24 and map in Fryxell, 1973), with *P. oxyphylla* var. *melanommata* occurring to the NW and *P. oxyphylla* var. *oxyphylla* to the SE of each other.

Additional illustrations: Cervantes Aceves (1992: fig. 65); Fryxell (1979: fig. 2B).

Specimens examined. MEXICO. Guanajuato: Mun. Yuriria, 11 km N of boundary between Michoacán and Guanajuato, *Kishler 90* (MEX). **Jalisco:** W of Juchitlán, *Cobián 223* (IBUG); 3 km W of Juchitlán, *P. A. Fryxell & Bates 2138* (BH, CAS, CHAPA, ENCB, MICH, MSC, NY); ca. 10 km N of Venustiano Carranza, on rd. to Tapalpa, *Lott & Magallanes 325* (MEXU); Mun. V. Carranza, 13.4 km N of V. Carranza, *Lott & Magallanes 777* (MEXU, MO, NY, TEX); 2 mi W of Juchitlán, *Moran 14725* (MICH, RSA, SD, UC); Lake Chapala, *E. W. Nelson 6528* (NY, US); Mun. Tamazula, km 5 on rd. from Soyatlán de afuera to Soyatlán de adentro, *Pérez de la Rosa 1691* (CTES, IBUG); between Atequiza and Chapala, *Rose & Painter 7611* (MO, NY, US); ca. 10 km N of V. Carranza, *Solheim et al. 981* (NY, WIS); 8 km NW of Tapalpa, *Soltero 311* (IBUG, NY). **México:** Mun. Ixtapan del Oro, 4 km W of the Valle del Bravo dam, *Koch & P. A. Fryxell 8322* (CHAPA, CTES, ENCB, F, MEXU, MICH, NY, RSA, TEX, XAL). **Michoacán:** 22 km S of Zitácuaro, *P. A. Fryxell 566* (BH, BM, CTES, NY); 20.5 km S of Zitácuaro, *P. A. Fryxell et al. 1649* (BH, BRIT, NY); Zitácuaro–Laureles, *Hinton et al. 13341* (ARIZ, CTES, K, LL, NY, OSH, PH, POM, US); Zitácuaro–El Sause, *Hinton et al. 13368* (ARIZ, MEXU, K, LL, NY, PH, RSA); 2 km N of Penjamillo, *A. Martínez 81* (IEB, NY); Penjamillo, *Pérez 1163* (IEB).

Pavonia subgen. Typhalea

Pavonia subgen. **Typhalea** (de Candolle) Ulbrich, Bot. Jahrb. Syst. **57:** 62. 1921. Basionym: *Pavonia* sect. *Typhalea* de Candolle, Prodr. **1:** 442. 1824. *Pavonia* subsect. *Typhalea* (de Candolle) Endlicher, Gen. Pl. 981. 1840. *Typhalea* (de Candolle) C. Presl, Abh. Königl. Böhm. Ges. Wiss. ser. 5, **3:** 449. 1845 (non *Typhalea* Necker, Elem. Bot. **2:** 412. 1790, illegit. art. 32.8 *ex* Britton & Wilson, Bot. Porto Rico **5:** 560. 1924). Type. *Urena typhalea* Linnaeus (≡ *Pavonia typhalea* (Linnaeus) Cavanilles = *P. fruticosa* (Miller) Fawcett & Rendle).

Synonym: *Typhalea* Necker *ex* Britton & Wilson, Bot. Porto Rico 5: 560. 1924. Type. *Hibiscus spinifex* Linnaeus (≡ *Pavonia spinifex* (Linnaeus) Cavanilles).

Perennial procumbent to erect herbs, subshrubs, or shubs, variously pubescent; flowers solitary or in definite inflorescences (racemes, panicles, heads); corolla yellow, white, or rose; fruits with indehiscent mericarps, these usually with 1 subapical and 2 lateral spines (rarely 1- or 2-spined), the spines erect or divergent, usually retrorsely barbed, the barbs sometimes obscure.

Synopsis of *Pavonia* subgen. *Typhalea*

1. **Pavonia** sect. **Typhalea.** Type species: *P. typhalea* (Linnaeus) Cavanilles (9 species).
2. **Pavonia** sect. **Urenoideae** St.-Hilaire. Lectotype species: *P. sepium* St.-Hilaire (20 species).
3. **Pavonia** sect. **Diathericae** Fryxell. Type species: *P. kearneyi* D. M. Bates (1 species).
4. **Pavonia** sect. **Carcerariae** Fryxell. Type species: *P. martii* Colla (3 species).
5. **Pavonia** sect. **Varians** Fryxell. Type species: *P. varians* Moricand (1 species).

This group of species has been discussed or treated by Garcke (1881), Gürke (1892b), Ulbrich (1921), Kearney (1951, 1954, 1958), Monteiro (1961), and Krapovickas (1982b), among others. A few workers (Britton & Wilson, 1924; Williams & Cheesman, 1929; Monteiro, 1961) treat this group in generic rank, but most retain it in infrageneric rank as a section or subgenus. Krapovickas (1982b: 282) summarized the nomenclatural aspects of these various views.

There have also been varying views on the species to be included in this taxon. Kearney (1954, 1958) included certain species in his keys under *Pavonia* sect. *Typhalea* that in footnotes he acknowledged to be better placed in other sections. Krapovickas (1982b) specifically excluded *P. varians*, *P. martii*, *P. arachnoidea*, *P. linearis*, and *P. submutica* from *P.* sect. *Typhalea* on the basis of carpel morphology. He defines the section on the basis of the fact that the central spine of the mericarp is subapical rather than apical. I agree with this definition, but I believe that, in the species named, the spine is indeed subapical and that these species should therefore be included (see Fig. 73), except *P. linearis,* which is here transferred to *P.* subgen. *Asterochlamys* (see following paragraphs).

Of the remaining species, *Pavonia varians* is a special case, inasmuch as it does not appear to have a strong affinity with any other American species. Ulbrich (1921) suggested that *P. varians* might be allied to the Indian *P. ceratocarpa* Dalzell *ex* Masters, but such an alliance seems unlikely on both phytogeographic and morphological grounds. There is some resemblance in mericarp morphology between *P. varians* and *P. submutica* (see Fig. 73), but other morphological distinctions, such as leaf and calyx form, support the isolation of *P. varians*. The best solution seems to be to recognize this isolation by placing *P. varians* in its own section.

Two species from Cuba (*Pavonia linearis* and *P. intermixta*) were explicitly included by Kearney (1954) in *P.* sect. *Typhalea*, and they were implicitly so included by Sauget and Liogier (1953). An examination of the morphology of the fruits, however, suggests that they are better included in *P.* sect. *Asterochlamys.* The mericarp morphology is variable (see Fig. 53), and sometimes there are two short barbed spines on the lateral margins, but a central (subapical or terminal) spine is always lacking. Even if these lateral spines are present, the overall conformation of the mericarp, including the tendency toward loculicidal dehiscence, matches well that of other species of *P.* sect. *Asterochlamys*, and the two Cuban species (here considered conspecific) are included in that section.

Another species of doubtful affinity is *Pavonia kearneyi.* It clearly has no close allies, but Bates (1966) hinted at a placement in *P.* sect. *Typhalea*, even though he did not then have access to information on fruit morphology. The subsequent availability of a fruiting specimen reveals that the fruit morphology (see Fig. 87) is as distinctive as is that of the rest of the plant. Each mericarp has two long, stout spines that radiate outward. The spines of adjacent mericarps are connivent, so that the fruit as a whole is 5-spined. The subapical (or terminal) spine is lacking, and the lateral spines are minutely and obscurely barbed toward the apex. The stout spines with minute barbs (see Fig. 87) are similar to those found in *P. flavispina* and *P. varians* (see Figs. 73, 89), but no suggestion of relationship is implied by this comparison. In truth, *P. kearneyi* can only be considered taxonomically isolated. It is here included in *P.* subgen. *Typhalea* but segregated to the monotypic *P.* sect. *Diathericae* in recognition of this isolation.

Several species of *Pavonia* subgen. *Typhalea* are characterized by having large appressed simple hairs on the upper leaf surface rather than having stellate-pubescent or glabrate surfaces. Since this characteristic is readily discerned, even in an imperfect specimen, and since the species exhibiting it are found in different sections of the subgenus, it is useful to discuss this group of species together. One species, *P. peruviana*, has terminal, leafless inflorescences (*P.* sect. *Typhalea*); the remainder of the species (*P. intermedia*, *P. oxyphyllaria*, *P. penduliflora*, *P. pseudotyphalea*, and *P. rudis*) have solitary, axillary flowers (*P.* sect. *Urenoideae*). These and other characters are presented in Table III, as a supplement to the key, to aid in identifying specimens. Additional discussion of the position of *P. intermedia* appears immediately after that taxon's description.

Previous attempts to discover the origin of the name *Typhalea* (e.g., Fryxell, 1988a) have not been successful. Fuertes (1989) honestly said "etymology not known." However, new information is now available that clarifies the origin of the name.

Table III

Comparison of the species of *Pavonia* subgen. *Typhalea* that have appressed simple hairs on the upper leaf surface

	P. intermedia	*P. oxyphyllaria*	*P. penduliflora*	*P. peruviana*	*P. pseudotyphalea*	*P. rudis*
Stem pubescence	Loosely stellate	Hispid and stellate	Coarsely stellate	Stellate or hirsute (simple hairs)	Roughly stellate, hairs antrorse	Stellate and simple hairs
Hair length (mm) on upper leaf surface	Ca. 1	1–2	0.5–0.7	0.5–2	Subglabrous	0.4–0.8
Stipule length (mm)	3–6	3–4	4–5	8–12	8–12	3–4
Disposition of flower or fruits	Intermediate	Axillary, solitary	Axillary, solitary	Racemose or capituliform	Axillary, solitary	Axillary, solitary
Internodes	Subequal	Subequal	Subequal	Foreshortened	Subequal	Subequal
Number of bracts	7–8	13–16	8–9	8–9	7–9	7–9
Involucel length (mm)	6–7	16–20	7–9	5–9	5–9	ca. 8
Corolla length (mm)	8–12	16–20	10–20	6–9	8–12	10
Corolla color	White	White	Red	White to pink	White	White
Fruit diam. (mm)	6	10–12	11–13	6–9	8–10	6–7
Distribution	Brazil	Costa Rica to Peru and Brazil	Central America	Ecuador and Peru to Central America	Colombia	Peru

The epithet enters the literature, seemingly independently, three times: *Typhalea* Necker 1790, nom. illeg., *Pavonia* sect. *Typhalea* de Candolle, 1824, both names implicitly or explicitly taken from Linnaeus's earlier binomial, *Urena typhalea* Linnaeus 1771, which is a synonym of *Pavonia fruticosa* (Miller) Fawcett & Rendle. Thus, Linnaeus's epithet is the ultimate origin of the name in the published literature and from it are derived the following binomials in *Pavonia*: *P. typhalea* (Linnaeus) Cavanilles, *P. typhaleoides* Humboldt, Bonpland & Kunth, and *P. pseudo-typhalea* Triana & Planchon. In an attempt to typify the Linnaean binomial *Urena typhalea*, the protologue was examined in detail. Linnaeus provided a lengthy (for him) and clear description of the species and a reference to "Allamand." This reference concerns a manuscript flora sent by the Swiss botanist Frédéric Allamand to Linnaeus, a manuscript now kept in the Linnean Herbarium (LINN). Consultation of the manuscript revealed that Allamand had proposed the generic name "Tryphalea" (and incidentally the specific epithet "aspera") for the plant that Linnaeus published as *Urena typhalea*. Thus, Linnaeus used Allamand's generic name as a specific epithet but altered its spelling by omitting the letter *r*, either deliberately or accidentally, and this spelling has been followed by all subsequent authors. One can argue that the change was not intentional (a typographical error) because there is no evident reason for the change and because the name appeared in Linnaeus's publication only once, not repeatedly. The matter is worth recounting because the derivation of the name now becomes clear. Attempts to derive the name from the Greek *typhe* or *typhos* (e.g., Fryxell, 1988a) have never been persuasive, although this root is surely the origin of the generic name *Typha* (Typhaceae). Rather, the name evidently derives from the Greek *tri* and *phales* or *phallos*, in clear reference to the three erect spines on each mericarp that characterize this species, and which, incidentally and appropriately, characterize nearly all of the species in the subgenus bearing the name (see Fig. 73). The same feature is the basis for the vernacular names "cadillo de tres pies" (*P. fruticosa*, *P. spinifex*), "carrapicho de tres pies" (*P. intermedia*), and "guisato tres puyas" (*P. schiedeana*).

Pavonia subgen. *Typhalea* sect. *Typhalea*

Erect subshrubs or shrubs, the leaves lanceolate or elliptic to oblanceolate, penninerved, the internodes often more or less foreshortened (sometimes the leaves almost in a rosette form); flowers in terminal leafless racemes or heads more or less above the leaves; involucellar bracts 5–10; corolla white or pink; mericarps each 3-spined (the 2 lateral spines suppressed in *P. stellata*).

Type species. *Pavonia typhalea* (Linnaeus) Cavanilles (= *P. fruticosa* (Miller) Fawcett & Rendle.)

Key to the species of *Pavonia* sect. *Typhalea*

1. Leaves entire or nearly so (sometimes with minute serrulations); inflorescence an open terminal panicle. . . . . 167. *P. integrifolia*
1. Leaves manifestly crenate, dentate, or serrate.
   2. Bracts of the involucel 4–6, each 2–4 mm wide; lateral spines of the mericarp ca. 1 mm apart at the base; dorsal wall of mericarp smooth, lustrous. . . . . 166. *P. fruticosa*
   2. Bracts of the involucel 7–12, each 0.5–2(–4) mm wide (if 2–4 mm wide, then 10 or more in number); lateral spines of the mericarp 1.5–4 mm apart at the base (or absent in *P. stellata*); dorsal wall of mericarp smooth or reticulately veined but not lustrous.
      3. Involucel shorter than calyx; inflorescence paniculate, the flowers (sometimes solitary) usually in groups of 3–6. . . . . 180. *P. leucantha*
      3. Involucel equaling or longer than calyx; inflorescences various.
         4. Mericarps with single erect spine 8–10 mm long, the 2 lateral spines suppressed. . . . . . . 173. *P. stellata*
         4. Mericarps with 3 erect or divaricate spines.
            5. Involucellar bracts ovate, 2–4 mm wide, often overlapping in fruit; stems with appressed 4-armed stellate hairs, the arms aligned with the stem axis; mericarp body 10–11 mm long. . . . . 168. *P. longitudinalis*
            5. Involucellar bracts linear, seldom >2 mm wide, not overlapping; stems otherwise pubescent; mericarp body 5–9 mm long.
               6. Upper leaf surface with appressed simple hairs 0.5–2 mm long.
                  7. Leaves distichous, asymmetrical; fruits 6 mm diam.; mericarps 5–6 mm long the central spine somewhat longer than body of mericarp; inflorescences axillary (sometimes terminal); Brazil. . . . . 178. *P. intermedia*

7. Leaves spirally disposed, symmetrical; fruits 6–9 mm diam.; mericarps 7–8 mm long, the central spine subequal to body of mericarp; inflorescences terminal, capituliform or racemose; Nicaragua to Peru. . . . . *171. P. peruviana*
6. Upper leaf surface with stellate hairs or subglabrous (if hairs simple, then <0.5 mm long).
 8. Involucel 8–10 mm long, gamophyllous basally for ⅓–½ of its length; mericarp body 6–9 mm long, with relatively prominent venation.
  9. Peduncle antrorsely hispid, the hairs simple, rigid; fruits reddish.. . . . . . . 170. *P. parva*
  9. Peduncles more or less hirsute, the hairs stellate or simple, neither rigid nor antrorse; fruits brownish. . . . . 165. *P. castaneifolia*
 8. Involucel 4–8 mm long, gamophyllous basally for ¼ of its length; mericarp body 5–6 mm long, smooth.
  10. Pedicels up to 12 mm long; central spine of mericarp usually 5 mm long or less, the lateral spines 3–4 mm apart at base. . . . . 172. *P. schiedeana*
  10. Pedicels 5–30 mm long; central spine of mericarp 9–10 mm long, the lateral spines 1.5–2 mm apart at base. . . . . 169. *P. nemoralis*

Note: *Pavonia intermedia* and *P. leucantha* are included in this key for sect. *Typhalea*, although they are classified in sect. *Urenoideae* and also keyed out there, for reasons noted in the discussion under *P. intermedia* (sp. 178).

165. Pavonia castaneifolia St.-Hilaire & Naudin, Ann. Sci. Nat. Bot., ser. 2, **18:** 44. 1842. *Malache castaneifolia* (St.-Hilaire & Naudin) Kuntze, Revis. Gen. Pl. **1:** 70. 1891. *Typhalea castaneifolia* (St.-Hilaire & Naudin) Britton *ex* Williams & Cheesman, Fl. Trin. Tobago **1:** 86. 1929. Type. Brazil. *Blanchet 33* (holotype, P as photo F-35468; isotype, CTES [fragment]). Figs. 72, 73

Pavonia longipes Standley, Publ. Field Mus. Nat. Hist., Bot. Ser. **18:** 672. 1937. Type. Panama. Bocas del Toro: Changuinola Valley, 29 Jan 1924, *Dunlop 408* (holotype, F as photos F-45080 = F-56236; isotype, US).

Pavonia longipes var. *hirsuta* Robyns, Ann. Missouri Bot. Gard. **52:** 542. 1965. Type. Panama. Darien: Near mouth of the Río Yapé, 12–14 Jul 1937, *P. H. Allen 362* (holotype, MO; isotype, F as photo F-56237).

Subshrubs 0.5–1 m tall, the stems minutely stellate-pubescent, the hairs often appressed. *Leaf* blades mostly 9–18(–28) × 2.5–7.5(–10) cm, elliptic to oblanceolate or obovate, sometimes subrhombic, penninerved, basally cuneate or narrowly truncate, serrate, acuminate, minutely stellate-pubescent, sometimes subglabrous on upper surface; petioles 0.5–7 cm long (shorter upward); stipules 6–16 mm long, narrowly lanceolate, erect. *Inflorescences* on long leafless peduncles, open racemes or condensed heads, the pedicels (or groups thereof) subtended by stipuliform bracts, each pedicel 0.5–1(–5) cm long; involucellar bracts 8–10, basally gamophyllous, 8–10(–14) x ca. 1(–3) mm, minutely pubescent, hirsute (hairs 1–2 mm long) or rarely glabrous; calyx 3–6 mm long, largely concealed by involucel, sparsely pubescent; corolla up to 12 mm long, white or pink. *Fruits* 6–9 mm diam., essentially glabrous, longitudinally costate, the mericarp body 6–9 × 3–4 mm, 3-spined, the spines 4–7 mm long, retrorsely barbed, the lateral spines 3–4 mm apart at base.

Distribution (Fig. 74). Widely distributed from Costa Rica to Brazil and Bolivia, often growing in the shaded understory of wet forests, sometimes in disturbed habitats. The type locality for *Pavonia longipes* is at 9°20'N, 82°30'W; that for *P. longipes* var. *hirsuta* is at 08°02'N, 77°40'W.

Representative specimens examined. **HONDURAS**. Wampusirpi, *Clewell 4563* (NY, TEFH).

NICARAGUA. Boaco: Cerro Mombachito, between Boaco and Camoapa, *Stevens et al. 14545* (MO, NY). **Chontales:** 2.8 km N of Cuapa, *Stevens 23382* (MO, NY). **Zelaya:** Río Punta Gorda, Atlanta, *Moreno & Sandino 12794* (MO); near Río Okanwas, 12 km E of Rosita, *Neill 4462* (MO, NY); Mun. Siuna, El Torno, *F. Ortiz 320* (MO).

COSTA RICA. Near San Ramón, *Brenes 14401* (K, NY). **Guanacaste:** Between Guayabo and La Unión, Bagaces, *Jiménez 1149* (F). **Heredia:** La Selva Biological Station, *Grayum 10007* (MO, NY). **Limón:** Cantón de Limón, Valle de la Estrella, *A. Fernández 1258* (MO, TEX); Tortuguero Park, Est. Agua Frío, *Robles 1143* (CR, MO, NY). **Puntarenas:** Finca Las Cruces, San Vito de Java, *Burch 4557* (F). **San José:** Nameless lake at the foot of Fila Lleskila Talamanca, *L. D. Gómez et al. 23113* (CR, NY).

PANAMA. Bocas del Toro: 15 km S of Changuinola, *Antonio 3179* (MO). **Canal Zone:** Barro Colorado Island, *Croat 6608* (F, MO). **Chiriquí:** Santa Clara region, 27 km NW of El Hato del Volcán, *Mori & Bolten 7149* (MO). **Colón:** Guanche River, *Sytsma 1679* (MO). **Darien:** Vic. of Pinogana, *P. H. Allen 931* (GH, MO, US); Tuquesa River, at Charco Peje, *Mori 7073* (MO, NY). **Los Santos:** 10 km SW of El Cortezo, *Hammel 5434* (MO, NY). **Panamá:** Near Torti Arriba, *Folsom 5118* (MO, pf); ca. 25 km NE of Cerro Azul

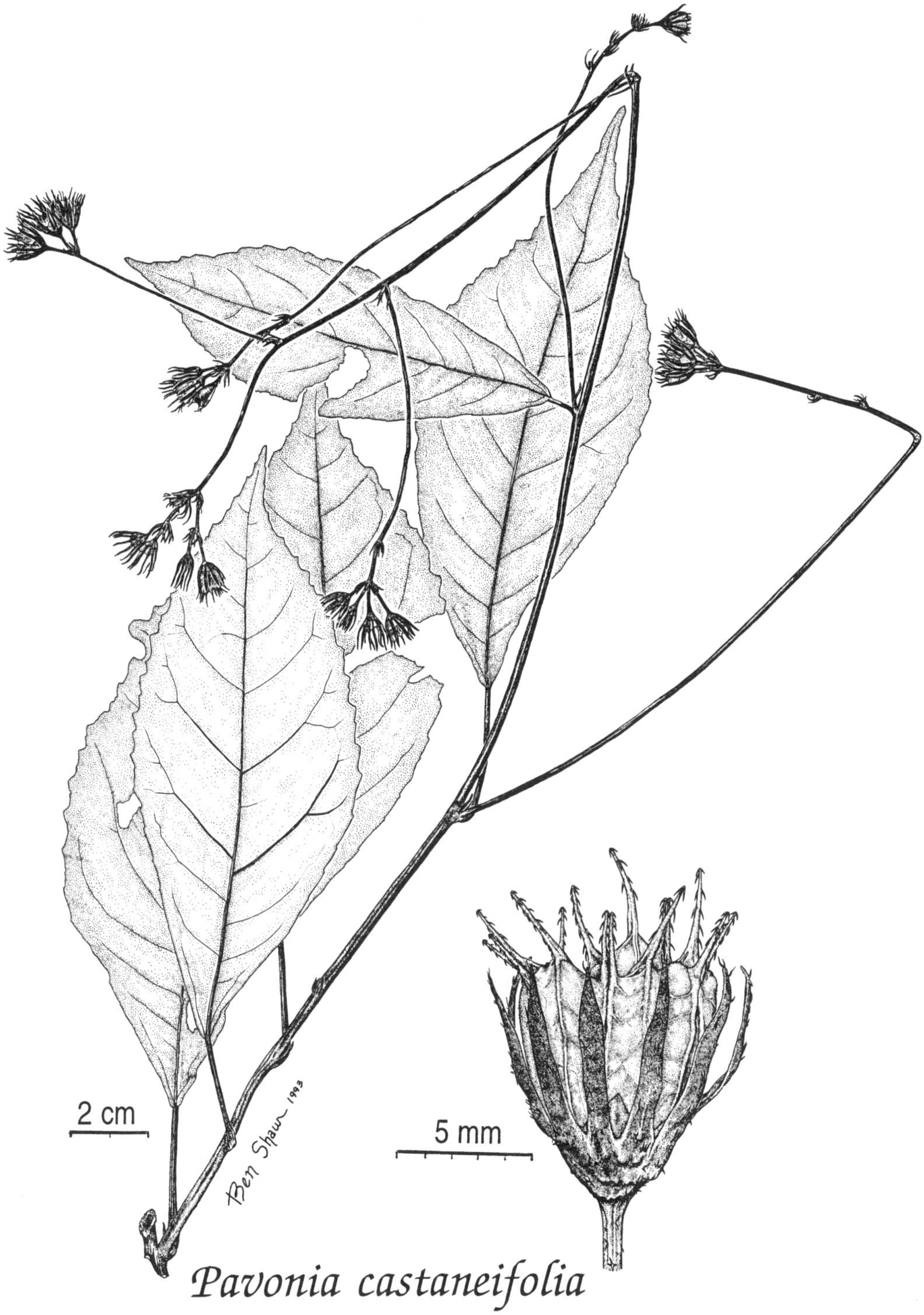

FIG. 72. *Pavonia castaneifolia*. Fruiting branch with enlargement of single fruit (bottom right) (*Cerón & Cerón 2112*).

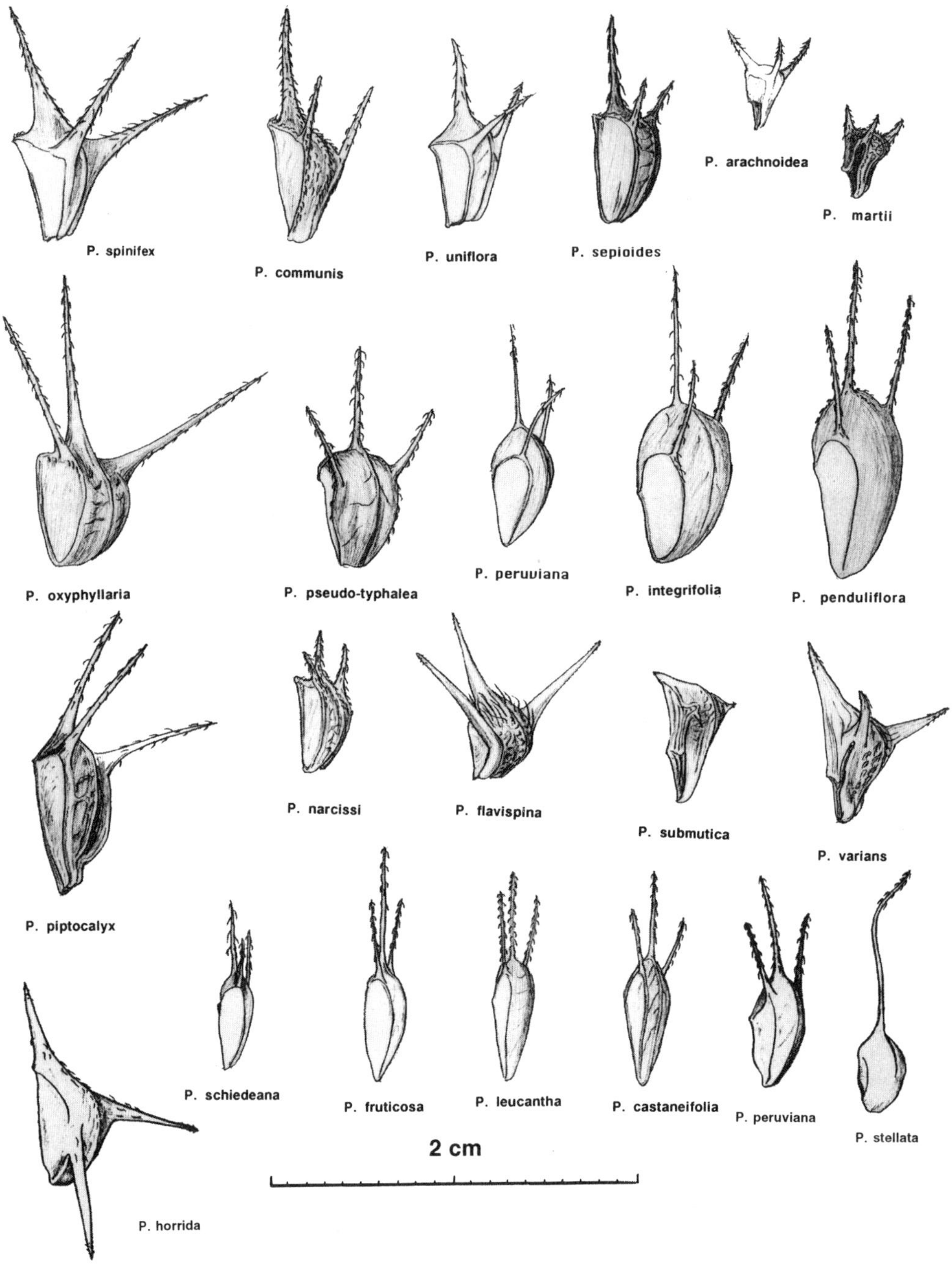

FIG. 73. *Pavonia* subgen. *Typhalea* mericarps in left lateral view. Top row (left to right): *P. spinifex* (*Britton & Wheeler 135*), *P. communis* (*Williams & Assis 6689*), *P. uniflora* (*Conzatti & González 26*), *P. sepioides* (*Hitchcock 20810*), *P. arachnoidea* (*Palmer 153*), *P. martii* (*Irwin et al. 23341*). Second row (left to right): *P. oxyphyllaria* (*Foster & d'Achille 10056*), *P. pseudo-typhalea* (*Lehman 5294*), *P. peruviana* (*Standley 42121*), *P. integrifolia* (*Dressler & Jones 146*), *P. penduliflora* (*Maas & Alpin 1481*). Third row (left to right): *P. piptocalyx* (*Rabelo 1415*), *P. narcissi* (*Balansa 1622*), *P. flavispina* (*Krapovickas & Cristóbal 28650*), *P. submutica* (*Harriman 14399*), *P. varians* (*Fryxell 1929*). Fourth row (left to right): *P. horrida* (*Krapovickas & Cristóbal 3772*), *P. schiedeana* (*Liebmann 1173*), *P. fruticosa* (*Proctor 28806*), *P. leucantha* (*Prance et al. 16790*), *P. castaneifolia* (*Killip & García 33567*), *P. peruviana* (*Williams & Assis 7015*), *P. stellata* (*Gardner 324*).

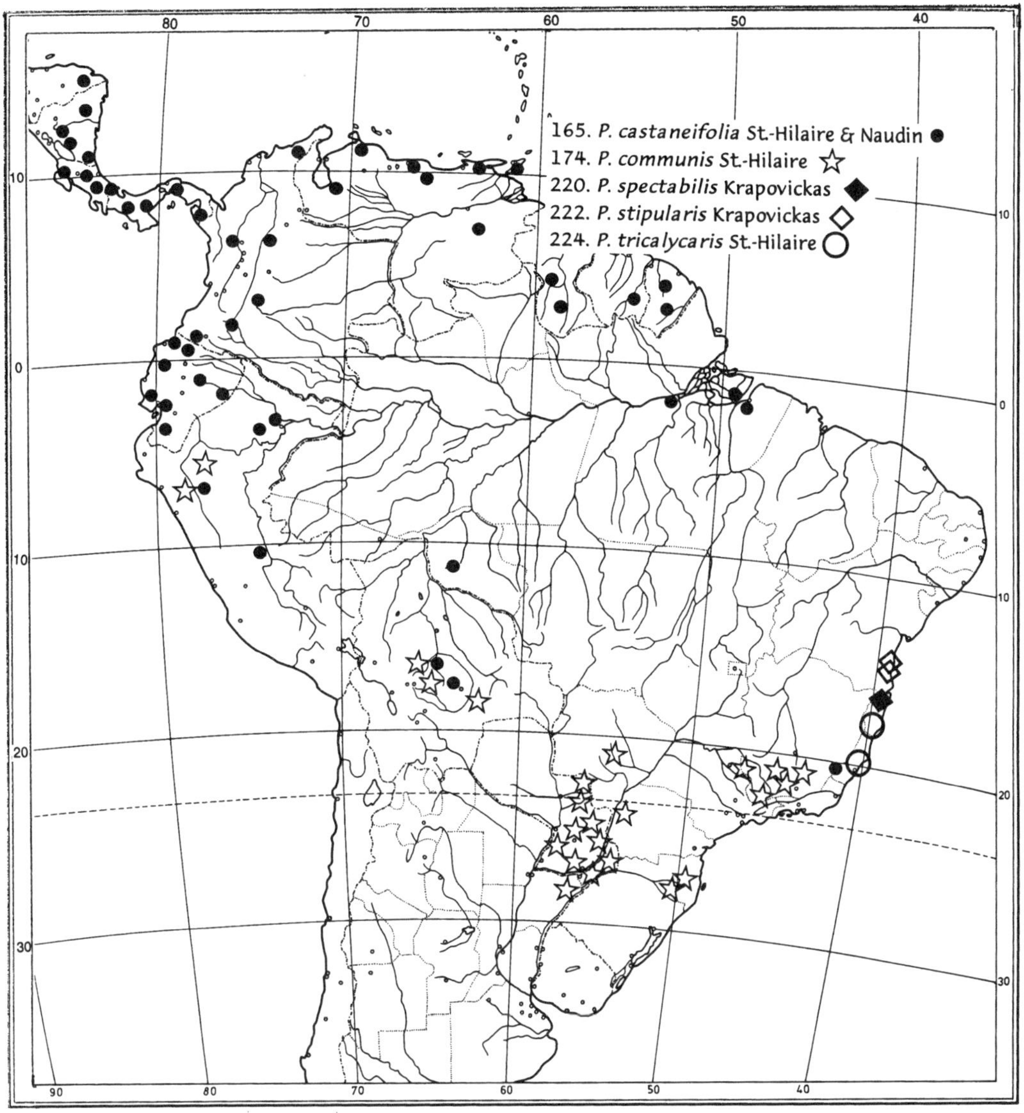

FIG. 74. Distribution map of five neotropical species of *Pavonia* subgen. *Typhalea* and subgen. *Goetheoides*.

on Piedras River, *Mori & Kallunki 3470* (MO, NY). **VERAGUAS:** Cerro Tute, W of Santa Fe, *McPherson 7167* (MO).

COLOMBIA. ANTIOQUÍA: 15 km W of Chigorodó, Río León, 15–20 km S of mouth, *Feddema 1873* (MICH, NY, S, US). **CHOCÓ:** Bahía Solano, *Espina et al. 2721* (NY). **MAGDALENA:** Upper Buritaca River, *Madriñán & Barbosa 401* (MO); Mun. Fundación, 3 leagues N of Santa Rosa, *Romero Castañeda 11174* (MO, NY). **META:** Buenavista, valley of Tigre River, *Fosberg 19538* (CAS, US). **NARIÑO:** Across San Juan River from Maldonado, Ecuador, *Boeke 853* (CTES, NY). **PUTUMAYO:** Mun. Mocoa, corregimiento de San Antonio, *J. L. Fernández et al. 10712* (COL, TEX). **SANTANDER:** Vic. of Puerto Berrio, on Carare River near Puerto Araujo, *Haught 1601* (F). **VALLE:** Zarzal, near Bugalagrande, *Arbeláez & Cuatrecasas 6369* (F, US); near Cauca, *Holton 745* (syntype excl. of *Pavonia peruviana* Gürke: G as photo F-23705, PH, NY); El Tambo, *von Sneidern 435* (F, NY, S). **VAUPÉS:** Guayabero River, *Cuatrecasas 7563* (US).

VENEZUELA. BARINAS: Mun. Ciudad Bolivia, Ticoporo Forest Reserve (08°N, 70°40′W), *A. Castillo 2278* (CTES, MO). **BOLÍVAR:** Paragua River, Casabe Island, *Killip 37306* (US); above the Para Falls, *Steyermark et al. 117190* (CTES); Cerro Coroba, *Wurdack 361* (F, K, NY, US). **FALCÓN:** Sierra San Luis, above La Chapa, *Burandt & van der Werff V0761* (NY, UCO); in pass from San Joaquin to Cueva Coycoy, *P. A. Fryxell et al. 4304* (CTES, NY).

Federal District: El Aguacatal, *Benítez de Rojas 2186* (F, MY); valley SE of Colonia Tovar, *Fendler 116* (GH, NY). **Monagas:** Distr. Caripe, Cerro de los Rastrojos, *Huber & Canales 6331* (CTES, NY).

TRINIDAD. Vic. of Tabaquite, *N. L. Britton et al. 2601* (GH, NY, US); heights of Aripo, *Broadway 9791* (F, GH, NY, US).

GUYANA. Kanuku Mtns., Maipaima, on Tsikoma Creek, *Jansen-Jacobs et al. 948* (K, MO, NY, U, US); Surama village, *McDowell et al. 1890* (NY, U, US).

SURINAM. Near Pommier, along the Surinam River, *P. A. Florschütz & J. Florschütz 282* (U); Kayserberg-airstrip, on the Zuid River, *Kramer & Hekking 2984* (U).

FRENCH GUIANA. Mont Galbao, *J.-J. de Granville et al. 8598* (MO, NY, US); Mont Atachi Bacca, Ininí region, *de Granville et al. 10488* (MO, NY, U); Camopi River, *Oldeman & Sastre 132* (CTES, NY).

ECUADOR. Esmeraldas: Quinindé (Rosa Zárate), *Asplund 16315* (S); Zapallo Grande, *Barfod et al. 48204* (AAU). **Carchí:** Between Chical and Peña Blanca (on Colombia border), *A. H. Gentry & Shupp 26460* (MO, NY, SEL). **Cotopaxi:** La Mana Cantón, 13 km N of La Mana, *Croat 73752* (MO, TEX). **Guayas:** Palme, between Balzar and Quevedo, *Fagerlind & Wibom 343* (S). **Loja:** Alamor–Cazaderos rd. just W of El Limo, *Harling & Andersson 22332* (GB, NY, QCA). **Los Ríos:** Hacienda Clementina on Pita River, *Asplund 5309* (S); 12 km E of San Carlos, *Sparre 19428* (S). **Manabí:** Vic. of Agua Blanca, ca. 15 km from Puerto López, *Hekker & Hekking 10244* (NY, QCA, U). **Napo:** Jatun Sacha Biological Reserve, *C. E. Cerón et al. 2112* (MO, NY, TEX); Cantón Putumayo, Aguarico River, Dureño, *Plowman et al. 4010* (GH, S). **Napo-Pastaza:** Near Canelos, *Mexia 6899* (US). **Pastaza:** 1 km E of Topo between Baños and Mera, *Palacios et al. 206* (MO, NY, QAME, QCNE); Sarayaquillo, ca. 10 km E of Sarayacu, *Lugo 4020* (GB, MO). **Pichincha:** 11 km W of Tandapi, *A. H. Gentry et al. 12115* (AAU, MO, NY, QCA); Maquicupuna Biological Reserve, *C. E. Cerón 12406* (MO, QCNE). **Santiago-Zamora:** Opposite Chupiantza, *Camp E-1044* (CAS, NY). **Tungurahua:** Baños–Puyo rd. near Río Negro, *A. H. Gentry & Bonifaz 28745* (MO, NY); vic. of Margarjitas River on Canelos trail, *Penland & Summers 161* (COLO).

PERU. Amazonas: Village of Kusu, Numpatakai River (tributary of Cenepa), *Berlin 924* (GH). **Huánuco:** Tingo María, *Asplund 12939* (S, US). **Loreto:** Santa María, 25 km below Yurimaguas on Huallaga River, *Allard 22477* (US); Prov. Maynas, 2 km W of Indiana on Amazon River, *A. H. Gentry et al. 55941* (MO); hwy. from Iquitos to Nauta, *R. Vásquez et al. 11935* (MO, NY, TEX). **Pasco:** Prov. Oxapampa, Palcazu Valley, *D. N. Smith 3992* (K, MO, NY). **San Martín:** San Roque, *L. Williams 7637* (F).

BRAZIL. Amazonas: Mun. Borba, across from Canumã, *Hill et al. 12864* (F, K, NY, US). **Espírito Santo:** Mun. Cachoeira do Itapemirim, Estação Experimental do Norte, *Kallunki et al. 613* (MO, NY, TEX). **Pará:** Mocoões River, 45 min. by boat below Frances, *H. T. Beck et al. 390* (INPA, MG, NY, TEX); Maicuru River, Igarapé do Mutum, *Jangoux & Ribeiro 1560* (NY). **Rondônia:** Vida Nova, *Duarte 7175* (CTES); rd. from São Luiz to Maloca Tupari-Buyitará (Territ. Guaporé), drainage of the Rio Branco, tributary of the Guaporé, *Scolnik & Luti 732* (NY).

BOLIVIA. Cochabamba: Valle de Sacta Project Station, *D. N. Smith et al. 12972* (LPB, MO, NY). **La Paz:** Prov. Nor Yungas, at the bridge over the Beni River, 9 km toward Litoral, *S. Beck 13363* (CTES).

Local name. Wíawía (Peru).

Pavonia castaneifolia is closely related to *P. fruticosa*, *P. nemoralis*, *P. parva*, *P. peruviana*, *P. schiedeana*, and *P. stellata*. It differs from *P. fruticosa* in having more (8–11), narrower involucellar bracts, prominently veined (vs. smooth) carpels, and less crowded spines at the apex of the mericarps. It differs from *P. nemoralis* in having prominently veined (vs. smooth) carpels and less crowded spines at the apex of the mericarps but shares the relatively large number of narrower involucellar bracts. It differs from *P. peruviana* in lacking appressed simple hairs on the upper leaf surface but shares the prominently veined carpels. It differs from *P. schiedeana* in having larger flowers and fruits and in having prominently veined (vs. smooth) carpels. It differs from *P. stellata* most obviously in that the latter has only a solitary spine (rather than 3) at the apex of the carpel. It differs from *P. parva* by the characters given in the key.

166. Pavonia fruticosa (Miller) Fawcett & Rendle, Fl. Jamaica **5:** 130. 1926. *Sida fruticosa* Miller, Gard. Dict., ed. 8, 1768. *Typhalea fruticosa* (Miller) Britton, Sci. Surv. Porto Rico, Virgin Islds. **5:** 560. 1924. Type. In cultivation, *P. Miller s.n.* (holotype, BM-as photo NY-257). Fig. 73

Urena typhalea Linnaeus, Mant. Pl. 258. 1771. *Pavonia typhalea* (Linnaeus) Cavanilles, Diss. **3:** 134. 1787. *Pavonia typhalea* var. *cavanillesii* Triana & Planchon, Ann. Sc. Nat. Bot., ser. 4, **17:** 159. 1862 [≡ var. *typhalea*]. *Pavonia fruticosa* var. *typhalea* (Linnaeus) Stehlé, Bull. Soc. Bot. France **92:** 263. 1945. Type. "Habitat in Surinam, Jamaic." (specimen unknown).

Hibiscus salicifolius Linnaeus, Pl. Surin. 12. 1775. *Urena leptocarpa* Linnaeus fil., Suppl. 308. 1781, nom. superfl. *Pavonia leptocarpa* (Linnaeus f.) Cavanilles, Diss. **3:** 351. 1787. *Diplopenta leptocarpa* (Linnaeus fil.) Alefeld, Oesterr. Bot. Ztschr. **13:** 11. 1862. *Malache leptocarpa* (Linnaeus fil.) Kuntze, Revis. Gen. Pl. **1:** 70. 1891. Type. Surinam. *Dahlberg s.n.* (lectotype, LINN-875.42 as microfiche; isotypes, LINN-875.41 as microfiche, S as microfiche).

Hibiscus guianensis Aublet, Hist. Pl. Guiane. **2:** 708. 1775. *Pavonia guianensis* (Aublet) Hochreutiner, Rev. Hibiscus 151. 1900. Type. Guyana. "ad marginas lucosum sylvarum insule Cay," *Richard s.n.* (P –n.v., as photo CTES).

Pavonia typhaleoides Humboldt, Bonpland & Kunth, Nov. Gen. Sp. **5:** 279 (folio ed., p. 217). 1822. *Malache typhaleoides* (Humboldt, Bonland & Kunth) Kuntze, Revis. Gen. Pl. **1:** 71. 1891. Type. "Novo-Granatensium, Santa Ana," *Humboldt & Bonpland s.n.* (holotype, P-HBK as microfiche and as photo CTES).

Pavonia glomerata Casaretto, Nov. Stirp. Bras. 38. 1842 (fide Gürke). Type. Brazil. Bahia: In hills near Cachoeira and S. Amaro, *Casaretto s.n.* (holotype, TO–n.v.).

Pavonia surinamensis Miquel *in* Hoeven & DeVriese, Tijdschr. Nat. Geschied. **10:** 77. 1843. Type. Surinam. *Focke 441* (holotype, U; isotype, U).

Pavonia brachypoda Turczaninow, Bull. Soc. Imp. Naturalistes Moscou **36:** 563. 1858. Type. Jamaica. near St. Marys, May 1840, (*ex* herb. Graham) *McNab s.n.* (holotype, KW; isotype, E). Fryxell & Krapovickas (1990) incorrectly stated that the type of *P. brachypoda* Turczaninow was collected by Graham and the specimen was unknown. In fact, the holotype is in KW and is cited correctly above.

Shrubs 1–1.5 m tall, the stems minutely stellate-pubescent, the hairs appressed. *Leaf* blades mostly 10–22 × 2–10 cm, ovate, lance-ovate, elliptic, or oblanceolate, penninerved or basally 3-nerved, truncate to cuneate, serrate (at least distally), more or less acuminate, minutely stellate-pubescent to glabrate; petioles 1–6 cm long (reduced upward); stipules 8–9 mm long, lanceolate-acuminate. *Inflorescences* long-pedunculate, capituliform, terminal, crowded and subtended by several ovate-acuminate bracts 5–10 mm long, the flowers 8–15 per head, short-pedicellate to subsessile; involucellar bracts 4–6, 7–8 × 2–4 mm, basally connate; calyx 5–6 mm long; corolla ca. 1 cm long, white fading pink. *Fruits* 5–6 mm diam., the mericarp body 5–6 mm long, glabrous, longitudinally costate (otherwise smooth and often shiny, with a distinctive pale, yellowish tan color), the mericarps 3-spined, the spines erect, 4–6 mm long, retrorsely barbed, crowded (the lateral spines ca. 1 mm apart at base). Chromosome number, $2n = 28$ (Ford, 1938; Skovsted, 1941 [identity not verified]).

Distribution (Fig. 75). Widely distributed from Costa Rica and the West Indies to Brazil and Bolivia. Giménez et al. (1993) discuss the unpublished manuscript "Flora de Guatemala" by Mociño, in which the name "Vrena typhalaca" appears (op. cit., p. 28). This seems to be a clear reference to the Linnaean binomial *Urena typhalea* and further appears to be a northward extension of this species to Guatemala. However, no other evidence is available for the occurrence of this species N of Costa Rica, and this is undoubtedly a case of misidentification, probably in reference to the common and widespread *Pavonia schiedeana* (q.v.). The type locality for *P. glomerata* is at 12°33′S, 38°52′W.

Illustrations. Cavanilles (1788: t. 197); Gürke (1892b: pl. 87, fig. 1); Robyns (1965: fig. 6E).

Representative specimens examined. **COSTA RICA.** PUNTARENAS: Corcovado National Park, *Delprete 5119* (MO, TEX). SAN JOSÉ: Vic. El General, *Skutch 4342* (F, GH, S).

PANAMA. CANAL ZONE: Between locks and Fort Sherman, *Croat 15368* (MO, US); 2–3 mi N of Gamboa, *Knapp 1050* (MO, NY); near Petitpie River, rd. from Gatún to Fort Sherman, *Mori 6927* (MO, NY). CHIRIQUÍ: Boquete, *Ebinger 758* (MO); San Bartolo Arriba, W of Puerto Armuelles, *Croat 26723* (MO). COCLÉ: Vic. of El Valle, *Bartlett & Lasser 16667* (MO). COLÓN: Ft. Sherman, 2 mi SW of Ft. San Lorenzo, *Antonio 1957* (MO, NY); near bridge over Guanche River, *Antonio 4798* (MO, NY); behind Portobello, *Croat & Porter 15605* (MO). DARIEN: Vic. of Canglon, *Antonio 4521* (CTES, MO, NY); Pucra to Pucra River, *Duke 5367* (GH, MO); Choco village, Piji Vasal, *Folsom 4567* (MO, NY); village of Mannene, *Kirkbride & Bristan 1574* (MO, NY); vic. of Cana, *Stern et al. 469* (GH, MO, US). PANAMÁ: Panama City on rd. to Tocumen, *W. G. D'Arcy & J. J. D'Arcy 6111* (GH, LL, MO, NY); rd. between Panamá and Chepo, *Dodge et al. 16668* (K, MO, U); 6 mi E of Chepo, *Duke 4084* (GH, MO); Río Tataré, *Woodson & Schery 1014* (MO, NY). SAN BLAS: SE of Puerto Obaldia, *Croat 16743* (MO); Aligandai River, *Duke 10846* (MO). VERAGUAS: Montijo, *Cisneros 28* (MO).

CUBA. "Cuba orientali," *Wright 33* (GH, MO, NY, PH, S, US); La Luisa in Monte Yoro, *Wright 2063* (GH, MA, MO). CIENFUEGOS: Limones, Soledad, *Jack 5464* (A); Hanabanilla Falls, *N. L. Britton et al. 4868* (NY); Mina Carlota, SE of Cumanayagua, Sierra de San Juan, *Hodge & Howard 4497* (CAS, GH, NY); San Blas, La Sierra, *Jack 5916* (A, POM). LA HABANA: Calvario, *Bro. León 1457* (NY); Anafe, near Ariguanabo, *Ekman 1035* (S). PINAR DEL RÍO: Sierra del Rosario, Loma del Salon, *Acevedo et al. 5663* (NY, US); next to Soroa, 4–5 km N, *Fernández-Casas & Morales 10745* (MA, MO, NY); vic. of Sumidero, *Shafer 13404* (GH, MO, NY, US); Herradura, *Van Hermann 286* (F, NY). SANTIAGO DE CUBA: Ensenada de Mora, *N. L. Britton et al. 12987* (NY); valley of the Bayamita River, N slope of Sierra Maestra, *Maxon 3934* (F, MO, NY, US); forest about Paso Estancia, *Shafer 1569* (F, NY, US).

JAMAICA. Vic. of Mandeville, *Crawford 749* (NY, PH); near Port Antonio, *Fredholm 3110* (NY, US); Mansfield, near Bath, *Maxon 1848* (US). ST. THOMAS: Along Sulphur River above Bath Fountain, *Proctor 28806* (LL, NY, U).

HAITI. Massif de la Selle, Port-au-Prince, N of l'Hôpital, *Ekman 8180* (K, NY, S, US); vic. of St. Michel de l'Atalaye, Dept. du Nord, *E. C. Leonard 7561* (CM, MO); vic. Kalacroix, Dept. de l'Artibonite, *E. C. Leonard 7975* (NY, US); Bayeux, near Port Mergot, *Nash 70* (F, NY, US).

DOMINICAN REPUBLIC. Barahona, *Fuertes 618* (A, GH, K); 6 mi NW of Paraíso, *S. A. Thompson et al. 7531* (CM). MONTE CRISTI: Lagunas de Cenobí, Distr. Sabaneta, *Valeur 39* (POM, NY, S, US). SAMANÁ: Vic. of Sánchez, Samaná Peninsula, *Abbott 45* (US); 8 km W of Boca del Infierno, *Zanoni et al. 34128* (JBSD, NY, TEX). SANTO DOMINGO: Near Mamayes River, *von Eggers 1519b* (S). SAN CRISTÓBAL: Arroyo del Oro, *Mejía et al. 797* (JBSD, NY, TEX). SAN JUAN: 9.4 km N of Cano Creek, *García & Pimentel 1193* (JBSD, MO, NY, TEX). SANTIAGO: San José de las Matas, *Valeur 1004* (C, K, LL–2, US). LA VEGA: Cotuy, *Abbott 841* (GH); vic. of Piedra Blanca, *Allard 13832* (S, US).

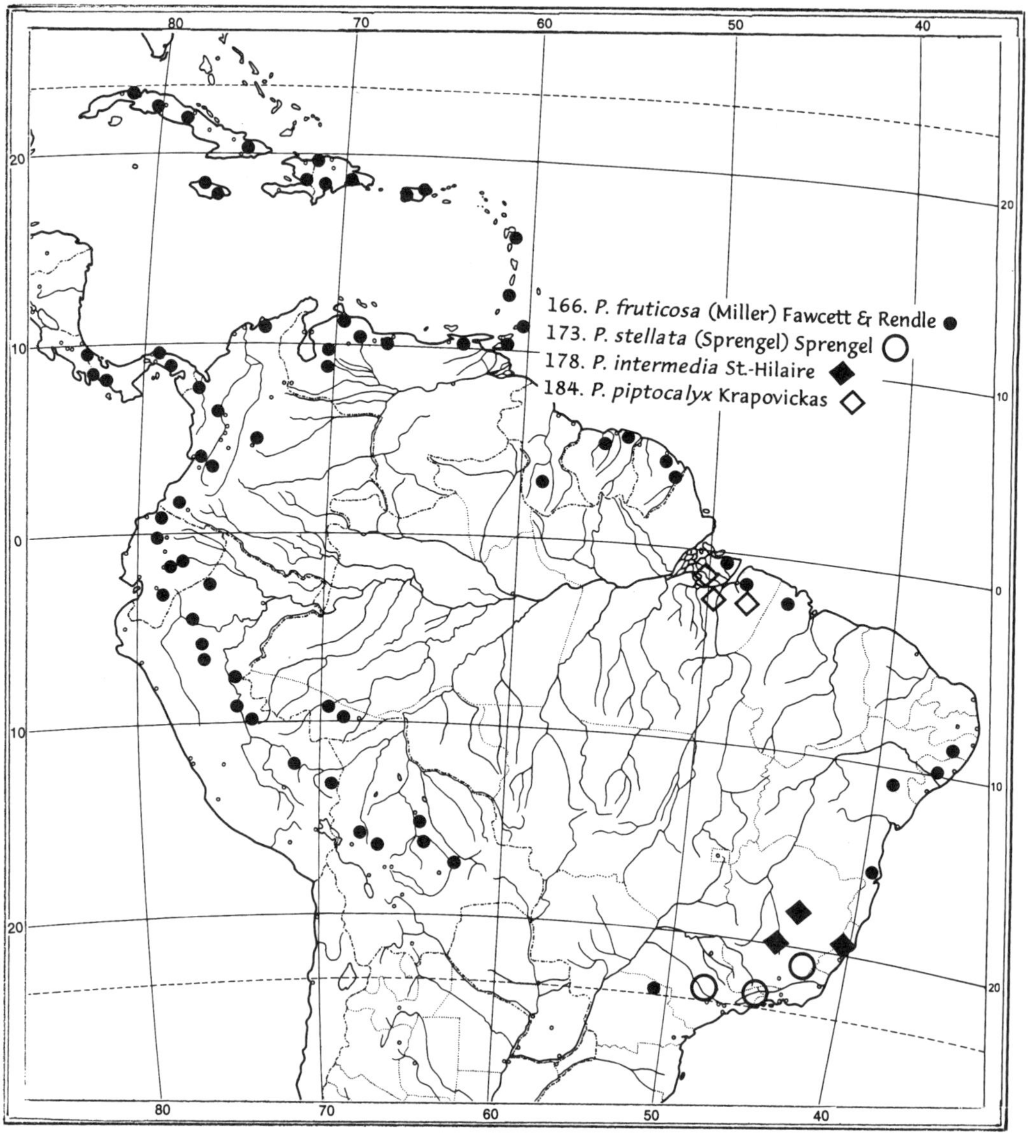

FIG. 75. Distribution map of four neotropical species of *Pavonia* subgen. *Typhalea*.

PUERTO RICO. Caribbean National Forest, *Boom 7051* (NY, TEX); Lares, *N. L. Britton et al. 2738* (NY, US); near Utuado, *N. L. Britton & Cowell 373* (NY, US); Yauco, *Garber 53* (GH, NY, PH); near Mayaguez, *Heller 4442* (A, E, F, GH, MO, NY, PH, POM, US); Sierra de Naguabo, Icaco River and adjacent hills, *Shafer 3473* (NY, US); near Mayaguez, *Sintenis 55* (C, F, GH, S, US).

WINDWARD ISLANDS. DOMINICA: Grande Savane between Salisbury and Batali River, *Ernst 2123* (TEX, US); near Sindicate, *Whitefoord 3945* (A, US). **ST. VINCENT:** Vic. of Kingston, *Morton 4893* (GH, US); without locality, *H. H. Smith & G. W. Smith 1359* (GH, NY).

COLOMBIA. Gorgona Island, *Harcourt 537* (K). **ANTIOQUÍA:** 3 km W of Santa Fé de Antioquía, *Barkley et al. 279* (S, US); Mun. Turbo, 11 km from Currubao, *Callejas et al. 4994* (HUA, MO, NY, TEX). **BOLÍVAR:** Quimari, *von Sneidern 5813* (A, LL, NY, S). **BOYACÁ:** Region of Mt. Chapon, NW of Bogotá, *Lawrance 181* (A, F, K, MO, NY, S, TEX, US). **CALDAS:** Sta. Cecilia, Pueblo Rico, *von Sneidern 5130* (A, CAS, F, LL, MO, NY, S, US). **CHOCÓ:** Mun. Pizarro, km 21–24 of hwy. from Pie de Pepé to Berrecul, *Espina 1890* (CHOCO, MO, NY); San Juan River, below Noanamá, *Forero et al. 4858* (COL, MO, NY); W of Unguía, base of Serranía del Darien, *A. H. Gentry & Aguirre 15231*

(MO, NY); Mun. Riosucio, Tilupo, *Romero-Castañeda 6259* (MO, NY). **Cundinamarca:** W of Guaduas, hacienda "Palmilla," *García-Barriga 12352* (US); Guaduas, Alto de Ficalito, *Uribe 2375* (NY). **Los Llanos:** Meta River, María, *Cuatrecasas 3769* (F, US). **Meta:** Villavicencio, *Alston 7621* (GH, S, US). **Magdalena:** Santa Marta, *H. H. Smith 846* (CAS, CM, F, GH, K, MO, NY, PH, S, US); Rincón Hondo, Magdalena Valley, *C. Allen 255* (K, MO). **Nariño:** Mun. Tumaco, Vereda la Sirena, *Madrigal & Preciado 836* (MO); Mun. Tumaco, between La Sirena and Palambi, *Madrigal & Caicedo 617* (HUA, NY, TEX); Ricáurte, *von Sneidern A521* (A, LL, NY, S). **Norte de Santander:** Between La Quebrada de Gibraltar and La Palma, *Cuatrecasas 13250* (US). **Santander:** Jordan, 10 km SSE of Landazari, *Ewan 15670* (CAS, US); Jordan, 50 km N of Velez, *Fassett 25268* (CAS, US). **Tolima:** La Trinidad, Libano, *Pennell 3342* (NY, US). **Valle:** Hills near Alcalá, *Cuatrecasas 22868* (F, NY, S, US); Mun. Zarzal, between Caserio La Paila and Zarzal, *A. H. Gentry 54130* (MO, NY); Mun. Cartago, rd. from Piedras de Moler to Coloradas, *Silverstone-Sopkin 1013* (MO, NY, TEX).

VENEZUELA. Aragua: La Victoria, *Holt 264* (CM, MO); near Rancho Grande, *Killip & Lasser 37787* (US); Ocumare, *Krapovickas 15583* (CTES, NY). **Barinas:** Near Barinitas, *Breteler 4175* (F, MO, NY, S, U, US); 15 km NW of Barinitas, *Aymard & Moran 2972* (CTES, PORT). **Bolívar:** Las Trincheras, Caura River, *Burandt & Clark V1216* (CTES, MO, NY). **T.F. Delta Amacuro:** Between Tucupita and Los Güires, *G. Agostini & T. de Agostini 1618* (NY, TEX, U, US, VEN). **Distrito Federal:** Near Tovar, *Fendler 117* (GH, MO, PH, US); Caruao Valley, *Pittier 11836* (A, NY, US). **Monagas:** Caripe, *Trujillo 9455* (MY, NY). **Falcón:** Sierra de San Luis, above Santa María, *Wingfield & Lemus 7714* (CORO, NY). **Miranda:** Banks of the "Los Chorros" Creek, *Tamayo 192* (US). **Portuguesa:** Caserío Villa Rosa, 20 km E of Biscucuy, *Aymard et al. 3643* (CTES); Dept. Araure, Camburito, *Ortega & Aymard 1662* (PORT, NY). **Sucre:** Between Quebrada Seca and Bohordal, *Burandt & Clark V1228* (MO, NY); 25 km NW of Irapa, *Dumont et al. 7408* (NY). **Tachira:** Hacienda Santa Teresa, old hwy. from San Cristóbal to Rubio, *Bunting & Medina 36* (NY). **Trujillo:** Buena Vista–Monte Carmelo, *Pérez-Levis 41* (F). **Yaracuy:** 5 km S of Bella Vista, *Agostini et al. 1770* (NY, VEN); 25 km NW of San Felipe, *Aymard et al. 2649* (CTES, PORT); slopes of Cerro Negro, vic. of San Felipe, *Trujillo 4284* (MY, U).

TRINIDAD. Dabadie, *N. L. Britton 2487* (GH, NY, US); Tamana Rd., *Kalloo B500* (NY). **Tobago:** Parlatuvier, *Sandwith 1885* (K, NY, U).

GUYANA. Akyma, Demerara River above Wismar, *Hitchcock 17445* (GH, NY, S, US); 1 km S of Tipu village, *McDowell & Gopaul 2584* (NY, TEX, U); on the upper Essequibo, *Schomburgk 137* (GH, K, MO, OXF, US).

SURINAM. Ca. 10 km S of Paramaribo, *Kramer & Hekking 2615* (C, LL); Paramaribo, near Agric. Exper. Sta., *Lanjouw 5* (NY, U); N of Charlesburg, *Stahel 35* (NY, U).

FRENCH GUIANA. Cayenne Island rd. from Matoury to Trio, *Granville 4993* (CTES, U); Cayenne, 1859, *Sagot s.n.* (U); Approuague River, Pierrette, *Sastre 6356* (CTES, U).

ECUADOR. El Oro: Rd. from Piñas to Santa Rosa, above El Placer, *Harling et al. 15505* (GB, NY). **Esmeraldas:** Cupa, *Poveda 26* (QCA, NY). **Los Ríos:** Montalvo, 40 km E of Babahoyo, *Holm-Nielsen et al. 2703* (AAU, CTES, F, GB, MO, NY, S). **Morona-Santiago:** 10 km S of Sucua, Upano River, *P. Gómez 415* (NY, TEX); vic. of Pan de Azúcar, E of Indanza, *Harling 26008* (GB, NY). **Napo:** Lower Aguarico River, *Harling et al. 7387* (AAU, GB, QCA, NY, US); between Coca and Armenia Vieja, *Harling & Andersson 11990* (AAU, GB, NY, TEX); Jatun Sacha Biological Station, *Bennett & Alarcón 4472* (NY, TEX). **Napo-Pastaza:** Tena, *Asplund 9007* (CAS, K, NY, S, TEX, US). **Pichincha:** Along the Toachi River, *Játiva & Epling 349* (AAU, MO, NY, UC); rd. from Aloag to Santo Domingo, Alluriquin, 600 m, *Sparre 14809* (S, US); Corazón Volcano, *Zak 1499* (MO, NY). **Santiago-Zamora:** Near Chupiantza, Valley of the Upano River, *Camp E-1418* (CAS, NY); Zamora to Yanzasa, *Mathias & Taylor 5227* (F, MO). **Zamora-Chinchipe:** N side of Palanda River at crossing with Zumba Rd., *Harling & Andersson 21272* (GB, MO, NY, QCA).

PERU. Amazonas: Huampami, 54 km E of Chávez Valdivia, *Ancuash 1470* (MO, NY); Cenepa River, Chigkan Island, *Kayap 755* (MO, U). **Cuzco:** Prov. Carabaya, Ollachea, Hda. Quillabamba, *Vargas 17542* (US); Sahuayaca, *Vargas 6308* (F). **Huánuco:** Rd. to Monosn, near Tingo María, *Ferreyra 12664* (US); Pachitea, Puerto Inca, *Foster 8833* (MO); Leoncio Prado, across Huallaga River from Tingo María, *A. H. Gentry et al. 15931* (AMAZ, F, MO, NY). **Junín:** Pinedo River, N of La Merced, *Killip & A. C. Smith 23582* (NY, US); Puerto Yessup, *Killip & A. C. Smith 26226* (NY, US). **Loreto:** Peña Negra, rd. from Iquitos past Quisto Cocha, *Davidson 5345* (MO, NY, RSA, US); Prov. Alto Amazonas, Yurimaguas, *A. H. Gentry et al. 52119* (MO, NY); Maucallacta, Paranapura River, *Klug 3953* (CAS, F, GH, K, MO, NY, S); Washintsa, *W. H. Lewis & M. Elvin-Lewis 11529* (MO, NY); Shucushuyacu, *R. Vázquez & N. Jaramillo 2378* (MO, NY). **Madre de Dios:** Prov. Manu, Cerro de Pantiacolla, *Foster et al. 11042* (F, NY); 39 km SW of Puerto Maldonado, *S. F. Smith et al. 1534* (K, US). **San Martín:** Pongo de Cainarachi, *Klug 2732* (F, GH, MO, NY, S); Prov. Mariscal Cáceres, Santa Rosa de Mishollo, *Schunke-Vigo 6818* (AMAZ, MO, NY, US); Puerto Pizana (Huallaga River), *Schunke-Vigo 4619* (F, GH, K, MO, NY, US). **Ucayalí:** Contamana, Maquia River (07°20′S, 75°05′W), *R. Vásquez & N. Jaramillo 1679* (MO, NY).

BRAZIL. Acre: 22 km from Rio Branco, Porto Velho Rd., *de Albuquerque et al. 1299* (GH, MO, NY, TEX); Mun. Sena Madureira, 10 km above Sena Madureira, Capoeira, *Prance et al. 7804* (CTES, F, GH, K, NY, S, U, US). **Alagoas:** Mun. Marechal Deodoro, APA de Santa Rita, Oiteiro, *Esteves & Lyra-Lemos 2043* (K). **Amazonas:** Esperança, *Murça-Pires & Black 904* (GH); banks of Jarí River, *N. T. Silva 1554* (CTES, K, NY, S, US). **Bahia:** Hwy. from Itajuípe to Ubaitaba, 12 km N of Barro Central, *Belém & Magalhães 885* (NY); Mun. Ilhéus, CEPEC area, km 22 on hwy. from Ilhéus to Itabuna, *Hage & Brito 636* (CTES); Uruçuca, Pancada Formosa, *Krapovickas 12763* (BAA, CTES, NY, UC). **Goiás:** Without locality, *Heringer et al. 16767* (CTES, MO). **Maranhão:** Alzilândia, Pindaré

River, *Jangoux & Bahia 339* (NY). **Pará:** Mun. Barcarena, Onças Island, *A. B. Anderson et al. 1075* (K, NY, US); Belém, *Archer 7842* (F, K, NY, S, US); Pará, *C. F. Baker 288* (K, POM, S, U); Itaituba, Tapajos National Park, *M. G. Silva & Rosario 3747* (MG, NY, TEX). **Paraná:** Near Caioba, *Lindeman & de Haas 3814* (U). **Pernambuco:** Recife, Dois Irmãos, *Krapovickas et al. 38079* (CTES, G, IPA, NY). **Rondônia:** Rd. from Porto Velho to Cuiabá, Trecho Jarú–Ouro Preto, *Freitas et al. 142* (MO).

BOLIVIA. Yapacauí, 1892, *Kuntze s.n.* (US); Cachuela Esperanza, *White 1799A* (K). **Beni:** Prov. Ballivián, Espiritu (Yacuma River), *S. Beck 5377* (CTES, US); Prov. Yacuma, Beni Biological Station, *Moraes 926* (US). **Cochabamba:** Near Chimore, 143 km NE of Cochabamba, *Eyerdam 24700* (F). **La Paz:** Yungas, *Bang 545* (E, F, GH, MO, NY, PH, US); Milluguaya in N Yungas, *Buchtien 270* (E, F, GH, NY); Polo-Polo near Coroico, N Yungas, *Buchtien 6023* (MO, US); Prov. Murillo, vic. of Cahua Hydroelectric Plant, *Solomon 12981* (MO, NY). **Pando:** Nicolás Suárez, near Cobija, *Fernández-Casas 8097* (MO, NY). **Santa Cruz:** Hacienda Simaco, on the rd. to Tipuani, *Buchtien 5458* (E, F, GH, MO, S, US); Gral. Saavedra, *Krapovickas & Schinini 31495* (C, CTES, F, G, SI); Mapiri, *Rusby 1487* (GH, NY, PH, TEX, US).

Local names. Abrazo (Colombia), anamú (Colombia, Puerto Rico), cadillo (Dominican Republic), cadillo de cacao (Dominican Republic), cadillo de perro (Dominican Republic), cadillo de tres pies (Dominican Republic), cadillo pequeño (Puerto Rico), carrapicho de brejo (Brazil: Pará), chicraten nupa (Peru, *Jívaro*), chucchuc pango (Ecuador, *Quechua*), escobilla (Colômbia), embiandé (Colombia), fierébrae (Surinam, Arawak), hoja de paludismo (Ecuador), imbiandi (Colombia), ismiandé (Colombia), kromantie (Surinam), malva carrapicho (Brazil), mapa huasca (Peru, Quecha), misnaga (Panamá), pega pega (Panamá), piojo de vaca (Peru), polatatai (Surinam), rosita quihua (Peru), tabadillato (Panamá, Chocó), tábano (Cuba), weawea (Peru), zavoine (Haiti), zelosa (Colombia).

Pavonia fruticosa is distinctive for its few (4–6) broad involucellar bracts, for its smooth (often yellowish and lustrous) carpels, and for its crowded spines at the apex of the mericarp, a character that it shares with *P. nemoralis*.

167. Pavonia integrifolia Standley, Publ. Field Columbian Mus., Bot. Ser. **8:** 24. 1930. Type. Mexico. Veracruz: Zacuapán, Apr 1927, *Purpus 11118* (holotype, F as photo F-56166; probable isotypes, MO, NY). Note: The holotype at F is dated Apr 1927; the isotype at MO is dated Apr 1928; the isotype at NY is dated May 1928; collection numbers agree. Fig. 73

Subshrubs 0.5–1 m tall, the stems with some stellate hairs, the nodes very short (5 mm or less). *Leaf* blades oblanceolate, up to 22 × 8 cm, basally truncate, essentially entire (or with minute serrulations), acuminate, slightly discolorous, 3-nerved from the base but appearing penninerved, glabrate above, nearly so beneath, the margins hispid-ciliate; petioles 2–5 cm long, sparsely stellate-pubescent to glabrate; stipules 10–14 mm long, subulate, erect. *Inflorescence* a leafless open raceme or panicle, each node with 2–3 stipuliform bractlets; pedicels 2–10 cm long, sparsely stellate-pubescent, articulated 6–10 mm below the flowers; involucellar bracts ca. 9, triangular, 6–9 mm long, basally connate forming a cup, the sinuses broadly rounded, sparsely stellate-pubescent; calyx subequal to involucel, pubescent; corolla 12–20 mm long, pink; staminal column 7–8 mm long, glabrous, pallid, the filaments 1–2 mm long. *Fruits* 9–13 mm diam.; mericarp body 7–8 mm high (excluding spines), straw-colored, essentially glabrous, 3-spined, the spines erect, 4–7 mm long, retrorsely barbed.

Distribution (Fig. 76). From S Mexico (Veracruz and Chiapas), Guatemala, and Honduras, in shady understory of wet forest at 300–1500 m elevation.

Illustration. Fryxell (1979: fig. 2D).

Specimens examined. **MEXICO. Chiapas:** Mun. Ocosingo, ca. 45 km E of Ocosingo, *Dressler 1560* (GH, NY, US); 3 km NW of Pueblo Nuevo Solistahuacán, *Thorne & Lathrop 46097* (RSA). **Veracruz:** San Andrés Tuxtla, near Cerro Tapalcapan and Cerro Mastagaga, *Dressler & Q. Jones 146* (CTES [fragment], GH, MEXU, NY, US); El Mirador, *Purpus 10768* (S, US), *Purpus 14128* (A); Zacuapan, *Purpus 15256* (UC).

GUATEMALA. Alta Verapaz: Near Tactic, *Standley 90441* (F, US). **Baja Verapaz:** 3 km N of Niño Perdido, rd. to Cobán, *R. Jones 2025* (CTES, MEXU, NY, TEX, US); Niño Perdido, on Quebrada Seca, *Lundell & Contreras 21032* (LL–2); Niño Perdido, Quebrada La Lima, *Lundell & Contreras 21182* (LL–2, MO, NY, UCR). **Huehuetenango:** Cerro Chiblac, between Ixcan and Finca San Rafael, *Steyermark 49157a* (F), *Steyermark 49454* (F); Cerro Negro, 2 mi E of Las Palmas, Sierra de los Cuchumatanes, *Steyermark 51686* (F, NY).

HONDURAS. New El Rincón, 10 mi W of Siguatepeque, *Yuncker 6062* (F, NY).

Pavonia integrifolia stands out for its entire (or nearly so) leaf margins and its relatively large fruits.

168. Pavonia longitudinalis Fryxell sp. nov. Type. Peru. Loreto: Prov. de Requena, Cocha Iricahua, left bank of the Ucayali River, below Jenaro Herrera; herb 80–100 cm tall; flower pale rose; in "Restinga," forming an understory in a periodically inundated

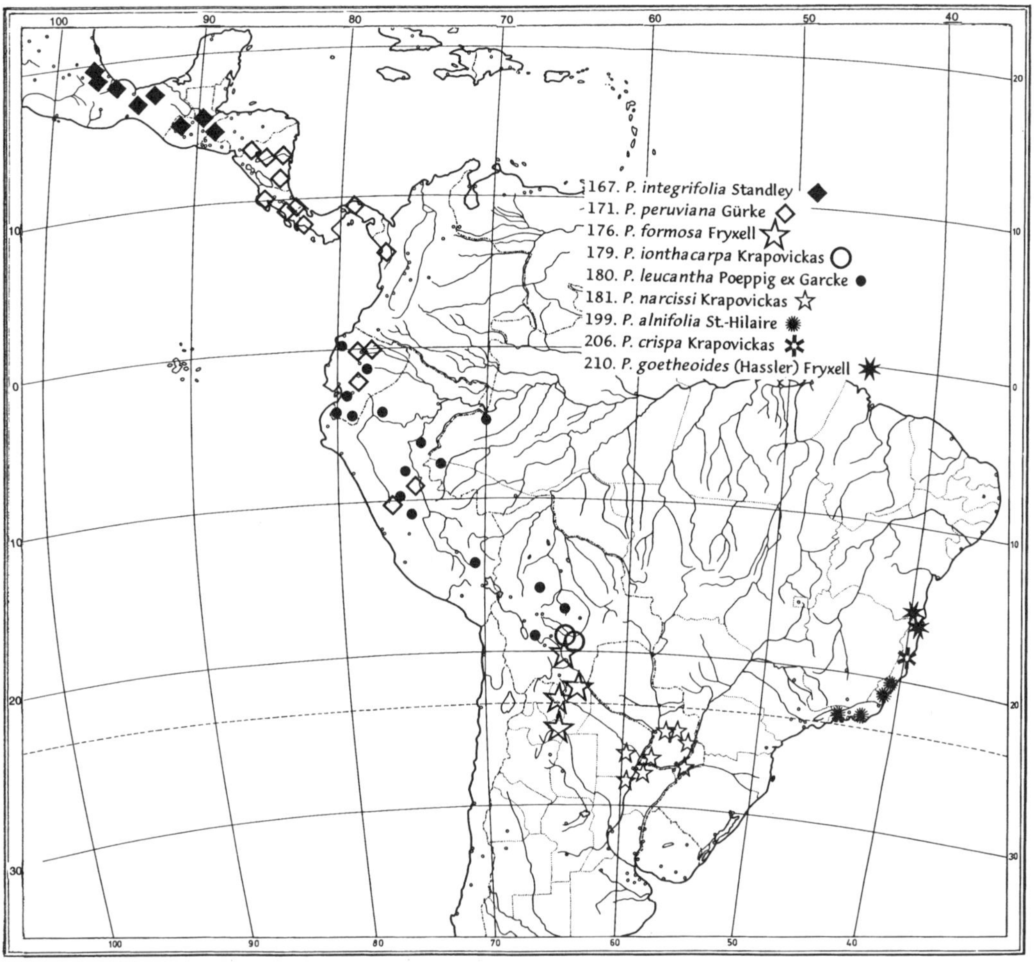

FIG. 76. Distribution map of nine neotropical species of *Pavonia* subgen. *Typhalea* and *P.* subgen. *Goetheoides*.

area; 180 m; 17 Feb 1982, *Encarnación 1290* (holotype, CTES).

Suffrutices caulibus pilis adpressis 4-brachiatis, brachiis longitudinaliter positis cum axe caulis (2 antrorsis et 2 retrorsis); bracteis involucellorum 10–12 ovatis ca. 4 mm latis plus minusve imbricatis; mericarpiis ca. 11 mm longis costis lateralis duabus, spinis tribus apicalis erectisque, spinis costisque retrorsum barbatis.

Subshrub to 1 m tall, the stems densely pubescent with appressed stellate hairs, the hairs usually 4-armed, with the arms aligned longitudinally, parallel with the stem axis. *Leaf* blades broadly oblanceolate, to 21 × 8.5 cm, basally cuneate, apically acuminate, the margin subentire proximally to shallowly dentate distally, penninerved, the upper surface subglabrous but with widely scattered simple or bifurcate hairs (rarely stellate) 0.5 mm long, especially toward the base, the lower surface stellate-pubescent, the hairs <1 mm diam., the pubescence of the nerves similar to that of the stem; petioles up to 4 cm long, with pubescence like stem; stipules lanceolate, 11 × 2 mm, erect, ascending-pubescent. *Inflorescence* capituliform, the peduncle with scattered appressed 3–4-armed stellate hairs similar to those of the stem but sparser; pedicels 2–4 mm long. *Flowers* and fruits crowded; involucel 8 mm long, with appressed bifurcate and stellate hairs, the bractlets 10–12, connate in basal half, ovate and somewhat overlapping, 2–4 mm wide, acuminate; calyx ca. 5 mm long, wholly concealed by involucel; corolla light rose. *Fruits* glabrous, the mericarp body ca. 11 × 3 mm, carinate and with 2 subparallel lateral ribs, with 3 apical retrorsely barbed spines, the central spine erect, 7 mm long, the lateral spines erect, 3 mm apart at base, 4–5 mm long, the retrorse spines extending down the lateral ribs of the mericarp body.

Distribution. At present known only from Peru in the province of Loreto at low elevation.

Pavonia longitudinalis is characterized by having appressed, longitudinally oriented stem pubescence (whence the specific epithet) similar to that found, for example, in the relatively unrelated *Sida ciliaris* Linnaeus, *Hibiscus coulteri* A. Gray, and *Malvastrum coromandelianum* (Linnaeus) Garcke, among others. Why these species share this distinctive character is not known. The new species is also distinguished by broad involucellar bracts, a condition also found in *P. fruticosa*, but the number of bracts is approximately double that found in *P. fruticosa*; because of the resulting crowding, the bracts are often overlapping. The fruit morphology of the new species is also quite different from that of *P. fruticosa*.

169. Pavonia nemoralis St.-Hilaire & Naudin, Ann. Sci. Nat. Bot., ser. 2, **18:** 43. 1842. *Pavonia typhalea* var. *nemoralis* (St.-Hilaire & Naudin) Triana & Planchon, Ann. Sci. Nat. Bot., ser. 4, **17:** 159. 1862. *Typhalea nemoralis* (St.-Hilaire & Naudin) Monteiro, Anais XII Reun. Soc. Bot. Bras. 30. 1961. Type. Brazil. Rio de Janeiro: *Gaudichaud 935* (lectotype, here designated, P as photo CTES; isolectotype, CTES). Fig. 77

Pavonia brachysepala St.-Hilaire & Naudin, Ann. Sci. Nat. Bot. ser. 2, **18:** 44. 1842. Type. Brazil. In Organ Mtns., *Gardner 324* (lectotype, P as photo CTES; isolectotype, CTES [fragment]).

Subshrubs 0.5–1 m tall, the stems scabridulous, the hairs stellate, 3–6-armed, 1 mm diam. *Leaf* blades elliptic to oblanceolate, mostly 9–20 × 4–9 cm, basally truncate to cuneate, serrate, acute or acuminate, penninerved or basally 3(–5)-nerved, somewhat discolorous, with minute (0.5 mm diam.) stellate hairs scattered evenly over both surfaces; petioles 1–6(–9) cm long, stellate-scabridulous; stipules subulate, 8 mm long, deciduous. *Inflorescence* terminal, with elongated peduncle (sometimes branched) rising above the leaves, racemose to subcapituliform; pedicels 0.5–3 cm long, the lower scattered, the upper crowded to subumbelliform; involucellar bracts 7–9, lance-linear, basally connate, 6–8 × 1–2 mm, minutely and sparsely stellate-pubescent, marginally ciliate (cilia 1 mm long); calyx 2–3 mm long, sparsely pubescent; corolla 7–10 mm long, pink, externally pubescent, the petals apparently narrow (2 mm wide?); androecium included, the anthers subsessile at apex. *Fruits* 5–6 mm diam., glabrous; mericarp body 5–6 mm long, smooth, the 3 distal spines erect and parallel, retrorsely barbed, the central spine 9–10 mm long, with a dorsoventrally broadened base, the lateral spines 6 mm long, 1.5–2 mm apart at base.

Distribution (Fig. 78). Largely confined to Brazil (Guanabara, Minas Gerais, Paraná, Rio de Janeiro, Rio Grande do Sul, Santa Catarina, and São Paulo) and with three collections known from Paraguay. It grows in shaded forests at 200–1400 m elevation.

Representative specimens examined. BRAZIL. GUANABARA: Pai Ricardo, *Duarte 5510* (CTES, G, MO, U, US). **MINAS GERAIS:** Near Lagoa Santa, *Warming 1318* (C–2); Caldas, *Regnell III-172* (C, F, S–4, U, US). **PARANÁ:** Cubatão, *Dusén 13666* (F, NY, S); Usina Electrica Marumbi, *Hatschbach et al. 13429* (CTES, K, U); Mun. Antonio, Caixa d'Agua, *Hatschbach et al. 13525* (CTES, U); Mun. Bocaiuva do Sul, Sesmaria, Capivari River, *Hatschbach 20929* (C, CTES, F); Mun. Cerro Azul, Barra of the Bom Sucesso River, *Hatschbach 33755* (C, CTES). **RIO DE JANEIRO:** Angra dos Reis, *Castellanos 25536* (CTES); Rezende, Palmital River, *Castellanos 25699* (CTES); Corcovado, *Dusén 1622* (S). **RIO GRANDE DO SUL:** Quinta near Rio Grande, *Malme 1586* (S–2); Quadros Lake, near Osorio, *Rambo 49736* (CTES, S). **SANTA CATARINA:** Penha, *Krapovickas & Cristóbal 39605* (C, CTES, K, MBM, MICH, NY, UC); rd. from Santo Amaro de Imperatriz to Fazenda Jomar, *Krapovickas & Cristóbal 44825* (CTES, TEX); Mata San Pedro, *Reitz 3507* (CAS, CTES, NY, S, US). **SÃO PAULO:** Butantan, *Gehrt 8280* (NY); Lapa, *Holway 1940* (GH, US); between S. João da Boa Vista and S. Pedro, *Mosén 1116* (S); Juruagua, *Usteri s.n.* (NY, PENN).

PARAGUAY. In vic. of Caaguazú, *Hassler 8876* (NY); near Igatimí, *Hassler 5632* (K, MO, NY, S); Villarica, *Jörgensen 4476* (A, C, F, NY, PH, S, US).

Pavonia nemoralis was treated as a synonym of *P. rosea* by Gürke (1892b), and this view has been followed by most subsequent authors. Consequently, many of the above-cited specimens of *P. nemoralis* may be found in herbaria annotated by me as a new species. When I examined the appropriate type material, I recognized that this is the same as the species earlier described by St.-Hilaire, and is distinct.

Pavonia nemoralis is similar to *P. fruticosa* in many respects, especially in having the 3 erect spines of the mericarp very crowded at the apex of each mericarp, the 2 lateral spines being less than 2 mm apart at the base, and in having smooth dorsal walls on the mericarps. It differs from *P. fruticosa* most notably in the more numerous, narrower involucellar bracts and in the fact that the central spine is longer (up to 10 mm) and prominently buttressed (dorsoventrally) at the base, but also in other characters including a smaller calyx and longer pedicels. A third species, *P. stellata*, is probably also allied to these two, although the three are widely separated in the preceding key. It may be

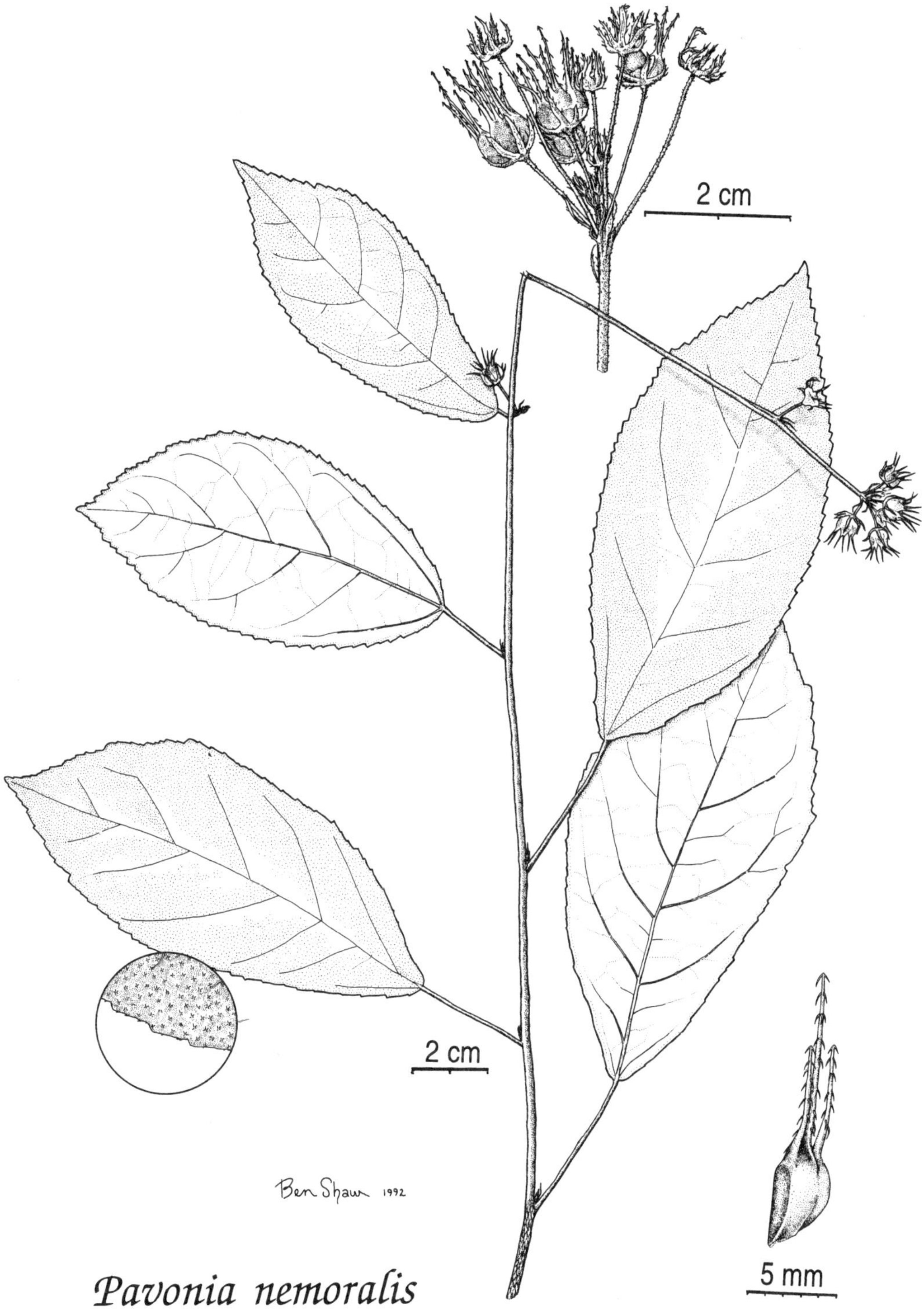

FIG. 77. *Pavonia nemoralis*. Fruiting branch with enlargements of fruiting inflorescence (top), mericarp (bottom right), and leaf margin (bottom left) (*Holway & Holway 1940* [branch], *Duarte 5510* [inflorescence]).

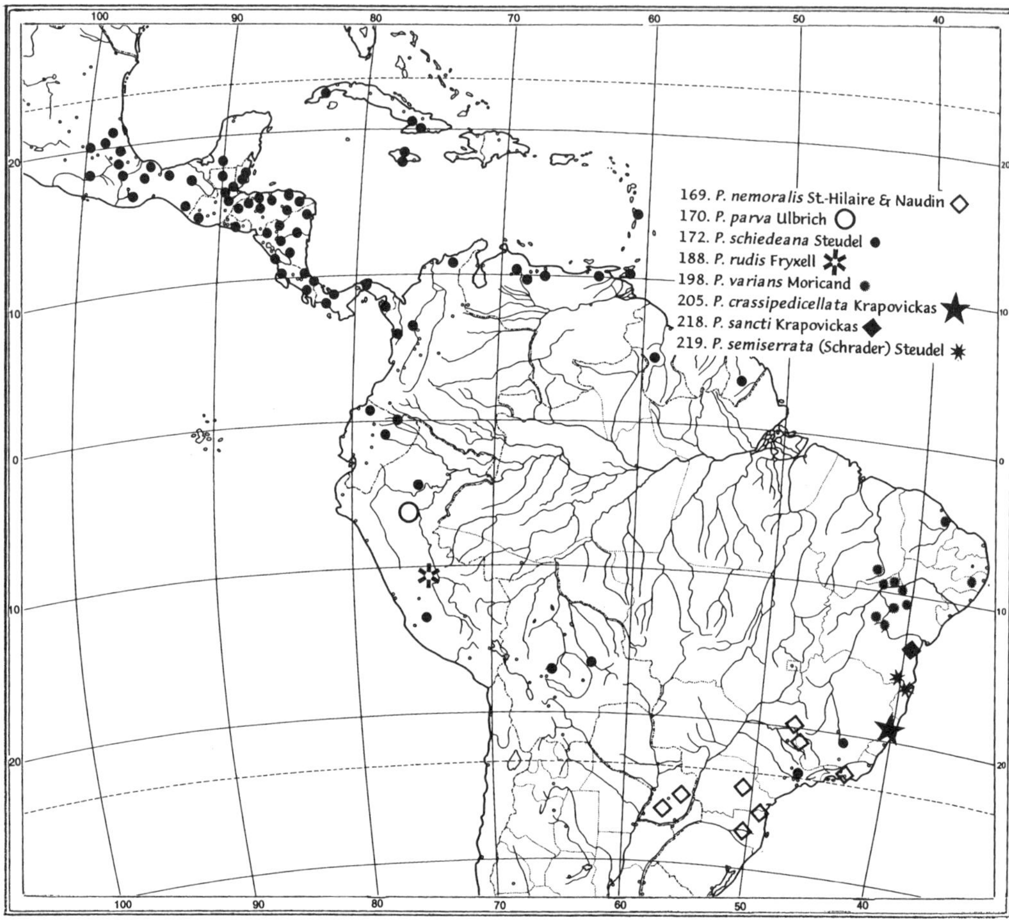

FIG. 78. Distribution map of eight neotropical species of *Pavonia* subgen. *Typhalea* and *P.* subgen. *Goetheoides*.

suggested that the solitary spine on the mericarp of *P. stellata* resulted from the apical crowding of the 3 spines being carried to the extreme of fusion into a single spine. This interpretation appears to be borne out by the observation (Krapovickas, pers. comm., based on *Gardner 324*) that sometimes the mericarps of *P. stellata* have 3 spines. However, whether this resulted from the Gardner collections being a mixed gathering of more than one species or represents variability within a population remains to be learned from further study of the relevant specimens and from field study at the type locality. The view is adopted here is that *Gardner 324* was a mixed gathering and represents two distinct species, following the implicit intepretation of Gürke (1892b), who cited *Gardner 324* under both species. Therefore, *P. brachysepala* is lectotypified with the Paris duplicate, and the name is treated as a synonym of *P. nemoralis*, whereas *P. begoniaefolia* is lectotypified with the G duplicate and treated as a synonym of *P. stellata* (q.v.). All other duplicates seen of *Gardner 324* (see citations under *P. stellata*) have only a single spine per mericarp. Moreover, the single spine usually found in *P. stellata* is buttressed at the base similar to that of the central spine of *P. nemoralis* and is of comparable length.

170. Pavonia parva Ulbrich, Notizbl. Bot. Gart. Berlin **6:** 327. 1915. *Pavonia nana* Ulbrich, Verh. Bot. Verein Brandenburg **51:** 85. 1909 (non *Pavonia nana* R. E. Fries, 1907). Type. Peru. Loreto: Tarapoto, mtn. forest near S. Antonio de Cumbaso, Mar. 1903 (fl. fr.), *Ule 6861* (holotype, B as photo F-9458; isotypes, CORD, CTES [fragment], MG).

Subshrubs 20 cm tall, the stems minutely puberulent. *Leaf* blades elliptic to oblanceolate, up to 16 × 7 cm, narrowly cordate basally, coarsely crenate, acute, pennınerved, discolorous, the upper surface subglabrous, the lower surface densely pubescent with minute stellate hairs; petioles 2–4 cm long, minutely puberulent; stipules lanceolate, erect, 7 × 1 mm, obscurely pubescent. *Inflorescences* on axillary peduncles, racemose or paniculate to subcapituliform; peduncles antrorsely pubescent with rigid simple hairs 1 mm long; pedicels to 5 mm long, densely antrorsely hispid; involucel of 10–12 bractlets, connate in basal half, 7–8 mm long, antrorsely hispid; calyx subequal to involucel, antrorsely hispid; corolla rose, ca. 8 mm long. *Fruits* glabrous, reddish; mericarps carinate and reticulate-veined, the body ca. 6 mm long, with 3 widely spaced spines, the central spine 4–5 mm long, erect, the lateral spines 3–4 mm apart at the base, somewhat divaricate.

Distribution (Fig. 78). Known only from the type collection.

The distinctive pubescence found on the peduncles, the short stature, and the reddish fruits combine to set *Pavonia parva* apart. Reddish fruits are also found in *P. penduliflora* (Standley) Standley. The photo of the holotype includes a drawing (presumably by Ulbrich) of a floral dissection.

171. Pavonia peruviana Gürke in Martius, Fl. Bras. **12(3):** 487. 1892. *Typhalea peruviana* (Gürke) Monteiro, Anais XII Reun. Soc. Bot. Bras. 30. 1961. Type. Peru. Huánuco: Near Pampayaco, *Poeppig 1584* (lectotype chosen by Macbride, 1956, F; isolectotype, OXF). The excluded syntype is *Holton 745* from Cauca, Colombia, which represents *P. castaneifolia* (q.v.). Fig. 73

Pavonia maxonii (Standley) Standley, Publ. Field Mus. Nat. Hist., Bot. Ser. **4:** 231. 1929. *Malache maxonii* Standley, Contr. U.S. Natl. Herb. **18:** 116. 1916. *Pavonia rosea* var. *maxonii* (Standley) Robyns, Ann. Missouri Bot. Gard. **52:** 543. 1965. Type. Panama. Chiriquí: along the Ladrillo River and vic., above El Boquete, 17–19 Mar 1911, *Maxon 5394* (holotype, US; isotypes, NY, US).

Shrubs 0.5–1(–3) m tall, the stems stellate-pubescent or often coarsely hirsute with spreading simple and bifurcate (occasionally stellate) hairs 1–2 mm long. *Leaf* blades mostly 8–18(–21) × 3.5–10(–15) cm, broadly elliptic or oblanceolate (i.e. widest at or above the middle, often lobulate), penninerved, truncate to cuneate, serrate-crenate, more or less acuminate, the upper surface with appressed simple hairs 0.5–2 mm long, the lower surface with scattered simple and stellate hairs often >1 mm long; petioles 1–5 cm long (shorter upward), the uppermost leaves sometimes subsessile; stipules 8–12 mm long, narrowly falcate, ciliate. *Inflorescences* capituliform or racemose (but the lowermost flowers often solitary), rising above the leaves, the heads terminal, crowded or open, the pedicels 1–9 cm long, subtended by several ciliate, lance-linear bracts 4–7 mm long; the 8–9 involucellar bracts 5–9 × 1–1.5 mm, basally connate, obscurely pubescent or prominently hirsute; calyx 2.5–6 mm long; corolla 6–9 mm long, white to pink. *Fruits* 6–9 mm diam., subglabrous, longitudinally costate, the mericarps each 3-spined, the spines erect, retrorsely barbed, the central spine 7 mm long, the lateral spines 4–5 mm long, 3–3.5 mm apart at base.

Distribution (Fig. 76). From Nicaragua to Peru, growing in shaded forests at 1200–1700 m elevation. The type locality for *Pavonia maxonii* is at 8°48′N, 82°30′W.

Illustration. Gürke (1892b: pl. 88).

Representative specimens examined. **NICARAGUA. CHONTALES:** Ca. 2 km N of Santo Domingo, *Grijalva & Ríos 3454* (HNMN, NY). **ESTELÍ:** Laguna de Miraflores, *Grijalva 964* (HNMN, MO, NY); 22 km N of Estelí, *Moreno 11035* (MO, NY). **JINOTEGA:** Finca Santa Elena, *Sandino 1050* (HNMN, NY). **MATAGALPA:** Above Santa María de Ostuno, *D'Arcy 10471* (MO, NY); Aranjuez, *Moreno 16325* (MO, NY). **RÍO SAN JUAN:** Sábalo, *Araquistain 3200* (HNMN, MO, NY). **ZELAYA:** Near Siuna, *Seymour 3102* (MO, NY); between Toro Bayo and Esquipulas, *Shank & Molina 4603* (F).

COSTA RICA. ALAJUELA: Guadalupe de Zarcero, *A. Smith 693* (A, F, MO, NY). **CARTAGO:** Turrialba, along Reventazon River, *Lent 687* (F, LL, MO); 10 air mi S of Cartago, *Liesner & Judziewicz 14605* (MO); vic. of Orosí, *Standley 39860* (F, US). **GUANACASTE:** Rincón de la Vieja National Park, *Rivera 520* (CR, MO, NY); near Monteverde, *Weston 5071* (F, MO, NY, US). **PUNTARENAS:** Cantón de Buenos Aires Olán, *Chacón 419* (CR, MO); near Coton, 23 km N of La Unión (on Panama border), *Croat 26625* (MO, NY); Corcovado National Park, *Liesner 2858* (MO, NY). **SAN JOSÉ:** Cantón de Turrubares, SW spur of Cerro Turrubares, *Grayum et al. 10541* (MO, TEX); Cantón de Acosta, Cuenca del Río Tabarcia, on the W slope of Cerro Cedral, *Morales 2759* (MO, TEX); Cerros de Turrubares, *Zuñiga et al. 156* (CR, MO).

PANAMA. CHIRIQUÍ: Chiriquí Viejo Volcano, *P. H. Allen 979* (MO, US); NW side of Cerro Pando, *Croat 15979* (F, GH, MO); Boquete Distr., Bajo Mono, *M. E. Davidson 503* (A, F, MO); 13 km by rd. S of Sereno River, *Hensold 1023* (F, MO, NY). **COCLÉ:** 3 km E of El Valle, *Hammel 4795* (MO); 5 mi N of El Valle de Antón, *Luteyn 1183* (MO). **DARIEN:** Cerro Pirre, *Folsom 4439* (MO, NY); Cerro Tacarcuna, *A. H. Gentry & Mori 14089* (MO). **PANAMA:** Serranía de Majé, S of Ipetí, *Huft et al. 1656* (MO); Cerro Campana, *Mori & Kallunki 2455* (MO, NY). **VERAGUAS:** 6–7 km W of Santa Fé, *M. Nee 9761* (MO); Alto de Piedra, *Tam 18* (MO).

COLOMBIA. CHOCÓ: Río Imgará, hwy. from San Juan del Palmar to Novita, *Forero et al. 2107* (MO, NY).

ECUADOR. **Esmeraldas:** Río Santiago at Concepción, *Holm-Nielsen et al. 26003* (AAU); ca. 30 km S of Esmeraldas, *Maas et al. 2943* (U). **Los Ríos:** Juaneche Forest between Mocachi and Palenque on the Estero Peñafiel, *Dodson et al. 8795* (MO, SEL); Río Palenque Field Station, *Løjtnant & Molau 15790* (AAU). **Manabí:** El Recreo, *von Eggers 15023* (US). **Morona-Santiago:** 25 km SE of Logrono, *Madison & Coleman 2559* (GH, SEL). **Napo:** 2 km W of Archidona, *Holm-Nielsen & Jeppesen 1031* (AAU, QCA, S). **Napo-Pastaza:** Mera, *Asplund 18319* (S); 4 km E of Puyo, *Prescott 446* (NY). **Pastaza:** 5 km NE of Mera, hwy. to Ansu River, *Neill et al. 5893* (MO, NY, TEX). **Santiago-Zamora:** Zamora to Yanzasa, *Mathias & D. Taylor 5221* (F, MO). **Tungurahua:** Between Baños and Cashurco, *Hitchcock 21779* (GH, NY, US); between Mapoto River and Margaritas River, *Penland & Summers 162* (COLO, GH); 2 km E of Topo River, ca. 31 km E of Baños on rd. to Puyo, *Stein 2976* (MO, NY).

PERU. Without locality, *Matthews s.n.* (OXF). **Huánuco:** Hwy. from Tingo María to Pucallapa, *Rimachi 5007* (K, NY). **San Martín:** San Roque, *L. Williams 7735* (F).

The distinction between *Pavonia maxonii* and *P. peruviana* is somewhat problematical, and the two species are not here maintained as distinct. The characters that have been used to distinguish them are not always clearly visible in an imperfect specimen, nor are they always clear-cut, even in complete specimens. Plants treated as *P. maxonii* tend to have more-open, racemose inflorescences (compared with closely capituliform in *P. peruviana*), larger fruits, shorter appressed hairs on the upper leaf surface, and differently shaped mericarps (rounded in *P. maxonii*, similar to those of *P. penduliflora*, although fruits are lacking in the type of *P. maxonii*; "rectangular" in *P. peruviana*, similar to those of *P. castaneifolia*). Further study is required to clarify the relationship between these species, but herein they are treated as a single species, albeit a variable one.

172. Pavonia schiedeana Steudel, Nom. Bot., ed. 2, **2:** 279. 1841. Basionym: *Pavonia rosea* Schlechtendal, Linnaea **11:** 355. 1837 (non *Pavonia rosea* Wallich ex Moris, 1833). *Malache rosea* (Schlechtendal) Kuntze, Revis. Gen. Pl. **1:** 71. 1891. Type. Mexico. Veracruz: Near Hacienda de la Laguna, *Schiede s.n.* (lectotype, here designated, GOET; isotypes, B as photo F-9467, BM as photo CTES, P as photo CTES, U). Fig. 73

Pavonia lappacea Casaretto, Nov. Stirp. Bras. 37. 1842. Type. Brazil. São Paulo: in the island S. Sebastiani, *Casaretto s.n.* (holotype, TO?–n.v.).

Hibiscus umbellatus Sessé & Mociño, Fl. Mexic., ed. 2, 160. 1894. Type. Mexico. "in montibus Totonacis," *Sessé & Mociño* (specimen unknown).

Pavonia silvatica Diehls, Biblioth. Bot. **29 (Heft 116)**: 106. 1937. Type. Ecuador. Tungurahua: "Tal des Río Pastaza bei Río Negro," *Diels 976* (specimen unknown).

Pavonia fruticosa var. *angustiloba* Stehlé, Bull. Soc. Bot. France **92:** 264. 1945. Type. Martinique. *H. Stehlé & M. Stehlé 5375* (isotype, US).

Pavonia rosea var. *rigida* R. E. Fries, Kongl. Svenska Vetenskapsakad. Handl., ser. 3, **24(2):** 25. 1947. Type. Cuba. near Canto River, *Ekman 2785* (holotype, S).

Pavonia umbrosa R. E. Fries, Kongl. Svenska Vetenskapsakad. Handl., ser. 3, **24(2)**: 25. t. 3, f. 1. 1947. *Typhalea umbrosa* (R. E. Fries) Monteiro, Anais XII Reun. Soc. Bot. Bras. 31. 1961. Type. Bolivia. S Yungas, Sirupaya near Yanacochi, *Buchtien 508* (holotype, S).

Subshrubs or shrubs to 2 m tall (to 4 m in *Wedel 149*), the stems minutely stellate-pubescent. *Leaf* blades mostly 6–21 × 2–10 cm, elliptic, oblong, or obovate, truncate, crenate-dentate (the margins often ciliate), acute, slightly discolorous, minutely stellate-pubescent; petioles ¼ the length of the blades or less; stipules 5–9 mm long, subulate. Peduncles axillary, sometimes exceeding the subtending leaves, bearing a dense headlike *inflorescence* that is actually a condensed raceme; pedicels up to 12 mm long but usually much shorter; involucellar bracts ca. 8, linear, 4–6 mm long, basally gamophyllous; calyx 4–5 mm long; corolla 6–9 mm long, pink; androecium subequal to petals. *Fruits* 6–8 mm diam.; mericarp body 5 mm high, smooth, each mericarp with 3 retrorsely barbed spines 2.5–5 mm long.

Distribution (Fig. 78). Widely distributed from central Mexico and the West Indies to Brazil and Bolivia. It grows in shaded forests and also in disturbed habitats. The type locality for *Pavonia umbrosa* is at 16°22′S, 67°45′W.

Illustrations. Dodson & Gentry (1978: pl. 93D, as *P. rosea*); Fries (1947: pl. 3, fig. 1); Fryxell (1979: figs. 2E, 3A); Gürke (1892b: pl. 87, fig. 2, as *P. rosea*); Robyns (1965: fig. 6G, as *P. rosea*).

Representative specimens examined. **MEXICO.** **Campeche:** 10 km S of Candelaria, *E. Cabrera & H. de Cabrera 4779* (GH, MEXU, MO, NY). **Chiapas:** 20 km S of Palenque, rd. to Ococingo, *E. Cabrera & Canul 5003* (MEXU, NY, TEX); Mun. Cacahoatán, 1 km W of Santo Domingo, lower slopes of Tacaná Volcano, *P. A. Fryxell & Lott 3354* (CHAPA, MEXU, NY); 5 km S of Frontera Echeverría, *E. Martínez 9081* (GH, MEXU, MO, NY); Mun. Yajalón, Agua Fría, *Ton 6568* (MEXU, MO, NY). **Guerrero:** Distr. Galeana, Carrizo–Sto. Domingo, *Hinton et al. 14516* (ARIZ, LL, NY, PH, RSA, US). **Hidalgo:** Jacala, *Kenoyer 521* (F). **Oaxaca:** Mun. Candelaria Loxicha, hwy. from Pochutla to Oaxaca, 6.5 km S of Candelaria, *Koch, P. A. Fryxell & Wendt 79518* (CHAPA, ENCB, MEXU, NY); Mun. Juchitán, Los Angeles (17°22′N, 95°08′W), *C. Martínez 1070* (MEXU, MO, NY). **Puebla:** Mun. Cuetzalan, Finca los Flores, al SE de Xochical, *Tenorio et al. 13946* (MEXU, NY); Mun. San José Acateno, Jiliapa, *A. Ventura 11993* (CTES, ENCB, NY). **Querétaro:** Mun. Landa, 10 km SE of Agua Zarca, *Rzedowski 43887* (ENCB, IEB); Mun. Jalpan, 2–3 km E of La Parada, *Servín 1216* (IEB, NY). **San Luis Potosí:** On banks of Aztlam River, 2 km E of Alfredo M. Terrazas (ca. 20 mi N of Tamazunchale), *P. A. Fryxell 1075* (BH, CTES, MICH, NA, NY, TEX); lowlands, Micos, *Pringle 5035* (NY, US). **Tabasco:** Mun. El Centro, on Grijalva River, N of Villahermosa, *Gilly & E. Hernández 324* (BRIT, GH); Sabana de Huimanguillo, *Ricardez 272* (MEXU, NY). **Veracruz:** Near Orizaba, *Bourgeau 2721* (C, CTES, K, US); Mun. Coetzala, 3 km WNW of Cuichapa, 1.5 km SE of Coetzala (18°47′N, 96°54′W), *Diggs et al. 2730* (BRIT, F, MEXU, NY, XAL); near Lerdo de Tejada, *P. A. Fryxell 670* (BH, CTES, F, MEXU, MO, NY, TEX); "Los Tuxtlas" Tropical Biological Station, *P. A. Fryxell 2607* (BRIT, CAS, CHAPA, CTES, ENCB, NY); San Andres, rd. to Sierra de Zongolica, *Rosas 450* (A, C, F, GH, MEXU, NY); Mun. Tlaltetela, El Limón, *F. Ventura 12813* (CHAPA, CTES, ENCB, MEXU, NY).

GUATEMALA. Alta Verapaz: Chapultepec Farm, *Contreras 4811* (LL–2, MO); Telemán Panzós, *E. M. Martínez et al. 22757* (GH, MEXU, MO, NY); Cobán, *von Türckheim II-1243* (C, F, GH, MO, NY, S). **Chimaltenango:** Near Sibajá, *Standley 62265* (F). **Escuintla:** Below Las Lajas, *Standley 64780* (F). **Huehuetenango:** Jacaltenango–San Marcos, Canton Llano, *Boeke 176* (CTES, MO, NY); Sierra de los Cuchumatanes, 3 mi NW of Santa Cruz, Mun. Barrillas, *Breedlove 11649* (DS, F, LL). **Izabal:** Puerto Barrios, *Deam 33* (GH, NY); Los Amates, *Kellerman 7475* (BRIT, F, NY); Quiriguá, *Standley 24574* (GH, MO, NY); Cerro San Gil, along Frio River, *Steyermark 41613* (A, F). **Petén:** Tikal, around Aguada Seca, *Lundell 16074* (LL–2, MEXU, NY); 20 km E of Santa Elena, *Molina 15692* (F, NY, US). **Quetzaltenango:** Finca Azucena above Colomba, *Standley 68077* (F, NY). **Retalhuleu:** E of Chivolandia, *Standley 67020* (F). **Santa Rosa:** Near El Molino, *Standley 60722* (F); about Guazacopán, *Standley 78658* (F). **Sololá:** Primavera, *Shannon 424* (US).

BELIZE. Belize: Burrel Boom, *Dwyer 14944* (LL, MO, NY). **Cayo:** Along Sibun River just W of Hummingbird Hwy., *Gentle 8806* (F, LL–2, MO, S, US); Vaca Plateau, Macal River, *Balick et al. 2008* (NY, TEX). **Stann Creek:** 2 km N of Victoria Peak, *A. H. Gentry 7930* (F, MO–2); 5 mi S of Lynam Agric. Coll., *Spellman 1590* (E, MO). **Toledo:** San Miguel Village, *Arvigo et al. 569* (NY, TEX); Columbia Forest Station, *Dwyer 9942* (COLO, F, MO, NY); San José, 6.7 mi N of Columbia Forest, *Dwyer 11161* (C, MO).

HONDURAS. Vic. of Siguatepeque, *Yuncker et al. 5578* (F, GH, MO, NY, S, U). **Atlantida:** Lancetilla Valley, near Tela, *Standley 53659* (A, F, US); near La Ceiba, *Yuncker et al. 8322* (F, GH, K, MO, NY, S, TEX, US). **Colón:** *C. Nelson & J. R. Martínez 1283* (MO). **Comayagua:** La Libertad, *Edwards P-623* (A, F); El Cedral, Cordillera Montecillos, *Molina 7192* (F, US). **Copán:** 6 mi S of Nueva Arcadia, *Harmon & Fuentes 6435* (MO, NY); just N of Copán, *Poole & Watson 1170* (LL). **Cortés:** 5 km W of Puerto Cortés, *Blackmore & Chorley 3996* (BM, NY); Quebrada Agua Azul, *Molina 5532* (F). **Gracias a Díos:** Tonsin, *Barkley & Hernández 40637* (GH). **Olancho:** River bottom in village of Culmí, *C. Nelson & Romero 4679* (MO, NY, TEFH); Mosquitia coast, Playabilas near mouth of Río Plátano, *P. A. Fryxell 2800* (CTES, NY). **Santa Bárbara:** San Pedro Sula, *Thieme 5144* (GH, NY, US); El Mochito, *E. Vargas 698A* (MO).

EL SALVADOR. Ahuachapán: Rd. from San Francisco Menéndez to Tacuba, 1–3 mi above intersection with rd. to Cara Sucia River, *Croat 42100* (MO, NY); Sierra de Apaneca, near Finca Colima, *Standley 20108* (US). **La Libertad:** Finca El Paraíso, Jayaque, *Montalvo 4712* (MO). **San Salvador:** Near San Salvador, *Calderón 774* (GH, MO, NY, US). **Sonsonate:** Near Sonsonate, *Standley 22273* (GH, US); vic. of Izalco, *Standley 22231* (GH, NY, US).

NICARAGUA. Boaco: Cerro Alegre, *Moreno 24924* (MO, NY); 1–2 km SE of Santa María, *M. Nee et al. 28007* (MO, NY, TEX). **Chontales:** 17.4 km NE of Río Mayales at ford of El Bizcocho River, *Stevens 4044* (MO, NY); 2.8 km N of Cuapa, *Stevens 3638* (MO, NY, TEX). **Comarca del Cabo:** *Molina 15049* (F, NY, US), *Seymour 5621* (GH, MO, NY). **Estelí:** Estanzuela Falls, *Hall 7683* (GH, MO, NY); 21.5 km E of Estelí, *Moreno 15401* (MO). **Granada:** Mombacho Volcano, *Neill 875* (MO, NY). **Jinotega:** Hwy. from Matagalpa to Jinotega at km 1392, *Vega & Quezada 194* (HNMN, MO, NY); between Matagalpa and Jinotega, rd. to Fundadora and La Salvadora, *Vincelli 427* (MO, NY). **Matagalpa:** Sta. María de Ostuma, *Araquistain & Moreno 729* (MO, NY); La Verona, *Moreno 18245* (HNMN, NY); Cerro El Picacho, *Stevens 23105* (MO, NY); Matagalpa, *Zelaya 2320* (BRIT, F, GH, MO, NY). **Nueva Segovia:** 1.6 km W of Murra, rd. to El Jícaro, *Stevens 17563* (MO, NY). **Zelaya:** Mun. Siuna, Wany, *F. Ortíz 58* (HNMN, MO, NY); Corn Island, *Seymour 4311* (BRIT, F, GH, MO, NY, TEX, US).

COSTA RICA. Alajuela: San Isidro de San Ramón, *Brenes 23174* (NY); 3.5 km W of Fortuna, *R. J. Taylor & C. E. Taylor 11572* (US). **Cartago:** Cartago, *Cooper 5720* (F, GH, K, US); above Humo River, Pejivalle Farm, *Dodge & W. S. Thomas 4387* (GH, MO); slopes above Platanillo, *Wilbur & Stone 10601* (DUKE, GH, LL, NY, TEX, US). **Heredia:** Between Puerto Viejo and Guapiles, 7 km N of Buenos Aires, *Croat 68394* (MO, NY). **Guanacaste:** Cantón de Tilarán, 2 km N of Las Nubes, *Haber & Ivey 10151* (MO); El Arenal, *Standley & Valerio 45256* (US). **Limón:** S of Puerto Limón, Cieneguita, *Khan et al. 1218* (BM, NY); near Westfalia, *J. F. Utley & K. Utley 5460* (MO). **Puntarenas:** Curú, at mouth of Curú River, *Sanders et al. 17469* (TEX, UCR). **San José:** Below La Palma, NE of San Jerónimo, *Burger & Liesner 6247* (F, MO, US); Cuenca del Río Negro, *Morales 2812* (MO, TEX); vic. of El General, *Skutch 2726* (GH, MO, NY, S, US).

PANAMA. Bocas del Toro: Chiriquicito to 5 mi S along Río Guarumo, *W. H. Lewis et al. 2044* (GH, MO, US); Old Bank Island, vic. of Chiriquí Lagoon, *von Wedel 2003* (GH, MO, US). **Canal Zone:** Paraíso River, above East Paraíso, *Standley 29862* (US); Barro Colorado Island, *Standley 40852* (US). **Chiriquí:** Vic. of Santa Clara, *Croat 66250* (MO, NY); Cerro Pando, *W. G. D'Arcy & J. J. D'Arcy 6497* (F, GH, LL, NY). **Coclé:** Near Limón, *Folsom 5884* (MO, NY); El Valle de Anton, *W. H. Lewis et al. 1791* (GH, MO, US); La Mesa, 2 km NW of Cerro Pilón, *Sullivan 556* (MO, NY). **Darien:** Ca. 18 km E of Pucuro, *de Nevers et al.*

8389 (MO, NY). **Panamá:** Cerro Campana, *Hammel 3817* (MO, NY); vic. of Arenoso, *Seibert 633* (K, MO). **Veraguas:** 18 km W of Las Minas, *Hammel 4360* (MO, NY).

CUBA. Santiago de Cuba: Sierra del Cobre, *Ekman 7842* (S); Bayate, on banks of Jagua River, *Ekman 2538* (NY, S); Santa Elena Estate, El Cobre, *Clemente 6270* (GH). **Pinar del Río:** Near San Diego de los Baños, *Bro. León 5136* (GH).

JAMAICA. St. Ann: Rd. between York Castle and Corner Shop, *W. R. Anderson & Sternberg 3092* (GH, LL, US); above Bowden Pen, along footpath to Bath via Cuna Cuna Pass, *Crosby & W. R. Anderson 1041* (DUKE, LL); Castleton, *Harris 11875* (C, GH, MO, NY); foothills of John Crow Mtns., E of Seamen's Valley, Portland, *Maxon & Killip 222* (F, GH, NY, US); Orange River valley, near Montego Bay, *Maxon & Killip 1684* (F, GH, NY, US); St. Andrew, Hope River valley ca. 1 mi E of Mona, *Yuncker 17167* (F, NY, S).

WINDWARD ISLANDS. Martinique: Côte de Bellefontaine au Morne Vert, *H. Stehlé & M. Stehlé 4822* (US).

COLOMBIA. Antioquía: In valley of Anorí River between Dos Bocas and Anorí, *Denslow 2305* (MO). **Chocó:** behind Solano Bay, *A. H. Gentry & Fallen 17508* (MO, NY). **Magdalena:** Sierra Nevada de Santa Marta, *Kirkbride 2134* (NY, US).

VENEZUELA. Distrito Federal: Vic. of Macarao, ca. 20 km W of Caracas, *Lasseigne 4455* (ISC, NY); Gamboa, near Caracas, *Pittier 9618* (US). **Lara:** Vic. of Barquisimeto, *Saer 283* (NY). **Monagas:** Vic. of Cuevas del Guaracho, *Croat 54328* (MO). **Sucre:** El Guayabito, *Davidse & González 19112* (CTES, US); S of Fila La Baquete, *Davidse & González 19252* (CTES, US). **Yaracuy:** Chinacoa, *Gines 4340* (US).

TRINIDAD. Toward Nariva swamp, *Broadway 9393* (U); Mt. Aripo, N Range, *R. S. Cowan 1347* (GH, NY).

GUYANA. W extremity of Kanuku Mtns., in drainage of Takutu River, *A. C. Smith 3244* (NY, U).

FRENCH GUIANA. Saül, Monts La Fumée, *Mori & Boom 15132* (NY, TEX); Mont Atachi Bacca, Ininí region (03°33'N, 53°55'W), *Granville et al. 10896* (U).

ECUADOR. Bolívar: 3–5 km above Balzapamba, *Fosberg 22506* (CAS, US). **Los Ríos:** Río Palenque Field Station, *A. H. Gentry 9524* (MO, NY, QCA). **Manabí:** Montecristi, *Plowman & Alcorn 14344* (F, NY). **Napo:** Lower Aguarico River (above Puerto Loja military station), *Harling et al. 7397* (GB, NY).

PERU. Cuzco: Dep. Madre de Diós, Salvación, *Vargas 16918* (US). **Loreto:** Prov. Maynas, 28 de Julio (Itaya River), *R. Vásquez 2878* (MO, NY).

BRAZIL. Minas Gerais: 12 km E of Ponte Nova, *Irwin 2232* (NY, TEX, US); Fazenda de Aguada, *Mexia 5446* (A, CAS, F, GH, MO, NY, PH, S, US). **Paraná:** Morretes, *Dusén 14318* (CAS); Mun. Morretes, Est. Marumbi, *Oliveira 287* (MU, NY). **Rio Grande do Sul:** Quadros Lake, near Torres, *Rambo 46017* (CAS). **São Paulo:** 9 km N of center of São Paulo, *G. Eiten & L. Eiten 2438* (US); Ubatuba, *Viégas 3391* (CAS).

BOLIVIA. Santa Cruz: Prov. Guarayos, 4 km S of Perseverancia, *M. Nee 38715* (MO, NY, TEX).

Local names. Cadillo (Mexico: Oaxaca, Veracruz, Tabasco), chicojoos (Mexico: Querétaro), conger watchman (Jamaica), cuayó (Guatemala), diente de chucho (Guatemala), guisato tres puyas (Cuba), jinuqui (Mexico: Oaxaca, Zoque), mosote (Guatemala, Honduras), mozote (Belize, Guatemala, Honduras), mozote de caballo (Guatemala), mozote de perro (Nicaragua), mozotillo (Costa Rica), oreja de coche (Guatemala), oreja de conejo (Guatemala), pega pega (Mexico: Oaxaca), sak'al zak (Mexico: Chiapas, Tzeltal), ts'ik'i uxum (Mexico: San Luis Potosí), txikim txikam (Guatemala, Jacaltec).

Pavonia schiedeana is distinguished from other species of this section by its smaller flowers and fruits.

173. Pavonia stellata (Sprengel) Sprengel, Syst. Veg. **3:** 97. 1826. *Urena stellata* Sprengel, Neue Entd. **2:** 163. 1821. Type. Brazil. *Sellow s.n.* (isotype, US). Fig. 73

Pavonia monatherica Casaretto, Nov. Stirp. Brasil 37. 1842. Type. Brazil. Rio de Janeiro: Organ Mtns., *Casaretto [1056]* (isotype, G as photo F-23704).

Pavonia begoniaefolia Gardner *in* Hooker, J. Bot. **2:** 333. 1843. Type. Brazil. In Organ Mtns., dense virgin forests, 1837, *Gardner 324* (lectotype, here designated, G-Del as photo CTES; isolectotypes, E-2, GH, K, MO, NY-2, OXF). (See comments concerning the choice of lectotype under *P. nemoralis*.)

Pavonia paraibica Wawra, Oesterr. Bot. Ztschr. **32:** 39. 1882. Type. Brazil. Juiz de Fora, between Parahiba and Paraibuña (rivers), *Wawra II, 132* (holotype, W?–n.v.).

Shrubs 0.5 m tall, the stems with widely scattered 3–6-armed stellate hairs ca. 0.5 mm diam., and with longitudinal lines of dense smaller (somewhat recurved) hairs. *Leaf* blades elliptic or oblong-elliptic, mostly 7–14 × 2–5 cm, basally asymmetrical (distichous), serrate, acuminate, penninerved, slightly discolorous, with minute stellate hairs (mostly 4–5-armed) evenly scattered over upper surface, the lower surface with fewer hairs; petioles 4–12 mm long, densely stellate-pubescent, especially on adaxial side; stipules subulate, 5–7 mm long, erect. *Flowers* sometimes solitary in the leaf axils, more commonly in terminal corymbose inflorescences, the pedicels 2–5 mm long, stellate-pubescent; involucellar bracts 7–10, linear, 5–6 × 0.5–1 mm, ciliate; calyx ca. 2 mm long, glabrous; petals pink, 10–12 mm long; androecium ca. ½ length of petals, the anthers few, subsessile. *Fruits* 4–5 mm diam., glabrous; mericarps ca. 5 mm long, dorsally rounded, smooth, with single apical spine 8–10 mm long (retrorsely barbed) that is basally buttressed dorsoventrally, the 2 lateral spines suppressed.

Distribution (Fig. 75). From central Brazil, where it is apparently not common. The type locality for

Pavonia monatherica and for *P. begoniaefolia* is approximately 22°20–30′S, 42°15′–43°10′W; that for *P. paraibica* is at 21°45′S, 43°20′W.

Illustrations. Gürke (1892b: pl. 89, fig. 2, as *P. monatherica*); Wawra (1883, vol. 1: pl. 6).

Specimens examined. **BRAZIL. WITHOUT LOCALITY:** *Sellow 1934* (US). **MINAS GERAIS:** Carangola, Fazenda Santa Rita, *Leoni 606* (GFJP, NY). **SÃO PAULO:** Ubatuba, Picinguaba, Trilha do Corisco, *Esteves & Rossi 2620* (NY, SP).

Although Garcke (1853) concluded that *Urena stellata* Sprengel was a synonym of *Pavonia typhalea*, a disposition followed by Gürke (1892b), the description (and type) clearly indicate it to be distinct. The possible relationship of *P. stellata* with *P. fruticosa* and with *P. nemoralis* is discussed under the latter species. *Pavonia stellata* stands out from all other species of *P.* subgen. *Typhalea* in having only a solitary spine at the apex of the mericarp.

Pavonia subgen. *Typhalea* sect. *Urenoideae*

Pavonia sect. **Urenoideae** St.-Hilaire, Fl. Bras. Mer. **1:** 223. 1827.

Erect or decumbent subshrubs or shrubs, the leaves ovate, lanceolate, or elliptic (suborbicular in one species), palmately nerved or penninerved, with more or less equal-length internodes; involucellar bracts 5–16; flowers axillary, the corolla yellow or white (red in one species); mericarps each 3-spined.

Lectotype species, here designated. *Pavonia sepium* St.-Hilaire.

Key to the species of *Pavonia* sect. *Urenoideae*

1. Leaf blades reniform or suborbicular, obtuse; plants decumbent; corolla yellow. . . . . 187. *P. renifolia*
1. Leaf blades longer than wide, ovate, lanceolate, elliptic, or spatulate; plants erect; corolla yellow, white, or red.
  2. Fruiting pedicels to 10 cm long, often exceeding subtending leaves; petals 10 × 3 mm, white; leaves penninerved, the upper surface glabrous. . . . . 192. *P. stenopetala*
  2. Fruiting pedicels usually <6 cm long, or if >6 cm long then shorter than subtending leaves; petals usually larger and always wider (only slightly wider in *P. piptocalyx* but then pedicels only 1 cm long), yellow, white, or red; leaves penninerved or palmately nerved, the upper leaf surface usually more or less stellate-pubescent or with simple hairs (sometimes glabrous).
    3. Calyx 4–5 mm long, circumscissally dehiscent, falling with the white corolla; leaves penninerved; flowers axillary but sometimes densely aggregated at apex of stems. . . . . 184. *P. piptocalyx*
    3. Calyx 3–14 mm long, persistent; leaves penninerved or palmately nerved.
      4. Leaves lanceolate or elliptic, 2.5–4 times as long as wide, penninerved or palmately nerved; petals 0.8–2 cm long.
        5. Plants decumbent; spines of mericarps 8–9 mm long, obscurely barbed; corolla yellow. . . . . 175. *P. flavispina*
        5. Plants erect; spines of mericarps prominently barbed; corolla yellow, white, or (in *P. penduliflora*) red; leaves sometimes distichous.
          6. Hairs of upper leaf surface predominantly simple; corolla white.
            7. Stems hispid, the hairs 1–2 mm long, ferruginous; involucellar bracts 13–16, prominently ciliate (hairs 2–5 mm long); fruits 10–12 mm diam. . . . . 182. *P. oxyphyllaria*
            7. Stems pubescent, the hairs <1(–1.5) mm long, not ferruginous; involucellar bracts 7–9, short-pubescent (hairs 1.5 mm long or less); fruits 5–10 mm diam.
              8. Leaves more or less asymmetrical at base, densely pubescent; stipules 3–6 mm long; fruits 5–7 mm diam.
                9. Pedicels 3–5 cm long; petioles 2–5 mm long; involucel with rounded sinuses; mericarp body 7 mm high, the central spine shorter; Peru. . . . . 188. *P. rudis*
                9. Pedicels 1–3 cm long; petioles 4–10 mm long; involucel with acute to narrowly rounded sinuses; mericarp body 5–6 mm high, the central spine longer; Brazil. . . . . 178. *P. intermedia*
              8. Leaves basally symmetrical, sparsely pubescent except on margin; stipules 8–12 mm long; fruits 8–10 mm diam.; pedicels 2–11 cm long, antrorsely pubescent. . . . . 185. *P. pseudo-typhalea*
          6. Upper leaf surface stellate-pubescent to subglabrous; corolla yellow or white.
            10. Corolla white; involucel shorter than calyx; flowers often forming a terminal inflorescence. . . . . 180. *P. leucantha*

10. Corolla yellow; involucel equaling or exceeding calyx; flowers always solitary in the leaf axils.
    11. Lower leaf surface with tufts of hairs in axils of main veins and often of the secondary veins. . . . . 190. *P. sepium*
    11. Lower leaf surface lacking tufts in the vein axils, variously pubescent.
        12. Lower leaf surface heterotrichous, with scattered 3–4-armed spicules (1 mm or more diam.) and underlying finer pubescence; calyx and involucel prominently hispid-ciliate, the hairs 1–1.5 mm long; leaves narrowly elliptic or oblong, widest in the middle; Mexico. . . . . 193. *P. uniflora*
        12. Lower leaf surface uniformly stellate-pubescent (hairs <1 mm diam.); calyx and involucel sparsely hispid-ciliate, the hairs 0.5–1 mm long; leaves ovate-lanceolate or ovate-elliptic, widest below the middle; NW South America. 189. *P. sepioides*

4. Leaves ovate, 1–2.5 times as long as wide, palmately nerved; petals 2–4 cm long [unknown in *P. horrida*].
    13. Lateral spines of mericarp erect; stems with fragile arachnoid hairs; corolla yellow.
        14. Calyx 4–7 mm long; involucel 5–9 mm long, exceeding calyx; corolla 1.5–2 cm long; uppermost leaves often notably reduced; mericarp spines 2–3 mm long; NE Argentina, Paraguay. . . . . 181. *P. narcissi*
        14. Calyx 6–9 mm long; involucel 8–11 mm long, subequal to calyx; corolla 2–2.8 cm long; uppermost leaves only slightly reduced; mericarp spines 3–5 mm long; NW Argentina, S Bolivia. . . . . 176. *P. formosa*
    13. Lateral spines of mericarps radiate.
        15. Spines of mericarps obscurely barbed, 3–6 mm long; body of mericarp hirsute; corolla yellow.
            16. Calyx longer than involucel, puberulent, not ribbed; stems with fragile arachnoid hairs; spines of mericarps 5–6 mm long; SE Brazil. . . . . 177. *P. horrida*
            16. Calyx shorter than involucel, ciliate, basally 5-ribbed; stems stellate-hirsute; spines of mericarps 3 mm long; Bolivia. . . . . 179. *P. ionthacarpa*
        15. Spines of mericarps prominently barbed, 3–7 mm long; body of mericarp glabrous or retrorsely hispid.
            17. Corolla red, ca. 2 cm long; leaves ovate-elliptic to oblanceolate, distichous; hairs of upper leaf surface mostly bifurcate, sometimes simple; fruits 11–13 mm diam. . . . . 183. *P. penduliflora*
            17. Corolla yellow, 1–3 cm long; leaves ovate, spirally disposed; hairs of upper leaf surface simple or stellate (or absent); fruits 6–10 mm diam.
                18. Genitalia shorter than corolla; spines of mericarps 3–6 mm long; leaves about twice as long as wide; South America.
                    19. Pedicels 0.5–5.5 cm long, mostly axillary; petals 2–3.5 cm; mericarps retrorsely barbed on spines and main veins. . . . . 174. *P. communis*
                    19. Pedicels 0.5–3 cm long, often crowded apically in a congested inflorescence; petals 1–2 cm long; mericarps retrorsely barbed on spines, main veins, and all reticulae. . . . . 186. *P. ramosissima*
                18. Genitalia somewhat exserted from corolla; spines of mericarps 6–7 mm long; leaves ca. 1.5 times as long as wide; Florida and West Indies. . . . . 191. *P. spinifex*

174. Pavonia communis St.-Hilaire, Fl. Bras. Merid. **1:** 224. 1827. *Pavonia spinifex* subsp. *communis* (St.-Hilaire) Gürke *in* Martius, Fl. Bras. **12(3):** 481. 1892. Type. Brazil. Minas Gerais: Near Poso Alto, *St.-Hilaire 563* (holotype, P as photo CTES; isotypes, CTES, US–2). Fig. 73

Pavonia regnelliana Miquel, Linnaea **22:** 551. 1849. Type. Brazil. Minas Gerais: Caldas, (various dates), *Regnell I-19* (holotype, U; isotypes, C, F, NY, S–6, US).

Pavonia canaminensis Rusby, Mem. New York Bot. Gard. **7:** 297. 1927. Type. Bolivia. Canamina, July 1921, *Rusby 85* (holotype, NY). Annotated by T. H. Kearney: "seems very close to . . . *P. speciosa* H.B.K." but clearly to be placed here.

Pavonia urticifolia Arechavaleta, Anal. Mus. Nac. Montevid. **3:** 144. 1898. (Non *Pavonia urticifolia* C. Presl, 1835.) Type. Uruguay. "Orillas de montes ó selvas, en Tacuarembó," *Arechavaleta s.n.* (holotype, MVM?–n.v.).

Shrubs 1–2 m tall, sometimes scandent to 3 m, the stems stellate-pubescent, the hairs of variable size (i.e., heterotrichous). *Leaf* blades ovate, mostly 5–14 × 2–8 cm, truncate to subcordate, serrate, acute or acuminate, palmately 5-nerved, discolorous, sparsely to densely stellate-pubescent, more densely so beneath; petioles 0.3–2(–3) cm long; stipules filiform, 5–12 mm long. *Flowers* solitary in the leaf axils, the pedicels 0.5–5.5 cm long; involucellar bracts 5–7, linear-lanceolate, 7–12 × 1–2 mm, ciliate; calyx 6–11 mm long, ciliate; corolla 2–3.5 cm long, yellow, obscurely pubescent externally, bearded on margins of claw; staminal column 1 cm long, glabrous, the filaments 1–2 mm long; styles exserted from column for 5–6 mm, free for distal 2–3 mm. *Fruits* 8–10 mm diam., the mericarp body 7 mm long; mericarps strongly keeled, the keel retrorsely scabrid, rugose, 3-spined, the spines divergent, 3–6(–8) mm long, retrorsely barbed. Chromosome number, $2n$ = 112 (Krapovickas & Cristóbal, 1965, as *P. spinifex*).

Distribution (Fig. 74). Widely distributed in South America, from Peru to Uruguay and Argentina, in forest and shrubland, at 600–2500 m elevation. The type locality is at 22°12′S, 44°58′W; that for *Pavonia regnelliana* is at 21°55′S, 46°40′W; that for *P. urticifolia* is at 31°40′S, 56°00′W.

Illustrations. Gottsberger (1972: figs. 14–16); Gürke (1892b: pl. 85, as *P. spinifex* subsp. *communis*).

Representative specimens examined. **PERU. AMAZONAS:** Chachapoyas, Nogal Cucho, *C. P. Cowan et al. 4289* (TEX, USM). **CAJAMARCA:** Near San Miguel de Asunción, between Magdalena and Cajamarca, *Ferreyra 7072* (F). **JUNÍN:** Prov. of Tarma, 1.5 km below Huacapistana, *Ferreyra 310* (GH).

BRAZIL. MATO GROSSO DO SUL: Mun. Amambai, 10 km S of Cap. Bado, *Gregory et al. 9684* (CTES); 1 km from Mundo Novo, *Hatschbach et al. 58656* (CTES, MBM, TEX). **MINAS GERAIS:** Ouro Preto, base of Pico Itacolomi, *Arbo et al. 3990* (CTES, K, NY); Serra do Espinhaço, 3 km N of Mariana, *Irwin et al. 29711* (CTES, F, NY); trail Viçosa to São Geraldo, *Mexia 4108* (A, CAS, F, GH, K, MO, NY, PH, S, U, US). **PARANÁ:** Ponta Grossa, *Dusén 7821* (GH, MO, NY, S); 15 km from Curitiba, *A. H. Gentry & Zardini 49755* (MO, NY); Serra Capivari Grande, Mun. Campina Grande do Sul, *Hatschbach 26327* (CTES, NY); Jaguariahyva, *Jönsson 334a* (GH, NY, S); Cerro Azul, along rd. to Rio Branco do Sul, *Landrum 4074* (NY, TEX); Curitiba, *Lindeman & de Haas 480* (K, NY, U); Mun. Guarapuava, Palmeirinha, *Pedersen 10989* (C, CTES, MO, NY). **RIO GRANDE DO SUL:** Mun. Bom Jesus, Pelotas River, rd. from São Joaquim to Roncinha, *Krapovickas & Schinini 38336* (CTES, HAS, UC); Pinhal Quiemado, *J. Pereira 8560* (= *Pabst 7835*) (K, NY). **RIO DE JANEIRO:** Mun. Patido Alferes, 75 km N of Rio de Janeiro, *Mizoguchi 1073* (A, NY); Vassouras, Morro Azul, *Monteiro 2093* (CAS). **SANTA CATARINA:** 2 km E from the access to Rio do Sul, BR-470, *Krapovickas & Cristóbal 43569* (CHR, CTES, EACS, LG, LIL, MBM, NY, TEX); Passo das Contas-S. Joaquim, *Reitz 3315* (CAS, S). **SÃO PAULO:** Between Capao Bonito and Guapiara, *Krapovickas et al. 23259* (CTES, NY); Sorocaba, Brigadeiro Tobias, Sitio Arco Iris, *Mello-Silva & Cecantini 541* (NY, SPF); Mun. Cubitão, Paranapiacaba, *Pedersen 8799* (C–2, CTES); Campinas, Itapura, *Santoro 844* (CAS, S, US).

BOLIVIA. LA PAZ: Prov. Inquisivi, below Cajuata, *Dorr & Barnett 6838* (LPB, NY, TEX); Coripati, Yungas, *Bang 2067* (E–2, F, GH, MO, NY, PH, US); rd. from La Paz to Chulumani km 105, *Krapovickas 6884* (CTES, SI). **SANTA CRUZ:** Yungas de San Mateo, Comarapa, *J. Steinbach 8407* (A, E, GH, K, MO, NY, PH, S, U).

PARAGUAY. Villarica, *Jörgensen 4288* (A, CTES, F, GH, MO, NY, PH, S, US); Tati Yupí Biological Reserve, *Marmori 870* (CTES). **AMAMBAY:** Pedro Juan Caballero, *Schwarz 11889* (CTES); Capú Bary, *Schwarz 12154* (CTES). **CAAGUAZÚ:** 5 km E of Caaguazú, *Krapovickas et al. 12483* (CTES); 34 km W of Caaguazú, *Schinini 9190* (CTES). **CANENDIYÚ:** Near Igatymí, *Fernández-Casas & Molero 5939* (MO, NY). **GUAIRA:** Cordillera de Ybytyruzú, *Zardini & Aguayo 8578* (CTES, MO). **PARAGUARI:** 2 km N of Salto Cristal, *Zardini & R. Velásquez 10118* (CTES, MO).

ARGENTINA. CORRIENTES: Prov. Ituzaingó, rt. 38 and Aguapey River, *Cáceres et al. 355* (CTES, NY). **MISIONES:** Dept. Candelaria, Santa Ana, *Montes 1564* (CTES, LIL, TEX, US); Dept. San Pedro, Monte Carlo, *Schwindt 1115* (CAS, CTES, MO, NY, TEX); Dept. San Ignacio, San Ignacio, *Vanni et al. 1421* (CLEMS, CTES, F, LIL, NY).

URUGUAY. RIVERA: Bajada de Pena, *Del Puerto 6052* (CTES, F).

A group of five yellow-flowered species of *Pavonia* subgen. *Typhalea* sect. *Urenoideae* have frequently been confused in previous literature. Because they are not juxtaposed in the key, their distinctions merit additional discussion. The five are *P. communis*, *P. sepioides*, *P. sepium*, *P. spinifex*, and *P. uniflora*. As is indicated in the key and in Table IV, *P. communis* and *P. spinifex* are relatively large-flowered, and *P. sepioides*, *P. sepium* and *P. uniflora* are relatively small-flowered. *P. spinifex* may be distinguished by its broad, often cordate leaves, its usually exserted stigmas, and its Caribbean distribution. *P. uniflora* stands out for its narrow leaves, long mericarp spines, and Mexican distribution. The three South American species are distinguished by differences in corolla size, leaf size, and stem pubescence, *P. communis* having larger corollas, larger leaves, and denser pubescence than *P. sepioides* or *P. sepium*.

Gottsberger (1972) describes bee-pollination in *Pavonia communis*, as well as the floral phenology and other insect visitors.

175. Pavonia flavispina Miquel, Linnaea **22:** 551. 1849. *Typhalea flavispina* (Miquel) Monteiro, Anais XII Reun. Soc. Bot. Bras. 29. 1961. Type. Brazil. Minas Gerais: City of Caldas, *Regnell II-21* (holotype, U; isotypes, B, destroyed, as photo F-9443, F as photo F-56240, NY, S–4, US). Fig. 73

Table IV

Comparison of five yellow-flowered species of *Pavonia* sect. *Urenoideae* that are often confused

	P. communis	*P. sepium*	*P. spinifex*	*P. uniflora*	*P. sepioides*
L/W ratio of leaves	Ca. 2	Ca. 2	Ca. 1.5	2.5–4	Ca. 2
Pedicel length (cm)	0.5–5.5	1.5–6	1–2	1–3.5	1.5–6
Calyx length (mm)	6–11	5–6	8–11	6–9	5–9
Stem pubescence	Densely stellate, heterotrichous	Glabrescent, narrow lines of hairs	Sparse to glabrescent	Stellate, heterotrichous	Glabrescent, narrow lines of hairs
Corolla length (cm)	2–3.5	0.8–1.5	2–2.5	0.9–1.5	1.2–2.2
Staminal column length (mm)	10	3–6	Ca. 20	4–7	3–7
Stigma position	Included	Included	Exserted	Included	Included
Mericarp spine length (mm)	3–6	3–6	6–7	7–10	3–7
Chromosome number	$2n = 112$	$2n = 56$	—	—	—
Distribution	Peru and Brazil to Argentina	South America, S of 20° lat.	Antilles, Florida	Mexico	NW South America

Subshrubs, decumbent and rooting at the nodes, the stems with narrow lines of fine simple hairs 0.3–0.4 mm long, otherwise glabrous. *Leaf* blades mostly 2–6 × 1–3 cm, elliptic, truncate to cuneate (somewhat asymmetrical at base), serrate, acute, palmately 5-nerved, very sparsely and minutely pubescent to nearly glabrous; petioles 2–7(–15) mm long, densely pubescent on adaxial side; stipules 4–7 mm long, lance-linear. *Flowers* solitary in the leaf axils, the pedicels 0.5–2 cm long, slender; involucellar bracts 5, linear, 5–6 mm long (shorter than calyx), subglabrous; calyx 7–8 mm long, marginally ciliate but otherwise sparsely pubescent to nearly glabrous, divided nearly to the base; corolla 9–10 mm long, yellow; staminal column ca. 4 mm long, the filaments 1 mm; styles and stigmas exserted from the staminal column by 5 mm. *Fruits* 7 mm diam. (excluding spines), ascending-setose, rugose; mericarp body 5 mm tall, each mericarp 3-spined, the central spine erect, the lateral spines spreading, the spines stout, 5–9 mm long, with minute obscure barbs at tip.

Distribution (Fig. 79). From Brazil (Minas Gerais, Rio Grande do Sul, Santa Catarina) and Argentina (Misiones), in forests and on forest margins. The type locality is at 21°55′S, 46°40′W.

Illustrations. Gürke (1892b: pl. 86, fig. 2); Krapovickas (1982b: fig. 1E).

Specimens examined. **BRAZIL. MINAS GERAIS:** Sta. Rita de Caldas, *Krapovickas & Cristóbal 35498* (C, CTES, F, ICN, K, MBM, NY, US). **RIO GRANDE DO SUL:** Pelotas, *da Costa Secco 1281* (F); Mata do Sétimo Céu, Marcelino Ramos, *Jarenkow 699* (CTES). **SANTA CATARINA:** Araranguá, Camping Lago Dourado, *Krapovickas & Vanni 36973* (CTES).

ARGENTINA. MISIONES: San Igancio, Jesuit Ruins, *Esteves et al. 2703* (NY, SPF); Dept. San Ignacio, San Ignacio, *Krapovickas & Cristóbal 28650* (BAA, C, CTES, ENCB, F, G, LG, LIL, MO, NY, SI), *Krapovickas & Cristóbal 43733* (C, CTES), *Krapovickas 2302* (CTES, MBM, MO, S-2, SI); Dept. San Javier, San Javier, *Krapovickas & Cristóbal 28904* (CTES, G, LIL, NY, SI).

Local name. Rozeta de mato (Brazil: Río Grande do Sul).

Pavonia flavispina is distinguished by having plants that are decumbent and rooting at the nodes. The spines on the mericarps are only obscurely barbed, similar to those of *P. horrida* and *P. ionthacarpa*.

176. Pavonia formosa Fryxell sp. nov. Type. Bolivia. Prov. Tarija, Entre Ríos, *Krapovickas & Schinini 39069* (holotype, NY; isotypes, CTES, F, MICH, UC).

Frutices vel suffrutices, caulibus pube semi-arachnoidea; laminis foliorum ovatis, in axillis venarum principalium et secundarium caespitibus pilos tenuium; floribus axillaribus solitariis, corollis flavis; fructibus 9–10 mm diametro, mericarpiis valde reticulatis, spinis erectis tribus hamatis 3–5 mm longis, hamulis secus costas lateralas extensis.

Shrubs or subshrubs 0.5–2 m tall (or taller if supported on other vegetation), the stems sparsely stellate-pubescent with fragile semi-arachnoid hairs, sometimes also with a few small stellate spicules. *Leaf* blades ovate, up to 12 × 6 cm, dentate, narrowly acute to acuminate, palmately 5-nerved, the upper surface with sparse stellate hairs <0.5 mm diam., the lower surface with sparse to dense stellate hairs that are somewhat larger and with tufts of fine hairs (domatia?) more or less well developed in the axils of the principal and of the secondary veins; petioles to 4 cm long, the upper side with a dense line of fine hairs, otherwise with scattered, appressed 3–4-armed stellate spicules, these denser toward apex, and/or with arachnoid pubescence like stem; stipules 8–9 mm long, narrowly linear. *Flowers* solitary in the leaf axils, the pedicels 1–4 cm long, with even pubescence of fine recurved hairs and sometimes also simple hairs 1 mm long; involucellar bracts 6, ligulate to lanceolate, 8–11 × ca. 1.5 mm, more or less ciliate; calyx 6–9 mm long, whitish with green veins, ciliate, the hairs almost 2 mm long; corolla yellow, 20–28 mm long, the genitalia included. *Fruits* 9–10 mm diam.; mericarps strongly reticulate, the body ca. 5 mm tall, the 3 retrorsely barbed spines erect, 3–5 mm long, the barbs extending down the lateral ribs and less completely the central one, otherwise glabrous.

Distribution (Fig. 76). In S Bolivia and NW Argentina, more or less between 19°S and 23°S latitude, at approximately 64° W longitude, from 850 to 2200 m elevation. The type locality is at 21°32′S, 64°12′W.

Specimens examined. **WITHOUT LOCALITY:** "Herb. mart. Suppl. Nro. 52," *ex* Herb. J. Gay, *s. coll. s.n.* (K).

BOLIVIA. **SANTA CRUZ:** Prov. Cordillera, Camiri 75 km toward Monteagudo, *S. Beck & Liberman 9818* (CTES, US); Prov. Vallegrande, *Krapovickas 8781* (CTES); Prov. Vallegrande, Comarapa, *Krapovickas 8915* (CTES); border of Santa Cruz and Chuquisaca, Cuesta de Incahuasi, between Lagunillas and Monteagudo, *Krapovickas & Schinini 31349* (CTES); Prov. Cordillera, Alto Parapetí, *de Michel 165* (CTES). **TARIJA:** Prov. Acre, rd. from Emborozú to Mamora, 5 km from Mamora, *Cuezzo et al. 11609* (CTES); Prov. Acre, 85 km toward Bermejo, before La Mamora, *S. Beck 14102* (CTES, US).

ARGENTINA. **SALTA:** Oran, 5 km SE of San Pedrico, *Eyerdam & Beetle 22703* (K); Urundal, 25 Oct 1948, *Fries s.n.* (S); Orán, Pocitos, *Krapovickas 9522* (CTES, F, NY); Orán, *O'Donell 3118* (CTES, S).

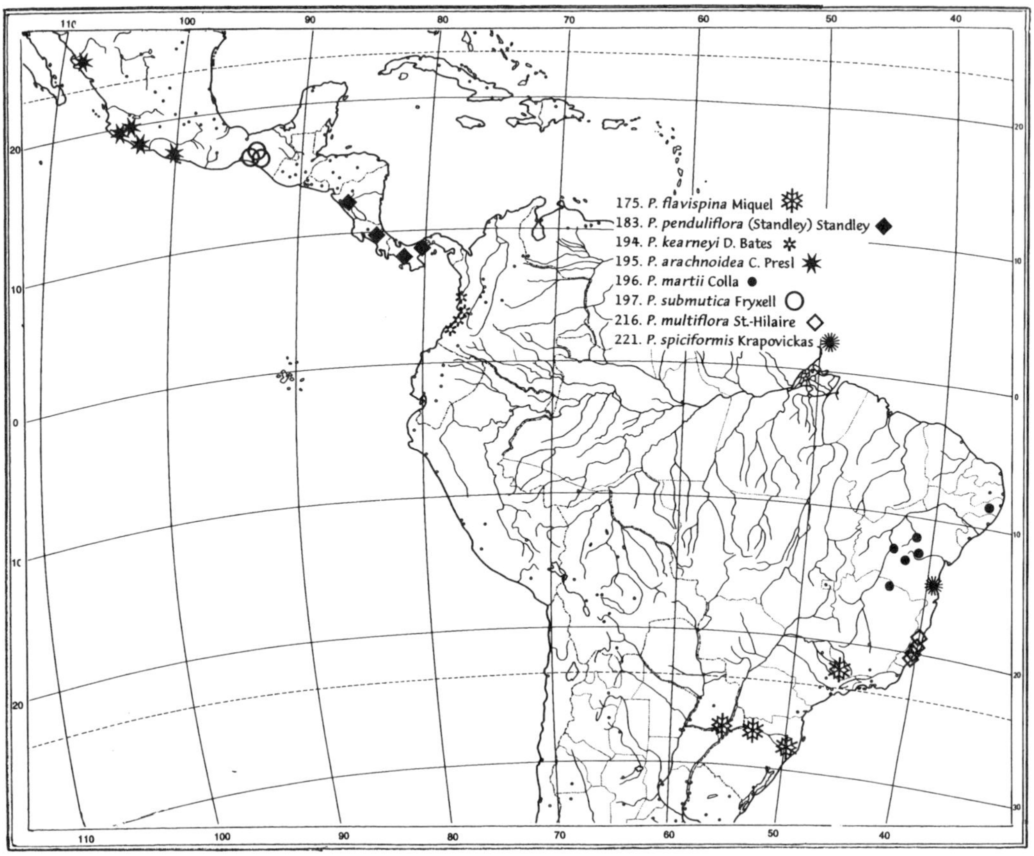

FIG. 79. Distribution map of eight neotropical species of *Pavonia* subgen. *Typhalea* and *P*. subgen. *Goetheoides*.

As indicated by its placement in the key, *Pavonia formosa* is most closely allied to *P. narcissi*, from which it is separated geographically and probably altitudinally as well as by the characters given in the key.

177. Pavonia horrida Krapovickas, Bol. Soc. Argent. Bot. **20:** 283. 1982. Type. Brazil. Santa Catarina: Laguna, *Pedersen 12655* (holotype, HBR–n.v.; isotypes, C, CTES, herb. Pedersen). Fig. 73

Shrubs ca. 1 m tall, the stems minutely arachnoid-pubescent. *Leaf* blades ovate to subelliptic, mostly 4–9 × 2.5–6 cm, truncate or subcordate, serrate, acute, palmately 5-nerved, minutely stellate-pubescent, sparsely so above; petioles 5–20 mm long; stipules subulate, 4–11 mm long. *Flowers* solitary in the leaf axils, the pedicels 7–14 mm long; involucellar bracts 5–6, linear-lanceolate, 7–9 × 1 mm or less, puberulent; calyx 8–12 mm long, hirsute with underlying puberulence; corolla (sometimes cleistogamous) 2–3 cm long, yellow, the genitalia ca. ½ the length of the corolla. *Fruits* 8–9 mm diam. excluding the spines, prominently hirsute, the mericarp body 6 mm long, prominently keeled and rugose, 3-spined, the spines stout, 5–6 mm long, obscurely and minutely barbed, the 5 apical spines erect and connivent, the lateral spines radiate and connivent in pairs, emerging between the calyx lobes.

Distribution (Fig. 48). Confined to the state of Santa Catarina, Brazil, in shrubland. The type locality is at 28°29′S, 48°45′W.

Illustration. Krapovickas (1982b: pl. 1, fig. 1F).

Specimens examined. **BRAZIL.** **Santa Catarina:** Mun. Laguna, Mar Grosso, *Hatschbach & Koczicki 27540* (BMB, CTES); Mun. Laguna, Caputera, *Hatschbach et al. 52362* (C, CTES); Garopaba, *Klein & Bresolin 8843* (CTES, HBR); Laguna, *Krapovickas & Cristóbal 37725* (CTES, F, GH, HAS, K, LIL, MBM, MEXU, MICH, NY, SI, US, WIS); Laguna, *Occhioni 4633* (CTES, RFA).

Pavonia horrida is characterized by very stout spines on the mericarps that are only obscurely and minutely barbed, in common with *P. flavispina* and *P.*

ionthacarpa. It is distinguished from *P. flavispina* by its erect habit and from *P. ionthacarpa* by the characters given in the key.

178. Pavonia intermedia St.-Hilaire, Fl. Bras. Merid. **1:** 223. 1827. Type. Brazil. Minas Gerais: Near Itajuru, not far from São Miguel de Mato, *St.-Hilaire 555* (holotype, P as photo F-35476; isotypes, P–2 as photos CTES).

Subshrubs ca. 1 m tall, the stems loosely stellate-pubescent, the hairs 0.5–1 mm long. *Leaf* blades lance-elliptic, mostly 4–13 × 2–4.5 cm, distichous, basally subtruncate to cuneate and more or less asymmetrically so, serrate, acute, penninerved or palmately 3-nerved, somewhat discolorous, the upper surface densely covered with appressed simple hairs ca. 1 mm long, the lower surface with simple and bifurcate (occasionally stellate) hairs 0.5–1 mm long (especially on veins); petioles 4–10 mm long; stipules 3–6 mm long, subulate. *Flowers* solitary or paired in the leaf axils or aggregated terminally or on lateral branches in subcapitate clusters, the pedicels 1–3 cm long, with pubescence similar to that of stem except more or less antrorsely oriented; involucellar bracts 7–8, narrowly linear with acute to narrowly rounded sinuses, 6–7 mm long, connate below, hirsute; calyx ca. 4 mm long, ciliate; corolla white, 8–12 mm long, minutely pubescent externally. *Fruits* 6 mm diam., glabrous, subglobose, the mericarp body 5–6 mm high, smooth, 3-spined, the spines 4–7 mm long, more or less erect, retrorsely barbed, the central spine dorsoventrally buttressed at base, the lateral spines 3 mm apart at base.

Distribution (Fig. 75). From Brazil (Bahia, Minas Gerais).

Specimens examined. BRAZIL. ACRE: Seringal Monte Alegre, ca. 40 mi S of Rio Branco, *Baldwin 3158* (US) [geographically doubtful]. **BAHIA:** Ilhéus, *Almeida & Santos 268* (CTES, NY); Mun. Maraú, 8 km before Uberitaba, *de Carvalho & Mori 107* (RB); Mun. Ilheus, km 22 on hwy. from Ilheus to Itabuna, *Hage 206* (CTES, NY, TEX). **MINAS GERAIS:** Viçosa, *Irwin 2652* (NY, TEX); Divino, Fazenda Barra de São Pedro, 700 m, Córrego São Pedro, *Leoni 738* (GFJP, unicate); Fanado (= Minas Novas), *von Martius s.n.* (M as photo CTES); Viçosa, Chacha Valley, *Mexia 4457* (A, CAS, F, GH, K, MO, NY, PH, RSA, S, TEX, U, US); Mun. of Serro, Mata da Mai d'Agua, *L. O. Williams & Assis 7015* (GH). **RIO DE JANEIRO:** Without locality, *Widgren 778* (S).

Local names. Carrapicho de tres pies (Brazil: Minas Gerais), vassoura do mata (Brazil: Bahia).

This species was described by St.-Hilaire as having "floribus subglomeratis" and was therefore named "intermedia," indicating its intermediacy between those species with solitary flowers and those with definite inflorescences. This is perhaps a matter of interpretation, but in my opinion the species accords better with the solitary-flowered species of *Pavonia* sect. *Urenoideae* than with *P.* sect. *Typhalea*. The slightly asymmetrically elliptic (distichous?) subsessile leaves, spaced at "normal" internode lengths agree with the condition found in *P. pseudo-typhalea* (for example) and not with that found in *P. peruviana*, in which the internodes are markedly shortened, and the leaf blades are symmetrically oblanceolate. The inflorescences in *P. intermedia* (when well developed) are axillary as well as terminal. In a somewhat different way, *P. leucantha* is also "intermediate," in that it has the leaf disposition characteristic of *P.* sect. *Urenoideae* but has flowers often aggregated into more or less definite inflorescences that basically are axillary cymes but eventually (in a well-grown specimen) become terminal panicles. Because of this combination of characters, both *P. intermedia* and *P. leucantha* are keyed out in the keys to both *P.* sect. *Typhalea* and *P.* sect. *Urenoideae*.

Pavonia intermedia probably has its nearest relationship to *P. rudis* and *P. pseudo-typhalea*, from which it can be distinguished by the characters given in the key.

179. Pavonia ionthacarpa Krapovickas, Bol. Soc. Argent. Bot. **20:** 287. 1982. Type. Bolivia. Santa Cruz: Prov. Cordillera, 5 km N of Yatarenda, *Krapovickas & Schinini 31481* (holotype, CTES; isotypes, MBM, NY, SI).

Shrubs 0.5–1 m tall, the stems hirsute with both stellate and simple hairs. *Leaf* blades ovate to subelliptic, mostly 4–7(–9) × 2–3.5(–5) cm, basally truncate to subcuneate, crenate-serrate, acute, palmately 5-nerved, discolorous, the upper surface minutely and sparsely pubescent (the hairs 0.5 mm long or less, simple, bifurcate, or stellate), the lower surface with 3–6-armed stellate hairs up to 1 mm long; petioles 5–10(–30) mm long; stipules subulate, 5–8 mm long. *Flowers* solitary in the leaf axils, the pedicels 1–2(–4) cm long; involucellar bracts 5–6, narrowly lanceolate, 9–11 × 1–1.5 mm, ciliate or hirsute; calyx 8–10 mm long, ciliate, pallid, the midrib prominent and raised; corolla 2–2.5 cm long, yellow, glabrous except pubescent on margins of claw; staminal column ca. ½ length of petals, pallid; styles emergent from the column, the stigmas villous. *Fruits* 7–8 mm diam. (excluding spines), hirsute; mericarps prominently reticulate-rugose, 3-spined, the spines 3 mm long, obsoletely barbed, the central spine erect, the lateral spines radiate.

Distribution (Fig. 76). Endemic to Bolivia. The type locality is at 19°12′S, 63°32′W.

Illustration. Krapovickas (1982b: fig. 1B, pl. 2).

Specimen examined. **BOLIVIA.** **SANTA CRUZ:** Comarapa, *Cárdenas 4624* (CAS).

Pavonia ionthacarpa is characterized by the spines of the mericarps being only obscurely and minutely barbed, in common with those of *P. flavispina* and *P. horrida*. It is distinguished from *P. flavispina* by its erect habit and from *P. horrida* by the characters given in the key.

180. Pavonia leucantha Poeppig *ex* Garcke, Jahrb. Bot. Gart. Mus. Berlin **1:** 211. 1881. *Typhalea leucantha* (Garcke) Monteiro, Anais XII Reun. Soc. Bot. Bras. 29. 1961. *Lassa leucantha* (Garcke) Kuntze, Revis. Gen. Pl. **3(2):** 19. 1898. Type. Peru ["Brazil" in original description]. Maynas, *Poeppig 2302* (holotype, B as photo F-9452; isotypes, F–3, OXF, P as photo CTES). Fig. 80

Subshrubs or shrubs 0.5–3 m tall, the stems stellate-hispidulous, the hairs more or less appressed. *Leaf* blades mostly 9–16 cm by up to 9 cm (usually <5 cm wide), oblong-lanceolate to oblanceolate, basally slightly asymmetrical, truncate to cuneate, serrulate, acuminate, penninerved or weakly 3-nerved, stellate-pubescent (sparsely so on upper surface); petioles 2–7 mm long; stipules ca. 2 mm long. *Flowers* sometimes solitary but usually on 3–6-flowered axillary or terminal peduncles, the individual pedicels 4–25 mm long; involucellar bracts 8–9, gamophyllous, ca. 4 mm long, stellate-pubescent; calyx 5–6 mm long (exceeding involucel), stellate-pubescent; corolla ca. 1 cm long, white. *Fruits* 5–6 mm diam., the mericarp body 6.5 mm high, glabrous, the mericarps 3-spined, the spines retrorsely barbed, erect, 4–5 mm long, the lateral spines ca. 2 mm apart at base.

Distribution (Fig. 76). From Ecuador to Bolivia and in westernmost Brazil, growing in wet forest (sometimes in disturbed areas) at 100–1100 m elevation. The type locality may be near Tabatinga at 04°17′S, 69°56′W.

Additional illustration. Gürke (1892b: pl. 89, fig. 1).

Representative specimens examined. **ECUADOR.** **ESMERALDAS:** Muisne Canton, 50 km SW of Atacames, *Croat 73130* (MO). **MORONA-SANTIAGO:** Rd. from Méndez to Chipana, 7–8 km SE of crossroad to Méndez, *Bohlin et al. 1512* (GB, NY); Centro Shuar Yukatais, *Bennett & Gómez 3616* (NY, TEX). **NAPO:** Jatun Sacha Biological Station, *C. E. Cerón & M. Cerón 4719* (MO, NY); Añangu, NW corner of Yasuní National Park, *Korning & Thomsen 47064* (QCA). **NAPO-PASTAZA:** Napo, *Asplund 9105* (CAS, K, NY, S, US); Tena, *Mexia 7233* (UC, US). **PASTAZA:** Landayacu River, 75 km E of Puyo, *Gudiño 1115* (MO, US). **PICHINCHA:** Along rd. from Aloag to Santo Domingo, at confluence of Pilatón and Toáchi Rivers, *Sparre 13797* (S). **SANTIAGO-ZAMORA:** W of Upano River, along Chupiangas River, *Prieto 23* (CAS, NY); Macas, *Asplund 19827* (NY, S). **ZAMORA-CHINCHIPE:** Near Zamora, *Freire-Fierro et al. 2491* (AAU, NY, QCA); along the Nangaritza River, 1 km upstream from Shaime, *van der Werff et al. 13124* (MO, TEX).

PERU. **AMAZONAS:** Valley of the Santiago River, Quebrada Caterpiza, *Tunqui 772* (CTES, MO, NY). **HUÁNUCO:** Prov. Leoncia Prado, Distr. Cueva de las Pavas, *C. Díaz & Baldeón 2266* (F, MEXU, MO, NY, TEX); vic. of Tingo María, Cueva de las Pavas, *Ferreyra 1610* (CTES, F, GH, US); Prov. Pachitea, Iparia National Forest, *Schunke-Vigo 1537* (F, NY, US). **LORETO:** Santa María, ca. 25 km below Yurimaguas on the Huallaga River, *Allard 22463* (US); Tigre River, *Ayala et al. 2673* (AMAZ, F, MO, NY); Miahuayacu, near Iquitos, *Klug 939* (F, NY, US); Prov. Maynas, Iquitos, Caserio Caracocha, *Revilla 270* (F, MO, NY, TEX). **MADRE DE DIÓS:** Prov. Manu, Cocha Cashu Station, *Foster & Terborgh 6130* (F, US); Prov. Manu, a few km downriver from Atalaya, *Foster & Baldeón 12863* (F, NY). **PASCO:** 3–7 km N of the Paucartambo bridge, *A. H. Gentry et al. 39846* (MO, NY). **SAN MARTÍN:** Zepalacio, near Moyobamba, *Klug 3540* (F, GH, K, MO, NY, S, US); Tarapoto, *Woytkowski 35069* (F, MO). **UCAYALÍ:** Prov. Coronel Portillo, von Humboldt National Forest, *A. H. Gentry et al. 31098* (MO).

BRAZIL. **ACRE:** Cruzeiro do Sul-Monte Belô, left bank of Moa River, *Marinho 181* (CTES). **AMAZONAS:** Village of Kusu, Numpatakai River (tributary of Cenepa), *Berlin 932* (GH); Aramaçá Island, almost opposite Tabatinga, *Prance et al. 16790* (MO, NY, RB, U, US).

BOLIVIA. **BENI:** Vic. of Reyes, "Rurrenabaque" on Beni River, *Fleischmann 523* (S–2); Prov. Yacuma, Beni Biological Station, *Villanueva & Foster 787* (CTES). **LA PAZ:** Prov. S Yungas, Quiquibey River, *S. Beck 8114* (CTES, US); Prov. S Yungas, Alto Beni 10 km from Palos Blancos toward Yucumo, *Seidel & Schulte 2197* (CTES, US); Prov. S Yungas, Monte Alro, *Vargas et al. 2066* (CTES, US). **SANTA CRUZ:** Mapiri region, San Carlos, near Sarampiuni, *Buchtien 875* (GH, US); Dept. Obispo, Santisteban, Montero, *Krapovickas 9120c* (CAS, US); Prov. Ichilo, Amboró National Park, along Saguayo River, *M. Nee 37311* (CTES, NY, TEX); Prov. Sara, Buena Vista, *J. Steinbach 5414* (F, G, GH, MO, NY).

Local names: Carrapicho (Peru), japajapa (Peru), wíawía (Peru).

Reasons for placing *Pavonia leucantha* in *P.* sect. *Urenoideae* rather than in *P.* sect. *Fruticosa* are discussed under *P. intermedia*. Both species, however, are included in the keys to both sections.

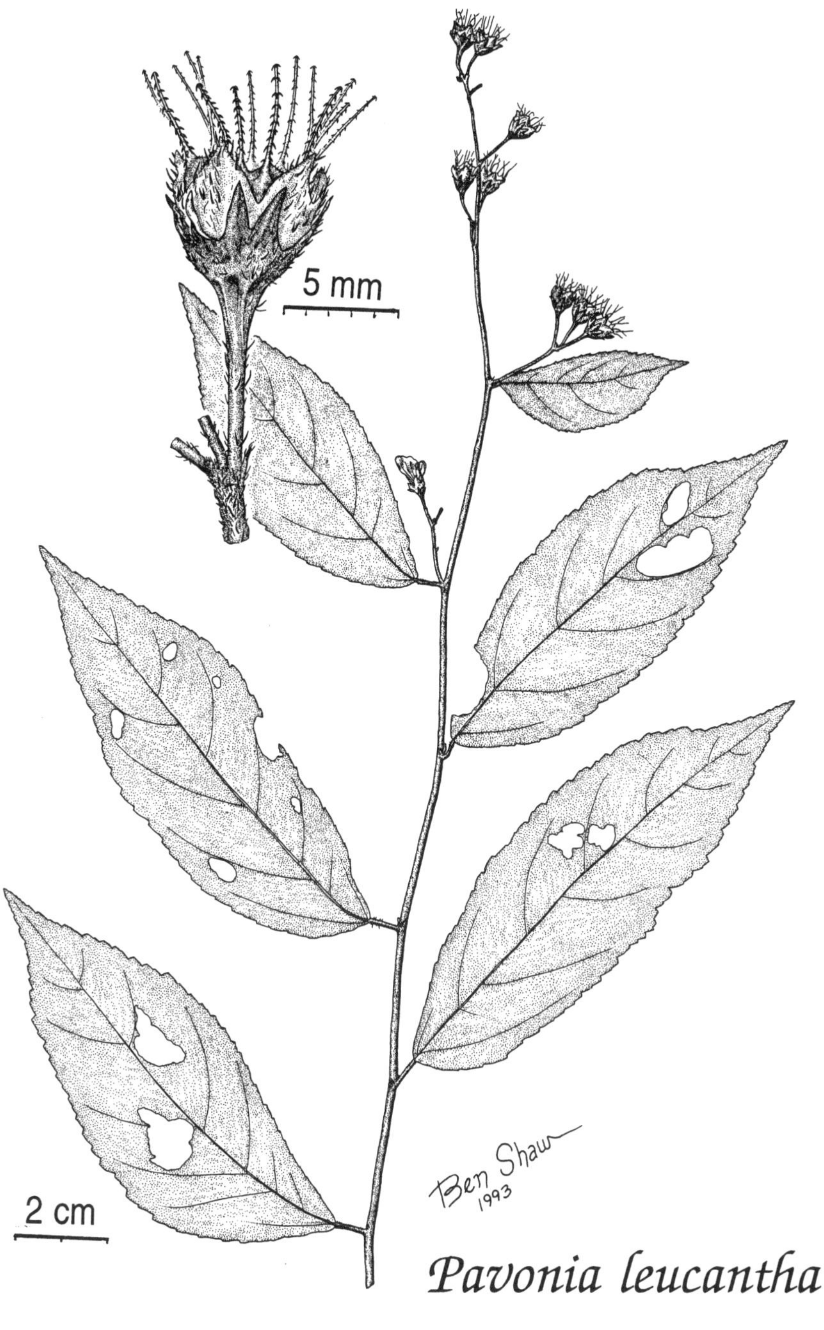

FIG. 80. *Pavonia leucantha.* Fruiting branch with enlargement of single fruit (*Mendenhall 12*).

Pavonia leucantha is the only species of *P.* subgen. *Typhalea* to have the involucel markedly shorter than the calyx.

181. Pavonia narcissi Krapovickas, Bol. Soc. Argent. Bot. **20:** 289. 1982. Type. Argentina. Corrientes: Dept. Itatí, Ensenada Grande, 26 Apr 1964, *Krapovickas*

& Cristóbal 11387 (holotype, CTES; isotypes, F as photo F-64928, G, SI). Fig. 73

Pavonia sepium var. *balansae* Gürke *in* Martius, Fl. Bras. **12(3)**: 482. 1892. *Pavonia communis* var. *balansae* (Gürke) Hassler, Ostenia 340. 1933 non *Pavonia balansae* Gürke, 1892. Type. Paraguay. Rengger, near Asunción, *Balansa 1622* (isotypes, CTES, G as photo F-23709, K, P–n.v.).

Shrubs 1.5 m tall, freely branched, the stems covered with minute arachnoid pubescence and with scattered larger stellate hairs. *Leaf* blades ovate, truncate, crenate-serrate, acute, mostly 3–9(–12) × 2–6(–10) cm, palmately 5-nerved, stellate-pubescent, more densely so beneath, sparsely so to glabrate (in age) above; petioles 0.5–2 cm long; stipules 3–8 mm long, linear-lanceolate. *Flowers* solitary in the leaf axils, the pedicels 2–4 cm long; involucellar bracts 6, linear-lanceolate, 5–9 mm long, minutely pubescent and sometimes ciliate; calyx 4–7 mm long, pubescent and sometimes ciliate; corolla 1.5–2 cm long, yellow; androecium 5–8 mm long; styles emergent from column, sparsely pubescent, the stigmas capitate, 0.6–0.8 mm diam., villous. *Fruits* ca. 6 mm diam., longitudinally ridged (retrorsely scabridulous on ridge) and rugose, the mericarp body 5 mm high, 3-spined, the spines erect, 2–3 mm long, retrorsely barbed.

Distribution (Fig. 76). From the valley of the Río Paraná in Argentina and Paraguay, in forests and on forest margins.

Illustrations. Krapovickas (1982b: fig. 1A, pl. 3).

Representative specimens examined. **BRAZIL.** Without locality, *Riedel 1334* (K).

PARAGUAY. Central Paraguay, without locality, *Morong 473B* (E, GH, K, MO, PH); in region of Lake Yparacay, *Hassler 11505* (A, C, E, F, GH, K, MO, NY, US). **CENTRAL:** Yaguarón, base of Mt. Oratorio, *Krapovickas et al. 12284* (CTES, G); Areguá, *Schinini & Mereles 24567* (CTES). **CORDILLERA:** 5 km N of Tobatí, *Krapovickas & Cristóbal 11467* (C, CTES). **GUAIRÁ:** Tororo, rd. to Cerro Polilla, *M. Ortiz 806* (CTES, MO); Cordillera de Ybytyruzú, rd. to Polilla, Cerro Acatí (25°55′S, 56°10′W), *Zardini & Velásquez 13588* (CTES, MO). **MISIONES:** 6 km S of Santa Rosa, rd. to General Delgado, *Arbo et al. 6128* (CTES, MO, TEX). **PARAGUARÍ:** Compañia Costa Segunda, Cerro Palacios, *Zardini 4526* (CTES, MO); Macizo Acahay, *Zardini & Cuevas 5362* (CTES, MO).

ARGENTINA. CHACO: Colonia Benítez, *Schulz 1868* (CTES); Dept. San Fernando, Antequara, La Mora, *Schulz 2084* (CTES). **CORRIENTES:** Dept. San Cosme, Paso de la Patria, *Krapovickas 43442* (AAU, BAA, C, CHR, CLEMS, CTES, EACS, F, FCQ, G, GB, HAS, K, LG, LIL, LPB, MICH, MO, NY, SI, TEX-2, TIAC, U, UB, WIS); Dept. Capital, Corrientes, Arroyo Límite, *Krapovickas & Cristóbal 11323* (C, CTES, ENCB, F, G, LIL, MBM, MEXU, MO, NY, WIS); Saladas, *Krapovickas & Cristóbal 11409* (BAB, CTES, ICN, WIS); Dept. General Paz, Ibahay, *Pedersen 11524* (C, CTES, K, MO); Dept. Empedrado, Estancia "Las Tres Marías," on the bank of Río Paraná, *Pedersen 11680* (C, CTES, NY); Dept. Esquina, Esquina, *Schinini & Carnevali 24953* (CTES, F, HAS, LIL, MICH, NY, QAS, SPF, USM). **MISIONES:** Dept. Cainguas, rt. 105, *Schwindt 4451* (CAS). **SANTA FÉ:** Dept. Gral Obligado, Los Laureles, *Blanchoud 2279* (CTES).

When in fruit, *Pavonia narcissi* is easily distinguished, but if fruits are lacking, it can sometimes be confused with other members of the "*P. sepium* complex" (see discussion under *P. sepium*). Its nearest relative is probably *P. formosa*, from which it is distinguished by the characters given in the key.

182. Pavonia oxyphyllaria J. D. Smith, Bot. Gaz. **23:** 237. 1897. Type. Costa Rica. Río Verde, Llanura de Santa Clara, *J. D. Smith 6449* (holotype, US). Fig. 73

Pavonia costaricensis Hochreutiner, Annuaire Conserv. Jard. Bot. Genève **10:** 18. 1906. Type. Costa Rica. hills of Zent, Atlantic slope, *Tonduz 14863* (holotype, G–n.v.; isotypes, B as photo F-9437, BM as photo CTES, K).

Malache fulva Standley, Contr. U.S. Natl. Herb. **18:** 115. 1916. *Pavonia fulva* (Standley) Standley, Field Mus. Nat. Hist., Bot. Ser. **4:** 231. 1929. Type. Costa Rica. Hacienda de Churipió [Chirripó?], Mar 1900, *Pittier 16080* (holotype, US; isotype, GH).

Shrubs 0.5–1(–2) m tall, the stems hispid (hairs ferruginous, 1–2 mm long) and with underlying finer pubescence. *Leaf* blades mostly 6–15 × 2–5 cm, narrowly elliptic (sometimes slightly asymmetrical), basally cuneate, serrulate, acuminate, penninerved, the upper surface with appressed simple hairs 1–2 mm long, those of the lower surface simple or bifurcate (rarely stellate); petioles 2–5(–12) mm long, the leaves often subsessile; stipules 3–4 mm long, subulate, hispid. *Flowers* solitary in the uppermost leaf axils, the pedicels 1–7 cm long, densely hispid, the hairs somewhat retrorse; involucellar bracts 13–16, filiform, 16–20 mm long, arcuate, copiously ciliate, the hairs 2–5 mm long, ferruginous; calyx ca. 5 mm long, stellate-pubescent and hispid; corolla subequal to involucel, white; staminal column ca. 4 mm long, glabrous, the anthers subsessile at apex. *Fruits* 10–12 mm diam., the mericarp body 6 mm high, glabrous, reticulate-veined, the mericarps each with 3 apical spines, the spines 6–10 mm long, divergent, retrorsely barbed.

Distribution (Fig. 81). From Costa Rica to Bolivia and Brazil, growing in tall rain forest at 200–1200 m elevation. The type locality for *Pavonia costaricensis*

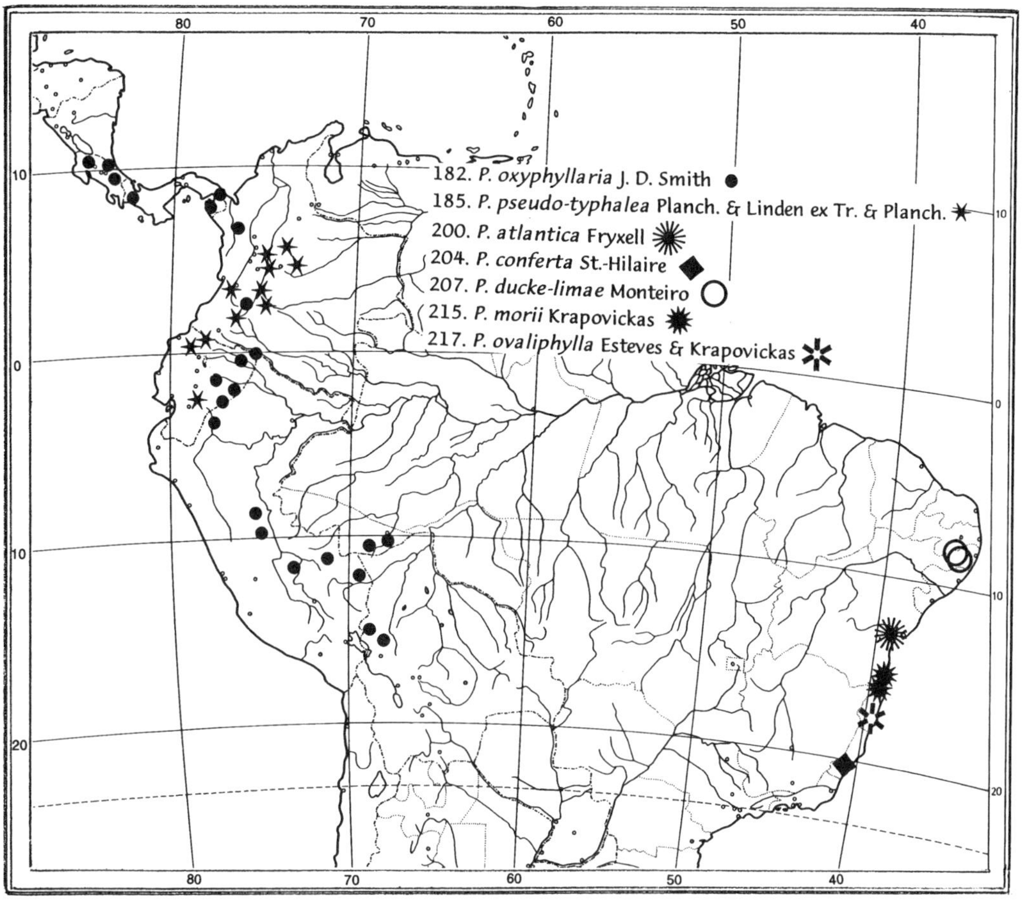

FIG. 81. Distribution map of seven neotropical species of *Pavonia* subgen. *Typhalea* and *P.* subgen. *Goetheoides*.

is at 10°00′N, 83°25′W; that for *Malache fulva* is at 09°45′N, 83°35′W.

Illustration. Robyns (1965: fig. 7).

Representative specimens examined. COSTA RICA. ALAJUELA: Santa María National Park, *Liesner 5062* (MO). HEREDIA: Tirimbina, *Proctor 32254* (LL–2). LIMÓN: Montana Andromeda, *Shank & Molina 4534* (F, GH, NY, US); vic. of Guapiles, *Standley 37120* (F, US). PUNTARENAS: 17 km N of San Vito, rd. to Potrero Grande, *Davidson 7152* (F, MO, NY, RSA). SAN JOSÉ: Vic. of El General, *Skutch 3865* (A, MO, NY, S, US).

PANAMA. BOCAS DEL TORO: Vic. of Chiriquí Lagoon, *von Wedel 1099* (GH, MO, US). CHIRIQUÍ: 2–8 km N of Cañas Gordas, *Busey 651* (F, MO). COCLÉ: La Mesa, *P. H. Allen 2742* (GH, MO, US). DARIEN: Trail SE of Manene to the Coasi River, *Hartman 12216* (MO, NY); on trail above Cana, *McPherson 15043* (MO, NY).

COLOMBIA. META: Mun. Mesetas, Vereda Gaviotas, *Callejas & Marulanda 5991* (MO); Sierra de la Macarena, Caño Entrada, *Philipson & Idrobo 1763* (F, US).

ECUADOR. MORONA-SANTIAGO: To the W of Macas, *M. A. Baker 6623* (MO, NY, QAME, QCNE, TEX); 20 km SE of San Juan Bosco, *A. H. Gentry et al. 30969* (MO). NAPO: Añangu, Yasuni National Park, *Argüello 577* (AAU, NY, QCA); Canton Aguarico, N bank of the Napo River, *Bensman 328* (QCNE, MO, WIS); 3 km E of the Caserío de Huamaní, N of hwy. from Hollin to Loreto, *Hurtado & Alvarado 385* (MO, NY); rd. from Hollin to Loreto between Guamaní and Pucuno Rivers, *Palacios 2195* (MO, NY). PICHINCHA: Santo Domingo de los Colorados, *C. E. Cerón & Benavides 6222* (MO, NY); 6 km WNW of P. Vicente Maldonado, *Harling & Andersson 23326* (GB, NY, QCA); along Toachi River, near Sto. Domingo, *Játiva & Epling 353* (NY, UC). SANTIAGO–ZAMORA: Shore of Pastaza River, opposite Mera, *Asplund 19252* (S).

PERU. AMAZONAS: Santiago River, 1 km from Caterpiza, *Huashikat 590* (MO, NY); vic. of community of Kusu, Numpatkin River, *Kayap 522* (GH). HUÁNUCO: Pachitea, Codo de Pozuzo, *Foster 9356* (MO, NY, TEX–2, USM); carretera marginal ca. 14 km from a point across Pachitea River from Puerto Inca, *Smith 1282* (MO). LORETO: Prov. Alto

Amazonas, Capahuari Sur, *R. Vázquez et al. 3078* (MO, NY). **Madre de Diós:** Manu National Park, *Foster & Baldeón 12778* (CTES, F, NY); near confluence of Tambopata and La Torre Rivers, *S. F. Smith et al. 516* (NY, US). **Pasco:** Prov. Oxapampa, Palcazu, Alto Isozacín River, Ozuz to Río Lobo, *Foster & d'Achille 10056* (F, NY). **San Martín:** Rd. to cacerio de Almendras, *Schunke-Vigo 3316* (F, US); rd. to Shunté, *Schunke-Vigo 3845* (F, US); Prov. Mariscal Cáceres, Dept. Tocache Nuevo, mouth of Mishollo River, *Schunke-Vigo 6375* (AMAZ, MO, NY, US).

BRAZIL. Acre: Mun. Rio Branco, rd. from Rio Branco to Porto Acre, km 20, *Cid & B. W. Nelson 2840* (MO, NY, TEX); 24 km from Rio Branco on rd. from Rio Branco to Porto Velho, *Lowrie et al. 419* (NY, TEX, US); Moa River, at Serra da Moa village, *Prance et al. 12503* (CM, NY, TEX); Acre River, *Ule 9588* (MG, NY).

BOLIVIA. La Paz: Prov. N. Yungas, Alto Beni, from San Antonio climbing to the "Tropical" concession, *Seidel & Hirschle 2624* (CTES, US).

Local name. Weawea (Peru).

Pavonia oxyphyllaria is readily distinguished by its ferruginous-hispid stems, its 13–16 prominently ciliate involucellar bracts, and its relatively large fruits.

183. Pavonia penduliflora (Standley) Standley, Publ. Field Mus. Nat. Hist., Bot. Ser. **4:** 231. 1929. *Malache penduliflora* Standley, Contr. U.S. Natl. Herb. **18:** 117. 1916. Type. Panama. Chiriquí: S slope of Cerro de la Horqueta, 1700 m, 17–19 Mar 1911, *Pittier 3188* (holotype, US; isotype, NY). Fig. 73

Shrubs 0.5–2.5 m tall, the stems coarsely stellate-pubescent. *Leaves* usually distichous, the blades ovate-elliptic or oblong to oblanceolate, mostly 5–20 × 2–7 cm, asymmetrically truncate or rounded basally, serrate, acuminate, palmately 3(–5)-nerved, stellate-pubescent beneath, sparsely pubescent above (the hairs sometimes simple, mostly bifurcate, 0.5–0.7 mm long) to glabrate; petioles 0.5–1(–2) cm long; stipules 4–5 mm long, subulate. *Flowers* solitary in the leaf axils, the pedicels 2–10 cm long, stellate-pubescent; involucellar bracts 8–9, linear, 7–9 × 1 mm, hirsute; calyx 6–9 mm long, hirsute; corolla 1–2 cm long, red, externally pubescent. *Fruits* 11–13 mm diam., reddish, glabrous, the mericarp body 7–10 mm high, smooth, 3-spined, the spines 5–8 mm long, retrorsely barbed, divergent, the lateral spines 4 mm apart at base.

Distribution (Fig. 79). Central America from Nicaragua to Panama, in rain forest and cloud forest at 600–2200 m elevation. The type locality is at 08°45′N, 82°26′W.

Illustration. Robyns (1965: fig. 6F).

Specimens examined. NICARAGUA. Jinotega: At foot of Yalí Volcano, E side, *Moreno 8002* (MO). **Matagalpa:** El Arenal, *Moreno 9575* (CTES, MO).

COSTA RICA. Alajuela: Cerro Pato de Gallo, San Rafael de San Ramón, *Carvajal 490* (MO). **Puntarenas:** Zona Protectora Las Tablas, slopes of Cerro Chai, *Almeda et al. 6642* (CAS, NY); at the Coton (Las Alturas) River, *Burger & Matta 4576* (F, MO, NY); foothills of Cordillera Talamanca, near Sitio Cotón, *Davidse 24562* (MO, TEX, U); Las Alturas Biological Station, *A. Fernández et al. 333* (MO, TEX); upper Burú River, *L. D. Gómez et al. 21422* (MO); above La Tigra, *Hazlett 5012* (F); Cantón de Coto Brus, Las Mellizas, *Herrera 3445* (CR, MO, NY); Las Alturas and vic., *Maas & McAlpin 1481* (F, U).

PANAMA. Canal Zone: Along Ladrillo River, *Pittier 3287* (US). **Chiriquí:** Cerro Horqueta, *Blum & Dwyer 2597* (MO, SCZ); E of Boquete on Cerro Azul, *Croat 26827* (MO); above Boquete, Monte Rey, *Croat & Porter 15670* (F, GH, MO); above Boquete, La Popa, *W. G. D'Arcy & J. J. D'Arcy 6400* (F, GH, MO); Boquete, *Davidson 54* (A, F, MO); Boquete, *Dwyer & Hayden 7642* (GH, MO); Cerro Horqueta, *Dwyer & Hayden 7696* (GH, MO, US); Cerro Horqueta, *Dwyer & Lallathin 8741* (MO); Boquete, *Folsom 2182* (MO, NY); Cerro Horqueta, Boquete region, *W. von Hagen & C. von Hagen 2041* (MO, NY); near Cerro Pando, *Hamilton et al. 830* (MO); 3.5 mi NE of Boquete, *Hammel 5686* (MO); 4.5 mi NE of Boquete, *Hammel 7537* (MO, NY); between Pinola and Quebrada Honda, *Kirkbride & Duke 878* (NY); 10 km N of Los Planes de Hornito, *Knapp & Vodicka 5609* (MO, NY); Boquete, *Bro. Maurice 877* (MO, US); above Boquete toward Cerro Punta, *de Nevers & McPherson 6806* (MO, NY); Palo Alto, just E of Boquete, *Stern et al. 1033* (GH, MO); vic. of Casita Alta, Chiriquí Volcano, *Woodson et al. 983* (A, F, MO, NY); vic. of Bajo Mona and Quebrada Chiquero, *Woodson & Schery 529* (GH, MO); vic. of Bajo Chorro, *Woodson & Schery 662* (MO); Bajo Chorro, *Woodson & Schery 705* (GH, MO).

Pavonia penduliflora is notable for its red corollas and for its relatively large (11–13 mm diam.) reddish fruits. The upper surface of the leaf is recognizable by its distinctive, small bifurcate hairs.

184. Pavonia piptocalyx Krapovickas, Bol. Soc. Argent. Bot. **20:** 293. 1982. Type. Brazil. Pará: Belém, banks of the Guamá River, floodable zone, 22 Jan 1981, *Krapovickas & Cristóbal 37174* (holotype, IAN; isotypes, C, CLEMS, CTES–2, E, F, G, GH, HAS, INPA, IPA, K, LIL, MBM, MICH, MO, NY, P, RB, SP, SI, TEX, U, US). Fig. 73

Shrubs (scandent?) 1.5–3 m tall, the stems with appressed stellate hairs 0.3 mm diam. *Leaf* blades narrowly elliptic, 5.5–12 × 1–4.5 cm (2.5–8 times as long as wide), cuneate, serrate, acuminate, penninerved, with minute appressed stellate hairs (or glabrate on upper surface); petioles 4–30 mm long; stipules linear-lanceolate, 6–8 mm long. *Flowers* solitary in the leaf axils but sometimes congested apically among the leaves, the pedicels 6–12 mm long; involucellar bracts 5, ligulate, 8–10 × 2–4 mm, sparsely and minutely pubescent; calyx 4–5 mm long, deciduous with the corolla after anthesis; petals ca. 10 × 3 mm, white; staminal column somewhat shorter than corolla, the anthers subessile. *Fruits* 8–10 mm diam., glabrous, basally constricted, the mericarp body 6–8 mm high, with strong lateral ribs, 3-spined, the spines divergent, 6–10 mm long, retrorsely barbed, the lateral spines 3–4 mm apart at base.

Distribution (Fig. 75). Apparently confined to the state of Pará, Brazil, where it occurs on riverbanks and seasonally flooded areas, presumably near sea level. The type locality is at 1°28′S, 48°30′W.

Illustrations. Krapovickas (1982b: fig. 1G–H, pl. 4).

Specimens examined. **BRAZIL.** Without locality, *Burchell 4713* (GH); Fazenda Macapá-AP, riverbanks, *Rabelo 1415* (HAMAB, NY, TEX). PARÁ: Pará, *Curran 11* (F, GH); basin of the Tocantins River, *Fróes 27071* (CTES); Belém, Guamá near Condor Airport, 15 Aug 1938, *Monteiro s.n.* (F); Jararaca, *Monteiro 153* (F); Belém, Guamá River, *Murça-Pires & Black 1841* (US); Marajó Island, Mun. Breves, across from Vila São Vicente, *Sobel et al. 4947A* (NY, TEX).

Pavonia piptocalyx is unique in the genus for its circumscissally dehiscent and deciduous calyx. It also has narrow, ligulate, white petals, like *P. stenopetala*, but its pedicels are much shorter than those of the latter species.

185. Pavonia pseudo-typhalea Planchon & Linden *ex* Triana & Planchon, Ann. Sci. Nat. Bot., ser. 4, **17:** 160. 1862. *Typhalea pseudo-typhalea* (Triana & Planchon) Monteiro, Anais XII Reun. Soc. Bot. Bras. 30. 1961. Type. Colombia. Mariqueta: forests of las Juntas, 1836, *Linden 936* (lectotype, here designated, G as photo F-23706; isolectotypes, BR, CTES [fragment], G, OXF, P as photo CTES). Fig. 73

Pavonia goudotiana Triana & Planchon, Ann. Sci. Nat. Bot., ser. 4, **17:** 160. 1862. *Typhalea goudotiana* (Triana & Planchon) Monteiro, Anais XII Reun. Soc. Bot. Bras. 29. 1961. Type. Colombia. Muso, *Goudot s.n.* (holotype, P as photo F-35473; isotype, CTES [fragment]).

Subshrubs 0.5–1 m tall, the stems roughly stellate-pubescent, the hairs somewhat antrorse. *Leaf* blades elliptic to oblanceolate, mostly 5–12 × 2–5 cm, basally rounded, truncate, or subcuneate, serrate, acuminate, scabridulous beneath (the hairs often bifurcate), nearly glabrous above except on midrib, or with appressed simple hairs, the margin often prominently short-ciliate or -setose; petioles 3–15(–40) mm long, with antrorse pubescence; stipules 8–12 mm long, subulate. *Flowers* solitary in the leaf axils, the pedicels 2–11 cm long, antrorsely pubescent; involucellar bracts 7–9, linear, 5–9 mm long (of variable length), connate below, ciliate; calyx 3–5 mm long, ciliate; corolla white, 8–12 mm long, minutely pubescent externally. *Fruits* 8–10 mm diam., glabrous, subglobose, the mericarp body 5 mm high, smooth, 3-spined, the spines 4–5 mm long, retrorsely barbed, more or less erect, the lateral spines 3.5 mm apart at base.

Distribution (Fig. 81). Fuertes (1989) states that this Colombian near-endemic (some collections are also known from Ecuador) occurs in the Depts. of Caldas, Cundinamarca, Nariño, Tolima, and Valle, in woodland understory and on roadsides, at elevations of 1750–2400 m, but *Schultes & Villarreal 5192* is from 1400 m elevation. The type locality is near 03°50′N, 76°05′W (?); that of *Pavonia goudotiana* is at 05°35′N, 74°07′W.

Illustration. Fuertes (1989: pl. 9).

Specimens examined. **COLOMBIA.** Ibagué, 1844, *Goudot s.n.* (syntype of *P. pseudo-typhalea*: CTES [fragment], P as photo CTES); Inza, Popayán, *Lehmann 5294* (F); highlands of Popayán, *Lehmann BT401* (GH). ANTIOQUÍA: Mun. Caramanta, 9.8 km from Caramanta toward Supia, *Betancur et al. 1128* (HUA, MO, NY); Mun. Jardín, 2 km N of Jardín, rd. to Morro Amarillo, *Callejas et al. 3958* (HUA, NY, TEX). CALDAS: Boquía River, Salento, *Killip & Hazen 8782* (GH, NY, PH, US). CAUCA: 5 km E of Timbio, *Grant & Drew 10675* (CAS, NY, TEX, US); La Cumbre, *Pennell & Killip 5731* (GH, NY, PH, US). CUNDINAMARCA: Pacho, mtns. above Patasía, *Uribe 5077* (US). HUILA: 32 km E of Neiva, *Fosberg 19749* (US); E of Neiva, *Rusby & Pennell 617* (NY, US); Villalobos River, vic. of Suazita River, *Schultes & Villarreal 5192* (GH). QUINDIO: Mun. Circasia, Bremen Forest Reserve, *Arbeláez et al. 1124* (CTES). RISARALDO: Mun. Sta. Rosa, Vereda El Crucero, *Idrobo et al. 9965* (U); Mun. Pereira, Ucumarí Regional Park, ca. 22 km ESE of Pereira, *Luteyn & Rangel 13087* (LSU, NY, TEX). TOLIMA: Mun. Santa Isabel, Vereda La Pava, *Idrobo et al. 10522* (U); "La Virginia," Libano, *Pennell 3252* (NY). VALLE: Mun. Cali, Villacarmelo, *I. Cabrera 6575* (MO); confluence of the Pichindecito and Picindé Rivers, *Cuatrecasas 18758* (F, US); Los Cárpatos, *Cuatrecasas 21631* (F, S, US); between Las Brizas and Albán, *Cuatrecasas 22169* (F, S, U, US); Mun. Cartago, Corregimento El Cairo, old rd. "Las Brisas" toward Argelia, *Delprete & Apreza 6304* (TEX); San Antonio Forest, W of Cali, near television tower, *A. H. Gentry et al. 48120* (MO, NY); El Silencio, Cordillera Occidental W of Yumbo, *A. H. Gentry et al. 65436* (LL, MO); Mun. Cali, Villacarmelo, *de Pascuas 330* (MO, NY).

ECUADOR. IMBABURA: Cotachachi, above Apuela, *Palacios & Tipaz 10046* (MO, QCNE, TEX); Cotacachi Cantón, Parroquia Apuela, Secto Cuellage, *Tipaz & Aulestia 1688* (MO, QCNE, TEX). ZAMORA-CHINCHIPE: Along rd. from Zamora to Loja, *Knight 385* (WIS).

Pavonia pseudo-typhalea is closely allied to *P. rudis* and possibly also to *P. intermedia*. These species may be distinguished by the characters given in the key.

186. Pavonia ramosissima (Arechavaleta) Fryxell & Krapovickas, comb. nov. Basionym: *Pavonia sepium* var. *ramosissimum* Arechavaleta, Anal. Mus. Nac. Montevideo **3:** 144. 1898. Type. Uruguay. "en parajes sombriós y húmedos," locality not stated, *Arechavaleta s.n.* (holotype, MVM?–n.v.). Fig. 82

Shrubs 0.5–2(–3) m tall, the stems densely stellate-pubescent. *Leaf* blades ovate-elliptic, up to 10(–12) × 5 cm, basally truncate or rounded, serrate, acute, palmately 5-nerved, somewhat discolorous, minutely stellate-pubescent above and beneath (the hairs beneath slightly larger but <0.5 mm diam.); petioles 3–12 mm long, densely stellate-pubescent; stipules 5–8 mm long, subulate-filiform. *Flowers* axillary but usually aggregated apically in a relatively congested inflorescence; pedicels 0.5–3 cm long with subarachnoid pubescence (finer than that of stem); involucellar bracts 5–6, lanceolate, 7 mm long, <1 mm wide, hirsute; calyx 5 mm long, ciliate (the hairs 1 mm long); petals yellow, 11–20 mm long, pubescent on margins of claw; staminal column 11 mm long, glabrous, the filaments 2–3 mm long; styles and stigmas emergent from column by 4–5 mm, almost equaling the petals. *Fruits* 6–7 mm diam.; mericarps 3-spined, the central spine erect, the lateral spines divergent, all 5 mm long, the spines retrorsely barbed, the mericarp body carinate and reticulate-nerved with retrorse barbs on all of the nerves.

Distribution (Fig. 48). In S Brazil (Paraná, Rio Grande do Sul, Santa Catarina) and N Uruguay. This species evidently differs ecologically from *Pavonia sepium*, prefering forest margins and spreading to roadsides, etc., rather than ocurring deep within the forests (Arechavaleta, 1898; Krapovickas, pers. comm.).

Specimens examined. BRAZIL. PARANÁ: Rio Branco do Sul, Itaperuçú, *Dombrowski 2447* (CTES); Quatro Barras, *Dombrowski & Kuniyoshi 2705* (CTES); Rio Branco do Sul, Madre, *Hatschbach 7666* (CTES); Mun. Colombo, Bacaitava, *Hatschbach 7781* (CTES); Capão Pomar, *Imaguire 3500* (CTES); Mun. Colombo, Ressaca, *Kummrow 886* (CTES). **RIO GRANDE DO SUL:** Mun. Farroupilha, *Aguiar & Martau 5523* (CTES); Planalto, *Bueno 4993* (CTES); Mun. Lagoa Vermelha, BR-285 km 61, *Krapovickas & Cristóbal 41933* (ASU, BAB, C, CTES, F, FCQ, G, GH, HAS, K, LIL, LPB, MBM, MICH, MO, NY, SGO, SI); Cruz Altinha BR-285 km 130, *Krapovickas & Vanni 37049* (CTES, HAS); Mun. Caxias do Sul, 10 km from Caxias do Sul, grounds of Hotel Samuara, *Pedersen 15929* (TEX); S. Fco de Paula, Vila Oliva, *Rambo 31098* (CTES). **SANTA CATARINA:** 13 km from Painel, rd. to Urupema, *Krapovickas & Cristóbal 41964* (CTES, MBM, UB); 16 km from BR-116, rd. to Campo Belo do Sul, *Krapovickas & Cristóbal 41973* (ASU, C, CTES, F, G, GH, HAS, K, LPB, MBM, MO, SGO, WIS); Pedras Brancas Park, 10 km SE of Lajes, *Krapovickas & Cristóbal 42079* (CTES, HAS); Fraiburgo (Reserva Renar), *Krapovickas & Cristóbal 43972* (CTES, MBM, MO); Mun. Urubici, 6 km N of Pericó, *Krapovickas & Schinini 38303* (CTES, HAS); Palmares, Campos Novos, *Reitz 6439* (CTES); Taquara Verde, Caçador, *Reitz & Klein 11769* (CTES).

The relatively crowded, terminal inflorescences are characteristic of *Pavonia ramisissima*. It is probably closely allied to *P. sepium* (q.v. for further discussion).

187. Pavonia renifolia Krapovickas, Bol. Soc. Argent. Bot. **20:** 296. 1982. Type. Brazil. Santa Catarina: Timbé do Sul, *Hagelund 11084* (holotype, CTES).

Prostrate *herbs* from a perennial root, the stems with narrow longitudinal lines of fine, dense pubescence, otherwise glabrous. *Leaf* blades reniform or suborbicular, cordate, coarsely crenate, obtuse, sparsely hirsute with simple hairs 1 mm long to glabrescent, palmately 7-nerved, 1.5–2 × 2–2.5 cm; petioles 5–12 mm long, with a narrow dense line of pubescence on adaxial side; stipules subulate, 3 mm long. *Flowers* solitary in the leaf axils, the pedicels 1–2 cm long, evenly short-pubescent; involucellar bracts 5–6, lanceolate, 5 × 1 mm, sparsely pubescent; calyx 6–7 mm long, subglabrous, deeply divided, the lobes lanceolate, distally ciliate; petals 17 mm long, yellow, the genitalia included. *Fruits* 6–7 mm diam. (excluding the spines), yellowish, hirsute (the hairs 1–1.5 mm long, sparse); mericarps with 3 stout spines 6–7 mm long, the central one erect and connivent with those of adjacent mericarps, the lateral ones spreading and also connivent with their adjacent partners, the fruit as a whole thus with 1 erect spine and 5 radiating spines, the spines very minutely and obscurely barbed at the tips.

Distribution (Fig. 48). Known from S Brazil in the state of Santa Catarina. The type locality is at 28°50′S, 49°52′W.

Illustrations. Krapovickas (1982b: fig. 1D, pl. 5).

Specimens examined. BRAZIL. SANTA CATARINA: Poco do Caixao, 5 km W of Timbe do Sul, *Krapovickas & Vanni 37004* (CTES); Mun. Araranguá, Jacinto Machado, *Reitz C88* (HBR).

Local name. Roseta (Brazil).

As a procumbent plant with reniform to suborbicular leaves, *Pavonia renifolia* is not to be confused with any other member of *P.* subgen. *Typhalea.*

188. Pavonia rudis Fryxell, sp. nov. Type. Peru. Pasco: Chontabamba Valley, 2–8 km W of Oxapampa, 1800–1850 m, remnants of natural vegetation along roadside, flowers white, 2 Feb 1983, *A. H. Gentry, D. Smith, R. Vásquez & León 39916* (holotype, MO; isotype, NY). Fig. 83

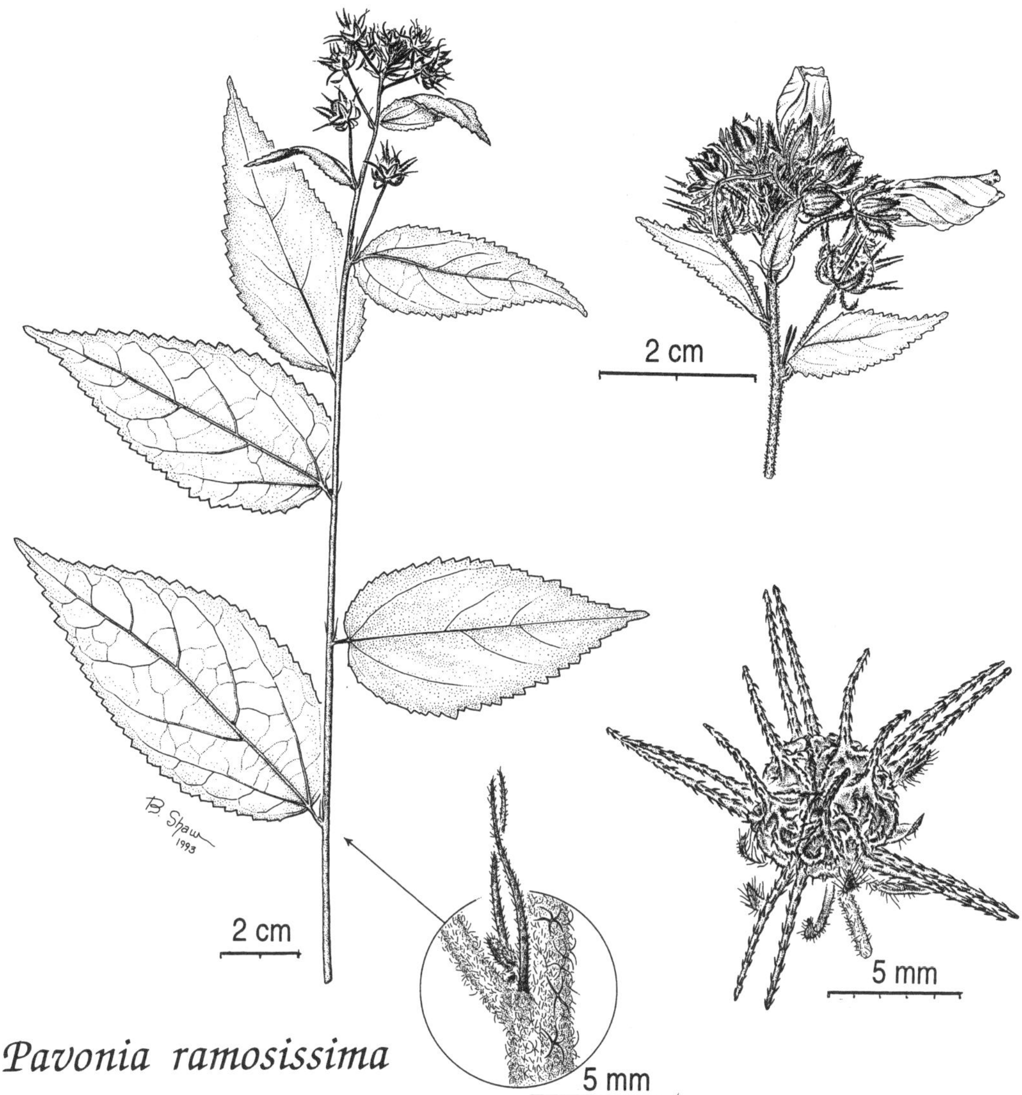

FIG. 82. *Pavonia ramosissima.* Fruiting branch (left) with enlargements of node and stipules (bottom center), of flowering and fruiting inflorescence (top right), and of mature fruit (bottom right) (*Krapovickas & Cristóbal 41933*).

Suffrutices caulibus pubescentibus, foliis subsessilibus, laminis supra pilis simplicibus vel bifurcatis adpressis, infra pilis stellatis; pedicellis solitariis 3–5 cm longis pilis plus minusve antrorsis; involucellis basaliter gamophyllis, distaliter 8–9 partibus anguste linearibus; corollis albis; mericarpiis laevibus, spinis tribus erectis 5–6 mm longis.

Subshrubs 1 m tall (?), the stems densely pubescent with a mixture of stellate and simple hairs hairs 0.5–1 mm long. *Leaf* blades 4–8 × 2–3.5 cm, lance elliptic, basally cuneate (somewhat asymmetrically so), serrate, acute, penninerved (3-nerved from the base, but the 2 lateral nerves reduced), the upper surface with appressed mostly simple but sometimes bifurcate hairs 0.4–0.8 mm long, the hairs on the principal nerves denser and slightly longer, the lower surface with stellate hairs 0.5–1 mm long, the principal nerves more densely pubescent; petioles 2–5 mm long, the leaves often subsessile, densely pubescent, the hairs 1 mm long; stipules subulate, erect, 3–4 mm long, pubescent. *Flowers* solitary (or paired) in the leaf axils, the pedicels 3–5 cm long, densely pubescent, the hairs stellate, bifurcate, or simple, variable

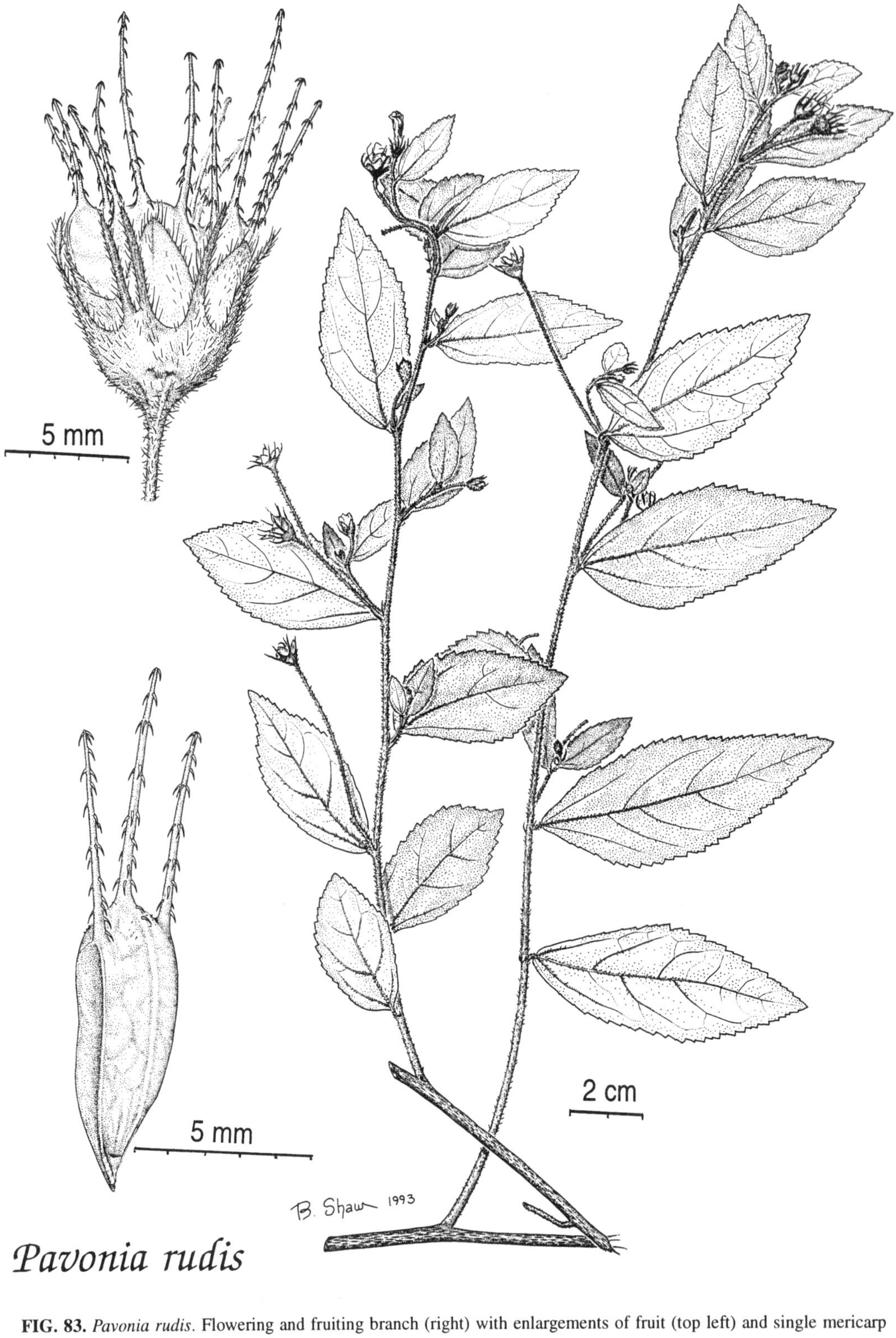

FIG. 83. *Pavonia rudis*. Flowering and fruiting branch (right) with enlargements of fruit (top left) and single mericarp (bottom left) (*Gentry et al. 39916*).

in size, the largest to 1 mm long, more or less antrorse; involucel ca. 8 mm long, cupuliform (i.e., gamophyllous in basal half), the 7–9 bractlets narrowly linear, separated by rounded sinuses, hirsute; calyx 5–6 mm long, ciliate, about half-divided; corolla 1 cm long, white. *Fruits* 6–7 mm diam.; mericarp body 7 mm tall, smooth, with minute glandular hairs apically, with 3 erect spines 5–6 mm long (the central spine the longest), retrorsely barbed, the lateral spines 2.5–3 mm apart at the base.

Distribution (Fig. 78). Known only from Peru in the provinces of Pasco and Junín. The type was collected at ca. 1800 m elevation. The type locality is at 10°31′S, 75°30′W.

Additional specimens examined: **PERU. JUNÍN:** Pichil Trail, Dos de Mayo, *Killip & A. C. Smith 25880* (NY, US).

Pavonia rudis is clearly allied to *P. pseudo-typhalea* from Colombia and perhaps also to *P. intermedia*. It can be distinguished from these by the characters given in the key.

189. Pavonia sepioides Fryxell & Krapovickas, sp. nov. Type. Colombia. Antioquía: Mun. Medellín, Vía Santa Helena, km 8–12, elev. 1520–1680 m, 26 Mar 1987, *Callejas & Escobar 3313* (holotype, TEX; isotypes, CAS, HUA, MO, NY, TEX).

Pavonia spinifex sensu Cavanilles, Diss. **3:** 133, t. 45, f. 2. 1787 (non *P. spinifex* [Linnaeus] Cavanilles). Although Cavanilles made the new combination in *Pavonia*, based on Linnaeus's plant, he drew his description and illustration from a cultivated plant raised from Peruvian seeds.

Pavonia sepium var. *macrocarpa* R. E. Fries, Kongl. Svenska Vetenskapsakad. Handl., ser. 3, **24(2):** 24. 1947. Type. Ecuador. At the village of Guápulo, near city of Quito, *Heilborn 321* (holotype, S; isotypes, NY, US). Fig. 73

Frutices, caulibus plus minusve glabrescentibus, caespitibus in axillis venarum principalium carentibus; laminis foliorum ovatis vel subellipticis, sparse stellato-pubescentibus; floribus solitariis in axillis foliorum, pedicellis 1.5–6 cm longis, bracteis involucellorum plerumque 6; corollis flavis, 12–22 mm longis; fructibus 7–9 mm diametro, mericarpiis spinis tribus hamatis divergentibus 3–7 mm longis.

Shrubs 1–2 m tall, the stems glabrescent or with narrow rows of more or less recurved hairs, sometimes with scattered large stellate hairs. *Leaf* blades mostly 5–10(–15) × 2–5(–6.5) cm, ovate to lanceolate (or subelliptic), usually widest below the middle, truncate or subcordate, serrate, acuminate, palmately 3–5-nerved, sparsely pubescent to glabrate, the lower surface with sparse uniform stellate hairs, lacking tufts of hairs in the nerve axils; petioles 4–15(–40) mm long, with small rigid stellate hairs; stipules 5–11 mm long, linear-lanceolate. *Flowers* solitary in the leaf axils, the pedicels 1.5–6 cm long, evenly pubescent with minute recurved hairs; involucellar bracts usually 6, lanceolate or oblanceolate, 7–11 mm long, sparsely pubescent; calyx 5–9 mm long, sparsely ciliate, sometimes basally pallid; corolla 12–22 mm long, yellow; staminal column 3–7 mm long, glabrous, pallid, the filaments 1–2 mm long; styles emergent from the column, pallid. *Fruits* 7–9 mm diam., the mericarp body 6 mm high, glabrous, reticulate-veined, the mericarps each with 3 apical spines, the spines 3–7 mm long, divergent, retrorsely barbed, the barbs sometimes extending onto the central costa of the fruit body.

Distribution (Fig. 84). From NW South America (Venezuela, Colombia, Ecuador, Perú), generally at relatively high elevation, 800–3000 m. The type locality of *Pavonia sepioides* is at 6°10′N, 75°20′W; that of *P. sepium* var. *macrocarps* is at 0°15′S, 78°59′W.

Illustrations. Fryxell (1992a: fig. 11, as *P. sepium*); Fuertes (1989: pl. 10, as *P. sepium*).

Representative specimens examined. COLOMBIA. ANTIOQUÍA: Mun. Fredonia, Vereda El Calvario, *Fonnegra et al. 2764* (NY, TEX). **BOYACÁ:** Soatá, *Cuatrecasas 1021* (F); from Santa Sofia to Villa de Leyva, 2 km from Santa Sofia, *Zuloaga & Londoño 4172* (MO, SI, US). **CUNDINAMARCA:** Between El Salto and El Colegio, *Cuatrecasas 8240* (F, US); San Bernardo, 10 km S of Arbeláez, *Fosberg 21002* (CAS, US); Las Musas, Canuelal Valley, 8 km W of Gutiérrez, 45 km S of Bogotá, *Grant & Giovanni 9668* (CAS, NY, TEX, US). **HUILA:** Valley of Fortalecillas River, 32–33 km E of Neiva, *Fosberg 19779* (CAS, US); Pitalito, Calamo, *Schultes & Villarreal 5137* (F, GH). **META:** Villavicencio Municipio Inspección de Policia de Servitá, *Quiñones 1343* (MO). **NARIÑO:** Dept. Pasto, across San Juan River from Maldonado, Ecuador, *Boeke 852* (CTES, NY); Aponte, *Ewan 16636* (NY, US); La Unión, *Fosberg 20449* (CAS, NY, US). **QUINDIO:** Mun. Salento, hwy. from Canaan to Sierra Moerna, *Arbeláez et al. 584* (CTES). **SANTANDER:** Between El Jaboncillo and Suratá, *Killip & A. C. Smith 16413* (A, GH, NY); vic. of Suratá, *Killip & A. C. Smith 16484* (A, GH, NY, US). **SUCRE:** Vic. of Carapas, *Badillo et al. 7301* (F, MY). **VALLE DEL CAUCA:** Quebrada de La Elvira, *Cuatrecasas 23781* (F, U, US); in shade, Popayan, *Hartweg 913* (K); Hondo River to Popayan, *Killip 8262* (GH, NY, PH); near Mares, hwy. from Cali to Buenaventura, *Killip et al. 39176* (F, NY); Popayan, *Lehmann 5520* (paratypes of var. *macrocarpa*: F, GH, NY, PH, S); Mun. La Elvira, Finca Zingara, *Luteyn et al. 12546* (NY, TEX); 21 km S of Popayan, 5 km S of Timbio, *Stein & McDade 3217* (K, MO, NY).

VENEZUELA. MÉRIDA: On mtn. rd. below Timotes, *Farenholtz 1560* (U); near El Paramito, *Lasser 1192* (US); near Mérida, *Pittier 12753* (A, NY, PH, US). **MIRANDA:** Knoop Park, Los Teques, *Aristeguieta 5511* (MO, NY, U). **SUCRE:** Cocollar, *Cumana 2614* (CTES). **MONAGAS:** Vic. of Cuevas del Guacharo, *Croat 54329* (MO), *Dumont et al. 7757* (NY). **TACHIRA:** At Delicias, *Steyermark & Liesner 118515* (MO).

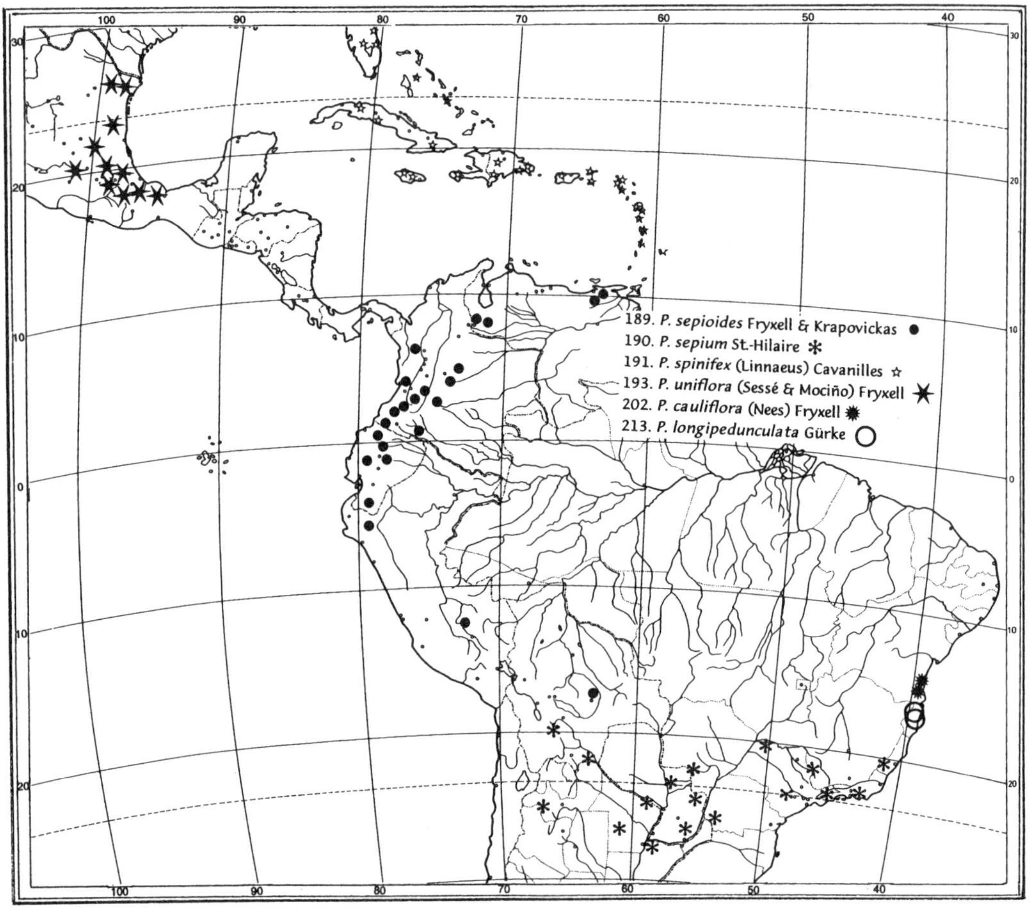

FIG. 84. Distribution map of six neotropical species of *Pavonia* sect. *Urenoideae* and *P.* subgen. *Goetheoides*.

ECUADOR. **CHIMBORAZO:** Huigra, *Hitchcock 20372* (GH, NY, US). **EL ORO:** Between Chaguarpamba and Balzas, *Harling et al. 20756* (GB, NY). **IMBABURA:** El Chota, *C. E. Cerón & M. Cerón 7037* (MO); E side of Ibarra, *Wiggins 10283* (CAS, DS, MO, NY). **LOJA:** Celica to Zapotillo rd., ca. 3 km below Pózul, *Harling & Andersson 18074* (F, GB, NY); km 12–15 Vilcabamba to Yangana rd., *Harling & Andersson 18459* (GB, QCA, NY); just W of Limo, *Harling & Andersson 22308* (GB, MO, NY, QCA). **PICHINCHA:** Tumbaco, *Asplund 6547* (paratypes of var. *macrocarpa*: S–2, US); Parroquía Calacali, Pululahua Geobotanical Reserve, *C. E. Cerón & M. Cerón 2744* (MO, NY); Otavalo to Malchinguí, *Hitchcock 20810* (GH, NY, US).

PERU. **CAJAMARCA:** Prov. Santa Cruz, Distr. Catache, ca. 1 km above Monte Seco on rd. to El Chorro, *Dillon et al. 4335* (F, NY, TEX); Prov Cajamarca, vic. of San Pablo, *Hutchison & Wright 5059* (BH, F, MO, NY, UC, US, USM); Prov. San Ygnacio, San Ygnacio to Nueva Esperanza, *A. López et al. 8991* (HUT, MO, NY, TEX). **CUZCO:** Prov. Convención, ca. 2 km S of Quillabamba, *Croat 50923* (MO, NY); Urubamba, Machupicchu ruins, 2000 m, *P. Núñez 7554* (MO). **HUÁNUCO:** Huánuco, *Asplund 13463* (S, US); Acomayo, between Tingo María and Huánuco, *Ferreyra 1829* (NY, US); 8 km N of Huánuco, *Stork & Horton 9411* (A, F, K). **JUNÍN:** Between Tarma and San Ramón, vic. of Huacapistana, *Croat 57651* (MO, NY). **PIURA:** Huancabamba, Distr. Chanchaque, *C. Díaz & Baldeón 2474* (F, MO, NY, TEX); Prov. Huancabamba, arriba de Palambla, *Ferreyra 10821* (CTES). **SAN MIGUEL:** Urubamba Valley, *Cook & Gilbert 1037* (US); Valley of the Urubamba, Machupicchu, *Herrera 3219* (F).

BOLIVIA. **SANTA CRUZ:** Prov. Florida, Quebrada El Durazno, 7 km NE of Mairana, *M. Nee 45332* (NY, TEX).

Local names. Anguía (Peru), cadillo (Colombia), chorro blanco (Peru), escoba (Colombia).

A collection from Peru (Prov. Cutervo, km 22 carr. Cutervo–Sócota, *Sánchez 2297*, CTES) is anomalous for having abundant long (2 mm) pilose hairs on stems, petioles, pedicels, and leaf veins. It also has relatively short (3–4 mm long) stout spines on the mericarps. For discussion of the relationship of *Pavonia sepioides* to other members of this complex, see discussion under *P. communis* and *P. sepium*, as well as Table IV.

190. Pavonia sepium St.-Hilaire, Fl. Bras. Merid. **1:** 225. 1827. *Malache sepium* (St.-Hilaire) Kuntze, Revis. Gen. Pl. **1:** 71. 1891. *Lassa sepium* (St.-Hilaire) Kuntze, Revis. Gen. Pl. **3(2):** 20. 1898. Type. Brazil. Near Sebastianopolis, *St.-Hilaire 50B* (holotype, P as photo CTES; isotype fragments, CTES, F).

Sida malvacea Vellozo, Fl. Flum. 277; vii. t. 13. 1825 [1829]. *Typhalea malvacea* (Vellozo) Monteiro, Anais XII Reun. Soc. Bot. Bras. 30. 1961. *Pavonia malvacea* (Vellozo) Krapovickas & Cristóbal, Lilloa **31:** 13. 1962. Type. Unknown.

Pavonia flava Spring *ex* Martius, Flora **20 (Beibl. 2):** 96. 1837. *Malache flava* (Spring) Kuntze, Revis. Gen. Pl. **1:** 70. 1891. Type. Brazil. Near Rio de Janeiro, *von Martius 95* (isotypes, BR, CTES, E, GH, MO, NY).

Pavonia flava var. *grandifolia* Spring *ex* Martius, Flora **22 (Beibl.):** 12. 1839; *Pavonia grandifolia* Spring *ex* Walpers, Repert. Bot. Syst. **1:** 300. 1842. Type. Brazil. Fields near João, *Luschnath 291* (holotype, BR as photo CTES; isotype, CTES [fragment]).

Pavonia gardneriana Turczaninow, Bull. Soc. Imp. Naturalistes Moscou **31:** 187. 1858. Type. Brazil. Piauí, Goiás, etc., *Gardner s.n.* (holotype, KW).

Pavonia sepium var. *silvatica* Arechavaleta, Anál. Mus. Nac. Montevideo **3:** 144. 1898. Type. Unknown [= *P.* sepium var. *sepium*].

Shrubs 1–2 m tall, the stems slender, glabrescent or with narrow rows of more or less recurved hairs. *Leaf* blades mostly 5–10(–15) × 2–5 cm, ovate to lanceolate (or subelliptic), widest below the middle, truncate or subcordate, serrate, acuminate, palmately 3–5-nerved, sparsely pubescent (especially on nerves), with tufts of fine hairs (domatia?) in the axils of the principal nerves beneath and often in the axils of the secondary nerves as well; petioles 4–15 mm long, pubescent distally and on adaxial side; stipules 5–11 mm long, linear-lanceolate. *Flowers* solitary in the leaf axils or on lateral axillary branchlets, the pedicels slender, 1.5–6 cm long, evenly and minutely pubescent; involucellar bracts usually 6, lanceolate or oblanceolate, 5–8 mm long, sparsely pubescent; calyx 5–6 mm long, sparsely ciliate, sometimes basally pallid; corolla 8–15(–20) mm long, yellow; staminal column 3–6 mm long, glabrous, pallid, the filaments ca. 1 mm long; styles emergent from the column, pallid. *Fruits* 6–8 mm diam., the mericarp body 5–6 mm high, glabrous, reticulate-veined, the mericarps each with 3 apical spines, the spines 3–6 mm long, divergent, retrorsely barbed, the barbs usually confined to the spines. Chromosome number, $2n$ = 56 (Krapovickas & Cristóbal, 1965, as *P. malvacea*); $2n$ = 112 (Skovsted, 1941, possibly a misidentification of *P. communis*?).

Distribution (Fig. 84). Found S of the 20° parallel of latitude in Brazil, Paraguay, Uruguay, Argentina, and Bolivia, generally at low elevation, but up to 800–1800 m in W Argentina and Bolivia. The type locality for *Pavonia sepium* is at 20°41′S, 49°56′W.

Illustrations. Gürke (1892b: pl. 86, fig. 1); Krapovickas (1953: fig. 107B).

Representative specimens examined. **BRAZIL. Espírito Santo:** Cachoeira do Itapemirim, rd. to Pedra de Itabira, *Esteves & Kameyama 2555* (NY, SP). **Guanabara:** Serra das Piabas, *D. Sucre & da Silva 8771* (RB); Mata da Tijuca, Açude da Solidão, *J. Pereira et al. 3873* (CTES, MO); Silvestre, rd. to Cristo Redentor, *J. Pereira & Duarte 4543* (MO). **Mato Grosso:** Between Ponta Poré and Amambay, *Gregory et al. 9666* (CTES, F). **Minas Gerais:** Sta. Rita de Caldas, Gruta Sta. Rita, *Krapovickas & Cristóbal 35497* (CTES, MBM); Divino, Fazenda Côrrego da Barra de São Pedro, *Leoni 740* (GFJP, NY); Caldas, *Regnell III-173* (C, NY, S-4, US). **Paraná:** 6 km SE of São Luis (between the Brazilian–Argentine border and the Iguazú River), *Davidse et al. 11208* (MO, NY); Ponta Grossa, *Dusén 7823* (C, CAS, LL, MO, S-2); Mun. Curitabá, Atuba River, *Hatschbach 22968* (CTES, NY); Mun. Palmeira, BR-277, descent to the Capivara River, *Hatschbach 47589* (CM, CTES, MO); Morro N of Barbaquá, ca. 85 km SW of Guarapuava, *Lindeman & de Haas 4960* (NY, U); Barique Park, Mun. Curitiba, *Paciornik 19* (CM, CTES, MU, NY); Iguazú Falls, *Schinini 8157* (CTES, NY); Obraje Lupíon, *Schwarz 7464* (CAS, CTES, TEX). **Rio de Janeiro:** Mun. Terra Boa, Ligeiro River, *Hatschbach 21546* (CTES); Organ Mtns., *Gardner 322* (GH); Corcovado, *Miers 3607* (E, US); Mun. Maricá, Itaipuaçu, Serra da Tiririca, *Pirani & Kallunki 3560* (NY, TEX). **Rio Grande do Sul:** Mun. Dos Irmãos, Feitoria River, *Bueno & Krapovickas 3602* (CTES, NY); Irai, Balneario Osvaldo Cruz, *Krapovickas & Cristóbal 43475* (CTES, TEX); Mun. Bagé, Passo dos Embarcaderos, 73 km NE of Bagé, bank of Camaquá River, *Lindeman et al. 7018* (CTES, U); Esteio, *Rambo 40642* (K, LIL, TEX, US); Vila Manresa near Porto Alegre, *Rambo 46058* (MO, US); Uruguayana, *Rosengurtt et al. 9484* (CTES, NY). **Santa Catarina:** Laguna, *Krapovickas & Cristóbal 37736* (ASU, BAB, CTES, F, G, GH, HAS, K, LIL, MBM, MICH, MO, UB, WIS); Fraiburgo, *Krapovickas & Cristóbal 43973* (ASU, CTES, HAS, LIL, MICH, MBM). **São Paulo:** Mun. Anhembi, Fazenda Barreiro Rico, *Kuhlmann 4520* (CTES, NY); Joanópolis, *Kuhlmann & Gonçalves 1355* (NY, SP); São Paulo, Cidade Universitária, *Motokane & T. R. S. Silva 17* (NY, SPF).

BOLIVIA. Chuquisaca: Prov. Tomina, Monteagudo 64 km toward Sucre, *S. Beck 6379* (CTES). **Santa Cruz:** Miske, on the rd. Vila Vila to Station Cruz, *Brooke 5896* (NY); Prov. Sara, *J. Steinbach 1855* (GH, NY, TEX). **Tarija:** Rt. from Tarija to Villa Montes, Entre Ríos, *Krapovickas et al. 19024* (CTES); Prov. Acre, 7.4 km E of Emboroza on rd. to Bermejo, *Solomon 10052* (CTES, MO, NY).

PARAGUAY. Cerros de Paraguary, *Hassler 6500* (C, GH, K, MO, NY, S); Villarica, *Jörgensen 3811* (A, C, CAS, F, MO, NY, PH, S, US). **Alto Paraná:** Puerto Presidente Stroessner, Technical Forestry School, km 12, *Fernández-Casas & Molero 5665* (MO, NY); Itabó Biological Reserve, trail to Tangará, 80 km N of Pto. Stroessner, *Krapovickas &*

Cristóbal 43200 (CTES, FCQ). **Amambay:** Chiriguelo, *Hahn et al. 1350* (MO, NY); 23 km W of P. J. Caballero (Chiriguelo), *Krapovickas et al. 14137* (C, CTES). **Caazapa:** Tavai, *Basualdo 2198* (CTES, MO). **Central:** Estero de Ypoá, 15 km SW of Nueva Italia, *Zardini & C. Velásquez 18797* (CTES, MO); San Antonio, on Paraguay River, *Zardini & E. Velázquez 22588* (MO). **Cordillera:** S of Tobatí, *Fiebrig 831* (A, CTES, E–3, F, K); 5 km N of Tobatí, *Krapovickas & Cristóbal 11469* (CTES, MEXU). **Guairá:** Colonia Independencia, *Hahn 2241* (CTES, MO, NY); Cordillera de Ybytyruzú, rd. to Cantera Jhú, *Zardini 14883* (CTES, MO). **Itapúa:** Salto Tembey, *Brunner et al. 885* (MO); between Capitán Meza and Puerto Triunfo, *Fernández-Casas & Molero 3706* (MO). **Misiones:** 10 km N of Ayolas, *Schinini & Vanni 25974* (CTES, G, MO, NY); Estancia La Soledad, 3 km S of Santiago, *Schinini & Vanni 26033* (CTES, G, MO). **Paraguarí:** Ybycu'í National Park, *Aguayo 37* (CTES, MO). **San Pedro:** Primavera, Alto Paraguay, *Woolston 1564* (C, K, NY, S).

ARGENTINA. Buenos Aires: Punta Lara, *Krapovickas 2886* (BRIT, CTES, F, LIL, MO, NY, US);Punta Lara, *Roivainen 13* (MO). **Catamarca:** Dept. El Alto, between El Alto and Súcuma, *Hunziker & Cocucci 17234* (CTES, MO, NY); Agua de Mato toward La Merced, *Hunziker 21953* (CTES). **Chaco:** Dept. 1° de Mayo, Guaycurú Creek, *Schulz 2173* (CTES, NY); Colonia Benítez, *Schulz 548* (CTES, NY). **Corrientes:** Rt. 14, km 720, *Huidobro 3770* (CAS, LIL, MO, S, TEX); Dept. Concepción, Carambola, *Pedersen 13368* (C–2, CTES, MO, NY); 17 km NW of San Carlos, *Tressens et al. 3891* (AAU, CHR, CLEMS, CTES, GB, LPB, MERL, NY, TIAC, U). **Entre Ríos:** Dept. Federación, Barra de Mocoretá, *Pedersen 8136* (C–2, CTES, herb. Pedersen); Dept. Uruguay, Estancia "La Selmira," *Pedersen 8206* (C, CTES, herb. Pedersen). **Formosa:** Guayendes, *Jörgensen 3166* (MO, US); Dept. Laishi, Herraura, *Petetin & Molina 1552* (CTES). **Jujuy:** Jujuy, *Cabrera 8183* (U); Dept. Ledesma, Sierra de Calilegua, *Venturi 5385* (US); Yala on the Yala lakes, *O'Donell 2872* (NY, S). **Misiones:** San Ignacio, Jesuit Ruins, *Esteves et al. 2704* (NY, SPF); Iguazú, Puerto Peninsula, *Krapovickas et al. 18338* (BAA, CTES, NY); San Javier, *Krapovickas & Cristóbal 28895* (CTES, G, NY, SI). **Salta:** 40 km N of Salta, *W. R. Anderson 12357* (CAS, CTES, MICH, NY); Dept. Anta, El Rey National Park, *A. H. Gentry et al. 51763* (CTES, MO, NY). **Tucumán:** Dept. Capital, vic. of San Miguel de Tucumán, *Spegazzini [BAB 15206]* (BAB, NY); El Timbó, *Venturi 2269* (G, US); Dept. Tafí, Yerba Buena, *Venturi 213* (CAS, F, NY, US).

URUGUAY. Flores: Rt. 12, km 13, Arias Creek, *Lavia et al. 19* (CTES, TEX). **Florida:** Casupá Creek, *Del Puerto 2252* (K); Est. Rincón de Santa Elena, Mansavillagra Creek, *Rosengurtt 5856* (MO, S). **Montevideo:** Miquelete, *Herter 2064* (F); Dept. Maldonado, Pan de Azucar Creek, *Krapovickas & Cristóbal 11169* (CTES, SI).

Gürke (1892b), like Cavanilles (1787), confounded *Pavonia sepium* and *P. spinifex* by citing South American specimens under *P. spinifex* var. *genuina*, a confusion that has been perpetuated by subsequent authors (e.g., Krapovickas, 1978b). The two species are not only allopatric but also morphologically distinct (Table IV) and need not be confused. However, the "*P. sepium* complex" presents problems that are only partially resolved here. *Pavonia sepium* s.str. is here seen to include only those plants occurring in southern Brazil, Uruguay, Paraguay, Argentina, and Bolivia, all south of the 20° parallel of latitude, generally at relatively low elevation, but up to 800–1800 m in western Argentina and Bolivia. It is clearly distinguished by the presence of tufts of hairs in the axils of the principal nerves on the underside of the leaf and often in the axils of the secondary nerves as well. Also found in this region are *P. ramosissima*, *P. narcissi*, and *P. formosa*, often occurring sympatrically with *P. sepium*. All of these species lack these tufts of hairs and are otherwise distinguishable from *P. sepium* (see key). In addition, *P. ramosissima* differs ecologically from *P. sepium* (A. Krapovickas, pers. comm.), in that *Pavonia sepium* occurs only deep within shaded woodlands, whereas, *P. ramosissima* is found on woodland borders in more open habitats. The other elements in this complex are *P. communis* (of a different ploidy level and readily distinguished morphologically), *P. spinifex* (allopatric and readily distinguished morphologically, Table IV), and the preceding species, *P. sepioides*, which is allopatric but has been freely confused with *P. sepium* in the past.

Stadler et al. (1987) have shown that *Pavonia sepium* (as *P. malvacea*) serves as a host of the cotton stainer (*Dysdercus albofasciatus* Berg), an important insect pest of cotton, and thus as a reservoir of the pest for reinfestation of the cotton crop in Argentina.

191. Pavonia spinifex (Linnaeus) Cavanilles, Diss. **3:** 133. 1787. *Hibiscus spinifex* Linnaeus, Syst. Nat., ed. x. 1149. 1759. *Pavonia spinifex* var. *ovalifolia* de Candolle, Prodr. 1: 443. 1824 [≡ var. *spinifex*]. *Typhalea spinifex* (Linnaeus) C. Presl, Bot. Bemerk. 19. 1844. *Lassa spinifex* (Linnaeus) Kuntze, Revis. Gen. Pl. **3(2):** 20. 1898. Type. Plumier, Pl. Amér., pl. 1 (1755). Figs. 73, 85

Pavonia aristata Cavanilles, Diss. **3:** 133, pl. 45, f. 3. 1787. Type. Cultivated Hort. Bot. Paris, 1786 (holotype, P-JU no. 12342A as microfiche).

Pavonia spinifex var. *grandiflora* de Candolle, Prodr. **1:** 443. 1824. Type. West Indies [erroneously attributed to Mexico by de Candolle], lectotype illustration, *Icones Florae Mexicanae s.n.* (Torner Collection acc. no. 6331.1592, Hunt Institute) (cf. copy at G as photo F-30487).

Shrubs 1–2 m tall, the stems sparsely and minutely stellate-pubescent to glabrate, with narrow longitudinal lines of minute recurved hairs. *Leaf* blades ovate,

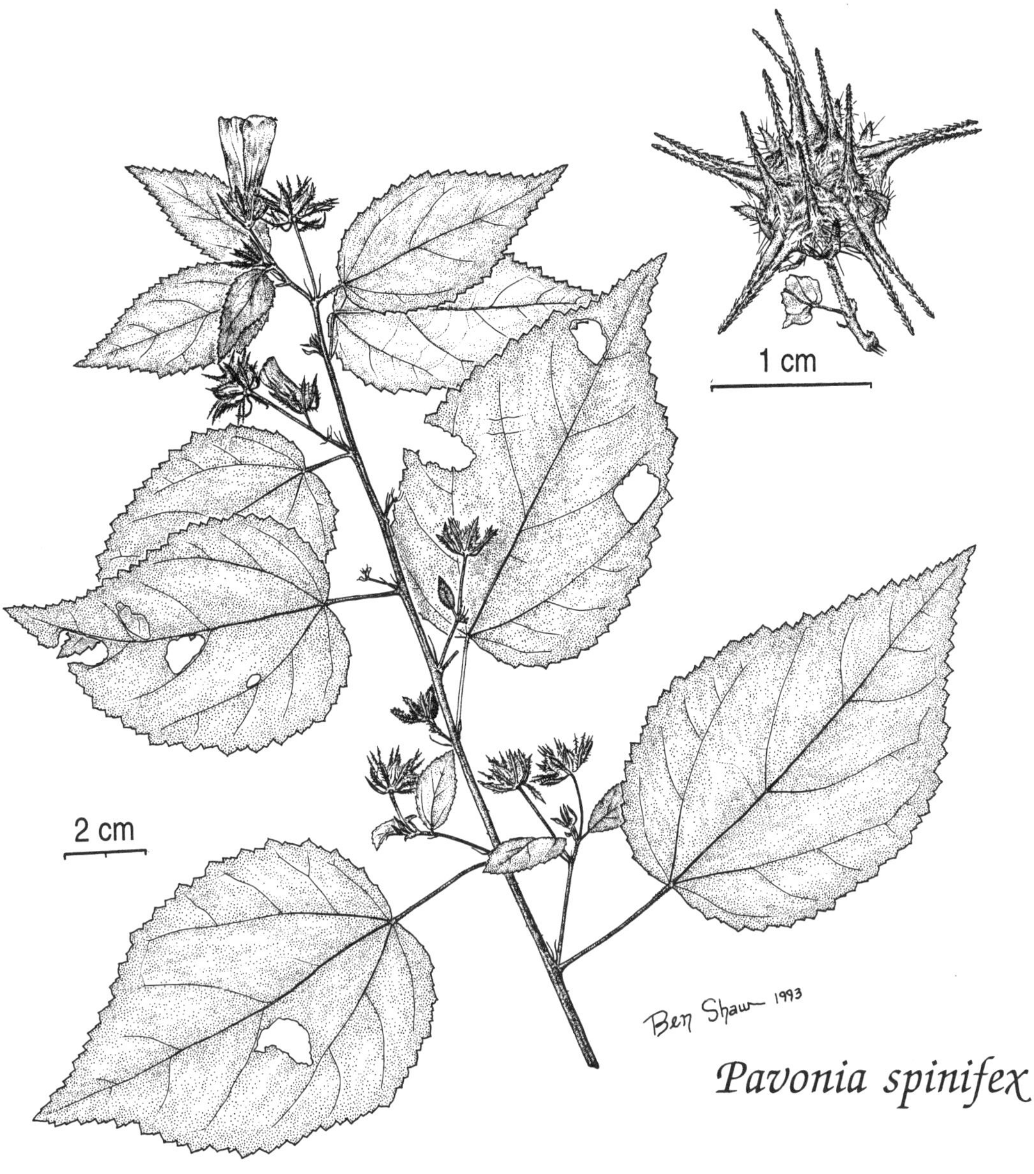

FIG. 85. *Pavonia spinifex*. Flowering and fruiting branch (left) with enlargements of mature fruit (upper right) (*Sintenis 145*).

mostly 4–12 × 2.5–8 cm, basally truncate to cordate, serrate, acute, sparsely pubescent (the hairs often simple) to glabrate above, stellate-pubescent beneath; petioles 0.5–3 cm long; stipules 5–10 mm long, subulate to filiform. *Flowers* solitary in the leaf axils, the pedicels 1–2 cm long; involucellar bracts 5–7, ligulate, 10–12 × 1 mm, ciliate; calyx 8–11 mm long, ciliate; corolla yellow, 2–2.5 cm long; genitalia subequal to the corolla to slightly exserted. *Fruits* 8–10 mm diam. (excluding spines), the mericarp body 6 mm high; mericarps 3-spined, the spines 6–7 mm long, retrorsely barbed, the central spine erect, the lateral spines radiate.

Distribution (Fig. 84). *Pavonia spinifex* occurs in the Greater and Lesser Antilles (with one record from Trinidad) and extends N to Florida and Georgia and to the Bahama Archipelago, growing in forest, shrubland, and savanna at 0–1000 m elevation.

Additional illustrations. Britton (1918: 237); Bot. Reg. 4: t. 339 (1819); Cavanilles (1787: pl. 45, fig. 2); Fawcett & Rendle (1926: fig. 50).

Representative specimens examined. UNITED STATES. FLORIDA: Vic. of Mayport and Jacksonville, 1870–1876, *Keeler s.n.* (NY); Merrits Island, *Curtiss 386* (CM, F, MICH, MO, NY, TEX). **Flagler Co.:** 1 mi S of St. John's county line at Marineland, *Perkins & Zomlefer 671* (RSA); ca. 9 mi S of New Smyrna Beach at Turtle Mound, *Ray 11188* (BRIT, USF); lime-sink near Pineola, *Small 8984* (NY). **Highlands Co.:** Ca. 6 mi W of Sebring, *Wilbur & Webster 2608* (BRIT, MICH); near Cape Canaveral, *Curtiss 5710* (K, MO, NY, POM). **GEORGIA:** Savannah, *Baldwin s.n.* (PH).

BERMUDA. Southampton Parish, Jan 1875, *Lefroy s.n.* (K).

BAHAMA ARCHIPELAGO. LONG ISLAND: Clarence Harbor, *Coker 495* (NY). **NEW PROVIDENCE:** Rd. to Fox Hills, *E. G. Britton 3385* (F, NY).

CUBA. CAMAGUEY: La Gloria, *Shafer 162* (BISH, F, GH, NY, US). **LA HABANA:** Near Havana, *Curtiss 563* (C, CAS, CM, E, F, GH, K, MO, NY, PH, US). **MATANZAS:** Tomory Valley at Matanzas, *Rugel 34* (BM, MO, NY, TEX); Santiago de las Vegas, *Van Hermann 3308* (F, GH, NY, POM). **LAS VILLAS:** Las Vegas de Mataguá, Trinidad Mtns., *Morton 10783* (US); Sierra de San Juan, above San Blas, *Morton 4038* (US). **SANTIAGO DE CUBA:** Finca Santa Elena, El Cobre, *Bro. Clemente 6282* (GH, NY); Finca Nancy, Jiguaní, *Bro. Clemente 7097* (GH, NY, US); Monte Verde, *Wright 34* (GH, K, NY, PH, S). **PINAR DEL RÍO:** *Guanaja 4271* (POM); Sierra de Anafe, *Wilson 11372* (NY, US). **SANCTI SPIRITUS:** Trinidad Mtns., *N. L. Britton & Wilson 5150* (F, NY). **CIENFUEGOS:** Cieneguita, *Combs 124* (F, GH, MO–2); Limones, Soledad, *Jack 6572* (A, CAS, DS, F).

JAMAICA. Porns, *Lloyd 1065* (F, MO). **ST. CATHERINE:** Mt. Rosser, *Adams 9902* (MO). **MANCHESTER:** Great Goat Island, Old Harbour Bay, *N. L. Britton & Hollick 1860* (NY); Lititz, Manchester, *Harris 12445* (CAS, F, GH, K, MO–2, NY, PH, TEX, US); vic. of Ewarton, *Maxon & Killip 377* (F, GH, US).

HAITI. Citadele de Roi Christophe, 5 km SE of Milot, *Bartlett 17381* (MICH, US); Massif de la Hotte, rd. Hab. Buttet to Etang-Ray, *Ekman 7158* (F, GH, K, LL, NY, S, US); Gonave Island, vic. of Anse Galette, *E. C. Leonard 3244* (PH, US); vic. of Mission, Fonds Varettes, *E. C. Leonard 3938* (GH, NY, TEX, US).

DOMINICAN REPUBLIC. ALTAGRACIA: Between Boca del Yuma town and El Caracol, *Zanoni et al. 10747* (JBSD, NY). **AZUA:** Jina Jaragua, N of Higüey, *Liogier 12348* (F, GH, NY, US); Santo Domingo, *von Türckheim 2506* (CAS, E, F, GH, MO, NY, S, U, US); Moncíón, *Valeur 244* (F, POM, S, US). **BARAHONA:** Barahona, *Fuertes 71* (CAS, E, NY, U, US); montiada Nueva, SE of Polo, *R. A. Howard & E. S. Howard 8609* (GH, NY, S, US). **INDEPENDENCIA:** 5–6 km NNW of Angel Felix Sierra de Neiba, *S. A. Thompson et al. 9753* (CM, NY); Samaná Peninsula, vic. of Samaná, *Abbott 345* (GH). **LA VEGA:** Near La Cienega, *Gastony et al. 420* (GH); La Culata de Constanza, *Zanoni & Boom 37110* (NY). **SAN JUAN:** Piedra del Aguacate to Río del Oro, *R. A. Howard & E. S. Howard 9444* (GH). **SANTIAGO:** 24–25 km N of Cabo Rojo, *S. A. Thompson et al. 7637* (CM, NY); Prov. Pedernales, 1.4 km N de Los Arroyos, *Zanoni et al. 27072* (NY). **SEIBO:** Between Punta Jicaco and Punta Rey, E of Jovero, *Abbott 2850* (A, GH, US); La Romana, *N. Taylor 386* (F, NY).

PUERTO RICO. Near Real, *Boom 9959* (NY, TEX); Island of Culebra, *N. L. Britton & Wheeler 135* (F, NY, US); Santurce, *A. A. Heller 4685* (A, BISH, E, F, GH, MO, NY, PH, POM, US); Arecibo, *A. A. Heller & E. G. Heller 344* (F, NY, US); Vieques Island, vic. of Isabel Segunda, *Shafer 2452* (CAS, NY, US); Mayagüez, *Sintenis 145* (C, GH, NY, US); near Maunabo, *Wagner 916* (A, U).

VIRGIN ISLANDS. ST. CROIX: Mt. Eagle, *Fosberg 59379* (CTES, OSH, NY, US); W end of Maroon Ridge, *Fosberg 60792* (NY, POM); Bassin Yard, *A. E. Ricksecker 96* (CTES, DS, F, GH, MO, NY, US); Bassin Yard, *J. J. Ricksecker 348* (E, F, MO). **ST. JOHN:** Dirt rd. to Bordeaux Mtn., *Acevedo Rodríguez 3124* (MO, NY, US); Coral Bay Quarter, *Acevedo Rodríguez & Siaca 3836* (MO, NY, US). **ST. THOMAS:** Villa Olga, *Croat 41* (MO); without locality, *von Eggers 109* (C–2, MO, S–2, U). **TORTOLA:** Pasea Hall Estate, *Fishlock 290* (GH, K, PH); rd. to Sea Cow Bay, *N. L. Britton & Shafer 691* (NY, US).

LEEWARD ISLANDS. ANTIGUA: Collins, *Box 953* (A, BM, NY, US); Halliday Hill, *Rose et al. 3331* (NY, US). **BARBADOS:** Sion Hill, St. James, *Bovell 552* (NY). **MONTSERRAT:** Windward Rd., W of Gages, *Shafer 149* (CM, F, NY, US). **SABA:** Lower part of the Bottom Mtn., *Boldingh 1401* (U); from Windwardside village to Mountaintop, *Howard 18154* (A, NY). **ST. BARTHELEMY:** L'Orient, *Le Gallo 353* (U), *Le Gallo 691* (A); Saint Caillon, *Le Gallo 2190* (A). **ST. EUSTATIUS:** The Quill, *Arnoldo 3196* (A, U). **ST. KITTS:** Without locality, 26 Feb 1885, *Kidder s.n.* (NY). **ST. MARTIN:** Mt. Paradis, *Boldingh 3188* (NY, U); Marigot Hill, *Boldingh 2650* (U).

WINDWARD ISLANDS. GUADELOUPE: Grande Savane, *Duss 2328* (F, GH, MO, NY, PH); Grand Terre, Le Moule, *Proctor 19905* (A, MO, US); Marie Galante, St. Louis River between Grand Bassin and Les Balisiers, *Proctor 21007* (A). **DOMINICA:** St. Joseph Parish, Coulibistri rd. up Batali River, *Hill et al. 25445* (NY). **MARTINIQUE:** Case Pilote, Carbet Valley, *Duss 843* (NY); au Carbet, 1787, *Isert s.n.* (C); Ceron à Grand Rivière, *H. Stehlé 2150* (NY). **ST. LUCIA:** Petit Piton, Soufriere, *Pierre 64* (A); slopes of Gros Piton, *Proctor 18073* (A, US).

TRINIDAD. "Flora Trinitatis No. 228" (E).

Local names. Abutilon espinoso (Puerto Rico), cadillo (Dominican Republic), cadillo amarillo (Dominican Republic), cadillo americano (Dominican Republic), cadillo de tres pies (Dominican Republic), cadillo espinoso (Puerto Rico), cousin (Dominican Republic), cousin piquant (Dominican Republic), gingerbush (Netherlands West Indies), grand cousin (Dominican Republic), majagüilla de costa (Cuba), majagüilla espinosa (Cuba), namble

(Dominican Republic), petit cousin (Dominican Republic), piquant cousin (Dominican Republic), spiny abutilon (Puerto Rico), spur bur (Jamaica).

When Cavanilles transferred *Hibiscus spinifex* L. to *Pavonia,* he cited it as from Peru, in reference to plants cultivated in the Paris botanical garden from seeds collected in "Maragnon, Peru," presumably by Dombey (P-JU no. 12341c, as microfiche). These plants are evidently representatives of *P. sepioides.* However, the name *P. spinifex* was based on *Hibiscus spinifex,* which Linnaeus published in the 10th edition of *Systema Naturae* (1759), although Cavanilles's reference is evidently to the 12th edition (1767) of this work, in which Linnaeus provides an expanded description. The 10th edition (the protologue) specifically cites Plumier's plate 1, which is the type. Plumier's drawings were based on West Indian plants, so it is clear that the name *Pavonia spinifex* applies to the West Indian species here described, despite Cavanilles's misunderstanding of its origins. De Candolle contributed to the misunderstanding when he published the name *P. spinifex* var. *grandiflora* (see above synonymy) and erroneously attributed it to Mexico. This geographic error was perpetuated by St.-Hilaire & Naudin (1842), Gürke (1892b), and other subsequent authors.

Stafleu and Cowan (1983: 301) note that Plumier's drawings were often based on specimens of Surian, and that the Surian herbarium is preserved at P. The basis for Plumier's drawing (his fig. 1) is a specimen of Surian preserved in the Jussieu herbarium (P-JU no. 12341b, seen as microfiche), but the existence of this specimen does not alter the fact that the drawing is the type. Moreover, the drawing, although highly stylized, is clearly identifiable with the Caribbean species, exhibiting the broadly ovate leaves, 3-spined mericarps, and exserted stigmas that are characteristic of it.

Gürke (1892b) divided *Pavonia spinifex* into two varieties, var. *genuina* (cited as from Peru, Panama, Florida, and the West Indies) and var. *communis* (with citation of Brazilian collections). Although his description of the former variety satisfactorily describes the Caribbean *P. spinifex* (including mention of the distinctively exserted stigmas), his inclusion of South American elements among his specimen citations has led subsequent authors to confound *P. spinifex* and *P. sepium.* These two species are not only allopatric but they are also distinct morphologically (see Table IV) and need not be confused.

Skovsted (1935) reported the chromosome number as $2n = 112$ and Davie (1933) reported $2n = 42$. Neither author states the origin of his material or cites voucher specimens except to cite various botanical gardens as seed sources, so it is uncertain what species they in fact counted. Since *Pavonia spinifex, P. sepium,* and *P. communis* have been freely confused in the literature, in the herbarium, and in botanical gardens, and since *P. communis* has been reported as $2n = 112$ (Krapovickas & Cristóbal, 1965), it is considered probable that Skovsted in fact had a plant of the latter species that he reported as *P. spinifex.* It is difficult to suggest what plant Davie may have studied. The only other count of $2n = 42$ is for *P. arachnoidea* (Bates, 1976), a species that is unlikely to be confused with *P. spinifex* and one that is not likely to be found in botanical garden collections.

192. Pavonia stenopetala Krapovickas, Bol. Soc. Argent. Bot. **20:** 298. 1982. Type. Brazil. Santa Catarina: Mun. Jacinto Machado, Sanga da Areia, *Reitz & Klein 9588* (holotype, HBR–n.v.; isotype, CTES).

Herbs <1 m tall (?), the stems with fine subarachnoid pubescence. *Leaf* blades elliptic, basally truncate or subcuneate, sharply serrate, acute or acuminate, penninerved, essentially glabrous, 6–9 × 2–3 cm; petioles 3–6 mm long, evenly stellate-pubescent; stipules 6–8 mm long, subulate or narrowly lanceolate, obscurely pubescent. *Flowers* solitary in the axils, the pedicels 2–3 cm long in flower, accrescent to 10 cm in fruit, with pubescence like stem but finer; involucellar bracts 6–7, ligulate, 4–5 mm long, obscurely pubescent, more or less reflexed; calyx 4 mm long, minutely pubescent, half-divided; petals 10 × 3 mm, white; staminal column ca. half length of petals, the anthers few (ca. 10), subsessile; styles emergent from the column, slender, the stigmas capitate, reddish. *Fruits* 8–9 mm diam. (excluding spines), glabrous; mericarps notably reticulate (ligneous?), with 3 stout spines ca. 1 cm long, the central spine erect, the lateral spines divergent, all retrorsely barbed toward apex.

Distribution (Fig. 51). Known from Brazil in the states of Paraná and Santa Catarina. The type locality is at 29°00′S, 49°47′W.

Illustrations. Krapovickas (1982b: fig. 1C, pl. 6).

Specimens examined. BRAZIL. PARANÁ: Mun. Cuatro Barras, Cuatro Barras, 23 km NE of Curitiba, *Dombrowski & Kuniyoshi 3383* (CTES).

Pavonia stenopetala is characterized by long fruiting pedicels (to 10 cm long) and narrow, ligulate, white petals. Its petals are like those of *P. piptocalyx,* but its pedicels are much longer.

193. Pavonia uniflora (Sessé & Mociño) Fryxell, Syst. Bot. **12:** 279. 1987. *Urena uniflora* Sessé & Mociño, Fl. Mexic., ed. 2, 158. 1894. *Hibiscus longifolius* Sessé & Mociño, Fl. Mexic., ed. 2, 161.

1894 (non *Pavonia longifolia* St.-Hilaire, 1827). Type. Mexico. Veracruz: Near Orizaba, *Sessé & Mociño 3532* (holotype, MA as photo F-46886; isotype, F). Figs. 73, 86

Pavonia spinifex var. *oblongifolia* de Candolle, Prodr. **1:** 443. 1824. *Pavonia oblongifolia* (de Candolle) Fryxell, Bol. Soc. Bot. Méx. **38:** 14. 1979. Type. *Icones Florae Mexicanae s.n.* (Torner Collection acc. no. 6331.818, Hunt Institute) (cf. copy at G, as photo F-30486).

Pavonia lanceolata Schlechtendal, Linnaea **11:** 356. 1837 (non *Pavonia lanceolata* Colla, 1833). *Malache lanceolata* (Schlechtendal) Kuntze, Revis. Gen. Pl. **1:** 70. 1891. Type. Mexico. [Veracruz:] Cuesta Grande, Chiconquiaco, 29 Sep [1829], *Schiede s.n.* (holotype, HAL).

Shrubs or subshrubs 0.5–1(–1.5) m tall, the stems sparsely to densely stellate-pubescent, the hairs variable in size, often twisted. *Leaf* blades mostly 4–9 cm long, 2.5–4 times as long as wide, elliptic, serrate, of variable pubescence (see discussion below); petioles to ⅓ the length of the blades (usually <¼, the leaves sometimes subsessile); stipules 3–7 mm long, lanceolate, erect. *Flowers* solitary in the leaf axils, the pedicels 1–3.5 cm long in flower to 6 cm in fruit, with minute recurved pubescence, sometimes also with a few larger hairs; involucellar bracts 6–8, linear-lanceolate, 6–9 mm long, hirsute; calyx subequal to involucel, hirsute; corolla 9–15 mm long, yellow, ciliate on margins of claw; staminal column ca. half length of petals, glabrous, pallid, the filaments 1–1.5 mm long; styles pallid, nearly glabrous, the stigmas prominently villous. *Fruits* 8–10 mm diam., excluding spines, yellowish, pubescent, the mericarp body ca. 5 mm long; mericarps reticulate-veined, 3-spined, the spines (3–)7–10 mm long, retrorsely barbed, divergent.

Distribution (Fig. 84). Endemic to E Mexico from Nuevo León and Tamaulipas to Oaxaca and Veracruz. The type locality is at 18°48′N, 97°06′W; that of *Pavonia lanceolata* is at 19°45′N, 96°48′W.

Additional illustrations. Fryxell (1979: figs. 2C, 3B; 1992b: fig. 17; 1993a: 110; 1993b: fig. 16).

Representative specimens examined. **MEXICO. GUANAJUATO:** El Guamuche, 12 km S of Xichú, *E. Ventura & E. López 7341* (IEB, NY). **HIDALGO:** San Bartolo Tutotepec, *Gimate-Leyva 783* (ENCB, NY). **NUEVO LEÓN:** Obispado, *Lacás 413* (F); Ramos River, 2 km S of Gral. Terán, *Rodríguez & Carranza 960* (ANSM, CTES, NY). **OAXACA:** Without locality, *Conzatti & V. González 26* (MO, US); San Pablo Huitzo, *Conzatti 1978* (F); Mun. San Miguel Chimalapa, 2.5 km N of Benito Juárez, valley of the Portamonedas River, *Maya 3822* (CHAPA, TEX). **PUEBLA:** Behind Cerro Colorado, between Nacozacalco and San Antonio Cañada, *Smith et al. 4076* (F, MEXU, US). **QUERÉTARO:** Mun. Landa, 2 km S of Matzacintla, *Carranza 910* (IEB, NY); Mun. Landa, 3 km NW of El Humo, *Rubio 1828* (IEB, NY, WIS); Mun. Jalpan, 2–3 km al E de La Parada, *Servín 430* (IEB, MO, NY). **SAN LUIS POTOSÍ:** Mun. Rioverde, San Martín, *R. Jones 148* (NY); Mun. Rayón, hwy. from Valles to Rioverde at K812, *P. A. Fryxell & W. R. Anderson 3620* (NY). **TAMAULIPAS:** Sierra de San Carlos, above El Rosario, vic. of Marmolejo, *Bartlett 10853* (LL, MICH, US); Hidalgo, Los Mimbres River, *Hinton et al. 24515* (TEX); Hidalgo, Divisadero, *Hinton et al. 24825* (TEX); 54 km N of González, *R. Jones & Treviño 173* (MO, NY); Rancho del Cielo, 11 km NW of Gómez Farías, *Torres & Hernández 3114* (MEXU, NY). **VERACRUZ:** 1 km E of Tenejapa, *Avendaño & Calzada 437* (F, NY, XAL); in Maltrata, *J. E. Fryxell 181* (F, NY, TEX); 2 km S of Huatusco, *P. A. Fryxell 2558* (BRIT, CHAPA, CTES, NY); Cerro San Miguel, 2 km W of Orizaba, *Nevling & Gómez-Pompa 2305* (A, C, F, MEXU, NY); Zacuapan, *Purpus 6337* (UC, US); near Jalapa, *Rose & Hay 6181* (NY, US); Teocelo, *F. Ventura 4448* (CTES, ENCB, LL, NY); Mun. Huatusco, Coscontla, *F. Ventura 5685* (CAS, ENCB, NY); Mun. San Juan Coscomatepec, *F. Ventura 8306* (CTES, ENCB, LL, NY); Mun. Jilotepec, La Concepción, *F. Ventura 11749* (CTES, ENCB, MEXU, NY); Mun. Naolinco, La Mesa, *F. Ventura 13150* (CHAPA, CTES, ENCB, MEXU, NY).

Some specimens (e.g. *Ventura 8306*; *Fryxell 2558*) have the upper leaf surface with long (1–2 mm) simple appressed hairs and the lower leaf surface with long simple and stellate hairs; other specimens (e.g. *Ventura 4448*, *11749*, *13150*; *Rodríguez & Carranza 960*) have simple or stellate hairs 0.5 mm long on the upper surface and uniform stellate hairs on the lower surface, lacking the large stellate spicules. These two extremes do not seem to differ in any other characters. Interestingly, the two branches on the type specimen (*Sessé & Mociño 3532*, MA) differ in type of pubescence, indicating it was either a mixed gathering or two gatherings later combined.

The similarities of *Pavonia uniflora* with four other species (*P. communis*, *P. sepioides*, *P. sepium*, and *P. spinifex*) are discussed under *P. communis* and are compared in Table IV. The typification and legitimacy of the name *Urena uniflora* are discussed by Fuertes and Fryxell (1993).

Pavonia subgen. *Typhalea* sect. *Diathericae*

Pavonia sect. **Diathericae** Fryxell, sect. nov. (from the Greek *di*, two, and *atheros*, spine or awn).

Suffrutices parvi in sitis salsuginosis habitat; foliis subsessilibus, laminis foliorum minute squamosis anguste spatulati-obtusis marginibus integris; pedicellis axillaribus recurvatis, floribus parvis; mericarpiis spinis duabus validis lateralibus quam corpum mericarpiorum duplo longioribus, spinis mericarpiorum contiguorum conniventibus.

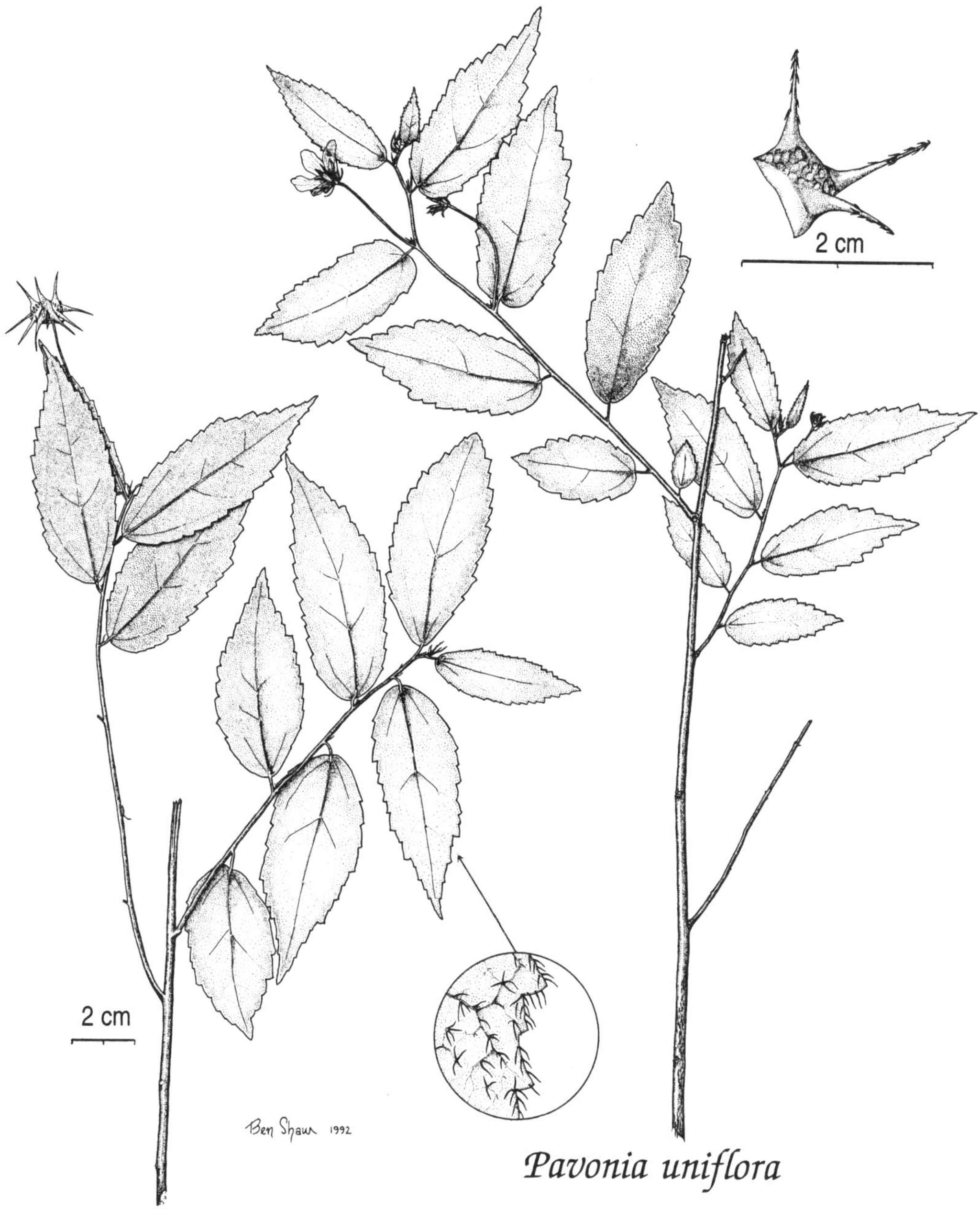

FIG. 86. *Pavonia uniflora.* Flowering branch (right) and fruiting branch (left) with enlargement of mericarp (above right) and of leaf margin (below center) (*Fryxell 2558*, *Ventura 8306*).

Small subshrubs of mangrove habitats, the herbage minutely squamose, the leaves narrowly spatulate-obtuse, entire; pedicels axillary, recurved, the flowers small; mericarps with 2 stout lateral spines twice the length of the mericarp body, the spines of adjacent mericarps connivent.

Type species. *Pavonia kearneyi* D. M. Bates

194. Pavonia kearneyi D. M. Bates, Kew Bull. **20:** 149. 1966. Type. Colombia. Valle del Cauca: at the mouth of the Saija River, Jun 1901, *Lehmann s.n.* (holotype, K–n.v.). Fig. 87

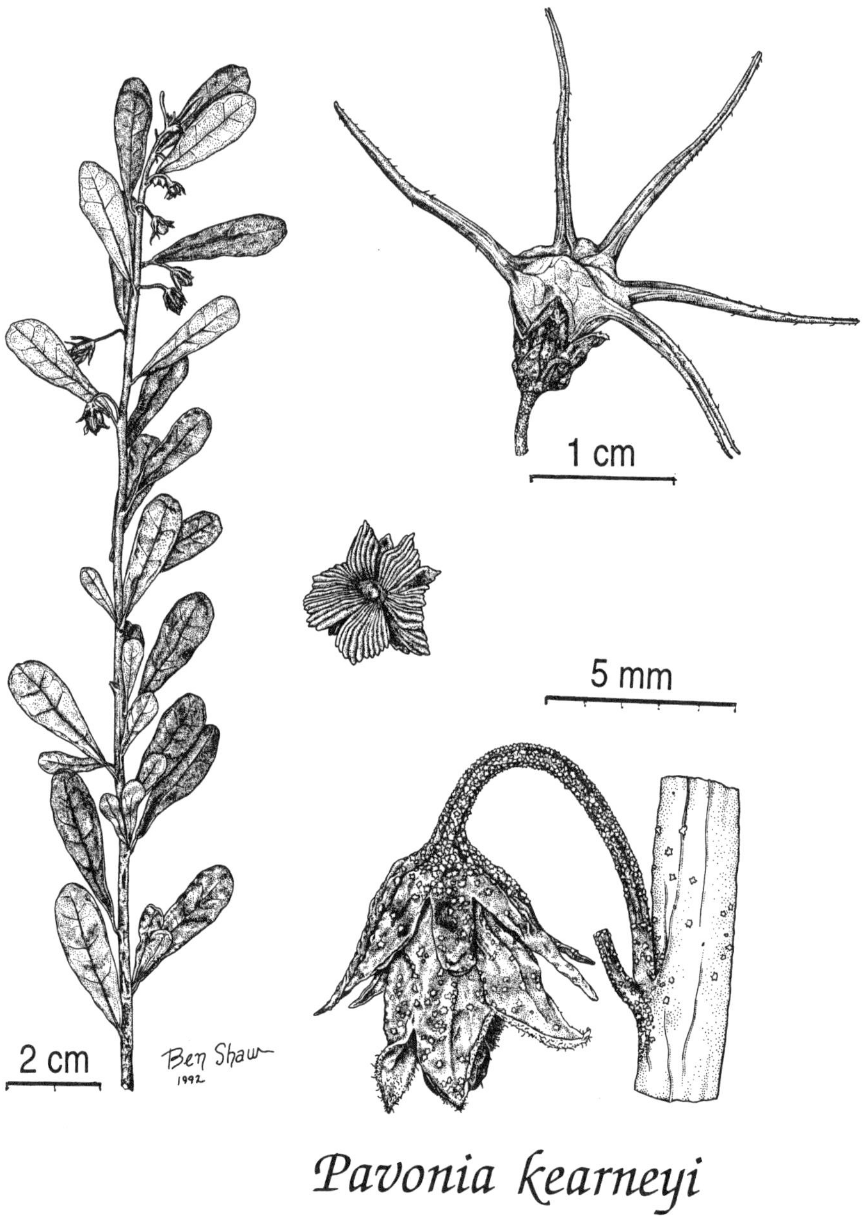

FIG. 87. *Pavonia kearneyi*. Flowering branch (left) with enlargements of fruit (upper right), flower (lower right), and scales (center right) (*Gentry 53854*, *Juncosa 2091*).

Subshrubs 0.5 m tall, the stems minutely and sparsely squamose to glabrate. *Leaf* blades narrowly spatulate, 20–40 × 5–12 mm (widest near apex), basally attenuate-cuneate, the margins entire (more or less undulate), obtuse, the midrib prominent (lateral nerves suppressed), minutely and sparsely squamose (hairs

0.1–0.2 mm diam.) to glabrate; petioles up to 2 mm long (the leaves subsessile); stipules vestigial. *Flowers* solitary in the leaf axils, the pedicels recurved, 4–8 mm long (to 12 mm in fruit); involucellar bracts 5–6, linear-lanceolate, 1–5 mm long, squamose; calyx 5–6 mm long, squamose or glabrate; corolla 8–11 mm long, white; gentitalia included, the anthers subsessile. *Fruits* 7 mm diam. (excluding spines), the mericarps 7 mm high, glabrous, 2-spined, the spines divergent, stout, 16–19 mm long, obscurely and minutely barbed toward apex, the spines of adjacent mericarps connivent.

Distribution (Fig. 79). *Pavonia kearneyi* occurs on the Pacific coast of Colombia, where it is sympatric with *P. rhizophorae* in mangrove habitats more or less at sea level. The type locality is at 02°50′N, 77°45′W.

Additional illustration. Bates (1966: fig. 1).

Specimens examined. **COLOMBIA. CHOCÓ:** Basin of San Juan River, Brazo Churimal, *Forero et al. 4398* (COL, MO); area of Baudó, on left bank of River Baudó, 1.5 km upstream from estuary, *Fuchs & Zanella 21847* (US); San Juan River delta, between Docordó and sea, *A. H. Gentry 53854* (MO, NY). VALLE: S of mouth of Yurumangui River, *A. H. Gentry & Juncosa 40588* (MO, NY); Mun. Buenaventura, Estero between Naya and Yurumangui Rivers, *Juncosa 2091* (MO, NY); Buenaventura, *Killip 11694* (GH, NY, PH, US); Punta Arenas, N shore of Buenaventura Bay, *Killip & Cuatrecasas 38613* (US).

Pavonia subgen. *Typhalea* sect. *Carcerariae*

Pavonia sect. **Carcerariae** Fryxell, sect. nov. (from the Latin *carcer*, jail or prison, in reference to the enclosing nature of the involucel).

Suffrutices vel frutices erecti; laminis foliorum ovatis palmatim nervatis; bracteis involucellorum 10–15 lineari-arcuatis setosis; floribus solitariis axillaribus (interdum aggregatis apicem versus) corollis luteis; mericarpiis spinis tribus brevis validis.

Erect subshrubs or shrubs, the leaves ovate, palmately nerved; involucellar bracts 10–15, linear-arcuate, setose; flowers solitary in the axils (sometimes aggregated apically), the corolla yellow; mericarps with 3 short, stout spines.

Type species. *Pavonia martii* Colla.

Key to the species of *Pavonia* sect. *Carcerariae*

1. Leaves <2 cm long; Brazil . . . . 196. *P. martii*
1. Leaves usually >2.5 cm long; Mexico.
    2. Pedicels subequal to involucel; calyx 10–18 mm long . . . . 197. *P. submutica*
    2. Pedicels 2 or more times as long as involucel; calyx 4–6 mm long . . . . 195. *P. arachnoidea*

195. Pavonia arachnoidea C. Presl, Reliq. Haenk. **2:** 129. 1835. *Malache arachnoidea* (C. Presl) Kuntze, Revis. Gen. Pl. **1:** 70. 1891. Type. Mexico. "in parte occidentali," *Haenke s.n.* (lectotype, PR-197900B; isotypes, BM, PR, W as photo F-32647). Fig. 73

Subshrubs or shrubs 0.5–1.5 m tall, the stems green (sometimes reddish), slender, pubescent with stellate and simple hispid hairs (1–2 mm long), minute stellate hairs, and longitudinal lines of recurved hairs. *Leaf* blades mostly 2–6 cm long, about twice as long as wide, ovate, truncate, irregularly serrate, acute or acuminate, sparsely and variably pubescent, the upper surface with the largest hairs simple (or bifurcate) and 1–1.5 mm long, the lower surface with 3–5-armed stellate hairs; petioles mostly ¼ the length of the blade or less; stipules 4–6 mm long, filiform. *Flowers* solitary in the leaf axils, the pedicels up to 6 cm long; involucellar bracts 10–15, filiform, 10–15 mm long, prominently hispid-ciliate, more or less arcuate; calyx 4–6 mm long, minutely stellate-pubescent; corolla 1.5–2.5 cm long, yellowish-white, glabrous except ciliate on claw; staminal column 5–6 mm long, glabrous, pallid, the filaments 1.5–2 mm long; styles exceeding the staminal column, connate for most of their length, pallid, glabrous, the stigmas prominently villous. *Fruits* ca. 6 mm diam., the mericarps reticulate-veined, 3-spined, the spines 1–2 mm long, retrorsely barbed, the mericarp body 3.5 mm long; seeds 2–2.5 mm long. Chromosome number, $2n = 42$ (Bates, 1976).

Distribution (Fig. 79). Endemic to W Mexico (Sinaloa to Guerrero), in shrubland and deciduous forest, at elevations of 300–400 m, flowering from Oct to Jan.

Illustrations. Cervantes Aceves (1992: fig. 61); Fryxell (1977: fig. 13D; 1979: fig. 3C).

Specimens examined. **MEXICO.** Without locality, *Sessé & Mociño 3537* (F, MA). GUERRERO: Acapulco, *Hancock 23* (K, NY); in 1894–1895, *E. Palmer 153* (CTES [fragment], F, K, MO, NY, U, UC as photo CTES, US). JALISCO: Rancho Tecuastitlán, some 18 km W of La Huerta, *Flores et al. 3021* (WIS); ca. 6 km NE of Nacastillo, *Iltis & M. Nee 1545* (CAS, ENCB, F, IBUG, MICH, NY, RSA, WIS); old rd. to Nacastillo, *Lott et al. 1675* (MEXU, NY); 7 mi S of La Huerta, *McVaugh 21029* (ENCB, MICH, US); Nacastillo, *Solís 1193* (CAS, MEXU). SINALOA: Between Rosario and Escuinapa, *P. A. Fryxell & Bates 2126* (BH, CAS, CTES, ENCB, NY).

196. Pavonia martii Colla, Herb. Pedem. **1:** 397. 1833. Type. Brazil, Valo, 1827, *von Martius s.n.* (holotype, TO). Figs. 73, 88

Pavonia luschnathiana Klotzsch *ex* Schlechtendal, Linnaea **14:** 301. 1840. *Malache luschnathiana* (Klotzsch *ex* Schlechtendal) Kuntze, Revis. Gen. Pl. **1:** 70. 1891. Type. Brazil. Bahia: Cruz de Casma, *Luschnath 167* (isotypes, BR, CTES [fragment], MO).

Pavonia microphylla Casaretto, Nov. Stirp. Bras. 38. 1842. *Typhalea microphylla* (Casaretto) Monteiro, Anais XII Reun. Soc. Bot. Bras. 30. 1961. Type. Brazil. Bahia: on the island of Itaparica and in hills around Cachoeira, *Casaretto 2058* (holotype, TO?–n.v.; isotypes, CTES [fragment], G as photo F-23703).

Shrubs 0.5–2 m tall, the stems covered with minute glandular hairs, dense recurved hairs, and sometimes setose with scattered hairs 2–3 mm long. *Leaf* blades ovate-oblong, 4–15 × 4–8 mm, subcordate, serrate, acute to obtuse, glandular-pubescent, setose on veins, with large 3–4-armed stellate hairs on lower surface; petioles 4–9 mm long; stipules subulate, 2 mm long. *Flowers* solitary in the leaf axils, the pedicels 1–4 cm long; involucellar bracts 10–11, linear, viscid and setose, 4–8 mm long; calyx 4 mm long, minutely pubescent on margin, sometimes also setose, otherwise subglabrous; corolla 1.5–2.5 cm long, yellow with small red center, externally pubescent, glabrous on margins of claw; staminal column 6–8 mm long, reddish, the anthers subsessile; styles exceeding the column, greenish, ca. ½ the length of the petals. *Fruits* ca. 7 mm diam., sparsely appressed-pubescent, the mericarps 3-spined, the spines stout, 3 mm long, divaricate, retrorsely barbed.

Distribution (Fig. 79). From Brazil in the states of Pernambuco, Bahia, and Minas Gerais, growing in cerrado and caatinga vegetation at 250–900 m elevation. The type locality for *Pavonia microphylla* is at 13°3′S, 38°46′W.

Additional illustration. Gürke (1892b: pl. 90, fig. 2, as *P. microphylla*).

Representative specimens examined. BRAZIL. BAHIA: 3 km S of Morro do Chapeu, rd. to Utinga, *Arbo et al. 5360* (CTES, K, NY, TEX); Poções, rd. to Bom Jesus da Serra, *Esteves & Kameyama 2542* (NY, SP); Mun. Campo Formoso, *Esteves & Lyra-Lemos 2530* (NY, SPF); 30 km E of Jacobina, *P. A. Fryxell 1947* (BH, CTES, NY, TEX); Serra de Contas just N of Livramento do Brumado, *Harley et al. 15331* (CTES, K, MO, NY, U, US); Brejões, 6 km S, *Hatschbach 48186* (C, CTES, US); 23 km E of Morro de Chapéu, Serra do Tombador, *Irwin et al. 30721* (CTES, F, NY); Jaguaquara, BR–4, *Krapovickas 12770* (CTES, LP, UC, WIS); 14 km E of the access to Lençois, *Krapovickas et al. 38819* (C, CEN, CTES, F, G, MICH, NY, UC); Mun. Maracas, km 2, hwy. from Maracas to Contendas do Sincorá, *Mattos Silva et al. 220* (CTES, NY); Mun. Lafayette Coutinho, *Mori & T. S. dos Santos 11822* (CTES, NY, RB); Mun. Jacobina, 10 km on the rd. from Jacobina to Morro do Chapéu, *Saunders & de Carvalho 3092* (CTES, NY, TEX). MINAS GERAIS: ca. 8 km W of Grão Mogol, *Irwin et al. 23341* (F, K, MO, NY, TEX, US); Mun. Grão Mogol, Serra do Catuny, *Markgraf et al. 3277* (F); Pedra Azul, Pedra da Conceição, *Trinta 770* (= *Fromm 1846*) (CTES, NY, TEX). PERNAMBUCO: São Caetano, *Andrade-Lima 66-4533* (IPA, NY); Gravatá, *Pickel 1164* (n.v.; cited by Carneiro de Lira, 1969).

Pavonia martii is notable for its consistently very small leaves.

197. Pavonia submutica Fryxell, Phytologia **37:** 297. 1977. Type. Mexico. Chiapas: 11 miles E of Tapanatepec, 20 Oct 1965, *Breedlove & Raven 13715* (holotype, DS; isotypes, F as photo F-64927, MEXU, NY). Fig. 73

Shrubs to 2 m tall, the stems stellate-pubescent. *Leaf* blades mostly 5–9 cm long, almost twice as long as wide, ovate, truncate to subcordate, finely serrate, acute, discolorous, stellate-pubescent; petioles ¼ the length of the blade or less; stipules 2–3 mm long, subulate. *Flowers* solitary in the leaf axils, often crowded apically, the pedicels 1–4 cm long, pubescent; involucellar bracts ca. 15, filiform, 20–30 mm long, arcuate, copiously ciliate (hairs 2–3 mm); calyx 10–18 mm long, pubescent; corolla yellow, 1.5–2.5 cm long, densely pubescent on margins of claw; staminal column 15–20 mm long, pallid, glabrous, the filaments 3–5 mm long; styles glabrous, slightly exceeding the androecium, the stigmas minutely pubescent. *Fruits* 8–10 mm diam., glabrous; mericarps smooth, with 1 apical and 2 lateral points (rudimentary spines), these sometimes retrorsely barbed.

Distribution (Fig. 79). Endemic to the Isthmus of Tehuantepec, in S Mexico in a limited area along the Chiapas-Oaxaca state boundary, in open shrubland at 600–900 m elevation. The type locality is at 16°25′N, 94°5′W.

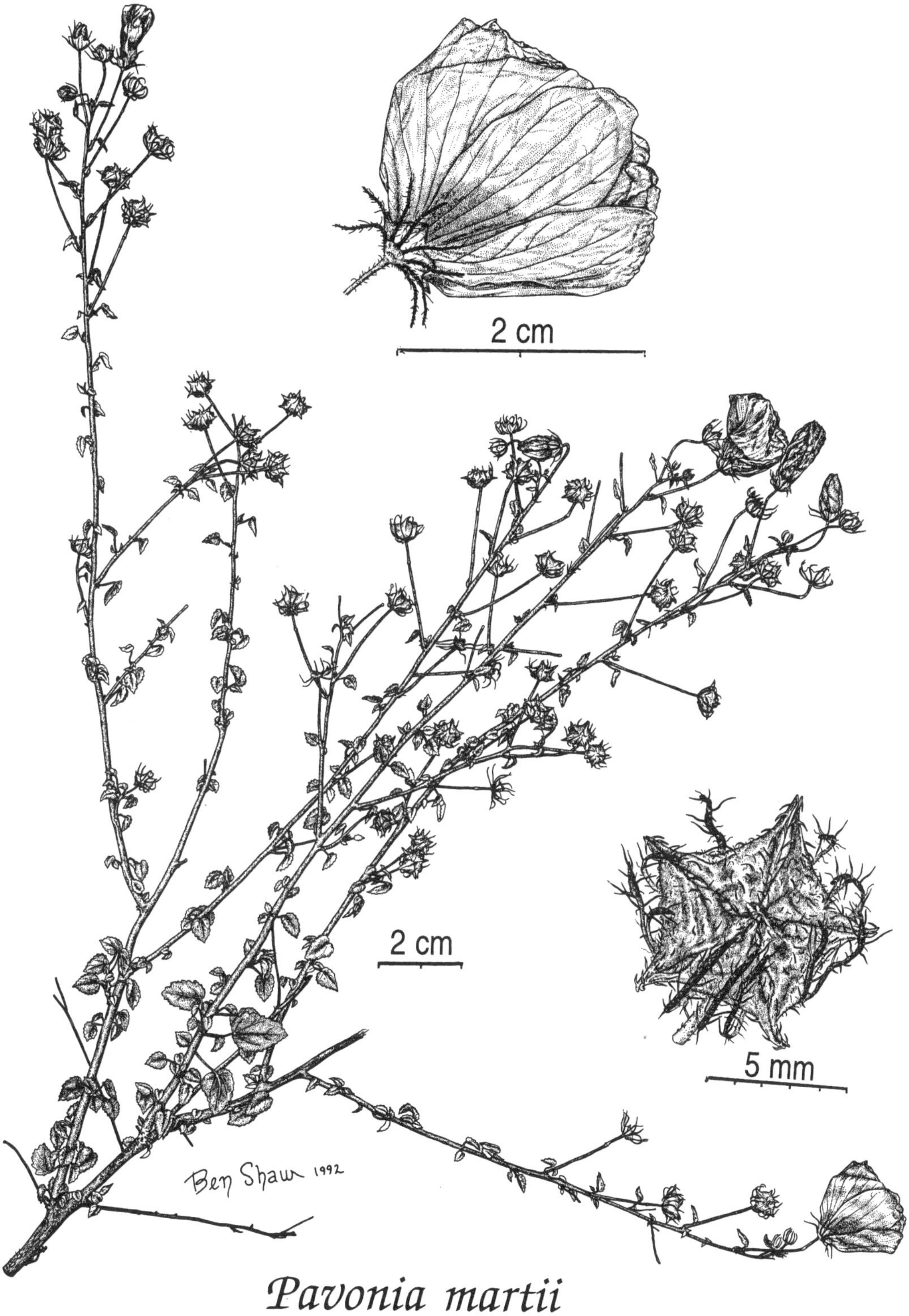

FIG. 88. *Pavonia martii*. Flowering and fruiting branch with enlargement of flower (above) and of fruit (lower right) (*Irwin et al. 23341*).

Illustrations. Fryxell (1977: figs. 10, 13C; 1979: fig. 3D; 1988a: fig. 85).

Specimens examined. MEXICO. CHIAPAS: 35 mi by rd. SW of Cintalapa on hwy. to Tehuantepec, *W. R. Anderson & Jessup 13546* (NY, TEX); SE of Tapanatepec, *Breedlove 49625* (CAS, MO); SE of Tapanatepec on Oaxacan border, *Breedlove & Almeda 60167* (C, CAS, GH, MO, NY, TEX); SE of Tapanatepec, *Daniel et al. 5866* (CAS, NY); Mun. Cintalapa de Figueroa, 3.5 mi NE of Oaxaca border, *P. A. Fryxell & Lott 3232* (BM, CAS, CHAPA, MEXU, MO, NY); foot of Cerro Ganeto, 20 km from Tapanatepec, *C. E. Smith & Ruíz 3237* (F, PH, US). **OAXACA:** On Hwy. 90, 0.5 mi W of Chiapas border, *P. A. Fryxell & Bates 902* (BH, NY); on rt. 90, 2 mi downslope from Chiapas border, *Harriman 14399* (CTES, NY, OSH); 19–21 km NE of Tapanatepec, *Koch et al. 73154* (CHAPA, ENCB, NY); Tres Cruces, Tehuantepec, 6 Dec 1969, *MacDougall s.n.* (NY).

The crowded flowers and fruits of *Pavonia submutica* and the morphology of its mericarps are distinctive.

Pavonia subgen. *Typhalea* sect. *Varians*

Pavonia sect. **Varians** Fryxell, sect. nov. (the name taken from that of the type species, *P. varians*).

Frutices laminis foliorum palmatim lobatis, lobis marginibus undulatis; floribus solitariis in axillis foliorum, involucello quam calyce dimidio longioribus vel minori; calycibus in alabastro plicatis-pentangularibus, lobis late cordatis; corollis flavis centro rubro; mericarpiis spinis tribus crassis, hamulis minutis fragilis, spinis apicalis conniventibus.

Shrubs with palmately lobed leaves, the lobes with undulate margins; flowers solitary in the axils, the involucel half the length of the calyx or less; calyx plicate-pentangular in bud, the lobes broadly cordate; corolla yellow with red center; mericarps 3-spined, the spines stout with very minute and fragile barbs, the apical spines connivent with those of other mericarps, the lateral spines radiate and connivent with those of adjacent mericarps, protruding between the lobes of the calyx.

Type species. *Pavonia varians* Moricand.

198. Pavonia varians Moricand, Pl. Nouv. Amér. 122, t. 74. 1844. Type. Brazil. Near "Japira" (Jacobina), *Blanchet 3148* (isotypes, CTES [fragment], F–3, K, P as photo CTES). Moricand cites the locality as "Japira" as does one of the F specimens, but the K specimen says "Jacobina." A second specimen at K (*Blanchet 3148*, from "Villa do Barra") is not a type. (Serra de Jacobina, not the city.) Figs. 73, 89

Pavonia cardiosepala Turczaninow, Bull Soc. Imp. Naturalistes Moscou **31:** 188. 1858, nom. superfl. *Malache cardiosepala* (Turczaninow) Kuntze, Revis. Gen. Pl. **1:** 70. 1891. Type. Brazil. Bahia: Autom, *Blanchet 3148* (holotype, KW; isotypes, CTES [fragment], F–3, K, P as photo CTES). The holotype gives "Autom" as the locality, the isotypes "Jacobina" or "Japira" (see above).

Shrubs 1–2 m tall, the stems densely stellate-tomentose. *Leaf* blades suborbicular in outline, angulate to deeply 5–7-lobed, at base deeply cordate, obscurely to prominently crenate or serrate, the margin undulate, the lobes acute or obtuse, palmately 7–9-nerved, densely and softly tomentose on both surfaces, up to 6 × 8 cm; petioles 1–4 cm long; stipules 2–6 mm long, subulate. *Flowers* solitary in the leaf axils (sometimes forming a terminal panicle), the pedicels 1.5–5.5 cm long, tomentose; involucellar bracts 8–10, subulate, 4–6 mm long (less than half the length of the calyx); calyx 12–16 mm long, tomentose, the lobes broadly cordate, 7–10 mm wide, 3-ribbed, plicate-pentangular in bud; corolla yellow, usually with a red center, 2–2.5 cm long; staminal column 12–15 mm long, pallid, the filaments 2 mm long; styles exserted from the column, the stigmas nearly equaling the petals. *Fruits* 9–10 mm diam., subconical, glabrous; mericarps 3-spined, the spines 2–4 mm long, stout, retrorsely barbed at apex (but barbs fragile and inconspicuous), the central spine erect and connivent with central spines of other mericarps, the lateral spines radiate between the lobes of the calyx.

Distribution (Fig. 78). Confined to central Brazil in the states of Bahia, Ceará, Pernambuco, and Piauí, at 500–800 m elevation.

Additional illustrations. Gürke (1892b: pl. 90, fig. 1); Moricand (1844: pl. 74).

Specimens examined. BRAZIL. BAHIA: Mun. Morpará, Mocambo Branco, *Bautista & Salgado 892* (RB); Serra de Santo Inacio, 18–30 km on the rd. from Xique-Xique to Santo Inacio, *de Carvalho & Saunders 2893* (TEX); hwy. from Irece to Xique Xique (BA-052) km 80, *Coradin et al. 6267* (CTES, K); 10 km from Bom Jesus da Lapa on the rd. to Ibotirama, *Ferruci et al. 986* (CTES, TEX); 97 km E of Ibotirama, *P. A. Fryxell 1929* (CAS, NY); 70 km W of Seabra, *P. A. Fryxell 1934* (NY); just S of São Inácio, *Harley et al. 19033* (K); Ibotirama, *Hatschbach 39522* (C, CTES); Ibotirama, *Krapovickas et al. 37892* (C, CEN, CTES, F, G, NY, UC); on the banks of São Francisco River near Joazeiro, *von Martius 2280* (M as photo CTES); Campo Formoso, 1943, *Mendonça s.n.* (CTES); near Joazeiro, *Pinto 11174* (CTES). **CEARÁ:** Aracati, *Fernandes 2799* (CTES). **PERNAMBUCO:** Buique, Fazenda Brejo de São José, *Andrade-Lima 55–2038* (F, IPA, NY); Araripina, Serra do Araripe, serrolandia, *Chiappeta et al. 535* (IPA, NY); Mun. Petrolina, rd. from Petrolina to Remanso, *Esteves & Lyra-Lemos 2220* (CTES); 40 km E of Remanso, *P. A. Fryxell 1945* (BH, CTES, NY); Petrolina, *Krapovickas 12903* (BAA, CTES, LP,

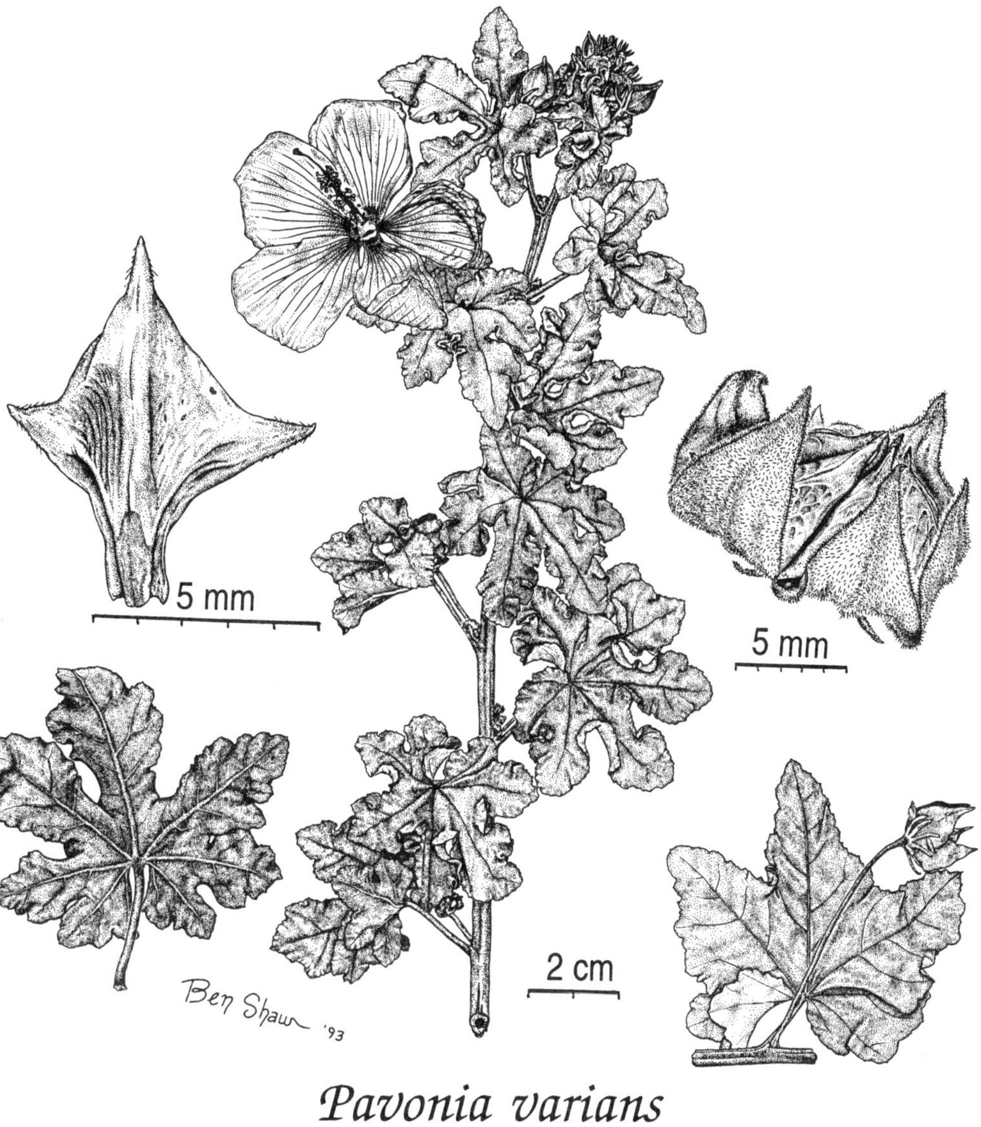

FIG. 89. *Pavonia varians.* Flowering branch (center) with leaf and flowering node (bottom) and enlargements of fruit (center right) and of single mericarp (center left) (*Fryxell 1929* [branch], *Krapovickas et al. 12903* [fruit], *Krapovickas et al. 37892* [mericarp]).

NY, P). PIAUÍ: Carnaiba, São Raimundo Nonato, *Fernandes 4111* (CTES).

Local name. Quati (Brazil: Pernambuco).

PAVONIA SUBGEN. GOETHEOIDES

Pavonia subgen. **Goetheoides** (Gürke) Ulbrich, Bot. Jahrb. Syst. **57:** 66. 1921. Basionym: *Pavonia* sect. *Goetheoides* Gürke *in* Martius, Fl. Bras. **12(3):** 479. 1892. Type (ICBN, art. 22.5). *Pavonia goetheoides* (Hassler) Fryxell.

Schouwia Schrader, Goett. Gel. Anz. **1:** 717. (5 May) 1821, nom. illegit. (non *Schouwia* de Candolle, Mem. Mus. Hist. Nat. **7:** 244 [20 Apr 1821]). Type. *Schouwia semiserrata* Schrader (≡ *Pavonia semiserrata*).

Goethea Nees, Flora **4:** 304. (28 May) 1821. *Pavonia* sect. *Goethea* (Nees) Endlicher, Gen. Pl. 981. 1840. Lectotype, designated indirectly by Gürke, 1892b, *Goethea cauliflora* Nees. Gürke (loc. cit.) retained *G. cauliflora* in *Goethea* while excluding *G. semperflorens* Nees, thereby effectively choosing the former as lectotype.

Pavonia sect. *Wiotiae* E. Morren, Belgique Hort. **25:** 113. 1875. Type. *Pavonia wioti* E. Morren (= *P. multiflora* St. Hilaire).

Pavonia sect. *Tricalycaris* Gürke *in* Martius, Fl. Bras. **12(3)**: 479. 1892. *Triplochlamys* Ulbrich, Notizbl. Bot. Gart. Berlin **6:** 333. 1915. Type (ICBN, art. 22.5). *Pavonia tricalycaris* St.-Hilaire.

Blanchetiastrum Hassler, Feddes Repert. Spec. Nov. Regni Veg. **8:** 28. 1910. Type. *Blanchetiastrum goetheoides* Hassler (≡ *Pavonia goetheoides*).

Shrubs or small trees, the leaves elliptic, often large, penninerved, the midrib raised both above and beneath; stipules sometimes simple, sometimes of 2–3 elements; involucellar bracts 4–24, linear or foliaceous, sometimes brightly colored; flowers axillary or in terminal racemes; corolla (often unknown) subequal to calyx or twice length of calyx, the genitalia sometimes exserted; mericarps (often unknown) unornamented and shiny.

Type species. *Pavonia goetheoides* (Hassler) Fryxell.

The species concerned have been variously placed in the genera *Blanchetiastrum* Hassler, *Goethea* Nees (Garcke, 1861), *Triplochlamys* Ulbrich, and *Pavonia*. In *Pavonia*, they have been variously allotted to *P.* sect. *Malvaviscoides* (Walpers, 1842–1848); *P.* sect. *Eupavonia*, as a typographically distinguished but unnamed subgroup (Garcke, 1881); *P.* sect. *Goetheoides* Gürke (Esteves, 1996; Esteves & Krapovickas, 1995; Gürke, 1892b; Krapovickas, 1982a); and *P.* sect. *Tricalycaris* Gürke (Esteves, 1996; Gürke, 1892b).

Few authors have accepted the inclusion of *Goethea* in *Pavonia*, but Ulbrich (1921: 66) significantly says, "Diese Arten [sect. *Goetheoides*] bilden gewissermassen einen Übergang zur Gattung *Goethea*, der sie vielleicht besser zuzuzählen sind, wenn man sie nicht als eigene Gattung betrachten will" [These species (sect. *Goetheoides)* to a certain extent form a transition to the genus *Goethea*, which one can probably better account for when it is not understood as a distinct genus]. It is also worthy of note that Steudel (1840–1841) listed *Pavonia semiserrata* (Schrader) Steudel and *Goethea semperflorens* Nees in separate genera, two names that are now generally accepted as representing a single species (e.g., Krapovickas, 1982a). Similarly, *Blanchetiastrum* has not previously been submerged in *Pavonia*, although Hassler, in describing the genus, noted its close alliance to *Goethea*.

Despite the fact that the species under discussion have been variously allocated among four genera, it seems clear that all of these species represent a single lineage and thus constitute a natural (monophyletic) taxon, regardless of the rank accorded it. Experimental support for this view is found in the report by Cheek (1989), in which a hybrid between two of the concerned species is described and named *Pavonia* × *gledhillii* Cheek. The parentage of the cross was *Goethea* × *Triplochlamys*, and the resulting hybrid was classifiable as a *Pavonia* (according to criteria conventionally used to distinguish among these "genera"). These facts demonstrate both the genetic commonality of these taxa and the artificiality of the criteria used to separate *Triplochlamys* and *Goethea* from *Pavonia*. The geographic and ecological cohesiveness of this group of species, all of which occur in the Atlantic coastal forest of easternmost Brazil (see Figs. 11, 15, 16, 60, 74, 76, 78, 79, 81, 84), also supports the view that these species form a coherent group.

Cauliflory has customarily been used to define the genus *Goethea*, yet cauliflory is characteristic of only one of the three species usually included in the genus. Airy Shaw (1966) notes that there are multiple buds in each axil (which is not unusual in the Malvaceae) in *Goethea cauliflora* and that some of these are delayed in development and are activated after leaf abscission. Thus some flowers develop on leafless stems, and the adjective "cauliflorous" results. If this is regarded as an acceptable generic distinction, then *Goethea* must be treated as a monotypic genus. It does not seem to me to be a sufficient distinction by itself for recognition at generic rank.

The broad, colored, and persistent involucellar bracts are also used as supporting characters in defining *Goethea* and *Blanchetiastrum*. But broadly ovate bracts are found in *Pavonia latibracteolata*, placed by its describer in *P.* sect. *Goetheoides*; in *P. garckeana* Gürke, generally accepted in *P.* sect. *Asterochlamys*; and *P. grazielae* Krapovickas and *P. serrana* G. Esteves of *P.* sect. *Malvaviscoides*, these only distantly related to the group under discussion here. It is generally true in the Malvaceae that broad involucellar bracts are few, presumably as a result of physical constraints, although *P. tricalycaris* and *P. longitudinalis* provide exceptions to this rule. Examples of such a rule can be adduced in *Abelmoschus* (e.g., *A. manihot*: Fryxell, 1988a) and *Hibiscus* (e.g., *H. dimidiatus*: Fryxell, 1984a), as well as in *Pavonia*. Broad, ovate bracts, thus, are not unique to *Goethea* or *Blanchetiastrum* and are a weak character at best on which to base a generic distinction. The fact that these bracts are sometimes colored (and are thus presumably taking on the pollinator-attraction function of the corolla) is again not unique to *Goethea* or *Blanchetiastrum* but occurs in other species: e.g., *Pavonia glazioviana* Gürke and *P. decora* Fryxell of *P.* sect. *Malvaviscoides*. *Blanchetiastrum*'s most significant distinction is the greatly enlarged, colored involucel, which has taken over the showy role of the corolla to a greater degree than in any of the other species of this group; but it is a question only of degree not of generic distinction.

The defining character for *Triplochlamys* has been the biseriate involucel, made up of many elements. The number of elements in the involucel is highly variable in *Pavonia*, ranging from 4 in *P. grazielae*, *P. garckeana*, and *P. quadrifida* to more than 20 in *P. blanchetiana* and *P. multiflora*. Thus, the number alone is not unusual. The distinctive character here is that the involucellar bracts have arranged themselves in two series by being alternately erect and reflexed or (in *P. atlantica* and *P. ciliata*) being of two size classes. Nevertheless, they are (as in all *Pavonia* spp.) a single whorl of elements and are not two distinct morphological entities. In *P.* subgen. *Goetheoides*, four or five species exhibit the biseriate involucel (*P. atlantica*, *P. conferta*, *P. longifolia*, *P. multiflora*?, and *P. tricalycaris*), and *P. ciliata* has the elements of the involucel of variable length, which is perhaps an intermediate condition. The biseriate involucel is indeed a distinctive arrangement, perhaps resulting from crowding, and possibly with some adaptational value, but it seems to have less taxonomic significance than previously thought, especially in view of the hybridization results reported by Cheek (1989).

Other distinctive characters of this subgenus include the sometimes compound stipules, each made up of two, three, or even four unequal elements (*P. cauliflora*, *P. ciliata*, *P. crassipedicellata*, *P. ducke-limae*, *P. latibracteolata*, *P. makoyana*, *P. ovaliphylla*, and *P. spiciformis*). Fruits are evidently of distinctive morphology in this group (see Fig. 30) but are known for only a relatively few species. Most species have solitary, axillary flowers, but a few have racemose inflorescences rising above the leaves (*P. ducke-limae*, *P. spectabilis*, *P. spiciformis*, and sometimes so in *P. sancti*), or in *P. conferta* the inflorescence is congested into a headlike structure.

The species of *Pavonia* subgen. *Goetheoides* are geographically and ecologically confined to the narrow strip of wet forest along the Atlantic coast of Brazil, extending from Pernambuco south to Guanabara. Within this zone they are extremely rich in species and diversity. Indeed, the number of species described has been rapidly increasing in recent years as a result of intensive collecting in the area because of concern that the habitat is disappearing. Many (most?) of these species are of very limited geographic amplitude, and given the pace of habitat destruction, they can all be considered vulnerable to extinction in the near future. Indeed, many species may already be extinct and others will become so, without ever having been discovered and described. The present description of the group, thus, may well be considered as a memorial. To be sure, a few of the species of section *Goetheoides*, such as *P. multiflora* and *P.* × *gledhillii*, will continue in cultivation in tropical botanical gardens in various parts of the world to serve as reminders of what may well become lost.

Several of the species in this group were introduced into cultivation in European greenhouses in the early 19th century (cf. Cheek, 1989) and as a result of such cultivation were illustrated in some of the serials of botanical illustration of that century. The group as a whole, thus, is relatively well illustrated.

The name *Goetheoides* means "like the genus *Goethea*." The latter name commemorates Johann Wolfgang von Goethe (1749–1832), German poet, dramatist, and novelist, whose interests extended to the natural sciences, including botany.

Key to the species of *Pavonia* subgen. *Goetheoides*

1. Involucellar bracts >12.
    2. Involucellar bracts 10 mm wide. . . . . 224. *P. tricalycaris*
    2. Involucellar bracts 1–5 mm wide.
        3. Leaves no more than 15 × 3.5 cm.
            4. Pedicels 1–4 cm long; stipules simple, 8–9 mm long.
                5. Petals dark purple; upper leaf surface glabrate. . . . . 204. *P. conferta*
                5. Petals cream; upper leaf surface minutely pubescent, at least on veins. . . . . 200. *P. atlantica*
            4. Pedicels 12–16 cm long; stipules of 2 or 3 elements, 13–21 mm long. . . . . 203. *P. ciliata*
        3. Leaves larger (up to 26 × 8 cm).
            6. Flowers usually many per axil; petioles 2.5–12 cm long. . . . . 216. *P. multiflora*
            6. Flowers solitary; petioles 3–5 mm long. . . . . 212. *P. longifolia*
1. Involucellar bracts <12.
    7. Involucellar bracts 4–7.
        8. Involucellar bracts 5, gamophyllous in basal half; inflorescences spiciform. . . . . 220. *P. spectabilis*
        8. Involucellar bracts 4–7, essentially distinct; inflorescences otherwise (or if spiciform, involucellar bracts 7–8).

9. Stipules 8–15 mm wide; involucellar bracts 2.5–4 × 1–2.5 cm; pedicels 18–24 cm long, slender. . . . . . 210. *P. goetheoides*
9. Stipules 1–5 mm wide; involucellar bracts 1.2–3.5 × 0.2–1.9 cm; pedicels 1–30 cm long.
 10. Involucellar bracts 10–19 mm wide; pedicels 1–5 cm long; staminal column exserted; stipules sometimes divided into 2 or 3 elements.
  11. Corolla equaling or exceeding calyx; stipules of 2–3 elements. . . . . . 214. *P. makoyana*
  11. Corolla shorter than calyx.
   12. Pedicels 0.5–2 cm long, often nodding; leaves entire; stipules simple or rarely divided into 2 elements. . . . . . 202. *P. cauliflora*
   12. Pedicels ca. 1 cm long, erect; leaves more or less serrate; stipules simple. . . . . . 223. *P. strictiflora*
 10. Involucellar bracts 2–10 mm wide; pedicels 1–30 cm long; staminal column included or exserted; stipules usually simple (except in *P. latibracteolata* often divided into 2 elements).
  13. Pedicels 1–3.5 cm long; leaves 4–11 cm long.
   14. Stems evenly hirsute (hairs 1.5 mm long); leaves 4–9 cm long, more or less hirsute beneath; petioles <1 cm long, evenly hirsute; involucellar bracts 4–5, ca. 7 mm wide; corolla almost twice the length of the involucel; staminal column subequal to corolla. . . . . . 208. *P. froesii*
   14. Stems with narrow longitudinal lines of hairs 0.5 mm long, otherwise glabrous; leaves 8–11 cm long, glabrous; petioles usually >1 cm long, adaxially bearded; involucellar bracts 6, 2–4 mm wide; corolla subequal to involucel; staminal column exserted. . . . . . 219. *P. semiserrata*
  13. Pedicels 4–30 cm long; leaves often 11–23 cm long.
   15. Pedicels 4–6 cm long; stipules 9–15 mm long, often divided into 2 unequal elements.
    16. Stems hispid, the hairs 2–3 mm long; petioles uniformly hispid; involucellar bracts 5–7 mm wide, minutely glandular-pubescent. . . . . . 217. *P. ovaliphylla*
    16. Stems stellate-pubescent, the hairs 1 mm long or less; petioles bearded on adaxial side; involucellar bracts 8–9 mm wide, minutely stellate-pubescent. . . . . . 211. *P. latibracteolata*
   15. Pedicels 9–30 cm long; stipules 10–40 mm long, simple; involucel subequal to or exceeding calyx.
    17. Leaves obscurely crenate, to 14 cm long, nearly as wide, obtuse or subacute; involucellar bracts 4–10 mm wide. . . . . . 199. *P. alnifolia*
    17. Leaves entire, to 23 cm long, subobtuse to acuminate; involucellar bracts 4–10 mm wide.
     18. Plants scandent; leaves rounded-acute (sometimes retuse), twice as long as wide; petioles seemingly glabrous (minutely lepidote). . . . . 215. *P. morii*
     18. Plants erect shrubs; leaves acuminate, 3 times as long as wide; petioles minutely bearded on adaxial surface. . . . . . 213. *P. longipedunculata*
7. Involucellar bracts 7–11.
 19. Pedicels 5–15 cm long.
  20. Involucellar bracts 25–30 × 5–7 mm; calyx 16–17 mm long; pedicels stout (3–5 mm diam.); stipules of 3(–4) unequal elements; petioles and midrib evenly pubescent. . . . . . 205. *P. crassipedicellata*
  20. Involucellar bracts 16–25 × 1–2 mm; calyx 8–11 mm long; pedicels slender; stipules single; petioles pubescent.
   21. Petioles bearded only on adaxial surface; leaf blades mostly <10 cm long; stems sparsely pubescent to subglabrous; pedicels 12–15 cm long; involucel ca. 2 cm long; calyx sparsely ciliate to glabrous. . . . . . 201. *P. calyculosa*
   21. Petioles evenly stellate-pubescent; leaf blades mostly >10 cm long; stems densely stellate-pubescent; pedicels 5–9 cm long; involucel ca. 1.6 cm long; calyx stellate-pubescent and ciliate. . . . . . 222. *P. stipularis*
 19. Pedicels 0.5–4 cm long.
  22. Leaves entire, no more than 5 cm long; stipules 3–3.5 mm wide, obtuse. . . . . . 218. *P. sancti*
  22. Leaves entire or crenate-dentate, to 10–30 cm long; stipules 1–2.5 mm wide, acute.
   23. Calyx 18–23 mm long; stipules simple, 1.2–3 cm long; involucellar bracts 2.5–8 mm wide; flowers solitary.

24. Involucellar bracts with undulate margins; leaves prominently serrate, to 23 × 11 cm; petioles glabrous; pedicels 1 cm long. . . . . 206. *P. crispa*
24. Involucellar bracts flat; leaves subentire, to 13 × 4 cm; petioles short-pubescent adaxially; pedicels 1.5–3 cm long. . . . . 209. *P.* × *gledhillii*
23. Calyx 12–13 mm long; stipules of 2–3 elements, 0.6–1 cm long; involucellar bracts 1–2 mm wide; inflorescences spiciform.
25. Pedicels 20 mm or more long; androecium 5.5 cm long; leaves dimorphic (see Fig. 91), the larger vegetative leaves serrate. . . . . 207. *P. ducke-limae*
25. Pedicels 4–8 mm long; androecium 0.7 cm long; leaves entire, not dimorphic. . . . . 221. *P. spiciformis*

199. Pavonia alnifolia St.-Hilaire, Fl. Bras. Merid. **1:** 241. 1827. *Goethea alnifolia* (St.-Hilaire) Garcke, Jahrb. Berlin Bot. Gart. **1:** 222. 1881. Type. Brazil. Rio de Janeiro: Fazenda de Paraty, on banks of Lake Araruama, *St.-Hilaire 625* (holotype, P as photo F-35487; isotype fragments, CTES, F). Fig. 30

Shrubs or trees 2–4 m tall, the stems sparsely and obscurely lepidote to glabrate. *Leaf* blades broadly elliptic, mostly 6–14 × 4–9 cm, basally truncate, crenate to subentire, acute to obtuse, penninerved (or 3-nerved from the base), minutely stellate-pubescent (on youngest leaves) to glabrate, sometimes vernicose; petioles 1–2.5 cm long, pubescent on adaxial side; stipules ligulate or spatulate, 8–13 × 1–2 mm. *Flowers* solitary in the leaf axils, the pedicels 6–12 cm long; involucellar bracts 4–5, ovate, 10–19 × 5–10 mm, more or less undulate-revolute at base, acute, nearly glabrous or sparsely ciliate on margins, sometimes vernicose; calyx 10–12 mm long, externally nearly glabrous and sometimes vernicose, internally densely yellowish-puberulent; corolla tubular, 16–20 mm long, subequal to involucel or exceeding it, rose or yellowish with red center and red venation; genitalia subequal to corolla. *Fruits* ca. 8 mm diam., 7 mm long, glabrous.

Distribution (Fig. 76). Along the Atlantic coast of Brazil (Espírito Santo, Guanabara, and Rio de Janeiro) from 20°40′S to 23°S, sometimes cultivated. The type locality is at 22°52′S, 42°10′W.

Illustration. Gürke (1892b: pl. 103).

Specimens examined. **BRAZIL.** Near Taipú, *Riedel 1292* (K). **ESPÍRITO SANTO:** Itaoca-Itapemirim, *Gibram 70* (CVRD, NY); rd. from Guarapari to Maratiazes, km 30, *Occhioni 7363* (CTES, NY); Mun. Itapemirim, 3 km N of Marataízes, *Pirani et al. 3593* (NY, SPF); Guarapari, *de Souza 88* (CVRD, NY); Sôto, Itapemirim, *de Souza 230* (CVRD, NY). **GUANABARA:** In cultivation, Botanical Garden, *Carauta 1730* (CTES, F); Restinga da Marabaia, *Chaso 167* (CTES); Restinga de Jacarepaguá, *Duarte 6230* (F, NY, RB); Recreio dos Bandeirantes, *Duarte 9873* (K). **RIO DE JANEIRO:** Restinga do Grumari, *Esteves 2385* (CTES); Mun. Saquarema, Reserva Estadual de Jacarepiá, *Ferruci & Somner 947* (CTES, TEX); without locality, *Glaziou 8780* (CTES, K), *Glaziou 18140* (C, K), *Glaziou 37805* (C); Jacarepaguá Biological Reserve, *Lanna-Sobrinho 8727* (F); Mun. Maricá, Lake Maricá, *Marquete et al. 141* (RB); Jacarepaguá Biological Reserve, *Moreira 4376* (CTES); Cabo Frio, *Sagadas-Vianna et al. 1339* (US); Porto Negri, *Schott 146* (F).

The broadly elliptic, more or less obtuse leaf blades are distinctive of *Pavonia alnifolia*.

200. Pavonia atlantica Fryxell, sp. nov. Type. Brazil. Bahia: km 3–4, hwy. from Itaimbó to Itamari, 12 Nov 1971, *Santos 2126* (holotype, CTES).

Frutices caulibus pubescentibus pilis ferrugineis simplicibus atque stellatis; laminis foliorum ellipticis usque 10 cm longis subintegris obscure pubescentibus; stipulis lanceolatis ciliatis atque ferrugineo-puberulentibus; floribus solitariis in axillis supernis; bracteis involucellorum ca. 14, classium duarum magnitudinis alternantis; corollis quam involucellum parum excedentibus; fructus ignotis.

Shrubs 3 m tall, the stems hispid with ferruginous setae 1–1.5 mm long and with underlying pubescence of distinctive multi-armed stellate hairs (also ferruginous) 0.1–0.2 mm long. *Leaf* blades elliptic, 6–10 × 2–3.5 cm, basally truncate or minutely auriculate, obscurely serrate to subentire, acuminate, penninerved, obscurely pubescent above and beneath (especially on veins) and often hispid on midrib near base; petioles 3–5 mm long, with pubescence like stem; stipules erect, lanceolate, 9–14 × 1–2 mm, ciliate and with ferruginous stellate puberulence. *Flowers* solitary in the upper leaf axils, the pedicels 1–6 cm long, with pubescence like stem, articulate 8–9 mm below the flower and more densely pilose distal to articulation; involucellar bracts ca. 14, of two alternate size classes, the smaller ligulate, 7–10 × 1–1.5 mm, the larger lanceolate or spatulate, 15–18 × 2–5 mm, both densely stellate-pubescent and sparsely ciliate; calyx 8–10 mm long, stellate-pubescent and hispid-ciliate; petals slightly exceeding involucel ("flore creme"), ca. 7 mm wide, externally stellate-pubescent; androecium subequal to petals; gynoecium unknown. *Fruits* (immature) subglobose (?), glabrous and shiny; mericarps 8–9 mm tall, carinate but otherwise unornamented.

Distribution (Fig. 81). At present only known from the vicinity of Gandu in Bahia. The type locality is at 13°54′S, 39°40′W.

Specimens examined. BRAZIL. Bahia: Mun. Algodão, cerca de 3 km desta na direcão de Gandu, Fazenda São José, 21 Apr 1991, *Esteves & Lyra-Lemos 2548* (SP, TEX).

The involucel of two alternating size classes of bracts is distinctive in *Pavonia atlantica*, as is the ferruginous pubescence. The specific epithet notes the geographic distinctiveness of the distribution this species and of all of *P.* subgen. *Goetheoides*, in the Atlantic coastal forests of E Brazil.

201. Pavonia calyculosa St.-Hilaire & Naudin, Ann. Sci. Nat., ser. 2, **18:** 40. 1842. Type. Brazil. Bahia: 1835, *Blanchet 2117* (holotype, P as photo CTES; isotypes, B destroyed, as photo F-9434, CTES [fragment], F as photo F-56239, G, K, MO, NY, OXF).

Pavonia bahiensis Gürke *in* Martius, Fl. Bras. **12(3)**: 525. 1892, nom. superfl. Type. Brazil. Bahia: 1835, *Blanchet 2117* (holotype, B destroyed, as photo F-9434; isotypes, CTES [fragment], F as photo F-56239, G, K, MO, NY, OXF, P as photo CTES).

Shrubs or trees up to 6 m tall, the stems with narrow longitudinal lines of curled hairs 0.5–1 mm long, otherwise subglabrous. *Leaf* blades elliptic, mostly 4–11 × 2–5 cm, basally rounded or subtruncate, manifestly or obscurely serrate to entire, acute or acuminate, penninerved (the midrib raised above and beneath), glabrous above, very sparsely stellate-pubescent beneath especially on veins, sometimes vernicose beneath; petioles 7–11 mm long, densely bearded on adaxial surface, otherwise subglabrous; stipules 3–13 mm long, subulate, erect, deciduous. *Flowers* solitary in the leaf axils, the pedicels (7–)12–15 cm long, slender, sparsely pubescent, the hairs mostly simple, ca. 1 mm long, more abundant on adaxial surface and at distal end; involucellar bracts 8–10, linear-lanceolate to spatulate, more or less incurved, 17–23 × 1–2 mm, sometimes ciliate (densely so at juncture with pedicel), connate basally for 1–1.5 mm; calyx 9–12 mm long, sparsely ciliate, the lobes prominently 3-nerved; corolla 1.5 cm long, yellowish, ca. 7 mm wide distally, with long narrow claw; androecium subequal to corolla, the filaments 1–2 mm long. *Fruits* 8–9 mm diam., subglabrous; mericarps 5–6 mm high, smooth and rounded, obscurely crested apically.

Distribution (Fig. 14). The Atlantic coast of Brazil (Bahia, Espírito Santo) from 15°S to 20°S.

Illustration. Gürke (1892b: pl. 102, fig. 2, as *P. bahiensis*).

Specimens examined. BRAZIL. Bahia: Valley of Mucuri River, roadside of BR-101, *Belem 3879* (CTES, NY, TEX), *Belem 3882* (CTES, NY, S, US); Conquista, rd. to São Paulinho, *Fróes 12670/36* (A, CAS, NY); between Itambé and the Pardo River, *Fróes 20054* (NY, US), *Fróes 20069a* (NY, US); 87 (Mun. Nova Viçosa), river brush, *Hatschbach & J. M. Silva 48718* (C, CM, CTES, NY, US). **Espírito Santo:** Linhares, Companhia Vale do Rio Doce Forest Reserve, *Esteves & Kameyama 2615* (NY, SP); CVRD Forest Reserve, Linhares, Est. Gávea, km 16, *Farias 371* (CVRD, NY); CVRD Forest Reserve, Linhares, Est. CAINGA, ant 234, km 0,870, 16.vii.1980, *Folli 241* (CVRD, NY); Vale do Rio Doce, *Santos 946* (CTES).

Pavonia calyculosa has subglabrous foliage and a distinctively crinoid-like involucel.

202. Pavonia cauliflora (Nees) Fryxell, comb. nov. Basionym: *Goethea cauliflora* Nees, Flora **4:** 304. 1821. Type. Brazil. On the Ilheos River, *Wied-Neuwied 848* [Mart. Herb. Fl. Bras. no. 649] (holotype, M as photo F-20000; isotypes, G as photo CTES, NY).

Large *shrubs* or small trees (evidently), the stems obscurely pubescent to glabrate. *Leaf* blades lance-ovate, mostly 15–29 × 7–11 cm, basally truncate to subcuneate, entire (or very coarsely serrate), acute or acuminate, penninerved (i.e., basally 3-nerved), the midrib raised above and beneath (with 5–7 pairs of secondary nerves), glabrous above and beneath; petioles 4–6 cm long, obscurely pubescent adaxially (the hairs very small), otherwise glabrous; stipules 17–22 × 1–2 mm, ligulate, glabrous, rarely divided into 2 unequal elements. *Flowers* solitary or paired in the leaf axils, the pedicels 0.5–2 cm long, slender, usually of reduced lateral branchlets bearing stipuliform elements 4–6 mm long; involucellar bracts 4, ovate, 1.5–2 × 9–14 mm, entire, acute, subglabrous, reddish showing prominent venation and white margin; calyx included in (and largely concealed by) the involucel, 1 cm long, deeply divided, minutely stellate-pubescent; corolla subequal to calyx, "pale pink, white distally"; androecium exserted from involucel, the column white, the filaments 4–7 mm long, reddish; styles emergent from the staminal column, white. *Fruits* unknown.

Distribution (Fig. 84). The Atlantic coast of Brazil (Bahia, Rio de Janeiro), sometimes cultivated. The type locality is at 14°48′S, 39°05′W.

Illustration. Nees & Martius (1823: pl. 8, as *Goethea cauliflora*).

Specimens examined. IN CULTIVATION. United States: Hawaii: Waimea Botanic Garden, *s.coll., s.n.* (NY). **BRAZIL. Without locality:** *von Martius 649* (MO). **Bahia:** Mun. Ilhéus, city of Ilhéus, 3 km N of Rodoviaria,

Thomas et al. 10603 (NY, TEX), *Thomas et al. 10744* (NY, TEX). RIO DE JANEIRO: Tijuca Forest, *Glaziou 18140a* (CTES [fragment]). SANTA CATARINA: Corupá [cult.], *Reitz 6243* (CTES).

Pavonia cauliflora is distinctive for its broad involucellar bracts, its compound stipules (of 2–3 elements), and its cauliflorous habit. Its nearest relative appears to be *P. makoyana*, from which it is distinguished by its obscure corolla.

203. Pavonia ciliata G. Esteves & Krapovickas, Hoehnea **22:** 41. 1995. Type. Brazil. Bahia: Mun. Una, BA-265, 19–22 km from Una; S Bahian wet shrubland, 50–75 m alt., 25 Feb 1978, *Mori, Kallunki & Pennington 9279* (holotype, CEPEC–n.v.; isotypes, CTES, K, NY, RB–n.v.). Fig. 90

Shrubs 4 m tall, the stems densely hispid, the hairs simple, yellowish, 1.5–2 mm long. *Leaf* blades elliptic, mostly 6–12 × 2–3.5 cm, basally cuneate, distally serrate, acuminate, the upper surface with evenly distributed simple hairs that are erect or more or less appressed, the lower surface similar, except the hairs more variable in size and on average smaller, penninerved, the midrib raised both above and beneath, more prominently so beneath; petioles 5–10 mm long, hispid; stipules 13–21 mm long, erect, linear or falcate, hispid-ciliate, divided to base into 2 or 3 elements of unequal length, deciduous leaving separate scars (i.e., 4–6 stipular scars per node). *Flowers* solitary in the leaf axils, the pedicels 12–16 cm long, hispid, the hairs 1.5–2 mm long, yellowish, somewhat antrorse; involucellar bracts 12–14, subulate, of variable length, the longest 20 mm long, hispid-ciliate; calyx 9–11 mm long, densely hispid; corolla 12–13 mm long, white (drying yellowish), apparently glabrous or with a few stellate hairs externally; androecium slightly exserted, the column glabrous, subequal to corolla, the anthers golden; styles and stigmas exserted, the stigmas capitate. *Fruits* 9 mm diam., glabrous, yellowish, retuse; mericarps 9 mm tall, carinate, slightly reticulate-veined.

Distribution (Fig. 11). From the Atlantic coast of Bahia, Brazil. The type locality is at 15°20′S, 39°15′W.

Illustrations. Esteves & Krapovickas (1995: figs. 1–8).

Specimens examined. **BRAZIL.** BAHIA: Una, Fazenda São Rafael, Mata para Seringa, *Pinheiro 1324* (CEPEC, CTES, SPF).

Pavonia ciliata is distinguished by its hispid herbage, compound stipules, and relatively small leaves (15 cm or less in length).

204. Pavonia conferta St.-Hilaire, Fl. Bras. Merid. **1:** 238. 1827. *Triplochlamys conferta* (St.-Hilaire) Ulbrich, Notizbl. Bot. Gart. Mus. Berlin-Dahlem **6:** 334. 1915. Type. Brazil. Espírito Santo: Near Ponta da Fruta, *St.-Hilaire s.n.* (holotype, P as photo F-35469).

Shrubs 1–2 m tall, the stems minutely puberulent or glabrescent, with prominent leaf scars. *Leaves* congested apically, the blades lanceolate-acuminate, entire or obscurely denticulate, 8–15 × 1.5–3 cm, glabrate above, scabrid beneath; petioles 1–3 cm long, deflected, scabridulous; stipules linear-lanceolate, 8 mm long. *Flowers* congested apically, forming a terminal capitulum, the pedicels 8–13 mm long, minutely puberulent; involucellar bracts ca. 15, linear-acute, tomentose, slightly shorter than calyx, in two series, the bracts of the outer series fewer and shorter; calyx tomentose, 5-lobed, the lobes more or less lanceolate; petals erect, narrow, twice length of calyx (i.e., exserted), dark purple; staminal column exserted, the filaments emerging in upper portion of column; styles exserted from staminal column, glabrous, with capitellate stigmas. *Fruits* unknown.

Distribution (Fig. 81). Known only from the type collection. The type locality is at 20°33′S, 40°22′W.

No specimens of *Pavonia conferta* have been examined, other than the photograph of the type cited above. Hence, the above description is derived from the original description. Apparently its distinguishing feature is the congested, capituliform inflorescence, from which the specific epithet was derived.

205. Pavonia crassipedicellata Krapovickas, Anais XXXII Congr. Nac. Bot. [Brasil] p. 68. 1982. Type. Brazil. Espírito Santo: Mun. Linhares, Sooretama Biological Reserve, Mata de Cupido, *Sucre 8674* (holotype, RB; isotype, CTES). Fig. 30

Tree ca. 8 m tall (trunk 4 m tall, dbh 8 cm, diameter of crown ca. 7 m), the young stems densely stellate-pubescent. *Leaf* blades lance-oblong to obovate, 14–44 × 5–20 cm, basally narrowly cordate, serrate, acuminate, penninerved (the midrib raised above and beneath, densely pubescent), the upper surface with simple erect hairs ca. 0.5 mm long, the lower surface stellate-pubescent; petioles 1.5–3.5 cm long distally to 12 cm below; stipules divided into 3(–4) distinct unequal elements, the largest 2.5–3 × 4–6 mm, ligulate to falcate, pubescent. *Flowers* solitary in the leaf axils, the pedicels to 5.5 cm long, stout (3–5 mm diam.); involucellar bracts ca. 7, spatulate, 25–30 × 5–7 mm, with a stout midrib/stipe, densely pubescent; calyx 16–17 mm long, yellowish-pubescent;

FIG. 90. *Pavonia ciliata*. Flowering branch with enlargement of node (upper left) to show multiple stipules (*Mori et al. 9279*.)

corolla ca. 2 cm long, tubular, yellow-green, glabrous (fleshy?); genitalia exserted, subequal to involucel. *Fruits* 12–14 mm diam., glabrous and shiny, weakly 10-ribbed; mericarps 17 mm tall, much longer than the reduced column.

Distribution (Fig. 78). From the Atlantic coast of Espírito Santo, Brazil. The type locality is approximately at 19°0′S, 39°58′W.

Illustrations. Krapovickas (1982a: pl. 1, fig. 2).

Specimens examined. BRAZIL. Espírito Santo: Linhares, CVRD Forest Reserve, *Esteves & Pirani 2397* (SPF), *I. A. Silva 374* (CTES, CVRD, NY).

The unique feature of *Pavonia crassipedicelata*, as the name suggests, is the stout pedicel, which is 3–5 mm thick.

206. Pavonia crispa Krapovickas, Anais XXXVI Congr. Bras. Bot., Curitiba [1985], p. 285. 1990. Type. Brazil. Bahia: Km 6 of hwy. from Teixera de Freitas to Alcobaça, 9 Oct 1971, *Santos 2080* (holotype, CEPEC–n.v.; isotype, CTES).

Tree to 6 m tall, the stems essentially glabrous, somewhat sulcate. *Leaf* blades broadly elliptic, truncate (with minute auriculae at base), coarsely serrate (6–7 teeth in 5 cm), acute or short-acuminate, penninerved, essentially glabrous, to 23 × 11 cm; petioles 1.5–2 cm long, glabrous; stipules erect, 2–3 cm long, ligulate, glabrous. *Flowers* solitary in the leaf axils, the pedicels ca. 1 cm long; involucellar bracts ca. 8, 2 × ca. 5–7 mm, the margins prominently undulate; calyx subequal to involucel, both very minutely and obscurely lepidote; corolla and genitalia unknown, the buds rose (*ex* collector's label). *Fruits* 1.5 cm diam., glabrous, the mericarps ca. 1 cm tall, smooth, carinate, apically pointed.

Distribution (Fig. 76). Known only from the type collection. The type locality is at 17°34′S, 39°23′W.

Illustrations. Krapovickas (1982a: pl. 2, fig. 2).

The distinctive feature of *Pavonia crispa* is the involucel, the bracts of which have undulate margins

207. Pavonia ducke-limae Monteiro, Anais XXIII Congr. Nac. Bot. [Brasil], p. 127. 1972. Type. Brazil. Pernambuco: São Lourenço da Mata, beginning of the rd. to Aldeia, *Ducke & Andrade-Lima 7* (holotype, R-76224–n.v.). Figs. 30, 91

Tree 4 m tall, the stems minutely stellate-pubescent. *Leaves* dimorphic, those of vegetative branches ovate-elliptic, up to 30 × 15 cm, basally cordate, undulate-serrate, acute, penninerved, sparsely and minutely stellate-pubescent; leaves of flowering branches rotund, 2.5–3 cm long, about as wide, basally cordate, entire, obtuse; petioles up to 11 cm long on lower leaves, sparsely pubescent to subglabrous, the upper leaves subsessile; stipules often divided into 2–3 elements, erect, ligulate or falcate, 1.5–3 × 1–2 mm, ciliate. *Flowers* solitary in the uppermost axils, the pedicels 2–4 cm long, the inflorescence racemiform; involucellar bracts 7–8, subulate, 10 × 1–1.5 mm, minutely stellate-pubescent; calyx 12–14 mm long, more densely pubescent and of lighter coloration than involucel, sometimes also hirsute, deeply divided, the lobes somewhat ribbed, internally villous; corolla 1–2 cm long, tubular, red; staminal column exserted (4–5 times the length of the corolla), dark red, the filaments 3–4 mm long, principally distal, the column and filaments covered with peglike hairs; styles and stigmas exserted. *Fruits* 10–11 mm diam., glabrous; mericarps 8–9 mm high, smooth (mericarps ca. 3 times height of reduced column).

Distribution (Fig. 81). From the Atlantic coast of Pernambuco, Brazil between 07°50′S and 08°S, the northernmost representative of the subgenus. The type locality is at 07°57′S, 35°03′W.

Additional illustrations: Monteiro (1972: figs. 6, 7).

Specimens examined. BRAZIL. Pernambuco: Side of rd. to Aldeia, *Andrade-Lima 50-621* (CTES); 15 km N of Recife, *Krapovickas 12872* (CTES [incl. wood sample], F, ENCB, G, GB, HB, LIL, MBM, MO, NY, SI, UC, WIS); rd. from Aldeia to Pernambuco, *Leal & da Silva 308* (CTES, RB).

Local name. Esfola-bahinha (Brazil).

Pavonia ducke-limae is characterized by dimorphic leaves, a racemiform inflorescence, and prominently exserted genitalia, as well as in having compound stipules. Two other species of *P.* subgen. *Goetheoides* have spiciform inflorescences (rather than solitary, axillary flowers) and might be compared to this species. The present species differs from *P. spectabilis* in having more numerous involucellar bracts, longer pedicels, subglabrous petioles, and compound stipules. It differs from *P. spiciformis* in its longer pedicels and exserted genitalia.

208. Pavonia froesii Fryxell, sp. nov. Type. Brazil. Bahia: Between Salobinho and Itatinga, 4 Dec 1942, *Fróes 20055* (holotype, US; isotype, NY).

Frutices dense hirsuti, pilis 1.5 mm longis; laminis foliorum ellipticis 4–9 cm longis distaliter obscure crenatis, infra in costos dense hirsutis; stipulis 10–12 mm longis hirsutis lanceolatis vel aliquantum spatulatis; pedicellis 2.5 cm longis hirsutis; bracteis involucellorum 4–5 hirsutis 10 × 7 mm; calyce quam involucello parum breviore; petalis quam calyce

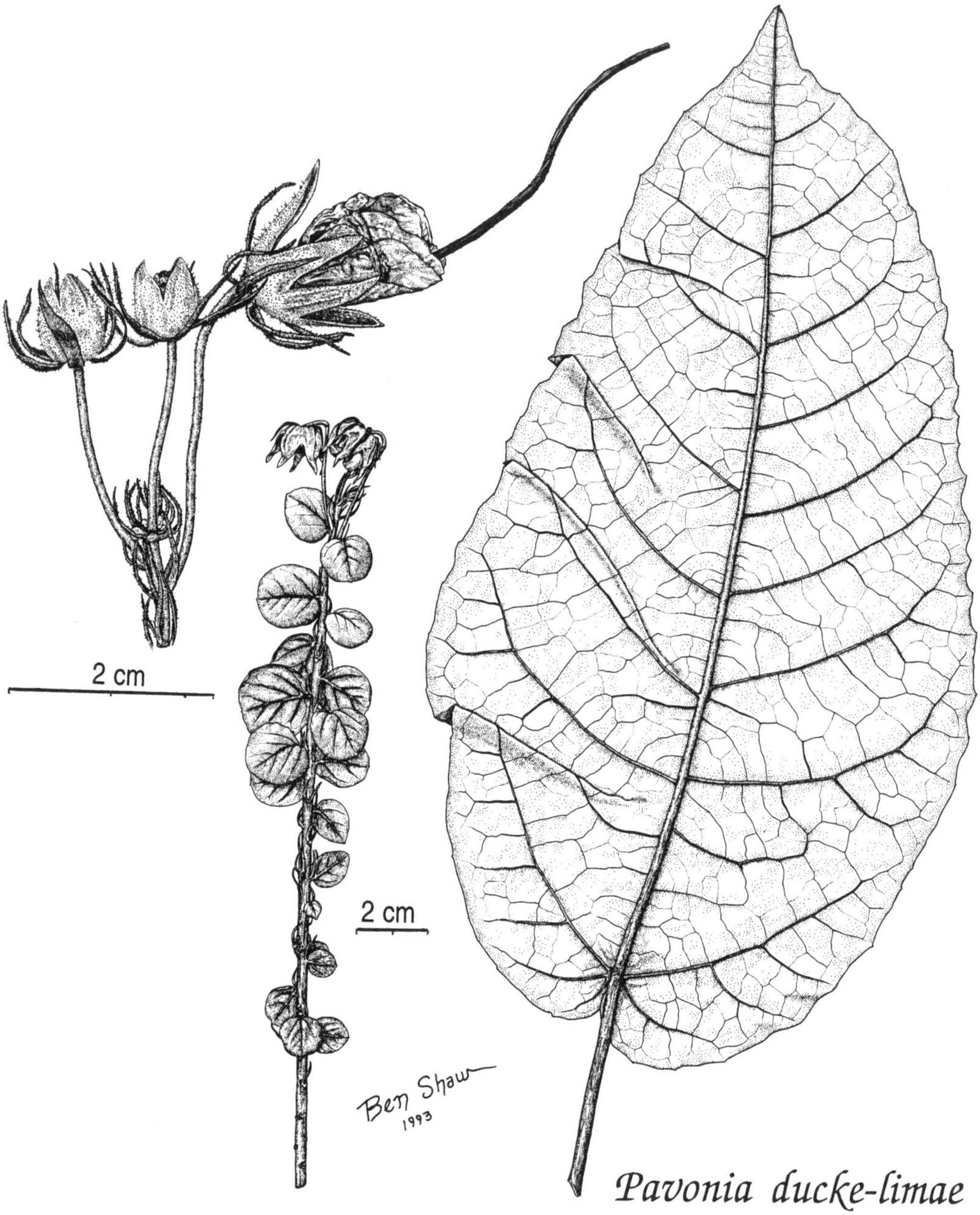

FIG. 91. *Pavonia ducke-limae*. Flowering branch (inflorescence) with reduced leaves, enlargement of a flower (at left) showing exserted staminal column (the stamens and styles broken off and missing in the available specimen), and a maximally developed vegetative leaf (right) (*Krapovickas 12872*).

duplo longioribus; androecio quam corolla subaequali; fructibus ignotis.

Shrubs with stems densely hirsute, the hairs 1.5 mm long. *Leaf* blades elliptic, 4–9 × 2.5–4 cm, basally rounded, obscurely crenate (at least distally), acute or acuminate, penninerved (or basally 3-nerved), the midrib raised above and beneath, somewhat discolorous, the upper surface very sparsely pubescent to glabrate, the lower surface hirsute (densely so on main veins), the hairs 1 mm long; petioles 5–9 mm long, evenly hirsute with pubescence like stem; stipules 10–12 mm long, linear-lanceolate to somewhat spatulate, hirsute, deciduous. *Flowers* (one seen) solitary in the leaf axils, the pedicels 2.5 cm long, densely hirsute, the hairs 1.5–2 mm long; involucel of 4 or 5 ovate bracts, each 10 × ca. 7 mm, hirsute; calyx slightly shorter than involucel, hirsute; petals ca. 18 mm long (ca. twice the length of the calyx), erect, apparently yellowish with darker veins; androecium subequal to corolla or slightly emergent. *Fruits* unknown.

Distribution (Fig. 15). Known only from the type collection from Bahia. The type locality is near Itatinguí (?) 15°13′S, 39°25′W (?).

Pavonia froesii is a small-leaved, hirsute species that is distinguished from *P. semiserrata* by pubescence characters. Its involucel includes only 4–5 bracts. The specific epithet is chosen to honor the collector of the type, Ricardo de Lemos Fróes (1891–?).

209. Pavonia × gledhillii Cheek, Kew Bull. **44:** 146. 1989. Type. From plants cultivated at the University of Bristol Botanic Garden, 1987, *Gledhill 1038* (holotype, K–n.v.; isotypes, BRIST–n.v., CTES–n.v., NY, RB–n.v.).

Subshrubs ca. 2 m tall, the stems minutely and inconspicuously pubescent. *Leaf* blades oblong-elliptic, mostly 10–13 × 3–4 cm, basally rounded-truncate, obscurely serrulate or subentire, acute, penninerved, obscurely and minutely pubescent to subglabrous; petioles 3–5 cm long, short-pubescent on adaxial surface, the hairs 0.3 mm long; stipules linear, 12–14 ×1.5 mm, ciliate, the hairs 0.5–1 mm long. *Flowers* solitary in the leaf axils, the pedicels 1.5–3 cm long, minutely pubescent; involucellar bracts 9–10 (?), lance-linear, 2 × 2.5–3.5 mm, ciliate (the hairs 1 mm long), reddish, basally connate; calyx ca. 18 mm long, densely puberulent; corolla 2.5–3 cm long, tubular, dark purplish; androecium exserted 1.5–2.5 cm beyond corolla, purplish, the filaments 3–5 mm long; styles emergent from the column, red or purplish, sparsely and minutely pubescent. *Fruits* unknown.

Distribution. This species is of hybrid origin, being the artificial hybrid of *Pavonia multiflora* and *P. makoyana*. It occurs only in cultivation.

Illustration. Cheek (1989: fig. 1).

Specimens examined. **IN CULTIVATION**. **UNITED STATES: California:** Los Angeles State and County Arboretum acc. no. 53-P-596, *Griffiths & Hawkinson 4646* (NY, US); S. California Horticultural Inst., *s.coll., s.n.* (RSA). **Missouri:** 1896, Missouri Botanical Garden (MO).

210. Pavonia goetheoides (Hassler) Fryxell, comb. nov. Basionym: *Blanchetiastrum goetheoides* Hassler, Feddes Repert. Spec. Nov. Regni Veg. **8:** 28. 1910. Type. Brazil. Ilhéus, *Blanchet s.n.* (holotype, G as photo F-23717).

Shrubs to 2 m tall, the younger stems minutely and obscurely sublepidote (the hairs 0.1 mm diam.) to glabrate. *Leaf* blades oblong-elliptic, mostly 9–19 × 3–7 cm, basally rounded or subcuneate (and minutely auriculate), entire, acuminate, penninerved, glabrous or subglabrous; petioles 0.5–2.5 cm long; stipules foliaceous, 1.5–2.5 × 0.8–1.5 cm, ovate-elliptic, sessile, basally cuneate to cordate-clasping, acute or obtuse, glabrous. *Flowers* solitary in the leaf axils, the pedicels 18–24 cm long, filiform (0.5 mm diam.), distally dilated forming 4 flanges or buttresses; involucellar bracts 4–5, foliaceous, red, 2.5–4 × 1–2.5 cm, ovate, entire, acute, glabrous; calyx 10–12 mm long, externally sparsely and minutely puberulent to subglabrous, internally densely puberulent; corolla subequal to calyx or somewhat shorter; staminal column 1–1.5 cm long, the filaments 6–9 mm long, erect, exserted from calyx/corolla. *Fruits* unknown.

Distribution (Fig. 76). From the Atlantic coast of Bahia, Brazil. The type locality is at 14°48′S, 39°3′W.

Illustration. Hassler (1910: 29, as *Blanchetiastrum goetheoides*).

Specimens examined. **BRAZIL. BAHIA:** Km 9 of new hwy. from São José to Una, *T. S. Santos 3105* (CTES, NY); Una, Unacau, *Sobral et al. 5793* (CTES, F, NY).

Pavonia goetheoides is notable for its prominent, foliaceous stipules.

211. Pavonia latibracteolata Krapovickas, Anais XXXII Congr. Nac. Bot. [Brasil], p. 70. 1972. Type. Brazil. Bahia: Hwy. from Porto Seguro to Eunápolis, *Pinheiro 1746* (holotype, CEPEC–n.v.; isotype, CTES).

Shrub 3 m tall, the stems densely yellowish-hirsute (when young) to minutely stellate-pubescent, the hairs ca. 0.5 mm long. *Leaf* blades elliptic, truncate, obscurely serrate, acuminate, 9–13 × 2–4 cm, penninerved, with scattered stellate pubescence beneath, nearly glabrous above; petioles 1–2 cm long, densely bearded on adaxial side, otherwise with scattered minute stellate pubescence; stipules divided into 2 unequal elements, the larger ca. 9 mm long, falcate, minutely pubescent, deciduous. *Flowers* solitary in the leaf axils, the pedicels to 5.5 cm long, densely yellowish-hirsute; involucellar bracts 5, broadly ovate, ca. 12 × 8–9 mm, minutely pubescent; calyx slightly exceeding the involucel, with the stellate pubescence coarser than that of the involucel, yellowish. (Note by Krapovickas on the isotype: "tubo de la corola 3 mm long., glabro;

pétalos 11 mm long. × 4 mm lat.; tubo estaminal 22 mm long., ápice 5-dentado; estambres numerosos, agrupados en el extremo del tubo; tubo y filamentos con pelos glandulares, breves, capitados—del holotipo, *Pinheiro 1746* [CEPEC].") *Fruits* unknown.

Distribution (Fig. 11). Known only from the type collection from Bahia. The type locality is approximately at 16°25′S, 39°05–30′W.

Illustration. Krapovickas (1982a: pl. 3, fig. 1).

212. Pavonia longifolia St.-Hilaire, Fl. Bras. Merid. **1:** 239. 1827. *Malvaviscus longifolius* (St.-Hilaire) Spach, Hist. Vég. Phan. **3:** 370. 1834 (non *Malvaviscus longifolius* Garcke, 1854). *Triplochlamys longifolia* (St.-Hilaire) Ulbrich, Notizbl. Bot. Gart. Mus. Berlin-Dahlem **6:** 334. 1915. Type. Brazil. Espírito Santo: Near the town of Villa da Victoria, *St.-Hilaire s.n.* (holotype, P as photo F-35477; isotype fragment, F).

Shrubs 1.5–2 m tall, the stems scurfy, rarely stellate-pubescent. *Leaf* blades elliptic to slightly oblanceolate, to 15–30 × 5–8 cm, basally cuneate, subentire or with a few denticulations near the apex, acuminate, penninerved (the midrib raised beneath), essentially glabrous (or with a few scabrid hairs beneath); petioles 3–5 cm long, scabrous, canaliculate; stipules linear-lanceolate, 15–20 × 1–2 mm, with scabrous stellate hairs, erect and rigid. *Flowers* solitary in the upper axils, the pedicels 1.5–3 cm long, pubescent with both stellate and simple hairs; involucel of ca. 16 bractlets in two series, the inner 8 erect, ca. 20 mm long, subspatulate, reddish, the outer bractlets half as long, apparently spreading; calyx 13 mm long, deeply divided, the lobes lanceolate, 3-nerved; petals subequal to or slightly longer than the involucel, externally greenish and stellate-pubescent. *Fruits* unknown (the ovary glabrous).

Distribution (Fig. 60). Known only from the type from Espírito Santo, Brazil. The type locality is at 20°20′S, 40°20′W. No specimens of *Pavonia longifolia* have been examined, except for the Field Museum photograph cited above. The description is modified from the original description, as is that of Gürke (1892b).

213. Pavonia longipedunculata Gürke *in* Martius, Fl. Bras. **12(3)**: 525. 1892. Type. Brazil. Bahia: *Blanchet 3949* (isotypes, C, CTES [fragment], F, K, MG, MO). Fig. 92

Shrubs 3–4 m tall, the stems minutely lepidote to glabrate. *Leaf* blades mostly 12–23 × 3–9 cm, narrowly elliptic to oblong-elliptic, basally truncate to cuneate, entire, acuminate, penninerved, sparsely and obscurely lepidote to glabrate; petioles 2–7 cm long, bearded in a narrow line on adaxial surface (the hairs 0.2–0.4 mm long); stipules lanceolate, 10–13(–26) × 1–1.5 mm, erect, with dense grayish puberulence, deciduous. *Flowers* solitary in the leaf axils, the pedicels 9–30 cm long; involucellar bracts 5–7, narrowly to broadly ovate, 12–18 × 3–7 mm, with minute sublepidote stellate hairs; calyx 10–12 mm long, densely yellowish-puberulent; corolla tubular, 13–15 mm long, yellowish-green, the genitalia slightly exserted. *Fruits* unknown.

Distribution (Fig. 84). From the Atlantic coast of Brazil (Bahia) from 16°30′S to 17°S.

Additional illustration. Gürke (1892b: pl. 102, fig. 1).

Specimens examined. **BRAZIL. BAHIA:** Mun. Itamarajú, Fazenda Pau-Brasil, entrance at km 5 of hwy. from Itamarajú to Eunápolis, *de Carvalho et al. 2035* (CTES, NY, TEX); Eunápolis, Itabela, Guaratinga, *T. S. Santos 902* (CTES, NY).

214. Pavonia makoyana E. Morren, Belgique Hort. **28:** 59. pl. 3. 1878. *Goethea makoyana* (E. Morren) J. D. Hooker, Bot. Mag. **105** (n.s. **35**): pl. 6427. 1879 (as "mackoyana"). Type. In cultivation *ex* Brazil (holotype, LG?–n.v.).

Pavonia rubriphylla Esteves, Bol. Bot. Univ. São Paulo **17:** [in press, 1999]. Type. Brazil. Bahia: Mun. Ibicarai, 7 Feb 1993, *Kallunki & Pirani 448* (holotype, SPF–n.v.; isotype, SP–n.v.).

Shrubs 1.5–3 m tall, the stems with whitish hairs 0.5–1 mm long and with an understory of minute stellate hairs. *Leaf* blades elliptic or oblong-elliptic, mostly 10–19 × 3–6 cm, basally cuneate, obscurely serrate to subentire, acuminate, penninerved, the midrib raised above and beneath, with minute stellate (sublepidote) pubescence to glabrate; petioles 0.5–3.5 cm long, adaxially pubescent; stipules single or often divided into 2–3 unequal elements, ligulate to falcate, 15–19 × 2–5 mm, erect, with minute stellate (sublepidote) hairs, sometimes also ciliate, the cilia 1–1.5 mm long. *Flowers* solitary in the leaf axils, often at terminal nodes above the leaves, the pedicels 2–3.5 cm long; involucellar bracts 4, broadly ovate, deeply cordate at base, 2.5–3.5 × 13–19 mm (enclosing and concealing the calyx), sparsely pubescent or sometimes ciliate, reddish; calyx 1.5–2.5 cm long, puberulent, deeply divided, the lobes with prominent midribs; corolla tubular, 2.5–3 cm long (subequal to involucel), red, externally pubescent; androecium exserted, reddish (at least distally), the filaments 3–7 mm long; styles emergent from the column. *Fruits* unknown.

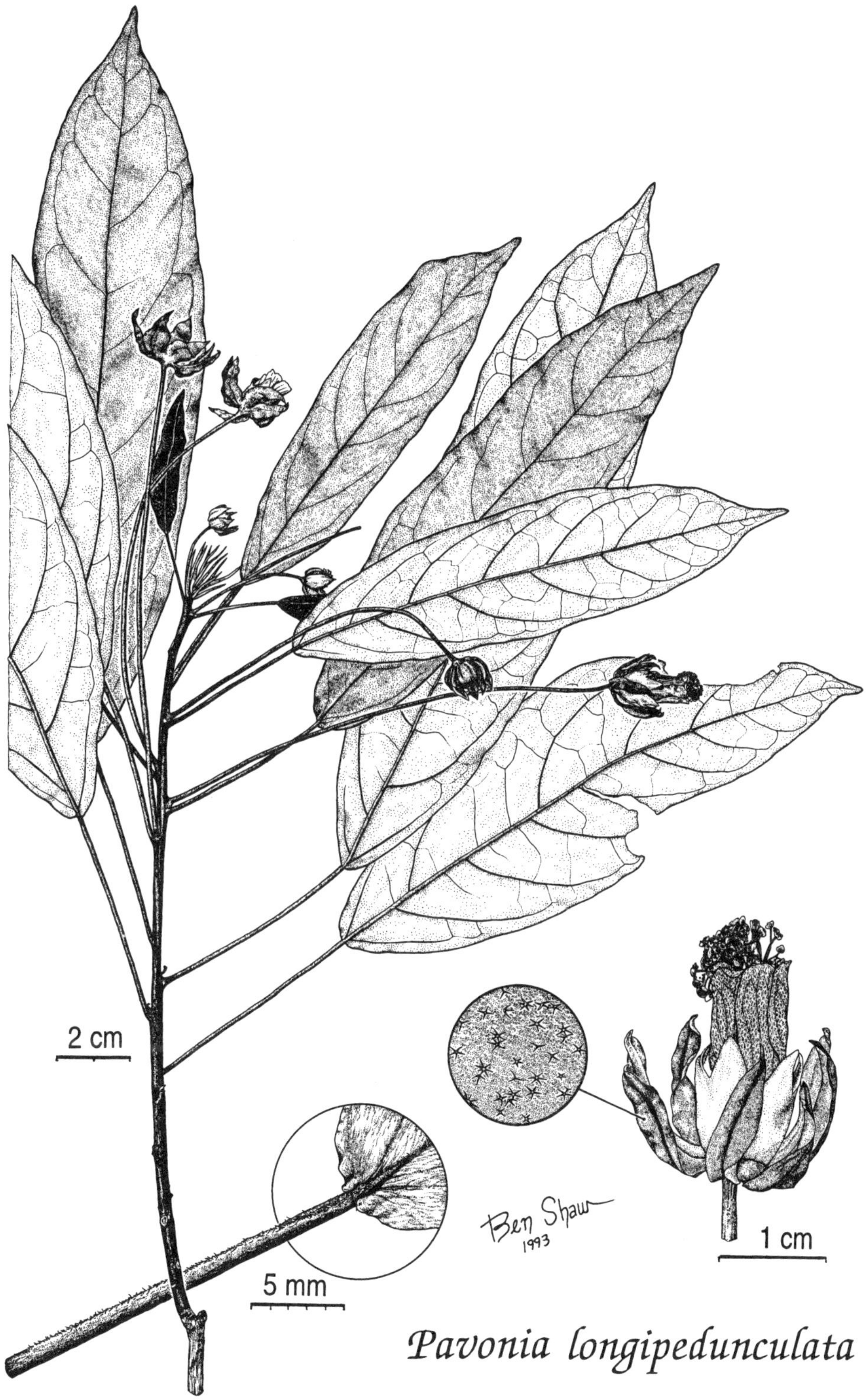

FIG. 92. *Pavonia longipedunculata*. Flowering branch with enlargement of flower (bottom right) and of leaf base (bottom center) (*de Carvalho et al. 2035*).

Distribution (Fig. 15). From the Atlantic coast of Brazil (Bahia, Espírito Santo, and Rio de Janeiro) from 16°30′S to 22°30′S, sometimes cultivated.

Illustrations. Gürke (1892b: pl. 105, as *Goethea makoyana*); J. Hooker (1879: pl. 6427, as *Goethea mackoyana*); Morren (1878: pl. 3), reproduced by Lebl (1878: pl. 22) and by Regel (1881: 279).

Specimens examined. **BRAZIL. BAHIA:** Eunapolis, rd. to Pôrto Seguro, *T. S. Santos 1674* (CTES, IPA, NY); 16 km W of Porto Seguro, *Santos et al. 367* (CTES). **ESPÍRITO SANTO:** Sta. Leopoldina, Ponte de Balanço, *Bittencourt 106* (CTES); Mun. Itaguassú, Limoeiro, Santa María, *Brade 18289* (CTES); Santa Thereza, *Duarte 9862* (CTES); Mun. Morro do Câco, near the city of Morro do Câco, *Gurken 1* (CTES); Mun. Domingo Martins, near Araguaia River, *Hatschbach & Silva 57991* (C, CTES); Mun. Santa Teresa, Lombardia Biological Reserve, Picada de Cachoeira, *Martinelli et al. 10946* (K, NY, RB); Santa Teresa, Santa Lúcia Biological Station, *Menandro 192* (CVRD, NY); Mun. Serra, rd. from Vitória to Jacaraípe, *Pirani et al. 2778* (NY, SPF, TEX); Mun. Domingos Martíns, *Sazima 14058* (NY, UEC). **RIO DE JANEIRO:** In the primeval forest of Macahé [Macaé], June 1832, *s.coll. 338* (NY, US, both *ex* LE).

Esteves and Fryxell (1999) distinguished *Pavonia rubriphylla* from *P. makoyana*, but the differences were largely details of pubescence and relative size of leaves, differences that now seem to fall within the range of infraspecific variation.

215. Pavonia morii Krapovickas, Anais XXXII Congr. Nac. Bot. [Brasil], p. 71. 1972. Type. Brazil. Bahia: Mun. Una, Hwy. BA-265 27 km from Una, *Mori, Kallunki & Pennington 9339* (holotype, CEPEC–n.v.; isotype, CTES).

Trees or scandent shrubs to 6 m tall, the stems sparsely and very minutely lepidote to glabrate. *Leaf* blades elliptic, 10–17 × 5–8 cm, basally truncate to cuneate, entire, acute or rounded-acute (sometimes retuse), penninerved (the midrib prominently raised beneath, scarcely raised above), sparsely lepidote to glabrate; petioles 2–9 cm long, sparsely lepidote to glabrate, not adaxially bearded; stipules erect, ligulate, 16–40 × 1–3 mm, minutely lepidote. *Flowers* solitary in the leaf axils, the pedicels 11–30 cm long, distally flared and sulcate; involucellar bracts 5, basally connate, 20–25 × 5–10 mm, ligulate to ovate, sparsely lepidote or almost glabrous; calyx 1.5 cm long, lepidote to almost glabrous; corolla 2 cm long, distally yellowish, but with a reddish cast where exposed, inconspicuous; androecium exserted 1.5 cm or more beyond the corolla, the column white, the anthers yellow. *Fruits* 11–12 mm diam., yellowish, glabrous, the mericarps smooth and shiny, with a medial ridge.

Distribution (Fig. 81). From the Atlantic coast of Bahia, Brazil. The type locality is at 15°12′S, 39°23′W.

Illustration. Krapovickas (1982a: pl. 4, fig. 1).

Specimens examined. **BRAZIL. BAHIA:** Mun. Una, BA-265, 27 km from Una, *Mori et al. 9340* (NY); km 28, hwy. from Olivença to Una (km 15, Una to Olivença), *dos Santos & Judziewicz 4083* (CTES, US); Mun. Una, Mico-leão Biological Reserve (IBAMA), entrance at km 46 of BA-001 from Ilhéus to Una, *de Carvalho et al. 6223* (CEPEC, NY).

216. Pavonia multiflora St.-Hilaire, Fl. Bras. Merid. **1:** 239. 1827. *Malvaviscus multiflorus* (St.-Hilaire) Spach, Hist. Vég. Phan. **3:** 370. 1834. *Triplochlamys multiflora* (St.-Hilaire) Ulbrich, Notizbl. Bot. Gart. Mus. Berlin-Dahlem **6:** 335. 1915. Type. Brazil. Espírito Santo: Near Fregesia da Serra, *St.-Hilaire s.n.* (holotype, P as photo CTES; isotypes, CTES [fragment], F). Figs. 30, 93

Pavonia lanceolata Colla, Herb. Ped. **1:** 398. 1833 (non *Pavonia lanceolata* Schlechtendal, 1837). Type. Brazil. Barra de Tuna, *Martius s.n.* (holotype, TO). See Fryxell (1976) for redescription and illustration.

Pavonia wioti E. Morren, Belgique Hort. **25:** 113, pl. 7. 1875. Type. The plate.

Shrubs 1.5–2.5 m tall, the stems sometimes prominently hirsute, usually with narrow longitudinal lines of fine curved hairs and otherwise glabrate. *Leaf* blades narrowly elliptic, mostly 12–26 × 2–8 cm, basally truncate to cuneate, obscurely to prominently serrate, acuminate, with appressed simple hairs to glabrous or subglabrous, penninerved (6–8 secondary nerves on each side), the midrib raised on both surfaces; petioles 2.5–12 cm long, sometimes prominently hirsute, usually minutely pubescent on adaxial surface and otherwise glabrous; stipules subulate, erect, 9–13 × 0.5–1 mm. *Flowers* solitary in the leaf axils or often in axillary fascicles of 2 or 3, the flowers often crowded apically because of shortened internodes and deciduous leaves, the leaf scars prominent, the pedicels up to 7 cm long, minutely pubescent; involucellar bracts 20–24, seemingly in two series, the outer spreading to reflexed, shorter than the arcuate to erect inner bracts, all narrowly ligulate, stellate-pubescent and ciliate, sometimes reddish, the longer bracts 12–20(–30) mm long; calyx 1–2 cm long, deeply divided, with stellate and glandular hairs, sometimes basally setose; corolla tubular, 2–2.5 cm long, barely exceeding calyx to twice length of calyx, externally stellate-pubescent, yellow (reddish in bud) or dark purple in cultivated (selected?) forms; androecium exserted, ca. twice length of corolla, pallid or purplish, the filaments 3–5 mm long; styles exserted, slender, the stigmas villous. *Fruits* 7–9 mm diam., short-beaked, glabrous, yellowish;

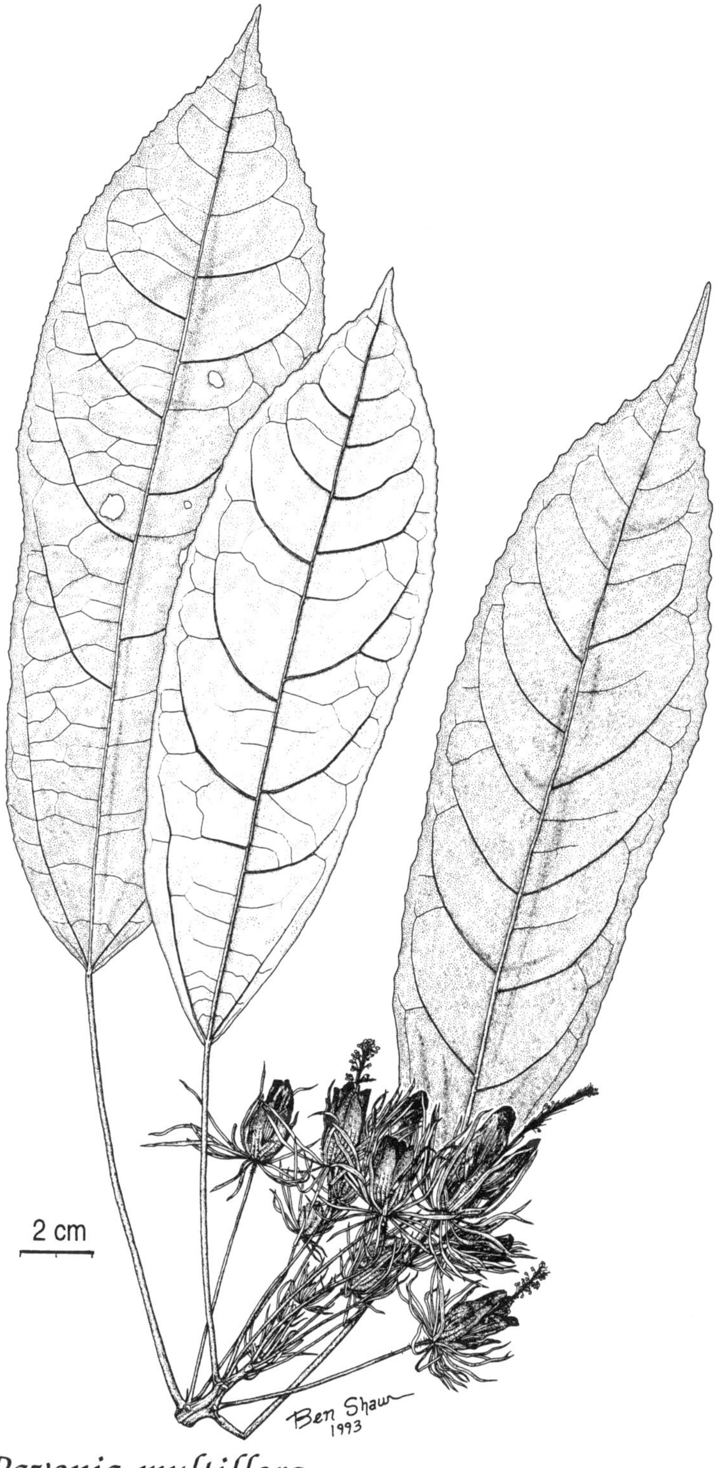

FIG. 93. *Pavonia multiflora.* Flowering branch (*Farias 113*, *Peixoto et al. 3051*).

mericarps 7–9 mm high, smooth. Chromosome number, $2n = 56$ (Bates, 1967).

Distribution (Fig. 79). From the Atlantic coast of Espírito Santo, Brazil, often cultivated in various parts of the world. The type locality is at 20°10′S, 40°18′W.

Additional illustrations: Floral Mag. n.s. pl. 276 (1877); Fryxell (1976: fig. 1); Hooker (1878: pl. 6398); Morren (1875: pl. 7); St.-Hilaire (1827: pl. 47); Vogel (1929: 494).

Specimens examined. IN CULTIVATION. UNITED STATES: Maryland: In greenhouse cultivation, College Park, *Hill 13325* (NY, TEX). **New York:** New York Botanical Garden Conservatory, original from Gould-1900, 23 Jul 1904, *Nash s.n.* (NY).

BRAZIL. ESPÍRITO SANTO: Sooretama Forest Reserve, *Belem 1502* (CTES); Linhares, Vale do Rio Doce, *Belem 3810* (MO, NY); Mun. Nova Venécia, Serra da Cima, *Duarte 4201* (CTES); Linhares, CVRD Forest Reserve, *Farias 113* (CTES, CVRD, NY), *Farias 397* (CVRD, NY); 20 km S of Linhares, *Hatschbach 46911* (CTES, US); Linhares, *Hatschbach et al. 58068* (C, CTES); Mun. Bananal, forest behind Fazenda Lagoa Nova, *Kallunki et al. 588* (MO, NY, TEX), *Kallunki et al. 691* (NY, TEX); Linhares, *Lino 96* (CTES); Mun. Colatina, Fazenda San Gabriel, *Lourteig 3218* (K, US); Linhares, *Martinelli & Soderstrom 9763* (US); Vila Velha, Convento da Penha, *Pereira 14230* (K, NY, TEX, UEC); Linhares, *A. Peixoto et al. 3051* (MO, NY, TEX); Linhares, *A. L. Peixoto & O. L. Peixoto 396* (MO); km 8 from Linhares to Vitória, *Pinheiro & Santos 2264* (CTES); Mun. Serra, perto de Vitória, Mestre Álvaro Biological Station, *Pirani et al. 183* (CTES); Mun. Serra, rd. from Vitória to Jacaraípe, *Pirani et al. 2784* (NY, SPF, TEX); Mun. Aracruz, Bairro Coqueiral, *Pirani et al. 3077* (NY, SPF); Mun. Nova Venécia, 1 km N of Córrego Boa Esperança, *Pirani et al. 3377* (NY, SPF); Mun. Marilândia, hwy. from Marilândia to the Bananal River, 11 km N of Marilândia, *Pirani et al. 3410* (NY, SPF); Linhares, *Queiroz 2470* (CVRD, pf); Linhares, *Santos 1463* (CTES); Linhares, vale do Río Dôce, km 6 of BR-101, S side, *T. S. Santos 1999* (CTES, NY); Vila Velha, *Sazima 14053* (NY, UEC).

Pavonia multiflora is relatively variable, especially in size and coloration of the corolla and involucel. It is quite floriferous and therefore sometimes is cultivated as an ornamental; some of the showier forms have probably been selected for this purpose. The large number of involucellar bracts that are presented in two series and are often reddish, together with the relatively crowded nature of the flowers, are distinctive features of this species.

217. Pavonia ovaliphylla G. Esteves & Krapovickas, Hoehnea **22:** 43. 1995. Type. Brazil. Bahia: Mun. Prado, rd. from Prado to São Francisco de Assis, BR-101, km 16, Fazenda São Antonio, 29 Mar 1989, *Mattos Silva, Hage & dos Santos 2638* (holotype, CEPEC; isotype, CTES).

Shrubs 1 m tall, the stems densely hispid, the hairs 2–3 mm long, spreading, tan. *Leaf* blades oblong to oblanceolate, 12–16 × 4–5 cm, basally narrowly truncate, entire or irregularly undulate, acuminate, penninerved, the midrib raised beneath and somewhat hispid near base, the blade otherwise glabrous; petioles 4–7 mm long, densely hispid like stem; stipules narrowly ligulate, sometimes double, 10–15 × 1.5–2 mm, attenuate, ciliate, persistent or tardily deciduous. *Flowers* solitary in the leaf axils, the pedicels 4–6 cm long, hispid like stem and with minute underlying glandular hairs; involucellar bracts 5–7, ovate to lanceolate, 12–15 × 5–7 mm, glandular-pubescent; calyx 7–8 mm long (concealed by the involucel), glandular pubescent; petals shorter than the calyx, 4–5 mm wide; staminal column subequal to petals, the filaments 2 mm long, distally inserted, purplish; styles emergent from the column by 5 mm, purplish, the stigmas capitate. *Fruits* unknown.

Distribution (Fig. 81). Known only from the type collection from Bahia. The type locality is at 17°22′S, 39°16′W.

Illustrations. Esteves & Krapovickas (1995: figs. 9–16).

Collectors' note: "Flores com brácteas, cálices e pétalas verdes, estames e anteras róseas."

Pavonia ovaliphylla is characterized by its hispid stems and petioles (hairs 2–3 mm long) and the glandular pubescence found on the pedicels and calyces.

218. Pavonia sancti Krapovickas, Anais XXXII Congr. Nac. Bot. [Brasil], p. 72. 1972. Type. Brazil. Bahia: Gandu, rd. to Algodão, *Santos 1173* (holotype, CEPEC–n.v.; isotype, CTES).

Tree to 8 m tall, the stems farinose-puberulent. *Leaf* blades elliptic, truncate, entire, acute or obtuse, penninerved, to 5 × 2.5 cm (reduced upward to stipuliform structures in the terminal inflorescence), essentially glabrous; petioles 3–6 mm long, with pubescence like stem; stipules broadly falcate, 10 × 3–3.5 mm, puberulent. *Flowers* solitary in the axils but grouped terminally more or less above the leaves; pedicels 1–2.5 cm long, with pubescence like that of the stem but sparser, medially articulated; involucellar bracts 6–8, lanceolate, 18–20 × 3 mm, puberulent; calyx 10–15 mm long, half-divided, puberulent; corolla subequal to involucel, greenish (fide collector); androecium slightly exserted. *Fruits* 12–13 mm diam., glabrous, smooth or weakly reticulate-veined, shiny, the mericarps 9 mm tall.

Distribution (Fig. 78). Known only from the type collection from the Atlantic coast of Bahia, sympatric with *Pavonia spiciformis*. The type locality is at 13°50′S, 39°35′W.

Illustrations. Krapovickas (1982a: pl. 5, fig. 2).

219. Pavonia semiserrata (Schrader) Steudel, Nom., ed. 2, **2:** 279. 1841. *Schouwia semiserrata* Schrader, Gött. Gel. Anz. 717. (5 May) 1821. *Lebretonia semiserrata* (Schrader) de Candolle, Prodr. **1:** 446. 1824. *Malache semiserrata* (Schrader) Kuntze, Revis. Gen. Pl. **1:** 71. 1891. Type. Unknown (GOET?).

Goethea semperflorens Nees, Flora **4:** 304. (21 May) 1821. *Pavonia semperflorens* (Nees) Garcke, Jahrb. Bot. Gart. Mus. Berlin **1:** 222. 1881. Type. Brazil. Barra de Vareda, *von Wied-Neuwied 40* (isotype, GOET). Note: Krapovickas (1982a) cites a BR specimen (*von Wied-Neuwied 114*, collected in 1827, seen as photo *ex* CTES) as holotype but gives no reason for considering it such. Since its collection date is later than the date of publication of the name, it cannot be type material. The GOET duplicate cited above bears Nees's binomial in Nees's handwriting.

Shrubs 2–8 m tall, the stems with narrow longitudinal lines of hairs (0.5 mm long or less), otherwise glabrous. *Leaf* blades oblong-elliptic, mostly 8–11 × 3–4 cm, basally rounded-truncate, crenate-serrate to subentire, acuminate, penninerved (the midrib raised above and beneath), glabrous; petioles 1.5–2.5 cm long, adaxially bearded; stipules spatulate, 16–20 × 2 mm, erect, subglabrous. *Flowers* solitary in the leaf axils, the pedicels 1–3.5 cm long, swollen and minutely stellate-pubescent distally; involucellar bracts 6, narrowly lance-ovate to ligulate, 12–15 × 2–4 mm, sparsely and obscurely stellate-pubescent; calyx 7–8 mm long (largely concealed by involucel), obscurely pubescent or subglabrous, somewhat vernicose; corolla tubular, subequal to involucel, appearing yellow in sicco ("vermelhas" on label); staminal column exserted, the filaments 3–5 mm long. *Fruits* 7–8 mm diam., glabrous, reticulately veined.

Distribution (Fig. 78). Easternmost Bahia, Brazil, in semideciduous forest at ca. 900 m elevation. The type locality of *Goethea semperflorens* is near the confluence of Riberão de Vareda and Río Pardo, 15°18′S, 40°54′W.

Illustration. Nees & Martius (1823: pl. 7).

Specimens examined. **BRAZIL. BAHIA:** Encruzilhada, exit to Itambé, *Belem 3672* (CTES, NY, US); without locality, *Rudolph 42* (S); Mun. Encruzilhada, perto de uma carvoeira, *Santos & Lima 152* (CTES).

220. Pavonia spectabilis Krapovickas, Anais XXXVI Congr. Bras. Bot., Curitiba [1985] 283. 1990. Type. Brazil. Bahia: Mun. Santa Cruz de Cabrália, hwy. from Porto Seguro to Santa Cruz de Cabrália, km 18, 17 Jun 1980, *Mattos-Silva & Brito 850* (holotype, CEPEC–n.v.; isotype, CTES).

Tree to 8 m tall, the stems densely stellate-pubescent, the hairs 0.3–0.4 mm long, persistent. *Leaf* blades elliptic to oblong, to 30 × 16 cm, truncate to subcordate, serrate, acute, penninerved (or basally 3–5-nerved, the lateral nerves reduced), with ca. 10 secondary nerves on each side, the midrib prominently raised beneath, slightly raised above, obscurely pubescent above, with minute stellate and simple hairs beneath; petioles 3–8 cm long, minutely and densely pubescent, the hairs finer than on the stem; stipules 12–14 mm (smaller in inflorescence) by 3–4 mm, triangular-acuminate. *Inflorescence* terminal, spiciform, branched, leafless, with very short internodes and persistent stipules; pedicels 10–12 mm long, densely farinose-puberulent; involucel 12–15 mm long, basally gamophyllous, the bracts ca. 5, triangular, 4–6 mm wide at the base, with minute stellate pubescence; calyx subequal to involucel, more densely pubescent than involucel, the hairs yellowish; petals 15–16 × 4–5 mm (exceeding the calyx by 8–9 mm), dark reddish in sicco; androecium subequal to corolla, the column puberulent, the anthers purplish, grouped terminally, the pollen yellow (spiny); styles and stigmas barely exceeding the androecium. *Fruits* 8–9 mm diam., glabrous, smooth or reticulate-veined, shiny, the mericarps 1 cm tall.

Distribution (Fig. 74). In S Bahia, Brazil, in the wet forest along the Atlantic coast growing more or less at sea level. The type locality is at 16°24′S, 39°04′W.

Illustration. Krapovickas (1990: fig. 2).

Pavonia spectabilis is one of the species of *P.* subgen. *Goetheoides* with spiciform or racemiform inflorescences, together with *P. ducke-limae* and *P. spiciformis*. It differs from both of the latter species in having densely pubescent petioles, simple (undivided) stipules, and fewer involucellar bracts.

221. Pavonia spiciformis Krapovickas, Anais XXXII Congr. Nac. Bot. [Brasil], p. 73. 1982. Type. Brazil. Bahia: Gandu, rd. to Algodão, *Santos 1179* (holotype, CEPEC–n.v.; isotype, CTES).

Shrubs to 3 m tall, the stems minutely stellate-pubescent to glabrescent, more or less sulcate. *Leaf* blades elliptic, truncate (and minutely auriculate at base), entire, rounded-acute, penninerved, the midrib slightly raised above, prominently raised beneath, to

31 × 11 cm, glabrous; petioles 6–9 cm long, with a few obscure stellate hairs on the adaxial side, otherwise glabrous; stipules single or divided into 2–3 elements, lanceolate or ligulate, persistent. *Inflorescences* terminal, spiciform, leafless, with short internodes; pedicels 4–8 mm long, densely puberulent; involucellar bracts ca. 8, lanceolate, 10 × 1.5–2 mm, minutely stellate-pubescent; calyx 12 mm long, more densely pubescent than the involucel, the hairs yellowish, ca. half-divided, the lobes more or less 3-ribbed; petals subequal to or slightly exceeding the calyx, cream-colored (fide collector); androecium subequal to corolla; styles emergent from corolla, with capitate stigmas. *Fruits* unknown.

Distribution (Fig. 79). Known only from the type collection from Bahia, Brazil, sympatric with *Pavonia sancti*. The type locality is at 13°50′S, 39°35′W.

Illustration. Krapovickas (1982a: pl. 6, fig. 1).

Pavonia spiciformis has spiciform inflorescences, as the names indicates, and in common with *P. spectabilis* and *P. ducke-limae*. It differs from the former in its more numerous involucellar bracts (ca. 8 vs. 5), its glabrous petioles, and its compound stipules. It differs from the latter in its shorter pedicels and its included staminal column.

222. Pavonia stipularis Krapovickas, Anais XXXII Congr. Nac. Bot. [Brasil], p. 74. pl. 7. 1982. Type. Brazil. Bahia: Mun. Itacaré, km 6 to the W of Itacaré, 12 Apr 1980, *Plowman, Mattos-Silva & dos Santos 10070* (holotype, CEPEC–n.v.; isotypes, CTES, F as photo F-57408, NY).

Small *tree* to 4 m tall, the branches with grayish bark, the young stems densely stellate-pubescent. *Leaf* blades elliptic, mostly 10–17 × 4–6.5 cm, basally cuneate, entire, acuminate, penninerved, the midrib raised above and beneath, with 6–8 secondary nerves on each side, glabrous above, sparsely stellate-pubescent beneath (primarily on nerves); petioles 1–3 cm long, evenly stellate-pubescent; stipules ligulate, 9–12 × 1.5 mm, erect, stellate-pubescent and ciliate. *Flowers* solitary in the leaf axils, the pedicels 5–9 cm long; involucellar bracts 9–11, ligulate, ca. 16 × 1.5 mm, densely stellate-pubescent and ciliate (cilia 1.5–2 mm long); calyx 10 mm long, with pubescence like involucel; corolla barely exceeding calyx, yellowish green. *Fruits* unknown.

Distribution (Fig. 74). From the Atlantic coast of Bahia, Brazil. The type locality is at 14°19′S, 39°3′W.

Illustration. Krapovickas (1982a: pl. 7, fig. 1).

Specimens examined. **BRAZIL. BAHIA:** On rd. from Ubaitaba to Itacaré, 4 km do loteamento da Marambaia, *Amorim et al. 416* (NY, TEX).

223. Pavonia strictiflora (W. J. Hooker) Fryxell, comb. nov. Basionym: *Goethea strictiflora* W. J. Hooker, Bot. Mag. **78** (n.s. 8): pl. 4677. 1852. Type. Unknown.

Large *shrubs* of unknown stature, with glabrate stems. *Leaf* blades to 23 × 10 cm, lance-ovate, essentially truncate basally, very coarsely dentate (the teeth 4–6 per 10 cm), especially distally, to subentire, apically acute or acuminate (the tip itself rounded), penninerved, the midrib slightly raised above, prominently raised beneath, with 5–7 secondary veins on each side, glabrous or glabrate above and beneath, nearly concolorous; petioles 1.5–3(–5) cm long, very minutely puberulent to glabrate; stipules not seen but leaving prominent scars. *Flowers* axillary, borne on short lateral spurs with crowded internodes, thus appearing fasciculate; pedicels ca. 1 cm long, slender; involucellar bracts 3–4, broadly ovate, 1.5–2 × 1–1.5 cm, subglabrous (brightly colored?); calyx 10–13 mm long (concealed by involucel), minutely and sparsely stellate pubescent (especially on lobes), ca. half-divided, the lobes ovate; petals subequal to or slightly exceeding the calyx (color?); staminal column glabrous, exserted, exceeding the involucel, the filaments 4–5 mm long; styles slender, exserted from androecium. *Fruits* unknown.

Distribution. Not certainly known.

Illustrations (all as *Goethea strictiflora*). Planchon (1852/1853: pl. 814); Gürke (1892b: pl. 105, fig. 2); W. Hooker (1852: pl. 4677); Lemaire (1854: pl. 365).

Specimens examined. **IN CULTIVATION: ENGLAND:** Hort. Kew 1856, *s.coll., s.n.* (K); Hort. Kew 1932, *Davie 21* (K). **UNITED STATES: Hawaii:** Waimea Botanical Garden, *s.coll., s.n.* (TEX).

BRAZIL. Esperanza, *Riedel 783* (A). **BAHIA:** Mun. Ibicaraí, Rod. Itabuna–Itapetinga, 5 km W of entrance to city of Ibivaraí, *Pirani et al. 2985* (NY, SPF).

The cultivated specimen dated 1856 was presumably gathered from the same plant upon which the name was based, but is not the type, since it was preserved four years after the name was published. The Bahia specimen (*J. R. Pirani et al. 2985*) is only provisionally included here; it has notably pubescent stems and is described in the collectors' note as "semidecumbente."

224. Pavonia tricalycaris St.-Hilaire, Fl. Bras. Merid. **1:** 240. 1827. *Triplochlamys tricalycaris* (St.-Hilaire) Ulbrich, Notizbl. Bot. Gart. Mus. Berlin-Dahlem **6:**

335. 1915. Type. Brazil. Espírito Santo: Near the city of Villa da Victoria, *St.-Hilaire s.n.* (holotype, P as photo F-35483).

Pavonia selloi Gürke *in* Martius, Fl. Bras. **12(3)**: 530, pl. 104. 1892. *Triplochlamys selloi* (Gürke) Ulbrich, Notizbl. Bot. Gart. Mus. Berlin-Dahlem **6:** 335. 1915. Type. Brazil. Espírito Santo: Victoria, *Sellow 352* (lectotype, here designated, US).

Shrubs 1–3 m tall, the stems sparsely and minutely lepidote (the hairs 0.1 mm diam.) to glabrate. *Leaf* blades elliptic, mostly 10–20 × 3–7 cm, basally cuneate, shallowly serrate (at least distally), acute or acuminate, penninerved, the midrib raised above and beneath, minutely and very sparsely stellate-pubescent to subglabrous; petioles 0.5–4.5 cm long, adaxially bearded, sometimes obscurely so; stipules subulate, curved, 12–13 mm long, ciliate toward base. *Flowers* solitary in the leaf axils, the pedicels 1–2 cm long, densely pilose, the hairs 2 mm long; involucellar bracts ca. 14, red, dimorphic, the smaller ones reflexed, 10–15 × 8–10 mm, the longer ones erect, 35 × 10–16 mm, both forms abaxially stellate-pubescent (the hairs 0.1–0.2 mm diam.), adaxially hirsute with simple hairs 1–2 mm long, both ovate-acuminate; calyx and rest of flower concealed by involucel in available specimens. *Fruits* unknown.

Distribution (Fig. 74). From the Atlantic coast of Espírito Santo, Bahia, and Rio de Janeiro, Brazil, sometimes cultivated. The type locality is at 20°11′S, 40°16′W.

Illustration. Gürke (1892b: pl. 104).

Specimens examined. BRAZIL. BAHIA: Mun. Alcobaça, km 8 of the hwy. from Teixera de Freitas to Caravelas, *Mattos-Silva et al. 557* (CTES); Teixera de Freitas, valley of the Alcobaça River, *Santos 2382* (CTES). **RIO DE JANEIRO:** *Glaziou 13540* (C, P as photo CTES).

Pavonia tricalycaris is distinguished by its numerous but wide, ovate-acuminate, red, dimorphic involucellar bracts.

Excluded Taxa and Doubtful and Invalidly Published Names

This list excludes names in *Pavonia* that refer to plants of the Old World.

Pavonia acuminata (Cavanilles) Sprengel, Syst. **3:** 98. 1826. Basionym: *Hibiscus acuminatus* Cavanilles Diss. **3:** 144, t. 50, f. 2. 1787. ≡ *Kosteletzkya acuminata* (Cavanilles) Britten = *Kosteletzkya virginica* (Linnaeus) A. Gray?

Pavonia berteri Sprengel, Syst. **3:** 99. 1826. Based on *Hibiscus pentaspermus* Bertero *ex* de Candolle, Prodr. **1:** 447. 1824. = *Kosteletzkya depressa* (Linnaeus) O. Blanchard, Fryxell & D. Bates.

Pavonia bracteosa Bentham, J. Bot. **4:** 118. 1842. Lectotype, here designated. Brazil. Ceará: Without locality, 1838, *Gardner 1458* (lectotype, F as photo F-56158; isolectotypes, GH, K, NY, OXF, US). = *Peltaea trinervis* (C. Presl) Krapovickas & Cristóbal; see Krapovickas & Cristóbal, 1965.

Pavonia brasiliensis (Linnaeus) Sprengel *ex* Gürke in Martius, Fl. Bras. **12(3):** 571. 1892. Basionym: *Hibiscus brasiliensis* Linnaeus, Sp. Pl., ed. 2, 977. 1763.

Linnaeus cited Plumier, Pl. Amer.: pl. 160, fig. 1 (1758), but he did so with a "?" so this figure is unsatisfactory as a type. No other type is available, so the name must be regarded as a nomen dubium.

Pavonia conzatii Greenmann 1913 *ex* Anonymous (1991: 46).

This name is perhaps meant to refer to *Malvaviscus conzattii* Greenman, Field Mus. Nat. Hist., Bot. Ser. **2:** 333. 1912; *Malache conzattii* Britton & Rose 1923 *ex* Anonymous is probably the same. Both are nomina nuda, and are in any case to be excluded from *Pavonia*.

Pavonia cordifolia Wright *ex* Grisebach, Cat. Pl. Cub. 27. 1866. *Malache mutanda* Kuntze, Revis. Gen. Pl. **1:** 70. 1891. Type. Cuba. *Wright 2061* (isotypes, GH, MA, MO, NY–2, S, US). = *Peltaea speciosa* (Humboldt, Bonpland & Kunth) Standley; see Krapovickas & Cristóbal, 1965.

Pavonia drummondii Torrey & Gray, Fl. N. Amer. **1:** 682. 1840. Type. Texas, *Drummond s.n.* (specimen not located). = *Malvaviscus drummondii* Torrey & Gray, but heterotypic.

Malvaviscus drummondii is based on: Texas: San Felipe de Austin, 1835, *Drummond 1* (holotype, GH; isotypes, BM, OXF); the type of *Pavonia drummondii* is unknown, but there is no basis in the protologue for concluding that Torrey and Gray were making a new combination, as was interpreted by Turner and Mendenhall (1993).

Pavonia edouardii Hochreutiner *ex* Chodat & Hassler, Bull. Herb. Boiss., sér. 2, **5:** 298. 1905. Types. Paraguay. In the region of the Corrientes River, *Hassler 4520* (syntypes, GH, NY); Paraguay. In the region of the Tapiraguay River, *Hassler 4368* (lectotype, here designated, NY; isolectotype, GH). ≡ *Peltaea edouardii* (Hochreutiner) Krapovickas & Cristóbal; see Krapovickas & Cristóbal, 1965.

Pavonia edouardii forma *fruticosa-erecta* Hochreutiner *ex* Chodat & Hassler, Bull. Herb. Boiss., sér. 2, **5:** 298. 1905. Types. Paraguay, *Hassler 7876* (syntype, NY); Paraguay. In the region of the upper Apa River, Dec 1901, *Hassler 8173* (syntypes, K, NY, S). = *Peltaea edouardii* (Hochreutiner) Krapovickas & Cristóbal.

Pavonia grisea St.-Hilaire & Naudin, Ann. Sci. Nat., ser. 2, **18:** 41. 1842. Type. Brazil. Goiás: Without locality, 1841, *Gardner 3020* (holotype, P as photo F-35486; isotypes, CTES [fragment], E, F, NY, OXF). = *Peltaea speciosa* (Humboldt, Bonpland & Kunth) Standley; see Krapovickas & Cristóbal, 1965.

Pavonia hirta Schweigger, Enum. Pl. Hart. Regiom. 52. 1812, nom. dub.

Pavonia hirta Klotzsch, Linnaea **14:** 301. 1840, nom. nud. (non Sprengel, 1826), based on: Brazil. Bahia: Cruz de Casma, *Luschnath 165* (lectotype, here designated, MO). ≡ *P. sulphurea* Steudel, nom. nov. sine descr. = *P. cancellata.*

Pavonia hirta Sprengel, Syst. Veg. **3:** 99. 1826, based on: Montevideo, *Sellow s.n.* = *P. glechomoides* fide Index Kewensis.

Pavonia hispida Sprengel, Syst. Veg. **3:** 99. 1826. = *Kosteletzkya* sp.

Pavonia hispida Wright *in* Sauvalle, Anal. Acad. Ci. Habana **5:** 239. 1868 (non Sprengel, 1826). Type. Cuba. 1865, *Wright s.n.* (GH). = *Malachra radiata* (Linnaeus) Linnaeus.

Pavonia hostmannii Miquel, Linnaea **22:** 469. 1849. *Pavonia speciosa* var. *hostmannii* (Miquel) Gürke *in* Martius, Fl. Bras. **12(3):** 495. 1892. (≡ var. *speciosa*). Type. Surinam. *Hostmann 1288* (holotype, U; isotypes, CTES [fragment], G as photo F-23713, GH, S). = *Peltaea speciosa* (Humboldt, Bonpland & Kunth) Standley; see Krapovickas & Cristóbal, 1965.

Pavonia involucrata Klotzsch, Linnaea **14:** 301. 1840. Type. Brazil. Bahia: Cruz de Casma, *Luschnath 166* (lectotype, here designated, MO). = *Peltaea trinervis* (C. Presl) Krapovickas & Cristóbal; see Krapovickas & Cristóbal, 1965.

Pavonia julianae Uittien, Rec. Trav. Bot. Néerl. **33:** 770. 1936. Type. Surinam. Boven Sipaliwini, *Rombouts 430* (holotype, U; isotype, U). = *Peltaea riedelii* (Gürke) Standley; see Krapovickas & Cristóbal, 1965.

Pavonia lindmanii Gürke *ex* Ulbrich, Bot. Jahrb. Syst. **57:** 64. 1921, in syn. = *Pavonia hassleriana* Chodat.

Pavonia mexicana Humboldt, Bonpland & Kunth, Nov. Gen. Sp. **5:** 284 (folio ed., p. 221). 1822. *Kosteletzkya mexicana* (Humboldt, Bonpland & Kunth) Garcke, Jahrb. Königl. Bot. Gart. Berlin **1:** 201. 1881. *Malache mexicana* (Humboldt, Bonpland & Kunth) Kuntze, Revis. Gen Pl. **1:** 70. 1891. Type. Mexico. Guerrero: "Venta de Acahuizotla," *Humboldt & Bonpland 3911* (holotype, P-HBK as microfiche; isotype, B-W no. 12860). sp. dub.; see Fryxell, 1988a: 327, 457.

Pavonia minutiflora Duchassaing *ex* Linden & Planchon, nom. nud. = *Peltaea ovata* (C. Presl) Standley; see Krapovickas & Cristóbal, 1965.

Pavonia nigrobracteata (Rusby) Uittien, Rec. Trav. Bot. Néerl. **33:** 772. 1936. *Peltobractea nigrobracteata* Rusby, Mem. New York Bot. Gard. **7:** 298. 1927. Type. Bolivia. Near Lake Rogagua, *Cárdenas 1654* (holotype, NY; isotypes, CTES [fragment], US). = *Peltaea riedelii* (Gürke) Standley; see Krapovickas & Cristóbal, 1965.

Pavonia nudicaulis St.-Hilaire, Fl. Bras. Mer. **1:** 231. 1827. Type. Brazil. Puto River, *St.-Hilaire s.n.* (P as photo F-35478). ≡ *Peltaea nudicaulis* (St.-Hilaire) Krapovickas & Cristóbal; see Krapovickas & Cristóbal, 1965.

Pavonia palmata (Ulbrich) Turner & Mendenhall, Ann. Missouri Bot. Gard. **80:** 457. 1993. Basionym: *Malvaviscus palmatus* Ulbrich, Verh. Bot. Verein Prov. Brandenburg **51:** 89. 1909. Type. Brazil. "Minas Gerais": on the upper Juruá River, 25 Apr 1901 (fl. fr.), *Ule 5443* (isotypes, CTES [fr. fragment], G as photo CTES, MG as photo CTES). ≡ *Malvaviscus palmatus* Ulbrich.

Turner and Mendenhall (1993) excluded this species from *Malvaviscus* and transferred it to *Pavonia* on the basis of the absence of auricles on the petals and secondarily on the basis of the distinctive deeply lobed leaves, which are very different from those of other species of *Malvaviscus*. The photograph of the isotype at G indicates that the stamens have the characteristic retrorse position found in *Malvaviscus*. It is difficult to be certain of the condition of the fruit (*Ule 5443*, CTES) in the dried state, which does not clearly indicate whether it was fleshy when fresh. However, two collections of this species have recently come to light, evidently collected in the vicinity of the type locality, on the upper Río Juruá in the state of Acre (*not* Minas Gerais, as stated by Ulbrich in the protologue). These collections (*Daly et al. 7566*, NY, collected in 1992 at 08°18′S, 72°44′W; *Silveira et al. 477*, NY, TEX, collected in 1993 at 09°12′S, 72°44′W) lack fruits, but clearly represent the same species that Ulbrich illustrated in the original publication and that is represented by the photographs of the isotype cited above. Dissection of a flower bud of *Daly et al. 7566* clearly revealed the presence of auricles at the base of the petals, fully characteristic of those of *Malvaviscus*, and the open flower has the characteristic retrorse stamens found in all species of *Malvaviscus*. Thus I believe that the species is properly retained in the genus *Malvaviscus* (as *Malvaviscus palmatus*, and not *Pavonia*), even though there are no fruits on the recent collections. Schery (1942) retained it in *Malvaviscus* with some doubts (not having seen a specimen), noting the possibility that it "may be" a *Pavonia* but "had best remain as *Malvaviscus palmatus*."

Pavonia pellita Humboldt, Bonpland & Kunth, Nov. Gen. Sp. **5:** 282 (folio ed., p. 219). 1822. *Pavonia speciosa* var. *pellita* (Humboldt, Bonpland & Kunth) Gürke *in* Martius, Fl. Bras. **12(3):** 495. 1892. Type. Near Maypure, on the Orinoco River, *Humboldt & Bonpland 879* (holotype, P-HBK as microfiche). = *Peltaea* sp., possibly *Peltaea speciosa*.

Pavonia perini Perini *ex* Index Kewensis, Suppl. **3:** 130. 1908, originally published as *Pavonia* var. *perini* Perini *ex* Knight, as an alternative name for *Canhamo braziliensis* Perini *ex* Knight, The Brazilian Review **1905:** 5. 1905, nom. illegit., apparently = *Hibiscus cannabinus* Linnaeus. (See Perini, 1905.) The name "Canhamo" was probably not intended as a generic name, although it is listed as such in Index Kewensis (Suppl. III: 34) and in Airy Shaw, 1966; but not in Hutchinson, 1967; Mabberley, 1987; or Wielgorskaya, 1995.

Pavonia pilosa Willdenow *in* Link, Enum. Pl. Hort. Reg. Berol. **2:** 213. 1822. = *Kosteletzkya* sp.

Pavonia poeppigii (Sprengel) Schlechtendal, Linnaea **11:** 357. 1837. Basionym: *Achania poeppigii* Sprengel. Type. Cuba. 1824, *Poeppig s.n.* (lectotype, here designated, MO). ≡ *Hibiscus poeppigii* (Sprengel) Garcke.

Pavonia polymorpha St.-Hilaire, Fl. Bras. Mer. **1:** 232. 1827. *Pavonia speciosa* subsp. *polymorpha* (St.-Hilaire) Gürke *in* Martius, Fl. Bras. **12(3):** 495. 1892. ≡ *Peltaea polymorpha* (St.-Hilaire) Krapovickas & Cristóbal; see Krapovickas & Cristóbal, 1965.

Pavonia preslii Standley, J. Wash. Acad. Sci. **17:** 168. 1927. = *Peltaea ovata* (C. Presl) Standley; see Krapovickas & Cristóbal, 1965.

Pavonia riedelii Gürke *in* Martius, Fl. Bras. **12(3)**: 493. 1892. Type. Brazil. Cuyabá, *Riedel s.n.* (probable type, OXF). ≡ *Peltaea riedelii* (Gürke) Standley; see Krapovickas & Cristóbal, 1965.

Pavonia scabra Bentham *ex* Garcke, Eichler Jahrb. Berlin **1:** 221. 1881, pro syn. (non *Pavonia scabra* C. Presl). Type. Guyana. Berbice, *Schomburgk 253* (isotypes, GH–2, OXF). = *Peltaea speciosa* (Humboldt, Bonpland & Kunth) Standley; see Krapovickas & Cristóbal, 1965.

Pavonia schrankii var. *angustifolia* Ulbrich, Feddes Repert. Spec. Nov. Regni Veg. **13:** 516. 1915 (non R. E. Fries, 1908). Five specimens were cited by Ulbrich, including the lectotype of the earlier homonym, *P. schrankii* var. *angustifolia* R. E. Fries (Lindman A1553). Thus, Ulbrich's name is both a later homonym and a nom. superfl. = *Pavonia angustipetala* Krapovickas & Cristóbal.

Pavonia sempervirens Ruiz & Pavón, Syst. Veg. Fl. Peruv. Chil. 253. 1798. = *Laurelia aromatica* L. [Monimiaceae].

Pavonia sessiliflora Humboldt, Bonpland & Kunth, Nov. Gen. Sp. **5:** 281 (folio ed., p. 218). 1822. *Lebretonia sessiliflora* (Humboldt, Bonpland & Kunth) Britton *ex* Williams & Cheesman, Fl. Trin. & Tobago **1(2)**: 86. 1929. Type. Novae Granatae, near Santa Ana, *Humboldt & Bonpland s.n.* (holotype, P-HBK as microfiche). ≡ *Peltaea sessiliflora* (Humboldt, Bonpland & Kunth) Standley; see Krapovickas & Cristóbal, 1965.

Pavonia sessiliflora var. *acutifolia* Gürke *in* Martius, Fl. Bras. **12(3)**: 493. 1892. Type. Brazil. Rio de Janeiro, *Glaziou 7879* (lectotype, chosen by Krapovickas & Cristóbal, 1996, P–n.v.; isolectotype, C as photo F-21601); Rio de Janeiro, *Riedel 257* (syntypes, E, GH, US). = *Peltaea obsita* (Martius *ex* Colla) Krapovickas & Cristóbal; see Krapovickas & Cristóbal, 1996.

Pavonia sessiliflora forma *ecostata* Hassler, Feddes Report. Spec. Nov. Regni Veg. **7:** 378. 1909. Type. Paraguay. Caballero Cué, between the Apa River and the Aquidaban River, *Fiebrig 4769* (isotypes, M as photo CTES, CTES). = *Peltaea obsita* (Martius *ex* Colla) Krapovickas & Cristóbal; see Krapovickas & Cristóbal, 1996.

Pavonia sessiliflora var. *mattogrossensis* E. G. Baker, J. Bot. **63:** 239. 1925. Type. Brazil. Mato Grosso: "campo cerrado," *Kuhlmann 2143* (holotype, K). = *Peltaea* sp. cf. *P. riedelii* (Gürke) Standley

Pavonia sessiliflora var. *obtusifolia* Gürke *in* Martius, Fl. Bras. **12(3)**: 493. 1892. Type. Venezuela. near Tovar, 1854–1855, *Fendler 91* (syntype, GH). = *Peltaea trinervis* (C. Presl) Krapovickas & Cristóbal; see Krapovickas & Cristóbal, 1965.

Pavonia speciosa Humboldt, Bonpland & Kunth, Nov. Gen. Sp. **5:** 281 (folio ed., p. 219). 1822. Type. Between the Orinoco River and Trapiche D. Felix Farreras, near Angostura, *Humboldt & Bonpland s.n.* (specimen unknown). ≡ *Peltaea speciosa* (Humboldt, Bonpland & Kunth) Standley; see Krapovickas & Cristóbal, 1965.

Pavonia speciosa var. *sellowii* Gürke *in* Martius, Fl. Bras. **12(3)**: 495. 1892. Type. *Sellow B1921, c.1417* (syntypes, specimens not found). = *Peltaea speciosa* (Humboldt, Bonpland & Kunth) Standley.

Pavonia speciosa subsp. *stenochlamys* Gürke *in* Martius, Fl. Bras. **12(3)**: 495. 1892.

Gürke described *Pavonia speciosa* subsp. *stenochlamys* without citing any specimens but included two varieties, var. *sellowii* Gürke and var. *pellita* (Humboldt, Bonpland & Kunth) Gürke. Since *P. speciosa* var. *pellita* is a new combination, the name *P. speciosa* var. *sellowii* should be *P. speciosa* var. *stenochlamys*, typified with a collection of Sellow. = *Peltaea* sp.

Pavonia spiralis Cavanilles, Icones **5:** 20, t. 434. 1799. Type. Ins. Taboga, *L. Née s.n.* (holotype, MA-#476077, cited and described by Garilleti, 1993). = *Malvaviscus arboreus* Cavanilles.

Pavonia subpandurata Wright *ex* Grisebach, Cat. Pl. Cub. 26. 1866. *Pavonia speciosa* var. *subpandurata* (Wright) Gürke *in* Martius, Fl. Bras. **12(3)**: 495. 1892. Type. Cuba. *Wright 2060* (isotypes, BM, GH, MA, MO, NY–2, S–2, UC, US). ≡ *Peltaea subpandurata* (Wright) Krapovickas & Cristóbal; see Krapovickas & Cristóbal, 1965.

Pavonia subpandurata var. *flava* Helwig, Feddes Repert. Spec. Nov. Regni Veg. **24:** 235. 1928. Type. Cuba. Pinar del Río: in pinelands between Laguna Santa Barbara and Laguna Alcatraz Chico, 17 Nov 1923, *Ekman 18119* (isotype, NY). = *Peltaea subpandurata* (Wright) Krapovickas & Cristóbal.

Pavonia sulphurea Steudel, Nomencl. Bot., ed. 2, **2:** 279. 1841, nom. illegit. *Malache sulphurea* (Steudel) Kuntze, Revis. Gen. Pl. **1:** 70. 1891. ≡ *Pavonia hirta* Klotzsch (q.v.), non Sprengel.

Pavonia surumuensis Ulbrich, Notizbl. Bot. Gart. Mus. Berlin-Dahlem **6:** 325. 1915. Type. Brazil. Rio Branco, Surumu, on the Igarape Burete near Serra de Pracauá, *Ule 7980* (isotype, MG). = *Peltaea aliculata* Krapovickas & Cristóbal, which becomes a synonym of *Peltaea surumuensis* (Ulbrich) Krapovickas & Cristóbal, Bonplandia **7:** 55. 1993.

Pavonia trinervis (C. Presl) Robyns, Ann. Missouri Bot. Gard. **53:** 114. 1966. ≡ *Peltaea trinervis* (C. Presl) Krapovickas & Cristóbal.

Pavonia ulbrichiana Kearney, Madroño **12:** 116. 1953. Type. Brazil. Maranhão-Pará: *Snethlage 170* (holotype, F-693728). = *Hibiscus dimidiatus* Schrank.

Pavonia urticaefolia C. Presl, Reliq. Haenk. **2:** 128. 1835. Type. Mexico. "in terris mexicanis occidentalibus," *Haenke s.n.* (holotype, PR-197902). ≡ *Malvaviscus urticaefolius* (C. Presl) Fryxell.

Pavonia viscida Klotzsch *in* Schomburgk, Reise Brit. Guiana **3:** 1170. 1849, nom. nud., based on: Guyana. On the Suruma River, *Schomburgk 764* (specimen unknown). = *Pavonia geminiflora* Moricand.

Pavonia warmingiana Gürke *in* Martius, Fl. Bras. **12(3)**: 488. 1892. *Typhalea warmingiana* (Gürke) Monteiro, Anais XII Reun. Soc. Bot. Bras. 31. 1961. Type. Brazil. near Castelnuovo, *Riedel 422* (holotype, B destroyed, as photo F-9475; isotypes, CTES [fragment], P–n.v.), sp. dub.

This species has appressed simple hairs on the upper leaf surface, like those compared in Table III. Of these species, it conforms geographically (type locality at 14°38′S, 39°10′W) only to *Pavonia intermedia*, but does not conform to it morphologically. Neither can it be included in any of the other species in this group, on the basis of available information. It is therefore treated as a species dubia.

Acknowledgments

I am grateful to the curators and administrators of the herbaria cited herein for hospitality during study visits to their institutions and for their willingness to lend specimens in their care when I was unable to make visits or needed to make more intensive study of particular specimens than was possible during a short visit. These institutions include: A, ARIZ, BISH, BR, BRIT, C, CAS, CHAPA, CM, COLO, CORO, CTES, CVRD, E, F, G, GH, IEB, IPA, K, LL, LSU, MEXU, MICH, MO, MU, MY, NY, OSH, OXF, PENN, PH, POM, PORT, RB, RSA, S, SP, SPF, TEX, U, UCR, US, VT, and WIS (abbreviations according to Holmgren et al., 1990). In addition, I appreciate the special effort made by Troels Myndel Pedersen, of Estancia Santa Teresa, Paraguay, who made it possible for me to examine the specimens of *Pavonia* kept in his private herbarium.

I appreciate the support offered me by the Commission for Educational Exchange between the United States and the Argentine Republic for the opportunity to do research as a Fulbright Scholar (April through June 1993) at the Instituto de Botánica del Nordeste in Corrientes, Argentina, part of which time was devoted to this project. I equally appreciate the kindness and cooperation of the herbarium staff in Corrientes in pursuing my work there, especially of Ing. Antonio Krapovickas and Dra. Carmen L. Cristóbal, who were generous in sharing their knowledge of *Pavonia* as well as in providing other support. The enormous erudition and copious records concerning South American botany in general and Malvaceae in particular accumulated by them were generously put at my disposal. As is apparent from the number of CTES citations in this study, no revision of neotropical Malvaceae can be considered complete without consultation of the CTES herbarium!

This study was supported in part by Grant DEB-9420233 to J. C. La Duke and P. A. Fryxell, which is gratefully acknowledged.

I thank those individuals who have participated in various field explorations of the neotropics with me, some of which have yielded materials relevant to this study, especially William R. Anderson, David M. Bates, Orland J. Blanchard Jr., Charles L. Burandt Jr., Clark P. Cowan, Joan E. Fryxell, Steven R. Hill, Stephen D. Koch, Emily J. Lott, Meghan Mendenhall, Jesús Valdés, Thomas L. Wendt, and Robert C. Wingfield. I am equally grateful to colleagues who have directed specimens of Malvaceae (including *Pavonia*) to my attention over many years, including Laurence J. Dorr, Antonio Krapovickas, Jerzy Rzedowski, and numerous others.

The skill of the illustrator whose drawings grace this work is much appreciated; Benjamin P. Shaw produced all of the illustrations (Figs. 1, 3, 6, 7, 9, 12–14, 17, 19, 21, 23, 25–27, 29, 31, 32, 35, 38, 41, 43, 46, 50, 52, 56, 59, 61–64, 67–72, 77, 80, 82, 83, 85–93), except the maps and mericarp drawings (Figs. 4, 30, 36, 53, 73), which I prepared.

Thanks to Gerleni Lopes Esteves (SP) for continuing discussions about *Pavonia*; to Lyn A. Craven (CANB) for providing information on the typification of *Greevesia cleisocalyx*; to Charles E. Jarvis (BM) for comments on the typification of *Urena typhalea* and for elucidating the origin of the name "Typhalea"; to Javier Fuertes (MA) for information on type specimens kept at Madrid; to K. N. Gandhi (GH) for discussions of nomenclatural problems; to Laurence J. Dorr (US) for help in seeking a type for *Pavonia jonesii*, which was subsequently found at MO; and to Billie L. Turner (TEX) for bringing to my attention the problem presented by *Pavonia palmata*. I am also grateful to John F. Reed (NY), Barney Lipscomb (BRIT), and Ted Delavoryas (University of Texas) for help in running down various difficult-to-locate literature references, and to James L. Luteyn (NY) for advice on bringing this manuscript into conformity with the Flora Neotropica series. José Panero and Tom Wendt reviewed and corrected the Spanish abstract. I am especially grateful to reviewers William Alverson, Laurence J. Dorr, and Piero Delprete for their efforts to improve the manuscript.

Literature Cited

Acevedo-Rodríguez, P. (and collaborators). 1996. Flora of St. John, U.S. Virgin Islands. Mem. New York Bot. Gard. **78:** 1–581. (Malvaceae by P. A. Fryxell, pp. 297–310.)

Adams, C. D. 1972. Flowering plants of Jamaica. University of the West Indies, Mona, Jamaica.

Airy Shaw, H. K. 1966. Willis' dictionary of flowering plants and ferns. 7th ed. Cambridge University Press, Cambridge.

Andrews, J. 1992. American wildflower florilegium. University of North Texas Press, Denton.

Anonymous. 1991. Homenaje al Prof. Conzatti. Instituto de Biología, Universidad Autónoma Nacional de México.

Arechavaleta, J. 1898. Malvaceae. *In* Flora Uruguaya. Anal. Mus. Nac. Montevideo **3:** 112–153.

Baillon, H. 1866–1895. Histoire des Plantes. 13 vols. (Malvaceae in vol. 4, 1872–1873).

Bates, D. M. 1966. A new *Pavonia* (Malvaceae) from Colombia. Kew Bull. **20:** 149–152.

———. 1967. Chromosome numbers in the Malvales. I. Gentes Herb. **10:** 39–46.

———. 1976. Chromosome numbers in the Malvales. III. Miscellaneous counts from the Byttneriaceae and Malvaceae. Gentes Herb. **11:** 143–150.

Bentham, G. & J. D. Hooker. 1863. Genera Plantarum. Reeve & Co., London. 3 vols.

Bertoni, M. S. 1911. Contribuciones preliminares al estudio sistemático, biológico y económico de las plantas del Paraguay. II. Sobre fibras paraguayos: una nueva planta de fibras, *Pavonia restiaria*. Anales Cient. Paraguayos **9:** 1–5.

Borssum Waalkes, J. van. 1966. Malesian Malvaceae revised. Blumea **14:** 1–213.

Britton, N. L. 1908. Studies of West Indian plants. The genus *Malache*. Bull. Torrey Bot. Club **35:** 344–345.

———. 1918. The flora of Bermuda (illustrated). New York.

——— **& C. F. Millspaugh.** 1920. The Bahama flora. New York.

——— **& P. Wilson.** 1924. Malvaceae. Pages 545–568 *in* Botany of Porto Rico and the Virgin Islands. Vol. 5.

Candolle, A. P. de. 1824. *Pavonia, Malvaviscus* (in part), and *Lebretonia*. Pages 442–446 *in* Prodromus Systematis Naturalis Regni Vegetabilis. Vol. 1. Treuttel and Würtz, Paris.

Carneiro de Lira, O. 1969. Malvaceae de Pernambuco—II. Gênero *Pavonia*. Anais XX Congr. Nac. Soc. Bot. Bras., pp. 227–230.

Cavanilles, A. J. 1787. Dissertatio Botanico. 3. Madrid.

Cervantes Aceves, N. 1992. La familia Malváceae en el estado de Jalisco. Pages 1–393 *in* Colección flora de Jalisco. No. 3. Universidad de Guadalajara, Guadalajara, Mexico.

Chaudhuri, S. K. 1965. Pollen morphological studies of the order Malvales II. Bull. Bot. Soc. Bengal **19:** 147–158.

Chauhan, M. P. S. 1987. Seed structure in *Pavonia* species (Malvaceae). Geophytology **17:** 265–269.

Cheek, M. 1989. A new name for a South American *Pavonia* (Malvaceae). Kew Bull. **44:** 145–149.

Coe, F. G. & G. J. Anderson. 1996. Ethnobotany of the Garífuna of eastern Nicaragua. Econ. Bot. **50:** 71–107.

Correll, D. S. & H. B. Correll. 1982. Flora of the Bahama Archipelago. J. Cramer, Vaduz.

Curtis, W. 1848. *Lopimia malacophylla*. Curtis's Bot. Mag. **74:** pl. 4365.

Davie, J. H. 1933. Cytological studies in the Malvaceae and certain related families. J. Genet. **28:** 33–67.

Davis, H. B. 1936. Life and work of Cyrus Guernsey Pringle. University of Vermont, Burlington.

Dietrich, D. 1847. *Pavonia*. Pages 822–825 *in* Synopsis Plantarum. Vol. 4. Weimar.

Dodson, C. H. & A. H. Gentry. 1978. Flora of the Río Palenque Science Center, Los Ríos, Ecuador. Selbyana **4:** 1–628.

Don, G. 1831. *Pavonia*. Pages 472–475 *in* General history of the dichlamydeous plants. Vol. 1. London.

Ekman, E. L. 1909. Beiträge zur Columniferenflora von Misiones. Arkiv Bot. **9(4):** 25–39.

Esteves, G. L. 1994. *Pavonia serrana* (Malvaceae), a new species from the state of Minas Gerais, Brazil. Novon **4:** 100–102.

———. 1996. Sistemática de *Pavonia* Cav. (Malvaceae), com base nas espécies das regiões nordeste e sudeste do Brasil. Ph.D. dissertation, Universidade de São Paulo. São Paulo, Brasil.

———. 1999. Delimitação, classificação infragenerica e novos táxons de *Pavonia* Cav. (Malvaceae). Bol. Bot. Univ. São Paulo **17:** in press.

——— **& P. A. Fryxell.** 1999. Espécies novas de *Pavonia* Cav. (Malvaceae) da região nordeste do Brasil. Hoehnea [in press].

——— **& A. Krapovickas.** 1995. Duas novas espécies de *Pavonia* Cav. (Malvaceae) do sul da Bahia, Brasil. Hoehnea **22:** 41–45.

Fawcett, W. & A. B. Rendle. 1926. Malvaceae. Pages 92–149 *in* Flora of Jamaica. Vol. 5. London.

Fernández, A. 1974. Recuentos cromosómicos en Malváceas. Bol. Soc. Argent. Bot. **15:** 403–410.

———. 1981. Recuentos cromosómicos en Malvales. Bonplandia **5:** 63–71.

Ford, C. E. 1938. A contribution to a cytogenetical survey of the Malvaceae. Genetica **20:** 431–453.

Fries, R. E. 1908. Studien über die amerikanische Columniferenflora. Kongl. Svenska Vetenskapsakad. Handl. **42(12):** 1–67 + 7 plates.

———. 1947. Zur Kentniss der süd- und zentralamerikanischen Malvaceenflora. Kongl. Svenska Vetenskapsakad. Handl., ser. 3, **24(2):** 19–24 + 9 plates.

Fryxell, P. A. 1973. New species and other notes in the Malvaceae. Brittonia **25:** 77–85.

———. 1976. On some Brazilian Malvaceae of Martius, published by Colla in 1833. Taxon **25:** 589–593.

———. 1977. New species of Malvaceae from Mexico and Brazil. Phytologia **37:** 285–316.

———. 1979. Una revisión del género *Pavonia* en México. Bol. Soc. Bot. México **38:** 7–34.

———. 1980. Three new species of Mexican Malvaceae. Phytologia **46:** 391–398.

———. 1984a. A new species of *Hibiscus* from the Amazon Basin. Supl. Acta Amazonica **14(1/2):** 101–103.

———. 1984b. Four new species of Mexican Malvaceae. Syst. Bot. **9:** 415–422.

———. 1988a. Malvaceae of Mexico. Syst. Bot. Monogr. **25:** 1–522.

———. 1988b. The genus *Pavonia* Cav. (Malvaceae: Malvavisceae) in Australia. Nuytsia **6:** 305–308.

———. 1989. Malvaceae. Pages 199–263 *in* R. A. Howard, Flora of the Lesser Antilles. Vol. 5.

———. 1990. New species of Malvaceae from South America. Contr. Univ. Michigan Herb. **17:** 163–172.

———. 1992a. 118. Malvaceae. Pages 1–141 *in* G. Harling & L. Andersson (eds.), Flora of Ecuador, No. 44.

———. 1992b. Malvaceae. Pages 1–255 *in* Flora de Veracruz Fasc. 68.

———. 1993a. Malvaceae. Pages 1–175 *in* Flora del Bajío y de regiones adyacentes. Fasc. 16.

———. 1993b. Malvaceae. Pages 1–175 *in* Flora del Valle de Tehuacán-Cuicatlán. Fasc. 1.

——— **& L. A. Craven.** 1989. *Urena* (Malvaceae) in Australia. Austral. Syst. Bot. **2:** 455–460.

——— **& J. Fuertes.** 1991. Cuban plants of Charles Wright in the Madrid herbarium. Taxon **40:** 597–599.

——— **& S. D. Koch.** 1991. *Pavonia ecostata* (Malvaceae), a new species from Jalisco, Mexico. Syst. Bot. **43:** 24–26.

——— **& A. Krapovickas.** 1990. The Malvaceae published by Turczaninow. Contr. Univ. Michigan Herb. **17:** 173–182.

——— **& D. M. Stelly.** 1993. Additional chromosome counts in the Malvaceae. Sida **15:** 639–647.

Fuertes, J. 1989. Malvaceae. Pages 1–86 *in* Flora de la Real Expedición Botánica del Nuevo Reyno de Granada (1783–1816) Promovida y Dirigida por José Celestino Mutis. Vol. 25(2). Ediciones de Cultura Hispanica, Madrid.

——— **& P. A. Fryxell.** 1993. Types of New World Malvaceae of Lagasca, Ortega, and Sessé & Mociño. Taxon **42:** 345–354.

Garcke, A. 1850. Über *Asterochlaena*, eine neue Gattung der Malvaceen. Bot. Zeitung (Berlin) **8:** 666–669.

———. 1853. Ueber einige im Prodromus von De Candolle falsch untergebrachten Pflanzen. Bot. Zeitung **11:** 841–849.

———. 1861. Die Gattung *Goethea*. Bonplandia **9:** 17–18.

———. 1881. Ueber die Gattung *Pavonia*. Jahrb. Bot. Gart. Berlin **1:** 198–223.

———. 1896. Über einige Malvaceen Gattungen. Bot. Jahrb. Syst. **21:** 379–401.

Garilleti, R. 1993. Herbarium Cavanillesianum. Fontqueria **38:** [1–5] 6–248.

Gentry, H. S. 1946. Notes on the vegetation of Sierra Suratato in northern Sinaloa. Bull. Torrey Bot. Club **73:** 451–462.

Giménez, M. A., J. M. Losa, & J. L. Valverde. 1993. José Mariano Mociño's "Flora de Guatemala" in the Real Jardín Botánico de Madrid. (R. Losa, trans.) Huntia **9:** 21–30.

Gottsberger, G. 1972. Blütenbiologische Beobachtungen an brasilianischen Malvaceen II. Oesterr. Bot. Z. **120:** 439–509.

Gray, A. 1849. Malvaceae. Pages 43–82 + plates 116–133 *in* The genera of the plants of the United States illustrated. (I. Sprague, illustr.) Vol. 2. New York.

Grisebach, A. H. R. [1859–]1864. Flora of the British West Indian Islands. Lovell, Reeve & Co., London.

Gürke, M. 1892a. Beiträge zur Systematik der Malvaceen. II. Die Gattung *Urena*. Bot. Jahrb. Syst. **16:** 361–385.

———. 1892b. Malvaceae II. Pages 457–624 + plates 81–114 *in* Martius, Flora Brasiliensis. Vol. 12(3).

Hashmi, S. H. 1970. The palynology of the Malvaceae of Texas. Ph.D. dissertation. Texas A&M University, College Station.

Hassler, E. 1907. Plantae paraguarienses novae vel minus cognitae. Bull. Herb. Boissier ser. 2, **7:** 718–740.

———. 1909. Novitates paraguarienses. Malvaceae austroamericanae. Feddes Repert. Spec. Nov. Regni Veg. **7:** 69–78, 369–383.

———. 1910. Novitates paraguarienses. Malvaceaeaustroamericanae. Feddes Repert. Spec. Nov. Regni Veg. **8:** 28–31, 34–47, 113–132.

Heel, W. A. van. 1978. Morphology of the pistil in Malvaceae–Ureneae. Blumea **24:** 123–127.

Hitchcock, A. S. 1893. List of plants collected in the Bahamas, Jamaica and Grand Cayman. Ann. Rep. Missouri Bot. Gard. **4:** 47–179 + plates 10–14.

Hochreutiner, B. P. G. 1901. Le genre *Urena* L. Annuaire Conserv. Jard. Bot. Genève **5:** 131–146.

———. 1931. Le genre *Asterochlaena* Garcke, dont-il être conservé? Candollea **4:** 313–315.

Holmgren, P. K., N. H. Holmgren & L. C. Barnett. 1990. Index Herbariorum. Pt. 1. The herbaria of the world. Ed. 8. New York. (Plus supplements in Taxon **40:** 687–692. 1991; **42:** 489–505. 1993; **43:** 305–328. 1995; **45:** 373–389. 1996; **46:** 567–591. 1997; **47:** 503–514. 1998.)

Hooker, J. D. 1878. *Pavonia multiflora*, native of South Brazil. Curtis's Bot. Mag., ser. 3, **34:** pl. 6398.

———. 1879. *Goethea mackoyana*. Curtis's Bot. Mag., ser. 3, **35:** pl. 6427.

Hooker, W. J. 1852. *Goethea strictiflora*. Curtis's Bot. Mag., ser. 3, **8:** pl. 4677.

Hutchinson, J. 1967. Family 119, Malvaceae. Pages 536–567 *in* The Genera of Flowering Plants. Vol. 2. Oxford University Press, Oxford.

Johnson, C. D. 1983. Ecosystematics of *Acanthoscelides* (Coleoptera: Bruchidae) of southern Mexico and Central America. Misc. Publ. Entomol. Soc. Amer. **56:** 1–370.

Kearney, T. H. 1951. The American genera of Malvaceae. Amer. Midl. Naturalist **46:** 93–131.

———. 1954. A tentative key to the North American species of *Pavonia* Cav. Leafl. W. Bot. **7:** 122–130.

———. 1958. A tentative key to the South American species of *Pavonia* Cav. Leafl. W. Bot. **8:** 225–246.

Krapovickas, A. 1953. Malvaceae. Pages 303–312 *in* A. L. Cabrera, Manual de la flora de los Alrededores de Buenos Aires. Buenos Aires.

———. 1965. Malvaceae. Pages 169–220 *in* A. L. Cabrera, Flora de la Provincia de Buenos Aires. Vol. 4.

———. 1977. Sinopsis de la sección *Lebretonia* del género *Pavonia* (Malvaceae). Pages 307–322 *in* Trabalh. XXVI Congr. Nac. Bot. Brasil.

———. 1978a. Una nueva espécie de *Pavonia* (Malvaceae) del estado de Minas Gerais (Brazil). Rev. Brasil. Bot. **1:** 89–91.

———. 1978b. Malvaceae. Pages 589–598 *in* J. A. Steyermark & O. Huber, Flora del Avila. Ministerio del Ambiente y de los Recursos Renovables, Caracas, Venezuela.

———. 1982a. Novedades en *Pavonia* Cav. secc. *Goetheoides* Gürke (Malvaceae). Pages 67–84 *in* Anais XXXII Congr. Nac. Bot. [Brasil].

———. 1982b. Novedades en *Pavonia* Cav. secc. *Typhalea* DC. (Malvaceae). Bol. Soc. Argent. Bot. **20:** 281–301.

———. 1983. Notas sobre Malváceas, IV. Bonplandia **5:** 257–273.

———. 1990. Dos especies nuevas de *Pavonia* Cav. (Malvaceae) de Brasil. Pages 281–286 *in* Anais XXXVI Congr. Bras. Bot., Curitiba [1985].

——— **& C. L. Cristóbal.** 1962. Notas sobre la sección *Lebretonia*, *Pavonia* (Malvaceae) y revisión de las especies argentinas. Lilloa **31:** 5–74.

——— & ———. 1965. Revisión del género *Peltaea* (Malvaceae). Kurtziana **2:** 135–216.

——— & ———. 1993. Novedades en Malvaceas sudamericanas. Bonplandia **7:** 45–56.

——— & ———. 1996. *Peltaea obsita*. Bonplandia **9:** 148.

La Duke, J. C. & J. Doebley. 1995. A chloroplast DNA based phylogeny of the Malvaceae. Syst. Bot. **20:** 259–271.

Lebl [J. E. Lebel ?]. 1878. *Pavonia Makoyana*. Ill. Gart.-Zeitung **22:** 169, pl. 22.

Lemaire, C. 1854. *Goethea strictiflora*. Jard. Fleur. **4:** pl. 365.

Lentz, D. L. 1993. Medicinal and other economic plants of the Paya of Honduras. Econ. Bot. **47:** 358–370.

Link, J. H. F. 1821–1822. Enumeratio Plantarum. G. Reimer, Berlin. 2 vols.

——— **& C. F. Otto.** 1822. Icones Plantarum Select. Fasc. 5. Berlin.

Liogier, A. H. 1981. Antillean studies. I. Flora of Hispaniola, part 1. Phytologia Mem. **3:** 1–218.

Maas, P. J. M. & L. Y. Th. Westra. 1993. Neotropical plant families. Koeltz Scientific Books, Koenigstein.

Mabberley, D. J. 1987. The plant-book. Cambridge University Press, Cambridge.

Macbride, J. F. 1956. Malvaceae [Flora of Peru]. Publ. Field Mus. Nat. Hist., Bot. Ser. **13 (pt. 3A, no. 2):** 442–593.

Mattei, G. E. 1921. Un nuovo genere di Piante dedicato al Biologo italiano Enrico Marconi. L'Umbria Medica **2(14):** 214–216.

McDade, L. A. & P. Davidar. 1984. Determinants of fruit and seed set in *Pavonia dasypetala* (Malvaceae). Oecologia **64:** 61–67.

McKinley, D. 1992. Adrien Lebreton, S.J. (1662–1736): A search for the identity of a neglected botanist in early Martinique. Huntia **8:** 155–162.

Meehan, T. 1892. *Pavonia wrightii*. Meehan's Monthly **2:** 178–179 + pl. 12.

Mitchell, A. S. 1982. Economic aspects of the Malvaceae in Australia. Econ. Bot. **36:** 313–322.

Monteiro Filho, H. da C. 1961. *Typhalea* Necker (Malvaceae). Pages 28–31 *in* Anais XII Reun. Soc. Bot. Brasil.

———. 1969. Malvaceae brasilienses novae vel criticae. II. Rev. Faculdade Ciénc. Lisboa ser. 2, **16C:** 15–36 + 6 plates

———. 1972. Malvaceae brasilienses novae vel criticae. IV. Pages 115–135 *in* Anais Soc. Bot. Brasil, XXIII Congr. Nac. Bot.

Mori, S. A. & J. L. Brown. 1998. Epizoochorus dispersal by barbs, hooks, and spines in a lowland moist forest in central French Guiana. Brittonia **50:** 165–173.

Moricand, S. 1844. *Pavonia*. Pages 119–125 (plates 72–74) *in* Plantes nouvelles d'Amérique. Genève.

Morren, É. 1875. Description du *Pavonia wioti* Éd. Mn., nouvel arbuste du Brésil. Belgique Hort. **25:** 113–114.

———. 1878. Description du *Pavonia makoyana*, nouvel arbuste brésilien a fleurs ornementales. Belgique Hort. **28:** 59–60.

Nees von Esenbeck, T. F. L. & C. F. P. von Martius. 1823. *Goethea*. Nova Acta Phys.-Med. Akad. Caes. Leop.-Carol. Nat. Cur. **11:** 91–102.

Nokes, J. 1986. How to grow native plants of Texas and the Southwest. Texas Monthly Press, Austin.

Paul, T. K. & M. P. Nayar. 1988. Malvaceae. Pages 64–233 *in* Fascicles of the Flora of India. Fasc. 19.

Perini, V. A. de. 1905. *Canhamo braziliensis* Perini or "Brazilian Linen." Prospectus and explanatory notice regarding the cultivation of the above plant and its application to the textile and paper making industries. The Brazilian Review **1905:** 1–13 + several illustrations and an addendum by John Knight, who excerpted Perini's "Prospectus."

Philcox, D. 1997. Malvaceae. Pages 287–360 *in* M. D. Dassanayake (ed.), A revised handbook to the flora of Ceylon. Vol. 11. Amerind Publishing, New Delhi.

Planchon, J. E. [as "J. E. P."]. 1848. *Lopimia malacophylla*. Fl. Serres Jard. Eur., plate 350.

———. 1852/1853. *Pavonia strictiflora*. Fl. Serres Jard. Eur., pl. 814.
Porsch, O. 1929. Vogelblumenstudien II. Jahrb. Wiss. Bot. **70:** 181–277.
Powell, A. M. 1988. Trees and shrubs of Trans-Pecos Texas. Big Bend Natural History Association, Big Bend National Park, TX.
Rathcke, B. J. 1996. Preliminary observations on plant reproductive biology in mangrove communities on San Salvador Island, Bahamas. Pages 87–96 *in* D.C. Edwards and P. J. Godfrey (eds.), Proc. Sixth Symp. Nat. Hist. Bahamas.
———. 1999. Bird pollination of the endemic swamp-bush, *Pavonia bahamensis* Hitchc. (Malvaceae): The risk of specialization. Pages 105–110 *in* Proc. Seventh Symp. Nat. Hist. Bahamas.
Regel, E. 1881. Neue und empfehlenswerthe Pflanzen. Gartenflora **30:** 277–286.
Reichenbach, H. G. L. 1827–1830. Iconographia botanica exotica. Leipzig. 3 vols.
Richard, A. 1845. Historia Física, Política y Natural de la Isla de Cuba . . . Botánica. Paris.
Robyns, A. 1965. Malvaceae. *In* Flora of Panama. Ann. Missouri Bot. Gard. **52:** 497–578.
Roubik, D. W. 1982. The ecological impact of nectar-robbing bees and pollinating hummingbirds on a tropical shrub. Ecology **63:** 354–360.
St.-Hilaire, A. 1824. Plantes usuelles des Brasiliens. Paris.
———. 1827. Malvaceae–Malveae. Pages 173–257 + plates 33–50 *in* Flora Brasiliae Meridionalis. Paris. vol. 1.
——— **& C. Naudin.** 1842. Malvaceae. *In* Revue de la flore du Brésil Meridional. Ann. Sci. Nat. Bot. sér. 2, **18:** 31–54.
Salgado-Labouriau, M. L. & A. B. Guzman. 1967. Pollen grains of plants of the "Cerrado" XVI. Malvaceae. Bull. Mus. Paraense Emilio Goeldi, Nov. Ser. Bot. **25:** 1–13.
Sauget, J. S. & E. E. Liogier. 1953. Malvaceae. Pages 239–275 *in* Flora de Cuba. Vol. 3.
Sazima, M. 1981. Polinização de duas espécies de *Pavonia* (Malvaceae) por beija-flores, na Serra do Cipó, Minas Gerais. Rev. Brasil. Biol. **41:** 733–737.
Schery, R. W. 1942. Monograph of *Malvaviscus*. Ann. Missouri Bot. Gard. **29:** 183–244.
Sivarajan, V. V. & A. K. Pradeep. 1996. Malvaceae of southern peninsular India: A taxonomic monograph. Daya Publishing, Delhi.
Skovsted, A. 1935. Chromosome numbers in the Malvaceae. Intl. J. Genet. **31:** 263–296.
———. 1941. Chromosome numbers in the Malvaceae. II. Compt. Rend. Trav. Lab. Carlsberg, sèr. Physiol. **23:** 195–241.
Spjut, R. W. 1994. A systematic treatment of fruit types. Mem. New York Bot. Gard. **70:** 1–182.
Sprengel, K. P. J. 1826. *Pavonia*. Pages 97–100 *in* Linnaeus, Systema Vegetabilium. Ed. 16. Vol. 3. Gottingen.
Stadler, T., C. Meré & H. L. Cappozzo. 1987. La bionomía de *Dysdercus albofasciatus* Berg, 1878 (Hemiptera: Pyrrhocoridae), plaga de algodón: Su ciclo de vida, alimentación, estrategias adaptivas y enemigos naturales. Bol. San. Veg. Plagas **13:** 143–159.
Stafleu, F. A. & R. Cowan. 1979. Taxonomic literature—II. Vol. 2. Regnum Veg. **98:** 1–991.
——— & ———. 1983. Taxonomic literature—II. Vol. 4. Regnum Veg. **110:** 1–1214.
Steudel, E. T. 1840–1841. Nomenclator Botanicus. Ed. 2. Stuttgart & Tübingen. 2 vols. (Facsimile edition 1980, Delhi.)
Steyermark, J. 1968. *Pavonia imatacensis*. Acta Bot. Venez. **3:** 181–183.
Sweet, R. 1839. Hortus Brittanicus. Ed. 3, ed. G. Don. London: James Ridgway.
Trew, C. J. 1750–1773. Plantae Selectae. Nürnberg. 10 parts.
Turner, B. L. & M. G. Mendenhall. 1993. A revision of *Malvaviscus* (Malvaceae). Ann. Missouri Bot. Gard. **80:** 439–457.
Ulbrich, E. 1915. Malvaceae. *In* R. Pilger, Plantae Uleanae novae vel minus cognitae. Notizbl. Bot. Gart. Berlin **6:** 316–335.
———. 1921. Monographie der Afrikanischen *Pavonia*-Arten nebst Übersicht über die ganze Gattung. Bot. Jahrb. Syst. **57:** 54–184.
Vogel, ?. 1929. [*Pavonia multiflora*]. Gartenwelt **33:** pagination unknown.
Vollesen, K. 1995. 82. Malvaceae. *In* S. Edwards, M. Tadesse & I. Hedberg (eds.), Flora of Ethiopia and Eritrea **2(2):** 190–256. [*Pavonia* on pp. 224–236.]
Walpers, W. G. 1842–1848. Repertorium Botanices Systematicae. F. Hofmeister, Leipzig. 6 vols.
Wawra, H. 1883–1888. Itinera Principum S. Coburgi. Vienna. 2 vols.
Wielgorskaya, T. 1995. Dictionary of generic names of seed plants. Colombia University Press, New York.
Willdenow, C. L. 1800. *Pavonia*. Pages 833–839 *in* Linnaeus, Species Plantarum. Ed. 4. Vol. 3. Berlin.
Williams, L. O. & E. E. Cheesman. 1929. Malvales. *In* Flora of Trinidad and Tobago **1(2):** 72–112.
Zanoni, T. A. 1997. The herbarium of Paul A. Fryxell (pf) and U.S. Department of Agriculture, College Station, Texas, integrated into the New York Botanical Garden Herbarium (NY). Brittonia **49:** 197–203.

Numerical List of Taxa

Pavonia subgen. **Pavonia** sect. **Pavonia**
1. *P. corymbosa* (Swartz) Willdenow
2. *P. cristobaliae* Fryxell
3. *P. gracilis* R. E. Fries
4. *P. harleyi* Krapovickas
5. *P. heterotricha* Fryxell
6. *P. insperabilis* Fryxell
7. *P. laetevirens* R. E. Fries
8. *P. laxifolia* St.-Hilaire
9. *P. opulifolia* S. Moore
10. *P. paniculata* Cavanilles
11. *P. patuliloba* Hochreutiner
12. *P. pseudolaxifolia* Fryxell
13. *P. restiaria* Bertoni
14. *P. subhastata* Triana & Planchon
15. *P. vitifolia* Hochreutiner *ex* Chodat & Hassler

Pavonia subgen. **Pavonia** sect. **Lopimia**
16. *P. almasana* Ulbrich
17. *P. angustifolia* Bentham
18. *P. aschersoniana* Gürke
19. *P. aschersonioides* Fryxell
20. *P. aspera* Hassler
21. *P. blanchetiana* Miquel

22. *P. dasypetala* Turczaninow
23. *P. grandiflora* St.-Hilaire
24. *P. hirticalyx* Fryxell
25. *P. imatacensis* Steyermark
26. *P. immitis* Fryxell
27. *P. longitricha* Fryxell
28. *P. luetzelburgii* Ulbrich
29. *P. macrostyla* Gürke
30. *P. malacophylla* (Link & Otto) Garcke
31. *P. piauhyensis* Ulbrich
32. *P. pohlii* Gürke
33. *P. pterocarpa* R. E. Fries
34. *P. rosa-campestris* St.-Hilaire
35. *P. schwackei* Gürke
36. *P. setifer* (C. Presl) Fryxell
37. *P. spinistipula* Gürke
38. *P. windischii* Krapovickas

Pavonia subgen. **Pavonia** sect. **Cancellaria**

39. *P. cancellata* (Linnaeus) Cavanilles
40. *P. cochensis* J. R. Johnston
41. *P. humifusa* St.-Hilaire
42. *P. repens* Fryxell

Pavonia subgen. **Pavonia** sect. **Malvaviscoides**

43. *P. decora* Fryxell
44. *P. erythrolema* Gürke
45. *P. glazioviana* Gürke
46. *P. glaziovii* (Ulbrich) Fryxell
47. *P. grazielae* Krapovickas
48. *P. malvaviscoides* St.-Hilaire
49. *P. montana* Garcke *ex* Gürke
50. *P. paraënsis* Fryxell
51. *P. serrana* G. Esteves
52. *P. tiliifolia* (Ulbrich) Fryxell
53. *P. viscosa* St.-Hilaire
54. *P. zehntneri* Ulbrich

Pavonia subgen. **Pavonia** sect. **Lebretonia** subsect. **Lebretonia**

55. *P. angustipetala* Krapovickas & Cristóbal
56. *P. latifolia* (Nees & Martius) Sprengel
57. *P. ramboi* Krapovickas & Cristóbal
58. *P. reitzii* Krapovickas & Cristóbal
59. *P. rubra* Krapovickas & Cristóbal
60. *P. schrankii* Sprengel

Pavonia subgen. **Pavonia** sect. **Lebretonia** subsect. **Exsertae**

61. *P. achanioides* Grisebach
62. *P. aurantia* Liogier
63. *P. calcicola* (Britton) Ekman
64. *P. coccinea* Cavanilles
65. *P. commutata* Garcke
66. *P. cryptocalyx* Urban
67. *P. ekmanii* Helwig
68. *P. heterostemon* Urban *in* Urban & Helwig
69. *P. hotteana* Helwig
70. *P. lasiopetala* Scheele
71. *P. leiocarpa* Urban
72. *P. mogotensis* (Carabia) Fryxell
73. *P. pabstii* Krapovickas & Cristóbal
74. *P. punctata* Urban

Pavonia subgen. **Pavonia** sect. **Lebretonia** subsect. **Cordifoliae**

75. *P. fryxellii* Krapovickas
76. *P. gentryi* Fryxell
77. *P. macdougallii* Fryxell
78. *P. mattogrossensis* R. E. Fries
79. *P. rojasii* Hassler
80. *P. sapucayensis* Hassler
81. *P. venusta* Fryxell

Pavonia subgen. **Pavonia** sect. **Lebretonia** subsect. **Hastifoliae**

82. *P. auriglooba* Krapovickas & Cristóbal
83. *P. belophylla* Hochreutiner
84. *P. betonicaefolia* C. Presl
85. *P. bullulata* Hochreutiner *ex* Chodat & Hassler
86. *P. cryptica* Krapovickas & Cristóbal
87. *P. cymbalaria* St.-Hilaire & Naudin
88. *P. distinguenda* St.-Hilaire & Naudin
89. *P. dusenii* Krapovickas
90. *P. eurychlamys* Ulbrich
91. *P. friesii* Krapovickas
92. *P. glechomoides* St.-Hilaire
93. *P. glutinosa* Krapovickas & Cristóbal
94. *P. guerkeana* R. E. Fries
95. *P. hastata* Cavanilles
96. *P. hatschbachii* Krapovickas
97. *P. hoehnei* E. G. Baker
98. *P. kleinii* Krapovickas & Cristóbal
99. *P. lanata* R. E. Fries
100. *P. missionum* Ekman
101. *P. nana* R. E. Fries
102. *P. nepetifolia* (Standley) Standley
103. *P. orientalis* Krapovickas
104. *P. pedersenii* Fryxell
105. *P. prionophylla* R. E. Fries
106. *P. pulchra* Hochreutiner
107. *P. reticulata* Garcke
108. *P. revoluta* Krapovickas & Cristóbal
109. *P. rhodantha* Hochreutiner
110. *P. rosengurttii* Krapovickas & Cristóbal
111. *P. sagittata* St.-Hilaire
112. *P. schininii* Krapovickas & Cristóbal
113. *P. viscidula* St.-Hilaire & Naudin
114. *P. xanthogloea* Ekman

Pavonia subgen. **Asterochlamys** sect. **Asterochlamys**

115. *P. apiculata* R. E. Fries
116. *P. argentina* Gürke
117. *P. aurelii* Fryxell
118. *P. balansae* Gürke
119. *P. biflora* Fryxell
120. *P. cracens* Fryxell & Esteves
121. *P. fonsecana* (Standley) Fryxell
122. *P. garckeana* Gürke
123. *P. geminiflora* Moricand
124. *P. hassleriana* Chodat
125. *P. hexaphylla* (S. Moore) Krapovickas
126. *P. hieronymi* Gürke
127. *P. immaculata* R. E. Fries
128. *P. intermixta* A. Richard
129. *P. krapovickasii* Fryxell
130. *P. mollis* Humboldt, Bonpland & Kunth
131. *P. monticola* Fryxell
132. *P. mutisii* Humboldt, Bonpland & Kunth
133. *P. neei* Fryxell
134. *P. platyloba* R. E. Fries
135. *P. purpusii* Brandegee
136. *P. quadrifida* Fryxell
137. *P. rupestris* (Hassler) Krapovickas & Cristóbal

138. *P. sidifolia* Humboldt, Bonpland & Kunth
139. *P. spuria* Fryxell
140. *P. subrotunda* St.-Hilaire & Naudin
141. *P. vannii* Fryxell
142. *P. vinosa* Esteves
143. *P. yatarendana* Krapovickas & Cristóbal

Pavonia subgen. **Asterochlamys** sect. **Albae**
144. *P. alba* Seemann
145. *P. falconensis* Fryxell
146. *P. hirtiflora* Bentham

Pavonia subgen. **Malache** sect. **Malache**
147. *P. bahamensis* Hitchcock
148. *P. paludicola* Nicolson
149. *P. rhizophorae* Killip *ex* Kearney
150. *P. troyana* (Urban) Urban

Pavonia subgen. **Malache** sect. **Collicolae** subsect. **Collicolae**
151. *P. candida* (de Candolle) Fryxell
152. *P. chlorantha* (Kunth) Fryxell
153. *P. firmiflora* Schery
154. *P. paneroi* Fryxell
155. *P. pleuranthera* (de Candolle) Fryxell

Pavonia subgen. **Malache** sect. **Collicolae** subsect. **Rubriflorae**
156. *P. anisaster* (Standley) Fryxell
157. *P. baumliana* Fryxell
158. *P. durangensis* Fryxell
159. *P. nayarensis* Fryxell

Pavonia subgen. **Malache** sect. **Collicolae** subsect. **Conditiflorae**
160. *P. alia* Fryxell
161. *P. clarkii* Fryxell
162. *P. ecostata* Fryxell & S. Koch
163. *P. pulidoae* Fryxell

Pavonia subgen. **Malache** sect. **Laminares**
164. *P. oxyphylla* (de Candolle) Fryxell
  a. var. *oxyphylla*
  b. var. *melanommata* (B. L. Robinson & Seaton) Fryxell

Pavonia subgen. **Typhalea** sect. **Typhalea**
165. *P. castaneifolia* St.-Hilaire & Naudin
166. *P. fruticosa* (Miller) Fawcett & Rendle
167. *P. integrifolia* Standley
168. *P. longitudinalis* Fryxell
169. *P. nemoralis* St.-Hilaire & Naudin
170. *P. parva* Ulbrich
171. *P. peruviana* Gürke
172. *P. schiedeana* Steudel
173. *P. stellata* (Sprengel) Sprengel

Pavonia subgen. **Typhalea** sect. **Urenoideae**
174. *P. communis* St.-Hilaire
175. *P. flavispina* Miquel
176. *P. formosa* Fryxell
177. *P. horrida* Krapovickas
178. *P. intermedia* St.-Hilaire
179. *P. ionthacarpa* Krapovickas
180. *P. leucantha* Poeppig *ex* Garcke
181. *P. narcissi* Krapovickas
182. *P. oxyphyllaria* J. D. Smith
183. *P. penduliflora* (Standley) Standley
184. *P. piptocalyx* Krapovickas
185. *P. pseudo-typhalea* Planchon & Linden *ex* Triana & Planchon
186. *P. ramosissima* (Arechavaleta) Fryxell & Krapovickas
187. *P. renifolia* Krapovickas
188. *P. rudis* Fryxell
189. *P. sepioides* Fryxell & Krapovickas
190. *P. sepium* St.-Hilaire
191. *P. spinifex* (Linnaeus) Cavanilles
192. *P. stenopetala* Krapovickas
193. *P. uniflora* (Sessé & Mociño) Fryxell

Pavonia subgen. **Typhalea** sect. **Diathericae**
194. *P. kearneyi* D. M. Bates

Pavonia subgen. **Typhalea** sect. **Carcerariae**
195. *P. arachnoidea* C. Presl
196. *P. martii* Colla
197. *P. submutica* Fryxell

Pavonia subgen. **Typhalea** sect. **Varians**
198. *P. varians* Moricand

Pavonia subgen. **Goetheoides**
199. *P. alnifolia* St.-Hilaire
200. *P. atlantica* Fryxell
201. *P. calyculosa* St.-Hilaire & Naudin
202. *P. cauliflora* (Nees) Fryxell
203. *P. ciliata* G. Esteves & Krapovickas
204. *P. conferta* St.-Hilaire
205. *P. crassipedicellata* Krapovickas
206. *P. crispa* Krapovickas
207. *P. ducke-limae* Monteiro
208. *P. froesii* Fryxell
209. *P. ×gledhillii* Cheek
210. *P. goetheoides* (Hassler) Fryxell
211. *P. latibracteolata* Krapovickas
212. *P. longifolia* St.-Hilaire
213. *P. longipedunculata* Gürke
214. *P. makoyana* E. Morren
215. *P. morii* Krapovickas
216. *P. multiflora* St.-Hilaire
217. *P. ovaliphylla* G. Esteves & Krapovickas
218. *P. sancti* Krapovickas
219. *P. semiserrata* (Schrader) Steudel
220. *P. spectabilis* Krapovickas
221. *P. spiciformis* Krapovickas
222. *P. stipularis* Krapovickas
223. *P. strictiflora* (J. D. Hooker) Fryxell
224. *P. tricalycaris* St.-Hilaire

List of Exsiccatae

Abbott, W. L., 45 (166); 345 (191); 609 (1); 841 (166); 1183 (64); 2850 (191); 2852 (148).
Acevedo, P., 3124 (191); *s.n.* (92).
Acevedo, P. & A. Siaca, 3836 (191).
Acevedo, P. et al., 4507 (126); 5663 (166).
Acuña, *s.n.* (148).
Acuña & Díaz-Barreto, 17504 (66).
Adams, C. D., 9902 (191).

Agostini, G. & T. de Agostini, 1618 (166).
Agostini, G. et al., 1770 (166); 2005 (39).
Aguayo, A., 37 (190).
Aguiar & Martau, 5523 (186).
Aguilar, J. I., 657 (95); 1735 (144).
Ahumada, O., 1945 (84); 5280 (90).
Ahumada, O. & A. Schinini, 3926 (84); 3942 (88); 4093 (114).
Ahumada, O. et al., 1404 (84).
Alain (Bro.), 2605 (128); see also A. H. Liogier.
Alain (Bro.) & López, 4211 (68).
Alain (Bro.) & C. V. Morton, 5187 (68).
Albuquerque, B. W. de et al., 1299 (166).
Allard, H. A., 13832 (166); 22463 (180); 22477 (165).
Allen, C., 122 (138); 255 (166); 534 (138); 554, 819 (144); 846 (10); 911 (31).
Allen, P. H., 362, 931 (165); 979 (171); 2253 (10); 2742 (182); 4221 (10).
Almeda, F., 2541 (155).
Almeda, F. et al., 3126 (39); 6642 (183).
Almeida, J. & T. S. Santos, 268 (178).
Alston, A. H. G., 5334, 5909, 6398, 6747 (10); 7621 (166); 8646 (149).
Amorim, A. M. et al., 416 (222); 687A (31); 979 (21).
Ancuash, E., 1470 (166).
Anderson, A. B. et al., 1075 (166).
Anderson, W. R., 12097 (10); 12357 (190).
Anderson, W. R. & S. L. Jessup, 13546 (197).
Anderson, W. R. & G. D. McPherson, 12097 (15).
Anderson, W. R. & D. C. Sternberg, 3092 (172); 3228 (150).
Anderson, W. R. et al., 6226, 6234, 6666, 6741 (26); 6779, 6888 (129); 7757 (26); 7988 (32); 8193 (34); 8292 (32); 8524 (53); 8746 (111); 8779 (30); 8834, 9017 (122); 9090 (39); 9102 (122); 9132 (138); 9273 (39); 9354 (26); 9403 (19); 9854 (125); 10204 (30); 35930 (48); 36793 (41); 36883 (125); 37162 (138).
Andrade-Lima, D., 49–253 (45); 50–621 (207); 55–2038 (198); 62–4061 (21); 66–4533 (196); 67–5074 (30); 68–5344 (21); 72–6988 (49).
Andrade-Lima, D. & de Ataide Sil, 74–6527 (21).
André, E., 1492 (130).
Angélica, *s.n.* (45).
Anisits, J. D., 2335, 2511 (134); 2639 (115); 2747, 2751 (116); 2893 (134).
Antonio, T., 1957 (166); 3179 (165); 4521, 4798 (166).
Apolinar-María, M., 272 (10).
Appun, C., 2311 (123).
Araquistain, M., 3200 (171).
Araquistain, M. & P. P. Moreno, 729 (172).
Araujo, 4440 (30).
Araujo, F. S. de, 46 (45).
Arbeláez, E. P. & J. Cuatrecasas, 6369 (165).
Arbeláez, E. P. et al., 584 (189); 1124 (185).
Arbo, M. M. & A. Schinini, 2481, 2498 (92).
Arbo, M. M. et al., 828 (140); 933 (86); 1597 (138); 1715 (124); 2879 (140); 3113 (34); 3355 (23); 3609, 3636 (26); 3990 (174); 4020 (49); 4311 (53); 4349 (35); 4532 (39); 5169 (119); 5360 (196); 5443 (41); 5667 (37); 5804 (28); 6079 (137); 6128 (181).
Archer, W. A., 4972 (95); 7641 (30); 7842 (166).
Archer, W. A. & T. Rojas, 4759 (126); 4862 (140).
Arechavaleta, J., 991 (88); *s.n.* (174, 186).
Argüelles, E., 1504, 1812, 1927 (151).
Argüello, A., 577 (182).
Aristeguieta, L., 5465 (10); 5511 (189).
Aristeguieta, L. & G. Agostini, 4118 (22).
Aristeguieta, L. & Montoya, 2045 (148).
Aristeguieta, L. et al., 7238, 12501 (148).
Arnoldo, 3196 (191).
Arrillaga et al., 2689 (95).
Arsène, G., 6192 (70); 10621 (151).
Arvigo, R. et al., 569 (172); 612 (148).
Asplund, E., 5309 (165); 5882 (132); 6547 (189); 7641 (10); 9007 (166); 9105 (180); 12167 (1); 12939 (165); 12994 (1); 13463 (189); 13470 (10); 16315 (165); 16675 (132); 18183 (10); 18319 (171); 19252 (182); 19827 (180).
Avendaño, S. & J. I. Calzada, 437 (193).
Ayala, F. et al., 2673 (180).
Ayala, M. G. & R. Ayala, 209 (75).
Ayers, T. J., 115 (155).
Aymard, G. C. & R. V. Moran, 2972 (166).
Aymard, G. C. & F. Ortega, 2325 (39); 4939 (144).
Aymard, G. C. & B. Stergios, 3287 (17).
Aymard, G. C. et al., 2649, 3643 (166).
Bacigalupo, N. M. et al., 611 (84); 750 (92); 992, 9684 (84).
Badcock, W. J., 236 (126); 412 (116).
Badillo, V. et al., 7301 (189).
Badini, 19970 (53).
Bailleti, 274 (116).
Baker, C. F., 288 (166); 3466 (128).
Baker, M. A., 6623 (182).
Balansa, B., 1612 (118); 1622 (181); 4385 (118).
Baldwin, 3158 (178); *s.n.* (191).
Balick, M. J. et al., 1537 (30); 2008 (172).
Balls, E. K., 5926 (92).
Bang, M., 409 (10); 545 (166); 1053 (116); 1456 (30); 1496 (10); 2067 (174).
Barbosa, 5449 (32).
Barclay et al., 3459 (130).
Barclay, G. W., 1045 (30).
Barfod, A., 41413 (130).
Barfod, A. et al., 48204 (165).
Barkley, F. A. & Hernández, 40637 (172).
Barkley, F. A. & C. Nelson, 40665 (148).
Barkley, F. A. et al., 279 (166).
Barreto & A. C. Brade, 1291 (48).
Barreto, 2953 (111); 7946 (53); 7965 (122); 8020, 8024 (8); 9918 (48); 11122 (91); 17504 (66).
Barroso, G. M., *s.n.* (47).
Barroso, G. M. et al., 607 (26).
Bartlett, H. H., 10853 (193); 17381 (191); 20111, 20759 (82); 20796 (92); 21094 (82).
Bartlett, H. H. & T. Lasser, 16667 (166).
Bastos, N. C., 317 (21).
Basualdo, I., 2198 (190).
Bauml, J. A. & Voss, 2020 (157).
Bautista & Salgado, 882 (45); 892 (198).
Beck, H. T. et al., 390 (165).
Beck, S., 3298 (17); 5251 (9); 5377 (166); 6379 (190); 6506 (141); 7509 (10); 8114 (180); 9909 (17); 11462 (116); 13363 (165); 14102 (176).
Beck, S. & Liberman, 9818 (176).
Beechey, F. W., *s.n.* (155).
Belém, R. P., 1502 (216); 3672 (219); 3810 (216); 3879, 3882 (201).
Belém, R. P. & M. Magalhães, 885 (166).
Belo, 62 (30).
Belshaw, C. M., 3074 (10); 3357 (138).
Benítez de Rojas, C., 638 (39); 2186 (165).
Bennett, B. & R. Alarcón, 4472 (166).

Bennett, B. & P. Gómez, 3616 (180).
Bensman, R., 328 (182).
Berlin, B., 924 (165); 932 (180).
Bernardi, L., 19408 (78).
Berro, 4779 (107).
Bertoni, M. S., 2681, 4211 (13); 5870 (15); 6104 (13); 98416 (93).
Betancur, J. et al., 1128 (185).
Bettfreund, C., 96 (140).
Bezerra, 446 (142).
Bigelow, J. M., *s.n.* (70).
Bittencourt, 106 (214).
Black, G. A., 48–2926 (1).
Black, G. A. & Fernandes, 1280 (123).
Blackmore, S. & M. Chorley, 3996 (172); 4113 (135).
Blackmore, S. & G. L. A. Heath, 1663 (22).
Blanchet, J. S., 33 (165); 313 (21); 944, 1334 (123); 2117 (201); 2843 (45); 3148 (198); 3149 (41); 3662 (21); 3949 (213); *s.n.* (210).
Blanchoud, 2115 (84); 2279 (181); 2369 (84).
Blum, K. E. & J. D. Dwyer, 2597 (183).
Blydenstein, J., 1860 (30).
Bockerman, 137 (30).
Boeke, J. D., 176 (172); 852 (189); 853 (165).
Boelcke, O., 861 (95).
Boelcke, O. et al., 1586 (114).
Bohlin, J.-E. et al., 1512 (180).
Boldingh, I., 1401, 2650, 3188 (191).
Boom, B. M., 7051 (166); 8002 (148); 9959 (191).
Boom, B. M. & M. Grillo, 6501 (30).
Bornmüller, 341 (55); 462 (82).
Bourgeau, E., 1513 (10); 2721 (172); 2906 (193).
Boutin, F. & F. Brandt, 2909 (81).
Bovell, J. R., 552 (191).
Box, H. E., 953 (191).
Brace, L. J. K., 4048, 4334, 4622, 4763 (147).
Brade, A. C., 5972 (94); 7364 (97); 13014 (60); 18289 (214); 21186 (98).
Brainard, E. M., 69 (70).
Breedlove, D. E., 9432 (10); 11649 (172); 21598 (138); 23879 (10); 36689 (138); 37625 (135); 49625 (197); 56434, 56982 (10); 61882 (152).
Breedlove, D. E. & F. Almeda, 56982, 57190 (10); 60167 (197); 60336 (10).
Breedlove, D. E. & P. H. Raven, 13715 (197); 13716 (131).
Breedlove, D. E. & R. F. Thorne, 20685 (131).
Brenes, A. M., 14401 (165); 17051 (22); 23174 (172).
Breteler, F. J., 4175 (166); 4672 (148).
Bridarolli, 3225 (126).
Brinker et al., 5508 (91); 5510 (89).
Britton, E. G., 2920 (148); 3385 (191).
Britton, N. L., 515 (150); 2487 (166).
Britton, N. L. & J. F. Cowell, 373 (166); 9836 (148); 12811 (66).
Britton, N. L. & C. A. Hollick, 1860 (191); 2015 (148).
Britton, N. L. & C. F. Millspaugh, 6145, 6145bis, 6187 (147).
Britton, N. L. & J. A. Shafer, 691 (191).
Britton, N. L. & W. M. Wheeler, 135 (191).
Britton, N. L. & P. Wilson, 4534 (148); 5150 (191); 14796 (30).
Britton, N. L. et al., 198 (148); 2601 (165); 2738, 4868 (166); 5905 (63); 6517, 6973, 6976 (128); 9627 (30); 10060 (128); 12987 (166); 14350 (128).
Broadway, W. E., 4572 (148); 9393 (172); 9791 (165); *s.n.* (148).
Brooke, W. M. A., 5756 (91); 5896 (190); 5956 (86).
Brown, 1626 (90).
Brunner, D. R. et al., 885 (190).
Buch, W., 182 (74).
Buchtien, O., 80 (10); 270 (166); 508 (172); 875 (180); 3844 (8); 5458 (166); 5459 (10); 6023 (166).
Buckley, S. B., *s.n.* (95).
Bueno, O. L., 484 (84); 1251, 1793, 2823 (91); 2883 (57); 2893 (107); 3258 (114); 3260 (101); 3394 (91); 4726 (107); 4993 (186); 5234 (105); 5381 (114); 5431 (88).
Bueno, O. L. & A. Krapovickas, 3602 (190); 3641 (57).
Bueno, O. L. & A. Pott, 33 (115).
Bueno, O. L. et al., 663 (114); 664 (105); 667 (95); 3982 (100).
Bullock, S. H., 986 (75).
Bunting, G., 4273 (30).
Bunting, G. & L. Licht, 1296 (148).
Bunting, G. & R. D. Medina, 36 (166).
Burandt , C. L. & W. E. Clark, V1216, V1228 (166).
Burandt, C. L. & H. van der Werff, V0761 (165).
Burch, D., 2530 (71); 4557 (165).
Burchell, W. J., 4713 (184); 5547 (34); 6992 (125).
Burger, W. C. & R. Liesner, 6247 (172).
Burger, W. C. & G. Matta, 4576 (183).
Burkhart, 7311 (82); 27009 (11).
Burkhart et al., 30530 (126).
Busey, P., 651 (182).
Busey, P. & T. B. Croat, 253 (22).
Buttura, 426 (119).
Cabral, E. et al., 183 (140).
Cabrera, A. L., 2093 (114); 2523 (84); 4031 (86); 6348 (95); 8183 (190).
Cabrera, A. L. & H. A. Fabris, 16194, 19918 (116).
Cabrera et al., 13778 (126); 19092, 22246, 27329 (92); 28624 (84).
Cabrera, E. & H. de Cabrera, 4779 (172).
Cabrera, E. & O. Canul, 5003 (172).
Cabrera, E. et al., 7699 (10).
Cabrera, I., 6575 (185).
Cáceres, S. et al., 355 (174).
Calderón, S., 774 (172).
Callejas, R. & O. Escobar, 3313 (189).
Callejas, R. & O. Marulanda, 5991 (182).
Callejas, R. et al., 3958 (185); 4994 (166).
Calvert, D., 115 (138).
Camp, W. H., E-1044 (165); E-1418 (166).
Campêlo, C. R., 781 (15); 1317 (10).
Campêlo & Campêlo, 572 (138).
Carabia, J. P., 3088 (72).
Carauta, 1730 (199).
Cárdenas, M., 22 (15); 1341, 1347 (116); 2578 (126); 3778 (125); 4543 (39); 4623 (95); 4624 (179); 4767 (116); 5033 (15).
Carlson, M., 3105 (36).
Carnevali, R., 1668 (140).
Caro, 3434 (82).
Carranza, E., 910 (193).
Carreira et al., 822 (122).
Carsuer & Studhalter, 4618 (70).
Carvajal, A., 490 (183).
Carvalho, A. M. de & S. A. Mori, 107 (178).
Carvalho, A. M. de & T. Plowman, 1567 (44).
Carvalho, A. M. de & J. G. Saunders, 2855 (28); 2893 (198); 2921 (4).
Carvalho, A. M. de & W. W. Thomas, 3008 (28).

Carvalho, A. M. de et al., 2035 (213); 2625 (37); 3780 (44); 4531 (39); 6144 (28); 6223 (215).
Casaretto, J., 1056 (173); 2058 (196); *s.n.* (39, 172, 166).
Castañeda, 1469 (130).
Castellanos, 3383 (100); 24770 (94); 25536, 25699 (169).
Castillo, A., 2278 (165).
Castro et al., 1082 (45).
Cavalcanti, P. B. et al., *s.n.* [SPF 39755] (24).
Cerón, C. E., 12406 (165).
Cerón, C. E. & G. Benavides, 6222 (182).
Cerón, C. E. & M. Cerón, 2744 (189); 4719 (180); 7037 (189).
Cerón, C. E. et al., 2112 (165).
Cezar, 79 (32).
Chacón, 419 (171).
Chaso, 167 (199).
Chavarría, U., 136 (148).
Chávez, C., 8.9.9 (157).
Cházaro, M. & L. Robles, 3320 (75).
Cházaro, M., 3316 (30).
Chiang, F. et al., 7515 (70).
Chiappeta, A., 300 (30).
Chiappeta, A. et al., 535 (198).
Churchill, J. A., 7542560 (148).
Cid, C. A. & B. W. Nelson, 2840 (182).
Cid, C. A. et al., 4482 (34).
Cisneros, J., 28 (166).
Clark, O. M., 6801 (70); 6863 (161).
Claussen, P., 266 (49); *s.n.* (122).
Clemente (Bro.), 6270 (172); 6282, 7097 (191).
Clewell, A., 4563 (165).
Cobián, J. F., 223 (164b).
Cochrane, T. S. et al., 10845b (155).
Coker, D., 495 (191); *s.n.* (147).
Colella, M. et al., 1376 (71); 1492 (25).
Combs, R., 124 (191); 222 (148).
Commerson, P., 246, *s.n.* (95).
Conceição, 132 (137).
Contreras, E., 4811 (172).
Conzatti, C., 1978 (193); 3661–½ (39).
Conzatti, C. & V. González, 26 (193).
Cook, O. F. & G. B. Gilbert ,1037 (189).
Cooper, J. J., 5720 (172).
Coradin, L. & M. Cordeiro, 747 (39).
Coradin, L. et al., 5706 (34); 6053 (54); 6134 (4); 6267 (198); 6485 (4); 7413 (34).
Cordeiro, I. et al., 974 (119).
Correll, D. S., 43262, 43863, 44322 (147).
Correll, D. S. & G. R. Proctor, 48693 (147).
Correll, D. S. & D. Wasshausen, 46838 (147).
Cory, V. L., 41678, 49392 (70).
Costa Secco, da, 471, 1015 (103); 1281 (175).
Cowan, C. P. et al., 4289 (174); 5644 (75).
Cowan, R. S., 1347 (172); 38940 (39).
Crawford, 749 (166).
Cristóbal, C. L., 1339, 1425 (126).
Cristóbal, C. L. et al., 1553, 1603 (84); 1947 (92).
Croat, T. B., 41 (191); 6608 (165); 9827 (148); 12754 (22); 15368 (166); 15979 (171); 16743 (166); 23288 (148); 26625 (171); 26723 (166); 26827 (183); 38426 (130); 42100 (172); 45199 (155); 50923 (189); 51122 (138); 53960 (39); 54328 (172); 54329 (189); 54392 (39); 57651 (189); 66250, 68394 (172); 73130 (180); 73752 (165).
Croat, T. B. & D. P. Hannon, 65988 (10).
Croat, T. B. & D. M. Porter, 15605 (166); 15670 (183).
Crosby, M. R. & W. R. Anderson, 1041 (172); 1236 (150).
Cross, W. H., *s.n.* (119, 125).
Cruz, de la, 1432, 3005, 4094 (148).
Cuatrecasas, J., 1021 (189); 3747 (14); 3769 (166); 3949 (17); 4288, 4722 (14); 7563 (165); 8240 (189); 13250 (166); 14182,14517–A (14); 15917 (149); 16377 (10); 17234 (14); 18758 (185); 19722 (149); 21631, 22169 (185); 22868 (166); 23781 (189).
Cuatrecasas, J. et al., 12180 (10).
Cuezzo, A. R. & B. Balegno, 2122 (82).
Cuezzo, A. R. & A. de la Sota, 1604 (13).
Cuezzo, A. R. & T. Meyer, 5369 (126).
Cuezzo, A. R. et al., 11609 (176).
Cumana, 2614 (189).
Cuming, H., 1143, 1147 (30).
Curran, H. M., 11 (184).
Curran, H. M. & M. Haman, 788, 1305, 1340 (148).
Curtiss, A. H., 290 (128); 386 (191); 386* (148); 563, 5710 (191).
D'Arcy, W. G., 10471 (171).
D'Arcy, W. G. & J. J. D'Arcy, 6400 (183); 6497 (172); 6111 (166).
Dahlberg, C. G., *s.n.* (166).
Dalby, P., H22 (87).
Daly, D. C. et al., 451 (30); 1018 (50); D278 (39).
Damazio, L. B., 1772 (33).
Daniel, T. et al., 5866 (197).
Davidse, G., 3681 (39); 24562 (183).
Davidse, G. & W. G. D'Arcy, 10082 (22).
Davidse, G. & González, 19112, 19252 (172).
Davidse, G. & T. P. Ramamoorthy, 10520A (94).
Davidse, G. et al., 4523, 4620 (39); 11208 (190); 11281A (55); 11862 (28); 11863 (123).
Davidson, C., 54 (183); 5345 (166); 7152 (182).
Davidson, M. E., 503 (171).
Davie 21, (223).
Davis et al., 2406 (53).
Dawson, E. Y., 14425 (125).
Dawson, G., 47 (108); 644 (82); 3013 (92).
Dawson, G. & A. Celastremé, 2045 (92).
Deam, C. C., 33 (172); 358 (10); 6051 (148).
de la Sota: see Sota, A. de la
Del'Arco, M. R., 492 (31).
Del'Arco, M. R. & Castro, 1113 (45).
Del Puerto, 4 (95); 2252 (190); 2254 (103); 2257 (84); 3215 (88); 3892 (92); 5409 (84); 6052 (174).
Del Puerto & Marchesi, 6147 (110).
Delascio, J. F. & R. Liesner, 6855 (138).
Delprete, P. G., 5119 (166).
Delprete, P. G. & P. Apreza, 6304 (185).
Denslow, 2305 (172).
Desportes, J. B. R. P., *s.n.* (64).
Devia A., W. et al., 2929 (149).
Díaz & W. H. Hodge, 325 (10).
Díaz, C. & S. Baldeón, 2266 (180); 2474 (189).
Díaz-Barriga, H., 2477, 3479 (151).
Díaz-Luna, 3525 (75).
Diedrichs, E., 65 (39).
Diggs, G. et al., 2730 (172).
Digilio-Grassi, 2032 (82).
Dillon, M. O. et al., 1252 (10); 4335 (189).
Dlouhy & J. Fernández-Casas, 7335 (13).
Dodge, C. W. & A. A. Hunter, 8634 (144); 8646 (1).
Dodge, C. W. & W. S. Thomas, 4387 (172).
Dodge, C. W. et al., 6217 (39); 16668 (166).

Dodson, C. H. & P. M. Dodson, 11486 (132).
Dodson, C. H. & A. Embree, 13422 (10).
Dodson, C. H. & L. B. Thein, 1312 (132).
Dodson, C. H. et al., 8795 (171); 11041 (132).
Dombey, J., *s.n.* (10).
Dombrowski ,2447 (185).
Dombrowski & Kuniyoshi, 2705 (186); 3383 (192).
Dorr, L. J. & L. C. Barnett, 6838 (172); 7721 (10).
Dorr, L. J. et al., 2652 (102).
Dorrien Smith, G., 86 (125); 297 (129).
Dressler, R. L., 1560 (167); 1988 (95).
Dressler, R. L. & Q. Jones, 146 (167).
Drouet, F. E., 2170 (39); 2591 (123).
Dryander, E. I., 2597 (149).
Duarte, A. P., 607 (113); 2587 (48); 3026 (34); 4201 (216); 5510 (169); 7175 (165); 8148 (48); 9528 (41); 9652 (35); 9862 (214); 9873 (199); 10583 (37); 11662 (53).
Dubs, B., 172 (126); 2125 (115).
Ducke, A. & D. Andrade-Lima, 7 (207).
Dugand, A., 3129 (138); 3533 (130).
Duke, J. A., 4084 (166); 4677 (144); 5367 (166); 8465 (130); 8887 (148); 9751, 10552 (149); 10846 (166).
Dumont, K. et al., 7408 (166); 7757 (189).
Dunlop, 408 (165).
Duque, A. & C. Restrepo, 410 (14).
Duque-Jaramillo, J. M., 3674 (130).
Duré, R., 419 (137).
Dusén, P., 1622 (169); 2645 (55); 7163 (60); 7821 (174); 7823 (190); 8922 (94); 9358 (94); 9587 (107); 13666 (169); 14318 (172); 14322 (94); 16308 (60); 16641 (111); 16895 (107); 17804 (100); *s.n.* (88).
Duss, P. A., 841 (148); 843, 2328 (191); 2786 (148).
Dwyer, J. D., 2286 (149); 4154 (138); 9942, 11161 (172); 12895 (148); 14944 (172).
Dwyer, J. D. & Hayden, 7642, 7696 (183).
Dwyer, J. D. & Lallathin, 8741 (183).
Dwyer, J. D. & R. Liesner, 12308 (10).
Earle, F. S., 711 (128).
Ebinger, J. E., 758 (166).
Edwards, P-623 (172).
Edwards, J. B., P-821 (22).
Eggers, H. F. A. von, 109 (191); 1519b (166); 15023 (171); *s.n.* (148).
Ehrich, 113 (92).
Eiten et al., 2936 (30).
Eiten, G. & L. T. Eiten, 2438 (172); 4120 (39); 8975 (120); 9075 (17).
Ekman, E. L., 166 (55); 168 (95); 170, 171 (114); 172 (105); 173 (100); 174, 175 (8); 176 (140); 1002 (74); 1035 (166); 1349 (71); 2538, 2785 (172); 3042 (74); 3055 (64); 5395 (67); 5561 (1); 6635 (74); 7158 (191); 7183 (1); 7270 (67); 7842 (172); 8180 (166); 8370 (68); 8810 (74); 8952 (67); 8979 (66); 9164 (74); 9474, 10451 (148); 10465 (30); 10777, 10792 (69); 10807 (128); 10858 (69); 10877, 11437 (128); 11826 (71); 12090 (128); 12505 (30); 12530 (148); 12592 (64); 13705 (71); 14987 (10); 15334 (64); 16612 (74); 17982 (72); 18896 (63).
Elias, 651 (138); 1301 (148).
Emperaire, 2182, 2415, 2680 (45).
Emperaire & Campêlo, 2757 (31).
Emrich, 44786 (55).
Emygdio, L. et al., 3507 (53).
Encarnación, 1290 (168).
Epler, 70 (60).
Eriksen, 59252 (130).
Ernst, W., 2123 (166).
Escobedo, J. M., 1211 (151).
Espina, J., 1890 (166).
Espina, J. et al., 2715 (149); 2721 (165).
Esteves, G. L., 180 (39); 2316 (111); 2331A (122); 2358 (4); 2382 (16); 2384 (44); 2385 (199).
Esteves, G. L. & C. S. S. Barros, 2370 (29); 2571 (39); 2580 (30).
Esteves, G. L. & C. Kameyama, 2477 (125); 2503 (47); 2520 (41); 2522 (45); 2542 (196); 2555 (190); 2611 (30); 2615 (201).
Esteves, G. L. & R. P. Lyra-Lemos, 2043 (166); 2218 (41); 2220 (198); 2503 (45); 2515 (28); 2523 (4); 2530 (196); 2533 (37); 2535 (44); 2544 (45); 2546 (29); 2548 (200).
Esteves, G. L. & J. R. Pirani, 2397 (205).
Esteves, G. L. & L. Rossi, 2620 (173).
Esteves, G. L. et al., 2703 (175); 2704 (190); 13315 (47); 13343 (34).
Ewan, J. A., 15670 (166); 16636 (189).
Eyerdam, W. J., 335, 560 (67); 24700 (166); 24908 (126).
Eyerdam, W. J. & A. A. Beetle, 22703 (176).
Fagerlind, F. & Wibom, 343 (165).
Fanshawe, D. B., 5060 (148).
Farenholtz, 1560 (189).
Farias, G. L., 113 (216); 371 (201); 397 (216).
Fassett, 25268 (166).
Feddema, C., 303, 337 (153); 678, 820 (156); 1873 (165); 2030 (148).
Fendler, A., 90 (10); 99 (39); 116 (165); 117 (166); 1932 (138).
Fernandes, 29, 2428 (123); 2788 (10); 2799, 4111 (198); 6186 (31).
Fernandes & Martins, *s.n.* (45).
Fernandes et al., 2895 (30).
Fernández, A., 1258 (165).
Fernández, A. & E. Pérez-Arbeláez, 496 (14).
Fernández, A. et al., 333 (183).
Fernández, J. L. et al., 10712 (165).
Fernández-Casas, J., 8097 (166).
Fernández-Casas, J. & J. Molero, 3706 (190); 4084 (39); 5665 (190); 5939 (174).
Fernández-Casas, J. & Morales, 10745 (166).
Ferrari, G., 162 (10).
Ferraro et al., 4136 (82).
Ferreira, A. R., 99 (37).
Ferreyra, R., 310 (174); 1610 (180); 1829 (189); 4234 (1); 7072 (174); 10821 (189); 12664 (166); 17288 (130); 17306 (138).
Ferrucci, M. S. & G. V. Somner, 947 (199).
Ferrucci, M. S. et al., 208 (94); 346 (60); 359 (100); 559 (13); 986 (198).
Fiebrig, K., 513 (140); 692 (124); 758 (138); 812 (95); 831 (190); 1253 (15); 1434 (126); 1466 (118); 2535 (92); 3057 (116); 4021, 4056 (126); 4197 (118); 4418 (78); 4434 (118); 4781 (39); 4859 (15); 5913 (114); 6083 (106).
Filgueras, T. de S., 428 (26).
Fishlock, M. C., 290 (191).
Fleischmann, 523 (180).
Flores et al., 3021 (195).
Flores, A., 1990 (172); 2075 (159); 2486 (158).
Flores, A. & N. Cervantes, 208 (155).
Flores, G. & P. Tenorio, 1302, 1445 (159).
Flores, G. et al., 897 (155); 1131, 1595, 1661 (159).
Florschütz, P. A. & J. Florschütz, 282 (164).

Focke, H. C., 441 (166).
Folli, D. A., 241 (201).
Folsom, J. P., 2182 (183); 4439 (171); 4567 (166); 5118 (165); 5884 (172).
Fonnegra, R. et al., 2764 (189).
Forero, E., 730 (1).
Forero, E. et al., 2107 (171); 4186 (1); 4398 (194); 4409 (149); 4858 (166).
Fortunato et al., 2484 (9).
Fosberg, F. R., 19477 (144); 19538 (165); 19749 (185); 19779 (189); 19808 (130); 20449, 21002 (189); 21998 (130); 22506 (172); 29004 (1); 45222 (39); 55631 (22); 59379, 60792 (191).
Foster, R. B., 8833 (166); 9356 (182); 9850 (1).
Foster, R. B. & S. Baldeón, 12778 (182); 12863 (180).
Foster, R. B. & B. d'Achille, 10056 (182).
Foster, R. B. & Terborgh, 6130 (180).
Foster, R. B. et al., 11042 (166).
Franceschi, 282 (95).
Fredholm, A. J. A., 3110 (166); 3151 (148).
Freire-Fierro, A. et al., 1736 (28); 2491 (180).
Freitas et al., 142 (166).
Fries, R. E., *s.n.* (176).
Fromm, E., 1846 (196); 2119 (138).
Fróes, R. de L., 11889 (39); 12633, 12634 (29); 12670/36 (201); 19991 (44); 20054 (201); 20055 (208); 20069a (201); 20145 (4); 20184 (97); 27067 (10); 27071 (184).
Fruchard, *s.n.* (92).
Fryxell, J. E., 181 (193); *s.n.* (70).
Fryxell, P. A., 566 (164b); 607 (164a); 670 (172); 756, 760 (75); 1075 (172); 1145 (75); 1165 (138); 1187 (135); 1913 (32); 1914 (34); 1925 (39); 1926 (138); 1927 (39); 1929 (198); 1932 (138); 1934 (198); 1935 (28); 1943 (45); 1944 (41); 1945 (198); 1946 (39); 1947 (196); 1948 (138); 1956 (28); 1958 (41); 2558 (193); 2607 (172); 2675 (70); 2800 (172); 2805 (148); 4406 (138); 4943 (155).
Fryxell, P. A. & W. R. Anderson, 3620 (193).
Fryxell, P. A. & D. M. Bates, 891 (10); 902 (197); 913 (75); 2126 (195); 2138 (164b); 2163, 2172 (164a).
Fryxell, P. A. & C. L. Burandt, 4324 (10).
Fryxell, P. A. & E. J. Lott, 3232 (197); 3233 (131); 3354 (172); 3383 (77).
Fryxell, P. A. et al., 1649 (164b); 1651 (164a); 4304 (165); 4351 (39); 4356 (138).
Fuchs, H. P. & L. Zanella, 21720 (1); 21847 (194); 22068 (149).
Fuertes, M. D., 16 (148); 71 (191); 373 (64); 618 (166).
Funck, N. & L. J. Schlim, 191 (10); 1271 (22).
Furlan, A. et al., 1677 (16); 1959 (4); 4642 (48); 7551 (4).
Galeotti, H. G., 4092 (10).
Gallinal et al., B-2352 (110).
Garber, A. P., 53 (166); *s.n.* (148).
García, R. G. & J. D. Pimentel, 1193 (166).
García-Barriga, H., 10749, 12017 (130); 12036 (22); 12352 (166).
Gardner, G., 322 (190); 324-pars (169); 324-pars (173); 932 (30); 1244 (41); 1457 (138); 1458; 2048 (42); 2049 (39); 3020 (32); 3591 (138); 3593 (30); 3599 (138); 4430 (34); 4432 (10); 4433 (18); *s.n.* (190).
Garnier, A., 94 (39).
Gasparini, *s.n.* (9).
Gastony et al., 420 (191).
Gates, B. & G. Estabrook, 116 (26).
Gaudichaud, C., 935 (169); *s.n.* (82, 88, 95).
Gaumer & W. Clark, 6 (193).
Gay, C., *s.n.* (87).
Gehrt, A., 8280 (169); 8327 (8); SP-3208 (97).
Gentle, P. H., 1338, 3816 (148); 8806 (172).
Gentry, A. H., 2413 (22); 6690 (1); 7930, 9524 (172); 22165 (1); 53854 (194); 54130 (166).
Gentry, A. H. & L. E. Aguirre, 15231 (166).
Gentry, A. H. & C. Bonifaz, 28745 (165).
Gentry, A. H. & M. Fallen, 17508 (172).
Gentry, A. H. & E. Forero, 7380 (149).
Gentry, A. H. & R. B. Foster, 75485 (41).
Gentry, A. H. & A. Juncosa, 40588 (194).
Gentry, A. H. & S. A. Mori, 14089 (171).
Gentry, A. H. & G. T. Prance, 16411 (10).
Gentry, A. H. & G. Shupp, 26460 (165).
Gentry, A. H. & D. N. Smith, 44956 (138).
Gentry, A. H. & T. Zanoni, 50560 (64).
Gentry, A. H. & E. M. Zardini, 49755 (174).
Gentry, A. H. et al., 7559 (148); 12115 (165); 15931 (166); 30969 (182); 31098 (180); 37637 (10); 37696 (130); 39846 (180); 39916 (180); 40414 (149); 41500 (10); 48120 (185); 50103, 50176 (138); 51763 (190); 52119 (166); 53412 (149); 55941 (165); 65436 (185).
Gentry, H. S., 7279, 7287 (76); 19502 (153).
Geoffrey, 35 (108).
Ghiesbreght, A. B., 214 (164a).
Gibbs, P. E. et al., 1980 (10); 5251 (53).
Gibert, E. J., 126, 351 (84); 691 (97); 954 (92); 1038 (137).
Giberti & Ahumada, 40 (15).
Gibram, M., 70 (199).
Gillis, W. T., 9839 (148); 10614, 11840, 12300, 13165 (147).
Gilly, C. L. & E. Hernández, 324 (172).
Gimate-Leyva, J., 783 (193).
Gines, 77 (10); 3151, 3193 (39); 4340 (172).
Ginzbarg, S. & P. Cézar, 732 (26).
Giulietti, A. M. et al., 1407 (28); 6861 (16); 36377 (29).
Glaziou, A., 8780 (199); 10278 (30); 12448 (138); 13540 (224); 13543 (52); 13544 (18); 13545 (122); 13548 (39); 14507 (53); 15836 (105); 15842 (46); 18140 (199); 18140a (202); 18885a (35); 20200 (53); 20710, 20710bis (34); 20713 (3); 20715 (32); 37805 (199).
Gledhill, D., 1038 (209).
Goldman, E. A., 1037 (138).
Gomes, 463 (125).
Gómez, A. et al., 419 (149).
Gómez, L. D. et al., 21422 (183); 23113 (165).
Gómez, P., 415 (166).
González, M. & J. Rzedowski ,1596, 1848 (158).
González, S., 128 (151).
González-Ortega, J., 68 (155).
Goodland, R., 1026 (39).
Goodspeed, T. H., 17241 (108).
Gottsberger, G., 11–14272 (55); 11–16271 (122); 12–25171 (94); 14–7471 (60).
Gottsberger, I. S. & G. Gottsberger, 1004 (119).
Goudot, J., *s.n.* (185).
Grant, M. L. & W. B. Drew, 10675 (185).
Grant, M. L. & F. Giovanni, 9668 (189).
Granville, J. J. de, 4993 (166).
Granville, J. J. de et al., 8598, 10488 (165); 10896 (172).
Gray, A., *s.n.* (95).
Grayum, M., 10007 (165).
Grayum, M. & P. Sleeper, 3660 (148).
Grayum, M. et al., 10541 (171).

Gregory, W. C. et al., 9595 (13); 9601 (106); 9611 (83); 9612 (85); 9617 (125); 9660 (95); 9666 (190); 9671 (109); 9684 (174); 9737 (122); 9760 (119); 9787 (111); 9808 (83); 9950 (15); 9972 (125); 10004 (7); 10107 (17); 10145 (23); 10192 (32); 10423 (34); 10527 (111); 10529 (30); 10594 (138).
Griffiths, A. & W. D. Hawkinson, 4646 (209).
Grijalva, 964 (171); 2350 (22).
Grijalva, A. & D. Ríos, 3454 (171).
Grijalva, A. & Vanegas, 4090 (39).
Grondona, E., 2643 (87).
Grondona, E. & G. Dawson, 6367 (87).
Groth, B. H. A., 125 (70).
Grotta, A. S., 5823 (8).
Guanaja, 4271 (191).
Gudiño, E., 1115 (180).
Guillén & Choré, 1849 (30).
Guillén & Coria, 1561 (17).
Gurgel, 49 (107).
Gurken, 1 (214).
Gutiérrez, 312 (82).
Gutiérrez et al., 572 (17); 900 (38).
Haber, W. A. & C. Ivey, 10151 (172).
Haber, W. A. & W. Zuchowski, 8903 (39).
Hadač, E., 225, 446, 1914 (128).
Haenke, T., *s.n.* (10, 36, 84, 164a, 195).
Hage, J. L., 206 (178).
Hage, J. L. & Brito, 636 (166).
Hagelund, 11084 (187).
Hagen, W. von & C. von Hagen, 2041 (183).
Hahn, W., 764 (138); 1569 (137); 2241 (190).
Hahn, W. et al., 887, 998 (95); 1350 (190).
Halàcz, *s.n.* (151).
Hall, J. S., 7683 (172).
Hallman, 10 (14).
Hambleton, 9 (95).
Hamilton et al., 830 (183).
Hammel, B., 3817, 4360 (172); 4582 (148); 4795 (171); 5434 (165); 5686, 7537 (183).
Hammel, B. & W. G. D'Arcy, 5060 (148).
Hancock, 23 (195).
Harcourt, W., 537 (166).
Harley, R. M., 19136 (39).
Harley, R. M. et al., 15181 (29); 15331 (196); 15619 (39); 15997 (4); 16058 (28); 16256 (28); 16511 (37); 16559 (4); 17025 (54); 18936 (45); 19033 (198); 19136 (39); 19510, 19947 (29); 20646 (4); 22346 (28); 22933 (4); 25062 (47); 25200 (39); 25366, 25889 (29); 26451 (54); 26540 (16); 26942, 26953, 51207 (29).
Harling, G., 1670 (10); 26008 (166).
Harling, G. & L. Andersson, 11990 (166); 18074, 18459 (189); 21077 (132); 21272 (166); 22308 (189); 23326 (182); 22332 (165).
Harling, G. et al., 7387 (166); 7397 (172); 15505 (166); 20756 (189).
Harmon, W. E. & J. A. Fuentes, 6435 (172).
Harriman, N. A., 14399 (197); 14612 (10).
Harris, W. E., 8629 (148); 9457, 10868, 10964 (150); 11875 (172); 12364 (150); 12445 (191); 12775 (150).
Hartman, R. L., 12216 (182).
Hartweg, K. T., 23 (146); 913 (189); 914 (130).
Hassler, E., 1922 (124); 1930 (80); 2747 (11); 3154 (138); 3978 (137); 4384 (106); 4385 (109); 4602 (83); 5632 (169); 5684 (95); 5906 (140); 6143 (137); 6500 (190); 6785 (124); 6872 (137); 6987 (85); 7305 (11); 7568 (126); 7801 (15); 8757 (85); 8760 (95); 8823 (15); 8876 (169); 9091 (106); 9467 (20); 9894, 10073 (106); 10998, 11070 (134); 11223 (80); 11250 (3); 11392 (106); 11459 (138); 11505 (181); 11951 (3); 11953 (106); 12070 (80).
Hatschbach, G., 135, 507 (60); 1194 (89); 2798 (107); 3947 (60); 3996 (89); 7316 (99); 7666, 7781 (186); 7802 (89); 7889 (13); 8550 (99); 8788 (107); 8870 (100); 8883 (65); 9759 (138); 11217 (13); 12858 (65); 13218 (94); 14337 (9); 15411 (96); 15439 (89); 15444 (107); 15456 (94); 15544 (83); 15984 (94); 16181 (65); 19870 (107); 20618 (83); 20929 (169); 21148 (94); 21546 (190); 22264, 22727 (65); 22831 (99); 22910 (95); 22937 (107); 22968 (190); 23559 (119); 23694 (125); 23721 (138); 24253 (111); 26227 (94); 26327 (174); 27925 (53); 28137 (96); 29486 (17); 29494 (134); 29923 (111); 30131 (48); 30220 (34); 30728 (96); 32350 (34); 32858 (60); 33755 (169); 34048 (125); 34727 (125); 34729 (122); 35660 (60); 36964 (32); 36991 (30); 38024 (125); 38639 (138); 38649 (78); 39368 (27); 39448 (125); 39522 (198); 39528 (45); 35660 (60); 39684 (28); 41073 (60); 41298 (47); 41310 (122); 41432 (94); 41523 (47); 42831 (122); 42918 (47); 43047 (107); 43531 (65); 43814 (32); 43833 (122); 44113 (120); 44245 (28); 44452 (55); 44843 (95); 45224 (15); 46553 (16); 46753 (30); 46911 (216); 47404 (29); 47517 (28); 47575 (94); 47589 (190); 48186 (196); 48244 (4).
Hatschbach, G. & Ahumada, 31316 (111).
Hatschbach, G. & Cervi, 51394 (39).
Hatschbach, G. & Fontella, 20813 (99).
Hatschbach, G. & O. Guimarães, 18486 (65); 20569 (100); 21945 (17); 24848 (8); 55105 (138).
Hatschbach, G. & H. Haas, 13946 (94).
Hatschbach, G. & Koczicki, 27540 (177).
Hatschbach, G. & R. Kummrow, 34957 (122); 38240 (23); 46700 (30); 47897 (28).
Hatschbach, G. & Manosso, 51514 (65).
Hatschbach, G. & Nicolack, 53415 (29); 53932 (26); 54231 (18).
Hatschbach, G. & O. S. Ribas, 52562 (100); 54087 (8); 55240 (122).
Hatschbach, G. & J. M. Silva, 48718 (201); 49204 (118); 49432 (30); 50020 (54); 50073 (44); 50154 (45); 50219 (48); 54705 (26); 57991 (214).
Hatschbach, G. & L. B. Smith, 16036 (60).
Hatschbach, G. & F. J. Zelma, 49102 (134); 52015 (47).
Hatschbach, G. et al., 12206 (107); 13376 (100); 13429, 13525 (169); 23721 (138); 28283 (89); 36281 (23); 52362 (177); 52430 (137); 52517 (134); 53024 (53); 55000 (122); 56052 (39); 56822 (29); 56833 (54); 56914 (45); 58068 (216); 58656 (174); 58784 (134); 58799 (126); 58951 (115); 65151 (54).
Haught, O., 132 (70); 1601 (165); 2127 (130); 2294 (138); 2461 (22); 3649 (138); 3792 (10).
Havard, V., 132 (70).
Hayes, S., 482 (30).
Hazlett, 5012 (183).
Heilborn, O., 321 (189).
Hekker, F. & W. Hekking, 10244 (165).
Hekking, W., 1078 (39).
Heller, A. A., 4442 (166); 4570 (148); 4685 (191).
Heller, A. A. & E. G. Heller, 344 (191); 824 (148).
Henrickson, J., 11507 (70).
Hensold, N., 1023 (171).
Hensold, N. et al., 30121 (47); *s.n.* [CFCR 2841] (53).
Heriberto (Bro.), 443 (144).
Heringer, E. P., 3689, 6993, 7405 (111); 7988/250 (32); 8551/745, 8679/873 (34); 8894/1088 (122); 11465, 12827

(129); 13097 (23); 14315 (119); 14453, 15418 (32); 15554 (23); 16043 (30); 16934-A (32).
Heringer, E. P. & Eiten, 14184 (125).
Heringer, E. P. et al., 416 (23); 489 (30); 1118 (23); 1535 (30); 2129, 2464 (34); 4396 (23); 4564 (30); 5471 (122); 6641 (119); 16767 (166); 17415 (34).
Heringer-Salles, A. E., 92 (23).
Hernández, R. & Y. Vázquez de Hernández, 748 (151).
Herrera, G., 3219 (189); 3445 (183).
Herrera, G. et al., 4964 (10).
Herter, W. G., 253 (95); 253a (91); 711 (92); 2064 (190); 2502 (95); 11456 (92); 16983 (103).
Hespenheide, H. A. et al., 1143 (150).
Heyde, E. T. & E. Lux, 6070 (10).
Hieronymus, G., 662 (116); 839 (108).
Hieronymus, G. & P. G. Lorentz, 237 (126).
Hill, S. R., 13206 (39); 13325 (216); 27299 (17).
Hill, S. R. et al., 12864 (165); 17608 (39); 25445 (191).
Hind, D. J. N. & R. F. Queiroz, 50073 (29).
Hinton, G. B. et al., 1923, 3680, 3892, 5326, 7091, 7218, 8351 (152); 8579 (164a); 11197 (30); 12717 (155); 13341, 13368 (164b); 14516 (172); 16899 (70); 18881, 23405 (102); 24515, 24825 (193).
Hitchcock, *s.n.* (147).
Hitchcock, A. S., 16707 (148); 17095 (39); 17445 (166); 20181, 20248 (10); 20372, 20810 (189); 21779 (171).
Hjerting, J. P. & K. Rahn, 3054 (92).
Hodge, W. H. & B. T. Hodge, 3744 (148).
Hodge, W. H. & R. A. Howard, 4497 (166).
Hoehne, F. C., 5513 (138); 6134 (97); 10735, 20058 (8); SPF-10546 (111).
Hoenicka, H., 48 (10).
Holdridge, L. R., 890 (74).
Holm-Nielsen, L. & S. Jeppesen, 1031 (171).
Holm-Nielsen, L. et al., 2703 (166); 26003 (171).
Holst, B. K. et al., 1964 (39); 2274 (138).
Holt, E. G., 264 (166).
Holt, E. G. & W. Gehringer, 157 (39).
Holton, I. F., 745 (165).
Holway, E. W. D., 49 (164a); 1940 (169).
Hopkins, M. J. G. et al., 76 (30); 115 (125); 537 (17).
Howard, R. A., 12131 (64); 14112 (150); 18154 (191).
Howard, R. A. & E. S. Howard, 8609 (191); 9437 (71); 9444 (191).
Howard, R. A. & G. R. Proctor, 14416 (150).
Huashikat, V., 590 (182).
Huber, O., 236 (39); 1195 (148).
Huber, O. & H. Canales, 6331 (165).
Huft, M. et al., 1656 (171).
Huidobro, A. M. R., 1348 (92); 1732 (87); 3770 (190); 4954 (114); 4956 (100); 5350 (100).
Humboldt, A. von & A. Bonpland, 1077 (138); *s.n.* (130, 166).
Hunt & J. E. Ramos, 6228 (19).
Hunt, 5493 (32); 6100 (3); 6131 (136).
Hunter, A. A. & P. H. Allen, 757 (144).
Hunziker, A. T. & A. E. Cocucci, 17234 (190).
Hunziker, 1746 (84); 5300 (126); 5999 (108); 7149 (92); 10688 (116); 13311 (108); 15404 (116); 18775 (92); 21953 (190).
Hunziker & Di Fulvio, 16927 (116).
Hunziker et al., 13950 (116); 21751 (108).
Hurtado, F. & A. Alvarado, 385 (182).
Hutchison, P. C. & J. K. Wright, 3597 (126); 5059 (189); 5858 (126).
Idrobo, J. M. et al., 9965, 10522 (185).
Iltis, H. H. & T. S. Cochrane, 267 (151).
Iltis, H. H. & M. Nee, 1545 (195).
Insfran, 445 (84).
Imaguirre, 3500 (186).
Irwin, H. S., 612, 1019 (39); 2232 (172); 2652 (178).
Irwin, H. S. & T. Soderstrom, 5594 (34).
Irwin, H. S. et al., 7967, 8098 (34); 8494 (30); 10437 (111); 12308 (26); 13037 (32); 14897 (129); 14974 (125); 15585 (30); 16270 (34); 16310 (30); 16485 (34); 16982 (125); 17552 (136); 17865 (19); 17984 (30); 18242 (12); 18999 (32); 20317 (48); 20603, 20693 (53); 23333 (122); 23341 (196); 23654 (39); 23958 (111); 24671 (26); 24792 (23); 25670 (34); 26416 (119); 26532, 26698 (23); 27556 (125); 28003 (30); 28571, 29345 (49); 29670 (53); 29711 (174); 30721 (196); 30793 (54); 30864 (39); 31528 (125); 31529 (129); 31807, 32781 (26).
Isabelle, A., *s.n.* (140).
Isert, P. E., *s.n.* (148, 191).
Izaguirre, P. et al., 9741 (92).
Jack, J. G., 5464, 5916 (166); 6572 (191).
Jangoux, J. & R. P. Bahia, 339 (166).
Jangoux, J. & B. G. S. Ribeiro, 1560 (165).
Jansen-Jacobs, M. J. et al., 63 (17); 948 (165); 1335 (17); 2262 (39); 2753 (123).
Jaramillo, R. et al., 8266 (14).
Jardim, 1892 (115).
Jarenkow, 699 (175).
Játiva, C. & C. Epling, 7 (10); 349 (166); 353 (182); 2159 (1).
Jeffrey & Trujillo, 2268 (144).
Jermy, G., 852 (70).
Jiménez, A., 3754 (22).
Jiménez, J., 1149 (165); 1288 (148); 1572, 2156 (71); 2317 (64); 2686 (10); 3347 (39); 3747 (64).
Johnson, 1235 (148).
Johnston, I. M., 552, 840 (30).
Johnston, J. R., 18 (40).
Johnston, M. C. et al., 11940B (70).
Jones, 4468 (95).
Jones, M. E., 28141 (70).
Jones, R., 148 (193); 2025 (167).
Jones, R. & R. Treviño, 173 (193).
Jones, S. & G. Jones, 6464 (70); 5396 (135).
Jonker-Verhoef, A. M. E. & F. P. Jonker, 510 (148).
Jönsson, G., 334a (174).
Jörgensen, P., 2308 (11); 2326 (73); 3163 (84); 3166 (190); 3808 (95); 3811 (190); 3812 (138); 4288 (174); 4476 (169); 4846 (20); 4850 (106).
Juncosa, A., 2090 (149); 2091 (194).
Kalbreyer, W., 251 (130).
Kalloo, M. B., B500 (166).
Kallunki, J. A. & J. R. Pirani, 448 (214).
Kallunki, J. A. et al., 588 (216); 613 (165); 691 (216).
Kappler, A., 1693 (39).
Kauffmann, 12 (30).
Kayap, 522 (182); 755 (166).
Keeler, *s.n.* (191).
Kellerman, W. A., 5091, 7290 (135); 7475 (172).
Kellogg, *s.n.* (95).
Kenoyer, L. A., 521 (172).
Kerr, J. G., 66 (84).
Khan, R. et al., 1023 (39); 1218 (172).
Kidder, N. T., *s.n.* (191).
Kiehl, 4064 (111).
Kiesling & Meglioli, 6661 (116).

Kiesling et al., 3770 (92).
Killeen, T. et al., 4260 (126); 5434 (38).
Killip, E. P., 5237 (149); 8262 (189); 11692 (149); 11694 (194); 33007, 34964 (149); 37306 (165); 43553 (148); 43946a, 44122, 44626, 44879, 45026, 45145, 45751 (128).
Killip, E. P. & J. Cuatrecasas, 38613 (194); 38636, 38834, 39131 (149).
Killip, E. P. & T. E. Hazen, 8782 (185).
Killip, E. P. & T. Lasser, 37787 (166).
Killip, E. P. & A. C. Smith, 14491 (138); 15470 (130); 16413, 16484 (189); 20872 (132); 22554 (10); 23582 (166); 25880 (188); 26226 (166).
Killip, E. P. et al., 38377 (14); 39176 (189); 39915 (130).
King, R. M., 4414 (10).
Kirkbride, J. H., 2134 (172).
Kirkbride, J. H. & N. Bristan, 1574 (166).
Kirkbride, J. H. & J. A. Duke, 878 (183).
Kishler, J., 90 (164b); 802 (151).
Klein, R. M., 4471 (107); 5074 (105).
Klein, R. M. & Bresolin, 8843 (177).
Klug, G., 939 (180); 2044 (1); 2732 (166); 3538 (130); 3540 (180); 3953 (166); 4008 (6); 4221 (130).
Knapp, S. D., 1050 (166); 2260 (10).
Knapp, S. D. & J. Mallet, 2965 (30); 7018 (6).
Knapp, S. D. & M. Vodicka, 5609 (183).
Knight, 385 (185).
Koch, S. D. & P. A. Fryxell, 8322 (164b); 77392 (151); 78338 (75); 78356 (138); 78361 (75); 89198 (162).
Koch, S. D. et al., 7986 (164a); 73154 (197); 78255 (138); 78258 (139); 79484 (10); 79511 (30); 79518 (172); 79520 (30); 87242 (75).
Korning & Thomsen, 47064 (180).
Kramer, K. U. & W. Hekking, 2615 (166); 2984 (165).
Krapovickas, A., 1567, 1796 (116); 2174 (87); 2302 (175); 2354, 2389 (13); 2886 (190); 2892 (114); 2956 (87); 3268 (95); 3389 (87); 4398 (84); 4404 (110); 4802 (84); 4804 (93); 4808 (84); 4811 (93); 4832, 4833 (103); 4848 (110); 4850 (88); 4856, 4861 (103); 6574 (82); 6576 (95); 6627 (82); 6884 (174); 8031 (39); 8781, 8915 (176); 8968 (95); 9047 (5); 9119 (78); 9120c (180); 9189 (39); 9522 (176); 12763 (166); 12770 (196); 12789 (39); 12850 (28); 12862 (42); 12872 (207); 12901 (41); 12903 (198); 13108 (80); 14323 (84); 14388 (10); 14398 (111); 14427 (126); 14493 (88); 14494 (94); 15583 (166); 16171 (103); 29785 (15); 29789 (127); 29807 (7); 29917, 29937 (17); 34643 (3); 41872 (141); 43442 (181); 44410 (80).
Krapovickas, A. & C. L. Cristóbal, 11126 (82); 11127 (92); 11133 (95); 11144 (103); 11151 (88); 11156 (87); 11157 (92); 11165 (95); 11167 (87); 11169 (190); 11170 (103); 11173 (87); 11176, 11177, 11178, 11183, 11184, 11185 (103); 11187 (92); 11189 (91); 11195 (95); 11202 (84); 11204 (82); 11207 (92); 11209 (101); 11225 (84); 11229 (88); 11231 (93); 11261 (126); 11299 (84); 11323 (181); 11338 (95); 11387 (181); 11393 (126); 11409 (181); 11413 (95); 11414 (86); 11467 (181); 11469 (190); 11473 (137); 11661 (140); 11759 (126); 12051 (100); 13171 (80); 13218 (95); 13257, 13299 (95); 13672 (13); 14758 (82); 15337 (126); 16164 (103); 16292 (82); 21661 (95); 24531 (126); 28650 (175); 28833 (114); 28881 (105); 28895 (190); 28897 (114); 28904 (175); 28972 (114); 33479 (60); 33550 (111); 33554 (8); 33655 (99); 33656 (60); 33703 (94); 33723, 33732 (82); 34173 (92); 34190 (114); 34196 (91); 34323 (95); 34325 (83); 34363 (10); 34443 (2); 34451 (7); 34494 (119); 34548 (39); 35349 (111); 35355 (94); 35497 (190); 35498 (175); 37174 (184); 37575 (94); 37576 (105); 37624 (88); 37633 (57); 37725 (177); 37736 (190); 37807 (89); 37892 (190); 38385 (140); 38388 (82); 38391 (95); 38392 (88); 38408 (98); 38493 (91); 39590 (60); 39597 (65); 39605 (169); 39629 (94); 39722 (89); 40296 (92); 40761 (60); 40993 (13); 41933 (186); 41956 (89); 41964, 41973 (186); 41978 (94); 41980 (107); 42079 (186); 42755 (125); 42812 (122); 42821 (111); 42843 (122); 42863 (41); 42866 (138); 42905 (39); 43043, 43056 (115); 43128 (122); 43200 (190); 43436 (84); 43475 (190); 43569 (174); 43572 (89); 43704 (55); 43721 (89); 43733 (175); 43956 (107) 43957 (65); 43972 (186); 43973 (190); 43980 (94); 44006 (100); 44212 (141); 44455 (140); 44513 (95); 44588 (124); 44659 (91); 44661 (114); 44665 (91); 44726 (82); 44825 (169); 44875 (95); 44982 (138); 46534 (126).
Krapovickas, A. & Fortunato, 43921 (41).
Krapovickas, A. & A. Schinini, 28578 (15); 30389 (95); 30429 (126); 31349 (176); 31481 (179); 31482 (143); 31495 (166); 31548 (41); 31564 (138); 31586 (3); 32265 (122); 32389 (39); 32566 (109); 32595 (39); 32753, 34708 (126); 34802 (138); 34871 (17); 34884 (138); 34928 (9); 34929 (138); 35104 (30); 35499 (8); 36134, 36147 (3); 36364 (39); 36433 (119); 36456 (122); 36511 (41); 36657 (122); 38179 (86); 38185 (94); 38223 (91); 38224 (107); 38225 (100); 38226 (94); 38227 (82); 38228 (105); 38254 (105); 38263 (89); 38279 (65); 38283 (89); 38303 (186); 38305 (107); 38329 (105); 38335 (89); 38336 (174); 38342 (55); 38380 (85); 38835 (90); 38838 (92); 39047 (126); 39069 (176); 39217, 39296 (116); 39349 (88).
Krapovickas, A. & R. O. Vanni, 36781 (114); 36782 (100); 36872 (94); 36886 (89); 36887 (94); 36910 (107); 36973 (175); 37004 (187); 37011 (107); 37012 (89); 37013 (88); 37034 (100); 37036 (94); 37049 (186); 37060 (107); 37086 (91).
Krapovickas, A. et al., 12284 (181); 12483 (174); 12520 (95); 13053, 13834 (11); 14076 (83); 14137 (190); 14152 (83); 14306, 14311 (15); 14823 (114); 14835 (100); 15763 (13); 16681 (92); 16720 (84); 16865 (95); 16997 (100); 17189 (114); 18338 (190); 18680 (9); 19024 (190); 19069 (92); 19303 (126); 20305 (84); 21039 (114); 21113 (95); 21599 (88); 22830 (82); 22938 (88); 22939 (110); 22986 (88); 22992 (105); 22998 (94); 22999 (105); 23002 (107); 23036 (89); 23050 (94); 23052 (89); 23053 (60); 23085 (94); 23136 (60); 23259 (174); 23275, 23384 (94); 24061 (95); 24805 (86); 25861 (114); 27924, 28370, 28503 (126); 32997 (138); 33037 (119); 33107 (125); 33366 (111); 37462 (126); 37837 (129); 37872 (39); 37875 (138); 37881 (122); 37892 (198); 38079 (166); 38760 (39); 38819 (196); 38823 (28); 40111 (122); 40117 (3); 40133 (34); 40143 (30); 40149 (34); 40203 (126); 40219 (3); 40222 (138); 40228 (34); 40229 (129); 45526, 45881 (138).
Krukoff, B. A., 1085 (39).
Kuhlmann, M., 130 (94); 751 (10); 4520 (190).
Kuhlmann, M. & A. Gehrt, SP-40080 (60).
Kuhlmann, M. & P. Gonçalves, 1355 (190).
Kummrow, R., 886 (186); 891, 1285 (94); 2196 (55).
Kunth, K. S., 1627 (152).
Kuntze, O., *s.n.* (3, 82, 87, 111, 122, 166).
Lacás, 413 (193).
Lamotte, 288 (53).
Landrum, L., 4063 (94); 4074 (174).
Lane, M. & M. Leidig, 2888 (70).

Niederlein, 40 (88).
Noblick, L. R., 1770 (4).
Noblick, L. R. & Pinto, 2831, 2876 (4).
Northrup, J. I. & A. R. Northrup, 736 (148).
Novarra, 5162 (90).
Nunes & Martins, 5824 (142).
Núñez, P., 7554 (189).
Núñez, P. & M. Cárdenas, 10327 (10).
O'Donell, C. A., 2872 (190); 3118 (176).
O'Donell, C. A. & J. M. Rodríguez, 515, 992 (82).
Occhioni, E. M., 4633 (177); 7372, 7946 (41).
Occhioni, P., 7363 (199).
Oldeman, R., B-850 (148).
Oldeman, R. & C. H. L. Sastre, 132 (165).
Oldenburger, F. H. F. & V. V. Mecenas, 1501 (34).
Olea, 159 (92).
Oliveira, 287 (172).
Oliveira & W. R. Anderson, 478 (26).
Oliveira, F. C. A. et al., 371 (23).
Ortega, F. & G. C. Aymard, 1662 (166).
Ortíz, *s.n.* (151).
Ortíz, F., 58 (172); 320 (165).
Ortíz, M., 806 (181).
Osten, C., 4299 (101); 5189 (82); 5297 (95); 16983 (87); 21816b (88).
Pabst, G., 6233 (73); 7632 (65); 7834 (105); 7835 (174); 8014 (60); 8562 (44).
Paciornik, E. F., 19 (190).
Paciornik, E. F. & Oliveira, 335 (94).
Palacios, W. & G. Tipaz, 10046 (185).
Palacios, W., 2195 (182).
Palacios, W. et al., 206 (165).
Palacios-Cuezzo, 447, 1184 (88).
Palmer, *s.n.* (11).
Palmer, E., 48 (151); 89, 90 (70); 153 (195); 579 (102); 686 (151); 1835, 1990 (155).
Palmer, W. & J. H. Riley, 257 (61); 941 (128); 1031 (148).
Panero, J. L. et al., 5311 (154).
Pascuas, N. B. de, 330 (185).
Paula, de & Conceição, 1568 (9).
Pedersen, T. M., 1020 (86); 1059 (87); 1279 (126); 3004 (140); 3118 (126); 3122 (138); 3214 (95); 4076 (126); 4791 (84); 5325, 5859 (140); 7172 (84); 7225 (95); 8136, 8206 (190); 8268, 8753 (95); 8799 (174); 8999 (95); 9410 (138); 10141 (15); 10824 (116); 10989 (174); 11364 (84); 11524 (181); 11638 (82); 11660 (114); 11665 (82); 11680 (181); 12002 (86); 12013 (126); 12049 (125); 12127 (15); 12542 (84); 12655 (177); 12806, 12868 (116); 13172 (87); 13368 (190); 13407 (84); 13724 (89); 13842, 13861 (91); 14650 (39); 15643 (95); 15902 (104); 15921 (105); 15929 (186); 15951 (94); 16022 (138); *s.n.* (92, 108).
Peixoto, A. et al., 3051 (216).
Peixoto, A. L. & O. L. Peixoto, 396 (216).
Peña, M. & R. B. Foster ,226 (38).
Penland, C. W. & R. H. Summers, 161 (165); 162 (171).
Pennell, F. W., 2708 (10); 2805, 2862 (22); 3252 (185); 3342 (166); 10871 (130).
Pennell, F. W. & E. P. Killip, 5731 (185); 6245 (10).
Perani et al., 12864 (47).
Pereira, 397 (125); 2042 (4).
Pereira et al., 251 (17).
Pereira, B. A. S., 231 (39).
Pereira, E., 2503 (60); 2923 (53); 5142 (94); 5351 (13); 6406 (73); 8357 (65); 8559 (105); 8739 (60); 9011 (32); 9673 (44).
Pereira, J., 2042 (4); 5052 (39); 8560 (174).
Pereira, J. & A. P. Duarte, 4543 (190).
Pereira, J. et al., 3873 (190).
Pereira, O. J., 14230 (216).
Perkins, 1377 (150).
Pérez, B., 504 (137).
Pérez, E., 1163 (164b).
Pérez, L. A., 1361 (75).
Pérez de la Rosa, 1691 (164b).
Pérez de Molas, L. et al., 16 (137).
Pérez-Levis, 41 (166).
Perkins & W. Zomlefer, 671 (191).
Pertusi, 147 (92).
Petetin & Molina, 1552 (190).
Philcox, D. & Fereira, 4044 (125).
Philipson & J. M. Idrobo, 1763 (182).
Phillippe et al., 21398 (17).
Phillips, B. R., 184 (74).
Pickel. B. J., 107 (39); 1164 (196); 2415 (30); 2634 (10).
Pickersgill, B. et al., RU72-188 (39).
Pierotti, S. A., 20 (115); 5048 (82); 6193 (86); *s.n.* (116).
Pierre, 64 (191).
Pimentel, J. D. & R. G. García, 607 (148).
Pimentel, J. D. et al., 363 (74).
Pinheiro, R. S., 1324 (203); 1746 (211).
Pinheiro & Santos, 2264 (216).
Pinheiros & G. L. Esteves, 8571 (48).
Pinto, G., 351 (123); 22/83 (21); 67/86 (138); 314/81(4); 11174 (198); 20978 (4).
Piper, 5557 (30).
Pirani, J. R. & J. A. Kallunki, 3560 (190).
Pirani, J. R. et al., 183 (216); 1849 (26); 1897 (28); 2073 (30); 2079 (4); 2778 (214); 2783 (30); 2784 (216); 2985 (223); 3077, 3377, 3410 (216); 3593 (199); 12864 (47); 12985 (18); SPF- 39645 (47).
Pittier, H. F., 211 (135); 616 (14); 2140 (10); 2273 (22); 2548 (144); 3188, 3287 (183); 4168 (148); 4448 (144); 5535 (1); 6084 (39); 6879 (10); 6907 (144); 7439 (39); 8181 (138); 8284 (10); 8681 (39); 9145 (10); 9618 (172); 10643, 10715 (138); 11836 (166); 12687 (148); 12753 (189); 12864 (47); 16080 (182); 16243 (10).
Pittier, H. F. & A. Tonduz, 9091 (10).
Plowman, T., 5562 (126).
Plowman, T. & J. B. Alcorn, 14344 (172).
Plowman, T. et al., 4010 (165); 10070 (222).
Poeppig, E. F., 1053 (10); 1584 (171); 2302 (180).
Pohl, J. E., 1160, 1303 (125); 2611 (32); 2667 (125); 2928 (111); 3064 (39); 3344 (122).
Poole, J. M. & W. A. Watson, 1170 (172).
Pott, A., 3061 (2).
Pott, A. & G. Eiten, 194 (134).
Pott, A. & Ferreira, 2648B (2).
Pott, A. & C. C. de F. Tavares, 3195 (9).
Pott, A. et al., 388 (17); 2796 (127); 4070, 5610, 5661 (9); 7794 (134).
Poveda, D., 26 (166).
Powell, 971 (148).
Prance, G. T. & N. T. Silva, 58188 (34).
Prance, G. T., et al. 5587 (30); 7804 (166); 8311 (9); 10336 (1); 12503 (182); 16708 (1); 16790 (180); 18861 (34); 26096 (125).
Prescott, 446 (171).
Prieto, F., 23 (180).
Pringle, C. G., 27 (151); 1944, 2686 (70); 4132 (151); 4343 (164b); 5035 (172); 5447 (153); 8718 (164a); 13698 (102); 13750 (70).

Santos, E. B. dos et al., 367 (214).
Santos, T. S., 902 (213); 1674 (214); 1755 (30); 1999 (216); 3105 (210).
Saravia, C. et al., 11603 (116).
Saravia-Toledo, C., 2683 (141).
Sastre, C., 6356 (166).
Saunders, J. G. & A. M. de Carvalho, 3092 (196); 3102 (28).
Sayago, 23 (108).
Sazima, M., 14053 (216); 14058 (214); *s.n.* (48, 53).
Schaller, G., 98 (115).
Schessl, M., 91/1–3 (19).
Schiede, C. J. W., *s.n.* (172, 193).
Schinini, A., 4230 (124); 4291, 4299 (15); 4369 (124); 4579 (95); 4856 (126); 5430, 5540 (84); 5590 (114); 6172 (138); 6813 (114); 8157 (190); 8844 (140); 9190 (174); 19600 (126); 26696 (84).
Schinini, A. & O. Ahumada, 15987 (84).
Schinini, A. & E. Bordas, 15102 (141); 17853 (137); 17854 (116); 20153 (117); 20183 (39); 20499 (83).
Schinini, A. & Caballero Marmori, 27162 (112).
Schinini, A. & R. Carnevali, 24953 (181).
Schinini, A. & A. Fernández, 6023 (100); 6113, 6120 (13).
Schinini, A. & F. Mereles, 24567 (181).
Schinini, A. & R. Palacios, 25890 (126); 25894 (15); 25906 (126).
Schinini, A. & R. O. Vanni, 22494 (92); 25974 (190); 25985 (13); 26033 (190).
Schinini, A. et al., 8239, 8348 (86); 17365 (93); 17599 (82); 24326 (15); 27757 (140); 29343 (138).
Schipp, W. A., 635 (148).
Schlim, L. J., 49 (130).
Schomburgk, R., 137 (166); 468 (123); 731 (17).
Schott, A., 146 (199).
Schreiter, R., 11463 (116).
Schultes, R. E. & M. Villarreal, 5137 (189); 5192 (185).
Schulz, A. G., 21 (108); 26 (80); 38–43 (70); 543 (80); 547 (95); 548 (190); 563 (80); 894, 894a (126); 1868 (181); 2079, 2080 (80); 2084 (181); 2097 (126); 2173 (190); 2174, 2190 (80); 4030 (95); 6311 (108); 6616 (92); 6851 (80); 7004 (105); 7797 (95); 7908 (15); 8044, 8808 (80); 9101 (95); 9205 (138); 12342 (15); 17664, 17770 (84).
Schunck, 199 (90); 7908 (116).
Schunke-Vigo, J. M., 1537 (180); 3316, 3845 (182); 4619 (166); 4971 (6); 6375 (182); 6818 (166); 6989 (6); 10501 (10).
Schwacke, C. A. W., 8329 (35).
Schwarz, G. J., 444 (86); 1985 (100); 2209, 2258 (114); 3631 (88); 3880 (95); 5175 (114); 5637 (59); 6948, 7079 (13); 7464 (190); 7691, 8011 (114); 9137 (140); 10183 (86); 11811 (79); 11889, 12154 (174); 12214 (15).
Schwindt, E., 31 (13); 1115 (174); 1426 (13); 3186 (100); 4451 (181).
Scolnik, R. & R. Luti, 732 (165).
Seemann, B. C., 90 (144); 472 (10).
Sehnem, A., 3566 (105).
Seibert, R. J., 633 (172).
Seidel & Hirschle, 2624 (182).
Seidel & Schulte, 2197 (180).
Seidel, 157 (138).
Seidenschwarz, F. G., 273/1 (1).
Seler, G. E. & C. Seler, 3767 (10); 4227 (164a).
Sellow, F., 204, 271, 352 (224); 356 (60); 1426 (49); 1622 (88); 1934 (173); 3919 (107); 3935 (65); *s.n.* (65, 92, 99, 103, 105, 173).
Servín, B., 430 (193); 1216 (172).
Sessé, M. & J. M. Mociño, 3532 (193); 3537 (195); 3548 (152); 3564 (164a).
Seymour, F. C., 3102 (171); 4311, 5621 (172).
Shafer, J. A., 149, 162 (191); 521 (128); 594 (148); 1569 (166); 2452 (191); 3473 (166); 4336, 8342 (148); 10483, 11740 (128); 13404 (166).
Shank, P. J. & A. Molina, 4534 (182); 4603 (171).
Shannon, W. C., 424 (172).
Shattuck, O. E., 731 (10).
Shepherd, G. J. et al., 3694 (26).
Silva, A. P., 9 (33).
Silva, I. A., 374 (205).
Silva, J. M. & G. Hatschbach, 513 (65).
Silva, M. G. & C. Rosario, 3747 (166); 5226 (50).
Silva, N. T. da, 92 (1); 1554 (166); 2142 (15); 4880 (17).
Silverstone-Sopkin, P., 1013 (166); 2063 (14).
Silverstone-Sopkin, P. & F. J. Roldán, 2412 (10).
Simão-Bianchini, et al. 12995 (47).
Sintenis, P., 55 (166); 145 (191); 798 (148); 1081 (10).
Skutch, A. F., 2726 (172); 2926 (10); 3865 (182); 3900 (22); 4063 (10); 4342 (166).
Small, J. K., 7483 (148); 8984 (191).
Small, J. K. & G. K. Small, 4598 (148).
Smith, 1282 (182).
Smith et al., 4076 (193); 9385 (105).
Smith, A., 693 (171).
Smith, A. C., 2387 (39); 2400 (17); 3244 (172).
Smith, C. E. & Ruiz, 3237 (197).
Smith, C. E. & H. M. Smith, 3296 (22).
Smith, C. L., 6 (1).
Smith, D. N., 3992 (165).
Smith, D. N. et al., 12972 (165).
Smith, G. L., 10457 (148).
Smith, H. H., 492 (130); 495 (10); 846 (166); 2019 (148).
Smith, H. H. & G. W. Smith, 1359 (166).
Smith, J. D., 6449 (182); 6452 (1).
Smith, L. B. & R. M. Klein, 7808 (98); 8279 (55); 10720, 10779, 10952 (94); 11265 (98); 11972 (91); 12052 (65).
Smith, L. B. & R. Reitz, 10393 (98).
Smith, S. F. et al., 516 (182); 1534 (166).
Sneidern, K. von, 435 (165); 5130, 5813, A521 (166).
Snethlage, E, H., 158 (30).
Sobel, G. L. et al., 4947A (184).
Sobral & Silva, 4942 (89); 5887 (44).
Sobral et al., 3451 (84); 5793 (210).
Solheim, S. & S. Reisfield, 1470, 1543 (10).
Solheim, S. et al., 981 (164b).
Solís, 1193 (195).
Solomon, J. C., 6208 (30); 10052 (190); 10201 (92); 10956, 11007 (108); 12981 (166); 13715 (8); 18621 (126).
Solomon, J. C. et al., 12000 (10).
Soltero, R., 311 (164b).
Soria, N., 702 (13); 2545 (95); 3733 (125).
Sota, A. de la, 1479, 1577, 1781 (82); 2268 (116); 2676 (90).
Soto-Núñez et al., 12193, 12548 (36).
Souza, V. de, 88, 230 (199).
Sparre, B., 1240 (92); 13797 (180); 14809 (166); 18341 (149); 19428 (165).
Sparre, B. & F. B. Vervoorst, 2149 (95).
Spegazzini, C. L., 5556 (87); BAB15206 (190).
Spellman, D. L., 1590 (172).
Sperling, C. R. et al., 5765, 6100 (30).

Spooner, D. M. & W. García-Fernández, 6625 (126).
Spruce, R., 1617 (15).
Stahel, G., 35 (166); 37 (148).
Standley, P. C., 20108 (172); 20677 (121); 22231, 22273, 24574 (172); 26846 (144); 29862 (172); 37120 (182); 39860 (171); 40852 (172); 53568 (148); 53659 (172); 60628 (10); 60722, 62265, 64780, 67020, 68077 (172); 72173 (148); 78658 (172); 87369 (10); 90441 (167).
Standley, P. C. & Valerio, 45256 (172).
Stannard, B. L. et al., 51836 (16).
Stehlé, H., 302 (148); 2150 (191).
Stehlé, H. & M. Stehlé, 4822, 5375 (172).
Stein, B. A., 2976 (171).
Stein, B. A. & L. McDade, 3217 (189).
Steinbach, J., 1435 (138); 1642 (126); 1855 (190); 2090 (3); 2718 (138); 5258 (116); 5414 (180); 5730 (3); 6303 (41); 8407 (174).
Steinbach, R. F., 330 (138); 390 (10); 788 (138).
Stephens, S. G., M2 (39).
Stergios, B., 3287 (17).
Stergios, B., et al. 2833 (17).
Stern, W. L., et al. 469 (166); 1033 (183).
Stevens, W. D., 1407 (135); 3638, 4044 (172); 7611a (30); 10796, 11191 (39); 17563, 23105 (172); 23382 (165).
Stevens, W. D. et al., 14545 (165).
Stewart, R. M., 45 (70).
Steyermark, J. A., 41613 (172); 44045 (135); 49157a, 49454, 51686 (167); 53696 (10); 54081 (132); 57695 (39); 88149 (25); 88708 (39); 98274 (30).
Steyermark, J. A. & P. H. Allen, 16778 (22).
Steyermark, J. A. & V. C. Espinoza, 108753 (10).
Steyermark, J. A. & A. González, 113575 (145).
Steyermark, J. A. & R. Liesner, 118515 (189).
Steyermark, J. A. & B. Manara, 110449 (148).
Steyermark, J. A. et al., 108414 (148); 114280 (39); 114292, 114700 (148); 117190 (165); 117904 (25); 127138, 127187 (144).
Stork, H. E., 3200 (148).
Stork, H. E. & O. B. Horton, 9411 (189).
Straw, R. M., 2469 (10).
Strudwick, J. J. & G. L. Sobel, 4334 (30).
Stuckert, 1247 (108).
Sucre, 8674 (205).
Sucre, D. & da Silva, 8771 (190).
Sullivan, G. A., 556 (172).
Swallen, J., 4504 (39).
Swartz, O., *s.n.* (1).
Sytsma, K., 1679 (165); 2030 (22).
Sytsma, K. & W. G. D'Arcy, 3182 (30).
Tam, M., 18 (171).
Tamayo, 192 (166).
Tate, G. H. H., 122 (39).
Taylor, 120 (102).
Taylor, E. L. et al., E1283 (39).
Taylor, M., 307 (70).
Taylor, N., 386 (191).
Taylor, N. P. et al., 1600 (28).
Taylor, R. J. & C. E. Taylor, 11572 (172).
Teixeira, L. O. A. et al., 1117 (30).
Téllez, O., 9489, 10136 (155).
Téllez, O. & G. Flores, 11673, 11842 (155).
Téllez, O. & Miller, 10459 (156).
Téllez, O. et al., 12144 (159).
Tenorio, P. & G. Flores, 16080, 16192, 16396, 16512 (159).
Tenorio, P. et al., 9596 (158); 9961 (160); 13946 (172); 14905 (148).
Tharp, B. C. et al., 46281 (70).
Thieme, C., 5144 (172).
Thomas, W. et al., 9687 (4).
Thomas, W. W. et al., 10603, 10744 (202).
Thompson, S. A. et al., 7531 (166); 7637, 9753 (191).
Thompson, W. M., 268 (70).
Thorne, R. F. & E. Lathrop, 46097 (167).
Thorne, R. F. & G. R. Proctor, 48095 (150).
Tipaz, G. & C. Aulestia, 1688 (185).
Tiritan, O., 555 (111).
Toledo, M. et al., 1 (38).
Ton, A. S., 5293 (10); 6568 (172); 7095, 7229 (10).
Tonduz, A., 14863 (182).
Toro, R. A., 772 (130).
Torres, 1992 (77).
Torres & Campos, 14056 (77).
Torres, R. & L. Cortes, 9266 (75).
Torres, R. & H. Hernández, 3114 (193).
Torres, R. & P. Tenorio, 3733 (160).
Traill, J. W. H., 51 (39).
Tressens, S. G. & A. Schinini, 1304 (15).
Tressens, S. G. et al., 1848 (59); 2019 (140); 2060 (86); 2400 (92); 3286 (95); 3432 (13); 3448 (138); 3891 (190).
Triana, J. J., *s.n.* (130, 185); *s.n.* [3115] (14).
Trinta, Z. A., 770 (196); 1043, 1065 (138); 1142 (103).
Troetsch, 14 (144).
Troncoso & N. M. Bacigalupo, 2591 (92).
Trujillo, B., 2960 (10); 4284, 9455 (166).
Tsugaru, S. & Y. Sano, B-1183, B-1227 (39).
Tucker, J. M., 704 (10).
Tunqui, S., 772 (180).
Tún Ortíz, 1668 (10).
Türckheim, H. von, 341a (10); 2506 (191); 2621 (148); 3189 (71); II-1243 (172).
Tweedie, J., *s.n.* (108).
Tyson, E. L. et al., 3178 (10).
Ule, E. H. G., 35 (44); 5529 (1); 6861 (170); 7462 (31); 9588 (182).
Uribe, L., 2375 (166); 3941 (22); 5077 (185).
Usteri, *s.n.* (169).
Utley, J. F. & K. Utely, 5460 (172).
Valeur, E. J., 39 (166); 244 (191); 1004 (166).
Válio, 61 (39).
Van Hermann, H. A., 286 (166); 3308 (191).
Vanni, R. O. & S. Cáceres, 680 (100); 688 (95).
Vanni. R. O. & A. Radovancich, 992 (95); 1037 (140).
Vanni, R. O. et al., 569 (105); 572 (100); 1421 (174); 1948 (140); 2198, 2470 (141).
Vargas, 6308 (166); 16918 (172); 17542 (166).
Vargas et al., 2066 (180).
Vargas, C., 10594 (10); 14493 (30).
Vargas, C. & R. B. Foster, 841 (30).
Vargas, E., 698A (172).
Vargas, I. G., 642 (1).
Vasconcellos, 65 (10).
Vauthier, 491 (111); 492 (60); *s.n.* (113).
Vázquez, J. A., 1169 (155).
Vázquez, R., 2878 (172).
Vázquez, R. & N. Jaramillo, 902 (1); 1679, 2378 (166).
Vázquez, R. et al., 3078 (182); 11935 (1645).
Vega, S. & B. Quezada, 194 (172).
Ventenat, E. P., *s.n.* (1).

Index of Local Names

anamú (*P. fruticosa*, Colombia, Puerto Rico)
anguía (*P. sepioides*, Peru)
Bahama swamp bush (*P. bahamensis*, Bahamas)
barba-de-bode (*P. cancellata*, Brazil: Piauí)
barba-de-veado (*P. cancellata*, Brazil: Piauí)
caballo usa amarilla (*P. paniculata*, Peru: *Quechua*)
cadillo (*P. fruticosa*, Dominican Republic; *P. schiedeana*, Mexico: Oaxaca, Veracruz, Tabasco; *P. sepioides*, Colombia; *P. spinifex*, Dominican Republic)
cadillo altea (*P. paniculata*, Puerto Rico)
cadillo amarillo (*P. spinifex*, Dominican Republic)
cadillo americano (*P. spinifex*, Dominican Republic)
cadillo anaranjada (*P. paniculata*, Puerto Rico)
cadillo de cacao (*P. fruticosa*, Dominican Republic)
cadillo de ciénaga (*P. paludicola*, Puerto Rico)
cadillo de perro (*P. fruticosa*, Dominican Republic)
cadillo de tres pies (*P. fruticosa*, Dominican Republic; *P. spinifex*, Dominican Republic)
cadillo espinoso (*P. spinifex*, Puerto Rico)
cadillo pequeño (*P. fruticosa*, Puerto Rico)
cadillo pequeque (*P. fruticosa*, Puerto Rico)
camilia (*P. dasypetala*, Panama)
carrapicho (*P. leucantha*, Peru)
carrapicho de brejo (*P. fruticosa*, Brazil: Pará)
carrapicho de tres pies (*P. intermedia*, Brazil: Minas Gerais)
caye moho (*P. paludicola*, Belize)
chicojoos (*P. schiedeana*, Mexico: Querétaro)
chicraten nupa (*P. fruticosa*, Peru: *Jívaro*)
chorro blanco (*P. sepioides*, Peru)
chucchuc panga (*P. fruticosa*, Ecuador: *Quichua*)
conger watchman (*P. schiedeana*, Jamaica)
Congo mahoe (*P. troyana*, Jamaica)
cotton (*P. paludicola*, Belize)
cousin (*P. spinifex*, Dominican Republic)
cousin piquant (*P. spinifex*, Dominican Republic)
cuayó (*P. schiedeana*, Guatemala)
diente de chucho (*P. schiedeana*, Guatemala)
embiandé (*P. fruticosa*, Colombia)
escoba (*P. sepioides*, Colombia)
escobilla (*P. fruticosa*, Colombia)
escotadura rosella (*P. hastata*, Argentina: Misiones)
esfola-bahinha (*P. ducke-limae*, Brazil)
fierebérae (*P. fruticosa*, Surinam: *Arawak*)
gingerbush (*P. spinifex*, Netherlands West Indies)
grand cousin (*P. spinifex*, Dominican Republic)
guácima (*P. malacophylla*, Brazil: Pará)
guisato tres puyas (*P. schiedeana*, Cuba)
hoja de paludismo (*P. fruticosa*, Ecuador)
imbiandi (*P. fruticosa*, Colombia)
ismiandé (*P. fruticosa*, Colombia)
japajapa (*P. leucantha*, Peru)
jinuqui (*P. schiedeana*, Mexico: Oaxaca: *Zoque*)
kroomantie (*P. fruticosa*, Surinam)
lirio (*P. candida*, Mexico: Sinaloa)
mahout (*P. paludicola*, Dominican Republic)
majagua (*P. paludicola*, Venezuela)
majagüilla (*P. achanioides*, Cuba; *P. intermixta*, Cuba; *P. paludicola*, Cuba, Dominican Republic; *P. paniculata*, Dominican Republic, Mexico: Oaxaca)
majagüilla de costa (*P. spinifex*, Cuba)
majagüilla espinosa (*P. spinifex*, Cuba)
malva (*P. malacophylla*, Brazil: Paraiba)
malva carrapicho (*P. fruticosa*, Brazil)
malva chanana (*P. cancellata*, Brazil: Pará)
malva del monte (*P. paniculata*, Peru)
malva grande (*P. sidifolia*, Bolivia)
malva rastrera (*P. humifusa*, Bolivia)
malvarisco (*P. blanchetiana*, Brazil)
malva-rosa (*P. blanchetiana*, Brazil)
malva rosada (*P. malacophylla*, Brazil: Pará)
malva veludo (*P. malacophylla*, Brazil: Maranhão)
malvilla (*P. setifer*, Mexico: Guerrero)
mangrove mallow (*P. paludicola*, U.S.A.: Florida)
manzanilla (*P. malacophylla*, Mexico: Tabasco)
mapa huasca (*P. fruticosa*, Peru: *Quechua*)
misnaga (*P. fruticosa*, Panamá)
monacillo (*P. candida*, Mexico: Jalisco)
mosote (*P. schiedeana*, Guatemala, Honduras)
mozote (*P. schiedeana*, Belize, Guatemala, Honduras)
mozote de caballo (*P. schiedeana*, Guatemala)
mozote de perro (*P. schiedeana*, Nicaragua)
mozotillo (*P. schiedeana*, Costa Rica)
namble (*P. spinifex*, Dominican Republic)
obeja-quivan (*P. mollis*, Peru)
oreja de coche (*P. schiedeana*, Guatemala)
oreja de conejo (*P. schiedeana*, Guatemala)
pega pega (*P. fruticosa*, Panamá; *P. schiedeana*, Mexico: Oaxaca)
petit cousin (*P. spinifex*, Dominican Republic)
piojo de vaca (*P. fruticosa*, Peru)
piquant cousin (*P. spinifex*, Dominican Republic)
polatatai (*P. fruticosa*, Surinam)
purma (*P. paniculata*, Peru)
quati (*P. varians*, Brazil: Pernambuco)
remela de Cochorro (*P. malacophylla*, Brazil: Maranhão)
rosalia (*P. dasypetala*, Honduras)
rosellina (*P. hastata*, Argentina: Misiones)
rose mallow (*P. lasiopetala*, U.S.A.: Texas)
roseta (*P. renifolia*, Brazil)
rosita quihua (*P. fruticosa*, Peru: *Quechua*)
rozeta de mato (*P. flavispina*, Brazil: Rio Grande do Sul)
sak'al zak (*P. schiedeana*, Mexico: Chiapas, Tzeltal)
saltwater moho (*P. paludicola*, Belize)
sea cotton (*P. paludicola*, Guyana)
smaller mahoe (*P. paludicola*, Jamaica)
spiny abutilon (*P. spinifex*, Puerto Rico)
spur bur (*P. spinifex*, Jamaica)
swamp bush (*P. paludicola*, Bahamas, Puerto Rico)
tabadillato (*P. fruticosa*, Panamá: Chocó)
tábano (*P. fruticosa*, Cuba)
tablón (*P. paniculata*, Colombia: Tolima)
tamborcito (*P. paniculata*, Colombia: Chocó)
tjawassie (*P. paludicola*, Surinam)
ts'ik'i uxum (*P. schiedeana*, Mexico: San Luis Potosí)
tumin chij ak' (*P. paniculata*, Mexico: Chiapas: *Tzeltal*)
txikim txikam (*P. schiedeana*, Guatemala: *Jacaltec*)
tzajal (*P. paniculata*, Mexico: Chiapas: *Tzeltal*)
tzajalzak (*P. paniculata*, Mexico: Chiapas: *Tzeltal*)
vassoura do mata (*P. gracilis*, Brazil: Minas Gerais; *P. intermedia*, Brazil: Bahia)
weawea (*P. fruticosa*, Peru; *P. oxyphyllaria*, Peru)
wíawía (*P. castaneifolia*, Peru; *P. leucantha*, Peru)
wild cotton (*P. paludicola*, Belize)
yahuballi (*P. paludicola*, Guyana)
yak'il ch'o (*P. paniculata*, Mexico: Chiapas: *Tzotzil*)
zavoine (*P. fruticosa*, Haiti)
zelosa (*P. fruticosa*, Colombia)

Index of Scientific Names

New names and combinations are in **boldface**; synonyms and nomina dubia are in *italics*. Page numbers in **boldface** indicate primary page references; page numbers with an asterisk (*) indicate pages with illustrations or maps.

PAUL A. FRYXELL

Paul A. Fryxell was born in Moline, Illinois, on 2 February 1927. He attended Moline public schools and, after military service, attended Augustana College (Illinois), graduating with a B.A. in 1949. His studies continued at Iowa State University (M.S., 1951; Ph.D., 1955). After employment with the New Mexico Agricultural Experiment Station (1952–1955) and the University of Wichita (1955–1957), he joined the Agricultural Research Service, U.S.D.A., with which agency he spent most of his career, located on the Texas A&M University campus (see Zanoni, 1997). He retired from this position in 1994 and now (in active retirement) is Adjunct Professor in the Botany Department of the University of Texas at Austin and is also an Honorary Curator at The New York Botanical Garden. His research interests have centered on the taxonomy of the neotropical Malvaceae, including work on the evolution, biodiversity, and taxonomy of *Gossypium*, the genus that includes the world's cotton crop. He has served as president of the American Society of Plant Taxonomists (1983–1984) and of the Society for Economic Botany (1988–1989), has held a Fulbright Scholar Award for study in Argentina (1993), is a Fellow of the American Association for the Advancement of Science and of the Texas Academy of Science, and is a member of the Commission of Flora Neotropica. Major publications include *The Natural History of the Cotton Tribe* (Texas A&M University Press, 1979), *The Malvaceae of Mexico* (Systematic Botany Monographs, vol. 25, 1988), and "The American Genera of Malvaceae" (Brittonia 49: 204–269. 1997). He is a contributor of treatments of the Malvaceae to numerous neotropical floristic works and has conducted fieldwork in the Neotropics, primarily in Mexico but also in parts of Central and South America, and also in tropical Australia.